T0304448

Fundamentals of Aircraft and Airship Design
Volume 2—Airship Design and Case Studies

Fundamentals of Aircraft and Airship Design

Volume 2—Airship Design and Case Studies

Grant E. Carichner
Lockheed Martin Aeronautics Company
Advanced Development Programs (Skunk Works)
Palmdale, California

Leland M. Nicolai
Lockheed Martin Aeronautics Company
Advanced Development Programs (Skunk Works)
Palmdale, California

AIAA EDUCATION SERIES
Joseph A. Schetz, Editor-in-Chief
Virginia Polytechnic Institute and State University
Blacksburg, Virginia

Published by
American Institute of Aeronautics and Astronautics, Inc.
1801 Alexander Bell Drive, Reston, VA 20191-4344

American Institute of Aeronautics and Astronautics, Inc., Reston, Virginia
1 2 3 4 5

Library of Congress Cataloging-in-Publication Data
Record on file.

Skunk Works® is a registered trademark of the Lockheed Martin Corporation.

Photographs of Lockheed Martin aircraft on the first page of Chapters 8 and 12 and Case Studies 2 and 5 as well as in Figures 1.15, 9.8, and 10.16 are provided courtesy of Lockheed Martin Corporation. Images on the first page of Case Study 8 and Appendix B as well as in Figures 8.8, 8.10, 10.12, 13.2, and 13.8–13.11 are from the private collection of the authors. Photographs on the first page of Chapters 1, 2, 4, 6, 7, Appendices A, E, F, and Figure 1.8 are from the private collection of Rich Van Treuren. All photographs in each case study and contributed chapter are from the contributing author unless otherwise specified. Unattributed photographs are held in the public domain and are from either the U.S. Department of Defense or Wikipedia.

Data and information appearing in this book are for informational purposes only. The authors and the publisher are not responsible for any injury or damage resulting from use or reliance, nor do they warrant that use or reliance will be free from privately owned rights.

ISBN: 978-1-60086-898-6

To my wife Deborah,
who's patience and understanding allowed me
the time needed to complete this book.

Also, I would like to express my appreciation for
the commitment by Lee and his pursuit of excellence
that made this a better book.

G. C.

To my children—
Jeffrey Stephan, Debra Leigh Nicolai-Moon, Noelle Michelle,
Jeffrey Ray—
and to Carolyn, my best friend and wife of 55 years.

L. M. N.

FOREWORD

The second volume of this two-volume aircraft and airship design textbook and reference takes the reader on a unique time-travel journey. The authors, Grant E. Carichner and Leland M. Nicolai, both possess a rare multifaceted portfolio of flight vehicle system integration skills. This expertise has been proven not only to develop competitive flight vehicle systems and explore prospective technologies, but also to support future competitive corporate strategies. This book is surprising at first because of its focus on a flight vehicle type thought extinct since the traumatic Hindenburg disaster on May 6, 1937. In *Fundamentals of Aircraft and Airship Design, Volume 2—Airship Design and Case Studies*, the authors thoroughly explore airship transportation potential while providing the reader with a pragmatic engineering approach to quantifying such a vehicle in the systems context. This professional forecasting capability renders this book unique.

At the heart of this time-travel journey is the enduring quest for human mobility. The Montgolfier brothers initiated this movement when they sent the first balloon aloft in 1783. The sky was limitless from that moment onward, and inventors began developing steerable airships. The first generation of ellipsoidal nonrigid airships was followed by semirigid airships incorporating fixed keels. The evolving rigid airships pioneered by Count Zeppelin resulted in majestic and capable aircraft that were able to carry large civil and military payloads. At the turn of the last century, Zeppelins conquered the skies by establishing the world's first passenger-carrying airline and providing nonstop transatlantic air service. As time and technology progressed, the airship era was brought to an untimely end, surviving only within small and specialized markets today.

This volume invites the reader to experience traditional airship designs and operational concepts virtually by using the Lockheed Martin Skunk Works mindset. The book provides unique insight into the workings of this famed advanced design group because the reader has the unique opportunity to observe and learn from two Skunk Works practitioners, both of whom have had long careers at Lockheed Martin Aeronautics. The classical airship characteristics are systematically quantified. Past and present system choices are detailed correctly using the language of advanced design project engineers. The authors are able to assemble solid arguments and convincing proof that confirm the relegation of the classical airship mode of transportation to today's niche markets.

In the best tradition of Lockheed Martin's Skunk Works, the authors have identified in advance inherent drawbacks of the evolutionary pattern and eventual stagnation of the classical airship concept from the past to the present time. Based on this knowledge, they turn their forecasting proficiency to explore the unfolded potential in a modern airship implementation—the *hybrid airship*. At this point, this author duo succeeds in developing a strong argument justifying the continuation of the airship bloodline with the conception of the hybrid airship type.

The last century has been characterized by the speed seekers who shaped aeronautics. This thirst for high-speed transportation is leveling out naturally owing to modern environmental concerns, energy efficiency requirements, and cost pressures. In this context, the primary focus of this book is to provide the tools and mindset that are required for the quantification of a next-generation hybrid airship completely capable of transporting large payloads over long distances at low speeds, while requiring minimal infrastructure and securing mandatory safety performance even in demanding operating conditions. Such future designs will benefit from a synergy of several technological spirals observed today: mixed (aerostatic + aerodynamic) lift configurations, lightweight composite materials, efficient and ecology-friendly power plants, smart avionics, including, where needed, unmanned control, satellite positioning, networked operations, etc. Interestingly, the hybrid airship mode of transportation naturally complements the surface transportation analog: the well-established railroad and cargo ship systems and infrastructure. Given the impact of present day low-speed railroad and cargo ship transportation modes, the operational and commercial potential of the hybrid airship becomes apparent. The hybrid airship presents a technically feasible and experimentally proven concept with operational characteristics superior to the classical airship and even to surface and aircraft transportation in sparsely populated and remote areas.

With the availability of this text, it is up to the next generation of strategic planners and financial decision makers to incorporate this new concept and efficient mode of carriage into future multimodal transportation infrastructures and operations. In my opinion, this text provides a well-balanced analytical framework required for new design quantification. The authors' approach follows the best-practice flight vehicle design development methodologies and analytics known to engineers and technologists.

Leland Nicolai and Grant Carichner have succeeded in providing a cutting-edge two-volume aircraft design text and reference addressing probably the most productive modes of air transportation: fixed wing aircraft and the promising low-speed hybrid cargo airship. Flight vehicle design decision making cannot rest exclusively on capable flight vehicle analytical product development methodology tools; this volume uniquely complements flight vehicle design principles with carefully selected flight vehicle case studies detailing project engineer lessons learned by renowned designers. Because design is a form of organized ideation, it is of significance for the student and practitioner alike to systematically complement formal analytical quantification proficiency with project engineer lessons learned. It is the combination of analytical expertise and open-ended problem-solving skills that instill the "right stuff" in any flight vehicle design team.

In *Fundamentals of Aircraft and Airship Design, Volume 2—Airship Design and Case Studies*, Carichner and Nicolai have carefully selected a cross section of case studies throughout aviation industry. These case studies present a romance of successes and failures, studies of the ability of people in groups, fused together to merge into a balance-compromised designed-to-mission flight vehicles limited by time pressures, cost constraints, and the challenge of producing machines that often have never before existed. No textbook would be complete without telling the tales of these groups in order "to realize the immense potentials of an effort shared." As such, these development accounts are produced from the perspective of manager, integrator, and technologist. These design perspectives offer a message that benefits the student and practitioner alike. A genuine tour de design is complete with this second volume—an important and unique contribution to aeronautics.

Bernd Chudoba
Associate Professor
Director, Aerospace Vehicle Design
(AVD) Laboratory
Mechanical and Aerospace
Engineering
The University of Texas
at Arlington
February 2013

CONTENTS

LIST OF COLOR PLATES

PREFACE

This book is Volume 2 of the two-volume set *Fundamentals of Aircraft and Airship Design*. It contains everything needed to perform a conceptual design of an airship. As different as an airship is from winged aircraft the governing equations of motion are identical to those for winged aircraft. Even though most of an airship's lift force comes from a buoyant gas, the presence of aerodynamic and/or vectored propulsive forces and moments are essential for normal mission stability and control. Thus, there are many similarities between designing an airship and a winged aircraft. This work is the collaborative result of two practicing aerospace engineers with more than 90 years of air vehicle design experience. They have been designing airships since 1998 and have taken that experience and put it on paper in the form of this book.

As with Volume 1, the discussions in Volume 2 address the science, art, and state of mind of airship design; all three are needed for a successful design. The science is the compliance with the laws of physics, the analysis methodology, and the mechanics of designing the craft. The art of design is the beauty and timeless elegance of the craft and is captured in the history, lessons learned, the facts and stories (that appear in blue boxes), nine case studies, and a four-color section found at the back of the book. The state of mind is the passion and yearning for the unachievable that the designer brings to the game. The design of an airship or aircraft must be a love affair between the designer and his design.

As airships regain some of their lost prominence, there is an increased interest in designing new and improved versions of old designs as well as creating new and exciting buoyant vehicles. Much of this is due to improvements in the envelope fabric, landing gears (air cushion landing systems), vectored thrust, and hybrid configurations having more efficient lift generation. There are a few airship books available but none of them address the step-by-step process of designing a modern airship.

Both Volumes 1 and 2 are aimed at upper-level undergraduate and graduate students as well as practicing engineers. The books have a comprehensive treatment of the conceptual design phase that starts with the consideration of the user's needs to the decision to iterate the design one more time. Volume 2 is complete in that the reader should not have to go outside the text for additional information. It has a comprehensive set of appendixes that present general data in a central location.

The text is organized in two parts with the first 10 chapters giving background information on aerostatics, aerodynamics, performance, propulsion, weights, materials, structures, stability, and control. Chapters 11 and 12 then discuss the specific design of a conventional airship (body of revolution) and a hybrid airship respectively. The last technical chapter discusses balloon design and operation and was written by Rodger Farley of NASA.

The second part of the book presents nine case studies of air vehicles that have influenced the art and science of design. The case studies cover military to commercial to private sector, incompressible to hypersonic speeds, and hydrocarbon-powered to man-powered to no power.

Using design examples throughout the book, the authors guide the reader's journey through the design process as it would happen in the actual design environment. By sharing their unique background, the authors give practical guidance that can be used directly in an airship design project. Students and practicing engineers alike will find *Fundamentals of Aircraft and Airship Design: Volume 2—Airship Design and Case Studies* a perfect complement to aircraft design found in Volume 1.

A special thanks to our AIAA editor, Pat DuMoulin, whose contributions have made this book much better.

Grant E. Carichner
Leland M. Nicolai
February 2013

Chapter 1 / Introduction

M.1 Airship (1918

- Balloon Development Over the Past 200 Years
- Helium Discovery and Impact
- Airship Structural Development
- Comparing Aircraft and Buoyant Vehicles
- Hybrid Airship Design
- Overall Design Requirements
- Airship Design Phases

The quote below reminds us that customer requirements can be flawed. The flawed requirements must be challenged and hopefully changed to present a credible set. History is filled with flawed requirements. An example of a flawed requirement is discussed in this chapter.

Mother Nature cannot be challenged . . . but man-made rules can and must be.

Clarence "Kelly" Johnson

1.1 Lighter Than Air . . . the Beginning

Since the dawn of humankind, we have wanted to soar and fly like the birds, but humans have only had this ability for a little more than 100 years. The mile works of the Wright brothers, Samuel Langley, Otto Lilienthal, Octave Chanute, and Sir George Cayley are well documented. All of these engineering giants contributed to successful powered manned flight. However, more than 100 years before the birth of aviation, a group of engineers, scientists, and experimentalists individually pursued manned and powered flight using a lifting gas, rather than wings, to generate lift.

Centuries earlier, man's pursuit of flight inadvertently started with Archimedes, who is credited with quantifying buoyancy. His discovery around 200 BC was not motivated by aeronautic or hydrodynamic thoughts or any other scientific endeavor. He was merely trying to come up with an accurate method to determine whether or not the King's gold had been alloyed with a base metal. It would be nearly 2000 years later that Archimedes' principle would be applied to manned flight.

The Chinese are credited with building the first man-made airborne objects: signaling lanterns in 200 AD. Figure 1.1 shows a re-creation of such

Figure 1.1 Chinese lantern: first man-made airborne objects.

a lantern, which used the principles of buoyancy to stay aloft. Although there is much debate if this was indeed the first man-made airborne object, it has also been suggested that centuries earlier the Nazca Indians of Peru used hot-air balloons to view their geometric land patterns from the air. However, the evidence is not definitive that the Nazcas made balloons. British balloonist Julian Nott built a hot-air balloon (see Fig. 1.2) using materials and weaving technology thought to be available to the Nazcas, and Nott successfully flew it. It is left to the reader to decide if the Nazcas of 500 AD could have accomplished this lighter-than-air feat.

More than 1000 years later in the late 1700s, balloon development heated up, and the evolution has many similarities to how the airplane would be developed 100 years later. Several balloon designers such as the Montgolfier brothers and Jacques Charles were furiously building lighter-than-air (LTA) machines, and in late 1783 the Montgolfier brothers' hot air balloon carried the first people. Since hot air is not an efficient lifting gas, the discovery by British scientist Henry Cavendish of hydrogen, which was recovered using a new sulfur process, was immediately embraced by nearly

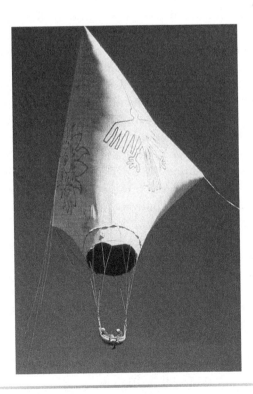

Figure 1.2 Demonstration that Nazca Indians of Peru had materials to make a balloon.

all serious balloon designers. Since hydrogen is so much less dense (by a factor of 14) than ambient air, it was a major improvement in lifting power compared to hot air.

Jacques Charles (of gas law fame) designed a new balloon that used hydrogen as the lifting gas in mid-1783 and was its first passenger just months later in December. Hydrogen would be the predominant lifting gas for all buoyant vehicles for the next 150 years! These efforts were followed closely by French mathematician and engineer Jean Baptiste Meusnier, whose design provided a modest ability to steer using an elongated shape, a rudder, and a source of propulsion. 1783 was indeed the year of the balloon! Figure 1.3 shows a collage of balloons from the late 1700s to modern day airships.

In the late 1700s many inventors attempted to develop an efficient means of flight. At first hot air and hydrogen-filled balloons were built and tested, but they were difficult to control, had a very low speed, and, in the case of hydrogen-filled balloons, used the dangerous gas. Based on these obstacles some engineers and scientists turned their energies to winged aircraft, which came with their own unique set of challenges.

Concurrently, there were other inventors and engineers that continued to design balloons that could also allow man to fly for long periods of time by using the property of gases that generated a buoyant force due to the density difference between a lifting gas (hot air or hydrogen) and ambient air. Years later, inventors such as German general and aircraft manufacturer Ferdinand von Zeppelin, who would come along in the early 1900s, significantly increased the aerodynamic performance of airships. Most of these new airship designs used hydrogen since it was much lower in density than heated air and much cheaper since no fuel had to be burned to maintain the density difference that generated the buoyant lifting force. *Buoyancy* is simply the total lifting force generated by the volume of contained lifting gas. Chapter 2 will develop the equations necessary to assess the aerostatic characteristics of a buoyant vehicle and the differing levels of buoyancy for various lifting gases.

Scientific studies of gas properties and engineering developments were performed that allowed for the later construction of successful balloons, dirigibles, blimps, and airships. For example, in the mid-1600s Evangelista Torricelli invented the barometer to measure atmospheric pressure and Robert Boyle (Boyle's gas law) started to weigh air. In 1766 Henry Cavendish isolated hydrogen, weighed it, and stated that hydrogen weighed about 1/14 as much as air. He called it phlogiston, a mystical fire element. Twenty-three years later, Frenchman Antoine Lavoisier properly recognized that it was an element and renamed it hydrogen.

The magical year for balloon development was 1783. In a small French village the Montgolfier brothers, Joseph and Etienne, noticed smoke rising,

Montgolfier balloon (hot air) - 1783

Jacques Charles hydrogen balloon - 1783

Meusnier's steerable hydrogen balloon - 1783

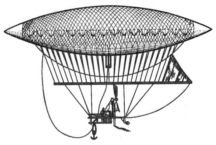

Henri Giffard blimp - 1852

Sentinel 1000 non-rigid airship - 1994

Zeppelin NT semi-rigid airship - 2001

Figure 1.3 Balloon development over the last 200+ years.

so they built a 35-ft balloon filled with smoke that ascended 6000 ft and then landed a mile away. Later in the year physicist Horace-Bénédict de Saussure explained that the balloon rose because the air was hot and not because of the smoke. (He would prove it two years later by putting thermometers inside and outside of a balloon.)

In mid-1783 in the big city, Parisians were not to be outdone by the Montgolfiers. They hired the distinguished physicist Jacques Charles to design a balloon. Using hydrogen the balloon rose to heights where it overpressured and burst. There was no pressure relief valve. Later in 1783 King Louis XVI invited the Mongolfiers to Versailles for a demonstration, so they built and flew a very successful balloon measuring 72 × 41 ft, and included a rooster, duck, and sheep among its passengers (the first living beings to fly!). They landed safely more than a mile away after soaring to 1500 ft. However, two years later while attempting to cross the English Channel, Pilâtre de Rozier became the first air fatality. His hydrogen-filled balloon had a metal vent valve that produced an electrostatic spark during venting causing the escaping gas to catch fire. Hydrogen valves were made of wood from then on. About 150 years later a similar electrostatic spark would also be responsible for the most famous airship disaster of all time, the Hindenburg.

It didn't take long for military minds to understand that these balloons could be used in battle. Soon, military observation balloons became popular, so the French formed a balloon corps that was organized and stationed at Chalais-Meudon, and was known as the Compagnie d'Aérostiers. This forward thinking was short-lived however, as Napoleon didn't include balloons in his military strategy and tactics, leading to the Aérostiers being disbanded. Napoleon's refusal to use balloons did not deter aviation progress, so in the 1880s, the world's largest hangar would be built at Chalais-Meudon to house Charles Renaud's revolutionary airship, La France. It was the first airship to be controllable during flight. Chalais-Meudon is now the world's oldest airbase.

After a flurry of balloon design activity in the late 1700s, not much progress was made due to the inability to supply power and forward speed to a balloon. Balloons were still at the mercy of the winds. It wasn't until 1853 that engineer Henri Giffard put a steam engine on a hydrogen-filled balloon, making it the world's first passenger-carrying dirigible. Although it could travel at 6 mph, it could not fly a circular route and return to the point from which it was launched. In the 1880s another power-generating system was integrated onto an airship when Charles Renaud introduced a battery-powered airship, La France, which could return to its launch point. It was the first aircraft of any type to do so. It would be another ten years before airships and manned gliders would have the gas engine as a source of propulsive power. Among legends such as Samuel Langley, the Wright

Brothers, and Count Zeppelin, it was the little-known engine designer, Charles Manly, who would deliver a useable engine for both aircraft and airships.

Armed with a reliable high power-to-weight ratio gasoline engine, developers such as Count Ferdinand von Zeppelin began their airship design efforts. Curiously, it was not until Zeppelin had retired from the military that he seriously started his rigid airship development. In his younger days, he served with the union army and saw firsthand the utility of balloons in military settings during the civil war. Intrigued by a balloon's operational possibilities, von Zeppelin studied the possibilities of building motorized balloons or airships.

In July 1900, against public opinion, von Zeppelin launched his first ship, which was 419-ft long. Powered by two 16 hp motors, it attained a speed of 18 mph. Realizing he needed a larger and more powerful ship, von Zeppelin started to raise funds for the project. Five years later he successfully flew a new Zeppelin ship to an altitude of more than 1600 ft. After several disasters the count was forced to go to the people for new investment. About $1.2 million was raised from public donations. This so touched von Zeppelin that he put the money in trust and founded the Zeppelin Endowment for the Propagation of Air Navigation. Since profits were to go back into aeronautics, the foundation as well as von Zeppelin, over time, became rich and powerful. The foundation designed and built its own motors, gears, hangar buildings, and performed its own research. Von Zeppelin had a special ability to attract gifted and clever people. Engineers such as Karl Maybach (motors), Dr. Karl Arnstein (mathematics), and Claude Dornier (design) became part of his foundation's team.

During the golden age of airships, which started around 1900, many designs were built and tested by famous inventors other than von Zeppelin. The very wealthy Alberto Santos-Dumont built nearly twenty balloons and airships in a few years' time (see Fig. 1.4). Shortly thereafter, he changed his focus to winged aircraft, which were much faster. Numerous other European inventors also tried their hand at designing balloons that could be controlled. But Count Ferdinand von Zeppelin's designs for balloons and airships were ultimately used extensively by Germany during World War I.

Technology was advancing quickly, so unstable spherical tethered balloons were soon being replaced by designs that would weathercock into the wind and not create too large a load on the ground tether. Military balloons and airships filled with hydrogen were doomed by the invention of the incendiary bullet and rapid development of the fighter plane. Even with these technological shifts, Zeppelin airships flew several million miles in support of fleet patrol, reconnaissance, and bombing operations during World War I. During this time numerous airships were built by Germany. Though they were greatly feared by people on the ground, they did

Figure 1.4   Alberto Santos-Dumont's ship.

relatively little damage and were responsible for relatively few deaths, because their role shifted to scouting and reconnaissance, such as to clear mines and to track submarines.

After World War I all of Europe was enchanted with airships and their potential. Under the guidance of Paul Jaray, Zeppelins were designed and built as commercial endeavors. Commercial airship designs were prevalent during the 1920s but it wasn't until the 1930s that these behemoths of the sky became extremely popular. Examples of Zeppelin airship designs are shown in Fig. 1.5. Jaray's design of the Bodensee (LZ120) became the prototype of the future large airships such as Graf Zeppelin (LZ127) and the Hindenburg (LZ129). Figure 1.5 compares early rigid airship designs to the much larger rigid Zeppelin designs.

The United States also became interested in airships in the early 1920s and purchased the semi-rigid ROMA airship from Italy in 1921. The ROMA crashed during a test flight the next year and became the last U.S. airship to use hydrogen (see Fig. 1.6). Over the next ten years the United States would build four large rigid airships named Shenandoah, Los Angeles, Macon, and Akron. The Shenandoah, Macon, and Akron would crash and the Los Angeles was retired after eight years of service. The

Shenandoah was the first U.S.-built rigid airship and it was the first airship to use helium, which was in very short supply. Helium supplies were so low that the Los Angeles could not be operated when the Shenandoah was operating.

There were numerous airship crashes in the early part of the 20th century and most were directly or indirectly caused by abnormal weather events. This slowed and for some countries stopped further development of lighter-than-air vehicles. While severe weather can still do great damage to an airship, it is no longer a significant threat because precise weather is known everywhere along the route. Current airships simply avoid, wait out, or out run any dangerous weather event.

During the 1920s and 1930s, hydrogen-filled Zeppelins were everywhere and were very

In the 1920s the famous airship designer, Charles Burgess, claimed that rigid designs were superior and became even better as airships got bigger. Of course, this has not turned out to be true. Burgess was basing his prediction on the facts that human labor was inexpensive at the time and no material could successfully contain lifting gases while under pressure. After 8 decades labor rates have skyrocketed and envelope materials have become nearly 30 times better making rigids too expensive and much heavier than a non-rigid equivalent.

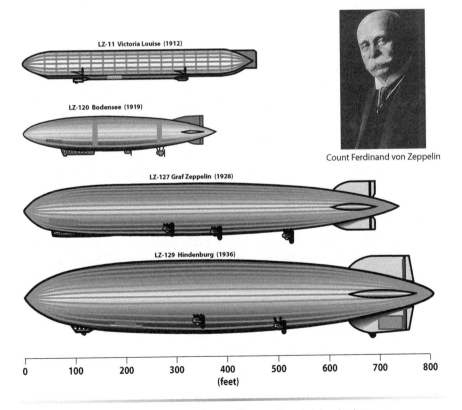

Count Ferdinand von Zeppelin

Figure 1.5 Comparison of Zeppelin airship designs.

Figure 1.6 ROMA airship, 1921. The last hydrogen-filled airship operated by the United States.

successful at carrying passengers around Europe and across the Atlantic Ocean and back. A Zeppelin airship was also the first to circumnavigate the globe. All of these Zeppelins used hydrogen since Germany had no supply of helium. The use of hydrogen instead of helium is what lead to the famous catastrophic Hindenburg crash. The sister ship Graf Zeppelin would be retired because of the dearth of helium supply. During the 1930s only Germany was building airships (all of which were rigid). After 1940 only the United States was building non-rigid airships and these were for military purposes. More than 150 non-rigid airships were built for the military and were used mainly for convoy escort and submarine scouting.

The U.S. airships were very successful in World War II and averaged more than 87% readiness . . . a very enviable record. An example of their value is illustrated by the fact that unescorted convoys had more than 500 ships sunk. Convoys escorted by airships had only one sinking occur out of more than 80,000 trips—an incredible record!

Appendix E is a compilation of airship characteristics. Most of the numbers seem correct but a few are suspect, so it's prudent to use for overall trends but one should be careful using data from a single airship. Weights, tail sizing, and ballonet sizing plots elsewhere in the book have come from this database. When viewing this data keep in mind that hull volumes for rigid designs did not equal the volume of the lifting gas. Rigids were filled with bags of lifting gas that were encased in an aerodynamically shaped hull. Hull volumes were 10%–20% larger than the lifting gas volume. However, the hull volume for non-rigids is within a few percent of the lifting gas volume because the envelope provides the aerodynamic shape and contains the lifting gas as well.

1.2 Helium

Helium manufacturing technology quickly responded to this scarcity with the large scale production of the non-flammable gas, helium, which came too late to be used in World War I. However, just twenty years later helium-filled airships became a military asset during World War II. The primary military role for airships was its unique ability to protect naval fleets from submarines. Another lesser known tactic was to use large numbers of small tethered balloons surrounding a battlefield (e.g. D-Day) to deter strafing and low level bombing by enemy warplanes. These were called barrage balloons.

French astronomer Jules Janssen originally identified helium in 1868 as he detected a yellow line in the sun's spectrum that had not been previously seen. English scientist Joseph Lockyer showed that the new line was an element not yet known on earth. They named this element helium from the word helios, the Greek word for sun.

Even though helium was discovered as a drilling byproduct in 1903, the first buoyant vehicle using helium did not appear until 1921. Helium was much more expensive than hydrogen so its use was not as widespread. It took the Hindenburg disaster to make all airships thereafter use helium as the lifting gas. The inert characteristic of helium was valuable to users because of its safety, but helium is more dense than hydrogen and thus not quite as good a lifting gas. Another drawback for helium is that its atom is smaller than the hydrogen molecule so it is harder to contain helium without some leakage. Typically, airship envelopes filled with helium have to be "topped off" every few weeks.

Helium is 1/7 the weight of air and has 7% less lifting capability when compared to hydrogen. However, design comparisons between helium and hydrogen airships will show that helium designs have 20%–30% more volume than hydrogen airships for the same buoyant lift. Some of this difference is explained by the 7% density difference between helium and the less dense hydrogen gas. However, most of the size difference comes from their ballonet sizes. Hydrogen is a relatively inexpensive gas especially when compared to helium. This means that venting some hydrogen to keep an airship's envelope from rupturing from overpressure is acceptable. This approach is not

Janssen was ridiculed for his claim that he had discovered a new element. Later in the year his discovery would be confirmed by Lockyer. Helium was the first element discovered in space before it was found on earth.

acceptable with airships having helium as the lifting gas. Therefore, a constant internal pressure must be maintained for all possible altitude and temperature excursions and it is the role of a ballonet system to maintain this constant internal pressure. Ballonet designs will be discussed in several of the following chapters.

1.3 Airship Structural Concepts—A Quick Overview

Historically, two structural concepts are prominent, namely *rigid* and *non-rigid*. While virtually all airship designs have been either rigid or non-rigid there is an intermediate concept called *semi-rigid*. Figure 1.7 illustrates

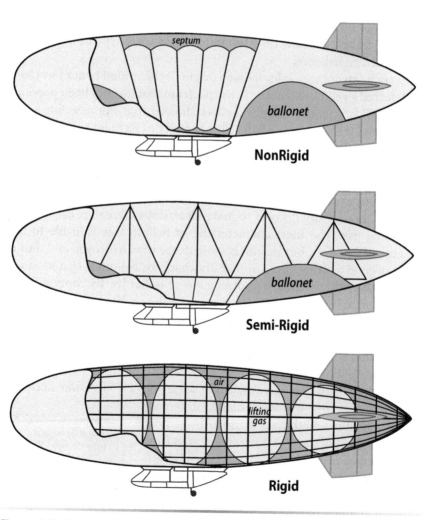

Figure 1.7   Comparison of structural concepts for body of revolution airship.

the general differences among these three approaches. A detailed discussion on airship structure is contained in Chapter 8 but a short overview is appropriate here.

In the late 1800s and early 1900s, virtually all airships were constructed as rigid designs. Early airship concepts had their designs limited by the ability of materials to safely contain the lifting gas under any pressure. Therefore, rigid shells were constructed and covered with material that would endure the rigors of flight but were also not under any pressure. These materials were simply attached to the outer structural frame. The buoyant lift was generated by gas cells containing hydrogen (and later by helium) that were attached to the bottom of the rigid frame. These gas cells were allowed to expand and contract as the lifting gas changed temperature and became fully expanded when the airship was at its maximum altitude. Figure 1.8 shows a rigid design in the process of being covered.

The intermediate structural approach, or *semi-rigid,* attempts to combine the best features of rigid and non-rigid designs. This concept adds a pressure-stabilized envelope with a modest structure running most of the length of the airship. Significant structural items such as engines and tails are attached to this internal structure. Although the Zeppelin NT-07 is

Figure 1.8 Applying the external shell fabric over a rigid structure.

such a design it has not proven to be commercially viable. As a result, there are very few successful semi-rigid designs.

Non-rigid designs were attempted in the early 1900s but these airships were small. No envelope materials were available in sufficient quantity or at a low enough cost that could contain either hydrogen or helium at pressures needed to structurally attach tails and resist puncture. Chapter 8 discusses the continuous improvements in fabrics over the last century that have made non-rigid designs the modern structural design of choice. Although rigid and semi-rigid designs are briefly discussed, this book concentrates on the non-rigid structural design approach. As seen later in Chapter 6 rigid designs are always heavier than non-rigid designs for a given volume.

Unlike the rigid design the non-rigid design uses a flexible bag called a ballonet inside the main envelope (shown in Fig. 1.7). The purpose of the ballonet is to maintain a constant pressure differential ($\Delta p = p_{amb} - p_{int\text{-}gas}$) across the envelope. For example, the interior gas may change its density due to temperature changes from the sun, or ambient conditions may change due to a change in altitude. If the pressure difference between external ambient conditions and internal conditions is held constant, the buoyant lift is also constant assuming perfect gas behavior, which means the envelope shape stays the same. The ballonet is normally filled with ambient air to compensate for the changing pressure of the internal lifting gas (nearly always helium) to maintain a constant ΔP between the outside air and the internal lifting gas. It functions as the pressure regulator for the airship's envelope. It is the constantly changing lifting gas properties of density, pressure, and temperature that create the basic need for ballonets. For airships operating with a constant ΔP across the envelope its internal gas density then varies inversely with absolute temperature (Boyle's Law). The ballonet constantly changes volume to maintain this constant delta pressure across the envelope as the lifting gas temperature changes. The lifting gas temperature is in turn a function of changes in either altitude, heating from solar radiation (superheat), or from directly heating the internal lifting gas.

Ballonets are commonly arranged to have one forward and one aft ballonet envelopes. This forward/aft design allows for air mass to inflate or deflate, in order to help with trimming the vehicle on the pitch axis. A rigid design does not need a ballonet as there is little pressure differential across the envelope, except at the nose. Internally, bags of lifting gas are attached to the rigid structure and have enough extra volume to operate at its maximum altitude. The envelope covering a rigid structure is for aerodynamic purposes only.

A prime example of the non-rigid structural concept is the ZP4K airship, which was produced by the Goodyear Aircraft Company of Akron,

Ohio in 1953 for the U.S. Navy. It was one of 134 K-class airships produced by Goodyear between 1938 and 1954 for ASW patrol and escort duties. The ability of the K-class airships to hover and operate at low altitudes and low speeds made them well suited to detect and destroy enemy submarines. The ZP4K descriptive arrangement is shown on Fig. 1.9 and has the following features:

Crew: 9–10

Maximum speed: 68 kt

Cruise speed: 50 kt

Range: 2205 nmile

Endurance: (2205/50) = 44 hr

Powerplant: (2) Pratt & Whitney R-1340 radial engines of 425 hp each

Buoyancy ratio = Static lift/Gross weight = 0.9

Propeller diameter = 11.5 ft

Propeller blade angle = −6 deg to +30 deg (small reverse thrust)

Sensors: ASG radar, sonobuoys, and magnetic anomaly detection (MAD)

Armament: (4) 350 lb Mk 47 depth charges & (1) 50 caliber machine gun

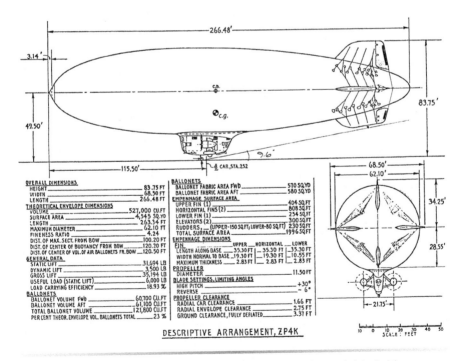

Figure 1.9 General arrangement of the ZP4K non-rigid airship, courtesy of Goodyear.

Figure 1.10 K-Class airship as escort during WWII. 134 K-Class airships were built.

The ZP4K was retired in 1959 after building 134 airships (see Fig. 1.10). It was also during this time that Goodyear came up with the term *dynamic lift* as a contrast to buoyant lift. While the term is popular within the airship community, we will use the term *aerodynamic lift* and *thrust vectored lift* in this book because that is a more accurate and literal description of what is happening It is also worth noting in Fig. 1.9 that the scrape angle is called out on the drawing. While generally not a critical design issue for conventional airships, scrape angle plays a more significant role in the design of hybrid airships. As we'll discuss and refer to later, hybrid airships are capable of generating more than 10% of its total weight with aerodynamic forces. The term *hybrid airship* does not have an exact definition. For this book the term *hybrid airship* means a buoyant air vehicle with takeoff $BR < 0.9$.

1.4 Failed Concepts and the Square-Cubed Law Revisited

Why is adding a small amount of volume for lifting gas to an existing airplane design such a bad idea? It is a temptation that must be resisted. Perhaps discussing the "square-cube law" (S-C) again (it first appears in

Volume I) can help to better understand the general limitations and benefits of buoyant vehicles as compared with wing aircraft. It was Galileo [3] who first stated "when an object undergoes a proportional increase in size, its new volume is proportional to the cube of the multiplier and its new surface area is proportional to the square of the multiplier." In physiological terms this means as an animal gets bigger its weight increases in proportion to its volume and its strength goes up by the cross-sectional area of its muscles. This explains why a flea can jump many times its height and an elephant cannot get all four feet off the ground even though the elephant is far stronger than the flea. As seen in Volume I, the S-C law limits airplane size as airplanes are photo-scaled to larger sizes since their weight increase is proportional to the volume (cube of the scaling factor). At the same time an airplane's ability to generate aerodynamic lift is proportional to the wing area (square of the scaling factor). For a given level of technologies, an aircraft still has a size limit to meet its requirements (weight, cost, performance). Beyond this limit any attempt to increase performance by simply photo-scaling the design would fail. But does this limitation apply to airships as well and, if so, is it the same?

Taking another look at the S-C law for buoyant vehicles shows some interesting contrasts. First of all, buoyant lift is proportional to volume, while most drag and weight are proportional to envelope surface area. This is just the reverse of physiological systems or winged aircraft. How does this distinction change the design approach to buoyant vehicles? The most dramatic change is the relationship between weight and drag. Based on the generalization of the S-C law, weight is less critical to the designer of a buoyant vehicle since volume can always be increased to generate lift faster than its weight increases (similar to sea ship design). In other words, buoyant vehicles become more efficient lift generators as their size increases *without limit*. Theoretically this is true but actual practice would place a limit on buoyant vehicle size based on structural design, manufacturing facilities, ground handling, launch and recovery, etc. Of course, actual magnitudes of forces such as drag must be accounted for as size increases, but this does not change the overall result that "bigger is always better and always more efficient" when generating buoyant lift. Some designers are uncomfortable with this result based on their aircraft design background and it takes a while for them to come to terms with this suspected heresy. For every pound of structural weight or payload that is added to a winged airplane, its takeoff weight will increase by 2–3 lb for transport aircraft and 3–5 lb for fighter aircraft. This is known either as the *growth factor* or the *multiplier effect* [see Eq. (1.1)].

Multiplier Effect = (Struct Wt_2 – Struct Wt_1)/(Base Weight Increase)   (1.1)

Relative to airplanes and using the S-C law, when it comes to adding weight we would expect different behavior for balloons and airships. To

Table 1.1 Sample Problem #1: Prolate Spheroid

Buoyancy = 0.065 lb/ft³		Payload change			Material density change		
Parameter	Base case	Parameter	+10% Payload	Change from base	Parameter	+10% Material density	Change from base
Envelope wt (lb)	6500	Envelope wt (lb)	7651	1.177	Envelope wt (lb)	8651	1.331
Payload wt (lb)	0	Payload wt (lb)	650	1.100	Payload wt (lb)	0	0
Lift force (lb)	6500	Lift force (lb)	8301	1.277	Lift force (lb)	8651	1.331
Volume (ft³)	100,000	Volume (ft³)	127,712	1.277	Volume (ft³)	133,100	1.331
Surface area (ft²)	12,297	Surface area (ft²)	14,475	1.177	Surface area (ft²)	14,879	1.210
Material (lb/ft²)	0.5286	Material (lb/ft²)	0.5286	1.000	Material (lb/ft²)	0.5814	1.100
Diameter (ft), a	20	Diameter (ft), a	21.699	1.085	Diameter (ft), a	22.000	1.100
Length (ft), b	59.68	Length (ft), b	64.75	1.085	Length (ft), b	65.65	1.100
Fineness ratio	2.984	Fineness ratio	2.984	1.000	Fineness ratio	2.984	1.000
		Multiplier effect =		1.77	*Multiplier effect =*		2.31

illustrate this difference, Sample Problem #1 in Table 1.1 compares three situations for a prolate spheroid filled with a lifting gas. The base case is for a 100,000 ft^3 prolate spheroid filled with helium. Spheroid sizes and weights will be recalculated for two design changes. The first change increases the payload by 10% of the envelope weight and the second change adds weight by increasing the envelope material density by 10%. At first these may seem the same but closer analysis shows that increasing material weight increases the airship size and total weight more than the 10% payload increase does. It is worth spending some time with this example until it is clear why these two changes give different results.

Example 1.1 Sample Calculation [See Eq. (1.1)]

Multiplier Effect = (7651 − 6500)/650 = 1.77

Look at Table 1.1 and notice that the envelope weight increases by 17.7% and 33.1% for these two cases that changed weight terms by 10% in two different ways. This results in a balloon or airship *multiplier effect* of 1.77 and 2.31, respectively. These numbers are lower than for aircraft but still not insignificant. The geometry changes associated with these 10% increases in weight only added 5–6 ft in length and 2 ft in diameter and there would be no theoretical limit to these size increases. All calculations keep the fineness ratio, *FR*, constant. If another parameter is kept constant, such as volume or length instead of *FR*, then these relationships would also change.

1.5 Comparison of Modes of Transportation

Since the 1990s there has been a renewed interest in airships, and in particular non-rigid airships. What are the reasons behind this renewed interest? Figure 1.11 presents a simplistic view of the relative efficiencies of buoyant vehicles vis-a-vis high speed winged aircraft. From Fig. 1.11 one concludes that for systems where high speed (>120 kt) is not important and/or long duration is important, there is no better solution than a buoyant vehicle. However, this benefit deteriorates rapidly with an increasing need for speed such that above about 150 kt the winged aircraft will always be a more efficient mode of transportation.

How these buoyant airships relate to other modes of transportation is shown in Fig. 1.12. It is clear from this figure that airships fill a void in the famous Gabrielli–von Kármán chart that relates the specific resistance (efficiency) to speed. The relationships postulated by Gabrielli and von Kármán [1] in the early 1950s show the *specific resistance* needed to move a weight at a certain speed using various modes of transport. It is traditionally shown with the three major transportation modes sea, land, and air.

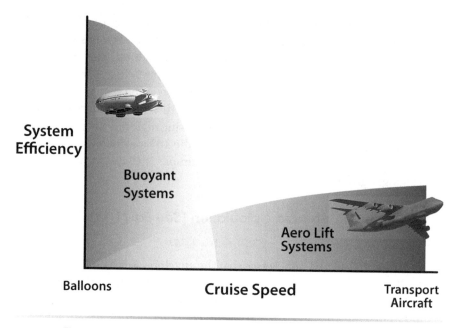

Figure 1.11   Comparison of relative efficiencies of aircraft vs buoyant vehicles.

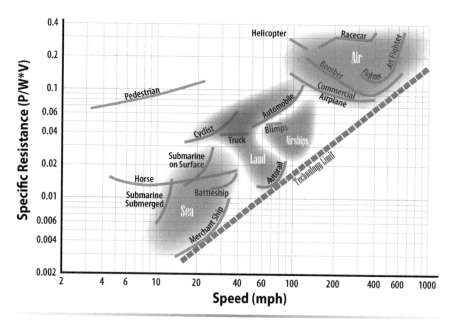

Figure 1.12   Gabrielli–von Kármán specific resistance data for transportation systems.

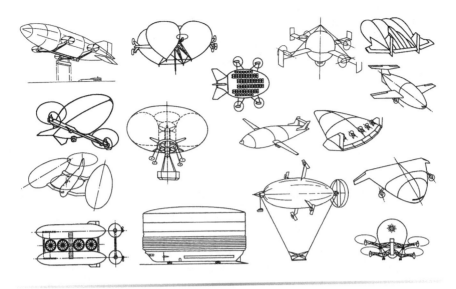

Figure 1.13 Failed attempts to blend buoyancy and fixed wing aircraft. [2]

However, notice that there is a region of operation between land and air that is uniquely filled in by airships and in particular hybrid airships. Even though buoyant vehicles are very efficient, adding buoyancy does not always result in a reasonable design. As a result, airplane designers, recognizing that a buoyant lifting gas offers efficiencies far beyond those of winged aircraft, have many failed attempts to integrate a lifting gas into an airplane. Why is this design road littered with failures such as in the examples of Fig. 1.13? The problem is that most "would be" designers fail to recognize that adding volume for buoyant lifting gases creates drag more rapidly than it creates lift until at least 50% of the lift force is generated by buoyancy. Most of the failed designs in Fig. 1.13 tried to add a modest amount of volume for a lifting gas to increase lift but it was at the expense of weight and drag. This is a losing proposition for modest volumes filled with a lifting gas.

1.6 The Hybrid Airship

Since the mid 1990s a new airship design has emerged that offers unique operational advantages over conventional airships. These new *hybrid airships* are dramatically different than a conventional ship and more capable than airship designs from 70 years ago, due to improved designs and much technological advancement. These improvements include a 20× improvement in envelope strength and improved digital flight controls, world-wide weather awareness, prediction, and dramatic improvements in landing systems. The bottom line is that hybrid airships are faster than trucks, trains, or sealift and have unique operational capabilities and flexibility not available in any fixed wing or rotary wing vehicle.

So, what is a hybrid airship and how will it make the world a better place? To answer this question, it is necessary to discuss buoyancy. Since Archimedes first postulated the laws of buoyancy nearly 2000 years ago, man has used this principle for transportation purposes. Most notably in ships, which transport most of the world's goods, it is buoyancy that allows flotation to occur. Similarly, buoyant air vehicles also float—in air, not water—with superior efficiency compared to airplanes.

The static or buoyant lift is constant and always "on." The dynamic lift (aero lift and thrust vectored lift) can be varied between on and off. Up to 40% hybrid airship liftcomes from aerodynamic lift, whereas conventional airships generate no more than 10% of their lift with aerodynamics.

When transporting goods, a hybrid's unique combination of buoyancy and varying aerodynamic lift makes it superior to its conventional airship siblings and airplane cousins. A conventional airship is nearly "neutrally buoyant" and as a result, it quickly becomes too light when offloading even modest payloads. A hybrid airship is designed to be partially buoyant, i.e. somewhat "heavier than air." This characteristic, together with its unique shape, permits it to generate larger variable amounts of aerodynamic lift. The ability to modulate this lift vastly increases a hybrid's operational flexibility and allows it to offload larger payloads without any loss of control.

A typical design layout for a hybrid airship is shown on Fig. 1.14 (see more details in Appendix D). Because of its multi-lobe shape, a hybrid airship can add features such as an Air Cushion Landing System (ACLS), which gives it a unique ability to take off, land, taxi, and park without substantial ground infrastructure. This ability is the brilliant result of the integration of hovercraft technology to the underside of an ellipsoidal or lobed shape. The capability of an ACLS was verified by Lockheed Martin Aeronautics in 2006 during the testing phase of a 3-lobe hybrid airship named P-791 shown in Fig. 1.15. With the ACLS system, a hybrid airship can easily travel to and from austere sites such as frigid Canadian mine sites, the Brazilian rain forest, Africa, China, and areas with deteriorating infrastructure such as Mexico. Yes, it can even operate from water.

The Northrop Grumman Corporation has a contract with the US Army for a long endurance multi-intelligence vehicle (LEMV), which is being designed and built by Hybrid Air Vehicles of the UK. The requirements for the LEMV are to operate unmanned for 21 days at 20,000 ft carrying a 2500 lb payload of sensors and communications equipment [6]. The LEMV is shown on the Chapter 6 title page. It first flew August 7, 2012, and is currently in flight test.

Over the last two decades natural and perceived operational issues have been explained, debunked, or overcome through judicious application of modern engineering and technology. It is a "green" machine that is very

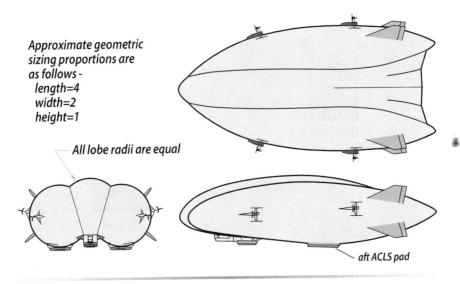

Approximate geometric
sizing proportions are
as follows -
 length=4
 width=2
 height=1

All lobe radii are equal

aft ACLS pad

Figure 1.14 Typical layout of a hybrid airship design.

fuel efficient, relative to an aircraft, while generating little noise—it is a good neighbor and requires little, if any, supporting infrastructure. The hybrid airship is also a friend to humanitarian and disaster relief missions. Its ability to efficiently and inexpensively deliver large payloads to isolated regions, quietly, and with little site preparation is unmatched by any of today's transportation systems. It is good for people, good for countries, and good for the world. The hybrid's time appears to have finally arrived.

Figure 1.15 P-791 hybrid airship demonstrator—Lockheed Martin.

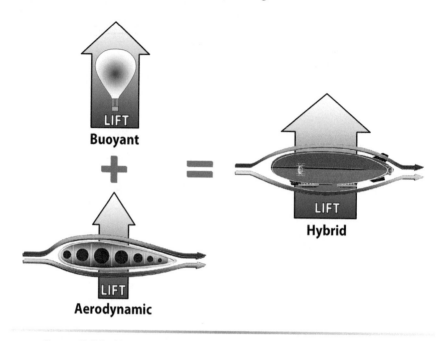

Figure 1.16   Hybrid airships combine aerodynamic and buoyant lift into one vehicle.

It is now time to review a few definitions that will be used throughout the book. First, the term, *hybrid airship* refers to any airship that is capable of generating more than 10% of its total weight with aerodynamic forces (see Fig. 1.16). Three other terms that we have used and will continue to refer to are also associated with the total lift forces and these are *buoyancy*, *buoyancy ratio (BR)*, and *heaviness*. *Buoyancy* is simply the total lifting force generated by the volume of contained lifting gas.

$$\text{Buoyancy ratio, } BR = (\text{buoyant lift, } L_{buoy})/(\text{total weight, } W_G) \quad (1.1)$$

$$\text{Heaviness} = \text{aerodynamic lift} = W_G\,(1 - BR) \quad (1.2)$$

Heaviness equals the amount of aerodynamic force during equilibrium flight or the airship's weight when on the ground at rest. Since fuel is burned during flight, the weight of an airship changes and therefore the *BR* changes. As for *BR*, only two conditions are of interest to the airship designer, the *BR@takeoff* and *BR@landing*. Discussions in Chapters 11 and 12 will provide more detail on their usefulness.

Compared to airplanes, hybrid airships are such poor generators of aerodynamic lift that very low values of *BR* result in too much drag-due-to-lift and/or the inability to generate sufficient aerodynamic lift during takeoff. For landing the lower the *BR@landing* the more weight (payload) that can be offloaded without immediately replacing it with new payload or ballast. Actually, offloading large payloads is simply a buoyancy control problem. Sometimes ballast/new payload must be onloaded at the same rate that the payload is offloaded. One of the most challenging problems for airship designers is controlling buoyancy when offloading payload. The unique ability of a hybrid airship to generate enough aerodynamic lift (which supports added payload or fuel during flight) reduces or eliminates ballast material. This design issue will be examined in greater detail in Chapters 11 and 12. Since low *BR@takeoff* also results in longer takeoff field lengths, bigger engines, and more fuel burned during the mission, a classic trade study is needed to identify the volume, fineness ratio, *AR*, *BR*, *T/W*, etc. The best combinations of these variables will vary with differing mission requirements.

Another source of vertical force that is easily integrated and adds flexibility into air vehicle designs is *thrust vectoring*, which results in unique designs that derive their vertical force from various combinations of three sources: aerodynamic lift, buoyant lift, and thrust vectoring. A Venn diagram (Fig. 1.17) shows air vehicle designs for all of the vertical force combinations. So far, the only vehicles that have successfully combined these three vertical forces have been hybrid airships.

For operators/owners, transportation system efficiency is important because it results in good productivity/throughput, yielding more revenue. An interesting comparison of transportation efficiencies is presented in

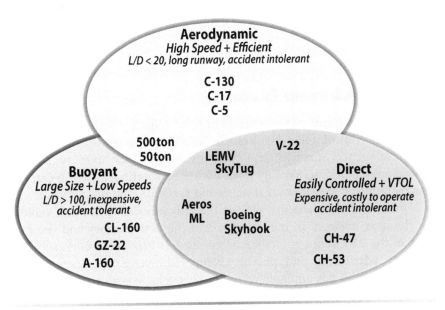

Figure 1.17 The three classes of lift generate the full spectrum of air vehicle designs.

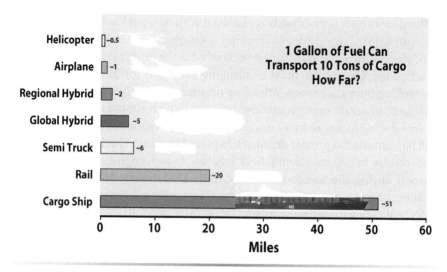

Figure 1.18 Comparison of efficiency between various modes of transport.

Fig. 1.18, which shows that a hybrid airship is not the most efficient of all known systems. If you narrow the comparison to air vehicles only, then Fig. 1.19 shows the superiority of airships and hybrid airships against their main air vehicle competition. However, other parameters are also important. Trucks and ships are more economical but suffer from longer delivery times or inability to travel from one point to another during or after a large weather event. What will be discussed in Chapter 12 is that the hybrid airship offers operational flexibility along with good efficiency that satisfies a unique market that cannot be served as well by any other transportation system.

1.7 Modern Airship Examples

A recent example of the interest in hybrid airship designs is the Long Endurance Multi-intelligence Vehicle (LEMV) program for high altitude ISR (Intelligence, Surveillance, Reconnaissance). A non-military use for a hybrid airship design (SkyTug) from Lockheed Martin focuses on carrying cargo and its operations over time should verify its commercial value.

However, hybrid airships are not the only innovative buoyant vehicles on the near horizon. In the latter part of the 20th century and the early part of the 21st century, airships were relegated to sightseeing, scientific research, advertising, and sporting events. However, a brand new buoyant vehicle program has emerged recently. This vehicle is a very high altitude airship named ISIS (Integrated Sensor Is Structure) that will be capable of station-keeping for a year or more above 60,000 ft by collecting solar energy from the sun. This energy is converted to electrical energy to power

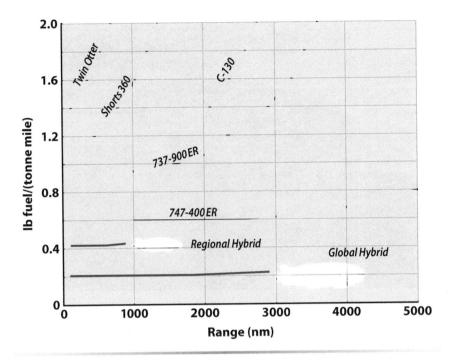

Figure 1.19   Performance comparison for various air vehicles.

electric motors and store excess solar energy in a storage device to power the vehicle through the night. An artist's rendition of ISIS is shown in Fig. 1.20. The solar power discussion and the Solar HALE example in Chapter 5 will expand upon this concept.

In 2005 a manned hybrid airship demonstrator was built to validate the hybrid airship concept. This demonstrator (P-791), shown earlier in Fig. 1.15, was built and first flown by Lockheed Martin in January 2005. Its primary purpose was to demonstrate that an Air Cushion Landing System (ACLS discussed in Chapters 9 and 12) would give the airship ground handling characteristics unlike any other airship (or airplane). The major success of this vehicle was instrumental in quieting the naysayers regarding the performance of the ACLS system on a modern airship and the possibilities of a hybrid airship for commercial and military use.

The term *hybrid airship* does not have an exact definition. For this book the term *hybrid airship* means a buoyant air vehicle with takeoff $BR < 0.9$. Other characteristics that are often associated with hybrid airships are listed below.

60% W_G < Buoyant Lift < 90% W_G

10% W_G < Aerodynamic Lift < 40% W_G (capable of modulation)

0% W_G ≤ Thrust Vectoring Lift < 15% W_G

Figure 1.20   ISIS high altitude, solar-powered airship-Lockheed Martin.

Multi-lobe configurations
Enhanced operations during landing, takeoff, and taxi.
Handling qualities more like an airplane than an airship
Rapid loading and unloading of large payloads
Reduced use of ballast weight
Operate from austere unimproved sites

Hybrid designs with $BR < 0.6$ should be reviewed closely as they are likely to have poor cruise performance due to large drag-due-to-lift and very poor takeoff performance due to a very limited takeoff C_L.

Specifically, how is a hybrid airship able to carry more payload when compared to a normal airship (body of revolution)? Figure 1.21 illustrates the hybrid advantage. This figure compares a hybrid to a conventional design operating on the same "out and back" route. Since the hybrid is able to generate much more lift it is able to carry significantly more payload. Both vehicles are assumed to be neutrally buoyant on their final landing for a consistent comparison. Remember that lift generated by a buoyant gas is very economical as it is purchased once and only requires modest topping off every month or so

Although it is not part of the traditional design effort for an airship the values of the infrared (IR) signatures are of interest to military users. From an IR standpoint the envelope is difficult to acquire as its temperature is the same as the surrounding air. The engines have thermal signatures but their low power levels make them hard to acquire as well.

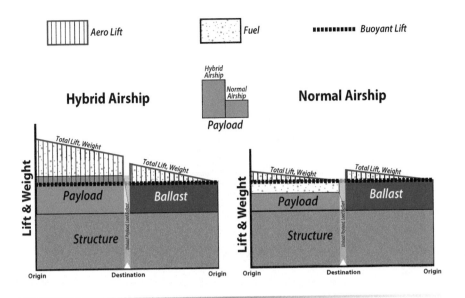

Figure 1.21 Advantage of a hybrid airship vs a normal airship in carrying payload.

to offset leakage. The disadvantage is that this buoyant lift cannot be turned off or modulated. As Fig. 1.21 shows, additional aerodynamic lift can reduce this problem but cannot eliminate it. The implications of using ballast by the hybrid airship designer will be discussed in detail in Chapter 12. It is true, however, that for unique routes which have similar payloads out and back ballast weight can be zero.

1.8 Airship Design

The design of an airship is a large undertaking by many talented engineers with expertise in the areas of aerodynamics, propulsion, materials, structures, flight control, performance, and weights. Specialists, who design such components as the crew station, landing interface system, interior layout, and equipment installation, must work together to produce the most efficient flight vehicle. It should be clear that the design process is a very involved integration effort, requiring the blending of many engineering disciplines and is essentially identical to that for a fixed-wing aircraft. However, modern day airship designers are not blessed with the rich databases available to airplane designers. If there is some data available it is probably based on very old experiments whose results are often inaccurate or suspect. This is especially true for the hybrid airship designer. What little design data there is for hybrid airships is simply not available to the general public as companies protect their private data.

Figure 1.22 Zeppelin LT doing scientific research in Botswana.

The key figure in the design process is the design team leader, IPT leader, or Chief Designer who acts as the main integrator and arbitrator. The Chief Designer is usually one who understands and appreciates all disciplines involved in the design process. He is often called upon to negotiate compromises among the design groups. For example, the propulsion group might propose thrust vectoring as a means of controlling the airship throughout its flight envelope. At the same time, the structures group might recommend a fineness ratio (*FR*) of 3.0 while the aerodynamics group might want the *FR* to be 4.5. The flight control group might propose putting an "X" tail on the design and the sensors group would prefer an inverted "Y" tail. The Chief Designer will bring about the best compromise among the design groups towards the design goal. An example of a modern airship design is the Zeppelin LT shown in Fig. 1.22.

1.8.1 Performance vs Cost

Airships will provide excellent cost vs performance. However, airships are only a realistic option when speed is relatively unimportant or not important at all. Mission scenarios requiring long duration are best satisfied by a buoyant vehicle such as an airship. Several studies over the last decade have shown that hybrid airships show significant benefits when delivering perishable payloads or operations involving austere sites. Other studies have also shown cost benefits for operations to and from mining sites such as in northern Canada. Cost will be discussed in more detail in Chapter 11, where it is shown that development costs for an airship program are about 10% of an equivalent aircraft development program.

1.8.2 There Is Never a Right Answer

In the design of an aircraft there is never a right answer—only a best answer at a given point in time [4]. The same can be said for an airship. This is because in aircraft and airship design a balance is always sought between competing requirements whose priorities change with time. An air vehicle might be designed to certain government technical and economic requirements, but if the government administration changes, then the requirements priority may change. The advice to the designer is to remain flexible and develop as robust a design as possible so that it will survive should the requirements change. The watchwords are *compromise, balance, and flexibility.*

1.9 Overall Design Requirements

Before an architect designs a building, he must first establish who and how many will occupy the building, what is its purpose, what is its scale, cost target, etc. The designer of an aircraft or airship must have similar requirements established before a design can proceed. The requirements define: (1) what mission will be performed, (2) how much it can cost, (3) how it should be maintained and supported, and (4) the schedule.

1.9.1 Mission Requirements

The mission requirements identify the following:

- Purpose: logistical transport, ISR (intelligence, surveillance and reconnaissance), etc.
- Manned or unmanned
- Payload: passengers, cargo, sensors, etc.
- Speed: cruise, maximum, loiter, landing, winds, etc.
- Range or radius
- Endurance or loiter (time-on-station)
- Takeoff area (vertical, short, conventional—takeoff and landing)
- Signature level: Not a design issue for airships—inherently low IR

1.9.2 Cost Requirements

The cost requirements for both aircraft and airships encompass the following:

- Development cost
- Acquisition cost
- O&S (operation and support) cost
- LCC (life cycle cost which is the sum of development, acquisition and O&S)
- CAIV (Cost As an Independent Variable) for government programs

1.9.3 Maintenance and Support Requirements

The maintenance and support requirements for aircraft and airships are as follows:

- Maintenance manhours per flight hour (MMH/FH, *maintenance index*)
- Ground support equipment (GSE)
- Maintenance levels (i.e. organizational, intermediate, and depot)
- Integrated Logistics Support (ILS) plan
- Contractor support or user support

Generally speaking maintenance and support is somewhat less costly for an airship than for an airplane. This is so because of the much higher flight hours (FH) giving a lower maintenance index and fewer landings and takeoffs. As a general rule the mission profiles are much less stressing than for aircraft.

1.9.4 Scheduling Requirements

The schedule requirements for aircraft and airships depend on the following:

- Development and test scheduling
- Product availability—when the airship should be available for deployment, Initial Operational Capability (IOC), to the warfighter or the commercial customer

1.9.5 Sources for Design Requirements

In the case of a commercial program, requirements are established by the airship designer based on input from potential users and customers. The airship company performs market analyses to determine what the public's needs or desires will be in the near future. Projections are made for future passenger travel or air freight needs. The commercial program starts up when a customer shows serious intent to buy the production airplane. A down payment usually entitles the customer to influence some of the requirements. Careful thought and research go into establishing the requirements because if they are inappropriate, then the airship (if it is fortunate enough to be built) may not find new customers or keep its initial customers.

Sometimes requirements are established by a military user such as the U.S. Air Force, U.S. Navy, U.S. Marines, U.S. Army, DARPA, etc. These requirements are usually developed to fill a military need (shortfall) or replace an obsolete system. These requirements are termed a "requirements pull" because the military need "pulls" the requirements.

Sometimes a new technology will push the requirements for a new aircraft (termed a "technology push"). The jet engine in the early 1940s, the stealth technology in the mid-1970s and the high energy airborne laser in the early 1990s are examples of technology push requirements which led to the XP-80, the Have Blue/F-117, and the YAL-1/ABL aircraft respectively.

The requirements usually come with a document called a Concept of Operations or ConOps for short. The ConOps describes how the airship will be deployed, operated, maintained, and supported . . . essentially all the information the designer needs to complete the design. The ConOps is helpful for a commercial airship but is essential for a military airship. For example the military airship designer needs to know if the threat defenses will be "up and in-place" or rolled back and what the maintenance concept will be.

1.9.6 The Need to Question the Requirements

When the requirements arrive, the designer MUST study them, understand them, evaluate them, and question them . . . and if necessary negotiate with the customer about their feasibility. Because, if the designer does not agree with the requirements he must walk away. This is very difficult to do. Disagreement with the fundamental requirements will sap the designer's passion and commitment, which are necessary to generate a successful conceptual design that will ultimately be selected to proceed into preliminary design.

Even when the customer tries very hard to generate credible requirements, as history shows, sometimes requirements are flawed. Some flawed requirements are discovered and changed, while others prevail and designs are produced. And some are ignored—this one is always risky. An example of a flawed requirement was the customer speed requirement of 132 kt for the Aerocraft Program conducted by Lockheed Martin in 1998–2000. The program evaluated numerous technical designs and their economies of transporting perishable goods across the Atlantic or Pacific Oceans. Realistic cost estimates ultimately showed that there was only a modest reduction in transportation costs compared to jet aircraft. Later designs operating at 80 kt or below showed significant cost reductions. Airship performance is very sensitive to maximum speeds so they should be selected with extreme care. The effects of speed on airship weight are discussed Chapters 6, 9, and 12.

1.9.7 Measure of Merit (MoM)

The airship design must meet (or exceed) the stated requirements in order to be acceptable to the customer. Meeting the requirements is a necessary condition for being a candidate to proceed to the next phase. If there is a requirement that the designer cannot meet or thinks is unrealistic, then he needs to petition the customer for a waiver.

The Measure of Merit, or MoM (sometimes called Figure of Merit), is similar to a requirement except that it is initially known only to the customer and is not overtly specified. The MoM is important to the customer and will be used by him as a "tie breaker" in his selection of the winning design. It is often said that meeting the requirements gets you invited to the dance . . . but meeting the MoM gets you out on the dance floor.

Since the MoM is initially unspecified, the designer (or someone in his design group) must do the homework to understand what the customer is really looking for. Sometimes the MoM is simply that the design must be aesthetically pleasing. But more often the MoM is more substantive and is learned by developing a close rapport with the customer. It goes without saying that developing a design to the wrong MoM will lose any contract.

1.10 Unmanned Aerial Vehicle (UAV)

The DoD defines a UAV as a powered unmanned aerial vehicle that uses aerodynamic forces to provide some control, can fly autonomously or be piloted remotely, is expendable or recoverable, and can carry a payload. With this definition aerostats and airships may be classified as UAVs, but an aerostat is similar to a balloon on a long tether. The design of unmanned and manned airships is the same in that they must obey the same laws of physics, but there the similarity ends. Each has advantages over the other. We should use unmanned airships where they have an advantage over their manned counterparts and vice versa. Since one of the main benefits of an airship is its long endurance, designing for unmanned operations is usually a winning strategy.

The main disadvantage of the unmanned airship system (and hence the manned airship advantage) is that it cannot think for itself and cope with unforeseen or dynamically changing events. No amount of autonomy and artificial intelligence can address all the uncertainties of war or unforeseen environmental events. Because of this shortcoming, unmanned systems will always have off-board human operators in the loop. This means that the unmanned system must have additional sensors and data-link capability onboard to make the off-board human operator aware of the situation at all times.

The advantages of an unmanned system are as follows:

1. The design of the unmanned system is not limited by the requirement to carry a human onboard and accommodate his frailties.
2. No human is at risk of capture.
3. No infrastructure is required to recover the crew if the airship crashes.
4. The unmanned airship does not need to fly to keep the unmanned system proficient.

1.10.1 Design Limitations—Human Operator

This feature has both pluses and minuses. On the plus side the unmanned airship does not have to accommodate a crew station. The elimination of the crew station also shortens the gondola a little thus reducing the empty weight. There is a modest cost reduction as well. Airship crew stations are not highly integrated into the design so there is little cost saving associated with designing the crew station into an underslung gondola or cargo bay.

Not having to "man-rate" the aircraft will simplify the design and development of the unmanned UAV somewhat. There will be a cost savings due to not having to man-rate the engine (engine testing), and the elimination of a crew escape system, and crew survival equipment design and test.

However, the down side to not having a human onboard is the requirement to recover pilot functionality by having an off-board operator that has complete situational awareness. This means increasing the software development for autonomous flight, adding sensors and data links, and of course adding a ground station to the overall system development cost. The consensus of knowledgeable aerospace professionals is that all of the plusses and minuses together will only give a modest cost reduction to the development and acquisition cost of an unmanned airship relative to a manned airship. Besides long endurance most advantages to unmanned airships tend to be political and will be discussed below.

1.10.2 Elimination of Search and Rescue

The elimination of the infrastructure to search for and rescue downed crewmen is a real opportunity for cost saving. In addition, the attention given a downed crew results in a significant resource shift away from combat operations. Since the crew has been eliminated from the airship, the political sensitivity of the unmanned mission is reduced as there is no crew to be held hostage and identified with a country (e.g. Gary Powers, 1960).

◢◣*1.10.3* Training and Proficiency

The fundamental premise is that the unmanned airship does not need to fly for the unmanned system to train or stay proficient. The human operator is off-board and trains by simulation. On the other hand, the ISR unmanned airship flies all the time since its flights are for the purpose of gathering continuous intelligence on target countries. Critics accept this premise but argue that the unmanned airship needs to fly during peacetime as well. As part of a combined arms team, the unmanned airships have to operate with the manned airships as the humans train. This argument fails when the capability of modern simulators is recognized. This notion of no (or at least minimal) peacetime flying presents a tremendous cost saving opportunity for unmanned airships and in general for any UAV.

◢◣*1.11* Specifications, Standards, and Regulations

The U. S. government regulates the operation of all airships in the United States by a system of specifications, standards, and regulations. An airship designer must not only meet (or exceed) the requirements discussed earlier, but must also comply with all the appropriate aircraft specifications, standards, and regulations if the aircraft is to be operated in the United States. The regulation of military aircraft is administered by the Department of Defense through the Department of Defense Specifications and Standards System (DODSSS) and of civil commercial aircraft by the Department of Transportation through the Federal Aviation Regulations (FAR).

Specifications are procurement documents which describe the essential and technical requirements for airship items, materials or services, including the procedures by which it will be determined that the requirements have been met. Standards establish engineering and technical limitations and applications for items, materials, processes, methods, designs, and engineering practices.

The documentation of design standards for airships is reported in the Federal Aviation Regulations FAA 8110-2 as well as CS-30T which was created by the European Aviation Safety Agency (EASA) and is temporarily recognized by the FAA. A full listing of potential DoD documents that may impact an airship's design are shown in Table 1.2.

It is a fact that the specifications and standards for aircraft are numerous. The number of regulations for airships is much less and often use modified versions of those for aircraft when appropriate. It has been asserted that military and FAA specifications and standards are excessive

Table 1.2 Partial Listing of Military Specifications and Standards—Aircraft Design

Doc number	Title
MIL-HDBK-1797	Flying Qualities of Piloted Airplanes (replaced MIL-F-8785C)
MIL-F-83300	Flying Qualities of Piloted V/STOL Aircraft
MIL-F-9490	Flight Control Sys-Design, Installation and Test of Piloted Aircraft
MIL-S-8369	Stall/Post-Stall/Spin Flight Test Demonstration Reqs for Airplanes
MIL-C-18244	Control and Stabilization Systems: Automatic, Piloted Aircraft
MIL-D-8708	Demonstration Requirements for Airplanes
MIL-I-8700	Installation and Test of Electronics Equipment in Aircraft
MIL-P-26366	Propellers, Type Test of
MIL-S-18471	Seat System, Ejectable, Aircraft
MIL-W-25140	Weight and Balance Control Data
MIL-STD-850	Aircrew Station Vision Req for Military Aircraft
MIL-STD-757	Reliability Evaluation from Demonstration Data
MIL-C-5011	Charts; Standard Aircraft Characteristics and Performance
MIL-STD-881	Work Breakdown Structure (WBS)
MIL-HDBK-516B	Airworthiness Certification – U.S. Tri-service
FAA 8110-2	Airship Design Criteria
CS-30T (EASA)	Certification Specifications for Transport Airships

and are part of the reason for the high cost of air vehicle systems. Companies spend considerable time and money in military "spec" compliance. The authors will remain neutral in this matter but suggest that the reader examine this issue and become involved.

1.12 Airship Design Phases

Design is the name given to the activities that create a new flight vehicle. It starts as a vision and finishes with the final inflation and integration of all major systems, subsystems, and components. It is the most important time in the life cycle of an airship as all its features both good and bad are locked in at this point. The design process is usually divided into the following three phases.

Conceptual Design
Preliminary Design
Detail Design

Although the specific activities during these three phases vary from one design group to another, they are generally formed as shown in Fig. 1.23.

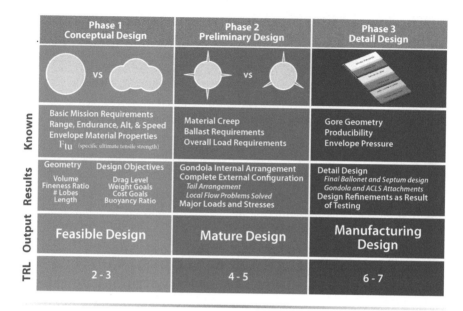

Phase 1 Conceptual Design	Phase 2 Preliminary Design	Phase 3 Detail Design
Known Basic Mission Requirements Range, Endurance, Alt, & Speed Envelope Material Properties F_{tu} (specific ultimate tensile strength)	Material Creep Ballast Requirements Overall Load Requirements	Gore Geometry Producibility Envelope Pressure
Output Results Geometry Design Objectives *Volume* *Drag Level* *Fineness Ratio* *Weight Goals* *# Lobes* *Cost Goals* *Length* *Buoyancy Ratio*	Gondola Internal Arrangement Complete External Configuration *Tail Arrangement* *Local Flow Problems Solved* Major Loads and Stresses	Detail Design *Final Ballonet and Septum design* *Gondola and ACLS Attachments* Design Refinements as Result of Testing
Output Results Feasible Design	Mature Design	Manufacturing Design
TRL 2 - 3	4 - 5	6 - 7

Figure 1.23 Comparison of the three design phases for an airship.

The TRL (Technology Readiness Level) used in Fig. 1.23 is an expression for the maturity of the technologies used in the three design phases [5].

1.12.1 Conceptual Design Phase

The conceptual design phase determines the feasibility of meeting the requirements with a credible airship design. The conceptual design process is shown schematically in Fig. 1.24. The general size and configuration of the airship along with a credible layout of the major systems are identified during this phase.

As we've discussed, the designer's first task is to study, evaluate, understand, question, and if necessary negotiate the requirements (or at least ask for a waiver). The requirements are flowed down to the design group in the System Requirements Document (SRD). The SRD lays out the ground rules for the design study along with information about the *Measures of Merit* (MoMs), program strategy, selection criteria, significant design decisions, and assumptions about technologies.

It is a good idea at the very beginning to have brainstorming sessions to identify all possible solutions to the design problem. These sessions need to be an open exploration of any and all concepts. Both left and right brain thinkers should attend as well as any person who will impact the design, e.g. engineers, maintenance, manufacturing, and cost personnel.

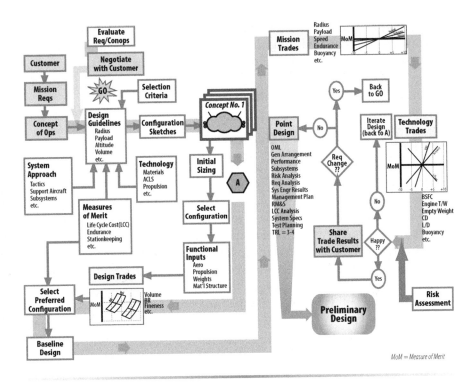

Figure 1.24 The conceptual design process.

Design trade studies are conducted around the more promising concepts using preliminary estimates of aerodynamics and weights to converge on the optimal volume, fineness ratio, and operational buoyancy range. Engine size, number of engines, and their locations are also evaluated for the best combination of cruise efficiency and thrust vectoring for control. Aerodynamic control surfaces are sized based upon the static stability and control considerations over the speed range from zero to the maximum. The performance requirements are varied (called mission trades) to determine the impact of each performance item on the airship volume, weight, and cost. This information is then shared with the customer to make sure he understands the penalty each requirement imparts to the design. The technologies being considered in the design are examined (called technology trades) and estimates made of their "maturity" (probability of success) and the consequence of their not meeting the required maturity level. The results of the technology trades form the design risk analysis.

The first look at cost and manufacturing is also made at this time. Only gross structural aspects are considered during the conceptual design phase as resources are usually limited and the design is changing often. The ability of the design to accomplish the given set of requirements is established during this phase, but the details of the configuration are subject to change. Most of the work done during this phase is done by a group of 15 to 40 people over a year. It should be emphasized that the cost of making a design change is small during conceptual design, is large during preliminary design, and is very large during detail design.

1.12.2 Preliminary Design Phase

The best configuration defined by the MoMs from the conceptual design phase is now fine tuned using wind tunnel parametric testing and computational fluid dynamics (CFD). This fine tuning is accomplished with a wind tunnel model capable of representing the general configuration with the provision for minor variations in body shape and tail arrangement. The engine is selected and the propellor is sized. Major loads and stresses in the envelope are determined along with considerable structural design.

Refined weight estimates are made and a more thorough performance analysis conducted. Dynamic stability and control analysis influences are determined and 6-DOF (degrees of freedom) simulations are conducted to establish flight control requirements and handling quality levels. The three trade studies (design, mission and technology) started in the conceptual design phase are continued but with more vigor.

The design is given serious manufacturing consideration with preliminary plans for jigs, tooling, and production breaks. Refined cost estimates are also made. Clearly the resources for the preliminary design phase are greater than the conceptual phase and typically number 100 or more people over about 1–2 years.

1.12.3 Detail Design Phase

In the detail design phase, the configuration is "frozen" and the decision has been made to build the airship. Detailed structural design is completed. All of the detail design and shop drawings of the joints, fittings, and attachments are accomplished. Interior layout is detailed as to location and mounting of equipment, hydraulic lines, ducting, control cables, and wiring bundles. Mockups are rarely required for the internal arrangements of airships due to the large volume available. The drawings for the jigs, tooling and other production fixtures are done at this time. A detailed cost estimate based upon Work Breakdown Structure (WBS) is made. All equipment and hardware items are specified. It is important that from this point on that the design changes be kept to a minimum because the cost of making a change is large once the drawing hits the shop floor. The next step is ordering all the equipment items (called Bill of Materials) and the fabrication and assembly of the prototype (usually at least two prototypes are built). Often, the fabrication of some components will be started during this phase as soon as their shop drawings are released. Usually there are a few items which are called *long lead items* that have to be ordered before detailed design starts to prevent schedules from being increased.

Figure 1.25 shows the three phases of design in a typical government program acquisition according to DoD 5000.1. The years shown are extremely optimistic since there are always breaks in the schedule while the government issues a Request for Proposal (RFP), industry submits proposals, the government evaluates the proposals, selects a winner and gets its funding in place. Commercial programs move much faster since the aircraft builder controls the tempo and funding of the program. Typical times from the decision to build the airship (Milestone 1 or B for the government and the start of preliminary design for commercial) to production is about four years for the government and three years for commercial. FAA certification can add a year to the commercial effort.

Figure 1.25 also shows the importance of the conceptual design phase in that over 70% of the design features that drive Life Cycle Cost (LCC) are selected during that phase.

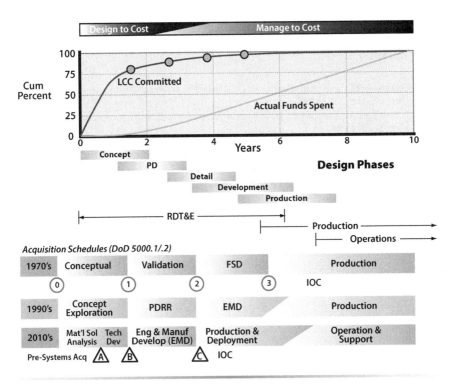

Figure 1.25 Design phases integrated into the entire government program.

1.13   Scope of the Text

This is the second volume of a two-volume set

Volume 1 —Aircraft Design

Volume 2 —Airship Design and Design Case Studies

Volume 1 covers the conceptual design phase of the aircraft design process. It is self contained as the chapters and appendices will lead the reader through one iteration of the conceptual design process. Volume 1 will give the reader an understanding and appreciation of how the different disciplines must blend together to produce an effective aircraft.

Volume 2 (this volume) is also written as a stand-alone volume; in some instances the text is based on Volume 1 but is rewritten for airship design. Virtually all text is new as it focuses on the unique design issues of airships, hybrid airships, and high altitude balloons. The second half of the book is comprised of nine sections discussing individual design case studies for both aircraft and airship programs. These case studies include the following air vehicles: 1) SR-71, 2) JSF/X-35, 3) Boeing 777, 4) Honda Jet, 5) Hybrid Airship, 6) Daedalus (human-powered aircraft), 7) Cessna 172, 8) T-46 Trainer, and 9) Hang Gliders. The authors of these case studies are highly regarded authors who have intimate knowledge of each vehicle through years of study or actually from working on the vehicle.

References

[1]   Gabrielli, G., and von Kármán, T., *What Price Speed*, Mechanical Engineering, Oct. 1950, pp. 775–781.

[2]   Khoury, G.A., and Gillett, J.D., *Airship Technology*, Cambridge Aerospace Series 10, New York, 1999.

[3]   Galilei, Galileo, *Two New Sciences* (1638); translated by Henry Crew and Alfonso de Salvio, McMillan, 1917.

[4]   Nicolai, L.M., and Carichner, G.E., *Fundamentals of Aircraft and Airship Design: Volume 1—Aircraft Design*, AIAA, Reston, VA, 2010.

[5]   Moorhouse, D.J., *Detailed Definitions and Guidance for Application of Technology Readiness Levels*, AIAA Journal of Aircraft, Vol. 39, No. 1, pp. 190–192, AIAA, Reston, VA, 2001.

[6]   Adams, Paul A, *Airship—The Journal of the Airship Association*, The Airship Association ltd., Wiltshire, UK, Dec. 2010.

- Classes of Buoyant Vehicles
- Characteristics of Air
- Laws of Physics that Govern the Behavior of Gases

When people think of airships, the Hindenburg disaster on May 6, 1937 usually comes to mind. This is unfortunate. The Hindenburg was a magnificent design and a superb engineering feat. In truth the disaster was the result of poor judgment on the part of German Chancellor Adolph Hitler (see box at end of chapter).

The mind is a fire to be kindled . . . not a vessel to be filled.

Plutarch

2.1 Introduction

A n airship is a unique flying machine that is very different from an aircraft with wings. Airships generate most of their lift using a lifting gas such as hydrogen or helium. Because bodies of revolution are inefficient producers of aerodynamic lift (very low L_{aero}/D), ellipsoid-shaped airships are designed such that most of their lift comes from a buoyant gas.

Historically, there have been many names given to buoyant vehicles that depend on their operations, structure, and speed. Fundamentally, there are three distinct classes of buoyant vehicles. First, there is the *balloon* that is untethered, free floating, and uncontrolled (Fig. 2.1). Next there is the tethered balloon shown in Fig. 2.2 that is known as an *aerostat*. The aerostat is fixed to the ground by a long cable that also serves as its power and communications backbone. The last category is the *airship* that has also been called a blimp or dirigible in the past. The terms blimp and dirigible are rarely used now. This book will concentrate on the fundamentals for designing all airships but will discuss in detail hybrid airship and balloon design in Chapters 12 and 13 respectively.

Aerostatics is the study of gases, their characteristics for changing pressure and/or temperature, and the lift they generate when enclosed. As

Figure 2.1   Scientific balloon launch.

Figure 2.2   Aerostat operating along southern U.S. border.

opposed to aerodynamics, which studies gases and bodies in motion, aerostatics concentrates on gases and bodies that do not move or move slowly. This chapter will present the principles of aerostatics and how they affect the behavior and performance of airships, aerostats, and balloons. All of the equations needed to calculate the buoyant lift of an airship under various temperature, pressure, and humidity conditions are developed.

Buoyant lift for any vehicle is generated by the density differences between the enclosed envelope volume of lifting gas and the outside ambient air. This vertical force (lift) is the result of a body submerged in a higher density fluid (liquid or gas) and was mathematically developed by Archimedes about 2300 years ago. Archimedes postulated and then later proved that the buoyant force depended only on the body's volume and the density difference between the submerged body and its surrounding liquid. The buoyant force was independent of shape!

Archimedes' principle states that *a body immersed in a fluid is buoyed up by a force equal to the weight of the displaced fluid. Independent of shape and composition, objects of equal volume in the same fluid experience equal buoyant forces.*

The revelation was that this simple relationship was independent of body shape! The derivation of the fundamental equation for calculating buoyant lift using the Navier-Stokes equation is shown as Eq. (2.1).

2.2 Aerostatics for Airships

Because the characteristics of air are important for determining airship buoyancy, it is important to know its composition and molecular weight. Although the composition of air varies slightly with altitude, the values shown in Table 2.1 will be assumed constant from sea level to 10,000 ft. Table 2.1 shows the gaseous content of air that is primarily made up of nitrogen and oxygen.

Table 2.1   Gaseous Content of Air [1]

Sea level with 0% air*–water vapor	
Nitrogen	78.00%
Oxygen	20.95%
Argon	0.91%
Carbon dioxide	0.03%
Hydrogen	0.01%
Other gases	0.10%

*The molecular weight for air is 28.8.

Hydrostatics is the study of fluids at rest. For this special case, $u = 0$ in the Navier-Stokes equations, the external force $F = g$ and the pressure gradient is just dp. The Navier-Stokes equations of motion are reduced to Eq. (2.1).

Example 2.1   Sample Buoyancy Calculation (See Fig. 2.3)

What is the buoyant force for 1 ft^3 of helium at sea level?

$$\frac{\partial u}{\partial t} + (u \bullet \nabla)u = -\frac{1}{\rho}\nabla p + F + \frac{\mu}{\rho}\nabla^2 u$$

$$0 = -\frac{1}{\rho}\nabla p + F$$

$$0 = -\frac{1}{\rho} + \frac{dp}{dz} + (g)$$

$$dp = \rho\, g\, dz$$

$$\text{Force} = p \times \text{area} = \rho\, g\, \underbrace{z \times \text{area}}_{\text{volume}} \qquad (2.1)$$

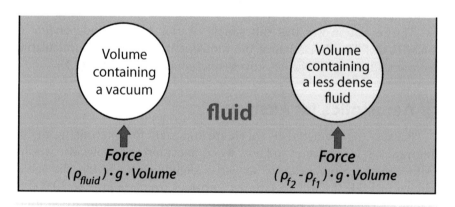

Figure 2.3   Buoyancy forces acting on a submerged body.

Force = $(\rho_{f2} - \rho_{f1}) \times g \times$ Volume

Force = $(0.002377 - 0.000327) \times 32.17 \times 1.0$ (note that density is in slugs/ft³)

Force = 0.0659 lb$_f$/ft³ (helium) or = 0.0711 lb$_f$/ft³ (hydrogen)

Force = 0.0646 lb$_f$/ft³ @ 98% purity for helium

All liquids and gases are fluids. For discussion purposes it can be assumed that gases and liquids are fluids that have significantly different densities. Gases are different from liquids in that they are compressible and change volume rapidly with changes in temperature or pressure. Liquids, on the other hand, change little over large changes in temperature or pressure. Compressibility has little effect on buoyancy and will not be taken into effect in ordinary airship design efforts. The change in density as temperature changes is very significant and is the primary gas property that governs airship design.

Density is the weight per unit volume or more accurately the force of gravity acting on the mass of a unit volume. Units are typically kilograms per cubic meter or pounds per cubic foot. A comparison of the densities of gases important to airship performance are shown in Table 2.2. Water is included only as a reference point. Specific gravities of gases use the value for dry air as the reference.

There are several laws of physics that govern the behavior of gases, which are summarized as follows. The most fundamental is Charles' Law as it governs how gases behave at constant pressure. Non-rigid airships are designed so that their internal gases are kept at a constant pressure using ballonets for controlling the pressure. The behavior and performance of aerostats and balloons are also a direct result of Charle's Law. Why modern airship designs are almost all non-rigid will be discussed in detail in Chapter 8.

- *Boyle's Law:* For a fixed amount of an ideal gas kept at a fixed temperature, P (pressure) and V (volume) are inversely proportional. Stated another way, the product of absolute pressure and volume is always constant. This law was named after chemist and physicist Robert Boyle, who published the original law in 1662 and appears as Eq. (2.2).

Table 2.2 Density of Various Gases and Fluids

Fluid	Density		Specific gravity
Water	62.428 lb/ft³	1000 kg/m³	1.0 (liquid)
Water vapor	0.05045 lb/ft³	0.80813 kg/m³	0.625
Air (dry)	0.08072 lb/ft³	1.293 kg/m³	1.0
Air (60% rh)	0.08047 lb/ft³	1.289 kg/m³	0.9969
Helium	0.01114 lb/ft³	0.1778 kg/m³	0.1375
Hydrogen	0.00562 lb/ft³	0.09002 kg/m³	0.0696

$$P\,V = \text{constant} \qquad (2.2)$$

- *Charles' Law:* At constant pressure, the volume of a given mass of an ideal gas increases or decreases by the same factor as its temperature on the absolute temperature scale. It is an experimental gas law that describes how gases tend to expand when heated. This law is based on unpublished work by Jacques Charles in the 1780s and appears as Eq. (2.3).

$$(T_2)(V_1) = (T_1)(V_2) \qquad (2.3)$$

- *Dalton's Law:* The pressure of a mixture of several gases in a given space equals the sum of the pressures that each gas would individually exert occupying that same space. This empirical law was observed by John Dalton in 1801 and appears as Eq. (2.4).

$$P_{total} = p_1 + p_2 + \cdots + p_n \qquad (2.4)$$

- *Joule's Law:* The internal energy of an ideal gas is independent of its volume and pressure, depending only on its temperature. This law is based on observations and studies performed in the 1840s by James Joule.
- *Pascal's Law:* The pressure of a fluid due to external pressure on the walls containing it is the same throughout the fluid and appears as Eq. (2.5).

$$(P_1)(V_1) = (P_2)(V_2) \qquad (2.5)$$

These basic gas laws govern the movement, behavior, and performance of airships and submarines. Submarine behavior and the governing equations of motion are similar to airships if one substitutes sea water for ambient air and air is the lifting gas instead of helium. Virtually all equations of motion can be used for either vehicle. One main difference is that as a submarine goes deeper the water density stays relatively constant (although the pressure changes significantly). Liquids do not behave exactly like gases so some motions of the submarine can be quite different from that of an airship. For instance, water is about 770 times more dense than air for a submarine and air is about 7.3 times more dense than helium for an airship. When the concept of added mass is discussed in Appendix C it will be noted that the displaced water for the internal volume of a submarine is far more significant than the displaced air for the internal volume of an airship. Hence, the added mass terms are larger for a submarine than for an airship.

Because lifting gases are responsible for creating most of the lift force, the equations that calculate buoyant lift must be developed. The basic equation in Eq. (2.1) assumes that the ambient outside air is dry, the lifting gas is pure, and the lifting gas occupies 100% of the volume inside its container (see Fig. 2.3).

The buoyant force is easily calculated using the equation developed from the Navier-Stokes equation in Eq. (2.1) and is generalized into Eq. (2.6).

This equation says that the buoyant force = volume × (density of ambient air–density of internal lifting gas).

$$\text{Buoyant Lift} = \text{Volume}\,(\rho_{air} - \rho_{gas}) \qquad (2.6)$$

where density, ρ, is in lb/ft^3 (if density is in $slug/ft^3$ then don't forget to multiply by $g = 32.174\ ft/s^2$).

This equation assumes that the air and lifting gas have the same temperature and that the pressure is constant. If the air and lifting gas temperatures are changed by the same amount, then there is no change in the buoyant lift.

$$L = V(\rho_a - \rho_g)$$
$$L = V(T/T_0)(\rho_a\,T_0/T - \rho_g\,T_0/T)$$
$$L = V(T/T_0)(\rho_a - \rho_g)(T_0/T)$$

This becomes the same as Eq. (2.6) and therefore has the same lift.

$$L = V(\rho_a - \rho_g)$$

Several sample problems will now be used to illustrate the variations in pressure, volume, density, and buoyant lift.

Sample Problem 2.1: Change in Volume Due to Change in Temperature

There is an elastic container whose initial volume is 1000 ft^3 and the internal gas temperature is 59°F. If the gas temperature increases by 20°F what is the volume after the gas expands? See Fig. 2.4 for an illustration of this problem and the governing perfect gas relationships.

The volume varies directly with the absolute temperature of the gas so,

Gas absolute temperature $= 459.7 + 59 = 518.7°R$

Heated gas absolute temperature $= 518.7°R + 20°F = 538.7°R$

New volume becomes $\quad V_2 = V_1 \times (T_1/T_2)$

$$= 1000 \times (538.7/518.7) = 1038.6\ ft^3$$

or, change in volume is $\Delta V = V_1 \times (\Delta T/T_1) = 1000 \times (20/518.7) = 38.6\ ft^3$

Sample Problem 2.2: Change in Density Due to Change in Temperature

If the 1000 ft^3 elastic container is filled with helium, what is the change in density when the temperature of the helium is increased from 59°F to 79°F?

Density of helium is 0.01114 lb/ft^3 @ 59°F (see Table 2.2)

New density at 79°F $= 0.01114 \times (518.7/538.7) = 0.01069\ lb/ft^3$

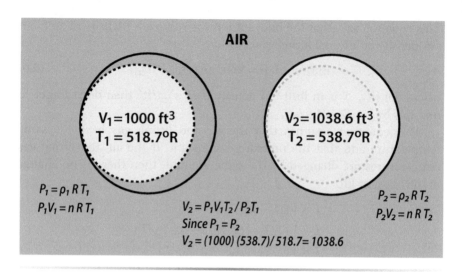

Figure 2.4   Effect of increasing the temperature of a lifting gas in an elastic container.

It is now time to expand Eq. (2.1) to put it into terms that are more easily known. Because density is not something that is easily measured, it is better to use parameters such as pressure and temperature that are known. Recognizing that

$$P = \rho RT \text{ and } \rho = P/RT \text{ and therefore } \rho_0 = P_0/RT_0$$

$$\text{Solving yields, } \quad \rho = P/[(P_0/\rho_0 T_0)T]$$

$$\text{Rearranging yields, } \quad \rho = (\rho_0 T_0/P_0)(P/T) \tag{2.7}$$

For convenience we now define the constant term $G = (\rho_0 T_0/P_0)$

Eq. (2.7) then becomes

$$\rho = GP/T \tag{2.8}$$

Table 2.3 shows values of G for various gases and units.
Using Eq. (2.5) the lift can be calculated in Eq. (2.9)

$$L = (G_a P/T - G_g P/T)V = VP/T(G_a - G_g) \tag{2.9}$$

Introducing the quantity *specific gravity,* $S = \rho_g/\rho_a$

$$L = V(\rho_a - \rho_g) = V(\rho_a - \rho_a S) = V\rho_a(1 - S) = [G_a PV/T](1 - S) \tag{2.10}$$

There is another correction to the buoyant lift equation that is straight-forward yet important. This is generally referred to as *superheat* and represents the higher temperatures of the internal gases relative to the

Table 2.3 Gas Properties for Airships

@ Sea level std day (59°F/15°C)				
Gas	Density (lb/ft^3)	Temperature (°R)	Pressure (lb/ft^2)	G ($\rho_o T_o / P_o$)
air	0.07648	518.7	2116.2	0.018746
He	0.01057	518.7	2116.2	0.002591
H$_2$	0.00533	518.7	2116.2	0.001306
@ 32°F/0°C				
Gas	Density (lb/ft^3)	Temperature (°R)	Pressure (in. Hg)	($\rho_o T_o / P_o$)
air	0.08066	491.7	29.92	1.3263
He	0.01114	491.7	29.92	0.18318
H$_2$	0.00562	491.7	29.92	0.09241

ambient air temperature. Superheat means that the internal lifting gas (and ballonet air) have higher temperatures due to radiation from the sun. Although the ambient air is also being radiated by the sun, its heat is dissipated by natural convection currents in the atmosphere. The internal gases are trapped inside the envelope and remain at a higher temperature for the same reason temperatures inside a greenhouse are higher than outside air. The change in lift due to a change in temperature (by superheat or any other means of changing the internal gas temperature) can be calculated using Eq. (2.10) resulting in Eq. (2.11).

$$\Delta L = G_a P V \Delta T_g / T_g^2 \qquad (2.11)$$

These rearrangements may initially seem pointless but are important because the resulting equations are in terms of quantities that are measureable such as pressure and temperature. Density is not easily measureable and so perfect gas relationships are substituted for density in most of the following equations.

Sample Problem 2.3: Change in Lift Due to Superheat

The envelope volume of an airship is 1,000,000 ft^3. The airship is brought out of the hangar and the sun raises its internal gas temperature by 20°F. The outside air is 60°F and its pressure is 2150 lb/ft^2 (30.40 in. Hg). The airship has ballonets whose total volume is 100,000 ft^3. How much does the lift increase due to the 20°F of superheat?

$$\Delta L = \rho_g V \Delta T_g / T = G_a V P \Delta T_g / T^2$$

$$\Delta L = 0.01875(2150)(1{,}000{,}000 - 100{,}000)(20)/519.7^2 = 2687 \text{ lb}$$

Introducing the quantity *fullness, F,* which indicates the fullness of the envelope, Eq. (2.10) can now be written as

$$L = [G_a PFV/T](1 - S) \quad \text{dry air, gas, and air at the same temperature}$$

where $F = 0.96$ means that the lifting gas can only fill 96% of the internal theoretical envelope volume.

Another effect on the lifting properties of an airship is the relative humidity of the air. Because the molecular weight of *water vapor,* gaseous H_2O, is 18 (2 + 16) and air has an average molecular weight of 28.8, water vapor weighs 18/28.8 or 5/8 as much as dry air alone. So, parts of the atmosphere that contain both air and water vapor weigh 1 + 5/8 as opposed to 1 + 1, which is 3/8 less than two portions of dry air. The term commonly used to designate how much water vapor is in the atmospheric air is called *relative humidity* and is defined as

Relative humidity = (vapor pressure-actual)/(vapor pressure-saturated)

Using the theory of gas partial pressures (Dalton's Law) the weight of a volume of wet air is equal to the weight of the dry air plus the weight of the water vapor within the same volume. Obviously wet air will weigh less than dry air because it is lighter (has lower molecular weight) than the dry air it has displaced. Therefore, within a constant volume, the pressure of dry air is $P\text{-}\omega$, where P is the atmospheric pressure and ω is the partial pressure of water vapor. This results in defining the density of dry air as Eq. (2.12).

$$\text{Density of dry air} = \rho_{da} = G_a(P - \omega)/T \tag{2.12}$$

Density of water vapor in same volume = $\rho_{wv} = G_a(\omega/T)(5/8)$.
The density of the dry air/water vapor combination in the volume is

$$\rho_a = \rho_{da} + \rho_{wv}$$
$$\rho_a = G_a(P\text{-}\omega)/T + G_a(\omega/T)(5/8)$$
$$\rho_a = G_a P/T\text{-}G_a\,\omega/T + G_a(\omega/T)(5/8)$$
$$\rho_a = G_a(P\text{-}\omega + 5/8\omega)/T$$
$$\rho_a = G_a(P\text{-}3/8\omega)/T \tag{2.13}$$

where ω is the pressure of water vapor ($\omega = 35.6$ lb/ft^3 at standard conditions).

Combining Eq. (2.13) with Eq. (2.10) yields Eq. (2.14),

$$L = [G_a FV(P\text{-}3/8\omega)/T](1 - S) \tag{2.14}$$

However, ambient air always has some moisture, there are always some impurities in the lifting gas, and the envelope has some internal volume (ballonets, etc.) that is not filled with lifting gas. Equation (2.14) incorporates all of these effects.

If we designate the amount of *purity* that a gas has as Y, then if the lifting gas is 98% pure, $Y = 0.98$. Eq. (2.15) is the result of modifying Eq. (2.14) for purity.

$$L = [G_a FYV(P\text{-}3/8\omega)/T](1 - S) \qquad\qquad (2.15)$$

Sample Problem 2.4: Lift Force for a Given Specific Gravity, Temperature, Pressure, and Humidity

A non-rigid airship with a volume of 1,000,000 ft^3 is filled with a lifting gas whose specific gravity is 0.213, temperature is 60°F, and ambient pressure is 2116 lb/ft^2. What is the lift force? Using Eq. (2.14) and Table 2.3 and $\omega = 0$ for dry air,

$$L = [G_a FYV(P\text{-}3/8\omega)/T](1 - S) = [0.1875(1,000,000)(2116)/(459.7 + 60)]$$
$$(1 - 0.213)$$
$$L = 60,081 \text{ lb}$$

Sample Problem 2.5: Effect of Humidity on Lift Force

What is the lift in Sample Problem 2.4 when the relative humidity (RH) is 60%?

$$L = [0.1875(1,000,000)(2116 - 0.375 \times 0.60 \times 35.6)/(519.7)](1 - 0.213)$$
$$L = 59,851 \text{ lb}$$

The lift has changed by only 0.3% for $RH = 60\%$. Because the change in lift is so small for relative humidity, its effect is usually ignored.

Sample Problem 2.6: Lift Force Due to Superheat

What is the new lift in Sample Problem 2.2.4 when its lifting gas is superheated by 30°F?

$$L = [0.1875(1,000,000)(2116)/(549.7 - 519.7)]0.213/(519.7 \times 549.7)$$
$$L = 60,969 \text{ lb}$$

The lift has changed by 1.3% for a 30°F superheat. A superheat of 30°F is considered to be near the maximum value that is experienced. This effect is small but should not be ignored.

The fundamental equation for buoyant lift was developed from the generalized Navier-Stokes equation in Eq. (2.1). Modifications to the

Navier-Stokes equation for zero velocity and the external force being caused only by gravity result in a very simple relationship for buoyant lift.

$$\text{Force} = (\rho_{f2} - \rho_{f1}) \times \text{Volume} \quad (\rho \text{ expressed in } lb_m/ft^3) \qquad (2.1a)$$

For simplicity and reduced confusion between pound-mass and pound-force units, it is suggested that when using English units for density that (slugs/ft³) be used rather than pound-mass/ft³. Remember that 1 slug = 32.174 lb_m. Equation (2.1a) then becomes

$$\text{Force} = (\rho_{f2} - \rho_{f1}) \times g \times \text{Volume} \quad (\rho \text{ expressed in slugs/ft}^3) \qquad (2.1b)$$

Care must also be taken to distinguish between the terms weight and mass. For airplanes weight and mass are essentially the same. For airships weight (because of buoyancy) is significantly different than mass. Figure 2.5 shows the changing weight and mass terms for various levels of buoyancy. This simple illustration shows that completely filling the sphere with a lifting gas such as helium reduces its weight by ~95% (4584/80,000) but only reduces its mass by 42% (2813/4863).

An object's buoyancy is always the result of the density differential between the enclosed liquid and the surrounding liquid. It does not matter

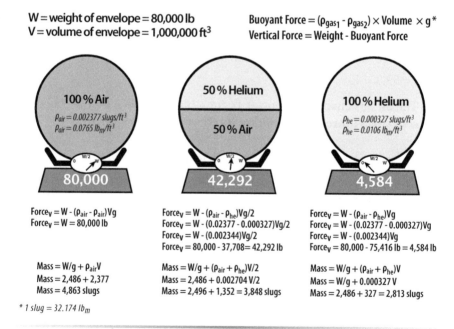

Figure 2.5   Weight, force, and mass comparison for three lifting gas combinations.

how this density difference comes about. Generally, it is the result of enclosing a gas that is less dense than air. However, the gases need not be different. Based on the perfect gas law, density is inversely proportional to temperature. This explains why a hot air balloon can generate buoyant lift. The heated air in the balloon is less dense than the cooler outside air. The greater the temperature difference the greater the buoyant lift. Figure 2.6 shows the buoyant lift achievable for a hot air balloon. All calculations assume that the sphere's volume is 1000 ft^3 and its envelope has no weight.

Two other buoyant lift systems are also illustrated in Fig. 2.6 showing relative lift capabilities for varying altitude. The first system, a closed container with no ballonet, plots the lift performance of the same 1000 ft^3 sphere as for the hot air balloon that is filled completely with helium. Notice that its lift performance varies a great deal with altitude. Although its buoyancy is the highest of all systems it is impractical as there is a significant pressure differential that increases with increasing altitude. Because typical envelope fabrics can only withstand 30–40 lb/ft^2 pressure airships would have little ability to fly at higher altitudes. A 40 lb/ft^2 pressure change is created by just a 500 ft change in altitude. At 5000 ft the pressure differential would be nearly 10× the value at 500 ft. The envelope thickness and therefore weight necessary to withstand these high pressure differentials would be prohibitive for normal airship operational altitudes of 4000 ft–10,000 ft.

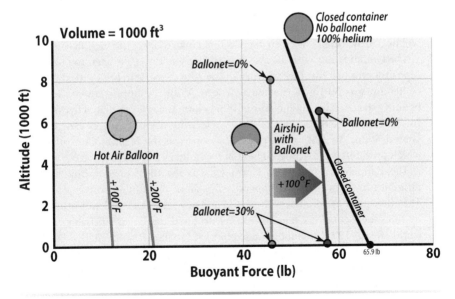

Figure 2.6 Buoyant forces for a 1000 ft^3 sphere—balloon, airship, closed container.

The third system shown in Fig. 2.6 shows the same size sphere that exhausts to the atmosphere and has a 70% helium fill and a 30% ballonet filled with air. Figure 2.6 shows that when the ballonet maintains a constant pressure differential across the envelope the buoyant lift force is constant with altitude. The reason for this slightly surprising result is shown below

$$p_2 - p_1 = \Delta p = \rho_2 R T_2 - \rho_1 R T_1 = \text{constant}$$

$$\Delta p = R(\rho_2 T_2 - \rho_1 T_1) = \text{constant}$$

$$RT(\rho_2 - \rho_1) = \text{constant \{when lifting gas temp = ambient temp\}}$$

$$(\rho_2 - \rho_1) = \text{constant}$$

. . . therefore, the buoyant force is constant

Hydrogen vs Helium

The central ingredient of an airship is its buoyant gas. There are only two practical lifting gas candidates, hydrogen and helium. Hydrogen has a density of 0.01114 lb/ft^3 and helium has a density of 0.00562 lb/ft^3 at 32°F. From Fig. 2.3 and Example 2.1 with the fluid being air at 60% relative humidity (density = 0.0805 lb/ft^3), hydrogen has 8% more buoyancy than helium. With 8% more lifting capability, hydrogen would be the preferred buoyant gas except for one very important characteristic: Hydrogen when mixed with air is an extremely flammable mixture. Hydrogen leaks can never be ruled out, so the use of hydrogen as the lifting gas in airships has been banned worldwide. Helium, on the other hand, is the lightest of the inert gases and will not burn. Thus, it is often used in fuel tanks to inert the fuel/air voids.

The United States with its vast helium fields in Texas has been the world's repository for helium since the turn of the 20th century. In the 1920s the cost of helium was 50 times that of hydrogen. With the improvements in the helium extraction technology the cost of helium today is 1/8 that of hydrogen. In the early part of the 20th century everyone used hydrogen except the United States. In 1927 the United States passed the Helium Control Act, which prohibited the export of helium for military purposes.

The Zeppelin Company in Germany was the most prolific builders of dirigibles in the world (see Appendix E)—all of them filled with hydrogen. They flew passengers all over the world and never had an accident until 1937. In 1933 the Zeppelin Company started to build the largest dirigible in the world, the LZ-129 Hindenburg (gas volume = 7 million ft^3 and length = 804 ft). The Zeppelin Company had been flying the LZ-127 Graf Zeppelin for 5 years on passenger routes and understood the danger of using hydrogen as the lifting gas. They considered the hydrogen airship to be a "ticking time bomb" and designed the Hindenburg for helium.

In 1933 Adolph Hitler became the Chancellor of Germany and immediately began using the Graf Zeppelin airship as a propaganda tool to trumpet the military supremacy of the Nazi party. Hitler made a poor judgment call and instructed propaganda minister Joseph Goebbels to make the Hindenburg ready for propaganda missions. This turn of events did not go unnoticed by President Roosevelt who could see war clouds looming over the horizon. He called up the Helium Control Act of 1927 and refused to sell helium to Germany. The Hindenburg was filled with hydrogen.

The German passenger airship LZ-129 Hindenburg began the second flight of its second year of commercial service in Frankfurt, Germany on May 3, 1937 and crossed the Atlantic without incident. It was raining as the Hindenburg was approaching the mooring mast at Lakehurst, NJ on May 6, 1937 and the air was rich with static electricity. As the wet handling ropes touched the ground, the circuit was completed and a spark ignited the hydrogen leakage in the aft part of the Hindenburg, resulting in the picture on the opening page of this chapter. Newsreel cameras rolled and radio announcer Herbert Morrison described the events for the first coast-to-coast news radio broadcast. As word of the disaster spread around the world, fueled by the horrifying newsreel coverage, the public lost faith in the technology . . . heralding the end of early commercial passenger airship flight.

The Graf Zeppelin II was the sister airship to the Hindenburg and was in production at the time of the Hindenburg disaster. The LZ 130 was quickly modified to use helium. However the United States still refused to sell any helium to Germany. The LZ 130 was converted back to hydrogen and very carefully operated in 1938 as a propaganda tool for the Third Reich. The majestic airship was scrapped in 1939 for aluminum to build Luftwaffe aircraft.

Reference

[1]   Khoury, G.A., *Airship Technology*, Cambridge University Press, 2004.

Chapter 3 Aerodynamics

- Importance of Lift
- Body Aerodynamics
- Lift Generation
- Streamlines
- Airship Drag
- Added Mass

The first civilian Zeppelin built after WW I was the LZ-120 Bodensee, which made its first flight on August 20, 1919. Bodensee's highly advanced and aerodynamically determined teardrop shape differed greatly from the thin, pencil-like shape of previous Zeppelins. Its shape was developed in the Göttingen wind tunnel.

Do not fudge your data . . . it may be right.
Wilbur Wright

3.1 Introduction

A erodynamics play a vital role in the behavior and performance of any air vehicle and lift is likely the most important of all aerodynamic parameters. For aircraft it is the wing that generates most of the lift and much of the drag. Because airships generate most of their lift using lifting gases instead of wings, there is no drag-due-to-lift that is created by the buoyant lift. When an aircraft flies it produces forces and moments (e.g. lift, drag, and pitching moment) around each of the three axes (x, y, z). Additionally, some derivatives with respect to angles are important, such as C_{L_α}, C_{m_α}, C_{n_β}, etc. These same forces and moments are produced when an airship flies as well. The difference is that the forces and moments are in different proportion and have differing importance for an airship compared to an aircraft. Actually, most of the aerodynamic forces and moments for an airship are much smaller due to its much slower speed and the fact that it does not have to generate all of its lift with wings.

When the airship flies at any angle of attack it generates aerodynamic lift and a drag-due-to-lift that is added to the buoyant lift and skin friction drag and zero lift pressure drag. Because bodies of revolution are inefficient at producing aerodynamic lift, airships generally fly at very low angles of attack and very close to neutral buoyancy. When bodies of revolution are discussed later it will be shown that their maximum aerodynamic L/D is usually less than three.

Other major contributors to airship behavior are the tails. They are wing-like surfaces generating aerodynamic forces enhanced by trailing edge control surfaces. Because tails are moment generators they should be able to create efficiently (without adding a lot of drag) up and down forces that produce stabilizing moments. The need to create both up and down forces means that the tails will likely be symmetric airfoil sections with a moveable control surface.

Because an airship's body dominates its aerodynamic characteristics, much of the discussion will be focused on optimizing body geometry for low drag or maximum buoyant lift-to-drag ratio. Tail arrangements and their sizing will be covered in Chapter 7.

3.2 Body Geometries

Historically, the standard approach to optimizing the design of an airship body was to use wind tunnel testing, which is a good means for accurately measuring aerodynamic forces and moments for any air vehicle. Figure 3.1 shows the Göttingen wind tunnel facility with the latest airship design in 1918, the Bodensee. Göttingen is best known as the home of Ludwig Prandtl, whose work in boundary layers and separated flow are legendary and still relevant today.

Figure 3.1 Göttingen wind tunnel model of LZ120 Bodensee airship (1918).

Because an airship has no wings, its wind tunnel models tend to be simpler and less expensive. However, it is also important to maximize the test Reynolds number (Re) by making the airship model as big as possible, which adds some cost. Tunnel test time for airship models is generally less than for aircraft because there is no wing and hence no trailing edge control surfaces. Wing-control surface combinations generally add significant test time to most aircraft wind tunnel tests.

Ludwig Prandtl had more than 80 doctoral students over his nearly 50-year career at Göttingen. A partial list of some of his most famous students includes Blasius, Schlichting, Tollmien, von Kármán, and Busemann.

Current body design techniques generally include some initial computational fluid dynamics (CFD) estimates that reduce the number of geometry variations to be tested to those with the highest performance. This technique can be extended to specialized CFD codes that specifically optimize airship bodies to minimize drag by defining shapes that result in large portions of the airship having laminar flow (see Fig. 3.2). Note that the airfoil section NACA Series 66 is used as a configuration starting point and how similar it is to the configuration that optimizes drag over the forebody. These body shaping techniques are very similar to designing sophisticated airfoil sections for a high performance aircraft wing.

Airship Body - Low Drag Configurations

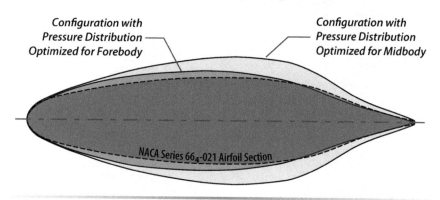

Configuration with Pressure Distribution Optimized for Forebody

Configuration with Pressure Distribution Optimized for Midbody

NACA Series 66_4-021 Airfoil Section

Figure 3.2 Comparison of shapes of low drag bodies designed using modern computer techniques (CFD). [1]

Optimizing aerodynamic performance (endurance, range, speed, etc.) for an airship is a complicated trade study that includes several new variables. The trade study process is virtually identical to that required for winged aircraft. These variables include volume (buoyant lift), body cross-section shape, envelope material properties, amount of buoyancy or buoyancy ratio (BR), and size of ballonets. Multidisciplinary optimization (MDO) tools are perfect for this type of trade study just as they are for aircraft trade studies.

Figure 3.3 shows ellipsoids that are the basis for many buoyant vehicle bodies. As discussed, these three ellipsoid shapes are applicable to specific types of airships or balloons. Spheres and oblate ellipsoids are used for balloons and will be discussed in Chapter 13. Prolate spheroids are very similar to most airship (and submarine) designs and their mathematical shapes makes them good candidates for analytic trade studies. Chapter 11 discusses designing specific airships that have nearly ellipsoid shapes. In Chapter 12 these ellipsoids are joined side by side to form a lobed airship body. The reasons why this lobed geometry is appropriate for a hybrid and not for a standard airship are also discussed. Appendix D presents the geometric and physical properties of ellipsoids and discusses the process of merging these shapes together to form multilobe hybrid configurations.

There are distinct geometric differences between a typically shaped body of revolution airship design and that of a hybrid airship design. Figure 3.4 compares the cross sections and planforms of the two designs. The symmetrical arrangement of circular arcs with the same radius is typical of the hybrid design philosophy. The arcs have the same radius to create the same level of stress throughout the envelope. It is important to have constant stress so the same material thickness can be used throughout. Because

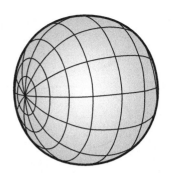

Sphere: All three axes a, b, and c are equal in length. It has the lowest ratio of surface area to volume and has equal fabric stresses everywhere. In terms of weight the sphere is the most efficient lifting shape. However, it can only function as a balloon that has no stationkeeping requirement. Its drag is high and it is at the mercy of the winds.

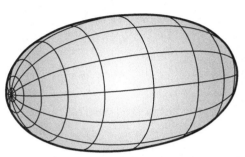

Prolate Spheroid: Major axis, a, is the longest and axes b and c are equal. Its surface area is somewhat larger than a sphere and stresses are different at every point on the envelope. This is the prototypical airship shape that is modified to address performance requirements. Its drag is much less than for a sphere. Its fineness ratio, a/b, is optimized for the best combination of drag, buoyant lift, and envelope weight. With tails and/or thrust vectoring this shape can be designed to stationkeep.

Oblate Spheroid: Major axis, a, is the shortest and axes b and c are equal. Like the prolate shape it has a high surface area to volume ratio. Interest in this shape is discussed in Chapter 13 on balloon design when the subject of "pumpkin shaped balloons" and their advantages are evaluated against standard spherical balloons.

Figure 3.3 Ellipsoidal bodies of revolution.

the radius of the circular arcs is obviously much smaller than the single circular shape (typically less than ½) the resulting stresses are also much lower.

The remainder of the discussion will focus on specific characteristics of bodies or revolution, i.e., spheres, spheroids, ellipsoids, and similar shapes. The flowfield of a body of revolution is in many respects similar to a wing. Air flows over an airship body and accelerates to about the midpoint then decelerates over the aft portions until the flow meets at the trailing edge/point. When there is a distinct trailing edge, as for a wing upper and lower surface, the flow adjusts so there is no pressure discontinuity. This

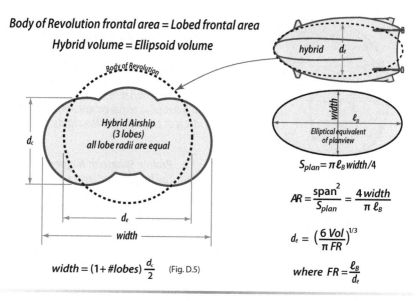

Figure 3.4 Comparison of ellipsoid airship to a hybrid airship with equal volume.

phenomenon is referred to as the Kutta condition. For a body of revolution there is no finite edge but the flow still cannot produce a pressure discontinuity at the trailing edge point. This acceleration of the air particles causes the static pressure on the surfaces to drop below the static pressure in the free stream. There is one streamline of air particles (called the dividing streamline) that slams into the body's nose and comes to a stop. This is referred to as the stagnation point and the total pressure at this location is equal to the free stream static pressure plus the dynamic pressure.

3.3 Body Aerodynamics

Lift, drag, and moment data for an airship are necessary ingredients for any design or performance analysis. The sign convention for these analyses and test data is shown in Fig. 3.5. The mean aerodynamic chord of a wing or tail, denoted by mac or $\bar{c}$, represents an average chord that, when multiplied by the average section moment coefficient, dynamic pressure, and reference area, gives the moment for the entire wing. It can be estimated for straight taper wings and tails by

$$\text{mac} = (2/3)C_R[(1 + \lambda + \lambda^2)/(1 + \lambda)]$$

where λ = taper ratio = C_T/C_R and C_T and C_R are the tip and root chords respectively.

In the following sections actual data is presented for various body shapes and flow conditions. Because airship drag is dominated by skin

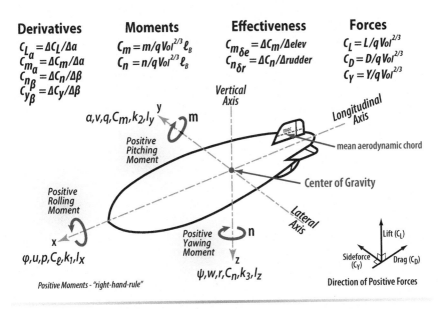

Figure 3.5 Airship sign convention.

friction, the Reynolds number will be very important in determining zero lift drag, coefficient C_{D_0}. Therefore, significant time is spent on methods for estimating skin friction drag and more importantly, how to reduce it. Actual data is also presented for various airship body geometries, and when possible, compared to estimated results. This chapter will focus on the fundamental aerodynamic concepts necessary to generate an acceptable database of lift, drag, and moment to be used for airship performance, stability, and control.

3.3.1 Reference Area

The aerodynamic forces and moments need to be non-dimensionalized into lift, drag, and moment coefficients. Dimensional analysis tells us that the lift and drag forces are made non-dimensional by dividing them by a pressure (the free stream dynamic pressure $q = \frac{1}{2}\rho V_\infty^2$ is the most obvious pressure) and a reference area. The moments are divided by q, reference area, and a characteristic dimension (wing span, mean aerodynamic chord, or airship length).

The reference area for aircraft is by international convention taken to be the planform area (S_{plan}). This is self evident because the wing is the main lift generating component and accounts for about half the zero lift drag due to the skin friction on its surface. The reference area for the airship was in debate for the first half of the 20th century and finally established by international convention to be $Vol^{2/3}$ where Vol is the volume of the airship envelope holding the lifting gas. This convention is not at all self evident because the airship does not generate much aerodynamic lift. The airship envelope surface area would be more in line because the airship drag is all aerodynamic and primarily due to skin friction. However determining the airship surface area requires considerable information about the airship configuration whereas the volume is easily determined by the equation [from Eq. (2.6)].

$$\text{Volume} = \text{weight} / 0.065\ \sigma$$

where 0.065 is the lifting capability for 98% pure helium gas and σ is the ratio of the densities between sea level and the maximum altitude for the airship for standard day conditions. So the $Vol^{2/3}$ is a matter of convenience and works well . . . except when you want to compare the aerodynamic forces and moments of airships with aircraft. For most airship configurations the exact planform area, S_{plan}, of the envelope is difficult to determine, however it can be approximated by a prolate spheroid (see Fig. 3.4). The relationship between S_{plan} and $Vol^{2/3}$ can be expressed as

$$S_{plan} = N_L\ Vol^{2/3} \tag{3.1}$$

where N_L depends upon the number of lobes in the configuration. N_L can be determined empirically from real airship/hybrid vehicles as

# lobes	N_L
1	2
2	2.25
3	2.4
4	2.5
5	2.54

Equation (3.1) can be used to convert airship/hybrid aero data from $Vol^{2/3}$ to S_{plan} for comparison with aircraft data.

An example of this conversion is as follows (note: the subscript V will denote reference to $Vol^{2/3}$ and subscript p for reference to S_{plan}.

For C_L, $C_{L\alpha}$, C_D, and C_M

$$L = C_{Lp}\ \mbox{q}\ S_{plan} = C_{Lv}\ \mbox{q}\ Vol^{2/3}$$

$$C_{Lp} = C_{Lv}\left(Vol^{2/3}/S_{plan}\right)$$

$$C_{Lv} = C_{Lp}N_L$$

For drag-due-to-lift factor $K = \Delta C_D/C_L^2$

$$D_L = K_P\ C_{L_p}^2\ \mbox{q}\ S_{plan} = K_V\ C_{L_V}^2\ \mbox{q}\ Vol^{2/3}$$

Using C_{L_P} mentioned previously,

$$\left(K_P / N_L^2\right) C_{L_V}^2 S_{plan} = K_V C_{L_V}^2 \, Vol^{2/3}$$

$$K_P = N_L^2 \left(Vol^{2/3} / S_{plan}\right) K_V = \left(N_L^2 / N_L\right) K_V$$

$$K_V = K_P / N_L$$

When reading C_{L_α} or K from Fig. 3.8 or Fig. 3.26 remember they are referenced to S_{plan}. Before using these values for airship design they must be re-referenced to $Vol^{2/3}$, which requires that they be converted using the C_{L_V} and K_V relationships derived in this section.

3.3.2 Aerodynamic Database

The design of an aircraft or airship should start with an examination of what equipment is already out there in the real world doing the mission that you are designing to. This examination can be a big help in getting your design started. Your clean sheet design must be better than what is already "on the shelf" otherwise you will not sell any new aircraft or airships because your price will not be competitive. Also, as your design progresses you should always compare your estimated aerodynamics, weight, performance, etc. with existing aircraft and airships as a "sanity check" of your estimations. If your takeoff gross weight ($TOGW$) is very much different than an existing aircraft/airship doing a similar mission you had better find out why . . . the answer is usually a flaw in your analysis. Appendix E contains data on existing airships for getting started and conducting "sanity checks."

Table 3.1 contains design and aerodynamic data (wind tunnel or flight test) on seven body of revolution (single lobe) airships, three multi-lobe hybrid airships, six lifting body research aircraft, and seven low aspect ratio ($AR < 4$) fighter type aircraft. The lifting body research configurations [9] were introduced into the data set to fill in the AR gap between airships and aircraft. The lifting bodies have shapes similar to the multi-lobe hybrids. They were designed to have blunt leading edges, providing a large bow shock to manage the high heating rate of reentry but have an aerodynamic L/D of 3–5 for good cross-range, down-range, and low speed handling characteristics.

All of the airships (single lobe and multi-lobe hybrids) in Table 3.1 have their aerodynamic data referenced to $Vol^{2/3}$ and the lifting body and fighter type aircraft are referenced to the conventional wing planform area S_{plan}. This mix of reference areas is necessary in order to maintain the purity of the airship/hybrid data because the conversion $S_{plan} = N_L Vol^{2/3}$ is empirical.

Figure 3.6 shows a representative vehicle for each of the four classes of air vehicles shown in Table 3.1.

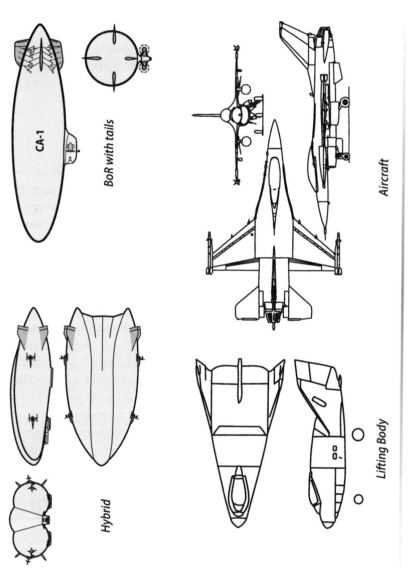

Figure 3.6   Representative vehicles from Table 3.1.

Table 3.1   Design and Aerodynamic Data Sheet

No.	Vehicle	Ref	AR	C_{L_α}	K	C_{D_0}	Ref
1	BoR, FR = 7.2	$Vol^{2/3}$	0.18	0.005	3.7	0.028*	6
2	BoR, FR = 6.0	$Vol^{2/3}$	0.21	0.006	2.9	0.030*	6
3	BoR, FR = 4.8	$Vol^{2/3}$	0.27	0.0088	1.7	0.0285*	6
4	BoR, FR = 3.6	$Vol^{2/3}$	0.35	0.007	1.3	0.031*	6
5a	USS Akron w/o tails, FR = 5.9	$Vol^{2/3}$	0.22	0.0063	2.8	0.019*	7
5b	USS Akron + tails, FR = 5.9	$Vol^{2/3}$	0.23	0.0125	1.24	0.025*	7
6a	ZP5K w/o tails, FR = 4.4	$Vol^{2/3}$	0.29	0.0066	2.0	0.0152*	12
6b	ZP5K + tails, FR = 4.4	$Vol^{2/3}$	0.3	0.0115	0.9	0.026*	12
7a	HALE w/o tails, FR = 3.2	$Vol^{2/3}$	0.4	0.008	1.15	0.016*	14
7b	HALE + tails, FR = 3.2	$Vol^{2/3}$	0.41	0.0122	0.55	0.024*	14
8	P-791	$Vol^{2/3}$	0.54	0.046	0.32	0.096	14
9	HA-1	$Vol^{2/3}$	0.60	0.045	0.28	0.033*	14
10	Aerocraft	$Vol^{2/3}$	0.46	0.027	0.46	0.032*	14
11	M2-F1	S_{plan}	0.65	0.0225	0.69	0.062	9
12	M2-F2	S_{plan}	0.712	0.0216	0.95	0.065	9
13	HL-10	S_{plan}	1.16	0.023	0.57	0.05	9
14	X-24A	S_{plan}	0.62	0.024	0.623	0.04	9
15	X-24B	S_{plan}	1.11	0.0217	0.5	0.025	9
16	Space Shuttle	S_{plan}	2.27	0.0437	0.33	0.061	9
17	SR-71	S_{plan}	1.72	0.04	0.3	0.006	13
18	F-117A	S_{plan}	2.06	0.05	0.33	0.0108	13
19	F-22A	S_{plan}	2.36	0.046	0.16	0.016	13
20	F-16C	S_{plan}	3.2	0.054	0.11	0.018	13
21	F-104C	S_{plan}	2.45	0.058	0.17	0.017	13
22	F-15E	S_{plan}	3.02	0.057	0.18	0.028	13
23	F-5E	S_{plan}	3.83	0.066	0.12	0.018	13

All $Re \geq 10^7$ and $M < 0.2$.

$AR = 4/\pi FR$ for bodies of revolution.

*Wind tunnel C_{D_0} value and models did not have full operational features such as lines, cooling drag, and landing gear.

The P-791 (see Fig. 1.15) was a demonstrator vehicle to examine integration of an ACLS and general hybrid configuration handling qualities. As such most of the wiring, cables, and attach fittings were external on the hull resulting in an unusually high C_{D_0} value.

3.3.3 The Generation of Lift

Both aircraft and airships generate lift and are thus able to move in three dimensions. Aircraft rely on aerodynamic lift whereas airships rely primarily on buoyant lift with some aerodynamic lift (sometimes called dynamic lift).

Because buoyant lift cannot be turned on or off, the airship relies on aerodynamic lift to modulate its total lifting force to account for ascent, descent, and fuel burn. The buoyancy ratio, BR, is defined as the buoyant lift divided by the total lift:

Buoyancy Ratio $= BR =$ (buoyant lift)/(buoyant + aerodynamic lift)

Airships typically have BRs of 90–95% and hybrids have 70–80% at takeoff.

Aerodynamic lift is generated by a fluid flowing over a body such that at angle-of-attack α the fluid flows faster over the top surface than the bottom surface. The energy contained in a fluid streamline is a constant and is a combination of pressure energy, kinetic energy, and thermal/internal energy. We will assume that the speed is less than 100 KEAS such that the thermal/internal energy does not change. Thus, as the speed increases the static pressure must decrease (Bernoulli's Theorem) resulting in a lift force (called circulation lift) normal to the free stream. This lift is expressed as

$$\text{Aerodynamic Lift} = C_L\, q\, S_{Ref} = C_{L\alpha}\, \alpha q\, S_{Ref} \qquad (3.2)$$

where $C_{L\alpha}$ is the lift curve slope and α is the angle of attack. The C_L is the non-dimensional lift coefficient referenced to S_{plan} or $Vol^{\frac{2}{3}}$.

Lift generation is illustrated in Fig. 3.7 showing the changing flow field around a 20% thick airfoil and body of revolution. There is one flow streamline (called the dividing streamline) that smashes into the airfoil and body of revolution at the stagnation point (point A).

In Fig. 3.7a one streamline goes over the top of the airfoil and one goes along the lower surface and they both meet (coalesce) at point C. The flow over the upper surface initially has to speed up going around the nose and then slow down as it approaches point C. As the flow speeds up the static pressure on the surface drops (denoted by −) and as it slows down the static pressure increases (denoted by +). The flow on the lower surface does just the opposite generating an increased pressure along the nose and a decreased pressure on the aft end as it accelerates around the aft end to meet the upper streamline at point C. The resultant summation of static pressures results in zero lift but a nose-up moment is generated. There is a drag force on the airfoil due to the skin friction.

The airfoil in Fig. 3.7b is the same as in Fig. 3.7a except that the aft end has been made sharp and approximates a NACA 0020 airfoil. Physically the

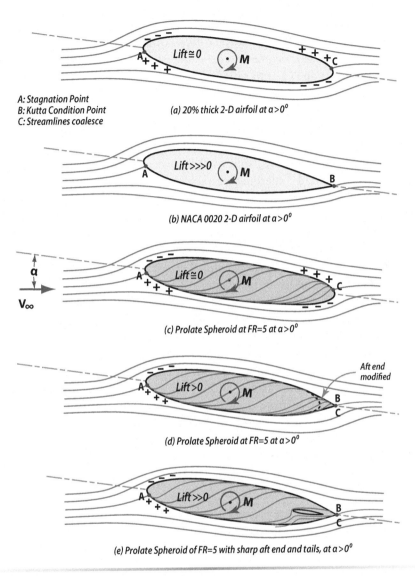

A: Stagnation Point
B: Kutta Condition Point
C: Streamlines coalesce

(a) 20% thick 2-D airfoil at α>0⁰

(b) NACA 0020 2-D airfoil at α>0⁰

(c) Prolate Spheroid at FR=5 at α>0⁰

Aft end modified

(d) Prolate Spheroid at FR=5 at α>0⁰

(e) Prolate Spheroid of FR=5 with sharp aft end and tails, at α>0⁰

Figure 3.7 The generation of lift on airfoils and bodies of revolution. (A is stagnation point where streamlines divide and B is point of Kutta condition.)

lower surface flow cannot negotiate the sharp trailing edge and flow forward to meet the upper surface flow. Consequently the upper and lower surface streamlines coalesce at the trailing edge (point B), resulting in the upper surface streamline having gone further and faster than the lower surface. A subtle note is that air molecules on adjacent streamlines at point A do not arrive at point B at the same time. The air molecule over the upper surface arrives at the trailing edge before the lower surface air molecule. This coalescing of the flows at the trailing edge is an important physical phenomena called the Kutta condition and results in a differential pressure between the top and bottom surface generating a lift force (and a nose-up moment). A necessary condition for generating lift on an airfoil or body is a sharp trailing edge (TE) to satisfy the Kutta condition. The modification of the airfoil in Fig. 3.7a by adding a sharp TE results in a very efficient airfoil for generating lift.

Figure 3.7c shows a prolate spheroid (body of revolution) of fineness ratio $FR = 5$ (20% thick) inclined at $\alpha > 0$ deg. Again, there is no lift generated (same as in Fig. 3.7a) due to the lower surface streamline flowing around the body and coalescing with the upper surface flow at point C. The blunt aft end is unable to support a Kutta condition. There is a moment generated due to the up force on the forebody and the down force on the aft body and there is a drag force as well.

Figure 3.7d shows the same body from Fig. 3.7c except a semi-sharp cone has been attached to the aft end. The semi-sharp cone on the aft end forces the upper and lower flow (single streamline) to coalesce at point C, which generates a small amount of lift (due to a weak Kutta condition). Because of the higher pressure on the lower surface there is a flow around the entire body reducing the pressure differential between the upper and lower surface that reduces the lift generated.

Figure 3.7e shows the same body from Fig. 3.7d except that horizontal tails have been attached to the aft end for pitch stability. The horizontal tails interrupt the flow from the lower surface to the upper surface at the aft end, which reinforces the Kutta condition and increases the lift generated. The tails also generate lift adding to the overall body lift and decreased nose-up pitching moment.

A useful expression for calculating C_{L_α} for aircraft is (from [10])

$$C_{L_\alpha} = \frac{2\pi AR}{2 + \sqrt{4 + AR^2 \beta^2 \left(1 + \dfrac{\tan^2 \Lambda}{\beta^2}\right)}} \tag{3.3}$$

where $\beta^2 = 1 - M^2$ and Λ is the sweep of the wing leading edge and $AR = (\text{span})^2/S_{plan}$. For speeds < 100 KEAS, $\beta^2 \approx 1$ and Eq. (3.3) becomes

$$C_{L_\alpha} = \frac{2\pi AR}{2 + \sqrt{4 + AR^2 \left(1 + \tan^2 \Lambda\right)}} \tag{3.4}$$

and for low sweep $\tan^2 \Lambda \approx 1$ and Eq. (3.4) becomes the familiar Helmbold equation (from [11])

$$\frac{dC_L}{d\alpha} = C_{L_\alpha} = \frac{2\pi AR}{2 + \sqrt{4 + AR^2}} \tag{3.5}$$

And finally for $AR < 1$, which is the case for bodies of revolution, the equation for C_{L_α} becomes the slender body expression

$$C_{L_\alpha} = \pi AR/2 \qquad (3.6)$$

The C_{L_α} vs AR for the data set of Table 3.1 is shown on Fig. 3.8. The C_{L_α} for airships and hybrids has been referenced to S_{plan} using Eq. (3.1) (dividing by N_L) so that all C_{L_α} are referenced to the same area. It is observed that the single-lobe body of revolution without tails is a poor lifting body. This is because the body of revolution typically does not have a sharp TE and the lower surface flow rolls around the aft end (from the high pressure to the low pressure regions) with the flows coalescing forward/upstream of the TE (see Fig. 3.7d). Lift efficiency improves somewhat when tails are added (see data points 5a to 5b, 6a to 6b, and 7a to 7b in Table 3.1) as the tails interrupt the lower surface flow around the aft end reinforcing the Kutta condition. However, C_{L_α} is still low.

Another approach to improving airship lift is by laying out a multi-lobe configuration that increases C_{L_α} by a factor of two to four over the single lobe body of revolution. This is due to the multi-lobe configuration having (1) an increased aspect ratio, (2) connecting the aft end of the outer lobes with a

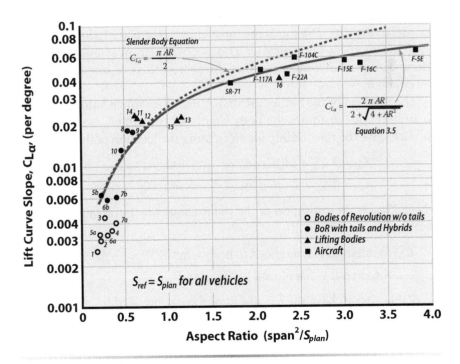

Figure 3.8 Lift curve slopes for the vehicles from Table 3.1.

sharp-edged piece of unpressurized fabric (as shown on Figs. 3.4, 3.6, and D.1 in Appendix D), and (3) possible vortex lift due to high leading edge sweep.

3.3.4 Body Reference Points

A design parameter that is unique to airships is called the *center of buoyancy* (c.b.). It refers to the x, y, z location on the airship where the buoyancy forces produce no moment about any of the three axes and is coincident with the centroid of the displaced fluid. Typical aircraft design efforts concentrate on maintaining a proper c.g. location that results in good performance and handling qualities. The c.b. is located at the centroid of the displaced fluid and adds another design parameter that must be positioned carefully. Its relationship to the c.g. must also be understood throughout the flight envelope. Generally, the c.b. should be slightly aft of the c.g. so that there is a slight nose-down moment from the buoyant gases.

Airships also have an *aerodynamic center* (a.c.) just like an airplane, and as such, their a.c. has a nominal position. For subsonic winged aircraft, the a.c. is usually located very close to $\bar{c}/4$ where $\bar{c}$ is the *mean aerodynamic chord* of the wing. The a.c. is that point on an aircraft, wing, airship, or body about which the pitching moment is independent of angle of attack (i.e., $dC_M/d\alpha = 0$). The a.c. is the most convenient place to locate the lift, drag, and moment of an airship. This is obvious from stability considerations because this results in $dC_{M_{ac}}/d\alpha = 0$ so there is one less term to worry about.

The position of the a.c. for an airship uses the same relationship as for an aircraft but substitutes the body length for the wing $\bar{c}$. Equations (3.7) and (3.8) give the relationship for calculating the location of the a.c.

$$X_{ac} = X_{mrp} + \ell \,(dC_M/dC_L) \qquad \{\text{airships}\} \qquad (3.7)$$

$$X_{ac} = X_{mrp} + \bar{c} \,(dC_M/dC_L) \qquad \{\text{aircraft}\} \qquad (3.8)$$

where X_{mrp} is the x location of the point about which the moments were measured

C_M for airships is referenced to body length

C_M for aircraft is referenced to the wing $\bar{c}$

Stable aircraft have the c.g. ahead of the a.c. but airships also have the c.b. to contend with as well. Because most of the lift is generated by the lifting gas that acts through the c.b., the c.b. should be very close to the c.g. or it will produce a constant longitudinal moment that may be hard to trim out. If this moment is too large then deflected control surfaces are necessary to trim out this moment in flight thereby producing trim drag that reduces performance. For very low speeds, aerodynamic forces are small so any moment must be reacted by vectored thrust. Figure 3.9 is a detailed diagram of the forces and moments that an airship experiences and their relative positions.

The location of the c.b. is invariant as it depends entirely on the centroid of the displaced fluid and does not depend on the arrangement of the internal gases. A common misconception is that when the ballonets fill with air, as in Fig. 2.5, the c.b. moves as well. This is not true. The location of the c.b. does not depend on the arrangement or distribution of the internal gases at all. This fact is illustrated in Fig. 3.10 where three conditions are compared for their c.b. and c.g. locations. The location of the c.b. is typically 45% of ℓ_B for most airship shapes (see Table 7.1).

3.3.5 Body Pressures

Because the forces and moments on the body dominate other parts of an airship, the body's pressure distribution and boundary layer character become major design goals. Figures 3.11–3.13 show these pressure variations as they vary with body shape, computational accuracy, and *Re*.

Figure 3.11, which compares the surface pressures on three different bodies of revolution, introduces the quantity, *pressure coefficient, C_p*, which is the non-dimensional coefficient defined in Eq. (3.9). Clearly, the ellipsoid and paraboloid shapes are quite different and therefore their pressure distributions would be expected to be different. Of more interest is the difference

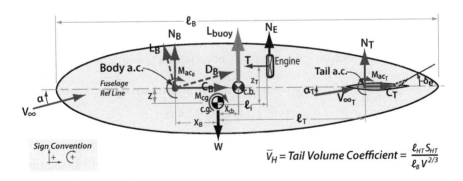

Figure 3.9 Forces and moments acting on a buoyant airship with an aft tail.

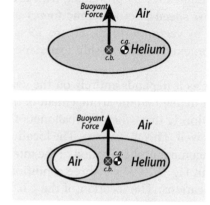

Surrounding fluid: homogeneous air
Internal gas: homogeneous helium

The location of the center of buoyancy (c.b.) is determined by the location of the centroid of the displaced surrounding fluid. Therefore, the c.b. is located at the centroid of the ellipse and a c.g. position is assumed.

Surrounding fluid: homogeneous air
Internal gases: non-homogeneous air+helium

The location of the center of buoyancy (c.b.) is determined by the location of the centroid of the displaced surrounding fluid and not the internal fluid distribution. Therefore, the c.b. is still located at the ellipse centroid but the c.g. moves forward.

Surrounding fluid: non-homogeneous fluid*
Internal gas: homogeneous helium

The location of the center of buoyancy (c.b.) is determined by the location of the centroid of the displaced surrounding fluid which is not homogeneous. Therefore, the c.b. is no longer located at the ellipse centroid but is shifted in the direction of the much denser fluid (water). The c.g. doesn't change.

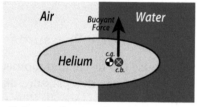

* Of course this fluid arrangement is impossible and is presented for instructional purposes only.

Figure 3.10   Effect of surrounding fluid properties on the location of the c.b.

in the pressure distribution of an actual fuselage shape and the similarly shaped ellipsoid. Notice how there are two distinguishing peaks in the C_p distribution for the fuselage shape that do not appear in either the plain ellipsoid or plain paraboloid shapes. These peaks are the result of a discontinuity in the curvatures (second derivative) as the ellipsoid nose fairs into the cylindrical section. The same explanation is made for the aft peak where the cylindrical section (curvature = 0) becomes parabolic (curvature > 0).

C_p = pressure difference from freestream/freestream dynamic pressure

$$C_p = (p - p_\infty)/\tfrac{1}{2}\rho_\infty V_\infty^2 = \Delta p/q \qquad (3.9)$$

Computational fluid dynamics (CFD) design codes are available for optimizing the shape of an airship's body. These codes vary from the simple potential flow solvers such as QuadPan to the full Navier-Stokes solvers such as CFD++. Figure 3.12 shows the results comparing inviscid vs turbulent and 2-D vs 3-D flowfields. Again notice how subtle changes in the area distribution of the two bodies shows large differences in the pressure distributions. However, comparing the axisymmetric body using an inviscid solution vs a turbulent solution shows little difference except for the aft closure region. This aft closure difference will not affect lift much but could be the reason for different values of drag between the two. Keep in mind that potential flow solvers inherently yield poor drag results regardless of the body shape.

Body pressures in Fig. 3.13 show the results of optimizing body shapes for various Re regions. This optimization consists of properly contouring the body to maintain a laminar boundary layer as far back as possible for each Re. Even though the boundary layer would transition to turbulent quickly on a flat plate, the laminar runs are maintained on the axisymetric body by the associated *favorable* ($dp/dx < 0$) pressure gradient on the forward portion of the 3-D body. Actual airship designs are capable of providing a proper area progression that can keep the boundary layer laminar for as much as a hundred feet or more. Notice in Fig. 3.13 that for each body the boundary layers turn turbulent the instant the pressure distribution becomes *adverse* ($dp/dx > 0$). Another difference among the three bodies in Fig. 3.13 is that they have different c.b. locations and different volumes. While volume is a fundamental optimizing parameter, c.b. location is a design parameter that impacts the handling qualities.

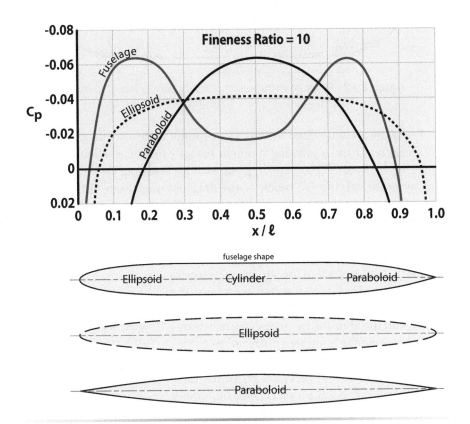

Figure 3.11 Pressure distribution comparison for three body shapes in potential flow. [2]

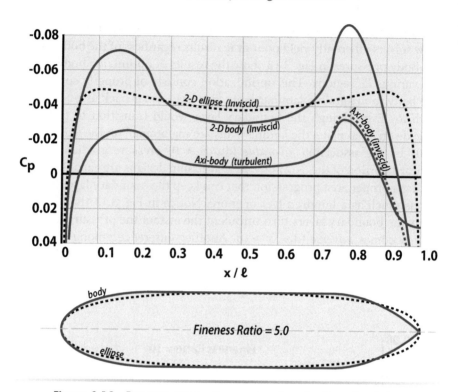

Figure 3.12 Pressure distribution comparison for three flowfields on *FR* = 5 bodies.

Real flow pressure distributions also produce a pressure drag parallel to the free stream and a pitching moment (usually measured about the body mid-length). Because pressure drag and skin friction drag vary in the opposite sense for varying *FR*, much of the drag optimization will trade these parameters to get the lowest zero lift drag coefficient C_{D_0} and still maintain significant buoyant lift created by a large volume of lifting gas. This optimization also recognizes that skin friction drag changes in two ways for varying *FR*. In one sense, for a given volume, the surface area increases for increasing *FR* that increases skin friction drag. However, skin friction drag also depends on where the boundary layer transitions from laminar to turbulent and this transition can be delayed by carefully designing the body area distribution.

Combining all of the contributing factors together, skin friction, pressure, pressure gradient, and volume, yields a total *L/D* for the airship that determines its range and endurance. Because volume is the means of generating buoyant lift from a lifting gas, the designer must seek the

"I am an old man now, and when I die and go to heaven there are two matters on which I hope for enlightenment. One is quantum electrodynamics, and the other is the turbulent motion of fluids. And about the former I am rather optimistic."

Werner Heisenberg (circa 1976)

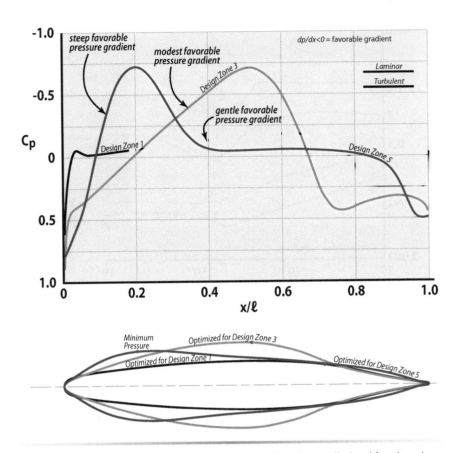

Figure 3.13 Inviscid pressure distributions of bodies optimized for drag in various regions. [4]

optimum combination of low drag body shape, boundary layer character, and volume to maximize performance. For buoyant vehicles that are required to perform stationkeeping, long endurance, or fly long distances the trade study to optimize total L/D is very important. Although total L/D does not appear in the range/endurance equations, the buoyant part of the lift term is very important from a design standpoint. If the buoyant lift is reduced then the aerodynamic lift would have to be increased, which would increase drag and thus reduce range/endurance for a given amount of fuel.

It is instructive to study the *drag area* results shown in Fig. 3.14. These generic 2-D bodies show major differences in total drag and how that drag uniquely varies with Re. In particular, note the drag variation of the ellipse. Because most airship bodies are very ellipse-like it is important to understand how drag changes with increasing Re. The $FR = 2$ ellipse in Fig. 3.14 will be shown to not be optimum when an airship is designed in Chapter 11.

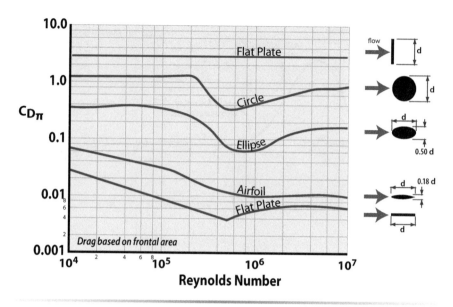

Figure 3.14   2-D drag coefficients variation with Reynolds number. [3]

3.4 Boundary Layers and Skin Friction Drag

Air molecules flow over a body in layers called streamlines. The air molecules in the streamline next to the body surface actually interact with the molecular structure of the surface and come to a stop. This is the "no slip condition" that makes up the foundation of boundary layer theory and is illustrated in Fig. 3.15. For streamlines away from the surface the air molecules are moving faster, which results in a varying velocity gradient dv/dz. At a distance δ (called the boundary layer thickness) from the surface the velocity gradient is zero. However, standard practice defines δ as the distance away from the body where the local streamline velocity is 0.99 of the freestream velocity. The thicknesses of the two forms of boundary layer can be estimated by Eqs. (3.10) and (3.11).

$$\text{Laminar thickness } \delta_L = 5.2x/Re_x^{0.5} \tag{3.10}$$

$$\text{Turbulent thickness } \delta_T = 0.37x/Re_x^{0.2} \tag{3.11}$$

The boundary layer is composed of many streamlines and can take one of three forms as shown in Fig. 3.15; (1) if the streamlines are smoothly flowing it is laminar, (2) if the streamlines are chaotic and vortical it is turbulent, and (3) if the streamlines are separated from the surface ($dv/dz \le 0$ at $z = 0$), it is called a separated boundary layer. The character of the separated boundary layer is such that the flow near the surface can actually reverse direction and flow upstream. The shearing action between the

Friction force ~ μ dv/dz (area) where μ is the fluid coefficient of viscosity and dv/dz is the velocity gradient evaluated at z = 0. This force acts parallel to the surface.
Boundary layer starts out laminar and transitions to turbulent at $R_{ex} \approx 5 \times 10^5$
where R_{ex} = local Reynold's Number = ρvx/μ
Laminar thickness $\delta_L = 5.2x/R_{ex}^{0.5}$ and turbulent thickness $\delta_L = 0.37x/R_{ex}^{0.2}$
Flow separates when dv/dz = 0 at the surface.

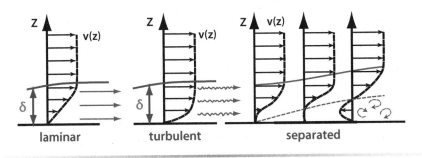

Figure 3.15 Boundary layer profiles and the resulting flat-plate skin friction coefficients.

streamlines creates a friction force in the streamline direction. At the surface ($z = 0$) this friction force is equal to $\mu dv/dz \times$ the surface area where μ is the fluid coefficient of viscosity and acts parallel to the surface. Notice that the velocity gradient dv/dz at the surface is smaller for the laminar boundary layer than the turbulent boundary layer, which results in a lower skin friction drag. Notice also that in the separated region $dv/dz = 0$ at the surface there is nearly zero skin friction drag but at the same time, there is a large increase in static pressure, pressure drag, and some loss of lift.

The character of the boundary layer is dependent upon a non-dimensional parameter called $Re_x = \rho Vx/\mu$, which is a ratio of the inertia forces in the boundary layer to the friction forces. For a flat plate the boundary layer starts out laminar and transitions to turbulent at a local $Re_x \approx 5 \times 10^5$. The laminar boundary layer is very delicate and will transition early if it encounters a disturbance or an increasing pressure gradient. Thus, the location of the maximum thickness is a good indicator of where the boundary layer transitions from laminar to turbulent because that is approximately where the pressure gradient (dp/dx) switches from negative to positive.

Ludwig Prandtl is considered by many to be the father of modern aerodynamics. His mathematical formulations created the foundation for subsonic and transonic analyses. But it was his work on characterizing the boundary layer and understanding its important role in defining drag, streamline bodies, and flow separation that is so important to the design of modern airships.

The averaged skin friction coefficient C_F acting on a square unit of an airship's surface is shown in Fig. 3.16 as a function of Reynolds number where the characteristic length is measured from the leading edge of a flat plate or the leading edge of a surface, i.e., the nose of the body. Note that several airships have been placed on this figure for reference purposes. The solid line on Fig. 3.16 is the fairing for a turbulent boundary layer and is referred to as the Schoenherr-von Kármán relationship. The dashed line at the left on Fig. 3.16 assumes the boundary layer is completely laminar and the transition line connecting them approximates the likely mixture of flow as the boundary layer transition progresses from laminar to fully turbulent flow. Using this data the airship skin friction drag force component is calculated as follows.

Skin friction drag force = C_F (surface area) (dynamic pressure)

In order to design a high performance airship it is necessary to understand the basics of the boundary layer. More than 100 years ago Ludwig Prandtl postulated the existence of the boundary layer and many of its governing equations. Because most of an airship's drag is due to skin friction and pressure and not drag-due-to-lift, it is important to understand the behaviors of these boundary layer forms.

It has been shown in earlier figures that body shapes that are not ellipsoidal can have significantly lower drag by shaping the area distribution to maintain as long a run of laminar flow as possible. However, long laminar runs on a flat plate are difficult once the Re approaches 1 million because there is no favorable pressure gradient (i.e., $dp/dx \geq 0$).

In a similar manner to airfoil design, airship bodies can be very carefully shaped to maximize the extent of the laminar flow region. Because the transition would happen quickly on a flat plate it is necessary to create an accelerated flow that results in a *favorable* pressure gradient to keep the boundary layer laminar for as long as possible. Luckily, airship forebodies are naturally shaped to create this natural *favorable* gradient. However, keeping the flow laminar aft of the maximum cross section is difficult. Designs of axisymmetric bodies are discussed in Chapter 11 and the flatter, more wing-like body shapes are presented in Chapter 12.

3.5 Airship Drag

3.5.1 Airship Drag Definitions

All possible types of drag are briefly defined in the following paragraphs. Even though they may not contribute to an airship's total drag, all terms are included for completeness.

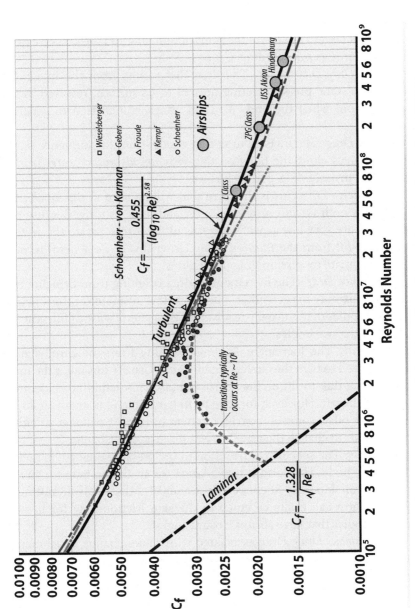

Figure 3.16 Skin friction on a flat plate vs Reynolds number showing various airships.

Skin Friction Drag: The drag on a body or component resulting from viscous shearing stresses over its entire wetted surface.

Pressure Drag (or Form Drag): The drag on a body or component resulting from the integrated effect of the static pressure acting normal to its surface resolved in the drag or wind axis direction.

Minimum Drag: The sum of the skin friction drag and the pressure drag. This is the same as zero lift drag, C_{D_0}.

Viscous Drag-Due-to-Lift: The drag that results from the integrated effect of the static pressure acting normal to its surface (resolved in the drag direction) when the body angle of attack is increased to generate lift.

Induced Drag: The inviscid drag that results from the influence of the trailing vortices along the body on the body *aerodynamic center* (sometimes called inviscid drag-due-to-lift).

Drag-Due-to-Lift: This is the term that will be used in this book as it accounts for both *viscous drag-due-to-lift* and *induced drag*. Because airship bodies are very low aspect ratio it is difficult to separate the *viscous drag-due-to-lift* from the *induced drag*. Using this more general term is more appropriate for airship drag discussions.

Interference Drag: The increment in drag resulting from bringing two bodies in proximity to each other. There is a flow mismatch at the junction of the two bodies that is usually resolved by the generous use of fillets at the junction of the two bodies.

Trim Drag: The increment in drag resulting from the aerodynamic forces required to trim the airship about its c.g. Usually this takes the form of added drag-due-to-lift on the horizontal tail. This definition is identical to that for aircraft. However, for long term trim situations on an airship it is possible to offset some of the trim moment by moving ballonet air either forward or aft to change the c.g. This was discussed in Chapter 2.

Base Drag: The specific contribution to the pressure drag attributed to a blunt after-body. This term is much larger for an airship than for an airplane as the airship body fineness ratio is about 1/2 that of an aircraft fuselage. This term can easily be equal to the skin friction drag for low *FR* bodies or bodies that have abrupt closure angles.

Miscellaneous Drag: Drag associated with cables, engine attachments, landing gear, Air Cushion Landing System pads (ACLS), propeller shrouds, antennas, sensors, and other protrusions external to the envelope.

3.5.2 Configuration Effects on Drag

All of the discussions about surface pressures and surface skin friction are necessary to provide an understanding of the zero lift drag of an airship. As a first step in calculating C_{D_0} of an airship, analyses showing the drag

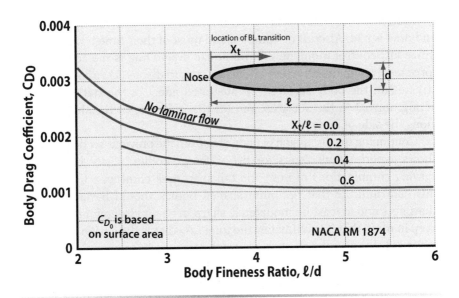

Figure 3.17 Body drag as a function of the location of boundary layer transition.

coefficient of ellipsoidal bodies of revolution are presented in Fig. 3.17 for various body FR and varying points on the body where the boundary layer transitions from laminar to turbulent. Notice that for $FR > 4$ there is very little pressure drag so most of the body drag is skin friction. As the FR is reduced to 3 and below the pressure drag term becomes more and more significant as it ultimately approaches $FR = 1$, which is a sphere. Figure 3.17 also shows that the no-laminar-flow (fully turbulent) body drag is reduced by half for the case where laminar flow is maintained to about 60% of the body length.

The aerodynamicist usually furnishes data in coefficient form that may show some variation with Re. Portions of this dataset consist of data showing the effects of compressibility and *Mach number* that are concerns to the aircraft designer but of no concern to airship designers. The coefficients of most interest are C_L, C_D, and C_M that are non-dimensionalized by area and length and by dynamic pressure $q = \frac{1}{2}\rho V_\infty^2$. Aerodynamic pitch axis characteristics of aircraft are referenced to wing area and its mean aerodynamic chord. For airships the area and length reference terms used in this book, unless specifically stated otherwise, are Volume$^{2/3}$ and body length, ℓ, respectively. Standard nomenclature in this book shows aerodynamic coefficients for three-dimensional bodies (bodies, tails, and combinations) as capital subscripts and for two-dimensional shapes as lowercase subscripts.

In airship design, lift is very good and mostly constant, moment is useful and drag is bad. Aircraft designers spend most of their time trying to maximize lift, control moment, and minimize drag. Drag is the aerodynamic force resolved in the direction of the free stream due to (1) viscous shearing stresses on the body surface, (2) integrated effect of the static pressures acting normal to the surfaces, and (3) the influence of the shed vortices along the side of body.

Continuing the discussion of body drag from the previous section, Fig. 3.18 shows experimental results for five different bodies. These bodies have varying combinations of nose and tail sharpness/bluntness. Results from this data indicate that optimum shaped bodies should have some nose sharpness and at least be moderately sharp in its aft closeout angle. Again, keep in mind that it is ultimately the total *L/D* that is the design measure of merit (MoM) so volume plays an important part in maximizing the total *L/D* parameter. Optimizing volume and body *FR* is the source of a very important trade study in airship design. In Chapter 12 the difficulties associated with sharp close-out angles will be discussed in detail.

It is also important to understand the main sources of airship body drag and their relative magnitudes. Figure 3.19 shows experimental data for

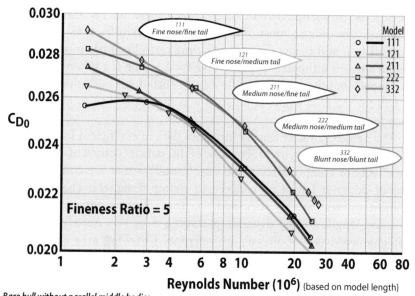

Bare hull without parallel middle bodies
Based on Volume 2/3

Figure 3.18 Effect of nose and tail bluntness on drag–prolate bodies. [5]

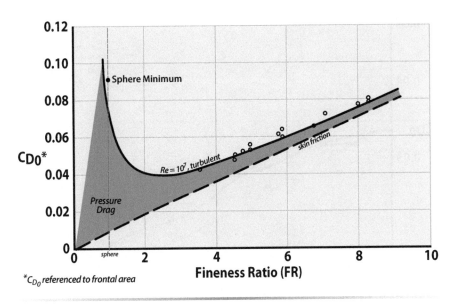

Figure 3.19 Drag coefficients of streamline bodies. [3]

various streamline bodies as their *FR* changes. Notice the skin friction line, which quantifies the contribution of fully turbulent skin friction drag. This highlights the dramatic change in pressure drag starting with *FR* = 1 for a sphere all the way to *FR* = 9. Pressure drag becomes fairly constant for *FR* > 5. This figure clearly shows how pressure drag dominates total drag for 1 < *FR* < 3. For bodies more slender than *FR* = 4 skin friction becomes an increasingly larger portion of total drag. Another example of body drag variation with both *FR* and *Re* is shown in Fig. 3.20 for several Goodyear Zeppelin shapes. The optimum *FR* depends on *Re* with the best *FR* being below 4 for *Re* ≈ 10^6 and the best *FR* at higher *Re* is approximately 6.

The pressure distribution is usually expressed as the surface pressure coefficient C_p defined in Eq. (3.9). There are, however, both skin friction and pressure drags. Most slender body drag is due to skin friction with the pressure drag term being rather small. For smaller body fineness ratios (*FR*) pressure drag becomes an increasingly larger portion of the total drag. At *FR* > 7 the pressure drag is no more than 5% of the total. At about *FR* < 3 the pressure drag becomes greater than skin friction drag.

3.5.3 Airship Zero Lift Drag

Both airships and aircraft display a parabolic behavior of C_D with C_L (see Fig. 3.21 and [10]). This behavior is expressed as

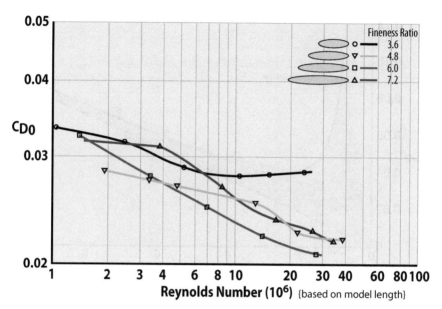

Bare hull without parallel middle bodies
Based on Volume $^{2/3}$

Figure 3.20 Effect of fineness ratio and Re on C_{D_0} (wind tunnel test on Goodyear Zeppelin designs). [6]

$$C_D = C_{D_0} + C_{D_L} = C_{D_0} + K'C_L^2 + K''\left(C_L - C_{L_{min}}\right)^2 \qquad (3.14)$$

where C_{D_L} is the drag-due-to-lift, K' and K'' are the inviscid and viscous drag-due-to-lift factors respectively, $C_{L_{min}}$ is the C_L for minimum C_D, and C_{D_0} is the zero lift drag coefficient. $C_{L_{min}} = 0$ for uncambered configurations, which is usually the case for airships and hybrids. For $C_{L_{min}} = 0$ Eq. (3.14) becomes

$$C_D = C_{D_0} + \left(K' + K''\right)C_L^2 = C_{D_0} + KC_L^2 \qquad (3.15)$$

where $K = K' + K''$ is defined as the total drag-due-to-lift factor and is computed as

$$K = \left(C_D - C_{D_0}\right)/C_L^2 = \delta C_D/C_L^2 \qquad (3.16)$$

This K is discussed in Sec. 3.5.4.

The zero lift drag coefficient is made up of the following four terms:

$$C_{D_0} = C_{D_P} + C_{D_F} + C_{D_{int}} + C_{D_{misc}} \qquad (3.17)$$

where C_{D_P} = pressure drag coefficient and is experimentally determined, estimated from Fig. 3.19, or assumed to be approximately 5%.

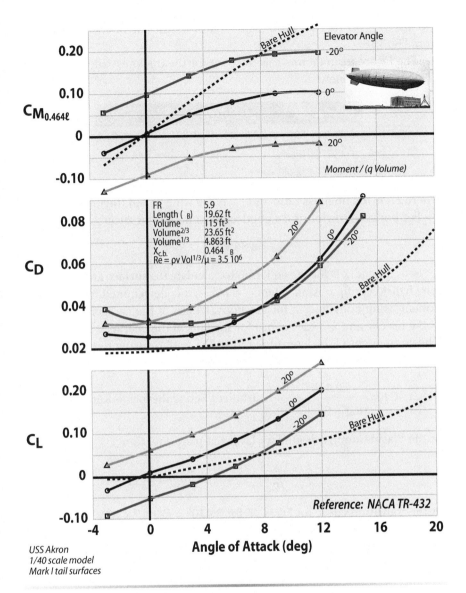

Figure 3.21 Effect of tails on aerodynamic characteristics for the USS Akron (ZRS-4). [7]

The C_{D_F} is the skin friction drag coefficient and is determined as follows (referenced to $Vol^{2/3}$):

$$C_{D_F} = \sum \text{component } C_{D_F} = \sum C_{f_{comp}} (FF)(S_{wet})_{comp} / Vol^{2/3} \quad (3.18)$$

for each component on the airship such as the body/envelope, the tails, the car/gondola, and engine nacelles. The S_{wet} is the wetted area of the component and the $C_{F_{comp}}$ is the flat plate skin friction coefficient of each component from Fig. 3.16.

The *FF* are form factors that account for the thickness and 3-D aspects of airship components:

Component	FF
Body/envelope, car/gondola	$1 + 1.5/(FR)^{3/2} + 7/(FR)^3$
Wings and tails	$1 + 1.2(t/c) + 100(t/c)^4$
Nacelles, smooth stores	$1 + 0.35/FR$

where *FR* is the fineness ratio and t/c is the wing or tail thickness ratio. For bodies of revolution a good approximation for surface area is

$$S_{wet} = 3.88 \, Vol^{2/3} \, (FR)^{1/3} \tag{3.19}$$

See Appendix D for the equation to calculate the surface area of a multi-lobe hybrid configuration. The $Re_\ell = \rho V_\infty \ell/\mu$ is determined for each component using body length for the body, length of the car/gondola, length of the nacelle and 2/3 root chord for the tails as the characteristic length ℓ. The component C_F is determined from Fig. 3.16 or using the Schoenherr-von Kármán turbulent boundary layer equation

$$C_f = \frac{0.455}{\left[\log_{10} Re_\ell\right]^{2.58}} \quad \text{(turbulent Schoenherr-von Kármán)} \tag{3.20}$$

or the Blasius laminar boundary layer flat plate equation

$$C_f = \frac{1.328}{\sqrt{Re_\ell}} \quad \text{(laminar flat plate)} \tag{3.21}$$

or the transition curve on Fig. 3.16 for $5 \times 10^5 < Re_\ell < 2 \times 10^6$.

Flat plate skin friction Eqs. (3.20) and (3.21) are plotted on Fig. 3.16. If it is assumed that boundary layers along a flat plate transition from laminar to turbulent at about a $Re_x = 5 \times 10^5 = \rho V x/\mu$ then the transition distance, x, would be about 1 ft from the leading edge of a flat plate for the following conditions {$x = (5 \times 10^5)(3.717 \times 10^{-7})/(0.002308)/(50 \times 1.689) = 0.95$ ft!} when traveling at a speed of 50 kt at a 1000 ft altitude. However, many airships have significant laminar runs that can extend to the maximum cross section and sometimes a little further aft. The calculation above showed transition occuring at 1 ft and yet actual transition may not occur for 100 ft or more! The difference is the pressure gradient. Flat plate skin friction coefficients, C_f, always assume no pressure gradient, $dp/dx = 0$. Gradual, well managed pressure gradients can usually maintain laminar flow on a body of revolution up to its maximum cross-sectional area.

An alternate method for determining the component C_{D_0} uses the zero lift drag area measured from existing airships. This alternate method can be used when the design is not mature or the information for the Re, wetted area, and FF is not available. The zero lift *drag area* is defined as *component drag area* (ft^2) = component zero lift drag/$q = C_{D_{comp}} \times$(component wetted area or frontal area) and is shown on Figs. 3.22 and 3.23 for most components on the airship. For example

$$\text{Envelope } C_{D_F} = (\textit{drag area} \text{ from Fig. 3.22})/Vol^{\frac{2}{3}} \qquad (3.22)$$

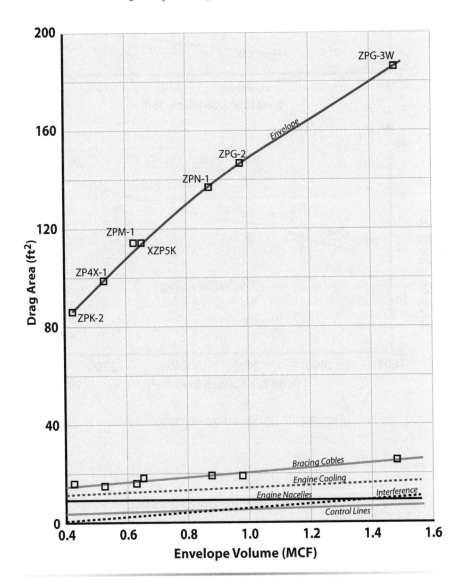

Figure 3.22 Airship component drags. [8]

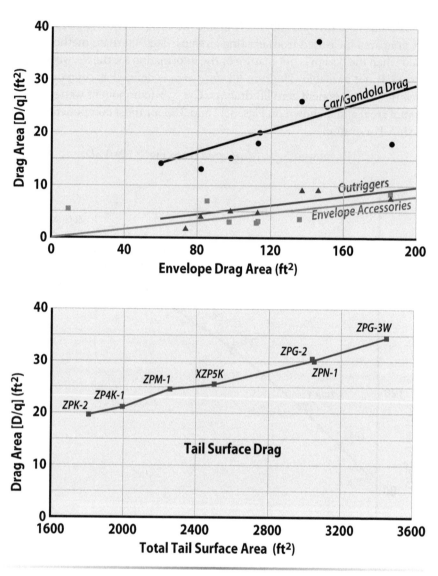

Figure 3.23 Airship drag components. [8]

The $C_{D_{misc}}$ is determined using the drag area values on Figs. 3.22 and 3.23 for the miscellaneous items on the airship such as the bracing cables, control lines, outriggers, landing gear, etc. The faired lines on Figs. 3.22 and 3.23 are curve fits through the data and the equations for these lines are given as follows.

Component	Equation for drag area	
Envelope	$C_DS_{env} = C_{Fenv} (FF)_{env} (S_{wet})_{env}$	(3.23)
Envelope accessories	$C_DS_{ea} = 0.04\ C_D\ S_{env}$	(3.24)
Tail surfaces	$C_DS_{ts} = 0.01\ (S_{plan})_{tails}$	(3.25)
Bracing cables	$C_DS_{bc} = 9.7 \times 10^{-6}\ Vol + 10.22$	(3.26)
Tail control lines	$C_DS_{cl} = 1.67 \times 10^{-6}\ Vol + 3.46$	(3.27)
Tail surface accessories	$C_DS_{tsa} = 7.0 \times 10^{-7}\ Vol + 0.625$	(3.28)
Car, gondola (+fairing)	$C_DS_{car} = 0.011\ (A_{frontal})_{car} + 0.6$	(3.29a)
Car, gondola (+fairing)	$C_DS_{car} = 0.108 C_D\ S_{env} + 7.7$	(3.29b)
Outriggers	$C_DS_{out} = 0.076(\#engine)(A_{frontal})_{out} + 3.3]$	(3.30a)
Outriggers	$C_DS_{out} = 0.044 C_D\ S_{env} + 0.92$	(3.30b)
Engine nacelle	$C_DS_{nac} = 4.25(\#engine)$	(3.31)
Engine cooling	$C_DS_{cool} = (\#engine)(2 \times 10^{-6}\ Vol + 4.1)$	(3.32)
Handling lines	$C_DS_{hl} = 7.9$	(3.33)
Interference	$C_DS_{mi} = 4.78 \times 10^{-6}\ Vol$	(3.34)
Fixed landing gear	$C_DS_{flg} = 1.76 \times 10^{-6}\ Vol + 4.68$	(3.35)
Retractable landing gear	$C_DS_{rlg} = 1.76 \times 10^{-6}\ Vol + 0.92$	(3.36)
Air cushion landing system	$C_DS_{acls} = 2.0$	(3.37)

In Eq. (3.25) the S_{plan} is the total planform area of the tails (not the wetted area). If the flight control system is a fly-by-wire and the tail rudders are actuated by servos (not externally mounted cables) then Eq. (3.27) is zero. If dimensions are available for the car/gondola and the frontal area can be estimated, use Eq. (3.29a) otherwise use Eq. (3.29b) [same for the outriggers in Eq. (3.30)]. Equation (3.37) uses empirical data from [14]. Find the best representation of the ACLS pad shape and use the $C_{D_p} \times$ the total frontal area of all ACLS pads. However, for the most part ACLS pads are retracted and faired for flight so drag is approximately zero.

The stabilizer and fins are constructed in one of two ways. When weight is critical (e.g., a high altitude long endurance airship) a fabric pressurized design is usually best. To keep the pressure low these pressurized tails look similar to air matresses with small radius curves stabilized by numerous internal septums. The other technique is to make extremely light space frame structure and cover it with material similar to that used for the envelope. Both tail designs are stabilized with guy wires at their tips. These guy wires have high drag as seen by their drag area given by Eq. (3.26). If the horizontal stabilizer/fin is cantilevered off the aft end with rigid structure then the bracing

The ZPG-3W is the last operational airship for the U.S. Navy and was decommissioned in 1962. It is the biggest non-rigid airship ever built. Four airships were built, all of which included a large internal radar antenna. Fifty years later an advanced airship design program called ISIS uses the same approach.

cables are not needed and $C_D S_{bc} = 0$. Notice that any external cables, control lines, or handling lines create a lot of drag.

Tail arrangements can also change the zero lift drag of an airship body. Figure 3.24 presents drag data for several tail configurations on the ZP5K airship. It also compares C_{D_0} and C_{L_α} for seven different tails to body alone data. Adding tails increases drag by ~50% but also increases C_{L_α} by more than double.

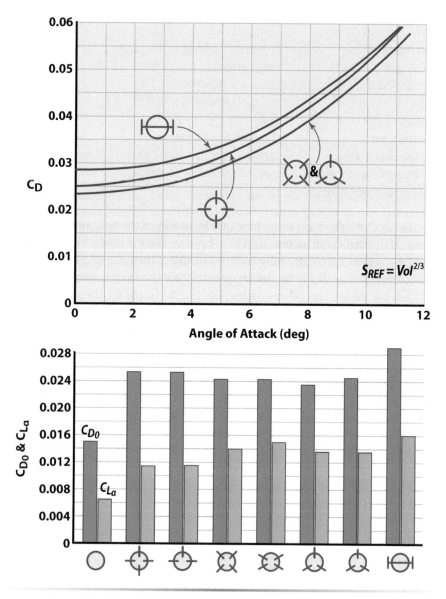

Figure 3.24   Drag and lift curve characteristics for ZP5K airship with different tails. [12]

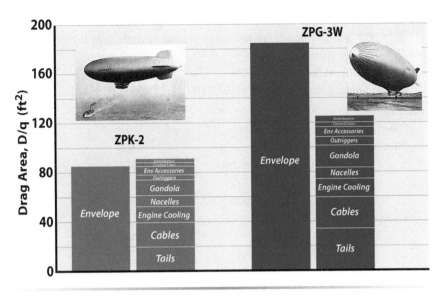

Figure 3.25   Zero lift drag area buildup for small and large airships. [8]

In Fig. 3.25 the drag area build-up is shown for a small airship ZPK-2 (volume = 425,000 ft^3) and a large airship ZPG-3W (volume = 1,465,000 ft^3). The body/envelope accounts for 50–60% of the total drag area with tails and cables accounting for approximately another 20%.

If the C_{D_0} for the operational airships and hybrids shown on Table 3.1 is referenced to S_{plan} by dividing their values by N_L it is observed that their C_{D_0} is similar to that of aircraft (C_{D_0} varying from 0.006 for the SR-71 to 0.018 for the F-16C). The C_{D_0} for lifting body configurations is larger (C_{D_0} = 0.04 – 0.062) because they were designed to have large C_{D_0} for the reasons mentioned earlier.

Sample Problem 3.1: Estimate the C_{D_0} for the ZP4K Shown in Fig. 1.9

The dimensions and performance of the ZP4K are as follows:

$$\text{Volume} = 527,000 \text{ ft}^3$$
$$Vol^{\frac{2}{3}} = 6518.7 \text{ ft}^2$$
$$\text{Length} = 263.34 \text{ ft}$$
$$\text{Maximum diameter} = 62.1 \text{ ft}$$
$$\text{Fineness ratio, } FR = 4.24$$

Surface area S_{wet} = 40,905 ft^2 [using Eq. (3.19) S_{wet} = 40,935 ft^2, which is very good agreement]

Assume maximum speed = 68 kt = 114 ft/s at 4000 ft (ρ = 0.00211 slug/ft^3 and μ = 3.657 × 10^{-7} slug/ft-s).

Envelope C_{D_0}:

$Re = \rho Vl/\mu = 17.3 \times 10^7$ (envelope is primarily turbulent flow)
$C_f = 0.00197$ [using Eq. (3.20)]
Envelope form factor $FF = 1.26$
Envelope drag area = drag/q = (0.00197)(1.26)(40,905) = 101.5 ft^2

Tail C_{D_0}:

From Fig. 1.15 the tail areas are:

Fins	upper	404
	horizontal	808
	lower	254
	subtotal	1466 ft^2
Rudders	upper	150
	lower	80
	subtotal	230 ft^2
Elevators (2)		300 ft^2
	Total tail plan form area =	1996 ft^2

Total tail wetted area $S_{wet} = 2(1996)(1.2) = 4790$ ft^2 where the 1.2 accounts for the increased area due to thickness. Using (2/3)(root chord) for the characteristic tail dimension gives a tail $Re = 18 \times 10^6$ (turbulent). Using Eq. (3.20) gives a tail $C_f = 0.0027$. Using a tail $t/c = 20\%$ and the wing/tail form factor equation gives $FF = 1.4$. The combined tail drag area is

$$\text{Tail drag area} = (0.0027)(1.4)(4790) = 18.1 \text{ ft}^2$$

The ZP4K drag area is the summation of all the components. Using the drag area from Eqs. (3.22–3.37) gives

Component	Drag area
Envelope ($C_D S_{env}$)	101.5
Envelope accessories	4.1
Tail surfaces	18.1
Bracing cables	15.3
Tail control lines	4.3
Tail surface accessories	1.0
Car/gondola	18.7
Outriggers	5.4
Engine nacelles (2)	8.5
Engine cooling	10.3
Handling lines	7.9
Retractable landing gear	1.85
Interference	2.5
Pressure drag (5% of envelope)	5.08
Total drag area	204.5 ft^2

The C_{D_0} of the ZP4K referenced to $Vol^{2/3}$ is $204.5/6518.7 = 0.031$. Notice that the envelope and accessories are 52% of the total C_{D_0}.

3.5.4 Drag-Due-to-Lift

The drag-due-to-lift coefficient C_{D_L} for uncambered airships and hybrids is composed of inviscid (induced) and viscous terms and is expressed as

$$C_{D_L} = K'C_L^2 + K''C_L^2 \tag{3.38}$$

where K' and K'' are the inviscid and viscous drag-due-to-lift factors respectively and are usually combined into the drag-due-to-lift factor $K = K' + K''$.

The inviscid $K'C_L^2$ term is due to the flow rolling around the wing tip or body edge from the high pressure to the low pressure region on the wing or body. The flow creates a vortex that trails behind the vehicle called a trailing vortex or wing vortex (the source of the destructive wake turbulence behind an aircraft that can cause a small aircraft trailing a large aircraft to be flipped upside down). This trailing vortex induces a downwash at the aerodynamic center of lift that results in an induced drag. This drag coefficient is expressed theoretically as

$$K'C_L^2 = C_L^2/\pi ARe \tag{3.39}$$

where the e is the *wing efficiency factor*. The e for airships is much less than 1 resulting in a large induced drag for airships. The e for aircraft is typically 0.5 to 1.0. The viscous drag-due-to-lift coefficient $K''C_L^2$ is due to the pressure drag on the hull of the airship or wing of the aircraft as lift is generated for $\alpha > 0$.

The combined drag-due-to-lift factor $K = \Delta C_D/C_L^2$ for the data set of aircraft and airships is determined from wind tunnel or flight test data and reported in Table 3.1. The K values for the airships and hybrids in Table 3.1 are multiplied by N_L (to reference them to S_{plan}), plotted vs AR, and shown on Fig. 3.26 along with the aircraft and lifting body values. Because K is dependent upon C_{L_α} the K for the body of revolution airships is extremely large, whereas the K for the multi-lobe hybrid airships is in line with that of the lifting bodies and fighter type aircraft having low AR.

3.5.5 Aerodynamic Data—Experimental

Figures 3.27–3.30 shows wind tunnel data for single lobe airships with and without tails. The data covers fineness ratios of 3.6 to 7.2. A body of $FR \leq 3.0$ is very bulbous and displays significant aft end flow separation and pressure drag (see Fig. 3.19). A body of $FR > 8$ is very long and slender, suffers severe body bending, and its skin friction drag is significantly greater than its pressure drag.

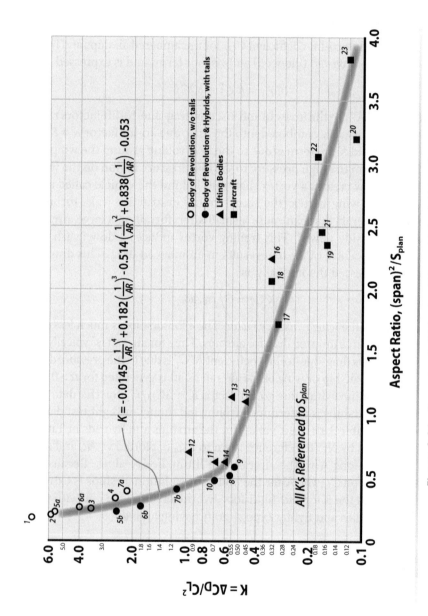

Figure 3.26   Drag-due-to-lift factor, K, for vehicles in Table 3.1.

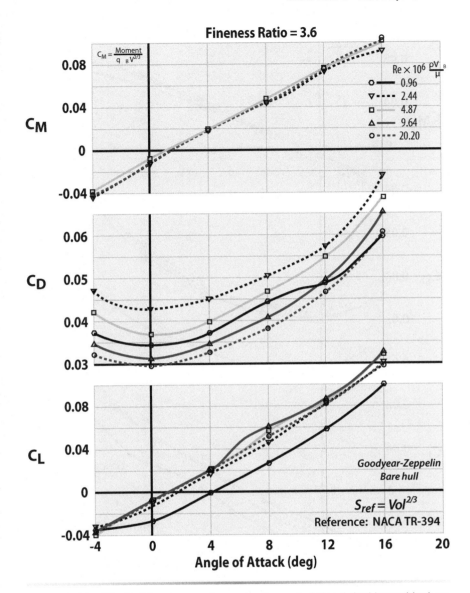

Figure 3.27 Effect of *Re* on aerodynamic characteristics (wind tunnel test on Goodyear Zeppelin designs). [6]

For slender bodies of revolution Fig. 3.31 shows the three primary aerodynamic quantities C_L, C_D, and C_M for a body (fuselage) with an ellipsoidal nose, followed by a constant cylindrical section, closed off by a parabolic shape. The slender body data shows symmetry about $\alpha = 0$ as it should. As alpha increases the efficiency of a slender body is very poor at low alphas and doesn't get much higher than $L/D = 2$ regardless of angle of attack.

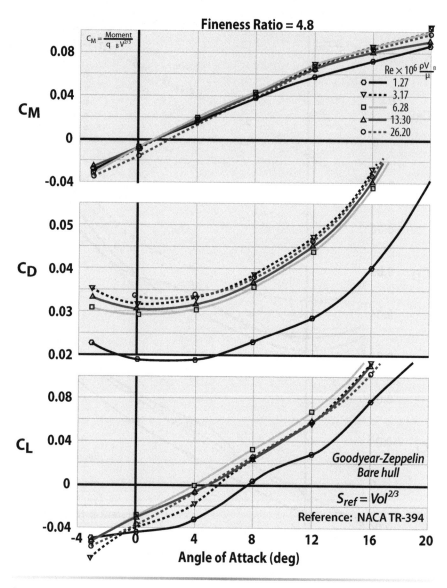

Figure 3.28   Effect of *Re* on aerodynamic characteristics (wind tunnel test on Goodyear Zeppelin designs). [6]

All of the experimental data display several important aspects. First, all of the drag polars are parabolic with C_L and symmetric about $C_L = 0$ when the tails are not deflected. Second, the body alone provides small amounts of lift (small C_{L_α}). The addition of tails gives a factor of two improvement in C_{L_α}. As shown in Fig. 3.24 the configuration of the tail has a strong influence on the C_{L_α} improvement. It is observed that after about $Re = 2 \times 10^6$ the *Re* has little effect on lift curve slope. However the *Re* continues to have a strong effect on C_{D_0}.

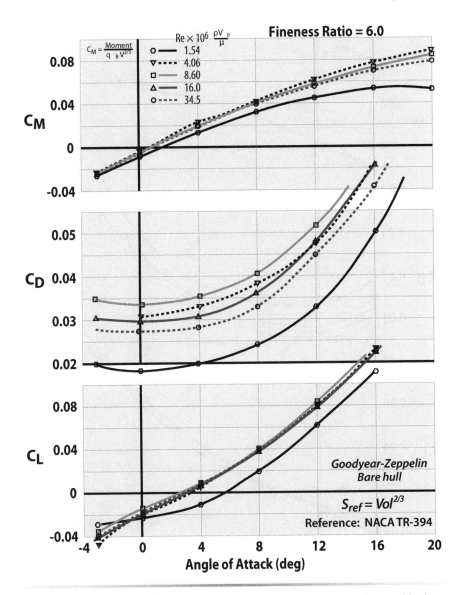

Figure 3.29 Effect of *Re* on aerodynamic characteristics (wind tunnel test on Goodyear Zeppelin designs). [6]

In Sec. 3.5.3 the C_{D_0} for the ZP4K was estimated to be 0.031. In Fig. 3.24 the C_{D_0} for the ZP5K is reported as 0.025. This does not suggest that the ZP5K is a more streamlined airship than the ZP4K. Figure 3.24 reports the results of a wind tunnel test of a 1/48 scale model of the ZP5K that did not have many of the features of an operational airship such as handling lines, envelope accessories, cooling drag, etc. An operational ZP5K has a $C_{D_0} = 0.03$.

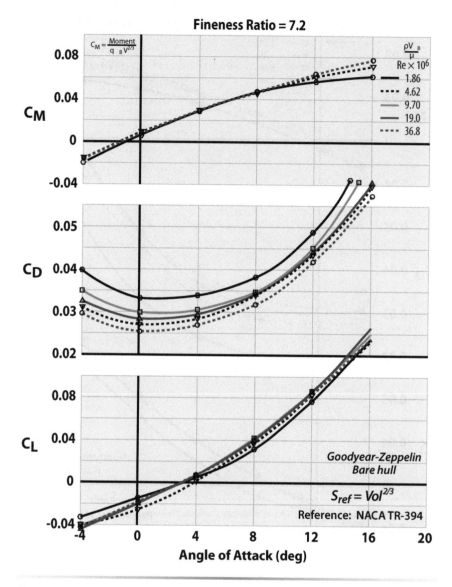

Figure 3.30 Effect of *Re* on aerodynamic characteristics (wind tunnel test on Goodyear Zeppelin designs). [6]

The USS Akron data in Fig. 3.21 demonstrates the following:

1. Drag polars are parabolic and do not exhibit any break for angles less than 15 deg.
2. Drag polars are symmetric when tails are not deflected.
3. The body provides small amounts of lift.
4. Adding tails more than doubles the amount of lift at a given alpha.
5. X_{ac} is located using Eq. (3.7) as follows.

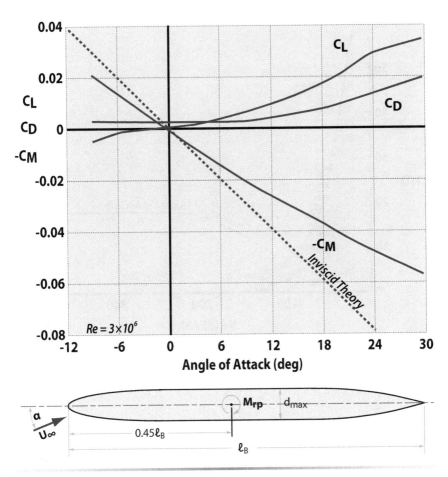

Figure 3.31 Aerodynamic characteristics of a slender body of revolution (fuselage). [2]

bare hull X_{ac} is ~$0.37\ell_B$ aft of the nose $\{0.464 - (0.10/0.25) \times 0.248 = 0.365\}$ + tails move X_{ac} to ~$0.44\ell_B$ aft of the nose $\{0.464 - (0.05/0.05) \times 0.248 = 0.44\}$

3.5.6 Total L/D

When performance comparisons are made between aircraft and airships it is useful to compare the total L/D of each vehicle. Figure 3.32 shows an overall comparison between an airship and a commercial transport. As was previously discussed speed is the important discriminator. One vehicle type is not the best at all speeds. When speed is less important the airship becomes the logical choice based on total L/D. Obviously, when the speed nears zero the total L/D becomes very large and essentially operates like a

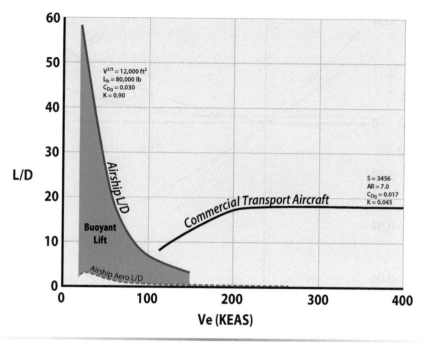

Figure 3.32 Total L/D comparison of an airship and a commercial transport.

balloon. Since there are always other requirements the vehicle with the highest L/D is not necessarily the best.

3.6 Added, Apparent, or Virtual Mass

One of the more misunderstood effects in all of aerodynamics is *added mass*. This term is sometimes also referred to as apparent mass or virtual mass. However, these terms do not accurately convey its true character because they imply that there is something mystical about them. In this book the term added mass will be used when discussing this effect. Another popular misconception is that added mass is only present for buoyant airships. This is not true. Added mass affects the motion of every object that is accelerating or decelerating in a surrounding fluid whether or not there is any appreciable buoyancy. However, the effects of added mass are only significant when the mass of the object is similar to that of the displaced surrounding fluid.

When a body accelerates, decelerates, or changes direction while moving in a fluid, it behaves as though it has more mass than it actually does. The apparent increase in mass and distribution of this added mass varies with the nature of the motion. A complete discussion of added mass appears in Appendix C and includes a complete development of the equations that are used to calculate the total force and moment that is consistent with a body's accelerated motion.

In Appendix C the buoyancy ratios of several objects are compared including an air bubble. Read Appendix C to find out why an air bubble in water has a buoyant force equivalent to ~770 g's and yet only experiences an acceleration of ~2 g's.

DELAG and the LZ-120 Bodensee

The world's first passenger airline, DELAG (German Airship Transportation Corporation Ltd.) was established in 1909 as an offshoot of the Zeppelin Company. While most of the early flights were sightseeing tours, the LZ-120 Bodensee began scheduled service in 1919 between Berlin and southern Germany. The flight from Berlin to Friedrichshafen took 4 to 9 hr compared to the 18 to 24 hr by rail. The Bodensee made 103 flights and carried almost 2500 passengers, 11,000 lb of mail and 6600 lb of cargo.

With its revolutionary design and four 245 hp Maybach engines, the LZ-120 Bodensee could reach a speed of 82 mph. The LZ-120's shape provided less drag, increased speed, and greater aerodynamic lift, and it became the basic model for the LZ-126 Los Angeles, LZ-127 Graf Zeppelin I, LZ-129 Hindenburg, and LZ-130 Graf Zeppelin II airships.

References

[1] Abbott, I.H., and von Doenhoff, A.E., *Theory of Wing Sections*, Dover, New York, 1959.

[2] Schlichting, H., and Truckenbrodt, E., *Aerodynamics of the Airplane*, McGraw-Hill, pp. 331–337, 1979.

[3] Hoerner, S.F., *Fluid Dynamic Drag: Practical Information of Aerodynamic Drag and Hydrodynamic Resistance*, Chapter 6, Midland Park, New Jersey, 1965.

[4] Lutz, T., and Wagner, S., *Drag Reduction and Shape Optimization of Airship Bodies*, AIAA Journal of Aircraft, Vol. 35/No. 3, May 1998.

[5] Abbott, I.H., *Fuselage Drag Tests in the Variable Density Wind Tunnel: Streamline Bodies of Revolution, Fineness Ratio of 5*, NACA TN-614, 1937.

[6] Abbott, I. H., *Airship Model Tests in the Variable Density Wind Tunnel*, NACA TR-394, 1931.

[7] Freeman, H.B., *Force Measurements on a 1/40 Scale Model of the U.S. Airship Akron*, NACA TR-432, 1933.

[8] Wright, J.M., and Adams, R.E., *An Empirical Method for Non-Rigid Airship Preliminary Drag Estimation*, AIAA 91-1277-CP, pp. 59–66, 1991.

[9] Saltzman, E.J., Wang, C.K., and Iliff, K.W., *Flight-Determined Subsonic Lift and Drag Characteristics of Seven Lifting-Body and Wing-Body Reentry Vehicle Configurations With Truncated Bases*, AIAA Paper 0383, 37th AIAA Aerospace Sciences Meeting, Reno NV, 11–14 Jan 1999.

[10]   Lowry, J.G., and Polhamus, E., *A Method for Predicting the Subsonic Lift Curve Slope for Straight Taper Wings*, NACA TR 3911, 1957.

[11]   Ashley, H., and Landahl, M., *"Aerodynamics of Wings and Bodies"*, Addison-Wesley, Reading MA, 1965.

[12]   Ross, S.A., and Liebert, H.R., *LTA Aerodynamics Handbook*, Goodyear Aircraft Corp., Akron, OH, 1954.

[13]   Nicolai, L.M., and Carichner, G.E., *"Fundamentals of Aircraft and Airship Design, Volume I—Aircraft Design"*, AIAA, Reston, VA, 2010.

[14]   Nicolai, L.M., and Carichner, G.E., *Airplanes and Airships . . . Evolutionary Cousins*, AIAA Paper 2012-1178, 50th AIAA Aeronautical Sciences Meeting, Nashville, TN, 12 Jan 2012.

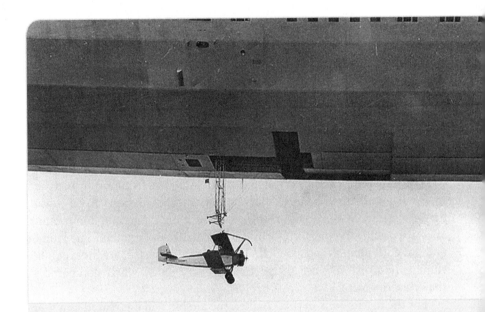

- Steady-state Performance Method
- Power Required
- Breguet Range and Endurance
- Payload-Range Curve
- Takeoff Performance
- Landing
- Field Length
- Turning

The USS Akron was fitted with a trapeze apparatus to launch/recover the Curtiss F-9C Sparrowhawk. The idea in 1920 was that airships could carry their own pursuit aircraft for protection from enemy aircraft. Flight trials were conducted but the idea never caught on. The same idea did not work for B-36 bombers, either.

He who forgets the mistakes of the past is destined to repeat them.
Anonymous

4.1 Introduction

S ince an airship is an air vehicle the laws of physics that govern the flight of a winged aircraft are exactly applicable. Similarly the governing equations of motion and the forces and moments are the same . . . with the understanding that the buoyant force is zero for the aircraft. Airship performance analysis of cruise range, loiter, landing, and takeoff is only dependent on the aerodynamic lift, aerodynamic drag, propulsive/reverse thrust, fuel consumption, and landing gear friction force. These forces are the same as for aircraft such that the airship performance equations are the same as for aircraft. The buoyant lift only enters into the analysis to define total lift [see Eq. (4.3)]. The buoyant lift can create a large pitching moment if the center of gravity (c.g.) and the center of buoyancy (c.b.) are very much misaligned (see Fig. 4.1).

This chapter considers steady-state and accelerated performance methods for airships. A large portion of an airship's mission can be considered as steady-state (equilibrium) because long range and long endurance are usually the most important performance parameters. The landing and takeoff phases and the climb-acceleration phase are not equilibrium conditions and are proportionally very small when compared to the cruise segment.

For the discussions in this chapter, the airship will be treated as a point mass system with translation and rotation degrees of freedom and subject to aerodynamic, propulsive, buoyant, and gravity forces. The force diagram for the airship is shown in Fig. 4.1 (x-z plane) where the lift and drag forces are normal and parallel to the free-stream velocity V_∞ respectively. See Fig. 3.4 for the sign convention in the x-y plane and y-z plane.

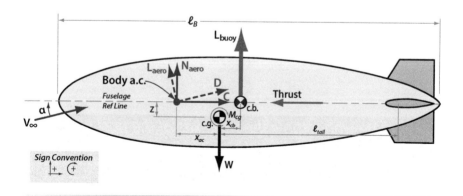

Figure 4.1   Forces and moments acting on a buoyant airship.

4.2 Level Unaccelerated Flight

During level unaccelerated flight the flight path angle γ is zero and all external forces acting on the aircraft are in balance. Thus, adding forces normal and parallel to V_∞ (the wind axis) yields the following

$$W = L_{aero} + L_{buoy} \cos \alpha + T \sin \alpha \qquad (4.1)$$

$$T \cos \alpha = D + L_{buoy} \sin \alpha \qquad (4.2)$$

Because α is usually small during most airship missions, Eqs. (4.1) and (4.2) can be expressed for equilibrium flight as

$$W = L = L_{aero} + L_{buoy} = C_{L_{aero}} q\, Vol^{2/3} + L_{buoy} \qquad (4.3)$$

$$T = D = C_D q\, Vol^{2/3} \qquad (4.4)$$

where $q = \tfrac{1}{2}\rho_\infty V_\infty^2$ is the dynamic pressure and $Vol^{2/3}$ is the reference area for C_L and C_D (airships use Volume$^{2/3}$, which is represented by $Vol^{2/3}$ throughout this book). It is important to recognize that

$$L_{buoy} \neq f(\alpha, V, h) = \text{constant}$$

and

$$L_{aero} = f(\alpha, V, h) = C_{L_{aero}} q\, Vol^{2/3}$$

It is also convenient to define two terms that will be used throughout the book. First, is the *buoyancy ratio*, or *BR*, and second is *heaviness*, W_H.

$$BR = \frac{L_{buoy}}{W} = \frac{L_{buoy}}{L_{aero} + L_{buoy}} \qquad (4.5)$$

$$W_H = \left(W - L_{buoy}\right) = W(1 - BR) = \frac{L_{buoy}}{BR}(1 - BR) = L_{aero} \qquad (4.6)$$

L_{buoy} cannot be "turned off," which is why airships with buoyancy ratios, *BR*, close to 1.0 must be able to carry ballast and be tied down during ground operations. This feature has been the major disadvantage of airships since their invention more than 200 years ago. On the other hand L_{aero} can be modulated or completely turned off by adjusting the angle of attack α, flight speed V, altitude h, or any combination of the three. Negative aerodynamic lift can even be obtained by flying at a negative angle of attack.

Designing the airship as a hybrid (*BR* < 1.0, discussed in Chapter 12) is a way of ameliorating the disadvantage mentioned earlier by being able to turn off/adjust part of the lift.

During ground operations $L_{aero} = 0$ and $W \geq L_{buoy}$. During cruise $W = L_{aero} + L_{buoy} = W_H + L_{buoy}$ and L_{aero} is varied to account for the fuel burned. During climb, $L > W$ by increasing α. During descent, $L < W$ by decreasing α. The α is controlled by the force on the horizontal tails. Climb and descent can also be modified by vectoring some of the engine thrust.

Remember the airship drag coefficient C_D (from Chapter 3) is

$$C_D = C_{D_0} + K'C_{L_{aero}}^2 + K''\left(C_{L_{aero}} - C_{L_{min}}\right)^2 \tag{3.14}$$

where K' and K'' are the inviscid and viscous drag-due-to-lift factors respectively.

Because conventional airships typically have low camber we can set $C_{L_{min}} \approx 0$. Thus, we can express the drag coefficient as

$$C_D = C_{D_0} + K\,C_{L_{aero}}^2 \tag{4.7}$$

where $K = K' + K''$ and C_{D_0} is primarily skin friction drag with some pressure drag on the aft body due to separation. There will be some interference drag and other miscellaneous drags as shown in Fig. 3.22. Figure 3.26 shows the influence of configuration on the drag-due-to-lift factor K for various aircraft and airship configurations.

4.3 Power Required

Because propulsion systems for airships do not involve jet engines and are only powered by piston and turboshaft engines driving propellers, thrust terms have little value and only power terms are used for airship performance. Therefore, starting with Eq. (4.7) the drag determines the thrust required T_R and is written as Eq. (4.8).

$$T_R = D = C_{D_0}\,q\,Vol^{2/3} + K\left(L_{aero}/q\,Vol^{2/3}\right)^2 q\,Vol^{2/3} \tag{4.8}$$

But power is what we are looking for so multiply Eq. (4.8) by velocity, V, to convert to a *power required* relationship as shown in Eq. (4.9). A typical plot of the zero lift drag and drag-due-to-lift contributions to power required for CA-1 and HA-1 configurations is illustrated in Fig. 4.2.

$$\text{Power} = V \times D = V \times C_{D_0}\,q\,Vol^{2/3} + V \times K\left(L_{aero}/q\,Vol^{2/3}\right)^2 q\,Vol^{2/3} \tag{4.9}$$

$$\{\text{zero lift term}\} \qquad \{\text{drag-due-to-lift term}\}$$

$$\text{Power} = V\left[\left(C_{D_0} + K\,C_{L_{aero}}^2\right)q\,Vol^{2/3}\right] \tag{4.10}$$

$$P_R = \frac{V\left[\left(C_{D_0} + K\,C_{L_{aero}}^2\right)q\,Vol^{2/3}\right]}{550\,\eta_p} \tag{4.11}$$

where 550 is the conversion from ft-lb/s to horsepower, V is the flight speed in fps, and η_p is the propeller efficiency.

The first term in Eq. (4.9) is the zero lift drag and the second term is the drag-due-to-lift during level unaccelerated flight. The zero lift drag is independent of the BR but the drag-due-to-lift term is not. Figure 4.2 shows calculations for both CA-1 and HA-1. Because long endurance and range are very important for an airship it is important to identify the operating conditions where maximum endurance and range occur. The minimum power required speed on Fig. 4.2 is also the minimum fuel flow point. For aircraft, flight below this speed is usually limited by either wing buffet or wing stall. However, neither of these conditions exists for airships. The maximum speed is at the intersection of the power available curve and the total power required curve for both aircraft and airships. Obviously, the maximum speeds for aircraft and airships are quite different.

Sample Problem 4.1: Airships CA-1 and HA-1 Power Required

This example will determine the power required for a conventional airship (CA-1) and a hybrid airship (HA-1) having the characteristics summarized in Appendix D and Table 4.1. The first airship CA-1 is a conventional body of revolution configuration with a fineness ratio $FR = 4.0$ and a BR @ takeoff of 0.91. The second airship HA-1 is a three lobe hybrid configuration with a BR @ takeoff of 0.75.

The one lobe body of revolution airship CA-1 is 91% buoyant allowing it to burn off its fuel and still have some heaviness (1000 lb) at landing. A $BR = 0.91$ equates to 9% aerodynamic lift for changing altitude and adjusting for fuel burn. CA-1 would need to be tied down during the unloading of its 20,000 lb payload at its destination because its heaviness is only 1000 lb at landing. It would then be loaded with a new payload of 20,000 lb and/or ballast and refueled for the return flight.

The hybrid airship HA-1 has a $BR = 0.75$ so that there is 25,000 lb ($0.25 \times 100,000$) of aerodynamic lift available for adjusting to fuel burn and unloading the payload. Its $C_{L_{aero}} = W(1-BR)/q$ $Vol^{2/3} = W_H/q\ Vol^{2/3}$ is determined for a range of flight speeds, V, and used in Eq. (4.11) to determine the power required shown in Fig. 4.2.

The power required (P_R) variation and the various operating points for both CA-1 and HA-1 are shown on Figs. 4.2a and 4.2b respectively.

Historically the term *dynamic lift* has been used instead of heaviness. The terms heaviness and aerodynamic lift will be used throughout this book as their names convey what they actually are. When there is no thrust, vectoring heaviness and aerodynamic lift are synonymous.

Table 4.1 Sample Problem 4.1

Config	TOGW	Fuel	C_{D_0}	K	C_{L_α}	η_p	BR
CA-1	56,000	5000	0.026	0.9	0.0115	0.65	0.91
HA-1	100,000	10,000	0.033	0.28	0.045	0.65	0.75

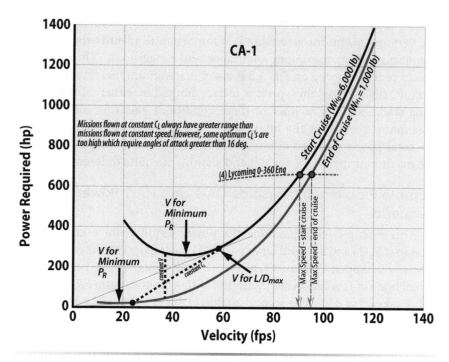

Figure 4.2a   CA-1 cruise power required for various speeds, $h = 4000$ ft (Sample Problem 4.1).

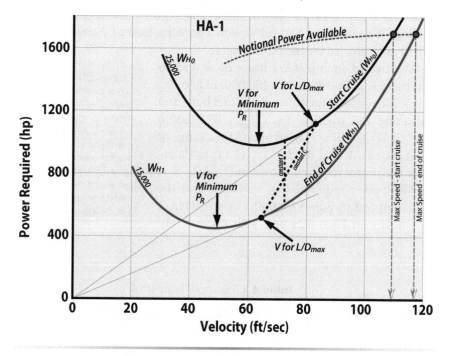

Figure 4.2b   HA-1 cruise power required for various speeds, $h = 4000$ ft (Sample Problem 4.1).

The minimum PR for CA-1 varies between 260 hp at 44 f/s and 18 hp at 18 f/s depending on the heaviness. For the HA-1 hybrid airship the P_R varies between 990 hp at 63 f/s and 460 hp at 49 f/s from start of mission to end of mission respectively.

The engine selection is a "designers choice" and depends upon how fast the airship needs to go and the number of engines. The engine is sized for the start of mission (maximum heaviness). The engines for this CA-1 airship example will be selected in Sample Problem 5.1.

From Fig. 4.2b it is observed that the minimum power for HA-1 is almost a factor of four greater than the minimum power for CA-1. Given that difference the question arises as to why anyone would design a hybrid airship that requires so much more power than a conventional airship. The answer is not found in the efficiency of the system but rather the flexibility, ease of operation, and superior ground handling. The CA-1 airship will need to be tied down during the unloading of its 20,000-lb payload and will need an equal weight of payload or ballast for the return trip. The HA-1 operates similar to an airplane on the ground and carries significantly more payload than a conventional airship. HA-1 does not need any unique infrastructure or ground support during unloading and loading of weight/ballast for the return trip. The benefits of these operational issues will be considered in more detail in Chapter 12 and form the basis for the benefit/penalty trade study performed in the conceptual stages of airship design.

4.4 Minimum Drag and Maximum L_{aero}/D

Continuing with Fig. 4.2 notice that it is possible to fly slower than the speed for minimum P_R and this is generally referred to as flying on the back side of the power curve. What this means is that when flying very slow increased speeds actually require less power. Referring to Fig. 4.2, as speed increases from the point of minimum power the speed for maximum L/D is attained that represents the flight condition for maximum range. It should be obvious from Fig. 4.2 that flying at maximum L/D as fuel burns off requires that the airship vary its speed to fly at the constant C_L for L/D_{max}. Subsequent discussions will compare flights at constant C_L vs flights at constant speed.

Power required curves are useful when analyzing reciprocating-engine propeller-driven airships. Reciprocating engine fuel flow rate is proportional to power output rather than thrust output. A useful conversion factor that the designer should remember is that horsepower equals 550 ft-lb/s.

It is worth looking at the equations for lift, drag, and power that often provide exact calculations for the speeds for minimum drag, minimum power, minimum fuel flow, maximum L/D, and maximum range and endurance. For these discussions the total drag coefficient for an airship is expressed in the same manner as for an aircraft [Eq. (4.7)]. However, the difference is that the approximate equality of lift and weight for equilibrium

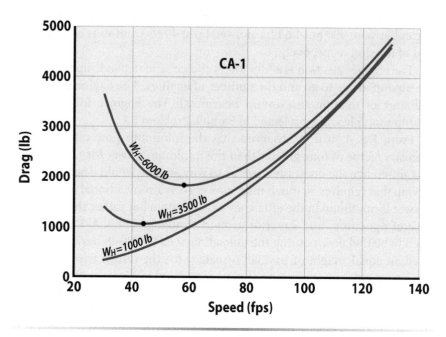

Figure 4.3a   Effect of speed on drag for CA-1 at various heavinesses.

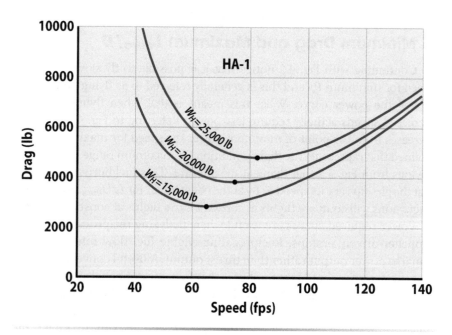

Figure 4.3b   Effect of speed on drag for HA-1 at various heavinesses.

flight typical of aircraft is not valid for airships. Because most of an airship's lift is produced by a buoyant gas there is no drag-due-to-lift produced by the buoyant lift force.

$$C_D = C_{D_0} + K C_{L_{aero}}^2 \qquad \text{(airships)} \qquad (4.7)$$

Equation (4.7) becomes a relationship for airship drag in Eq. (4.12) by multiplying Eq. (4.7) by $q\ Vol^{2/3}$.

$$\text{Drag} = D = \left(C_{D_0} + K C_{L_{aero}}^2\right) q\ Vol^{2/3} \qquad (4.12)$$

Although the speed for minimum drag (maximum L/D) is noted on Fig. 4.2 this speed is verified by the data plotted in Figs. 4.3 and 4.4. Here calculations are presented for CA-1 and HA-1 for both drag and L/D. The fact that the speed for minimum drag and the speed for maximum L/D are the same is shown in these figures. Equations will be developed later in the chapter that prove that the speed for maximum L/D and maximum range are also the same.

4.5 Breguet Range

Historically, the range and endurance of an aircraft were easily calculated using the Breguet equations whose genesis begins with Eq. (4.13a).

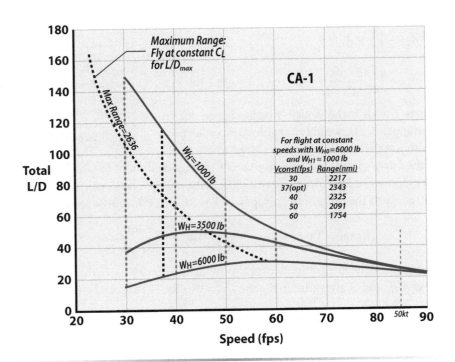

Figure 4.4a Effect of heaviness and speed on total L/D (CA-1).

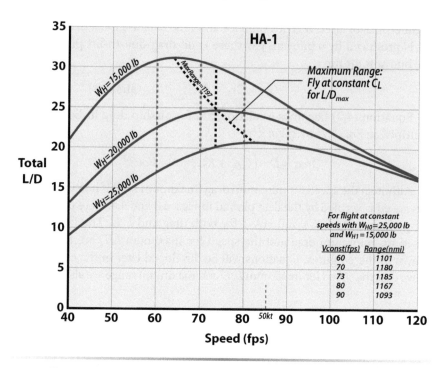

Figure 4.4b   Effect of heaviness and speed on total *L/D* (HA-1).

$$\text{Range} = \int_{t_i}^{t_f} V \, dt \qquad (4.13a)$$

$$\text{Range} = \int_{W_i}^{W_f} \frac{V}{dW/dt} \, dW \qquad (4.13b)$$

The Breguet range equation is named after Louis Charles Breguet, the record setting aviation designer and builder. Breguet pioneered the development of the helicopter and built the first piloted vertical ascent aircraft.

where dW/dt is the vehicle weight change due to burning fuel. The weight change for a propeller propulsion system is $dW/dt = (BSFC)(\text{hp})$ where $BSFC$ is the *brake specific fuel consumption* of the engine in lb of fuel/hr-hp and hp (horsepower) is hp $= (\text{drag})(\text{speed})/550\,\eta_p$. Also, $\eta_p = \text{propeller efficiency}$ and the 550 converts the lb-fps into horsepower. Equation (4.13b) can be rewritten as

$$\text{Range} = \int V \, dt = \int \frac{V}{BSFC \times P_R} \, dW = \int \frac{326\,\eta_p}{BSFC \times D} \, dW \qquad (4.14a)$$

and because the integration is over the initial heaviness W_{H_0} to the final heaviness W_{H_1} and at any point in flight $L_{aero} = W_H$ Eq. (4.14a) can be rewritten as

$$\text{Range (nmiles)} = 326 \int \frac{\eta_p}{BSFC} \frac{L_{aero}}{D} \frac{dW_H}{W_H} \qquad (4.14b)$$

Note that the heaviness term W_H has been substituted for weight in the above equation. From an aerodynamic point of view heaviness for an airship is the equivalent of weight for an airplane. Heaviness is the lift that must be generated by flying at an angle of attack and is the source of any drag-due-to-lift for the airship. As mentioned before the constant buoyant lift force generates no drag-due-to-lift.

4.5.1 Cruise Strategy #1

The airship has two possible cruise strategies. The first cruise strategy is at constant C_L that makes L_{aero}/D a constant over the mission. Because airships usually cruise at a constant altitude the speed must decrease as fuel is burned to keep C_L constant. The C_L is usually selected to fly at maximum $L_{aero}/D = 1/[4(C_{D_0})(K)]^{1/2}$, which is also a minimum drag flight.

If we assume C_L, η_p, and BSFC constant over the weight change (fuel burned), Eq. (4.14b) can be integrated to give an exact solution for cruise strategy #1

$$\text{Range (nmiles)} = \frac{326\eta_p}{BSFC}\frac{L_{aero}}{D}\ell_n\left[\frac{W_{H_0}}{W_{H_1}}\right] \qquad (4.15)$$

Equation (4.14a) can also be numerically integrated as follows:

$$\text{Range} = \sum\left[\bar{V}/(BSFC\times\overline{Power})\right](W_{H_{i+1}} - W_{H_i}) \qquad (4.16)$$

for $i = 0$ to n increments where $\bar{V}$ and $\overline{Power}$ are the average speed and power over the weight increment. The quantity $\bar{V}/(BSFC\times\overline{Power})$ is called the *range parameter* or *range factor*.

4.5.2 Cruise Strategy #2

The second cruise strategy is to fly at constant speed. For this strategy the C_L must decrease as fuel is burned to fly at constant altitude (constant q). Returning to Eq. (4.14a) and substituting $\left(C_{D_0} + KC_{L_{aero}}^2\right)q\,Vol^{2/3}$ for the drag term and $C_{L_{aero}} = W_H/q\,Vol^{2/3}$ gives

$$\text{Range(nmiles)} = 326\int\frac{\eta_p}{BSFC}\frac{dW_H}{q\,Vol^{2/3}\left[\left(C_{D_0} + K\dfrac{W_H^2}{q^2\,Vol^{4/9}}\right)\right]} \qquad (4.17)$$

Equation (4.17) can be integrated from the initial W_{H_0} to the final W_{H_1} for constant speed (q), BSFC, and η_p to give an exact solution for the cruise strategy #2 range.

Range

$$= \frac{326\eta_p}{BSFC}\frac{1}{\sqrt{K\,C_{D_0}}}\left[\tan^{-1}\left(\frac{W_{H_0}}{q\,Vol^{2/3}\sqrt{\frac{C_{D_0}}{K}}}\right) - \tan^{-1}\left(\frac{W_{H_1}}{q\,Vol^{2/3}\sqrt{\frac{C_{D_0}}{K}}}\right)\right] \qquad (4.18)$$

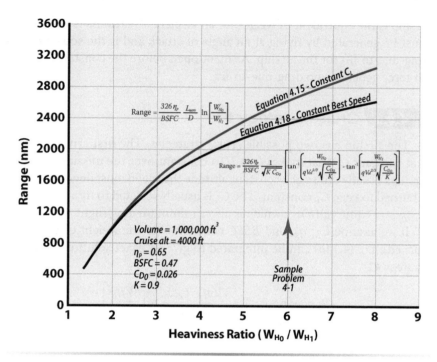

Figure 4.5 Range calculation comparison—for constant C_L and constant speed strategies (data for CA-1 Sample Problem 4.1).

The range for the two cruise strategies is compared on Fig. 4.5 using the data of Sample Problem 4.1. For a given amount of fuel (W_{H_0}/W_{H_1}) constant C_L flight [Eq. (4.15)] always gives an increased range over constant speed strategy [Eq. (4.18)] but takes longer because the speed decreases. The cruise strategy employed will depend upon the user requirements . . . maximum range or minimum time. It is important to remember that for constant C_L cruise Eq. (4.15) must be used and for constant speed cruise Eq. (4.18) must be used.

4.5.3 Cruise Strategy #3

Taking notice of Fig. 4.5 it might make sense to fly a combined cruise strategy: start the mission at a high value of W_{H_0}/W_{H_1} with a constant C_L cruise, then at a specified speed switch to a constant speed cruise for the remainder of the flight. This combined strategy would yield more range than a constant speed for the entire cruise distance and less cruise duration than for a constant C_L for the entire cruise distance. The CargoStar example in Sample Problem 6.1 will demonstrate this combined cruise strategy.

4.5.4 Optimal Flight at Constant C_L (Winged Aircraft)

For a winged aircraft we would find the $C_{L_{aero}}$ that minimizes an aircraft's total drag or maximizes its L/D. The assumption of flight at a constant C_L would be used to generate equations that specifically calculate optimum values of C_L and speeds. For subsonic incompressible flight the maximum aerodynamic L/D occurs at one C_L. However, this is not the case for airships. Flight is typically not at a constant C_L but is generally at constant speed and constant altitude, which means C_L varies over the flight. Optimum speeds for other important conditions such as minimum power or drag and maximum L/D are calculated the same regardless of whether flight is at constant C_L or at constant speed.

It must also be recognized that total L/D (which includes buoyant lift) is not a constant since the drag term varies with speed given a constant buoyant lift term. For an airship the L/D is a combination of the aerodynamic L/D (which doesn't vary with speed) and the buoyant L/D (which does vary with speed).

As a reminder the following equations have been developed in Reference 3 and many other textbooks on aerodynamic design. The equations will be presented without discussion. All of the equations use the standard approach for finding maxima and minima with differentiation and setting this differential equal to zero.

$$C_{L_{opt}} = \sqrt{\frac{C_{D_0}}{K}} \qquad \text{minimum drag and } L/D_{max} \qquad (4.19)$$

Substituting this value of $C_{L_{aero}}$ into Eq. (4.8) and dividing by C_L results in the expression for $(L_{aero}/D)_{max}$ below [Eq. (4.16)].

$$\left(L_{aero}/D\right)_{max} = \frac{1}{2\sqrt{C_{D_0} K}} \qquad (4.20)$$

It is also instructive to find the speed for maximum L_{aero}/D or minimum drag. This speed is presented with an expanded equation for $C_{L_{opt}}$, solving for speed, V, and then substituting Eq. (4.19).

$$V_{\left(L_{aero}/D\right)_{max}} = \sqrt{\frac{2\,L_{aero}}{\rho C_{L_{opt}} Vol^{2/3}}} = \sqrt{\frac{2\,W_H}{\rho C_{L_{opt}} Vol^{2/3}}} = \sqrt{\frac{2\,W_H}{\rho Vol^{2/3}}}\sqrt{\frac{K}{C_{D_0}}} \qquad (4.21)$$

where the *heaviness*, $W_H = L_{aero}$, ignores any contribution to vertical force (lift) from vectored thrust and assumes equilibrium flight. The term *heaviness* is often used to describe the weight state of an airship where it is defined as $W_H = W - L_{buoy}$. Recall that the definition of *weight* is $W = L_{aero} + L_{buoy} + T_{vec}$. In other words, *heaviness* is that portion of the weight that is not supported by the vertical buoyant force L_{buoy}. Technically speaking,

$W_H = L_{aero} + T_{vec}$ where $T_{vec} = T \sin\alpha$. For most cruise conditions the thrust vectoring force, T_{vec}, is assumed to be small and is ignored for discussion purposes in this book.

This process also leads to the speed for minimum P_R as

$$V_{min\,P_R} = \sqrt{\frac{2\,W_H}{\rho\,Vol^{2/3}}}\sqrt{\frac{K}{3C_{D_0}}} \qquad (4.22a)$$

$$C_{L_{min}\,P_R} = \sqrt{\frac{3C_{D_0}}{K}} \qquad (4.22b)$$

Notice that the speed for minimum P_R is 24% less than the speed for maximum L_{aero}/D or minimum drag in Eq. (4.21).

4.5.5 Optimal Flight at Constant Speed (Airships)

Since airships usually fly at constant speed rather than at constant C_L a different set of equations that calculate maximum L/D, minimum drag, minimum power required, and their associated speeds will now be developed. The technique is the same as in Sec. 4.5.4 except the differentiations are done with respect to velocity, V, instead of C_L.

It is possible to derive similar equations for the speeds where P_R is minimum, drag is minimum, L/D is maximum, and range is maximum. Unfortunately, a closed form solution for the speed where endurance is maximum does not exist. The following discussion develops the equations for minimum P_R, minimum drag, and maximum range (is shown to be the same speed as for maximum L/D).

As before, we start with an expression for drag such as Eq. (4.12) where the equation is then rewritten in terms of velocity as Eq. (4.23). This results in the following:

$$\text{Drag} = D = C_{D_0}\, q\, Vol^{2/3} + K\, W_H^2 / q\, Vol^{2/3} \qquad (4.12)$$

$$D = C_{D_0}\, \tfrac{1}{2}\rho V^2\, Vol^{2/3} + K\, W_H^2 / \tfrac{1}{2}\,\rho V^2\, Vol^{2/3} \qquad (4.23)$$

Applying the differential dD/dV and setting it equal to zero yields

$$V^4 = \frac{4KW_H^2}{\left(\rho Vol^{2/3}\right)^2 C_{D_0}} \qquad \text{speed for minimum drag} \qquad (4.24)$$

Notice that Eq. (4.24) is the same as Eq. (4.21) for constant C_L after recognizing that W_H for an airship is aerodynamically equivalent to W for an aircraft. The same technique is then used to find the speed for minimum power required. Equation (4.12) is multiplied by V to calculate P_R:

$$P_R = V \times D = V \times C_{D_0}\, \tfrac{1}{2}\rho V^2\, Vol^{2/3} + V \times K\, W_H^2 / \tfrac{1}{2}\,\rho V^2\, Vol^{2/3} \qquad (4.25)$$

Applying the differential dP_R/dV and setting it equal to zero yields the closed form solution for the speed at which minimum power required occurs:

$$V^4 = \frac{4KW_H^2}{3\left(\rho Vol^{2/3}\right)^2 C_{D_0}} \qquad \text{speed for minimum } P_R \qquad (4.26)$$

which is the same as Eq. (4.22a).

Given that the speed for minimum drag is the same regardless of whether C_L or V is constant, it's no surprise that Eq. (4.26) is also the same as the one for constant C_L [Eq. (4.22)]. When the best constant speed for a mission is needed for a given amount of fuel burned (*heaviness*$_0$ – *heaviness*$_1$) substitute the average $W_{H_0} W_{H_1}$ for the W_H^2 term in both Eqs. (4.24) and (4.26).

The last speed of importance is for maximum L_{aero}/D, which also happens to be the speed for maximum range. Taking the Breguet range equation derived for constant speed, Eq. (4.18), and expressing it in terms of V results in Eq. (4.27). Results of simplifying the equation in general terms and using the derivative in Eq. (4.28) gives the final form of the equation as Eq. (4.29).

Starting with Eq. (4.18),

$$\text{Range} = \frac{\eta_p}{BSFC} \frac{1}{\sqrt{K C_{D_0}}} \left[\tan^{-1}\left(\frac{W_{H_0}}{qVol^{2/3}\sqrt{\frac{C_{D_0}}{K}}}\right) - \tan^{-1}\left(\frac{W_{H_1}}{qVol^{2/3}\sqrt{\frac{C_{D_0}}{K}}}\right) \right]$$

$$(4.18)$$

which is equivalent to this general form for ease of integration

$$\text{Range} = A \left[\tan^{-1}\left(\frac{S_0}{V^2}\right) - \tan^{-1}\left(\frac{S_1}{V^2}\right) \right] \qquad (4.27)$$

Differentiating range with respect to speed, V, setting it equal to zero, and recognizing that

$$\frac{d}{dx} \tan^{-1} x = \frac{1}{1+x^2} \qquad (4.28)$$

yields the speed for maximum range for a constant speed mission,

$$V^4 = \frac{S_1^2 S_0 - S_0^2 S_1}{S_1 - S_0} = S_0 S_1 \qquad (4.29)$$

so, at a given heaviness, W_H, Eq. (4.29) becomes

$$V^4 = S_0^2 = \frac{4KW_H^2}{\left(\rho Vol^{2/3}\right)^2 C_{D_0}} \qquad \text{(at a given heaviness)} \qquad (4.30)$$

where

$$S = \frac{2W_H}{\rho Vol^{2/3}\sqrt{\frac{C_{D_0}}{K}}}$$

However, the best *constant* cruise speed for a given amount of fuel burned is a slight modification of Eq. (4.30).

$$V^4 = \frac{4KW_{H_0}W_{H_1}}{\left(\rho Vol^{2/3}\right)^2 C_{D_0}} \qquad \text{best constant speed for given fuel burned} \qquad (4.31)$$

Recall that the total *L/D* for the airship configuration is expressed as

$$L/D = \frac{L_{aero} + L_{buoy}}{Drag} \qquad (4.32)$$

$$L/D = \frac{C_{L_{aero}} q Vol^{2/3} + (BR)W}{\left(C_{D_0} + K C_{L_{aero}}^2\right) q Vol^{2/3}} \qquad (4.33)$$

A sample problem will help illustrate the relationships of various speeds to their specific performance related parameter.

Sample Problem 4.2: Calculate the Speeds for Minimum Drag, Minimum P_R, and Maximum *L/D* for the HA-1 Hybrid Airship Configuration

Volume = 1,000,000 ft^3; C_{D_0} = 0.033; K = 0.28; gross weight = 100,000 lb. For BR = 0.75, fuel = 10,000 lb; W_{H_0} = 25,000 lb; W_{H_1} = 15,000 lb; altitude = 4000 ft

$$V_{(min\ drag)} \ V^4 = \frac{4KW_H^2}{\left(\rho Vol^{2/3}\right)^2 C_{D_0}} \qquad \text{Eq. (4.24)}$$

$$V_{(min\ P_R)} \ V^4 = \frac{4KW_H^2}{3\left(\rho Vol^{2/3}\right)^2 C_{D_0}} \qquad \text{Eq. (4.26)}$$

$$V_{(L/D)_{max}} \ V^4 = \frac{4KW_{H_0}W_{H_1}}{\left(\rho Vol^{2/3}\right)^2 C_{D_0}} \qquad \text{Eq. (4.29)}$$

$V_{(min\ drag)} = [(4 \times 0.28 \times 25{,}000^2)/(0.00211 \times 10{,}000)^2/0.033)]^{1/4} = 83.1$ fps
$C_{L_{aero}}@\ D_{min} = (0.033/0.28)^{1/2} = 0.343$
$V_{(min\ PR)} = [(4 \times 0.28 \times 25{,}000^2)/(0.00211 \times 10{,}000)^2/0.033/3)]^{1/4} = 63.1$ fps
$C_{L_{aero}}@\ P_{R_{min}} = (3 \times 0.033/0.28)^{1/2} = 0.595$
$V_{(L/D)max} = (S_0\ S_1)^{1/4}\ [S_0 = 6903, S_1 = 4142] = 73.1$ fps (*best average speed*)
At W_{H_0},
 Min drag = $(0.033 + 0.28 \times 0.343^2)(0.5 \times 0.00211 \times 83.1^2) \times 10{,}000 = 4804$ lb
Max $L_{total}/D = (L_{aero} + L_{buoy})/\text{drag} = (25{,}000 + 75{,}000)/4804 = 20.8$

4.6 Endurance/Loiter

The expression for vehicle endurance or loiter is shown in Eq. (4.33a)

$$\text{Endurance} = \int dt = \frac{1}{dW/dt}\,dW \qquad (4.33a)$$

where dW/dt is the weight change due to fuel burned. Following the approach used in the range analysis (Sec. 4.5) where $dW/dt = (BSFC)$ (power) $= (BSFC)DV/550\eta_p$, Eq. (4.33a) becomes

$$\text{Endurance} = \int \frac{550\eta_p}{BSFC \times V \times D}\,dW \qquad (4.33b)$$

where 550 converts from hp to ft-lb/s.

4.6.1 Endurance/Loiter Strategy #1—Constant C_L

$$\text{Endurance (hr)} = \int \frac{550\eta_p}{BSFC}\frac{L_{aero}}{V \times D}\frac{dW}{W}$$

Expressing (4.33b) in terms of C_L and C_D requires a substitution for V.

$$\text{Endurance} = \int \frac{\eta_p}{BSFC}\frac{C_L^{3/2}}{C_D}\sqrt{\frac{\rho Vol^{2/3}}{2W}}\frac{dW}{W^{3/2}} \qquad (4.33c)$$

which is the same as Eq. (3.21) in [3].

If we recognize that $L_{aero} = W_H$ in Eq. (4.33c) and assume constant altitude and constant C_L we get

$$\text{Endurance} = \int dt = \int \frac{\eta_p}{BSFC \times V \times D}\,dW_H \qquad (4.34a)$$

$$\text{Endurance (hr)} = \frac{26.8\eta_p}{BSFC}\frac{C_L^{3/2}}{C_D}\sqrt{\frac{2\sigma Vol^{2/3}}{W_{H_0}}}\left[\left(\frac{W_{H_0}}{W_{H_1}}\right)^{1/2} - 1\right] \qquad (4.34b)$$

where $\sigma = \rho/\rho_{SL}$ and $BSFC = $ *brake specific fuel consumption* in lb of fuel per brake horsepower-hour. Because η_p and $BSFC$ are relatively constant with speed, it should be clear from Fig. 4.2 that maximum endurance for an airship will occur at minimum power required [Eq. (4.22)]. As fuel is burned and W_H decreases the speed must decrease in order to maintain constant altitude and C_L. The decreasing speed is generally not an issue for long duration missions.

4.6.2 Endurance/Loiter Strategy #2—Constant Speed

If we assume constant speed, η_p and $BSFC$ we can integrate Eq. (4.33b) getting

Endurance

$$= \frac{\eta_p}{V\,BSFC} \frac{1}{\sqrt{K\,C_{D_0}}} \left[\tan^{-1}\left(\frac{W_{H_0}}{q\,Vol^{2/3}\sqrt{\frac{C_{D_0}}{K}}} \right) - \tan^{-1}\left(\frac{W_{H_1}}{q\,Vol^{2/3}\sqrt{\frac{C_{D_0}}{K}}} \right) \right] \quad (4.35)$$

Since the speed is constant the C_L will decrease as fuel is burned for flight at a constant altitude. The two flight strategies are compared in Sample Problem 4.3.

The endurance equation is determined in the same fashion as was done for the range equation. Without discussion Eq. (4.34) presents the relationship for endurance assuming a constant C_L and altitude, which means that speed varies. However, should an airship choose to fly at a constant speed then it would use Eq. (4.35). Just as for range calculations, the propulsive efficiency, η_p, and brake specific fuel consumption, $BSFC$, are assumed to be constants.

Sample Problem 4.3: Compare the Range and Endurance for Both Constant Speed and Constant C_L for Both CA-1 and HA-1

Data

The data for CA-1 and HA-1 is shown in Table 4.1 and the vehicle configuration is in Appendix D.

Constant speed of $V = 50$ fps and fuel load $= 5000$ lb.

Using Eqs. (4.18) and (4.35) respectively the range and endurance are,

$$\text{Range} = A[\arctan(X_0) - \arctan(X_1)]$$
$$A = 326 \times 0.65/0.47/(0.9 \times 0.026)^{1/2} = 2947$$
$$X_0 = 6000/(0.5 \times 0.00211 \times 50^2)/10,000/(0.026/0.9)^{1/2} = 1.338$$
$$X_1 = 1000/(0.5 \times 0.00211 \times 50^2)/10,000/(0.026/0.9)^{1/2} = 0.223$$
$$\text{Range} = 2947\,[\arctan(1.338) - \arctan(0.223)] = 2092\ \text{nm}$$
$$\text{Endurance} = B[\arctan(X_0) - \arctan(X_1)]$$
$$B = 550 \times 0.65/0.47/50/(0.9 \times 0.026)^{1/2} = 99.6$$
$$\text{Endurance} = 99.6[\arctan(1.338) - \arctan(0.223)] = 70.7\ \text{hr}$$

If the *best constant speed* for maximum range is used instead of $V = 50$ fps this new speed is found from Eq. (4.29):

$$V^4 = \frac{4KW_{H_0}W_{H_1}}{\left(\rho\,Vol^{2/3}\right)^2 C_{D_0}}$$

$$V_{best} = [4(0.9)(6000)(1000)/((.00211 \times 10,000)^2)/(0.026)]^{1/4} = 37.0\ \text{fps}$$
$$X_0 = 6000/(0.5 \times 0.00211 \times 37^2)/10,000/(0.026/0.9)^{1/2} = 2.444$$
$$X_1 = 1000/(0.5 \times 0.00211 \times 37^2)/10,000/(0.026/0.9)^{1/2} = 0.407$$
$$\text{Range} = 2947\,[\arctan(2.444) - \arctan(0.407)] = 2345\ \text{nm}$$

If the previous problem is performed at a **constant** C_L instead of at a constant speed, the range using Eq. (4.15) becomes

Range $= 326(0.65)/(0.47)\ L/D_{max}\ \ln[6000/1000]$
$C_D = 2C_{D_0}$ at $C_{L_{opt}}$ so $L/D_{max} = C_{L_{opt}}/C_D = (0.026/0.9)^{1/2}/2(0.026) = 3.27$
Range $= 326(0.65)/(0.47)\ (3.27)\ \ln[6000/1000] = 2640$ nm

This represents the best range when cruise is at a constant C_L. This is the greatest range that can be achieved with the given fuel load. The only concern is that the angle of attack could be too high to achieve this optimum C_L. Although airships typically fly at constant speed operations a *best* constant speed may not be feasible because the angle of attack is just too high. A greater range results from flying at a constant 37 fps but the C_L at the start of cruise is 0.415, which requires an $\alpha = 36$ deg. This is clearly not practical. Actual operations must take into account what angles of attack are needed to achieve optimum performance. Angles of attack near minimum drag and minimum power required should be check using C_{L_α} calculated with Eqs. (3.5) or Eq. (3.6).

Constant C_L—Range and Endurance

$$\text{Range (nm)} = \frac{326\eta_p}{BSFC}\frac{L_{aero}}{D}\ln\left[\frac{W_{H_0}}{W_{H_1}}\right] \tag{4.15}$$

$$\text{Endurance (hr)} = \frac{26.8\eta_p}{BSFC}\frac{C_L^{3/2}}{C_D}\sqrt{\frac{2\sigma Vol^{2/3}}{W_{H_0}}}\left[\left(\frac{W_{H_0}}{W_{H_1}}\right)^{1/2}-1\right] \tag{4.34b}$$

Constant Speed—Range and Endurance

$$\text{Range} = \frac{326\eta_p}{BSFC}\frac{1}{\sqrt{K\,C_{D_0}}}\left[\tan^{-1}\left(\frac{W_{H_0}}{qVol^{2/3}\sqrt{\frac{C_{D_0}}{K}}}\right)-\tan^{-1}\left(\frac{W_{H_1}}{qVol^{2/3}\sqrt{\frac{C_{D_0}}{K}}}\right)\right] \tag{4.18}$$

$$\text{Endurance} = \frac{\eta_p}{V\,BSFC}\frac{1}{\sqrt{K\,C_{D_0}}}\left[\tan^{-1}\left(\frac{W_{H_0}}{qVol^{2/3}\sqrt{\frac{C_{D_0}}{K}}}\right)-\tan^{-1}\left(\frac{W_{H_1}}{qVol^{2/3}\sqrt{\frac{C_{D_0}}{K}}}\right)\right] \tag{4.35}$$

Cruise Strategy

Constant C_L cruise strategy would begin its cruise at the speed for optimum C_L at the initial heaviness

$$V^4 = \frac{4KW_H^2}{\left(\rho Vol^{2/3}\right)^2 C_{D_0}} \qquad \text{speed for minimum drag} \tag{4.24}$$

where

$$CL_{opt} = \sqrt{\frac{C_{D0}}{K}} \qquad \text{minimum drag and } L/D_{max} \qquad (4.19)$$

Constant speed cruise strategy would begin and maintain its cruise at the best speed determined by

$$V^4 = \frac{4KW_{H_0}W_{H_1}}{\left(\rho Vol^{2/3}\right)^2 C_{D0}} \qquad \text{best constant speed range} \qquad (4.29)$$

where this speed is constant throughout the entire flight.

Similarly the best constant speed for maximum endurance would be based on a slight modification to Eq. (4.26).

$$V^4 = \frac{4KW_{H_0}W_{H_1}}{3\left(\rho Vol^{2/3}\right)^2 C_{D0}} \qquad \text{best constant speed endurance}$$

However, maintaining constant speed for loiter missions is unusual and speeds for maximum duration would be used based on flying at optimum C_L using

$$CL_{opt} = \sqrt{\frac{3C_{D0}}{K}} \qquad (4.22b)$$

The instantaneous optimum speed throughout the mission is then easily calculated for the current heaviness, W_H.

A summary of the range and endurance capabilities for CA-1 and HA-1 are shown in Tables 4.2a and 4.2b and include performance restricted to a maximum angle of attack of 16 deg.

Conclusion

The CA-1 airship was not able to fly at the "Best Constant Speed" shown in Tables 4.2a and 4.2b because its low $C_{L_\alpha} = 0.0115$ per degree resulted in angles of attack of much greater than 16 deg. Airships and hybrids are usually

Table 4.2a CA-1 and HA-1 Optimum Speed and Range Comparison (α_{limit} = 16 deg)

Range

Config	W_{H_0} (lb)	W_{H_1} (lb)	Opt C_L	Initial speed (fps)	Final speed (fps)	Const C_L range (nm)	Best const speed (fps)	Const speed range (nm)	α_{limit} const speed (fps)	α_{limit} const speed range (nm)
CA-1	6000	1000	0.170	57.8	23.6	2640	37.0	2345	55.6	1905
HA-1	25,000	15,000	0.343	83.1	64.4	1198	73.1	1185	57.4	1063

Table 4.2b CA-1 and HA-1 Optimum Speed and Range Comparison (α_{limit} = 16 deg)

						Const C_L endur (hr)	Best const speed (fps)	Const speed endur (hr)	α_{limit} const speed (fps)	α_{limit} const speed endur (hr)
ndurance										
Config	W_{Ho} (lb)	W_{H1} (lb)	Opt C_L	Initial speed (fps)	Final speed (fps)					
CA-1	6000	1000	0.294	44.0	17.9	142	25.6	130	55.6	57.8
HA-1	25,000	15,000	0.595	63.1	48.9	31.6	54.5	31.3	57.4	31.2

given a free air $\alpha_{limit} \leq 16$ deg because their $C_{L\alpha}$ is starting to become non-linear past this α, handling qualities are sluggish, and operations are unacceptable by crew and onboard personnel. For this reason the constant speed range and endurance shown for CA-1 in the tables also include an "Alpha Limited Constant Speed" using α_{limit} = 16 deg. This was not a problem for HA-1 because its $C_{L\alpha}$ = 0.045 per degree kept the required $\alpha < 16$ deg.

4.7 Mission Performance

Because range and endurance are generally the most important design parameters for the airship designer, most design efforts try to maximize one or the other. However, there are other measures of performance that are as important or more important to the operator or owner of airships. For example, the payload vs range data shown in Fig. 4.6 is a standard Measure of Merit (MOM) for transport aircraft. This MOM is also important to an airship delivering cargo from one point to another. Although transport aircraft have significant performance advantages over an airship, when cost to deliver a pound of cargo over a given distance is compared the cost for an airship is somewhat lower than it is for an aircraft. This leads to a discussion of another parameter, namely, *fuel burned/payload/range*, which measures the efficiency of an air vehicle to transport goods. And finally, one last parameter similar to the previous one where operating cost substitutes for fuel burned and typically is expressed as ($\$\$/\text{ton-nmile}$). This parameter is often referred to as *productivity*. Because the operating costs are often not very well known in the early design stages this parameter is generally hard to accurately compute until the design has matured significantly.

4.7.1 Range-Payload

One of the most important performance curves used to design any aircraft or airship that transports cargo is the payload–range curve. An example of a hybrid airship performance presented in this manner is shown as Fig. 4.6. Three main points are better understood by examining this figure more closely. First, there is a maximum payload value that is the upper limit of all

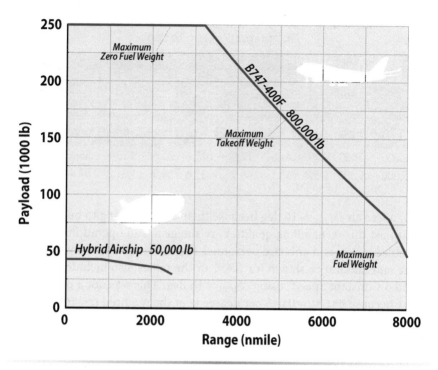

Figure 4.6   Range–payload comparison between a hybrid airship and a 747-400F.

range–payload curves and is 40,000 lb for the example hybrid airship shown in Fig. 4.6. This capability is contrasted with the 250,000-lb payload version of the B747-400F. Secondly, for a constant takeoff weight, as payload is reduced it is replaced with fuel that allows the airship to travel further until it hits the limit for maximum fuel. Finally, as payload is reduced even more the improvement in range continues but at a reduced rate as payload is reduced and not replaced by fuel. Range improvements are due to flying at reduced weight (heaviness for an airship) for the same maximum amount of fuel.

Most payload range data for aircraft is presented at its nominal cruise speed because time is an important aspect of its productivity. However, when speed is much less important the airship designer has a wider range of speeds that are acceptable. The dramatic impact of speed on airship performance is shown in Fig. 4.7. Although the differences between the curves are only 10 or 20 kt the percentage change is enormous. This is understandable because much of the power is used to overcome the zero lift drag that changes with the square of the velocity. This range improvement with lower speeds will continue until the speed for maximum L/D is reached (see Fig. 4.2).

Historically, helicopters have been used on short-range missions where speed is not important. Even though there are some slight speed advantages for a helicopter compared to an airship, the airship has a significantly better

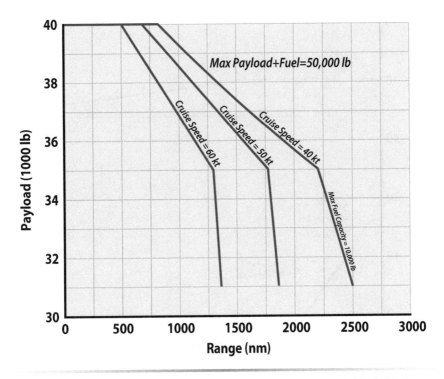

Figure 4.7 Effect of cruise speed on range–payload (hybrid airship).

$$/ton-mile value. Figure 4.8 compares the range–payload of a helicopter and a hybrid airship that shows how much more range–payload performance a hybrid airship has compared to a helicopter. Cost comparisons using $$/ton-mile as the measure would show helicopters to have significantly higher operating costs than that of an airship.

4.7.2 Mission Efficiency

The efficiency of an airship is measured in a number of ways but four of the most important MoMs are

1. Specific range/cruise efficiency (nmile/lb of fuel burned)
2. Endurance efficiency (lb of fuel/hr, i.e., fuel flow)
3. Mission efficiency (lb of fuel burned/ton of payload/nmile)
4. Mission productivity ($$/ton of payload/distance traveled)

The cruise and endurance efficiency parameters are important to most aircraft but the mission efficiency parameter is unique to cargo delivering systems. Just like aircraft, an airship's mission will determine the relative importance of each of these parameters and then they are prioritized or

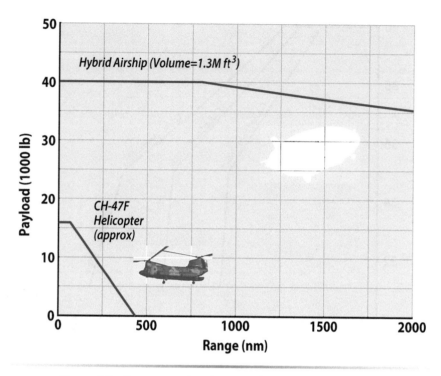

Figure 4.8 Range–payload comparison between a hybrid airship and a helicopter.

wieghted for trade studies. It is not possible to design an airship that is optimized for all four MOMs. Someone, usually the customer, will give the designer the relative importance of these parameters (weighting factors) so that an accurate trade study can be performed. Results of these trade studies will define the airship's volume, fineness ratio (FR), buoyancy ratio (BR), propulsion system type, engine size, tail sizes, body shape, etc.

Sometimes the efficiency of an air vehicle is more important than its actual performance. However, overall performance cannot be decoupled from flight efficiency. Figure 4.9 shows the *specific range* (nmile/lb of fuel burned) of a typical airship (ZP4K) for various speeds and values of heaviness. Obviously, flying at speeds where the specific range is maximum will result in maximum range. Also, note that the one engine operating specific range is somewhat better than for two engines. This increase in specific range is the result of one engine performing better (lower $BSFC$) at high power settings vs two engines operating at very low and inefficient power settings. It is common on airships to shut down one or more engines so that better $BSFC$ values can be obtained from the remaining engines. This is routine for aircraft as well when endurance is the most important mission parameter. For example, during long endurance missions the P-3 aircraft shuts down two of its four engines to obtain the best $BSFC$ from its two operating turboprop engines.

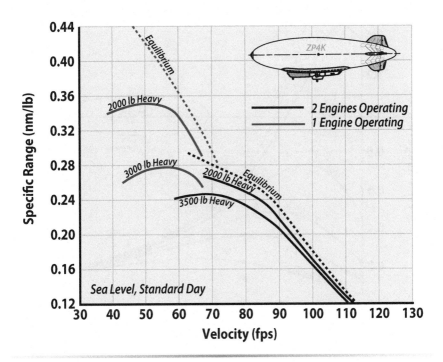

Figure 4.9 Specific range for an airship similar to the ZP4K.

Two more figures have been included as Figs. 4.10a and 4.10b, which summarize the efficiencies of both the CA-1 and HA-1 configurations. Referring to the vehicle characteristics in Appendix D should allow the reader to duplicate these calculations. Notice how the specific range becomes very large for low speeds and small values of heaviness. This is the direct result of the total L/D at these conditions and shown previously in Fig. 4.4. Figure 4.10 also has curves labeled "equilibrium" that show the extreme case of zero heaviness (no drag-due-to-lift) where all of the drag comes from zero lift drag, C_{D_0}.

As a basis for a final discussion of range and endurance efficiencies Figs. 4.11 and 4.12 show how a conventional airship (CA-1) compares to a hybrid airship (HA-1). It is clear that a conventional airship out performs its hybrid cousin casting doubt on the utility of a hybrid. Chapter 12 will discuss this issue in detail. At this point it is obvious that a hybrid airship's value is not based on its performance efficiency.

Mission efficiency is different from performance efficiency because it includes both cost and productivity. Although operators are interested in basic vehicle efficiencies, what is also important is how fast cargo can be moved from one point to another and the costs associated with moving it. Data presented in Fig. 4.13 shows how fuel burned/payload/range varies

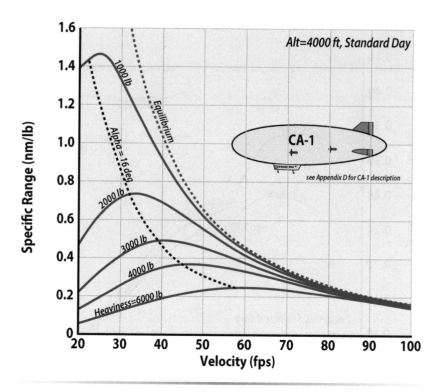

Figure 4.10a Specific range for CA-1 configuration.

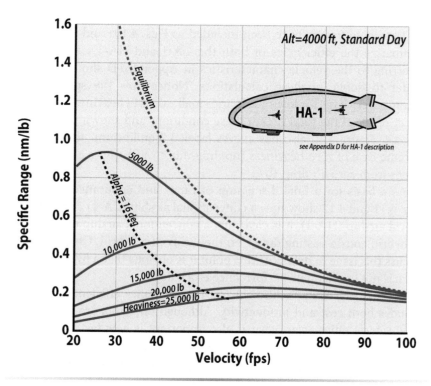

Figure 4.10b Specific range for HA-1 configuration.

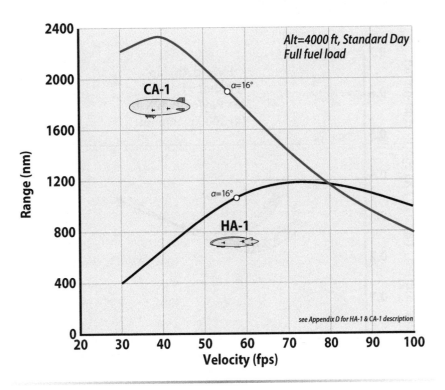

Figure 4.11 Range comparison for CA-1 vs HA-1 configurations.

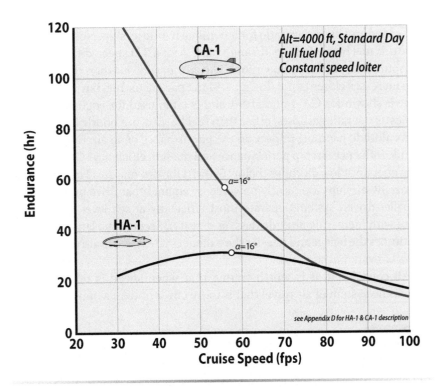

Figure 4.12 Endurance comparison for CA-1 vs HA-1 configurations.

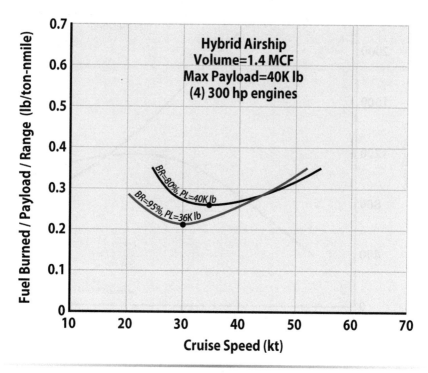

Figure 4.13 Effect of BR and payload size on productivity of a hybrid airship design.

with speed and buoyancy ratio for a nominal hybrid airship. Notice how sensitive this fuel burned/payload/range value is for a 5 kt speed difference from the optimum speed and how much better the value is when the buoyancy ratio increases closer to 1.0. In Fig. 4.14 the parameter fuel burned/payload/range is shown for CA-1 and HA-1 and is often used for engineering trade studies as operational costs, other than fuel burned, are poorly known.

As altitude has a big impact on the performance of an aircraft, how does altitude influence airship performance and mission efficiency? Generally, airships operate at low altitudes (below 5000 ft) for two reasons. First, and most important, airships are mostly powered by engines that drive propellers. All propeller-driven systems operate most efficiently at sea level and become worse with increasing altitude (less air density). Second, the larger the range of altitudes the bigger and heavier the ballonet system which also means there is less buoyancy at sea level.

All considered, it is worth stating that when speed is not important there is no system of air travel that is more efficient than a buoyant vehicle!

4.7.3 Ballast Requirements

One of the more difficult issues facing airship operators is the matter of requiring ballast in some instances to prevent operations at buoyancy

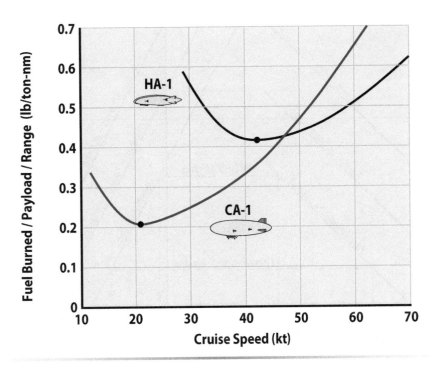

Figure 4.14 Comparison of productivity for CA-1 and HA-1 configurations.

ratios greater than 1 (heaviness < 0). For missions where fuel is burned off but no cargo is unloaded it is the role of the designer to make sure the airship can safely and efficiently operate at all *BR* associated with conditions from maximum fuel load to empty. However, this is not the most strenuous condition for the airship designer. Missions where significant cargo is unloaded at the destination and very little or no cargo is available for the return trip it is likely to require added weight in the form of ballast to keep the airship from a *BR* > 1 condition. What this ballast weight material is can be the topic of some interesting discussions but water is a popular choice. Further discussions are deferred to Chapter 12.

The criteria for adding ballast is simple . . . the airship has to have positive heaviness (*BR* < 1) at all flight conditions. After unloading payload at the destination the amount of ballast weight necessary depends on what heaviness can be supported by aerodynamic lift and vectored thrust during landing. An example that illustrates how the amount of ballast weight is determined appears as Fig. 4.15. In this example start with the value of the heaviness when the airship lands and move straight up to the amount of payload that is unloaded. From this point continue horizontally to the curves on the right and stop at the value of heaviness that is needed for the return flight. Continue down and read the amount of ballast that will be needed to meet the three requirements of heaviness at landing, unloaded payload, and heaviness at takeoff.

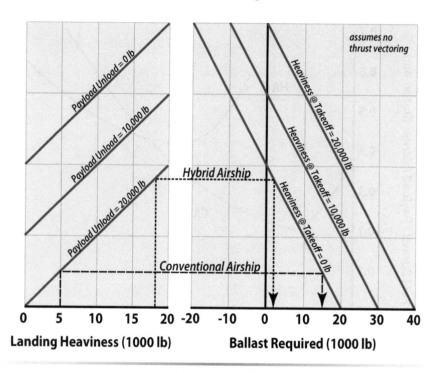

Figure 4.15 Ballast required for return flight (hybrid vs conventional airship).

Figure 4.15 also compares a conventional airship to a hybrid airship flying the same mission. In both cases the amount of payload unloaded is 20,000 lb and the heaviness at takeoff is zero. Notice how the hybrid is significantly better because it requires much less ballast. Because hybrids are much better at generating aerodynamic lift that they can modulate when the payload is onboard and then adjust the lift according to the heaviness on the return segment as fuel is burned off. Hybrid airship design specifics will be discussed in more detail in Chapter 12.

4.7.4 Buoyancy Control

Someday, optimal airship designs will not depend on ballast at all. Hybrid airships that burn off a lot of fuel for traveling long distances or staying aloft for long times should use another approach that doesn't require the use of ballast weight.

There are four candidate technologies that should be evaluated for their potential to reduce or eliminate the need for ballast weight.

1. Compressing helium within the envelope to reduce the buoyant lift to match the offloading of weight.
2. Compressing outside air that is brought onboard to increase an airship's heaviness.
3. Putting gaseous hydrogen in the ballonets and burning it as the ballonets reduce in size as the airship gains altitude. The ballonet volume is completely filled with much heavier air on landing.
4. Capturing water vapor from the engine's combustion products. This approach is very attractive as it offsets some of the fuel burned during the mission. These systems may get 20%–40% of the water available. One drawback is that this water is often contaminated with combustion products, which can make disposal difficult.

All of these techniques should be given some consideration during the airship design process. As technologies mature one or more of these processes will become the most popular means of eliminating the need for ballast weight.

4.8 Climb and Descent

In aircraft design, particularly for fighters, optimizing flight profiles is very important. For airships, it is far less interesting and only becomes important when clearing obstacles in and around the airport. Descent performance is centered around one issue, which is making sure the ballonet pumps can keep up with the most rapid descent possible. If the ballonet system is not capable then the airship could lose its rigidity since the lifting gas contracts as the airship descends to lower altitudes.

4.8.1 Rate of Climb

The rate of climb for an aircraft is given by $ROC = dh/dt = V \sin \gamma$ where γ is the climb angle and $\sin \gamma = [T \cos (\alpha + i_T) - D]/W_H$, which leads to

$$\frac{dh}{dt} = V \sin \gamma = \frac{V(T \cos(\alpha + i_T) - D)}{W_H} \tag{4.36}$$

where W_H is the airship heaviness, T = thrust, D = drag, α = aerodynamic angle of attack, and i_T = inclination angle of thrust.

Equation (4.24) is valid for all airship climb segments. Corrections to the ROC for accelerations during climb, necessary for many aircraft, are unnecessary for airship climb performance as accelerations are usually small.

For descents, thrust is assumed small so the flight angle, γ, is written as

$$\tan \gamma = (-D/W_H) = (D/L_{aero}) \tag{4.37}$$

For maximum range during the glide descent (i.e., stretching the glide) the aircraft should be flown at minimum γ which means, looking at Eq. (4.25), flying at the aerodynamic L/D_{max}. Therefore, the velocity for maximum gliding range is given by Eq. (4.30)

$$V^4 = \frac{4KW_H^2}{\left(\rho Vol^{2/3}\right)^2 C_{D_0}} \tag{4.30}$$

The condition for minimum rate of descent is different than the condition for maximum range. Assuming that γ is small such that $\sin\gamma \approx \tan\gamma$ the rate of descent, ROD, can be expressed as

$$ROD = V\tan\gamma = -\left(DV/W_H\right) = -\frac{C_D}{C_{L_{aero}}}\left[\frac{4KW_H^2}{\left(\rho Vol^{2/3}\right)^2 C_{D_0}}\right]^{\frac{1}{4}} \tag{4.38}$$

so that the velocity for minimum ROD is

$$V_{ROD_{min}} = \left[\frac{4KW_H^2}{3\left(\rho Vol^{2/3}\right)^2 C_{D_0}}\right]^{\frac{1}{4}} \tag{4.39}$$

which is about 24% less than the velocity for maximum gliding range. Notice that the velocity for minimum ROD is the same as for minimum power required, Eq. (4.26).

4.9 Takeoff

There is a significant difference between the importance of takeoff performance for an airplane and an airship. For airplanes, runways are often restricting because of their length, surrounding terrain, weather, altitude, etc. Many tests have to be performed and results verified to establish takeoff characteristics under all possible operational circumstances. The difficulty with taking off an airplane that depends on wing-generated lift is that the airplane has to accelerate to takeoff speeds to generate sufficient lift at an angle of attack. This is not the case for an airship. Much of an airship's lift comes from its internal lifting gas that typically is responsible for 70% or more of an airship's vertical force. Therefore, much of the takeoff field length needed to produce the 70% + lift does not apply to an airship. Thus, conventional airship field lengths tend to be much shorter than for comparably sized aircraft.

When an airship has a small heaviness it has the ability to operate in a Vertical Takeoff and Landing (VTOL) mode by vectoring its engine thrust. Because takeoff performance must be proven, vertical takeoff results are verified by an authorizing agency such as the FAA.

4.9.1 Ground Effects

For most airborne vehicles there is an increase in lift and a decrease in drag as the vehicle approaches the ground at heights less than its span. These aerodynamic changes are very sensitive to body/wing geometry and lift changes on the order of 10%–25% are typical. This phenomenon is the result of the ground interfering with the trailing legs of the horseshoe vortex system generated by the wing/body at an angle of attack. Ground effects are often analyzed by putting an image horseshoe vortex system of equal but opposite strength at the same distance below the ground that the wing is above the ground. Most modern CFD codes handle this situation easily. This is not to say that the CFD codes get accurate answers for vehicles flying near the ground. These codes are usually run using inviscid flow and there is rarely any test data to establish the validity of the theoretical answer. Occasionally, there is a database from a similar project that measured or calculated "in-ground" C_L, C_D, or C_M. Luckily for the designer in-ground effects do not influence air vehicle performance enough to make a difference in the overall design. The approach is simply one that statistically measures many near-ground test results and integrates them into an acceptable form used by operators of the vehicle.

For airships there is virtually no data showing the effect of the ground on aerodynamic characteristics so this becomes a statistical measurement as well. The in-ground effect is much less for bodies of revolution or aircraft with small aspect ratios. Because hybrid airship configurations have small aspect ratio bodies, they are capable of generating significant lift that will be affected by the proximity of the ground. However, the effect is still small and will be disregarded when designing hybrid airships in Chapters 6 and 12.

Chapter 10 in [3] discusses in more detail how ground effects change the aerodynamics of an aircraft. Some experimental data is also included. This discussion of in-ground behavior will give the reader some insight into the effects of the ground on aerodynamic properties. However, this treatment has little carryover to airships as bodies are very unlike lifting wings and airships spend little time flying close to the ground.

4.9.2 Takeoff Analysis

Takeoff field length is the maximum distance required for an airship to accelerate from $V = 0$ to clearing a 50-ft obstacle. Before developing the equations for making takeoff calculations a few definitions are helpful. First, Fig. 4.16 summarizes the fundamental forces associated with takeoff performance. Figure 4.17 shows a schematic of the takeoff problem and its various takeoff distance calculation segments. In general, the takeoff distance is the sum of the ground distance segments (S_G), rotation distance (S_R), and climb

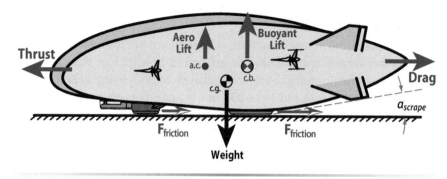

Figure 4.16   Force diagram during takeoff ground roll.

distance (S_{CL}). Equations for each of these segments are developed in the following sections. Special calculations are also made for losing an engine and either continuing the takeoff maneuver or coming to a complete stop. The engine failure speed that results in the accelerate-go distance being equal to the accelerate-stop distance is referred to as V_1 and results in a balanced field length. This is discussed in more detail in Sec. 4.11.

4.9.2.1   Available Takeoff C_L

The airship must accelerate down the runway to a takeoff speed V_{TO}, rotate about the main gear (or aft ACLS pads) and generate $L_{aero} = 1.2\,W_{H_0}$. This $V_{TO} = 1.1[2\,W_{H_0}/Vol^{2/3}\rho C_L]^{\frac{1}{2}}$ where C_L is the available lift coefficient. For aircraft this C_L is the maximum or stall lift coefficient. For airships it is

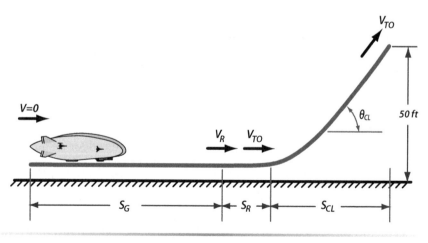

Figure 4.17   Schematic of an airship takeoff.

much less than the maximum lift coefficient. Airship C_L vs α data (see Figs. 3.21 and 3.27–3.31) show linear C_L up to about $\alpha = 16$ deg. Although airships have poor lift curve slopes, $C_{L\alpha}$, their α range is large. The available takeoff C_L is not limited by aerodynamic separation but rather by tail strike during rotation about the main gear or aft ACLS pad. Typical non-rigid airships (CA-1, Goodyear "K" class, Sentinel 1000, etc.) cannot rotate past $\alpha \sim 10$ deg before striking their tails. This limits the takeoff C_L (at $\alpha = 10$ deg) to a maximum value of 0.10 for bodies of revolution shapes or 0.45 for lobed hybrid shapes.

4.9.2.2 Ground Distance (S_G)

It is assumed that the airship accelerates to its takeoff velocity, V_{TO}, and at that speed the aircraft is rotated to an angle of attack whose maximum value is often determined by its physical scrape angle. The airship then lifts off and transitions from horizontal to climbing flight during the transition distance, S_{TR}. The airship must maintain its liftoff speed until its tail clears the 50-ft obstacle.

Equation (4.40) is the basic equation for calculating ground distance. Substituting the acceleration term from Eq. (4.41) into Eq. (4.40) gives the final relationship for ground distance and is shown as Eq. (4.42). Notice that the mass term (W/g) has been split into two parts to make sure that the mass of the gases in the ballonet and envelope are both included along with the $TOGW/g$.

$$S_G = \int_0^{V_{TO}} \frac{V \, dV}{a} = \frac{1}{2} \int_0^{V_{TO}} \frac{dV^2}{a} = \frac{1}{2} \frac{V_{TO}^2}{a_{(@\,0.707\,V_{TO})}} \tag{4.40}$$

$$a = \frac{g}{W}\left[T - D - F_f\right] = \frac{g}{W}\left[T - D - \mu\left(W_H - L_{aero}\right)\right] \tag{4.41}$$

$$S_G = \frac{\left(W/g + m_{gas}\right)V_{TO}^2}{2\left[T - D - \mu\left(W_H - L_{aero}\right)\right]_{@0.707\,V_{TO}}} \tag{4.42}$$

Equations (4.41) and (4.42) require the thrust at 0.707 V_{TO}. The takeoff acceleration varies from $V = 0$ to $V = V_{TO}$. The acceleration evaluated at $V = 0.707\ V_{TO}$ approximates the time-wise integration of the acceleration very well and is used in the takeoff analysis for convenience. The thrust available depends upon the engine/propeller combination and the speed for 0.707 V_{TO} and is discussed in Sec. 5.5. Equation (4.42) does not have any added mass terms and μ is the landing gear coefficient of friction for brakes off (see Table 4.3). If an ACLS is being used for the landing gear (see Chapter 9), the airship is in a hover mode and $\mu \approx 0$.

4.9.2.3 Rotation Distance (S_R)

Distance traveled during rotation is shown in Eq. (4.43) and is based on the time that is demonstrated during certification testing for agencies such

Table 4.3 Coefficients of Friction for Various Takeoff and Landing Surfaces

Coefficient of friction values		
Type of surface	Brakes off, average ground resistance coefficient	Brakes fully applied, average wheel braking coefficient
Concrete or macadam	0.015 to 0.04	0.3 to 0.6
Hard turf	0.05	0.4
Firm and dry dirt	0.04	0.30
Soft turf	0.07	0.5
Wet concrete	0.05	0.2
Wet grass	0.10	0.2
Snow or ice-covered field	0.01	0.07 to 0.10

as the FAA. This segment distance is generally computed by assuming that rotation takes an average amount of time at the V_R speed. Certifying agencies specify a fixed time increment of about 3 s for this action.

$$S_R = (\text{rotation time}) \times (\text{takeoff speed}) = t_{ROT} \times V_{TO} \quad (t_{ROT} \approx 3 \text{ s}) \qquad (4.43)$$

4.9.2.4 Climbout Distance (S_{CL})

The final segment where the airship climbs to an obstacle height of 50 ft has its distance calculated using Eq. (4.45) which is based on the climb angle defined in Eq. (4.44). This climbout is performed at the constant speed of V_{TO} as the vehicle is not allowed to decelerate from liftoff to the 50-ft obstacle.

$$\tan\theta_{CL} \approx \sin\theta_{CL} = \frac{V_{TO}(T-D)}{W_H} \qquad (4.44)$$

$$S_{CL} = \frac{50}{\tan\theta_{CL}} \qquad (4.45)$$

The final accelerate-go distance is the sum of the 3 segments shown in Eq. (4.46).

$$S_{total} = S_G + S_R + S_{CL} \qquad \text{accelerate-go} \qquad (4.46)$$

4.9.3 Takeoff Noise

Although noise can be a significant issue for jet-powered aircraft it is rarely a problem for the designers of propeller-powered aircraft. Because the power required by an airship is much less than a typical medium-sized

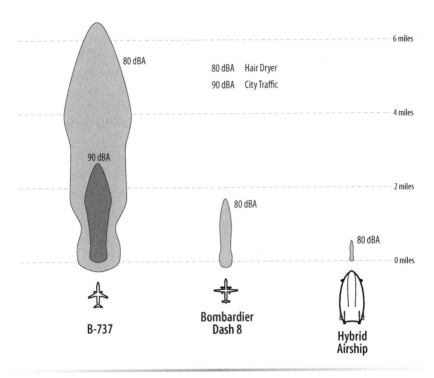

Figure 4.18 Airport noise profiles comparison—takeoff.

transport powered by propellers, its noise is far lower. A comparison of these three modes of transport is shown in Fig. 4.18. The data for the B-737 and Bombardier Dash 8 are measured and the airship noise profile is an analytical estimate.

However, it is not that simple. Noise calculations also take into account the amount of time (duration) for the noise source. Since the airship is typically flying somewhat slower than an airplane its noise value becomes worse than suggested by the relative values shown in Fig. 4.18.

4.10 Landing

Landing performance for an airship uses the same equations of motion as for airplanes and must be validated in a similar manner. The difference being that braking is usually much less important compared to the vectored or reverse thrust contributions. Typically, there is a constant speed approach at a constant angle, a flare just prior to touchdown, followed by a deceleration along the ground. Figure 4.19 shows a typical landing profile.

4.10.1 Landing Analysis

The landing distance is the horizontal distance required to clear a 50-ft obstacle, flare, free roll, and then decelerate to a complete stop. Figure 4.19 shows the schematic for the landing analysis. Similar to takeoff there are three segments that make up the landing distance. These segments are air distance, free roll, and deceleration. Similar to takeoff the three equations for these landing segments are presented without discussion.

The landing distance is the sum of the air distance (S_A), the free roll distance (S_{FR}), and the braking distance (S_B).

The air distance is computed by finding the total change in energy (KE + PE), which is equal to the retarding force × the air distance (S_A). Assume θ_{app} small such that $\cos\theta_{app} \approx 1$. Then,

$$m\left[\frac{V_{50}^2}{2} + 50g - \frac{V_{TD}^2}{2}\right] = F_R S_A$$

where m is the mass of the airship and its gases (4.47a)

$$S_A = \frac{m}{F_R}\left[\frac{V_{50}^2 - V_{TD}^2}{2} + 50g\right] = F_R S_A$$

where

$$F_R = \text{drag} + \text{thrust vector} \qquad (4.47b)$$

or

$$S_A = 50 / \tan\theta_{app} \qquad (4.47c)$$

where θ_{app} is typically 3 deg so, $S_A = 50 / 0.0524 = 954$ ft.

It is assumed that the velocity over the 50-ft obstacle is $V_{50} = 1.3\,V_L$ and the touchdown velocity is $V_{TD} = 1.15\,V_L$. The V_L is for the airship in its landing configuration, that is

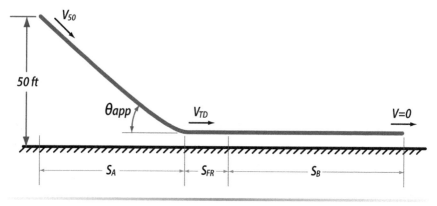

Figure 4.19   Schematic for landing performance analysis.

$$V_L = [2\,L_{aero}/Vol^{2/3}\rho\,C_L]^{1/2}$$
$$L_{aero} = W_{H_0} - \tfrac{1}{2}\,\text{fuel} = TOGW - \tfrac{1}{2}\,\text{fuel} - L_{buoy}$$
$$C_L = \text{lift coefficient at } \alpha = 10 \text{ deg (tail strike)}$$

The free roll distance during landing is assumed to be 3 s during which the airship

1. Chops power
2. Changes configuration to make $L_{aero} = 0$ to get maximum weight on wheels
3. Applies brakes
4. Applies reverse thrust

$$S_{FR} = 3\,V_{TD} \tag{4.48}$$

Also, substituting V_{TD} for V_1 and T_{rev} is reverse thrust in Eq. (4.36d) yields

$$S_B = \frac{W/g + m_{gas}}{\rho\,Vol^{2/3}\left(C_D - \mu_B\,C_{L_{aero}}\right)} \\ \ln\left[\mu_B W_H + T_{rev} + \frac{\rho}{2}Vol^{2/3}\left(C_D - \mu_B\,C_{L_{aero}}\right)V_{TD}^2\right] \tag{4.49}$$

which is the total braking distance starting at touchdown speed V_{TD}. The μ_B is the braking coefficient of friction and is determined from Table 4.3 for a conventional landing gear with brakes. If an ACLS is used the $\mu_B \approx 0$ because the ACLS is not used for braking as the ACLS curtain (see Chapter 9) would have to be replaced after several landings due to the abrasion of contacting the ground. Most airships use reverse thrust for braking where it is assumed that ~60% of the takeoff thrust is available to stop the airship.

Adding Eqs. (4.37), (4.38), and (4.39) together gives the landing distance in Eq. (4.40). A conservatism factor is always applied to these calculations. For the FAA, Landing Field Length = Calculated Landing Distance/0.6.

$$S_{LAND} = S_A + S_{FR} + S_B \tag{4.50}$$

4.11 Critical Field Length (Balanced Field Length)

Should the airship have an engine failure at a speed less than V_1 (decision speed) then the pilot is obligated to stop the airship rather than continue with its takeoff. Conversely, if the engine fails after V_1 then the pilot must continue the takeoff to become airborne. There are numerous options for computing the official takeoff performance and these options from [4] are summarized in Fig. 4.20. Similar to aircraft, the airship takeoff is the longest distance from these several options that have different assumptions and criteria.

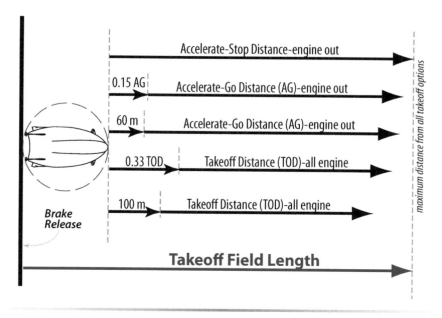

Figure 4.20   Takeoff field length definition from CS-30T. [4]

Basically there are two possibilities for taking off. First, the vehicle accelerates to its takeoff speed, V_{TO}, and continues to climb until it reaches the 50-ft obstacle. Second, the vehicle accelerates and during this acceleration one engine fails and the airship is braked to a stop. Which of these two scenarios applies depends on the speed at which the engine fails. There is a common term used for both aircraft and airships that is called the *decision speed*, V_1, which uniquely determines whether an air vehicle continues with its climbout or brakes to a stop. The decision speed is defined as that speed that makes the continuing distance from V_1 to the 50-ft obstacle equal to the distance from V_1 to a complete stop. Generally, V_1 is very slightly higher than the engine failure speed, V_{EF}, but it cannot be lower than V_{EF}. Figure 4.21 summarizes this calculation and defines all of the speeds.

There is a *free roll* segment that is added to the stopping calculation that accounts for the amount of time taken by the pilot before the brakes are applied. The generally accepted time for this action is 3 s. So free roll distance is computed using Eq. (4.47).

$$S_{FR} = 3\,V_{TO} \qquad (4.47)$$

Once the retarding force is applied (brakes or reverse engine thrust) the stopping distance is calculated using Eq. (4.48d). When discussing braking performance for a hybrid airship with an *air cushion landing*

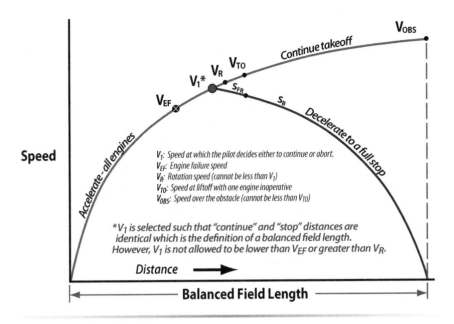

Figure 4.21 Schematic of balanced field length. [4]

system, ACLS, this calculation becomes more complex and is discussed further in Chapter 12.

$$S_B = \int dx = \int_{V_1}^0 \frac{V dV}{a} = \frac{1}{2}\int_{V_1}^0 \frac{dV^2}{a} \tag{4.51a}$$

$$-a = \frac{g}{W}(F_R + D) = \frac{g}{W}\left(\mu_B W_H + T + C_D\, q\, Vol^{2/3}\right) \tag{4.51b}$$

where F_R = braking force + retarding thrust = $\mu_B W_H + T$ where $T \approx 0.6$ (takeoff thrust)

$$S_B = \frac{W}{2g}\int_{V_1}^0 \frac{-d\left(V^2\right)}{\mu_B W_H + T + \frac{\rho}{2} Vol^{2/3} V^2 \left(C_D - \mu_B C_{L_{aero}}\right)} \tag{4.51c}$$

$$S_B = \frac{W/g + m_{gas}}{\rho\, Vol^{2/3}\left(C_D - \mu_B C_{L_{aero}}\right)} \\ \ell n\left[\mu_B W_H + T + \frac{\rho}{2} Vol^{2/3}\left(C_D - \mu_B C_{L_{aero}}\right)V_1^2\right] \tag{4.51d}$$

Equations (4.49) and (4.51) require thrust to be calculated and do not account for any added mass terms. As was done for Eq. (4.42), find the thrust at $0.707\, V_{TD}$ or $0.707\, V_1$. The thrust available depends upon the engine/propeller combination and the speed, which is discussed in Sec. 5.5.

4.12 Turning

An aircraft turns by banking and using a component of the wing lift force to turn the aircraft.

For some aircraft, turning performance is a requirement that can affect the design wing loading. For these aircraft turn rates are limited by the maximum load factor (g's) the airplane is designed for and the available engine power or thrust. For fighters, both the instantaneous turn rate and sustained turn rate are of interest. For commercial aircraft only the sustained turn rate is important. The same is true for airships.

For an airship the turning capability MOM is how long it takes to turn 360 deg. Turning an airship through a full circle is dependent on two basic parameters. The most important parameter is the speed during the turn and together with the turning coefficient, R/L establishes an airship's turning capabilities. This turning coefficient represents how tight the turn is relative to the size of the airship. Figures 4.22 and 4.23 show the relation of these parameters to the 360-deg turn time.

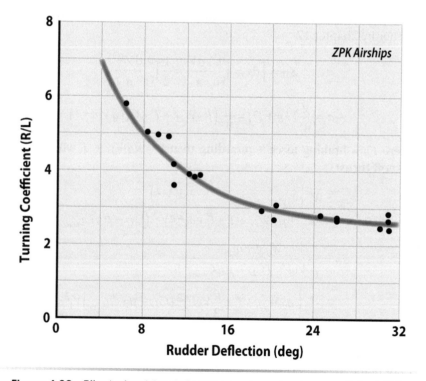

Figure 4.22 Effect of rudder deflection angle on turning coefficient. [2]

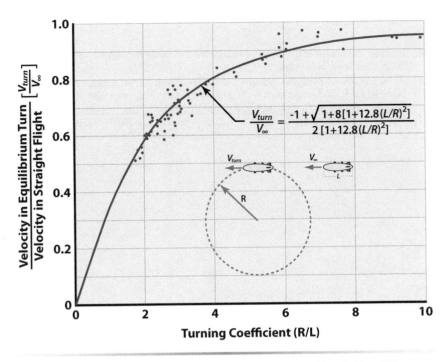

Figure 4.23 Turning performance for axisymmetric airships. [2]

A more generalized method of calculating the time to turn through 360 deg uses the empirical data from numerous airships. This technique determines the turning coefficient based on the available rudder deflection from Fig. 4.22. Using Fig. 4.23 the parameter V_{turn}/V_∞ is found next. Given the turning radius and the initial speed V_∞ yields V_{turn} that is then divided into the perimeter to get the time.

Sample Problem 4.4: Calculate the Time to Make a 360 Deg Turn

Assume a rudder deflection of 28 deg, $V_\infty = 60$ kt (101.3 f/s), and $R = 1000$ ft. From Fig. 4.22 at $\delta_R = 28$ deg read $R/L = 2.7$. Using Fig. 4.23, at $R/L = 2.7$ read $V_{turn}/V_\infty = 0.69$. $V_{turn} = 101.3 \times 0.69 = 69.9$ f/s, which results in $t_{360} = 2\pi R/V_{turn} = 89.9$ s. This turn would be considered to be a little slow as airship designers generally want a full circle turn to be performed in under a minute. Another way to look at this problem is to use the $R/L = 2.7$ and for an airship that is 200 ft long recognize that this example airship is capable of turning about a circle of $200 \times 2.7 = 540$ ft.

References

[1] Perkins, C.D., and Hage, R.E., *Airplane Performance Stability and Control*, John Wiley and Sons, New York, N. Y., 1949.

[2] Ross, S.A., et al, *Goodyear Aerodynamics Handbook*, Akron, Ohio, 1954.

[3] Nicolai, L.M., and Carichner, G.E., *Fundamentals of Aircraft and Airship Design, Volume I—Aircraft Design*, AIAA, Reston, VA, 2010.

[4] *Transport Category Airships*, CS-30T, European Aviation Safety Agency, Sept 2003.

Chapter 5 *Propulsion*

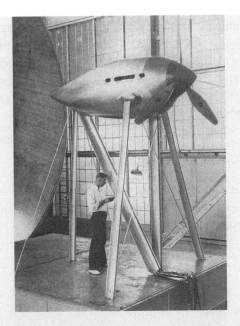

The picture shows a NACA test of a propeller in a wind tunnel. NACA conducted an aggressive program of propeller testing of the Clark Y and RAF-6 airfoils during the 1930s through the 1950s for the military and commercial sectors. This testing produced propeller performance maps and design charts that are still used today.

- Propeller Propulsion and Theory
- Blade Analysis
- Propeller Design
- Engine–Propeller Design and Interaction
- Propeller Systems
- Reciprocating Piston Engines
- Use of Vendor Charts
- Solar Airship System
- Solar Power

Propellers and mystery are synonymous.
Benjamin Carmina, Aviation, 1919

5.1 Introduction

The primary purpose of all aircraft propulsion devices is to supply a force (thrust) by imparting a change in momentum to a mass of fluid. The fluid may be air, air plus combustion products, or combustion products only. Isaac Newton's Second Law states that the force or thrust produced on a system is equal to the change in momentum of the system in unit time (dmV/dt). This fundamental principle is shown in Fig. 5.1 for a streamtube of air. The entrance conditions are denoted by the free stream symbol a and the exit conditions denoted by e. The mass flow rate of air through the stream tube is ρAV and has units of slugs/s. The streamtube boundaries are the fluid streamlines. The force or net thrust acting on the stream tube system is given by

$$T_{net} = (\dot{m}_{air} + \dot{m}_{fuel})V_e - \dot{m}_{air} V_a + P_e A_e - P_a A_a \qquad (5.1)$$

Notice that there may be a difference in the pressure and area at the entrance and exit such that a small pressure force would act on the system. Because the mass flow rate of the fuel added to the system is very small compared to the mass flow rate of the air, Eq. (5.1) is usually written

$$T_{net} = \dot{m}_{air}(V_e - V_a) + P_e A_e - P_a A_a \qquad (5.2)$$

Because of the low speed of airships (less than 100 KEAS) their principal propulsion device is the propeller.

Propellers are driven by reciprocating piston engines, gas turbines (turboshaft), or electric motors. A propeller operates by producing a relatively small change in velocity of a relatively large mass of air. Equations of motion will be developed later in this chapter that prove it is more efficient to move a large mass of fluid by a small increment in speed than a small mass by a large speed increase. Propellers are limited by the fact that their tip speeds

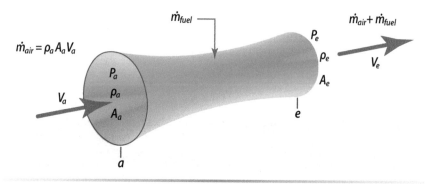

Figure 5.1   Momentum change on a fluid system.

must be much less than sonic due to the formation of shocks and therefore have a practical tip speed limit less than 500 kt (Mach = 0.75).

5.2 Why Propellers?

The open propeller, or airscrew, offers an efficient means of propulsion for airships that cannot be matched by the high exhaust velocity of a turbojet. Just as the turbofan engine is more efficient than a turbojet of the same thrust, a propeller is more efficient than either of them. The reason can be found from a brief look at Newton's Second Law in the form that a propulsion engineer would use.

$$T_{net} = \dot{m}_{air}(V_e - V_a) \qquad (5.3)$$

where, for this analysis, the pressure term is ignored and the resultant force in Eq. (5.3) is the thrust of the powerplant. A propeller achieves a specified level of thrust by giving a relatively small acceleration to a relatively large mass of air, while the turbofan and the turbojet each give a higher acceleration to a correspondingly smaller mass of air. From energy considerations, the powerplant producing the smallest change in kinetic energy will require the smallest expenditure of fuel, and thus the propeller provides the highest efficiency of the methods considered.

Another advantage of the propeller for an airship has to do with its ability to provide reverse thrust very quickly by simply changing its pitch angle. Because many airships have poor or no brakes they depend on the reverse thrust capability of their propeller propulsion system. Propeller reversing systems provide high deceleration for little weight or cost penalty.

5.3 Propeller Theory

The analysis of propeller performance can be accomplished using one or more of the following theories: momentum theory, blade element theory, and vortex theory. Each method has its own distinct advantages as well as shortcomings, yet all play an important role in providing an understanding of airscrew performance. The following discussion is intended to convey a general working knowledge of propeller theory, but for more details the reader is directed to [1]–[6]. References [1] and [3] cover the theory from a more practical point of view than the other references. References [2] and [4] are excellent from a theoretical perspective. References [5] and [6] present the theory from a helicopter point of view.

5.3.1 Momentum Theory

Any aerodynamic propulsive device produces a thrust by imparting a change in momentum flux to a specified mass of air (Newton's Second

Law). The basic momentum theory analyzes the effects of this change in momentum, the work done on the air, and the energy imparted to the air. Certain simplifying assumptions are made about the propeller and its surroundings in the development of this theory that divorce them from the real world, and yet the method remains a useful tool in calculating the maximum theoretical efficiency which a propeller can obtain.

The first assumption made by the momentum theory is that the propeller is replaced by an infinitesimally thin actuator disk which consists of an infinite number of blades. The disk is held to be uniformly loaded and is thus experiencing uniform flow and imparting a uniform acceleration to all of the air passing through it.

The actuator disk is further assumed to be surrounded by a sharply defined streamtube that divides the flow passing through the propeller and the surrounding air. Far upstream and downstream from the disk the walls of the streamtube are parallel, and the static pressure inside the streamtube at these points is equal to the freestream static pressure. Momentum theory deals with a working fluid (air in this case) that is inviscid and incompressible. As a consequence, the propeller does not impart any rotation to the air, and any profile losses from the blades of the propeller are ignored.

To an observer moving with the actuator, the air far upstream will be moving with the freestream velocity, V_∞ (Fig. 5.2). This air will be gradually

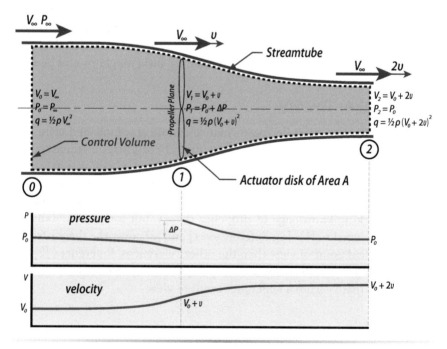

Figure 5.2 Propeller analysis by momentum theory.

accelerated until at station 1 (the propeller disk) $V_1 = V_0 + v$ where v is the induced velocity imparted to the air by the propeller. It can be shown at station 2, far downstream from the propeller that $V_2 = V_0 + 2v$. The net change in velocity through the control volume defined by the streamtube and planes perpendicular to the flow far upstream and far downstream is

$$(V_0 + 2v) - V_0 = 2v \tag{5.4}$$

and from continuity considerations for an incompressible fluid,

$$A_1 = A = 2A_2 \tag{5.5}$$

For steady flow the mass flux will be constant across every plane of the streamtube, which is perpendicular to the flow. Using the propeller as a reference plane,

$$\dot{m} = \rho A (V_0 + v) \tag{5.6}$$

The thrust, T, produced by the propeller will be

$$T = \Delta \text{ Momentum flux}$$
$$T = \rho A (V_0 + v)\, 2v \tag{5.7}$$

To produce this level of thrust, the propeller must supply energy to the slipstream. Because the theory ignores profile and rotational losses, this energy goes only to increasing the kinetic energy of the flow. The power required for this purpose, the induced power P_i, will equal the change in kinetic energy flux through the control volume and may be shown to be simply the product of the resultant thrust and the velocity at which the thrust is applied, or

$$P_i = T(V + v) \tag{5.8}$$

Equation (5.8) indicates that in order to minimize induced power requirements at a given thrust level and freestream velocity, the induced velocity must be kept as small as possible. Solving Eq. (5.7) for v (and remembering that $v > 0$ for a propeller) yields

$$v = -\frac{V}{2} + \left[\frac{V^2}{4} + \frac{T}{2\rho A}\right]^{1/2} \tag{5.9}$$

Two important conclusions are apparent from this expression. In order to minimize v (hence, P_i) at given values of V and T, the quantity T/A, the disk loading, must be minimized. Thus, within the limits of the assumptions made, it may be stated that the larger the propeller used to produce a given thrust, the smaller will be the power and energy requirements. The second result is that for a given thrust as the freestream velocity increases, the induced velocity will decrease. This is not to imply however, that the induced power requirement will decrease. For a given level of thrust, V will increase faster than v will decrease, and thus the required P_i will increase as free stream velocity increases. In practice, however, the thrust of a propeller will not remain constant with changing velocity, but the power of the engine turning it will, over moderate speed ranges, remain fixed. Because profile and rotational losses are being neglected, P_{avail} will remain constant and thrust will decrease with increasing velocity. This condition may be illustrated by combining Eqs. (5.8) and (5.9) to form an expression for P_i as a function of T and V. Solving for the static condition ($V = 0$ as designated by the o subscript) and for the condition of $V \neq 0$, and assuming that P_i is constant for all V, this finally gives the thrust ratio referenced to $V = 0$ as

$$\frac{T}{T_o} = \frac{2}{\dfrac{V}{v_o} + \left[\left(\dfrac{V}{v_o}\right)^2 + 4\left(\dfrac{T}{T_o}\right)\right]^{1/2}} \tag{5.10}$$

While a general solution for the thrust ratio $T/T_o = f(V/v_o)$ is not possible, the approximation

$$\frac{T}{T_o} \approx 1 - 0.32\frac{V}{v_o} \tag{5.11}$$

will hold for $V/v_o \ll 1$.

The theoretical power required by the propeller has been defined as the product $T(V + v)$. By defining the useful power output of the propeller as TV, it is possible to form an ideal efficiency. The following equations refer to Fig. 5.2.

$$\eta_i = \frac{\text{Output}}{\text{Input}} = \frac{\text{Thrust} \times \text{Velocity}}{\text{Total Work Done}} = \frac{T \times V}{\text{Kinetic Energy Increase}}$$

$$T = \dot{m}(V_2 - V_0) = 2\dot{m}v$$

$$\Delta KE = \tfrac{1}{2}\dot{m}\left(V_2^2 - V_0^2\right)$$

$$\eta_i = \frac{2\dot{m}vV_0}{\tfrac{1}{2}\dot{m}\left(V_2^2 - V_0^2\right)}$$

$$\eta_i = \frac{4vV_0}{(V_2 - V_0)(V_2 + V_0)} = \frac{4vV_0}{2v2(V_0 + v)}$$

$$\eta_i = \frac{V_0}{(V_0 + v)} \tag{5.12}$$

There are two results that are important. From Eq. (5.12) it is clear that for speeds near zero the efficiency factor, η_i, is close to zero. Another observation from Eq. (5.12) is that v should be low to obtain high efficiencies. For propellers, this means that it is more efficient to change a large area of air by a small v rather than changing a small area of air by a large v. Also note that η_i will increase with V_0. This concept of ideal efficiency is misleading for cases where $V/v < 1$. The use of the word "ideal" must again be emphasized as no real world losses are included in its calculation. It is also worth noting that high propeller efficiency and low power required both want small velocity changes over a large propeller area.

The momentum theory does not provide a means to predict propeller losses due to blade skin friction, rotational motion, or mutual blade interference, nor does it account for any geometry parameters other than propeller area. Although it is simple to apply, this theory must be combined with some other analytical tool in order to be of use to the designer.

5.3.2 Blade Element Theory

An airship propeller is nothing more than an airfoil rotating about a translating axis dividing a propeller blade into a number of chord-wise strips. It is possible to analyze the performance of the entire propeller by summing the contributions of all segments on all blades of the airscrew. This is essentially what is done by the blade element theory (or sometimes called strip theory).

In Fig. 5.3 a small element of the propeller blade is marked for consideration. This infinitesimal element is dr wide, has a chord c, and is located a distance r from the axis of rotation. The entire blade has a radius of R. A cross section of the blade element is shown in Fig. 5.4. The airfoil shape

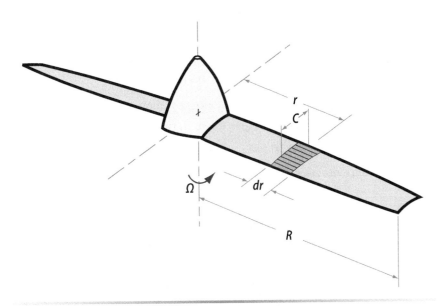

Figure 5.3 Propeller blade element.

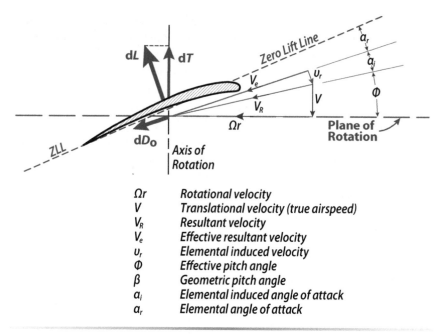

Ωr	Rotational velocity
V	Translational velocity (true airspeed)
V_R	Resultant velocity
V_e	Effective resultant velocity
v_r	Elemental induced velocity
Φ	Effective pitch angle
β	Geometric pitch angle
α_i	Elemental induced angle of attack
α_r	Elemental angle of attack

Figure 5.4 Forces, velocities, and angles for a blade element.

can be clearly seen, and many of the angular and velocity notations are analogous to those used in wing theory.

To simplify the development of the blade element theory, it is assumed that each element is subjected to two-dimensional flow only and that each element is independent of its neighbors.

The aerodynamic lift force produced by the dL, will be perpendicular to the effective velocity, V_e, and will be inclined from the axis of rotation by the angle $\phi + \alpha_i \approx \tan^{-1}[(V + v_r)/\Omega r]$. For freestream velocities experienced by airships it may be assumed that this angle is small, and

$$\sin(\phi + \alpha_i) \approx \phi + \alpha_i \text{ in radians}$$
$$\tan(\phi + \alpha_i) \approx \phi + \alpha_i \text{ in radians}$$
$$\cos(\phi + \alpha_i) \approx 1$$

Thus, the elemental thrust is

$$|dT| = |dL| \cos(\phi + \alpha_i) \approx |dL| \tag{5.13}$$

Similarly, the drag force opposing the rotation of the propeller element will consist of a drag component, dD_0, and a component due to the inclination of the lift force, the induced drag, dD_i:

$$|dD| = |dD_0| \cos(\phi + \alpha_i) + |dL| \sin(\phi + \alpha_i)$$
$$|dD| \approx |dD_0| + |dL|(\phi + \alpha_i) \tag{5.14}$$

It is now possible to express the thrust produced by a single element as

$$dT \approx dL = \text{dynamic pressure} \times \text{area} \times \text{lift coefficient}$$

$$dL = \left(\tfrac{1}{2} \rho V_e^2 \right) (c \, dr) \, c_\ell \tag{5.15}$$

where c_ℓ is the two-dimensional lift coefficient of the element. To determine the thrust of the propeller one must integrate this expression across the span of the blade and multiply by the number of blades (b):

$$T = b \int_0^R 0.5 \, \rho V_e^2 \, c \, c_\ell \, dr \tag{5.16}$$

Practical propeller blades do not run to the axis of rotation because some allowance must be made for a mounting hub and, possibly, a pitch changing mechanism. For this reason the inner limit of integration, r_i, is usually taken as $0.1R$. Similarly, some accounting must be made for losses caused by a decrease in effectiveness of outboard blade elements that results from the formation of a blade tip vortex. The outer integration limit is usually taken to BR where the empirically determined tip-loss-factor $B \approx 0.96$.

By ignoring compressibility effects, Eq. (5.13) becomes

$$T = 0.5 \rho \, b \int_{r_i}^{BR} V_e^2 \, c \, c_\ell \, dr \tag{5.17}$$

where V_e will vary with r, and c and c_ℓ may or may not be functions of radial position. Generally, $c = c(r)$ is specified, but in order to calculate the propulsive thrust, one must know $V_e = V_e(r)$ and $c_\ell = c_\ell(r)$.

From 5.4 it is obvious that

$$V_e \approx \left[(v_r + V)^2 + (\Omega r)^2 \right]^{1/2} \tag{5.18}$$

and the two-dimensional lift coefficient may be expressed as

$$c_\ell = a \, \alpha_r = a \left[\beta - (\phi + \alpha_i) \right] \approx a \left[\beta - \frac{V + v_r}{\Omega r} \right] \tag{5.19}$$

where $a = dc_\ell/d\alpha$. Due to variations in local Mach number across the blade span, a will vary with r. However, with little loss of accuracy, it may be assumed that a is a constant with a value appropriate for the conditions at $r = 0.75R$.

Equations (5.18) and (5.19) still cannot produce the key to solving for the thrust of the propeller until the local induced velocity, v_r, is known at every blade location. An expression for v_r can be obtained by employing simple momentum theory in an elemental approach. Figure 5.5 shows an actuator disk upon which an annulus dr wide and located a distance r from the center has been using the same logic as was used to develop Eq. (5.7), the differential thrust produced by this annulus will be

$$dT = \rho (2\pi r \, dr)(V + v_r) 2 v_r \tag{5.20}$$

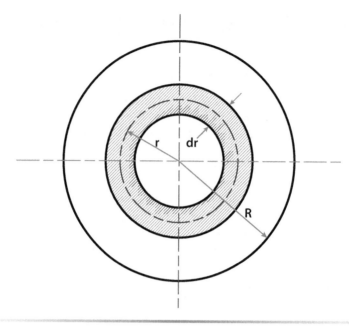

Figure 5.5 Annulus of an actuator disk.

From blade element considerations, the thrust generated by this same annulus will be the product of the thrust produced by a single element located a distance r from the axis of rotation [Eq. (5.15)] and the number of blades (b). With Eqs. (5.18) and (5.19) this becomes

$$dT = \frac{b}{2}\rho\left[(v_r+V)^2+(\Omega r)^2\right]c\,a\left[\beta-\frac{V+v_r}{\Omega r}\right]\tag{5.21}$$

Using Eqs. (5.20) and (5.21) and solving for v_r:

$$v_r = \left(\frac{V}{2}+\frac{bca\Omega}{16\pi}\right)\left\{-1+\left[1+\frac{2\Omega r\left(\beta-\dfrac{V}{\Omega r}\right)}{\dfrac{4\pi V^2}{bca\Omega}+V+\dfrac{bca\Omega}{16\pi}}\right]^{1/2}\right\}\tag{5.22}$$

which, within the limitations of the theory, will predict the induced velocity at a radial distance r of a propeller of known physical characteristics that is axially translating at a velocity, V.

Theoretically, it would be possible to introduce Eq. (5.21) into Eq. (5.22) and integrate the latter expression between appropriate limits to calculate the thrust of a propeller of arbitrary twist distribution. In practice, however, the resulting expression would prove extremely difficult to handle. Satisfactory results can be obtained by dividing the blade into a finite number of stations, calculating v_r and dT at each station, and finally computing r total thrust via graphical integration or some numerical technique such as Simpson's Rule.

The calculation of propeller thrust can be greatly simplified by the recognition that, as expressed by Eq. (5.22), the local induced velocity will be constant across the blade if the quantity $2\Omega r[\beta - (V/\Omega r)]$ is also a constant. (It can be shown [5] that constant v_r across the blade will require the minimum induced power for a given thrust and is thus desirable for reasons other than convenience of computation.) This may be accomplished by providing the blade with ideal twist such that for any element located at r, the geometric pitch angle is defined by

$$\beta = \frac{\beta_t R}{r} \qquad (5.23)$$

where β_t is the pitch of the tip section. This expression becomes unmanageable for $r \to 0$ as a result of the small angle assumption

$$(\phi + \alpha_i) \approx \tan^{-1}\left(\frac{V + v_r}{\Omega r}\right)$$

Practically, as $r \to 0$, $\beta \to \pi/2$. It must be noted that a unique twist distribution will be ideal only for a limited number of thrust and airspeed combinations. Because $T = f(\beta_t)$ for a given V, varying thrust levels will require variable β_t. However, because $\beta = \pi/2$ at $r = 0$ for all cases, the ideal twist distribution must be optimized for a single thrust/airspeed combination.

The blade element theory furnishes a method for approximating the total power requirements of the propeller by providing insight into the profile losses of the blade. From Fig. 5.5 it can be seen that the power required to rotate the propeller (and thus generate thrust) will be the power needed to overcome the forces in the plane of rotation. For a single infinitesimal element this is

$$dP = \Omega r dD_0 \cos(\phi + \alpha_i) + \Omega r dL \cos(\phi + \alpha_i) \qquad (5.24)$$

The term $dD_0 = \frac{1}{2}\rho V_e^2\, c\, c_{d_0}\, dr$ is the profile drag acting on the element, and thus the first series of terms in Eq. (5.24) may be thought of as the elemental profile power while the second group is the elemental induced power. Then,

$$dP = dP_0 + dP_i \qquad (5.25)$$

It must be noted that the induced power requirements are directly associated with the production of propeller thrust, and when the expression for dP_i is integrated across the blade radius provisions must be made for the loss of thrust at the tips. Profile losses, however, are present across the entire exposed radius of the blade. Thus, each of the terms in Eq. (5.25) must be integrated between separate limits:

$$P = 0.5\rho b\left[\int_{r_i}^{R}(\Omega r)^2 V_e\, c\, c_{d_0}\, dr + \int_{r_i}^{BR} V_e c(\beta\Omega r - V - v_r)(V + v_r)dr\right] \qquad (5.26)$$

This general equation is for modern propellers that employ ideal twist. Also, because most propellers are designed so that each section is operating at a low angle of attack, each element will also be functioning in the angle of attack region where the two-dimensional, incompressible profile drag coefficient, c_{d_0} is approximately constant, and for low speed application c_{d_0}, may be removed from the integral. This last statement is certainly not true for high speed propellers, however. As shown in Fig. 5.6, the resultant tip speed of a rotating blade is a function of rotational velocity and the true airspeed. At high flight speed and high propeller rpm (necessary for high thrust) the tip Mach number may approach or surpass the critical Mach number (~0.9) of the tip sections, and c_{d_0} will experience a drastic increase as $r \to R$. (For simplicity, skin friction, pressure, and wave drag effects are lumped together in c_{d_0}).

Equation (5.26) provides a key to understanding the rationale behind the selection of a certain propeller geometry to fulfill given design requirements. For low-to-moderate airspeeds where c_{d_0} will be constant, power requirements may be reduced by minimizing the blade chord toward the tip where dynamic pressure is greatest. However, this high dynamic pressure in the blade tip region is also responsible for the lion's share of the resulting thrust, and larger tip chords would be desirable from this standpoint. Some compromise must be reached, and the results are planforms of the type shown in Fig. 5.7b. Blade A is a type used on low speed general aviation craft. It features a circular or elliptical root section developing into an 8–12% thick section at the outer radii. Operating at rotational tip Mach numbers approaching 0.8, a propeller utilizing this blade can fly at airspeeds up to approximately

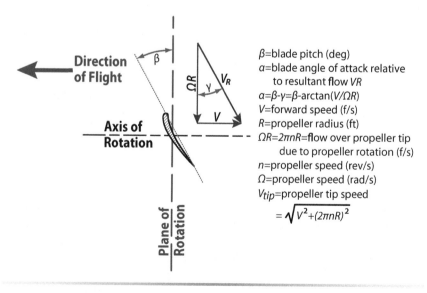

Figure 5.6   Resultant velocity at a propeller tip.

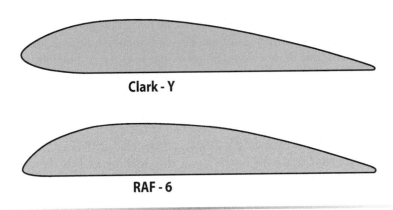

Figure 5.7a Typical propeller airfoil sections.

Mach = 0.4 before compressibility effects begin to be felt. Blade B exhibits a planform designed for use at high subsonic Mach numbers and features thin sections and reduced chord at the tip. In this way the drag effects of transonic tip conditions can be minimized. This class of propeller blade has not found widespread application because the speed range for which it is designed (Mach = 0.6–0.8) can be more efficiently handled by turbofan engines.

A practical blade planform for the middle subsonic range is the paddle blade design, blade C, which was used on the original C-130 and Electra aircraft. The wide tip chord of this blade would seem to produce higher compressibility losses, but, as demonstrated in [7], the opposite is true. The blade with a large chord at the tip will be more efficient than a tapered blade producing the same thrust at the same operating conditions because the tip

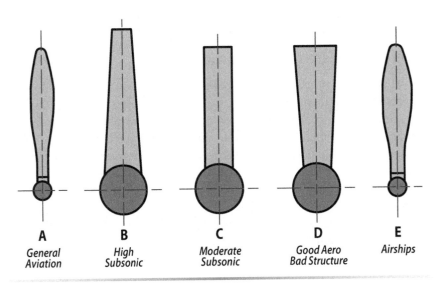

Figure 5.7b Typical propeller blade planforms.

sections of the wider, untapered blade will be operating at a lower c_ℓ and will have a higher critical Mach number. This argument would indicate that an even more efficient design would employ inverse taper as shown by blade D. While promising from an aerodynamic viewpoint, this approach has not been accorded wide acceptance because of structural difficulties.

Comparing blade E to aircraft propellers shows its unique geometry because airships have much smaller forward speeds, V, and lower rpm which means V_R is lower until getting near the blade tip. Generally speaking airship propellers have much lower disk loading and the blades have higher fineness ratios.

5.3.3 Vortex Theory

While providing a rapid method for the preliminary calculation of propeller performance, blade element theory does not provide the accuracy needed for detailed design work. Such factors as tip losses, three-dimensional effects, and mutual blade interference (which is significant) cannot be predicted by this method. For example, blade element theory indicates that a linear increase in thrust with no change in efficiency will result from adding blades to a propeller while, in fact, the most efficient propeller consists of a single blade with efficiency decreasing as the number of blades increase.

The third major branch of propeller theory, vortex theory, overcomes many of the limitations of the previous two methods and offers the capability for great accuracy. The equations required to implement this theory satisfactorily requires the use of high speed computers and sophisticated CFD codes. The details of the vortex theory are beyond the scope of this text and are more the tool of the propeller designer rather than the aircraft designer. The interested reader is referred to [1, 3, 4] and [8].

5.4 Propeller Design Parameters

While the previously discussed theoretical methods for propeller analysis provide convenient and relatively accurate schemes for predicting the performance of airscrews of known design, they would prove to be too cumbersome for application to preliminary design purposes. To establish the propeller design parameters required by the preliminary design process, various semi-empirical methods are employed. Reference 9, for example, provides rapid performance calculations for light aircraft propellers driven by engines of up to 300 hp and at flight speeds ranging up to 200 kt. Airships use scaled up versions of these general aviation propellers.

The task of identifying the characteristics of a propeller to meet a given set of performance specifications is essentially a two-part problem since it attempts to relate the horsepower available to the thrust provided by the propeller in the takeoff mode and in the cruise mode. Each of these segments requires independent methodology, and the results must be blended to provide continuous thrust output for a selected propeller from brake release through the speed limits of the airship.

At this stage of the design loop the drag characteristics of the airframe should be well established and should include a rough approximation of nacelle/shroud drag for the selected number of engines. The major design parameters to be determined at a given flight condition are the propeller diameter and the engine shaft horsepower required for that condition. All other parameters are defined by technology within rather narrow limits. If the propeller diameter is fixed by some structural consideration such as distance from the envelope then the resulting efficiency will be less than optimum, but the design process will be simplified as the required shaft horsepower can be calculated without iteration.

Certain definitions must be made at this point. As with most aerodynamic quantities, the thrust developed and power required by a propeller are conveniently expressed as non-dimensional coefficients in the following forms:

Power coefficient

$$C_P = \frac{P}{\rho n^3 D^5} \tag{5.27}$$

Thrust coefficient

$$C_T = \frac{T}{\rho n^2 D^4} \tag{5.28}$$

Speed-power coefficient

$$C_S = \left(\frac{\rho V^5}{P n^2}\right)^{1/5} \tag{5.29}$$

where n is the propeller rotational velocity in revolutions per second (rps), D is the diameter of the airscrew in feet, P is the engine/motor power in ft-lb/s, V is the flight speed in f/s, ρ is the density at altitude in slug/ft^3, and T is the thrust generated by the propeller in lb. Notice that C_S is independent of the propeller diameter D. The importance of this will be discussed in Sec. 5.5.2.

The ideal efficiency, η_i has been defined by Eq. (5.12) and should not be confused with another measure of effectiveness, the propulsive (or propeller) efficiency,

$$\eta_p = \frac{\text{Thrust Power Output}}{\text{Shaft Power Input}} = \frac{TV}{P} \tag{5.30}$$

This expression accounts for profile losses as well as induced losses and may be written as the product of an induced (or ideal) efficiency, η_i, and a profile efficiency, η_o. Thus,

$$\eta = \eta_i \eta_o \tag{5.31}$$

As with the ideal efficiency, the propeller efficiency will be zero under static condition.

Another useful parameter is the rotational tip speed of the propeller, V_{tip}, defined as

$$V_{tip} = \Omega R = \pi n D \tag{5.32}$$

The rotational tip speed has been given close consideration as a design point in recent years because of its importance in determining the operating noise level of the aircraft. Producing an aircraft with acceptable sideline noise levels is a major challenge to the designers of both civil and military STOL aircraft, and the reader is encouraged to consult [10] and [11] for further background on this problem. As a starting point 700–800 f/s is an upper limit for V_{tip}.

The ratio of the true airspeed, V, to the tip speed has proven to be a powerful propeller design variable in that it is related both to its efficiency and to its aerodynamic characteristics. This ratio is most often expressed as the proportional advance ratio

$$J = V/nD \tag{5.33}$$

Two more parameters are needed to completely define the propeller and its operating conditions. One is to establish the blade planform and the other to set the sectional lift characteristics. The latter condition was defined in the section on blade element theory as a two-dimensional lift coefficient, c_ℓ, which could vary across the blade span. In practical propeller designs the blade sectional camber is defined by the design lift coefficient, c_{ℓ_d}, and the camber for the entire blade is designated by specifying c_{ℓ_d} at $r = 0.7R$. Generally c_{ℓ_d} at $r = 0.7R$ will vary from 0.4 to 0.6 and minor excursions from the specified value at sections on either side of $r = 0.7R$ will have a negligible effect on the propeller performance.

The blade planform is expressed by the activity factor (AF), which represents the rated power absorption capability of all blade elements. Equation (5.26) indicates that the power absorbed by a blade element will be proportional to the area of the element times the cube of the velocity. By assuming $V_e \sim \Omega r$, the power may be expressed as

$$dP \alpha c(\Omega r)^3 dr \tag{5.34}$$

since at flight velocities $dP_o \gg dP_i$. This expression has been non-dimensionalized with V_{tip} and D to form a function of purely geometric properties and yet which reflects the relative ability of the blade to absorb power. The activity factor is conventionally defined as

$$AF = \frac{100,000}{16} \int_{0.15}^{1.0} \left(\frac{c}{D}\right)\left(\frac{r}{R}\right)^3 d\left(\frac{r}{R}\right) \tag{5.35}$$

for a single blade. The propeller *AF* is simply the blade *AF* times the number of blades (*b*). Values for blade activity factor are usually constrained by structural considerations to values between 80 and 180.

5.5 Shaft Engine Characteristics

In designing a propeller-driven airship, the designer must consider the propeller and its engine together as well as the location of the engine. Engine location is important as it affects weight, maintenance, vectoring, and many other issues (see Secs. 9.4 and 11.4.8).

Figure 5.8 shows the shaft horsepower-to-weight relationships for a spectrum of reciprocating piston and turboshaft engines. In most cases, the turbine engines include the weight of the reduction gearing required for their application as turboprop engines. The output of these engines is listed in terms of the shaft horsepower being produced to turn the propeller. This is not a complete picture in the case of the turboprop powerplant because a certain amount of residual jet thrust, T_J is also being generated.

The propeller tip speed is a function of both propeller diameter and shaft speed, n, thus the designer is concerned with the gear ratio between the power turbine or piston engine and the output shaft. Powerplant thrust changes are accomplished via simultaneous changes in fuel flow and propeller blade pitch. The engines could be designed to operate at a different output rpm with the attachment of a gearbox. Each turboprop or piston engine is evolved with a specific propeller in mind, thus the performance is based on the use of a standard reduction gear.

5.5.1 Propeller Suppliers

The design engineer will normally have available propeller operating curves supplied by propeller vendors. Propeller vendors in the United States are

Hartzell Propellers	One Propeller Place Piqua, OH 45356
McCauley Propellers	5800 East Pawnee Wichita, KS 67218
Sensenich Propellers	2008 Wood Court Plant City, FL 33563
Dowty Propellers	114 Powers Court Sterling, VA 20166-9321

5.5.2 Use of Vendor Propeller Charts

Propeller vendors will test their propeller designs in wind tunnels measuring T (thrust in lb), η_P (propulsive efficiency) for a given P (horsepower),

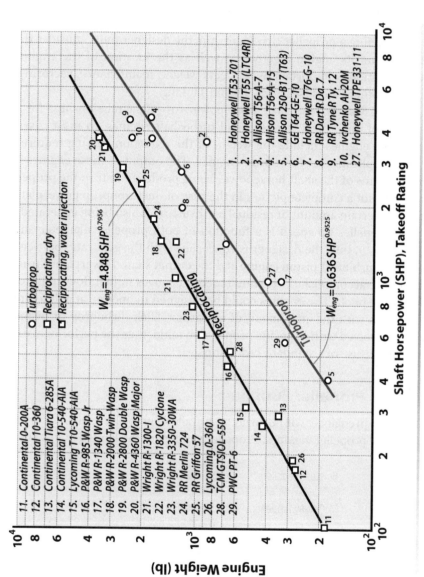

Figure 5.8 Weight vs shaft horsepower (shp), for turboprop and piston aircraft engines.

n (propeller speed in rps), D (prop diameter in feet), β (blade pitch angle in degrees), number of blades, and V (forward speed in ft/sec). This data will then be made available in coefficient form as C_P vs J and C_T vs J. This power coefficient and thrust coefficient are usually plotted together as shown on Fig. 5.9a for a three-bladed prop using a Clark Y airfoil section (shown in Fig. 5.7a). Figure 5.9a is called a propeller map or performance chart. Notice that at each value of $J = V/nD$ and blade pitch angle there is a unique combination of C_P Eq. (5.27) and C_T Eq. (5.28).

Because most propeller problems involve the determination of the propeller diameter D and blade angle β to give an acceptable propulsive efficiency η_P for a given flight condition (density and speed), a coefficient independent of D would be most useful. For such design purposes the speed-power coefficient C_S Eq. (5.29) was developed by modifying the C_P and presented as Fig. 5.9b for a three-bladed propeller with a Clark Y airfoil section. Figure 5.9b is a typical propeller design chart for determining D, β, and η_P for a required engine power, propeller speed, aircraft speed, and altitude.

The Clark Y and RAF-6 airfoil sections (see Fig. 5.7a) are the most popular airfoil sections used in modern day propellers. This is because there is an enormous amount of propeller data (i.e., performance charts like Fig. 5.9a and propeller design charts like Fig. 5.9b) available in the public domain as a result of the prolific propeller testing by NACA (National Advisory Committee for Aeronautics) in the 1930s through 1950s (see [1] and [12]). The modern day propellers that use high technology, laminar flow airfoils are only about 5% more efficient than the props of the 1940s.

Sample Problem 5.1: Engine and Propeller Selection for CA-1

We return to Sample Problem 4.1 and select the engine and propeller for the CA-1. If we assume the maximum speed to be approximately 90 f/s at 4000 ft (53 kt) there are several engine candidates available from Fig. 5.8. This figure gives the maximum *bhp* at SL and maximum rated rpm (2700–2800). Using Eq. (5.37a) to adjust for altitude = 4000 ft gives the following trade matrix:

Engine	bhp (SLS)	# of eng	Total bhp (4000 ft)	Max speed @start of mission (f/s)	Total engine weight (lb)
Wright R-3350-30	780	1	682	90	1050
PW R-985 Wasp	440	2	769	95	1340
Lycoming IO-540	305	3	800	96	1545
Lycoming O-360	185	4	647	88	1128

The single Wright R-3350 is not a good choice because of the possibility of losing one engine. The two, three, and four engine combinations are all

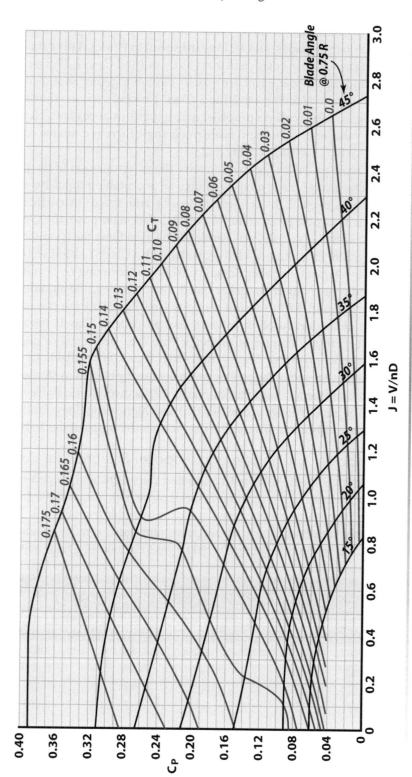

Figure 5.9a Performance chart for a 3 bladed propeller with a Clark Y airfoil section [1,12].

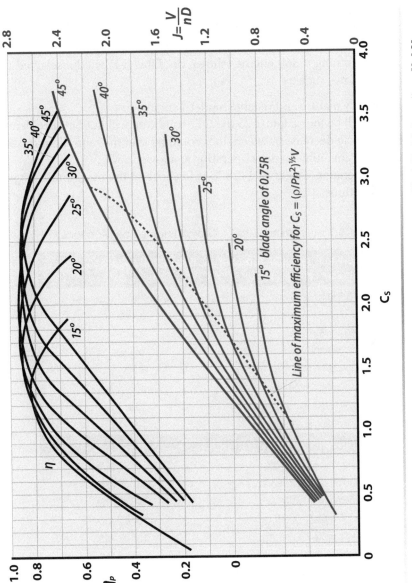

Figure 5.9b Design chart for a three-bladed propeller using a Clark-Y airfoil section. [1,12]

equally good with *BSFC* and total installed engine weight being major factors for selection.

As fuel is burned the engines would be throttled back to reduce the rpm and power available to keep $L = W$ for constant altitude. The CA-1 would be flown at constant speed, constant C_L or some combination for best range and endurance. From Fig. 4.2a the speed and power vary greatly from start of mission to the end. A propeller will need to be selected to give good performance (high η_P) over a broad range of hp and speed.

The Lycoming 0-360 engine (shown on Table 5.1) will be selected for several reasons, such as

1. It is a very mature and reliable modern engine.
2. There will be four engines so the OEI condition will not be a problem.
3. As the drag decreases, two engines could be shut down in flight to keep the two remaining engines operating at a lower *BSFC*.
4. Its total engine weight of 1128 lb is lower than the other multi-engine combinations.

Table 5.1 Lycoming 0-360-A Aircraft Engine Specifications [13]

Type—four cylinder, direct drive, horizontally opposed, wet sump, air cooled engine	
Weight, lb	282
Bore, in.	5.125
Stroke, in.	4.375
Displacement, in.^{3}	361
Compression ratio	8.5:1
Cylinder head temperature, max °F	500
Cylinder base temperature, max °F	325
Fuel: aviation grade, octane	100–130
Performance point	**hp**
Takeoff rating @ SLS, hp	185 @ 2900
Max rated @ 700 ft (28 in. of Hg), hp	180 @ 2700
Max rated @ 7000 ft, hp	143 @ 2700
Max rated @ 21,000 ft, hp	76 @ 2700
Cruise rpm @ 7000 ft, hp	135 @ 2450
Cruise rpm @ 21,000 ft, hp	74 @ 2450
Cruise rpm @7000 ft, hp	126 @ 2200
Cruise rpm @ 21,000 ft, hp	70 @ 2200
Cruise rpm @ 7000 ft, hp	106 @ 1800
Cruise rpm @ 21000 ft, hp	55 @ 1800
Cruise BSFC, lb/bhp-hr	0.47

Using Eq. (5.37a) the maximum rated power at 4000 ft and 2700 rpm is 162 hp for the Lycoming 0-360.

We will assume the cruise strategy for the CA-1 will be to fly a constant C_L cruise for $(L/D)_{max}$ down to 36 f/s and then a constant speed cruise to the end of the mission. This cruise strategy is shown on Fig. 4.2a. The propeller will be designed for $(L/D)_{max}$ (minimum drag) at the start of mission ($W_H = 6000$ lb) and $\eta_P = 0.65$ (to agree with Table 4.1). Then the η_P and propeller diameter will be checked for end of mission and maximum speed. The propeller will be a variable-pitch propeller.

From Fig. 4.2a the P_R for minimum drag at maximum W_H is 300 hp or 75 hp per engine at $V = 58$ f/s. From Eq. (5.37b) the engine rpm is 1250, however we will show that the propeller wants to operate at a slower speed for the low speed airships. Thus the engine will have a gear box that reduces the propeller speed to 600 rpm.

The data inputs for the baseline propeller are

$V = 58$ f/s
Power required $P_R = 75$ hp $= (75)(550) = 41{,}250$ ft-lb/s
$n = 600$ rpm $= 10$ rps
$\beta = 20$ deg (assumed)
$C_S = [\rho/P_R\, n^2]^{0.2}\, V = [0.00211/(41{,}250)(100)]^{0.2}\, (58) = 0.803$

From Fig. 5.9b

Advance Ratio $= J = V/nD = 0.54$
Blade airfoil: Clark Y
Number of blades $= 3$
Propeller efficiency $\eta_P = 0.65$
Diameter $D = 10.7$ ft

At the end of mission ($W_H = 1000$ lb) the speed is 36 f/s and the $P_R = 40$ hp. We will shut down two engines so that $P_R = 20$ hp for one engine at an rpm $= 333$. The data inputs into Fig. 5.9b are

$V = 36$ f/s
Power required $P_R = 20$ hp $= (20)(550) = 11{,}000$ ft-lb/s
$n = 333$ rpm $= 5.55$ rps
$\beta = 30$ deg (assumed)
$C_S = [\rho/P_R\, n^2]^{0.2}\, V = [0.00211/(11{,}000)(30.8)]^{0.2}\, (36) = 0.822$

From Fig. 5.9b

> Advance Ratio $= J = V/nD = 0.6$
> Blade airfoil: Clark Y
> Number of blades $= 3$
> Propeller efficiency $\eta_P = 0.55$
> Diameter $D = 10.8$ ft

The baseline propeller design checks out okay for the EOM design point except for a lower than desired η_P.

The data inputs for the maximum speed design point are

> $V = 88$ f/s at 4000 ft
> Power $= 162$ hp $= 89,100$ ft-lb/s at 2700 rpm

After some iteration the best propeller speed for this design point is propeller speed $= 700$ rpm $= 11.67$ rps.

> $C_S = 0.983$
> Blade angle $= 30$ deg
> $J = 0.73$
> $\eta_P = 0.63$
> Diameter $D = 10.3$ ft

The baseline 10.7-ft propeller checks out correctly at the three design points with the blade angle varying between 20 and 30 deg and the η_P dropping to about 0.55 at the end of the mission. The designer would probably elect to examine different airfoil shapes for the propeller and iterate on propeller speed n and blade angle β to improve the propeller efficiency. The propeller efficiency depends primarily on the blade airfoil, blade angle, airspeed, power, and propeller.

5.5.3 Use of Vendor Propeller Charts for Estimating Takeoff Thrust

At zero forward speed, the efficiency of a propeller is zero by definition, even though its thrust is not zero. In fact, for the same shaft power a propeller will produce more static thrust ($V = 0$) than it will with a forward speed. This can be explained using Fig. 5.6. The propeller develops the most thrust for a unique blade angle of attack α relative to the resultant flow V_R. This α is about 16 deg (depending on the blade airfoil section and the section lift characteristics). As the fixed pitch (β) propeller begins to move forward the blade angle of attack at the tip decreases as $\alpha = \beta - \gamma = \beta - \arctan(V/\pi n D)$ as shown on Fig. 5.6. At some forward speed V, the fixed pitch propeller blade $\alpha = 0$, and the prop would be slicing through the air over-speeding the engine and producing no thrust. This is the problem with fixed-pitch propellers. The variable-pitch propeller would increase its pitch angle automatically with forward speed (keeping its $\alpha \approx 16$ deg) producing thrust all the way out to its maximum speed.

Fortunately airships have low cruise speeds and a fixed-pitch propeller works quite well for takeoff and cruise. The following example will consider a three-blade, fixed-pitch propeller (Clark Y airfoil section and blade aspect ratio of 6) on an airship with a 100 ft/s cruise speed. An AR = blade length/blade chord = 6 propeller with a Clark Y airfoil section will stall at approximately 16 deg and $Re = 3 \times 10^6$. If we assume a blade angle $\beta = 20$ deg, prop speed $n = 20$ rps, and diameter = 14 ft, the prop tip speed and angle-of-attack at SL are

$$\text{Tip speed} = [100^2 + (\pi n D)^2]^{\frac{1}{2}} = [100^2 + (879)^2]^{\frac{1}{2}} = 885 \text{ ft/s (Mach} = 0.8$$
$$\text{so compressibility is not a concern)}$$

Tip α at $V = 100$ ft/s cruise speed is $\alpha = 20 - \arctan[100/\text{tip speed}] = 13.5$ deg

which is a good angle-of-attack for developing cruise thrust (tip is operating at approximately 84% of maximum C_L). At takeoff ($V = 0$) the propeller tip region would be stalled at $\alpha = 20$ deg but would still produce a static thrust at approximately 75% of maximum C_L. At a forward speed of 65 ft/s the propeller tip would be unstalled and operating at $\alpha \approx 15.8$ deg (α for maximum C_L) and producing maximum takeoff thrust.

For the same shaft power a variable-pitch propeller will produce the most thrust in zero forward velocity (i.e., its static thrust is greater than the thrust produced in forward flight). Figures 5.10 and 5.11 can be used to estimate the thrust available from a variable-pitch propeller at low forward speeds. The static thrust is first obtained from Fig. 5.11 and then reduced by a factor from Fig. 5.10 to get the thrust at 0.7 V_{TO} for a takeoff analysis. These charts only apply to a variable-pitch propeller that allows the engine to develop its rated power regardless of the forward speed.

The static thrust for a fixed-pitch propeller is obtained from Fig. 5.9a. The fixed-pitch angle β is selected (usually 15 to 20 deg). Then a C_P and C_T combination is located on the $V = 0$ axis of Fig. 5.9a.

Then using the value of C_P, power (ft-lb/s), prop diameter D (ft), and density the propeller speed n is determined from Eq. (5.27). Then solve for the static thrust $T_0 = C_T \rho n^2 D^4$. This static thrust should be less than the T_0 for the variable-pitch propeller case. The thrust at 0.7 V_{TO} can be determined the same way except locate the C_P and C_T combination on the $J = V/nD$ axis.

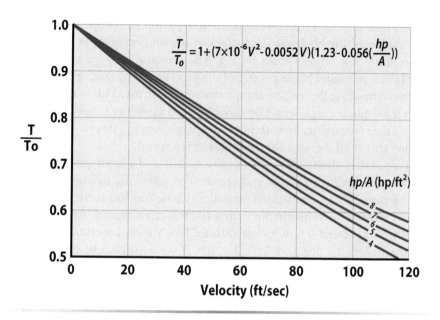

$$\frac{T}{T_o} = 1 + (7 \times 10^{-6} V^2 - 0.0052\, V)\left(1.23 - 0.056\left(\frac{hp}{A}\right)\right)$$

Figure 5.10   Decrease of thrust with velocity for different power loadings. [13]

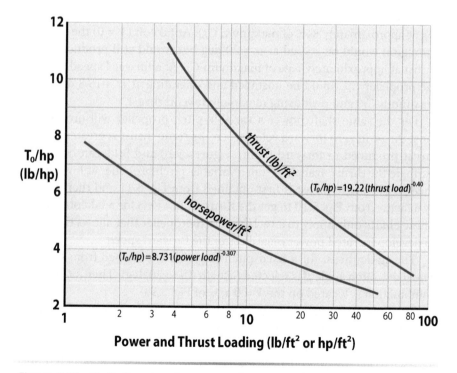

$(T_0/hp) = 19.22\,(\text{thrust load})^{-0.40}$

$(T_0/hp) = 8.731\,(\text{power load})^{-0.307}$

Figure 5.11   Static thrust and power performance of propellers or rotors. [13]

Sample Problem 5.2: Estimating the Takeoff Thrust for a Fixed- and Variable-Pitch Propeller

This example will estimate the takeoff thrust at $V = 0$ and 60 ft/s for a variable-pitch and fixed-pitch propeller for an airship with a three-bladed Clark Y propeller of $D = 14$ ft with power = 524 hp (each engine) at SL.

Variable pitch propeller: The blade angle β on a variable-pitch propeller is automatically changed in flight to operate at a resultant $\alpha \approx 16$ deg providing maximum thrust at all flight speeds V. From Fig. 5.11 the hp/prop disc area = 524/153.86 = 3.38, which gives $T_0/\text{hp} = 6.0$. The static thrust is $T_0 = (T_0/\text{hp})\,\text{hp} = 3144$ lb. At a forward speed of 60 ft/s, the curve fit equation for Fig. 5.10 gives a $T/T_0 = 0.70152$ or a $T = 2206$ lb.

Fixed pitch propeller: A fixed-blade angle $\beta = 20$ deg will be selected so that it produces good thrust at $V = 60$ ft/s. The $C_P = 0.09$ and $C_T = 0.1555$ lines merge at $\beta = 20$ deg on the $J = V/nD = 0$ axis. The C_P value is used to solve for the prop speed n:

$$n = [P/\rho\,D^5\,C_P]^{1/3} = [(524)(550)/(0.002377)(14)^5\,(0.09)]^{1/3} = 13.6 \text{ rps}$$

The static thrust is $T_0 = C_T\,\rho\,n^2\,D^4 = (0.1555)(0.002377)(184.43)(38416) = 2618$ lb. The thrust at a forward speed of 60 ft/s is determined using Fig. 5.9a for a $J = 60/(13.6)(14) = 0.315$. For $J = 0.315$ and $\beta = 20$ deg, $C_T = 0.14$ and $C_P = 0.088$. Solving C_P for n gives $n = 13.68$ rps. At 60 ft/s the prop is operating at $\alpha = 20 - \arctan[60/\pi\,(13.68)(14) = 14.3$ deg and $T = C_T\,\rho\,n^2\,D^4 = (0.14)(0.002377)(187.21)(14)^4 = 2393$ lb. It should be noted that the propeller blades are stalled at $V = 0$ but the prop is producing relatively good static thrust.

If we change the fixed pitch β to 15 deg we should improve the static thrust but will degrade the thrust at $V = 60$ ft/s because the resultant α will be decreased relative to the $\beta = 20$ deg case. For $J = 0$, $\beta = 15$ deg, $C_T = 0.14$ and $C_P = 0.06$. Solving C_P for prop speed gives $n = 15.55$ rps giving a static thrust $T_0 = C_T\,\rho\,n^2\,D^4 = (0.14)(0.002377)(241.67)(38416) = 3090$ lb (18% higher than the $\beta = 20$ deg case but close to the T_0 for the variable pitch prop). The thrust at $V = 60$ is determined along the $\beta = 15$ deg curve intersection with the $J = 60/(15.55)(14) = 0.275$ line giving $C_T = 0.103$ and $C_P = 0.056$. Solving C_P for n gives $n = 15.9$ rps and $T = (0.103)(0.002377)(252.81)(38416) = 2378$ lb. (within 1% of the $\beta = 20$ deg case). Clearly $\beta = 15$ deg is the better choice for the fixed-pitch blade angle.

5.6 Operation of Propeller Systems

The previous section discussed the analysis and design of propellers. This section will discuss the engines that drive the propeller. The engine provides a thrust power available equal to TV, which may be taken as the propeller output. The power input to the propeller from the engine shaft is the engine brake horsepower; thus the propeller efficiency is

$$\eta_p = \text{propeller thrust power/engine shaft brake horsepower} \quad (5.30)$$

In flight, the propeller accelerates a large mass of air rearward to a velocity only slightly greater than the flight speed, exhibiting efficiencies at normal flight speeds of between 85 and 90%. The lost horsepower appears mainly as unrecoverable kinetic energy of air in the slip stream.

The horsepower required for an aircraft to fly at a speed V is

$$\text{hp}_{\text{Req}} = DV/(550\eta_p) \quad (5.36)$$

where the 1/550 converts ft-lb/s to horsepower.

5.7 Reciprocating Piston Engines

The aircraft reciprocating piston engine uses the well-known four-cycle Otto cycle. An aircraft piston engine is similar to an automobile engine with a few differences. First, engine weight (hp/lb) is a major performance parameter. Most aircraft engines are air cooled for this very reason. Second, reliability is very important as a malfunction at any altitude is a serious situation. The current piston engines are well developed to give high performance (hp/lb), low brake specific fuel consumption $BSFC = $ (lb of fuel/hr)/brake horsepower, and high reliability.

Current piston and turboprop engine weights as a function of shaft horsepower are shown on Fig. 5.8. The hp/lb for the current spark-ignition piston engines varies from about 0.6 for the small engines (less than 600 hp) to almost 1.0 for the larger engines. The $BSFC$ for all the piston engines on Fig. 5.8 at SLS conditions varies from approximately 0.5 for the smaller engines (less than 400 bhp) to 0.42 for the larger engines. Most engines have a major overhaul recommended at 2000 hr. The engines have two spark plugs on each cylinder fired independently from engine driven magnetos.

The power output from a piston engine depends primarily on two parameters ... the engine rpm and the absolute pressure in the intake manifold. Maximum power is typically at 2800 rpm and SLS conditions of 59°F and 5.7 psia (30 in. of Hg).

Table 5.1 presents the specifications for the Lycoming 0-360-A aircraft engine. The 0-360-A is in the Piper Cherokee 180 and represents a very typical general aviation piston engine that could be used in the design of airships. Notice that it is designed to cruise at 66% of maximum power, which is an rpm range of 2200–2450. The maximum throttle performance degradation with altitude is linear from 700 ft (180 hp @ 2700 and 28 in. of Hg) to 21,000 ft (76 hp @ 2700 rpm). Cruise power is linear with altitude also.

A useful expression (from [15]) for the power loss (reduction in brake horsepower, bhp) with density altitude ρ is

$$bhp = bhp_{SL}\left(\frac{\rho}{\rho_{SL}} - \frac{1 - \dfrac{\rho}{\rho_{SL}}}{7.75} \right) \qquad (5.37a)$$

At a constant altitude the partial power performance is approximated by

$$bhp = bhp_{max\ rated}\,(rpm/rpm_{max\ rated}) \qquad (5.37b)$$

Piston engines are sometimes supercharged to increase SL power for air racing or to increase the operating altitude. Supercharging involves compressing the air entering the intake manifold by means of a compressor. In earlier piston engines, this compressor was driven by a gear train from the engine crankshaft. The more modern supercharged engines employ a turbine-driven compressor powered by the engine's exhaust and are called turbochargers. The advantage of the turbocharger over the gear-driven supercharger is two fold. First, the compressor does not extract power from the engine, but uses exhaust energy that would normally be wasted. Second, the turbocharger is able to provide sea-level-rated power up to much higher altitudes than the gear-driven type. Because the speed of airships is less important than for aircraft, adding heavy and expensive components to piston engines to increase power is not advisable. It is better to size the engine and number of engines to that for an existing off-the-shelf engine than to supercharge a smaller engine.

5.8 Turboprop Engines

Historically airships have been powered by reciprocating engines but another engine type can be used as well. Turboprop engines are used to power some aircraft where less speed and more range/endurance is part of the mission requirement. Compared to piston engines the benefits of a turboprop are lighter weight and less maintenance per flight hour but at the expense of fuel consumption.

A turboprop engine (sometimes called a turboshaft) is essentially a turbojet engine designed to drive a propeller. The turboprop is shown schematically in Fig. 5.12 and uses the basic gas generator section of a turbine engine. The propeller operates from the same shaft as the low spool compressor through reduction gearing. The hot gases are nearly fully expanded in the turbine first stage, which develops considerably more shaft power than required to drive the low spool compressor and accessories. The excess power is used to drive a conventional propeller equipped with a speed-regulated pitch control. The remainder of the hot gases are expanded through a nozzle thus providing jet

Turboprop

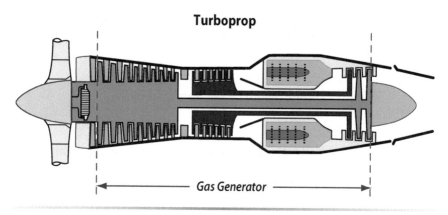

Figure 5.12   Schematic of typical engine showing basic gas generator core.

thrust. This engine retains the advantage of having a light weight and a low frontal area. In addition, it has a high efficiency at relatively low speeds.

The performance (hp/lb) of current turboprop engines is shown on Fig. 5.8. Turboprops are much lighter than an equivalent piston engine with hp/lb of approximately 2.2–2.4 for all engines. The shaft on a turbine engine typically rotates at 10,000 rpm, a speed much too high for propeller operation. In most cases, the weights shown on Fig. 5.8 for the turboprop includes the weight of the reduction gearing required for a propeller speed of approximately 2000 to 2700 rpm. The *BSFC* for turboprops is about 25% higher relative to a piston engine.

In a turboprop engine most of the power is extracted as shaft power to drive the propeller. However there is a residual energy that is expanded through the nozzle as jet thrust (T_J) that is not included in the listed shaft horsepower. To account for the power produced by this jet thrust an equivalent shaft horsepower (eshp) has been devised to account for the total power output of the engine. Using Eq. (5.36) the jet thrust is converted to a thrust horsepower (thp) by

$$thp = T_J \, V/(0.8)(550) \qquad (5.38)$$

where the 0.8 accounts for a conventionally assumed 80% propeller efficiency. With this expression *eshp* may be written

$$eshp = shp + T_J \, V/(0.8)(550) \qquad (5.39)$$

Notice that this relationship does not account for thrust horsepower under static conditions where $V = 0$. For such cases (and for $V < 100$ kt) another convention has been adopted to equate a given thrust level per horsepower. Some European turboprop companies use 2.6 lb of thrust per horsepower, but the usual equivalence is 2.5 lb of thrust equals one horsepower. Thus for $V < 100$ kt

$$eshp = shp + T_J/2.5 \qquad (5.40)$$

5.9 Electric Motors

Electric motors are simple and reliable (design life of 30,000 hr when operated at ~60% rated power). They have a specific power of approximately 0.27 hp/lb (0.2 kW/lb). The electric motors get their power from onboard APU (auxiliary power units, either piston or turboshaft engines driving electric generators), batteries, fuel cells, or solar cells (photovoltaic cells that convert incident solar energy into electricity).

For missions having several day/night (diurnal) cycles, the electric aircraft would have to seriously consider incorporating solar cells (photovoltaic). It would collect solar energy from the sun during the day and convert it to electricity through the photovoltaic action of solar cells. It would need to store energy in batteries or fuel cells to power the vehicle during the night. The solar cells would then recharge the batteries or fuel cells for the next nighttime operation by collecting excess power during the day. Theoretically this cycle could go on forever . . . however the batteries and fuel cells have finite recharging limits and performance degradation over time [14]. Table 5.2 contains data on electric motors, solar cells, batteries and fuel cells.

Table 5.2 Electric Airship System Data (2010)

Characteristic	Electric motor	Solar cell	Fuel cell	Batteries
Specific energy (kW-hr/lb)	0.2*	NA	0.89[†][††]	0.27[‡][††]
Design life	30,000 hrs	§	NA	300[¶]
Efficiency (%)**	97	28	55	90
Installed weight (lb/sq ft)	NA	0.1	NA	NA

* Weight includes motor, controller, and propeller. Increase weight by 25% for installation. Specific energy is kW/lb.

† H_2O regenerative fuel cell using proton exchange membrane technology. Increase weight by 25% for installation.

‡ Li-S batteries are projected to increase to 0.336 kW-hr/lb by 2015.

§ Solar cells degrade about 1.5% of power output per year.

¶ 300 full depth discharges in 2010. Decreasing the discharge to 50% would increase number of recharges to approximately 1000.

** Efficiency is energy out/energy in. Solar cell efficiency projected to increase to 32% and fuel cell to 65% by 2015.

†† Specific power based on discharge time.

5.10 Solar Power

The sun is a source of unlimited energy during the day. Every day it bathes the outer edge of the earth's atmosphere with 127 W/ft^2 of solar energy on average. The 127 W/ft^2 is termed the solar constant. The amount of solar energy received anywhere on the Earth at a point in time depends on the latitude Φ of the surface, the tilt (inclination) of the earth's spin axis as it orbits around the sun, and its position relative to the sun (time of day).

This dependence is shown in Fig. 5.13. The inclination of the earth to the orbital plane varies between $+23.5^0$ on 21 June and -23.5^0 on 21 December and is the reason the Earth has its four seasons. On 21 June the northern hemisphere is getting more solar energy and is enjoying summer while the southern hemisphere is getting less and is having winter. On 21 December the situation reverses. On 21 June the northern hemisphere has its longest day of the year and on 21 December the shortest.

The solar energy received on Earth is converted to useful electrical energy by the photovoltaic action of solar cells. The electrical energy per unit area available from a horizontal solar cell of efficiency η_{SC} at an altitude h and solar elevation angle θ is

$$\text{Power Available} = P_{Elect} = P_{Solar}\, \eta_{SC} \sin\theta \qquad (5.41)$$

where θ is the solar elevation angle of the sun above the horizon and $\sin\theta$ accounts for the presented area of the horizontal solar cell. The solar elevation angle is a complicated function of the latitude, inclination angle (time of year), and orientation to the sun (time of day). The best way to determine θ is to go to the NOAA (National Oceanic and Atmospheric Administration) Web site and use their solar position indicator.

P_{Solar} is the average solar radiation at altitude, h, and solar elevation angle θ. The Earth's atmospheric mass (AM) has a significant effect on the value of P_{Solar}. The water and ozone in the atmosphere absorbs and scatters the solar radiation. $P_{Solar} = 127$ W/ft^2 in space (outside the Earth's atmosphere at an altitude of approximately 320,000 ft or 53 nmiles) whereas $P_{Solar} = 96.5$ W/ft^2 on the Earth's surface and $\theta = 90$ deg having suffered a 24% energy loss due to atmospheric attenuation. The space condition is termed AM 0 and the condition on the Earth's surface and $\theta = 90$ deg is AM 1.0. Values for P_{Solar} at altitude h and solar elevation angle θ are given on Fig. 5.14 (essentially h and θ define the *slant range* through the atmosphere).

Even though the solar energy is limitless and free, it is small when compared to the energy available from burning hydrocarbon fuels (i.e., gasoline or JP-4). Because the power required increases by the cube of the speed and the available solar power is small, the speed of a solar airship is low—it will

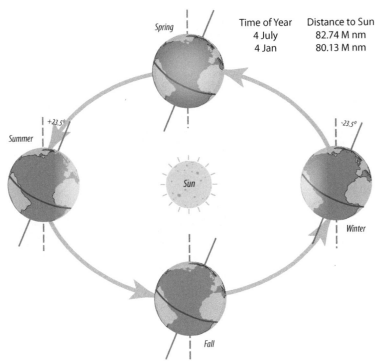

Time of Year	Distance to Sun
4 July	82.74 M nm
4 Jan	80.13 M nm

The solar energy incident on the earth changes due to its tilt (inclination) and orientation as it revolves around the sun. The latitude (location), earth's tilt (time of year), and position (time of day) relative to the sun results in different amounts of the sun's energy hitting the earth.

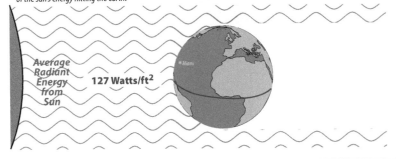

Figure 5.13 Solar energy radiated to Earth during the year.

always be less than 35 KEAS. This can be shown by setting power required = power available.

$$\text{Power required} = (\text{Drag})(\text{Velocity})/(\text{Propulsive efficiency})$$
$$= (\tfrac{1}{2}\,\rho\,C_D\,\text{Vol}^{\tfrac{2}{3}}\,V^2)(V)/\eta_{\text{prop}} \tag{5.42}$$
$$\text{Power available} = P_{Solar}\,\eta_{SC}\,S_{SC}\,\sin\theta \tag{5.43}$$

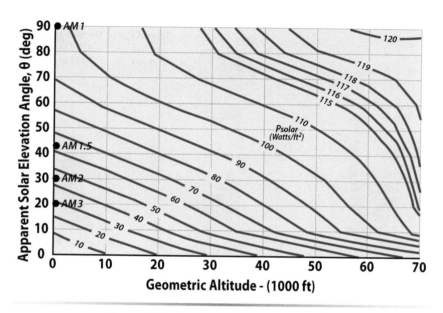

Figure 5.14   Direct clear-sky P_{Solar} (W/ft²).

Then assume typical values for the parameters and solve for V as follows:

Altitude = 60,000 ft	$\rho = 0.000224$ slugs/ft³
theta = 90 deg (optimistic)	$P_{Solar} = 120$ W/ft² (Fig. 5.14)
$\eta_{SC} = 31\%$ (optimistic)	35% in lab, 31% installed on wing
$\eta_{prop} = 0.81$ (optimistic)	Motor, propeller, and line losses
$Vol^{2/3} = S_{SC}$	Reasonable and it makes the math simpler
$C_D = 0.024$	HALE (Fig. 3.19)
Payload & vehicle power = 0	not realistic but makes the point
$V = 192$ f/s = 114 kt at 60,000 ft ~ 35 KEAS	and this is the best it can do

The electrical energy available from a solar cell of an assumed installed efficiency of 28.7% at Miami, FL (latitude $\Phi = +25°46'$) and Moscow ($\Phi = +55°45'$) is shown on Fig. 5.15 for 4 July and 5 January over a 24-hr period [15, 16]. The area under the curves is the total electrical energy per unit area [watt-hours per square foot (W-h/ft²) captured by horizontal solar cells during the daylight hours. Notice that Moscow and Miami have about the same total energy on 4 July even though Moscow is at a much higher latitude. The reason for this is that Moscow has more daylight hours than Miami (17 and 14 hr, respectively). This is not the case on 4 January.

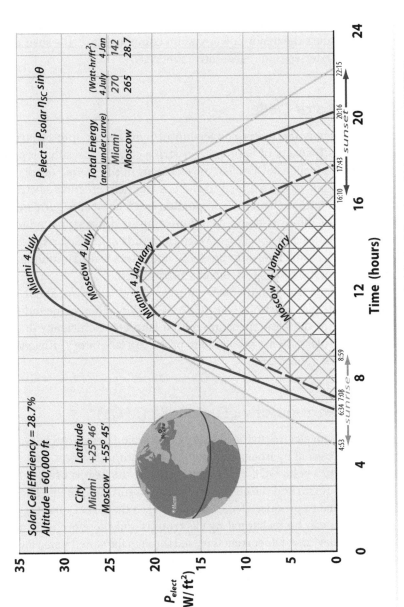

Figure 5.15 Comparison of electrical energy available for Miami and Moscow at 60,000 ft.

5.11 Solar-Powered Airship Subsystems

The sun is an unlimited source of energy . . . in the daytime. Solar-powered air vehicles must collect excess energy during the daytime and store it so that it can be use to power the vehicle during the night. This section will discuss the major power system components . . . the solar cells, rechargeable batteries, and regenerative fuel cells.

5.11.1 Solar Cells

Photovoltaics is the direct conversion of light energy into electricity. When photons (packets of light energy) strike certain specially prepared semiconductors (called solar cells), the electrons in the atomic outer rings loosen to flow out of the cell due to the electric field creating electricity. Thin wafers of certain elements/compounds (such as silicon, germanium, copper, indium, gallium arsenide, cadmium telluride) form semiconductors that capture photons in specific wavelength bands of the solar spectrum (UV, visible, IR). By stacking different semiconductors (called multijunction devices) more of the solar spectrum can be captured.

A single silicon semiconductor has a solar cell efficiency of about 20%. In 2008 the triple junction GaIn/GaAs/Ge thin film semiconductor captured most of the solar spectrum with a laboratory efficiency of 28%. Multijunction solar cells are projected to produce a laboratory efficiency of 32% in 2015 (see Table 5.2).

5.11.2 Regenerative Fuel Cells (RFCs)

A fuel cell is an electrochemical energy conversion device. A H_2/O_2 fuel cell converts external hydrogen and oxygen gas into water, heat, and electricity (about 0.75 volts DC). The water is stored, the heat is used to keep the equipment warm and the electricity is used to run equipment. The fuel cell will continue to produce electricity, as long as there is H_2 and O_2. The H_2/O_2 fuel cell can be recharged by electrolysis and therefore used as an energy storage device. A DC current is passed through the stored water. Hydrogen bubbles form at the anode (−) and oxygen bubbles at the cathode (+). The two gases are stored in separate tanks until needed to generate electricity. Currently the H_2/O_2 fuel cell has a specific energy of 0.89 kW-hr/lb and is projected to have a round-trip efficiency (energy out/energy in) of 65% in 2015 (see Table 5.2). The fuel cell power system is shown schematically in Fig. 5.16.

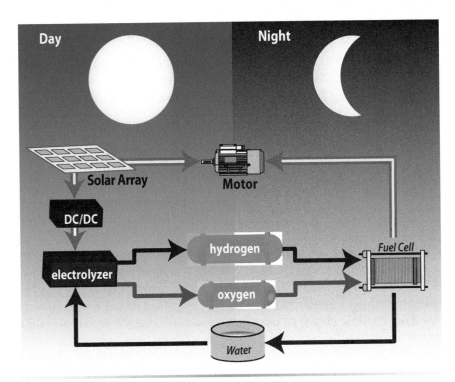

Figure 5.16 Day–night energy storage, solar, and fuel cells.

5.11.3 Rechargeable Batteries

A battery is an electrochemical energy conversion device with all the chemicals internal to the device. The battery can be made to be rechargeable and thereby used as an energy storage device similar to RFCs. The characteristics of a battery are given in Table 5.2.

The advantages of a battery relative to an RFC are

1. Simplicity – it is like a flashlight
2. Higher basic round-trip efficiency (i.e., 90% compared to 55%)

The disadvantages of a battery relative to an RFC are

1. Lower specific energy (i.e., 0.27 kW-hr/lb compared to 0.89). Not all of the reactants embodied in battery electrodes are accessible. Some cannot be reached and remain as useless weight after discharge.
2. Full depth charge/discharge cycle limits. The electrodes are consumed during the recharging process and the current cycle limit is about 300 cycles. Reducing the depth of discharge (i.e., partial discharge) increases the number of cycles.

5.12 Propulsion and Energy Storage System Sizing

This section will size the propulsion and energy storage system (solar cell area, batteries, RFCs, motors, and propellers) for a solar-powered airship conducting ISR over Miami, FL 24/7 for a duration of one year.

Sample Problem 5.3: Solar-Powered Airship Sizing

We will assume the HALE airship to have the following features:

Max altitude	65,000 ft
Station	60,000 ft

From Fig. 5.17 the winds aloft are the lowest around 65,000 ft with a maximum wind speed of 44 f/s (26 kt) and an average wind speed of 17 f/s (10 kt) over Miami on 4 Jan. The HALE propulsion system will be sized to counter the maximum winds on the critical winter day (4 Jan with its short daytime hours) to ensure the vehicle is not blown out of its ISR pattern box.

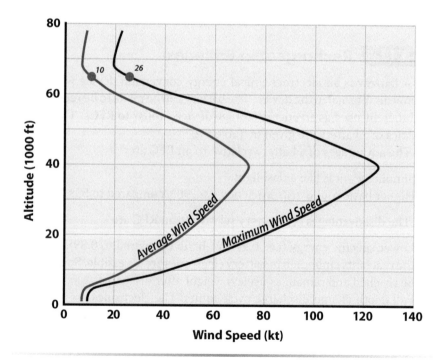

Figure 5.17   Variation of wind speed with altitude for Miami on 4 January. (NCAR data).

Assume the following weight and power for the HALE airship:

Item	Weight (lb)	Power (kW)
Payload	500	1
Vehicle	3400	4
Propulsion + power storage	800	TBD
Other/margin	300	1
Total	5000	TBD

The lifting capability of the Solar HALE airship using 98% pure helium is given by Eq. (2.6) as

$$\text{Weight} = \text{Buoyant Lift} = (\text{Volume})(0.065)\sigma \qquad (5.44)$$

where σ is the ratio of densities ($\sigma = 0.0754$ at 65,000 ft). Thus, the airship volume to lift 5000 lb to 65,000 ft with the ballonets empty (maximum altitude) is

Volume $= L_{buoy}/0.065 \ \sigma = 5000/(0.065)(0.074) = 1,040,000 \ \text{ft}^3$
$Vol^{2/3} = 10,264 \ \text{ft}^2$
Fineness Ratio = length/diameter = 3.2
Dynamic pressure, q, at 65,000 ft and 44 f/s = 0.17 lb/ft^2
Assume $C_D = 0.024$ (Table 3.1) and $\eta_p = 0.80$

The drag $= (0.024)(0.17)(10,264) = 42$ lb and the propulsion power required $P_R = (\text{drag})(\text{max speed})/550\eta_p = (42)(44)/(550)(0.80) = 4.2 \ \text{hp} = 3.1 \ \text{kW}$ and the total power required during the day is 9 kW.

We will refer to the all-electric aircraft database shown in Table 5.2 and assume a solar cell efficiency of 32% (note, these cells will be multijunction cells and will be expensive). Each solar cell generates 0.5 volt. The cells are connected together to form a blanket (typically 36 individual cells are connected together, generating 18 volts DC with a blanket-packing efficiency of ~96%). The blankets are connected together to form a solar array with an array electronics efficiency of 96%. The solar arrays are glued to the top of the HALE envelope. If the cells are going to be in service for long periods, the environment will degrade the cell efficiency by about 1.5% per year an effect called end-of-life efficiency. We will assume that 4 January occurs in the middle of the one year mission; then the cell degradation will be minimal during the period of short daylight hours. Thus, we will assume an installed solar cell efficiency $(32)(0.985)(0.985)(0.96)(0.96) = 28.7\%$ at the end of the second year.

At this point a decision needs to be made as to where the electrical energy comes from to power the HALE through the night when P_{Solar} is zero. The answer is that all daytime solar energy > 9 kW will be stored and used to power the HALE at night. The two candidate storage schemes are rechargeable batteries and *regenerative fuel cells* (RFCs ... usually H_2/O_2). Table 5.2 has design data on batteries and RFCs.

We will initially assume that the energy for nighttime operation will be stored in rechargeable batteries. While batteries have a much simpler installation than the RFCs, they have a limited life of approximately 300 full depth discharges (see Table 5.2), which would not work for a one year mission. However, if the discharges are less than full depth (which is the case for most of the year when the nighttime hours are much shorter than those for 4 January) the battery life is doubled.

The baseline *round-trip efficiency* (RTE) is 0.9. However there are line losses that need to be considered. First there is a transmission efficiency $\eta_{Trans} = 0.98$ going in and coming out of the battery. Then there is a power switch/step efficiency $\eta_{Switch} = 0.90$ going in and coming out of the battery. The line losses amount to $(0.98)(0.98)(0.9)(0.9) = 0.78$. Thus, the total RTE for the battery storage system is $\eta_{RTE} = (0.78)(0.9) = 0.7$. The continuous power that needs to be provided to the batteries is $9/0.7 = 13$ kW.

The solar cell sizing problem is one of collecting an excess amount of energy during the daylight hours, storing it in rechargeable batteries (or RFCs) and using it during the night to power the HALE vehicle. Figure 5.18 helps to understand the diurnal power balance.

We iterate on the Fig. 5.18 (a or b) Miami 4 January curve to find the CPL (daytime continuous power loading) that gives

$$A = 142 - (CPL)(\text{daytime hours}) = (CPL / 0.7)(\text{nighttime hours}) = R_{1+R^{2+margin}}$$

Using CPL $= 4.5$ W/ft$_2$ gives $A = 142 - (4.5)(9.3) = 49.5 + 43.8 + 6.9$

Thus $A = R_1 + R_2$ assuming a 7.4% margin. During the conceptual design stage it is always a good idea to carry a 6–10% margin. Notice that for all other days $A > R_1 + R_2$ and the airship will have the problem of rejecting the excess energy.

The daytime continuous power requirement is 9000 watts and the solar cells are providing 4.5 W/ft^2. Thus the solar cell area needs to be $9000/4.5 = 2000$ ft^2 of horizontal surface. Because the HALE $FR = 3.2$, the solar cell should be 80×25 ft. The solar cells are mounted on an insulated blanket to keep the rejected heat from conducting into the airship envelope. The blanket is glued to the top of the airship. The cell area would be adjusted upward on subsequent design iterations to account for the curvature of the surface. The airship would be pointed towards the sun as much as possible to minimize the cells deviation from a horizontal surface. The HALE configuration is shown on Fig. 5.19.

The weight of the batteries is based upon the discharge W-h. For the batteries the discharge power is 12.86 kW and the discharge time is 14.7 h or 189 kW-hr. From Table 5.2 the specific energy for batteries is 0.336 kW-hr/lb

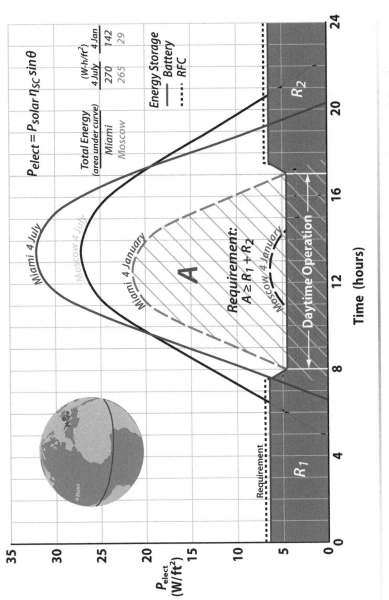

$P_{elect} = P_{solar} \, \eta_{SC} \sin\theta$

Total Energy	(W-h/ft²)	
(area under curve)	4 July	4 Jan
Miami	270	142
Moscow	265	29

Energy Storage
—— Battery
···· RFC

Miami 4 July

Moscow 4 July

Miami 4 January

Moscow 4 January

Requirement:
$A \geq R_1 + R_2$

A

Daytime Operation

R_1

R_2

Requirement

P_{elect}
(W/ft²)

Time (hours)

Figure 5.18a Diurnal energy balance example basics for stationkeeping over Miami.

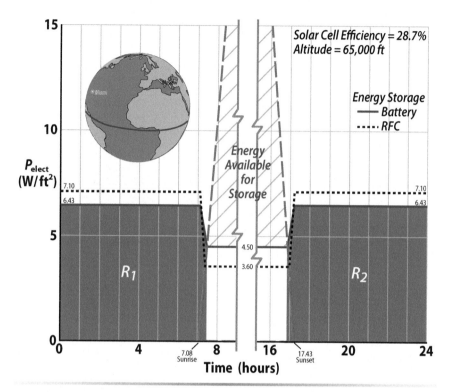

Figure 5.18b   Diurnal energy balance example for stationkeeping over Miami on 4 January. Data from Fig. 5.15 (cross-hatch is battery storage).

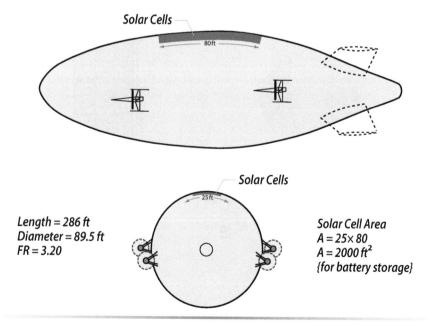

Figure 5.19   Solar High-Altitude-Long-Endurance (HALE) configuration.

(in 2015) giving a battery weight of 563 lb. There is a 15% installation factor for batteries. The weight of the installed solar cells is $(0.1)(2000) = 200$ lb.

The RFCs are sized in the same manner as the batteries. Using the 2015 value for the uninstalled RTE for the RFCs (Table 5.2) gives an installed RTE of $(0.65)(0.78) = 0.507$. Iterating on Fig. 5.18 indicates a daytime power loading of 3.6 W/ft^2 (nighttime of $3.6/0.507 = 7.1$), which gives a power balance of 107.8 W/ft^2 available and of 101 W/ft^2 required for a 6.7% margin. The RFC discharge power is $(29/0.507)(14.6) = 259$ kW-hr, which gives an installed RFC weight of $(259/0.89)(1.25) = 365$ lb. The required solar cell area is $9000/3.6 = 2500$ ft^2 with an installed weight of $(2500)(0.1) = 250$ lb. Table 5.3 shows a comparison of batteries and RFCs. At this point in the design of the Solar HALE the RFCs would appear to be a better storage selection because they are approximately 200 lb lighter than the batteries.

We will assume that the maximum power of 4.2 hp is provided by four electric motors of 1.0 hp each. The propeller will be designed using the following inputs:

Power $= 1.0$ hp $= (1.0)(550) = 550$ ft-lb/s
Maximum speed $= 44$ ft/s at 65,000 ft $(\rho = 0.000176$ slugs/ft$^3)$
Propeller speed $= 600$ rpm $= 10$ rps

Using the speed-power coefficient Eq. (5.29) gives

$$C_S = [(0.000176)/(550)(100)]^{0.2} (44) = 0.88$$

Table 5.3 Sizing Comparison for Battery and RFC Storage

Item	Battery	RFC
Critical Day	4 Jan	4 Jan
Round trip efficiency, η_{RTE} (%)	90	65
Installed η_{RTE} (%)	70	50.7
Daytime power loading on Fig. 5.18 (W/ft^2)	4.5	3.6
Nighttime power loading (W/ft^2)	6.43	7.1
Storage discharge time (hr)	14.7	14.6
Excess energy during daytime, A (W/ft^2)	100.2	107.8
Energy needed during nighttime, $R_1 + R_2$ (W/ft^2)	92.6	101
Continuous daytime power (kW)	28	28
Solar cell area (ft^2)	2000	2500
Discharge power (kW-h)	189	259
Battery weight (includes 15% install factor) (lb)	647	–
RFC weight (includes a 25% installation factor) (lb)	–	364
Installed solar cells (0.1 lb/ft^2) (lb)	200	250

From Fig. 5.9 (three blade, Clark Y airfoil propeller chart) we obtain

$\eta_p = 0.78$
Blade angle = 15 deg
$J = V/nD = 0.5$

where the $\eta_p = 0.78$ matches our assumed propeller efficiency and the advance ratio gives a propeller diameter of $D = 9.8$ ft.

All that remains in order to close the Solar HALE design is the sizing of the tail and check the assumed $TOGW = 5000$ lb. This will be done in Chapters 6 and 9 respectively.

5.13 Energy Balance Method

The energy balance method just discussed in the sizing of the Solar HALE airship is a very robust approach. The energy balance solution shown in Fig. 5.18 is independent of the airship size (volume) or the daytime power requirement. The solution is dependant only on the available solar energy shown in Fig. 5.15 (function of latitude and date) and the RTE of the storage device. Once the daytime power loading (W/ft^2) is determined, the airship size (volume) and required power can be changed very easily. The reason for this is the fact that the airship has so much surface area that an increase in the solar cell area [solar cell area = (daytime power requirement)/ (daytime power loading from Fig. 5.18)] does not usually require a volume change. Notice that this is not the case for a solar-powered aircraft where the solar cells are installed on the wing upper surface as the solar cell area increases the wing area has to increase, which is a major design change (see the Solar Snooper example in [16]). The size of the airship would be changed using Eq. (5.44) only if the component weight build-up did not meet the assumed weight. If the airship volume changes the energy balance solution is still the same.

References

[1] Nelson, W.C., *Airplane Propeller Principles*, John Wiley and Sons, NY, 1944.
[2] Theodorsen, Theodore, *Theory of Propellers*, McGraw-Hill, New York, 1948.
[3] Dommasch, Daniel O., *Elements of Propeller and Helicopter Aerodynamics*, Pitman Publishing Corp., New York, 1953.
[4] Glauert, H., "Airplane Propellers," *Aerodynamic Theory*, Vol. IV (William F. Durand, ed.), Dover Publications, Inc., New York, 1963.
[5] Gessow, Alfred, and Myers, Garry C., Jr., *Aerodynamics of the Helicopter*, Frederick Ungar Publishing Co., New York, 1952.
[6] Stepniewski, W.Z., *Introduction to Helicopter Aerodynamics*, Rotorcraft Publishing Committee, Morton, Pa., 1950.

[7] Stack, J. Draley, E.C., Delano, J.B., and Feldman, L., "Investigation of the NACA 4-(3)(08)-03 and NACA 4-(3)(08)-045 Two-Blade Propellers at Forward Mach Numbers to 0.725 to Determine the Effects of Compressibility and Solidity on Performance," NACA TR-999, 1950.

[8] Goldstein, S., "On the Vortex Theory of Screw Propellers," *Proceedings of the Royal Society*, A. 123,1929.

[9] Crigler, John L., and Jaquis, Robert E., "Propeller-Efficiency Charts for Light Airplanes," NACA TN 1338, 1947.

[10] Hubbard, H.H., "Propeller Noise Charts for Transport Airplanes," NACA TN 2968, 1953.

[11] Rosen, George, and Rohrbach, Carl, "The Quiet Propeller—A New Potential," AIAA Paper No. 69-1038, 1969.

[12] Hartman, E.P., and Biermann, David, "The Aerodynamic Characteristics of Full-Scale Propellers Having 2, 3, and 4 Blades of Clark-Y and R.A.F. 6 Airfoil Sections," NACA Report 640, Nov 1957.

[13] McCormick, Barnes, *Aerodynamic, Aeronautics and Flight Mechanics*, John Wiley and Sons, Inc., 1995.

[14] Mitlisky, F., Weisberg, A., Myers, B., *Regenerative Fuel Cells*, Lawrence Livermore National Laboratory, UCRL-JC-134540, June 1999, paper prepared for the U.S. DOE Hydrogen Program 1999 Annual Review Meeting, Lakewood, CO, May 4-6, 1999.

[15] Hill, P.G. and Peterson, C.R., *Mechanics and Thermodynamics of Propulsion*, Addison Wesley, Reading, MA, 1965.

[16] Nicolai, L.M. and Carichner, G.E., *Fundamentals of Aircraft and Airship Design, Volume I—Aircraft Design*, AIAA, Reston, VA, 2010.

[17] Youngblood, J.W., and Talay, T.A., "Solar Powered Airplane Design for Long Endurance, High Altitude Flight," Paper AIAA-82-0811, 18 May 1982.

[18] Youngblood, J.W., Talay, T.A., and Pegg, R.J. "Design of Long Endurance Unmanned Airplanes Incorporating Solar and Fuel Cell Propulsion," Paper AIAA-84-1430, 13 June 1984.

Chapter 6 Preliminary Weight Estimate

LE

- Estimating Empty Weight
- Estimating Fuel Weight
- Iterating for Gross Weight
- Hybrid Airship Sample Problem: CargoStar
- Cruise Strategies
- Propulsion Sizing
- Propeller Sizing
- Estimating Takeoff Distance

The Northrop Grumman/U.S. Army LEMV unmanned hybrid airship is 302 ft long, 84 ft in diameter, and has a volume of 1,350,000 ft^3. The LEMV mission is to carry 2750 lb of sensors to 20,000 ft and stay there for 21 days, providing tactical ISR.

KISS–Keep It Simple Stupid.
Skunk Works edict

6.1 Introduction

The design process begins with an estimate of the gross weight, W_G. The designer will know very little about the airship except its requirements—usually payload, range/endurance, and altitude. The designer must then decide on the

1. Airship configuration (body of revolution or a multi-lobe hybrid)
2. Propulsion system (reciprocating, turbine or solar)—need *BSFC* and hp
3. Structural arrangement (rigid, semi-rigid, or non-rigid)
4. Structural material
5. Maximum altitude (sizing the ballonets)
6. Buoyancy ratio (*BR*)
7. Fineness ratio–length/(equivalent diameter)
8. Manned/unmanned
9. Tail configuration
10. Mission flight strategy (constant C_L, constant speed, or some combination)

Getting started at this point is challenging because most designers must make assumptions based upon very little information. Perhaps the best rule here is to "assume something even if it's wrong" so that an estimate of W_G can be made.

It is important to remember that the conceptual design phase is a looping or iterative process, in which the assumptions are refined on subsequent passes as the design converges to a feasible baseline point (see Fig. 1.21).

6.2 Weights

The takeoff gross weight is defined as

$$\text{TOGW} = W_G = \text{payload} + \text{crew/equip} + W_{fuel} + \text{empty weight} \quad (6.1a)$$

The designer needs to pay particular attention to how payload is commonly defined as it may or may not include elements such as armament, accommodations, cargo, passengers, fuel (in the case of an aerial tanker). Sometimes the payload is expendable (i.e., bombs, missiles, bullets, tanker fuel, torpedos, sonobuoys, etc.) and other times it is not (i.e., ISR sensors, cargo, passengers, laser module, and gun). It is the designer's call as to whether the gun that shoots the bullets is bookkept as payload or part of the basic empty weight. But generally, payload is everything that is required to accomplish the mission. The payload and crew weight is usually defined by the customer.

Aircraft design text books [1, 2, 3] use the equation

$$\text{Fixed weight} = \text{payload} + \text{crew/equip} \qquad (6.2)$$

for preliminary sizing and defines $W_G = \text{Fixed weight} + W_{fuel} + W_E$

The empty weight (W_E) and fuel weight, W_{fuel}, are discussed in the next two sections.

Operating empty weight (W_{OE}) is another useful weight used by the mass properties and performance engineers in weight summaries

$$W_{OE} = W_E + \text{trapped fuel} + \text{crew} \qquad (6.3)$$

where trapped fuel is the fuel trapped in the fuel system lines and not available for the mission. Typically trapped fuel is less than 1% of the total fuel required for the mission and is often ignored in preliminary sizing.

The zero fuel weight W_{ZF} is defined as

$$W_{ZF} = W_{OE} + \text{payload} \qquad (6.4)$$

Using the definitions for W_{OE} and W_{ZF} Eq. (6.1a) is rewritten as

$$\text{TOGW} = W_G = \text{payload} + W_{fuel} + W_{OE} = W_{ZF} + W_{fuel} \qquad (6.1b)$$

Military [4] and commercial [5] performance analyses use the following weight definitions:

$$\textit{Commercial: Landing weight} = W_{OE} + W_{fuelres} \qquad (6.5a)$$
$$\text{Useful Load} = \text{payload} + W_{fuel} \qquad (6.5b)$$
$$\textit{Military: Landing weight} = W_G - \tfrac{1}{2} W_{fuel} \qquad (6.5c)$$
$$\text{Combat weight} = W_G - \tfrac{1}{2} W_{fuel} \qquad (6.5d)$$

6.3 Empty Weight

The empty weight W_E of the airship includes envelope fabric and structure, ballonets, propulsion system, gondola, tails, subsystems, avionics, and so on.

The designer will soon discover that this preliminary estimation of the aircraft empty weight is the weakest part of the conceptual design analysis and it has tremendous leverage on the airship gross weight. It is almost impossible to estimate with any degree of accuracy the empty weight of something that has not been built (usually with new subsystems and structural materials). However it is important not to get delayed by challenges such as this or the airship will never be designed.

Sometimes the empty weight is expressed as

$$W_E = W_{ME} + W_{FEQ} \qquad (6.6)$$

where $= W_{ME}$ is the manufacturers empty weight (sometimes called the AMPR weight or green weight) and is used in costing the manufacture of

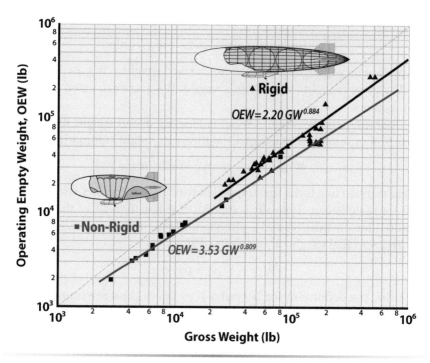

Figure 6.1   Operating empty weights of historical airships, rigid and non-rigid.

the airframe (Chapter 24 [1]). The W_{FEQ} is the weight of the procured items (such as engines, wheels, landing gear, avionics, instruments, environmental control system, and auxiliary power unit).

The operating empty weight is determined by using historical data and trends as shown on Fig. 6.1. Notice that the non-rigid data has a tight grouping around its trend line whereas the rigid data exhibits more scatter. There is a cluster of rigid airship data at $W_G = 1$–2×10^5 lb, which is well under the trend line. These were very lightweight and fragile German airships in the 1930s designed for reconnaissance at 20,000 ft. These designs were not included in the equation on Fig. 6.1. The W_{OE} will be refined by doing a weight build-up (discussed in Chapter 9). It is always a good idea to compare the W_{OE} estimate with real airship data in Appendix D.

6.4 Fuel Weight

The W_{fuel_A} is determined by assuming a gross weight, W_G and estimating the W_{OE} from Fig. 6.1. The fuel available is then

$$\text{Available fuel, } W_{fuel_A} = W_G - \text{Payload} - W_{OE} \qquad (6.7)$$

The range or endurance that can be achieved with the available fuel is determined and compared with the required range or endurance. If the comparison is close (within 5%) the assumed W_G is a good preliminary estimate of the W_G and the design is ready for refinement as shown on Fig. 1.21. If the comparison is not close the W_G needs to be increased—or decreased—and the analysis repeated.

The helium volume required is estimated knowing the amount of the W_G that needs to be lifted by the helium to the maximum altitude (where the ballonets are empty).

$$\text{Volume} = (BR)(W_G)/[(0.065)\ \sigma] \qquad (6.8)$$

where $\sigma = \rho/\rho_{SL}$ at the maximum altitude for the airship.

The helium volume is needed as it equals the volume of the hull (or envelope) and is used to calculate the airship dimensions and surface area for estimating the aerodynamics. This approach would not work for sizing a rigid airship since the hull volume is larger than the helium volume. The rigid airship hull is essentially a cage holding the helium balloons. Historical data for rigid airships indicates a hull volume approximately 10% larger than the helium/hydrogen volume and accounts for the rigid airship W_E being above the trend line for non-rigid airships on Fig. 6.1.

For a body of revolution (one-lobe conventional airship) configuration the designer must calculate the length l_B and diameter using the required volume and assumed fineness ratio FR.

For a hybrid configuration finding the length, width, and height is more involved. The length and width of the hybrid configuration are determined assuming a prolate spheroid having the same volume as the hybrid (see Fig. 3.4). Using volume $= \pi l_B (\text{diameter})^2/6$ and the assumed FR, it is possible to calculate the length and diameter of the prolate spheroid. The cross-section of the prolate spheroid is then flattened to an elliptical shape with the major (width/2) and minor (height/2) axis the same as the hybrid. Then calculate $S_{plan} = \pi (\text{width}/2)(\text{length}/2) = \pi (FR) (\text{width}/2)^2$ and aspect ratio $AR = (\text{width})^2/S_{plan}$. The envelope surface wetted area is estimated using Eqs. (D.3) and (D.4).

The tail area is estimated using the *tail volume coefficient* method of Chapter 7. Assume the tail moment arms $l_{HT} = l_{VT}$ both equal to 36–43% of the airship length l_B. The c.g. is typically at the same x-location as the c.b., which is located at ~0.45 l_B .

The C_{D_0} is estimated using the method discussed in Chapter 3, Sec. 3.5.3. Be sure to include all the miscellaneous drag items discussed in Eqs. (3.23) to (3.37). The drag-due-to-lift factor K is determined using Fig. 3.26. Be sure to reference the C_{D_0} and K to the same reference area . . . either $Vol^{2/3}$ or S_{plan}.

Finally calculate the range or endurance with the available fuel using the assumed flight strategy:

Constant C_L	
$C_L = (C_{D_0}/K)^{1/2}$	(4.19)
Range using Eq. (4.15) with maximum $L_{aero}/D = 1/(4\,C_{D_0}\,K)^{1/2}$	(4.20)
Endurance or cruise flight time using Eq. (4.34b)	
Constant speed: Range using Eq. (4.18)	
Endurance using Eq. (4.35)	

Compare the available range or endurance with the requirement. If the range or endurance is less than the requirement, the W_G will need to be increased and the fuel weight analysis repeated. Continue the iteration until the available range or endurance is within 5% of the requirement.

The analysis proceeds until it is clearly evident that the assumed W_G is not going to work. At that point the W_G is adjusted up or down and the analysis is started over. This iteration on W_G continues until the W_{OE} build-up (discussed in Chapter 9) plus the W_{fuel} and payload weight equals the assumed W_G [Eq. (6.1)] within 1%.

Notice that this sizing process could also start by assuming an initial volume and iterating on volume until there is design closure. The sample airship problems in Chapters 11 and 12 will be an example of preliminary sizing by initially assuming a volume.

Sample Problem 6.1: Estimate the Gross Weight of the CargoStar Vehicle

This problem will demonstrate the methodology for estimating the gross weight (W_G) of a hybrid airship designed for transporting cargo. The requirements for the CargoStar (Fig. 6.2) are a payload of 34,000 lb and range of 5000 nm. The assumptions are as follows:

Configuration:

3-lobe hybrid airship similar to HA-1 (see Fig. D.2)
Buoyancy Ratio (BR) = 0.75 @ takeoff
Propulsion – (4) IC engine/props at $BSFC = 0.47$ (see Table 5.1) and $\eta_P = 0.65$
Cruise speed = not specified (cruise can be constant V or C_L)
Cruise altitude = 4000 ft
Maximum speed = 65 kt @ SL
Maximum altitude = 12,000 ft (maximum altitude for manned airships without pressurization)
Helium @ 98% purity: Force = 0.0646 lbf/ft^3 (see Sec. 2.3)
Ballonet fraction = 0.45 (see Fig. 9.3, used "conservative" line)
Crew = 5
Fineness ratio, FR = length/(equivalent diameter) = 2.86

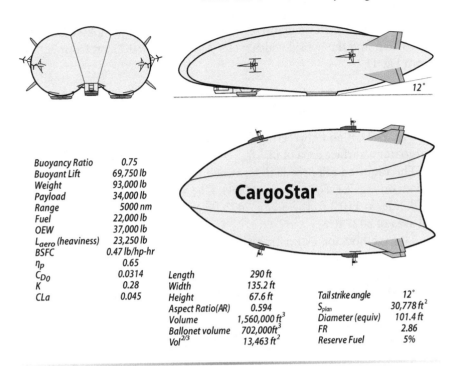

Buoyancy Ratio	0.75			
Buoyant Lift	69,750 lb			
Weight	93,000 lb			
Payload	34,000 lb			
Range	5000 nm			
Fuel	22,000 lb			
OEW	37,000 lb			
L_{aero} (heaviness)	23,250 lb			
BSFC	0.47 lb/hp-hr			
η_p	0.65			
C_{D0}	0.0314	Length	290 ft	
K	0.28	Width	135.2 ft	
CL_a	0.045	Height	67.6 ft	Tail strike angle	12°

CargoStar

Length	290 ft		
Width	135.2 ft		
Height	67.6 ft	Tail strike angle	12°
Aspect Ratio(AR)	0.594	S_{plan}	30,778 ft^2
Volume	1,560,000 ft^3	Diameter (equiv)	101.4 ft
Ballonet volume	702,000 ft^3	FR	2.86
$Vol^{2/3}$	13,463 ft^2	Reserve Fuel	5%

Figure 6.2 CargoStar—hybrid transport example (ref area = $Vol^{2/3}$).

Tail moment arm = 0.38 ℓ_B
Takeoff/landing distances = not specified
Reserve and trapped fuel = 5% and 1% respectively

The preliminary sizing starts by assuming a W_G = 93,000 lb. From Fig. 6.1 the W_{OE} = 37,000 lb, which gives an available fuel weight W_{fuel_A} = 93,000 − 37,000 − 34,000 = 22,000 lb. The L_{aero} will compensate for the fuel burned from 22,000 lb down to 1320 lb (5% reserve and 1% trapped fuel).

The sizing continues with determining the volume of CargoStar.

$$\text{Volume} = (BR)\,(W_G)/(0.0646\,\sigma)$$
$$= (0.75)(93,000)/(0.0646)(0.6932) = 1.56 \times 10^6 \text{ ft}^3$$
$$Vol^{2/3} = 13,463 \text{ ft}^2$$

Next we find the length and diameter of an equivalent body of revolution of the same volume and FR = length/diameter = 2.86. The volume = π (length)(diameter)^2/6 so that length = ℓ_B = 290 ft and diameter, d_e = 101.4 ft. Using the geometric condition given on Fig. D.2 that each lobe intersects the adjacent lobe by an amount equal to its radius, we use the relationship on Fig. D.3 that d_e/lobe diameter = d_e/d_c = 1.5. Thus the lobe diameter is d_c = 67.6 ft and the hybrid width = (1 + #lobes)d_c/2 = 135.2 ft from Fig. D.2. From Fig. D.2

$$S_{Plan} = \pi \ell_B \text{ (width)}/4 = 30{,}778 \text{ ft}^2$$

$$AR = 4 \text{ (width)}/(\pi \ell_B) = 0.594$$

Using Eq. (D.3) and correcting for a three-lobe perimeter (Fig. D.4) gives CargoStar a surface area of $(S_{wet})_{env}$ = (77,367)(1.081) = 83,634 ft².

The required tail area is determined from the tail volume coefficients shown on Fig. 7.1. For a C_{HT} = 0.0684, C_{VT} = 0.061 and a tail moment arm of (0.38)ℓ_B = 110.2 ft, the S_{HT} = 2423 ft² and S_{VT} = 2161 ft², giving a total tail area of 3427 ft².

The method for estimating the C_{D_0} and K follows the discussion in Chapter 3. The cruise speed is assumed to be 59 f/s (35 kt which is a realistic speed) at 4000 ft, resulting in a Re of 9.6×10^7, which gives a body turbulent C_f = 0.0021 from Eq. (3.20). The body form factor FF = 1.61 for the FR = 2.86. Using Eq. (3.23) and a body surface area of $(S_{wet})_{env}$ = 83,634 ft² and assuming 90% turbulent flow gives an envelope *drag area*, $C_D S_{env}$ = 252.6 ft². The remainder of the drag area components are determined using Eqs. (3.24) through (3.37). The drag build-up is as follows:

Component	Drag area (ft²)
Envelope	$C_D S_{env}$ = 252.6
Envelope accessories	$C_D S_{ea}$ = 10.2
Tail surfaces	$C_D S_{ts}$ = 32.47
Bracing cables	$C_D S_{bc}$ = 25.37
Tail control lines	$C_D S_{cl}$ = 0 (fly by wire)
Tail surface accessories	$C_D S_{tsa}$ = 1.72
Car/gondola (fairing incl)	$C_D S_{car}$ = 35.18
Outriggers	$C_D S_{out}$ = 12.11
Engine nacelle	$C_D S_{nac}$ = 17.0
Engine cooling	$C_D S_{cool}$ = 27.9
Handling lines	$C_D S_{hl}$ = 0 (has an ACLS)
Interference	$C_D S_{mi}$ = 6.5
Air cushion landing system	$C_D S_{acls}$ = 2.0 (retracted)
Total drag area = 423.0 ft²	
C_{D_0} (referenced to $Vol^{2/3}$) = 423.0/13,463 = 0.0314	

The drag-due-to-lift factor K is determined from Fig. 3.26. For an AR = 0.594 we obtain K = 0.67 (referenced to S_{plan}). Using the N_L = 2.4 determined in Sec. 3.3.1 gives a K (referenced to $Vol^{2/3}$) = 0.67/N_L = 0.28. The range will be determined using two different cruise strategies.

6.5 Cruise Strategy #1–Constant C_L at maximum aero L/D (V decreases)

This cruise strategy is at minimum drag (maximum aero L/D). As fuel is burned the CargoStar will slow down. The range equation for this strategy is

$$\text{Range(nm)} = \frac{326\eta_p}{BSFC} \frac{L_{aero}}{D} \ln\left[\frac{W_{H_0}}{W_{H_1}}\right] \tag{4.15}$$

which assumes the aero $L/D = 1/[4C_{D_0} K]^{\frac{1}{2}} = 5.33$ for a constant $C_L = (C_{D_0}/K)^{\frac{1}{2}} = 0.335$ and speed decreasing as fuel is burned. At the start of the mission $L_{aero} = (1 - BR)(W_G) = W_{H_0} = 23{,}250$ lb, and at the end of the mission after burning 20,680 lb of fuel (leaving 5% reserve and 1% trapped fuel) $L_{aero} = W_{H_1} = 23{,}250 - 20{,}680 = 2570$ lb giving $W_{H_0}/W_{H_1} = 9.047$. The range for this first set of assumptions (first iteration) is

$$\text{Range (nm)} = 326 \ (0.65)(5.33) \ln (9.047)/ (0.47) = 5292 \text{ nm}$$

and the speed varies from 41.4 kt (70 f/s) to 13.8 kt (23 f/s), which is close to the average speed of 35 kt used in the C_{D_0} calculation. Obviously, one of the disadvantages of flying at the optimum C_L is that end-of-cruise speeds can become unacceptably low.

At landing the CargoStar would weigh 72,320 lb and would be 2570 lb heavy. The hybrid airship would have to be ballasted or tied down as the payload is removed. In order to prevent this operational inconvenience the CargoStar would have to be designed with a $BR = W_{O_E}/W_G = 0.4$. However a $BR = 0.75$ is about the lower limit as it starts making takeoff a problem for the airship to generate $> 0.25 \ W_G$ in aerodynamic lift.

At a $W_G = 93{,}000$ lb the W_{fuel} is 22,000 lb giving a range of 5292 nm (~6% greater than the required range of 5000 nm). Thus, we will keep the 22,000 lb fuel weight for future calculations for cruises at conditions other than constant C_L. A $W_G = 93{,}000$ lb is a good preliminary estimate for the start of the design. Further iterations will refine the assumptions, which might cause the W_G to change up or down. The preliminary CargoStar design is shown on Fig. 6.2.

6.5.1 Cruise Strategy #2–Constant Speed (C_L Decreases)

With cruise strategy #2, flying the best constant speed for maximum range results in a flight speed of 39 f/s (23.1 kt) and a range of 4460 nm using Eq. (4.18). Since speed is held constant throughout the mission, C_L decreases as fuel is burned to keep $L = W$ at a constant altitude. While this cruise strategy avoids the low end of cruise speeds from maintaining a constant C_L it gives up 16% in range. If the mission requires flight at a constant speed then the W_G would have to be increased and the analysis repeated until the range is acceptably close to 5000 nm.

6.5.2 Cruise Strategy #3–Combined Constant C_L and Constant Speed

Cruise strategy #1 gives a larger range than strategy #2 for a given W_{H_0}/W_{H_1}. However, the endpoint speed of 14 kt might be unacceptable as the mission duration is 9.7 days.

A better cruise strategy might be to start the cruise at constant C_L and then shift to a constant speed cruise to reduce the cruise duration. For example assume that the CargoStar cruises at constant $C_L = 0.335$ for 18,000 lb of fuel ($W_{H_0} = 23,250 - 18,000 = 5250$ lb) or $W_{H_0}/W_{H_1} = 4.43$. The distance covered is 3576 nm, the endpoint speed is 20 kt, and [using the endurance Eq. (4.34b)] the cruise time is 5.34 days. Then the cruise strategy shifts to constant speed at 20 knt giving 1588 nm range [using Eq. (4.18)] over $1588/(20)(24) = 3.3$ days. The total range is 5164 nm over 8.55 days. This combined cruise strategy meets the 5000 nm range requirement at one less day cruise time than strategy #1. Table 6.1 gives the cruise range and time for various combinations of constant C_L and constant speed. The constant C_L cruise burning 16,000 lb of fuel followed by a constant speed cruise at 39 f/s is selected as the cruise profile since it meets the 5000 nm requirement (within 1%) and reduces the cruise duration by two days.

6.5.3 Propulsion Sizing

The propulsion for the airship is sized by either maximum speed or takeoff distance. For the CargoStar the takeoff distance is not specified but

Table 6.1 Trade Study of Combined Constant C_L and Speed Cruise Strategies for the CargoStar Example

Fuel burn @ const C_L (lb)	Const $C_L/\Delta V$ (f/s)	Const C_L range (nm)	Const C_L time (days)	Fuel burn @ const V (lb)	Const V speed (f/s)	Const V range (nm)	Const V time (days)	Total range (nm)	Total time (days)
20,680	70/23	5292	9.7	0	NA	NA	NA	5292	9.7
NA	NA	NA	NA	20,680	39	4460	8.05	4460	8.05
18,000	70/33	3576	5.34	2680	33	1588	3.3	5164	8.55
16,000	70/39	2800	3.8	4680	39	2142	3.86	4942	7.66
14,000	70/44	2215	2.8	6680	44	2472	3.96	4687	6.76

the maximum speed is 65 kt at SL ($q = 14.32$ lb/ft^2) which sizes the maximum power required. We obtain maximum $P_R = $ (drag)(velocity)/$550\eta_P$. At the start-of-mission the CargoStar is generating about 23,250 lb of aerodynamic lift for a $C_L = 0.12$ and a $C_D = 0.0314 + (0.28)(0.12)^2 = 0.0354$. The total drag is 6825 lb giving a maximum $P_R = 2096$ hp for $\eta_P = 0.65$ or 524 hp per engine at SL. From Fig. 5.8 the TCM GTSIOL-550 at 520 hp and 640 lb is available off-the-shelf (OTS) with a $BSFC = 0.47$. There is no OTS turboprop available for this horsepower. However the trend line indicates that a 520 hp turboprop could be developed (called a "rubber engine," one that changes its shape and component design for conceptual design sizing requirements) for a weight of approximately 250 lb and an estimated $BSFC = 0.67$. It remains as an exercise for the reader to show that the GTSIOL-550 spark ignition reciprocating engine is a better choice than a new turboprop engine (neglecting the development cost of several $B) even though its engine weight is 2.6 times greater. [Hint: Reduce the empty weight by the difference in engine weight. Put the weight difference into more fuel. Solve Eq. (4.15) for the larger $BSFC$ and W_{H_0}/W_{H_1}.]

The maximum speed requirement is a major design factor as it drives the size of the propulsion unit which impacts the empty weight. It also impacts the hull weight as the envelop fabric density is sized by the internal hull pressure which is a function of 1.2 q_{max} (Chapter 8). These effects will become clear when the airship weight is determined in Chapter 9.

6.5.4 Propeller Design

The propellers for the CargoStar need to deliver an acceptable propulsive efficiency η_P across the following wide set of conditions:

Case	Condition
1	Start of mission at constant C_L cruise at 70 f/s at 4000 ft
2	Start of constant speed cruise at 39 f/s at 4000 ft after burning 16,000 lb of fuel
3	End of mission at 39 f/s and 4000 ft after burning 20,680 lb of fuel
4	Maximum speed at SL of 65 kt

The propellers will be designed using the method discussed in Chapter 5, Sec. 5.7.1. The propellers will have three blades and a Clark Y airfoil section with performance given by Fig. 5.9. The performance "sweet spot" (minimum $BSFC$) for spark ignition reciprocating engines is an engine rpm in the range of 2000–2400. This rpm is much too fast for good propeller efficiency on a slow speed airship so the props will be geared down to propeller speeds of 1200 rpm or less.

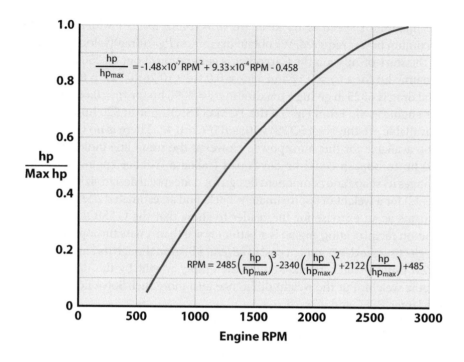

Figure 6.3   Typical power vs rpm for spark ignition reciprocating aircraft engines at SL.

The propellers will be designed (selection of blade angle β, propeller speed n, and diameter D) for case #1 and then checked for the other three cases for acceptable propulsive efficiency. At start of mission CargoStar is cruising at minimum drag (maximum L/D for cruise strategy #1) at 70 ft/s, 4000 ft. The drag $= L_{aero}/(\text{maximum } L/D) = 4362$ lb giving a total required power of 854 hp or 214 hp/engine. Using Eq. (5.37b) the maximum available power at 4000 ft for the GTSIOL-550 at 2800 rpm is 454 hp/engine or about twice the required. Thus, the engines will have to be throttled back to a lower rpm (this is the compromise that is always made when the engines are sized for conditions other than cruise). As shown in Fig. 6.3, the engine would need to be throttled back to about 1250 rpm to reduce the hp/engine to 214.

The data inputs into Fig. 5.9 are:

n = 600 rpm = 10 rps (assuming a gearbox ratio of 1:2.08)
P_R = (214)(550) = 117,727 ft-lb/s at 4000 ft
β = 20 deg
$C_S = [\rho/P_R\, n^2]^{0.2}\, V = [0.00211/(117,727)(10)^2]^{0.2}\,(70) = 0.787$

Table 6.2 Propeller Design Must Cover a Wide Set of Flight Conditions

Case	W_{H_0} (lb)	Speed/Alt (f/s & ft)	P_R/eng (hp)	Engine RPM	Prop RPM	C_S	η_P	J	Diam (ft)
1	23,250	70/4000	214	1250	600	0.787	0.66	0.5	14.0
2	7250	39/4000	74	825	400	0.64	0.60	0.4	14.6
3	2570	39/4000	45	725	350	0.744	0.65	0.45	14.9
4	23,250	110/SL	520	2800	1346	0.770	0.66	0.48	13.9

From Fig. 5.9

$\eta_P = 0.66$ which agrees with the assumed $\eta_P = 0.65$
$J = V/nD = 0.5$

giving a propeller diameter $D = 14$ ft

The other three cases were examined and their results are shown in Table 6.2. The 14-ft diameter propeller with a blade angle of 20 deg and a gearbox ratio of 1:2.08 will give acceptable propulsive efficiencies throughout the CargoStar flight regime.

6.5.5 Takeoff Distance

The takeoff distance is defined in Chapter 4, Sec. 4.9 as the distance needed for the airship to accelerate down the runway to a takeoff speed V_{TO}, rotate about the main gear (or aft ACLS pads), and generate $L_{aero} = 1.2 W_{H_0}$. This $V_{TO} = 1.1[2 W_{H_0}/Vol^{2/3}\rho \, C_L]^{1/2}$ where C_L is the available maximum lift coefficient usually limited by the α for tail strike. For CargoStar this α is 12 deg. Using $\alpha = 10$ deg the takeoff $C_L = (10)(0.045) = 0.45$ and $V_{TO} = 62.5$ f/s (37 kt). After liftoff the CargoStar would climb to 4000 ft and start its constant C_L or constant speed mission.

The ground roll distance is given by Eq. (4.42). For the CargoStar the retarding friction force is zero because the ACLS in hover mode gives a $\mu = 0$ (see Sec. 9.7). The ground roll is

$$S_G = \tfrac{1}{2}[\text{vehicle mass} + \text{helium mass} + \text{ballonet air mass}] \, V_{TO}^2/(T - D) \quad (6.4)$$

where the accelerating force $(T - D)$ is evaluated at 0.707 V_{TO} (43.8 f/s or $q = 2.27$ lb/ft^2).

The static thrust of each engine/prop is determined from Fig. 5.11. The engine hp/A = 3.4 is giving a T_0/hp = 6.0 and a static thrust of 3120 lb. This static thrust degrades with forward speed and using Fig. 5.10 we obtain 2523 lb at 43.8 f/s. The accelerating force at 43.8 f/s is

$$T - D = 4 \, (2523) - (0.0314)(2.27)(13,463) = 10,092 - 951 = 9141 \text{ lb}$$

The mass of the helium and air inside the CargoStar is determined using the densities in Table 2.2 (which are lb_f/ft^3). The volume of the ballonet (which is full of air at SL) is 702,000 ft^3. The mass of the helium is (0.01114) (858,000)/32.2 = 297 slugs. The mass of the air in the ballonet is (0.08072) (702,000)/32.2 = 1760 slugs. The total mass accelerated at takeoff is 93,000/32.2 + 297 + 1760 = 4945 slugs.

The takeoff ground roll distance using Eq. (6.4) is 1057 ft. A takeoff ground roll of ~2000 ft is considered the limit for remote area operation.

6.5.6 CargoStar Design Closure

The CargoStar example will be continued towards design closure by generating an empty weight build-up in Chapter 9 and comparing the build-up weight with the assumed W_{OE} of 37,000 lb.

References

[1] Nicolai, L.M., and Carichner, G.E., *Fundamentals of Aircraft and Airship Design, Volume I—Aircraft Design*, AIAA Education Series, AIAA, Reston, VA, 2010.

[2] Roskam, J., *Aircraft Design, Part 1*, Roskam Aviation and Engineering Corp., Lawrence, KS, 1985. [Available via www.darcorp.com]

[3] Raymer, D.P., *Aircraft Design: A Conceptual Approach*, AIAA Education Series, AIAA, Reston, VA, 1989.

[4] *Charts, Standard Aircraft Characteristics and Performance*, Military Specification MIL-C-5011A, 14 Feb 2003.

[5] *Airworthiness Standards: Part 23 and Part 25* Federal Aviation Regulation, Vol. 3, Department of Transportation, U.S. Government Printing Office, Washington D.C., 1996.

- Tail Volume Coefficient Method
- Vertical and Horizontal Tail Area Sizing
- Sample Problem: Initial Tail Sizing for the Solar HALE

More can be learned from failures than from successes. One should not fear failure but try to keep its cost low. Tolerate failure and, to learn from it, keep a list of lessons learned. Beware of the engineer who claims to never have had a failure . . . he is either not telling the truth or never takes any risk.

Good judgment comes from experience . . . and experience comes from bad judgment.

Anonymous

7.1 Introduction

At this point the airship has been sized (an envelope volume has been estimated) based upon an assumed takeoff gross weight ($TOGW$ = payload, empty weight, and fuel). A general configuration in terms of number of lobes, fineness ratio, and gondola/car size to house the payload has also been assumed. Before the design can further evolve, we need to do an aerodynamic and weight build-up in order to estimate the drag and fuel required to accomplish the mission, the empty weight and c.g. location. In order to do this it is required to assume an initial tail configuration and size. This initial tail sizing must be done wisely because the tail group will typically account for 10–14% of the empty weight and 20% of the airship C_{D_0}. Thus, it will have a large influence on the c.g. because of its location at the aft end of the vehicle. Once we have the fuel weight, empty weight, and c.g. location we can do a refined tail sizing using the discussion in Chapter 10.

What the tails look like and where they are located is a design decision. The designer needs to decide on the tail configuration (see Fig. 3.24), their location (the tail moment arm from the c.g.) and their size. The sizing of the tail surfaces is a lot of work. It requires a precise knowledge of the c.g. location and vehicle weight and is the subject of Chapter 10. Unfortunately we do not know the c.g. location at this point or its movement as fuel is burned but we do know that it is close to the c.b. At this point the tail surfaces are sized using a shortcut technique called the tail volume coefficient approach. It is based upon the observation that values of these volume coefficients are similar for like classes of aircraft (see Chapter 11 in [1]) and airships.

7.2 Initial Sizing of the Vertical Tail

The vertical tail area (fin plus rudder) provides static stability about the Z axis called weather cock stability (like a weather vane) and directional control through the deflection of the rudder. The static directional stability and control (S&C) criteria is very loose. The directional stability criteria is very simply stated as "the airship should exhibit good weathercock stability" but no level of damping is offered. The static directional control criteria is equally loose and is a legacy criteria from the World War II post war years when the U.S. Navy used airships for ASW and SAR (search and rescue). The Navy criteria was a 360-deg turn in one minute.

A convenient parameter to compare across classes of airships is the vertical tail volume coefficient:

$$C_{VT} = (\ell_{VT})(S_{VT})/[(\ell_B)Vol^{2/3}] \qquad (7.1)$$

where the ℓ_{VT} is the distance between the c.g. and the quarter chord of the vertical tail mean aerodynamic chord (mac) and ℓ_B is the airship body length.

As we don't know the location of the c.g. at this point we will assume it is located at the same fuselage station as the c.b.

Table 7.1 shows historical data of airship tail sizes. This data is plotted on Fig. 7.1 as vertical tail volume coefficient C_{VT} vs airship volume. The vertical tail area S_{VT} is determined using Eq. (7.1) and an appropriate value of C_{VT}. Notice that the c.b. is typically located at 45% of the airship length from the nose. The moment arm varies between 36%–43% of the body length for all of the airships in Table 7.1.

There are only two hybrid configurations in the data set, P-791 (Fig. 1.17) and Aerocraft (a Lockheed Martin three-lobe design with substantial analysis and wind tunnel testing). Notice that the vertical tail area is always less than or equal to the horizontal tail area.

7.3 Initial Sizing of the Horizontal Tail

The horizontal tail area (stabilizer plus elevator) provides static stability about the longitudinal axis called pitch stability and longitudinal control through the deflection of an elevator. As with the static directional case discussed earlier, the longitudinal S&C criteria is also very loose. The longitudinal stability criteria is very simply stated as "the airship should exhibit positive static pitch stability" but no level of damping is offered due to the fact that the time to double from upset is very large. It should be noted that the longitudinal situation is different from the directional case in that there are other mechanisms that can provide pitch stability and control. Most of the airships pitch stability comes from the fact that the c.g. is well below the c.b. giving a restoring moment as the nose or tail is pitched up (called pendulum stability). Also the pitch control from deflecting the elevator can be augmented by moving the c.g. fore and aft by transferring air from the fore and aft ballonets.

The horizontal tail volume coefficient is defined below as

$$C_{HT} = \ell_{HT}\, S_{HT} / [(\ell_B) Vol^{2/3}] \tag{7.2}$$

where ℓ_B is the airship length and ℓ_{HT} is the distance between the c.g. and the quarter chord of the horizontal tail mac. Most of the time ℓ_{HT} and ℓ_{VT} will have very similar values.

Figure 7.1 shows historical values for C_{HT} as a function of airship volume. The two mechanisms discussed previously (pendulum stability and moving air between the fore and aft ballonets) explain the flat character of the horizontal tail volume coefficient curve shown in Fig. 7.1. There is no explanation for the small tail shown for point 13 in Fig. 7.1. Data is shown but was not included in the curve fits.

The horizontal tail area S_{HT} is determined using Eq. (7.2) and an appropriate value of C_{HT}.

Table 7.1 Airship Data for Tail Volume Coefficients (from Appendix E)

#	Airship	Volume (ft³)	Length, ℓ_B (ft)	Width or dia (ft)	Moment arm (% ℓ_B)	Horizontal tail area (ft²)	Vertical tail area (ft²)	Hor Tail Vol Coeff C_{HT}	Vert Tail Vol Coeff C_{VT}
1	ZP4K	527,000	267	68.5	38	1108	888	0.063	0.0517
2	ZNP M-x	725,000	311	73.3	36	1280	1101	0.061	0.055
3	ZP2N-1	975,000	343	75.4	38	1511	1511	0.058	0.058
4	ZR-1	2,290,000	680	78.5	38	2870	2335	0.064	0.056
5	K-X	456,000	250	60	38	992	815	0.0632	0.0522
6	GZ-20	202,700	192	50	44	NA	280	–	0.036
7	M-2	648,000	302	69.5	36	1270	1130	0.0614	0.054
8	Skyship 600	235,400	194	50	37	NA	437	–	0.0424
9	Sentinel 1000	353,146	222	54.7	40	NA	732	–	0.0586
10	AeroCraft	40,500,000	894	476	34	28103	21252	0.08	0.061
11	ZRS-4	8,542,981	785	133	36	7170	6980	0.062	0.060
12	ZPN-1	875,000	321	73.5	38	1518	1518	0.063	0.063
13	P-791	120,000	120	58	40	177	143	0.029	0.0235
14	ISIS	5,858,597	510.5	159.4	40	5600	5600	0.069	0.069
15	R-31	1,610,000	615	65.5	40	2191	2060	0.0638	0.06
16	ZPK2	425,000	249	57.8	38	992	815	0.0664	0.055

Center of buoyancy (c.b.) assumed at 0.45ℓ from nose.

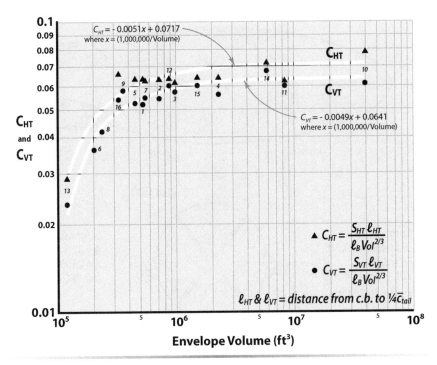

Figure 7.1 Initial tail sizing (see Table 7.1).

Sample Problem 7.1: Initial Tail Sizing for the Solar HALE

We return to the example of the Solar HALE in Chapter 5 and size the horizontal and vertical tails. Based on the data in Fig. 3.24, a cruciform tail was selected that gives an improved C_{L_α} (lower drag-due-to-lift) and lower C_{D_0}. Using the Solar HALE geometry shown in Fig. 7.2 we assume the ¼ chord of the tails mac to be located 100.1 ft behind the c.b. ($\ell_{VT} = \ell_{HT} = 0.35\,l_B$).

From Fig. 7.1 the value for C_{HT} will be estimated at 0.071 and C_{VT} at 0.063. Using Eqs. (7.1) and (7.2), the projected planform area for the horizontal and vertical tails is determined to be

$$S_{HT} = [C_{HT}\,\ell_B\,Vol^{2/3}]/\ell_{HT} = (0.071)(286)(10{,}264)/(100.1) = 2082 \text{ ft}^2$$
$$S_{VT} = [C_{VT}\,\ell_B\,Vol^{2/3}]/\ell_{VT} = (0.063)(286)(10{,}264)/(100.1) = 1848 \text{ ft}^2$$

The total tail exposed planform area (for all four tails) is 2784 ft² or 696 ft² for each tail. Because we are using a cruciform tail arrangement we need to bias the tail cant angle to give more projected area in the horizontal direction than the vertical direction, as shown on Fig. 7.3, which also

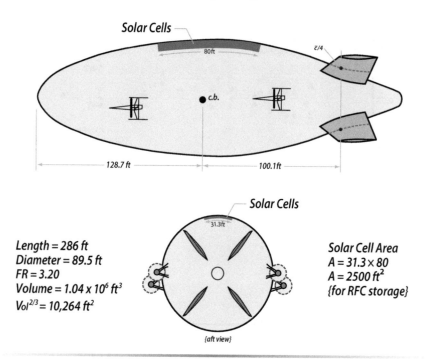

Figure 7.2 Solar High-Altitude-Long-Endurance (HALE) configuration.

displays the geometry selected for the tails. This tail area will be used to estimate the vehicle empty weight and c.g. location in Chapter 9. Finally the tail area and arrangement will be refined using the static S&C discussion of Chapter 10.

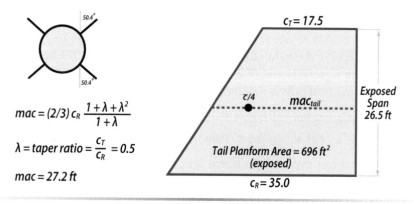

Figure 7.3 Tail geometry for Solar HALE.

Weather Sensitivity

Shortly after 1:30 p.m. on August 25, 1927, the USS Los Angeles (ZR-3) was moored on the high mast at Naval Air Station, Lakehurst, NJ. Suddenly a cold air front moved in and lifted the tail of the giant airship, causing it to rise before she could swing around the mast parallel to the new wind direction. There was a small riding-out crew onboard the ship. The change in attitude took seven to eight minutes so the crew had time to adjust their stance. The Los Angeles held the unflattering position shown on the first page of this chapter for almost thirty minutes before the tail started slowly to descend. None of the crew was hurt and the Los Angeles suffered only minor damage, but the incident underscored the vulnerability of airships to weather.

Reference

[1]   Nicolai, Leland M. and Carichner, Grant E., *Fundamentals of Aircraft and Airship Design, Volume I–Aircraft Design*, AIAA, Reston, VA, 2010.

Chapter 8 Structures and Materials

John Morehead

- Airship Materials
- Design Loads
- Stress Analysis
- Design Details
- Structural Testing

P-791, the first manned hybrid airship, takes shape at Lockheed Martin's Skunk Works. Instead of jigs and stands to hold an aircraft in place, airships and hybrids under construction float a few feet above the ground. For P-791, hold-down ropes were anchored to the hangar floor with electric winches for height adjustment.

> If black boxes survive air crashes—why don't they make the whole plane out of that stuff?
>
> George Carlin

8.1 Introduction

Airships are divided into three structural categories: rigid, non-rigid, and semi-rigid. Rigid airships transmit vehicle loads through a system of girders and wires that form an internal framework. For environmental and aerodynamic reasons, the rigid airship typically employs a tensioned fabric skin on the outside of the structure. Internally, the lifting gas is contained in separate gas cells, made from thin film or fabric. Non-rigid airships transmit vehicle loads through the pressure stabilized skin that form the ship's outer surface. This membrane fabric outer skin serves simultaneously as gas barrier, structural load path, and aerodynamic outer surface. In addition to the outer skin of the hull, non-rigid airships typically employ internal cables for suspending concentrated loads, such as a gondola. Semi-rigid designs combine the two concepts, using a sparse internal framework to address concentrated forces and moments, in addition to a pressurized hull forming the outer surface. Cutaway views of these three airship types are shown in Fig. 8.1.

All three structural types have been proposed for a wide range of Lighter-Than-Air (LTA) concepts. However most small airships are non-rigid, and all of the early 20th century giants were rigid. The fact that the large airships from history were rigid resulted from the lack of strong and reliable gas-tight fabric in that era. Rigid construction was the only option for high loads due to aerodynamic forces and moments in addition to con-centrated internal loads. Unfortunately many of these designs proved to be brittle and unforgiving, in some cases resulting in structural failures in flight. With the introduction of high-strength fabrics (many times stronger than titanium per pound) a new wave of airship designs has emerged that promises to be lighter, simpler to build, and, most important, more reliable and safe. This chapter primarily focuses on materials and design techniques associated with non-rigid pressure stabilized hull design. Refer to *Fundamentals of Aircraft and Airship Design—Volume I* for a complete treatment of metallic and composite aerospace structures.

8.2 Airship Materials

This section details the structural fibers, films, and fabrics used in airship design and construction. In recent decades, significant prog-ress has been made in synthetic materials resulting in dramatically improved strength to weight levels and durability. Modern gas tight fabrics are up to three times stronger than aerospace aluminum, as shown in Fig. 8.2.

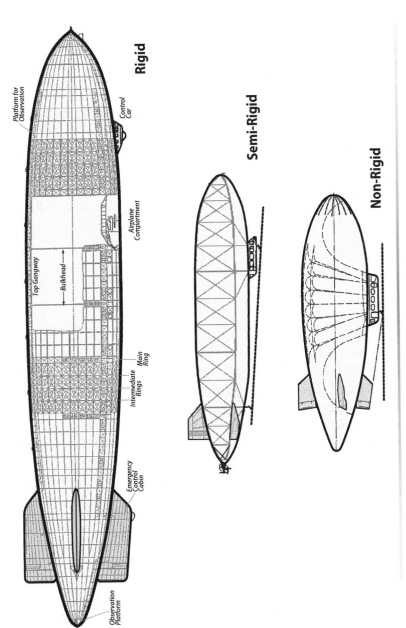

Figure 8.1 Airship structural categories.

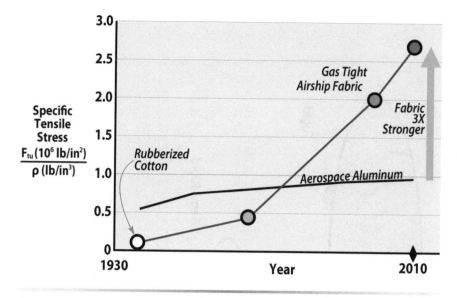

Figure 8.2   Material improvements for fabrics compared with aerospace aluminum.

8.2.1 Fibers

The choice of structural fiber is the first step in the development of high-strength fabrics. Early airship hulls were constructed with rubberized cotton cloth that had a breaking strength of 80 lb/in. This cotton based material load limit constrained non-rigid airships to approximately 200,000 ft³ or less in hull volume. With the advent of synthetic fibers such as Nylon (Polyamide) and Dacron (Polyester), engineers began to tailor the fiber properties to suit more demanding performance requirements. A variety of potential airship fibers are shown in Table 8.1; fiberglass and titanium are included for comparison.

Fiber strength is the breaking force recorded during a series of rapid pull tests. This value is often converted to units like "pounds per square inch" to match the way strength is documented for bulk materials. However, structural fibers are manufactured and used in bundles called yarns, which are available in a limited number of sizes. The smallest available yarn will affect how light a fabric can be designed. The unit of measure for yarn size is the denier, defined as the mass in grams per 9000 meters of yarn or filament. This unusual measurement has its origins in silk: a single strand of silk equals one denier. Typically available yarns range from 100 to 1000 deniers. Cost per pound for a given fiber increases as the denier

Table 8.1 Fiber Properties

Fiber	Type	Source	Density (lb/in.^{3})	Strength (10^3 psi)	Modulus (10^6 psi)	Strength to wt (10^6 in.)	Break strain (%)
Dacron	Polyester	DuPont	0.050	168	2	3.4	14
Nylon	Polyamide	DuPont	0.042	143	1	3.4	18
Kevlar	Aramid	DuPont	0.052	435	16	8.4	2
Vectran	Liquid Crystal Polymer	Kuraray	0.051	450	10	8.9	3
Spectra	UHMWPE (S-2000)	Honeywell	0.035	484	18	13.8	3
Carbon	PAN (T1000)	Toray	0.065	924	43	14.2	2
Zylon	PBO	Toyobo	0.056	840	26	15.1	3
Dyneema	UHMWPE	DSM	0.035	536	26	15.3	3–4
S-Glass	Fiberglass (S-2)	Owens Corning	0.090	665	13	7.4	5
Titanium	(Ref)		0.160	155	17	1.0	6

decreases, due to the increased difficulty in manufacturing and processing small yarns.

8.2.2 Films

Polymer films provide protection for the structural fibers from environmental factors such as sunlight exposure (including UV radiation), chemical interaction with the atmosphere, and severe weather. Films may also serve as a shear load path, increasing the overall stiffness and handling characteristics of the finished fabric. Table 8.2 lists several films used in the airship industry.

For low altitude airships and aerostats, a common choice for the outer surface is Tedlar due to its weatherability, chemical inertness, and hydrophobic properties. A Tedlar film thickness of 1 to 2 mils (1 to 2 oz/yd^2 areal density) is sufficient to protect the underlying structural fabric for a decade or longer. For the design of large non-rigid airships, a layer of Mylar (usually around 1 mil) is typically included to add shear stiffness and strength to the fabric. The two films combine to form a good gas barrier that is resilient to damage during the hull manufacturing process as well as during the service life of the airship. Stratospheric airships, being much more weight sensitive, cannot tolerate the

Table 8.2   Film Properties

Film	Type	Source	Density (lb/ in.^{3})	Strength (10^3 psi)	Modulus (10^6 psi)	Strength to wt (10^6 in.)	Break strain (%)
Dowlex	Linear Low Density Polyethylene	Dow	0.033	7	0.25	0.21	700
Nylon	Polyamide	DuPont	0.042	11	0.50	0.26	18
Tedlar	PVF – Floropolymer	DuPont	0.050	13	0.31	0.26	95
Mylar	Polyethylene terephthalate (PET)	DuPont Teijin	0.050	29	0.71	0.58	116
Kapton	Polyimide	DuPont	0.051	33	0.37	0.64	72
Teonex	Polyethylene naphthalate (PEN)	DuPont Teijin	0.049	39	0.83	0.79	88

2 to 3 oz/yd^2 required for a Tedlar/Mylar design. High altitude designs will typically use a thinner, lighter film (such as Teonex) that is less robust.

8.2.3 Fabrics

The most common process used in the construction of hull fabric consists of weaving the structural fibers, infusing the weave with resin adhesive, and then laminating on one or more films, illustrated in Fig. 8.3. The width of the resulting material (called the "fill" direction) is limited by the size of the equipment used at the weaver—typically 4 to 8 ft. The length of the material (called the "warp" or "roll" direction) can be very large, often exceeding 1000 ft. During the development of the fabric design, careful attention is paid to the arrangement of yarns in the warp and fill directions, to achieve strength levels and tear properties that meet minimum requirements. Due to the bi-axial nature of the weave in addition to yarn crimping and twisting during production, multiple iterations of the fabric design are usually needed during development.

Although traditional woven fabric is the standard material used in most LTA applications, it has some shortcomings: limited roll width, high initial stretch from yarn crimp, and overall low conversion of basic fiber strength to finished fabric strength. Recent industry efforts have focused on alternate weaving technologies (such as warp knit and multi-axial knit fabrics) and multi-layer uni-tape laminates. These approaches avoid the use of a traditional loom and allow for fabrics that are much wider and possibly less costly to produce. They also avoid crimping and bending of the main

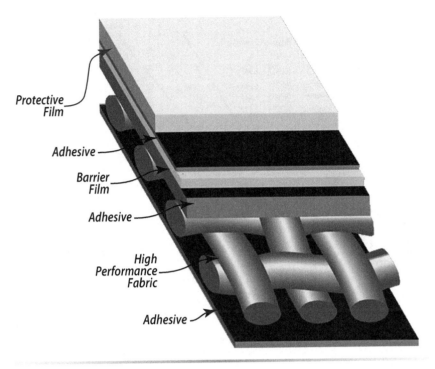

Figure 8.3 Woven materials layers.

structural fibers resulting in higher conversion of strength to the final product.

A selection of non-rigid airship fabrics used over the years is presented in Table 8.3 for reference. The Vectran and Dyneema fabrics listed are generic values for new applications, based on various approaches currently

Table 8.3 Hull Fabric Structural Properties

Fabric	Application	Strength (lb/in.)	Weight (oz/yd^2)	Specific strength (10^6 in.)
Cotton/Rubber	Circa 1920	80	14.0	0.1
Polyester/Neoprene	GZ 20	165	10.9	0.3
Polyester/Neoprene	ZPG 3W	320	16.5	0.4
Polyester/ Polyurethane	Skyship 600	210	11.5	0.4
Vectran/ Polyurethane	Low Altitude	650	11.0	1.2
Vectran (Laminated)	Low Altitude	740	7.6	2.0
Dyneema (Laminated)	Stratospheric	680	5.2	2.7

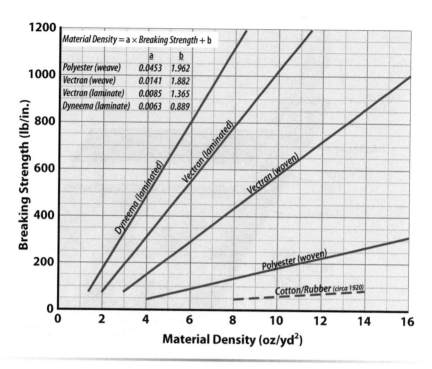

Figure 8.4 Hull fabric strength to weight.

in development. A graph of these materials is provided in Fig. 8.4 to illustrate general strength to weight characteristics over a range of capabilities. The higher strength materials, such as Vectran and Dyneema, are clearly the most attractive choices from a weight standpoint but consideration should be given to the fact that they are much more costly and are less forgiving due to the higher modulus. Dyneema also suffers from significant creep over a wide range of operating temperatures.

8.3 Structural Design Criteria

Airship structural design begins with definition of the structural design criteria and the certification basis—if applicable. The structural design criteria are the key parameters, such as design load factors, vehicle weights, speeds and altitudes, design life, factors of safety, and other operational considerations, that drive the design of the airship. The certification basis is the set of standards to be followed to obtain government approval for a type certificate. In the case of an experimental airship, while a type certificate is not necessary, the information contained in a standard certification basis is still valuable as guidance in the design process.

Prior to 1980, the only U.S. government agency that approved airship design and airworthiness was the Navy. In the preceding decades, they owned and operated nearly all airships in use, so consequently the FAA was not involved. That changed in the 1980s with the development of multiple commercial airships. With no military involvement with these new designs, the FAA established P-8110-2, Airship Design Criteria (ADC). Based primarily on Part 23 of the FAR, U.S. Navy detail design specifications of airships, and additional criteria developed by FAA/NASA, the ADC became the standard for type certification projects.

As interest grew in the development of larger more capable airships, European and U.S. officials embarked on the creation of CS 30T (Transport Airship Requirements [3]) in 2003. The transport category is defined as multi-engine propeller driven airships that have a capacity of 20 or more passengers (excluding crew), or a maximum take-off mass of 15,000 kg (33,000 lb) or more, or a design lifting gas volume of 20,000 m³ (700,000 ft³) or more, whichever is greatest. This section will primary present CS 30T requirements, although note that other methods may also be acceptable to the Aviation Authority, provided rational justification.

8.3.1 Limit and Ultimate Loads

Strength requirements are specified in terms of limit loads, the highest loads expected during the service life of the airship, and ultimate loads, which are limit loads multiplied by a factor of safety. Limit loads are determined by placing the airship in equilibrium, taking into account the lifting gas, air, and ground loads; inertia forces and moments; and, where applicable, virtual inertia as the effect of momentum changes in the surrounding air mass. By this definition, limit loads are both physically possible and expected to be observed at least once. Structural deformation must not be permanent or interfere with safe operation at limit. Ultimate loads, in contrast, may yield airship structure but must not initiate failure for at least 3 sec. The typical factor of safety is 1.5 for most structural components with some important exceptions. Most notably, non-rigid and semi-rigid airship envelopes must be designed with a minimum factor of safety of 4.0.

This historical factor-of-safety value of 4.0 has been part of airship design criteria for many decades, and since airship safety has been excellent during that time the FAA is not motivated to make a change. As a result, the envelope material will be loaded in service to a maximum of 25% of the original breaking strength. This large residual strength is considered necessary to account for end of life degradation, creep characteristics, material variability, and structural criticality. In the drive to minimize the empty weight of an airship design, designers often propose lowering the hull factor of safety—arguing that modern materials are more reliable. Although testing and operational experience may someday prove this assertion, until then all manned airships should use 4.0 factor of safety for ultimate hull strength.

8.3.2 Design Maneuvering Loads

Loads resulting from airship maneuvers listed in CS 30T (Table 8.4) must be evaluated, including control surfaces. Both steady state and transient effects should be calculated during checked (rapid sequence of positive and negative inputs) and unchecked (single direction movement of control surface) maneuvers. If thrust vectoring and/or mass-shifting is employed in the airship design, they must be included in the analysis.

Table 8.4 Design Maneuvering Loads*

| | | | | | | Control surface position | |
No	Condition	Speed	Weight	Attitude	Thrust direction	Rudder	Elevator
1	Level flight	VH	Wt	†	Forward	Neutral	†
2	Level flight	VH	Wt	†	Forward	Neutral	†
3	Nose down	VH	Wt	+30°	Forward	Neutral	†
4	Nose up	VH	Wt	−30°	Forward	Neutral	†
5	Descent & pull-up	VH	Wt	†	Forward	Neutral	†
6	Turn entry	VH	Wt	Horizontal	Forward	Full over	Neutral
7	Turn & reverse	VH	Wt	Horizontal	Forward	‡	Neutral
8	Dive entry	VH	Wt	Horizontal	Forward	Neutral	Full down
9	Climb entry	VH	Wt	Horizontal	Forward	Neutral	Full up
10	Turn & climb	VH	Wt	Horizontal	Forward	Full over	Full up
11	Turn & dive	VH	Wt	Horizontal	Forward	Full over	Full down
12	Turn	VH	Wt	Horizontal	Forward	Full over	Neutral
13	Turn recovery	VH	Wt	Horizontal	Forward	‡	Neutral
14	Turn recovery & climb	VH	Wt	Horizontal	Forward	‡	Full up
15	Turn recovery & dive	VH	Wt	Horizontal	Forward	‡	Full down
16	Light flight	VH	Wt	†	Forward	Neutral	†

* Velocity values must be determined after speed is stabilized, but not before a steady state condition.
† That necessary to produce maximum loading conditions.
‡ Full rudder must be applied followed by full reverse rudder after 75° of turn.

8.3.3 Gust and Turbulence Loads

A method for determining airship loads due to atmospheric gusts in level flight is presented below. CS 30T [3] requires the airship to withstand a 25 ft/s discrete gust, U_m, at flight speed V_H (maximum horizontal speed) and a 35 ft/s discrete gust at flight speed V_B (maximum gust operating or encounter speed). The definition of gust shapes and intensities may be calculated using Eq. (8.1)

$$u = (U_m/2)[1-\cos(\pi X/H)] \quad \text{(ft/s)} \tag{8.1}$$

where

U_m	gust velocity specified above (ft/s)
X	penetration distance (ft), $0 < X < 2H$
H	gust gradient length (ft), $\ell/4 < H < 800$ ft (244 m), (a sufficient number of gust gradient length must be investigated to find the critical response of the airship)
ℓ	length of the airship (ft)

Steady state loads and dynamic response must be taken into account for the airship, with control surfaces in their most critical position. The gusts must be evaluated in all directions relative to the vehicle: vertical, lateral, and axial. The maximum aerodynamic bending moment, M (ft-lb), applied to the envelope, is determined by Eq. (8.2) for airships with a minimum fineness ratio of 4. For a non-rigid design with a slightly lower fineness ratio, use the value of 4 for rough sizing.

$$M = 0.058 \, Vol \, (\ell/2)^{0.25} \, [1 + (FR\text{-}4) \, (0.5624 \, \ell^{0.02} - 0.5)] \, (qU_m/v) \quad \text{(ft-lb)} \tag{8.2}$$

where

FR	envelope fineness ratio, $FR \geq 4$
U_m	discrete gust velocity (25 ft/s @VH, 35 ft/s @ VB)
q	dynamic pressure (lb/ft^2) at the velocity V (ft/s) under consideration, $q = \rho \, V^2/2$
ℓ	length of the airship (ft)
D	maximum envelope diameter (ft)
ρ	density of air (slugs/ft^3)
V	airship equivalent speed (ft/s) (VH or VB)
Vol	total envelope volume (ft^3)

8.4 Initial Stress Analysis and Fabric Weight Estimates

Initial envelope stress analysis is performed using overall hull dimensions, maximum cross-section radius, and internal pressure. For a non-rigid airship, the internal pressure value, called super pressure P_{SP}, is driven by the top design speed. Typically for designs that have minimal nose stiffening, the super pressure must be greater than the maximum stagnation pressure in flight, $1.2 \times q_{max}$ usually used in early design. The envelope is treated as a membrane because the skin has essentially zero bending stiffness with the lifting gas super pressure providing enough pre-load in the skin to maintain tension in all directions to avoid wrinkles. In that regard, the lifting gas is viewed as part of the structure thereby carrying compressive loads throughout the airship. To calculate the envelope circumferential or "hoop stress" due to super pressure one can start with the cylindrical pressure vessel Eq. (8.3)

$$\sigma_{hoop} = (P_{SP} \times R)/t \tag{8.3}$$

However for membrane fabrics, the shell thickness t is unimportant so it is removed and the results in Eq. (8.4) are reported in load per unit length:

$$\text{Super Pressure Skin Load} = P_{SP} \times R \tag{8.4}$$

In this equation, the maximum radius R is used with the maximum anticipated super pressure (as measured at the low point of the hull). Example values for this calculation are: 23 ft radius (276 in.), 2.0 iwg pressure (0.072 psi), skin load equals 20 lb/in.

Next we look at the net buoyant force applied to the envelope skin. A column of lifting gas inside the airship weighs less than a column of air of the same height outside the vehicle—resulting in a pressure delta on the upper skin otherwise known as net buoyant lift. This pressure is calculated in Eq. (8.5)

$$\Delta p = (\rho_{atm} - \rho_{lifting\ gas})gh \tag{8.5}$$

The vertical distance from the hull's low point to its high point is used for h and the result Δp is the increment of additional pressure from buoyancy. Example values for this calculation are: 0.0635 lb/ft^3 for $\Delta \rho g$, 46 ft (552 in.) for h resulting in 0.020 lb/in.^{2} for Δp. Using the hoop stress equation above, we find that buoyancy contributes 5.5 lb/in. of additional tension in the skin at the top of the envelope. Notice for these example values the super pressure load is nearly four times greater than the buoyant pressure load.

The skin limit load acting of the envelope is expressed in Eq. (8.6a) as

$$\text{Envelope Skin Limit Load} = (P_{SP} + \Delta p)R \tag{8.6a}$$

where the pressures are in lb/in.^{2} and R is the envelope maximum radius in inches. The maximum envelope skin load is Eq. (8.6b)

$$\text{Maximum Skin Load} = (P_{SP} + \Delta p)R \times FS \tag{8.6b}$$

where $FS = 4.0$. This skin load is referred to as the *minimum acceptable breaking strength* for the fabric.

Most of the envelope skin load is the result of super pressure and lifting gas static pressure. Consequently a quick calculation of these two values will provide rough sizing for the required fabric strength. Aerodynamic and vehicle bending loads in conjunction with load alleviation and concentration due to hull shape deformations are usually smaller effects and are best evaluated using non-linear finite element modeling. Often the highly detailed stress analysis will fall within 10% to 20% of the initial rough estimate.

The fabric breaking strength from Eq. (8.6b) is used to determine the fabric density from Fig. 8.4 (note: 1 oz/yd^2 = 0.007 lb/ft^2). The weight of the envelope fabric is then calculated using Eq. (8.7).

$$W_{env} = \text{(fabric density)} \, (S_{wet})_{env} \, F_{MA} \, F_{AF} \qquad (8.7)$$

where $(S_{wet})_{env}$ is the envelope surface area [see Eq. (3.19)], F_{MA} is a manufacturing/assembly factor that accounts for joints, doublers, and load patches (use $F_{MA}=1.2$). F_{AF} accounts for envelope attach fittings and is equal to 1.26.

The ballonet fabric weight is determined assuming the ballonet as a sphere. The volume of two ballonets is determined as discussed in Chapter 2 or Sec. 9.3. The required radius of the sphere is determined from $Vol_{bal} = (4/3)\,\pi R^3$ and its surface area is $S_{bal} = 4\,\pi R^2$. Ballonet fabric strength is not based on the super pressure as it generally carries much lower loads than the hull. In practice ballonet fabrics range from 2–9 oz/yd^2 with 5 oz/yd^2 (0.035 lb/ft^2) used in early design. The ballonet fabric weight can be calculated using Eq. (8.8).

$$W_{bal} = 0.035 S_{bal} \qquad \text{(for } S_{bal} \text{ in ft}^2) \qquad (8.8)$$

8.5 Finite Element Modeling

Computer simulation is an essential feature of modern airship envelope design, with active roles at the shape definition, structural design, and fabrication stages.

Finite element modeling (FEM) provides verification of the equilibrium shape and stress distribution of the chosen design at reference operating conditions. Analysis of the structural performance of the envelope is undertaken for the full range of operating conditions including landing, with interest on both fabric stresses and deformed geometry. Computer-generated cutting patterns provide the fabrication geometry for the practical realization of the chosen envelope design.

With fabrics and cables as the principal structural materials, any analysis tools used must be readily able to cope with the non-linearities of on/off fabric wrinkling, cable slackening, large deformations, rigid body movement, and the potential for the development of local mechanisms. Together with the need for shape verification and patterning, these capabilities are not typically found in the general purpose FEM analysis packages used for more conventional aerospace design.

A finite element program suite developed by UK consulting engineering firm TENSYS has been used in recent years on several major projects including non-rigid hybrid airships and stratospheric super-pressure scientific balloons, see Fig. 8.5. Called "inTENS", the software relies on Dynamic Relaxation (DR) for the solution of these highly non-linear problems. This is an explicit dynamic analysis method, in which the central difference time stepping implementation of Newton's law of motion is automatically bought to a static solution through the application of kinetic damping.

The element displacements are the minimum number of geometric variables necessary to completely define the deformed configuration of that element, independent of rigid body motion, which is treated automatically by the DR process. A set of associated natural elements forces are determined from these basic displacements by means of the element natural stiffness relations.

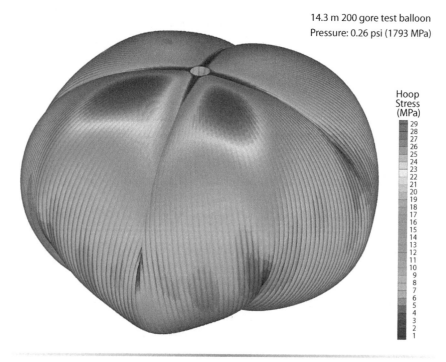

14.3 m 200 gore test balloon
Pressure: 0.26 psi (1793 MPa)

Hoop Stress (MPa)

29
28
27
26
25
24
23
22
21
20
19
18
17
16
15
14
13
12
11
10
9
8
7
6
5
4
3
2
1

Figure 8.5   Highly non-linear model result.

The iterative equations for the motion in direction x of any node i in space at time t are obtained directly from Newton ($F = ma$) and is expressed in matrix notation in Eq. (8.9)

$$R_{ix}^t = M_i A_{ix}^t \qquad (8.9)$$

which may be expressed in central difference form as Eq. (8.10)

$$R_{ix}^t = M_i \left(V_{ix}^{(t+\Delta t/2)} - V_{ix}^{(t-\Delta t/2)} \right) \Big/ \Delta t \qquad (8.10)$$

giving the recurrence relation for nodal velocities in Eq. (8.11)

$$V_{ix}^{(t+\Delta t/2)} = V_{ix}^{(t-\Delta t/2)} + \frac{\Delta t R_{ix}^t}{M_i} \qquad (8.11)$$

Where

R_{ix}^t is the x direction residual force matrix for node i at time t
V_{ix}^t is the associated node velocity matrix
M_i is the node mass matrix and Δt the time interval

The residual forces R_{ix}^t are computed for the then current node coordinates x_i^t. An updated set may now be calculated from the incremented node coordinates and is presented as Eq. (8.12):

$$x_i^{(t+\Delta t)} = x_i^t + \Delta t V_{ix}^{(t+\Delta t/2)} \qquad (8.12)$$

Similar recurrence relations apply to all unconstrained degrees of freedom of the structure. The stability of a time stepping dynamic analysis will depend upon the selection of a suitably small time increment, which can be shown to be a function of the local node relative stiffness. When only the final static solution is sought, one may use fictitious nodal masses that optimize convergence for a chosen time interval, which is typically taken as unity for computational efficiency.

The load extension response of the woven and coated fabrics used for airship construction can be represented within their working stress range by a linear elastic material model, with crimp interchange effects modeled by Poisson-type terms. The necessary material parameters are determined from biaxial tests cycled through varying warp to fill stress ratios within the anticipated working stress range.

The modeling of construction details such as tie-tabs and reinforcing patches requires multiple layers of elements with appropriate fiber orientations in order to represent the actual construction. Webbing belts are modeled with line elements within this assembly as necessary. Both stresses within the local assembly and those imposed on the main hull fabric are of

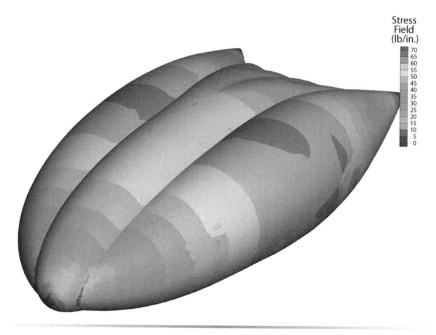

Figure 8.6 High resolution full vehicle model.

interest. Although useful results can be found from local models with single assemblies within a specified stress field, the complexity of boundary conditions associated with varying flight conditions are such that this level of detail should ideally be integrated into high resolution full vehicle models, as shown in Fig. 8.6.

8.6 Hull Joints and Assembly

An airship hull is usually constructed from strips of fabric cut from flat patterns, called gores, which are tapered on both ends. As the edge of one

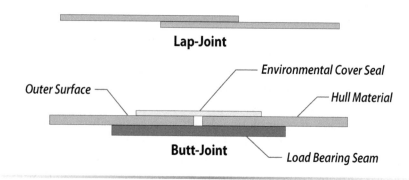

Figure 8.7a Hull joint geometry.

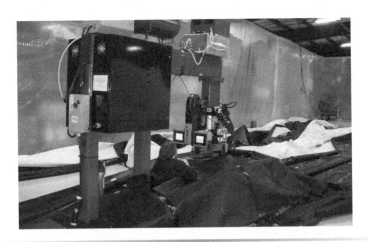

Figure 8.7b ILC Dover step sealing machine.

gore is joined to another, a surface with compound curvature is formed—much like the sections of a football. The gore-to-gore joint should be designed to be as strong as or stronger than the surrounding hull material while maintaining an appropriate stiffness to avoid banding or stress concentrations. For lightly loaded, simple hulls, an overlap joint is common. However, most LTA designs require a butt-joint seam construction, illustrated in Fig. 8.7a. The more complex butt-joint has several advantages over the simpler lap-joint: better environmental sealing to protect bonded surface, works with fabrics that have low bonding strength on outer surface (an issue with Tedlar), and lower joint eccentricity.

Tolerance buildup during gore seaming can have an impact on overall hull performance; stress concentrations and local defects will lower the theoretical strength. For low stiffness fabrics like those made from polyester, the material is able to stretch and redistribute loads thereby lessening the impact of stress concentrations. This allows for the joining of polyester gores to be done by visually aligning match marks as the joint is formed (typically with a heat press) as shown in Fig. 8.7b. However, higher stiffness material such as Vectran and Dyneema are much less forgiving and require tighter tolerances and greater care during assembly in order to retain the overall strength of the hull.

8.7 Structural Attachments and Load Paths

Because membrane surfaces cannot take compressive loads, introducing concentrated loads into a pressure stabilized hull skin can be challenging. First, the magnitudes of the loads at an interface are estimated and a determination is made as to whether or not the loads are reversible. If the load is always in tension (like the suspension cables for a gondola), then load patches for external locations and catenary curtains for internal locations are used to spread the load into the fabric. In cases where loads reverse or change direction dramatically, such as fins or engine pods, a direct attachment to the skin is necessary, often in combination with tension attachments. One type of direct attachment is a batten or lash tube (pictured in Fig. 8.8)—a rigid member positioned onto the surface of the envelope and held in place with a sleeve or rope. The batten provides a compressive load path normal to the hull surface (working against the internal gas pressure) and a load path tangential to the hull surface (working against the skin preload). Battens are significantly heavier per unit load carried than pure tension joints however, so the desire is to lay out primary load paths in a way to maximize the use of tension interfaces.

Internal load curtains, also known as septums, may be used to form multi-lobe shapes, such as the hull form of a hybrid airship or the airfoil contours of an inflated fin. For example, Fig. 8.9 shows a section cut from a hybrid design that aligns two internal septums with the load path required for carrying the payload weight. A good initial estimate of septum load can

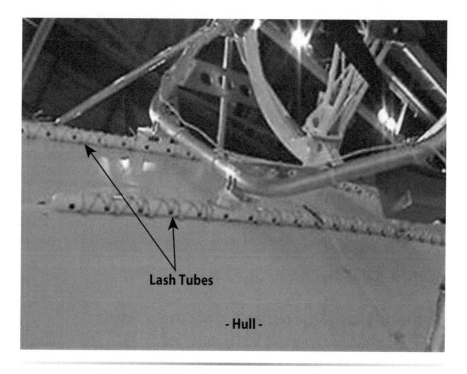

Figure 8.8   Hull surface mounting with lash tubes.

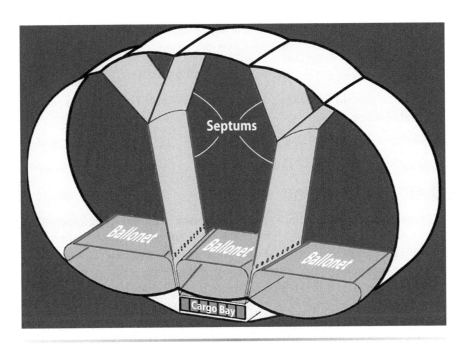

Figure 8.9 Section of typical hybrid airship hull showing cargo, ballonet, and septum layout.

be calculated using the internal hull pressure along with intersecting circular arcs (representing the outer surfaces) and straight lines connecting those intersection points. A joint where three load bearing fabrics come together is assumed to be in equilibrium; if two have known loads (the outer hull segments) and one is unknown (the septum), summing the known load components in the direction of the septum will reveal what the unknown load must be.

The septum/curtain fabric weight is determined based upon the loads in the envelope skin and the local angles and joints of the septum/curtain arrangement. A typical septum arrangement for a cargo carrying hybrid (such as CargoStar) is shown in Fig. 8.9. The typical curtain arrangement for a non-rigid airship is shown in Fig. 8.1. It is assumed for the initial design that the septum loads are 1.5 times that of the envelope maximum skin load [Eq. (8.7)] with $FS = 4.0$. The septum/curtain fabric density is determined from Fig. 8.4 using this minimum acceptable breaking strength. The septum/curtain fabric weight is estimated using Eq. (8.13).

$$W_{S/C} = (\text{surface area})(\text{fabric density}) \times F_J \qquad (8.13)$$

where $F_J = 1.06$ and accounts for the assembly joints. The hybrid septum fabric surface area is assumed to be a function of the envelope side projected area. Although septum surface area is highly dependent on the internal hull configuration, an initial estimate may be calculated for hybrids using

$$S_{Septum} = 0.75 \text{ (envelope side projected area)} \times 2 \qquad (8.14a)$$

Likewise, a curtain fabric estimate for standard non-rigid airships may be calculated using

$$S_{Curtain} = 0.20 \text{ (envelope side projected area)} \qquad (8.14b)$$

8.8 Structural Testing

Qualification of the airship material system is accomplished through a series of standardized tests. The exact make up of the Qualification Test Plan is dictated by the certification basis for civil projects or customer airworthiness requirements in the case of military projects. A sample test matrix is shown in Table 8.5 to illustrate the scope of testing required for each new material included in an airship design.

For material breaking strength, "allowable" values are derived from mean test data (coupon tests) reduced by a knock-down factor to account for batch to batch variability. The following formula is used to calculate the design allowable:

$$\text{Allowable} = m - K_b\, s$$

Where

m = data mean
s = standard deviation
K_b = a B-basis statistical factor representing 95% confidence that 90% of the samples will exceed the allowable (A-basis is rarely used)

Tear testing is also conducted at the coupon level. The cut slit tear test measures the ability of a material to resist further tearing after damage has occurred. A test coupon (4 in. wide by 6 in. long) is cut with a razor to produce a 1.25 in. long slit, as seen in Fig. 8.10. The material sample is loaded into a test apparatus then pulled at constant speed (12 in. per minute) while the tension load is recorded. This testing method is valuable for assessing relative tear performance among several materials but is limited to its applicability in predicting actual full-scale performance. A more realistic test involves large diameter inflated test articles in which cuts of varying lengths are used to determine the critical tear dimension (the length at which propagation starts for a given pressure and radius). These tests however are costly and time-consuming so the coupon level method remains the industry standard.

Table 8.5 Sample Test Matrix for Airship Hull Material (Courtesy ILC/Dover)

Test	Test method
Weight	FED-STD-191 TM5041
Bow and skewness	ASTM D 3882
Surface finish – interior	Visual inspection
Surface finish – exterior	Visual inspection
Water release – exterior	FED-STD-191 TM5504
Blocking at elevated temperature	FED-STD-191 TM5872
Surface polymer characterization	Infrared spectrophotometry
Tensile modulus	ASTM D 751
Breaking strength/elongation – strip method ultimate tensile	FED-STD-191 TM5102
Breaking strength/elongation – strip method, ultimate tensile after Wx exposure (QUV chamber)	FED-STD-191 TM5102
Seam tensile strength – heat seal	FED-STD-191 TM5102
Seam tensile strength at elevated temperature	
Heat seal	FED-STD-191 TM5102
Base cloth breaking strength – ravel strip method ultimate tensile	FED-STD-191 TM5104
Creep/hysteresis evaluation	Vendor test method
Tear strength – cut slit	MIL-C-21189 Para 10.2.4 FAA P-8110-2, Appendix A
Tear strength – tongue	FED-STD-191 TM5134
Coating adhesion – heat seal seam, back/structural tape	FED-STD-191 TM5970
Coating adhesion – heat seal seam, cover tape	FED-STD-191 TM5970
Coating adhesion – cement	FED-STD-191 TM5970
Film ply bond adhesion (dry)	FED-STD-191 TM5970
Film ply bond adhesion (elevated humidity)	FED-STD-191 TM5970
Seam deadload – elevated temp (underwater) heat seal	Vendor test method
Seam deadload – elevated temp (hot air) heat seal	Vendor test method
Seam deadload – elevated temp (underwater) cement	Vendor test method
Seam deadload – elevated temp (hot air) cement	Vendor test method
Cylinder deadload – elevated temp (underwater)	Vendor test method
Inflated cylinder flex testing	Vendor test method
Low temp flex	ASTM D 2136
Helium permeability	ASTM D 1434 or vendor test method
Helium permeability after Wx exposure (QUV chamber)	ASTM D 1434 or vendor test method
Seam helium permeability	ASTM D 1434 or vendor test method

Figure 8.10 Tear test method.

8.9 Weight Estimation—CargoStar Example

Using the CargoStar hybrid transport shown in Fig. 6.2, this section will walk through an example problem to determine the initial weight estimate for the envelope, ballonets, and septums. First, the maximum expected (limit) hoop direction skin load is calculated based on the largest lobe radius and hull super pressure. For an assumed V_{MO}, maximum operating speed of 65 kt, the airship will experience a sea level q equal to 14.4 lb/ft^2. With a factor of 1.2 to preclude nose wrinkling the resulting super pressure is 14.4 lb/ft^2 × 1.2 = 17.3 lb/ft^2 or 0.12 lb/in.^{2} Next, the pressure on the upper surface due to buoyancy is calculated by multiplying the vehicle height by the net gas lift, 70.8 ft × 0.0635 lb/ft^3 = 4.5 lb/ft^2 or 0.031 lb/in.^{2} Superimposing the two pressures results in 0.12 lb/in.^{2} + 0.031 lb/in.^{2} = 0.151 lb/in.^{2} Finally, the hoop skin load is determined by multiplying combined pressure and radius together, 0.151 lb/in.^{2} × 35 ft × 12 in./ft = 63.4 lb/in. Now that the limit load is known, use $FS = 4.0$ to find the minimum acceptable breaking strength for the fabric, 63.4 lb/in. × 4.0 = 253.4 lb/in. Using Fig. 8.4 to determine the material weight for a typical Vectran weave yields a value of 5.4 oz/yd^2 or 0.0378 lb/ft^2 for the hull fabric. The envelope weight equals the total surface area multiplied by the material weight, a seam factor to capture joint, doubler, and load patch weight and an attach fitting factor to capture integral load bearing attachments and gas tight ring sets for hull penetrations, $W_{Hull} = 74,100$ ft^2 × 0.0378 lb/ft^2 × 1.20 × 1.26 = 4235 lb.

Next, the septum material weight is determined based on the loads in the outer skin and the local angles at the septum "Y" joint. The particular geometry used for the CargoStar hull generates septum loads that are 1.5 times that of the hull so the minimum acceptable breaking strength is calculated by 253.4 lb/in. × 1.5 = 380 lb/in. Using Fig. 8.4 to determine the septum material weight for the same Vectran weave yields a value of 7.3 oz/yd^2 or 0.0511 lb/ft^2. The septum weight equals the total surface area, determined from airship CAD model, multiplied by the material weight and a joint factor, $W_{Septum} = 24,400$ ft^2 × 0.0511 lb/ft^2 × 1.06 = 1322 lb.

Lastly, the ballonet fabric weight is calculated from the surface area of a sphere sized to contain the required volume, in this case 45% of the total hull volume, 0.45 × 1,560,000 ft^3 = 702,000 ft^3. The radius of such a sphere is determined using $Vol = (4/3)\,\pi r^3$, resulting in $r = 55.1$ ft. The sphere surface area is $A_{bal} = 4\,\pi r^2 = 38,189$ ft^2. A spherical shape would not actually be used for the ballonet because it is too large to even fit. However, the surface area corresponds closely with the more complicated geometry that will eventually be developed later in the design process. The ballonet fabric strength is not sized based on super pressure; it generally carries much lower loads as compared to the hull. In practice, fabrics ranging from 2 oz/yd^2 to 9 oz/yd^2 have been used, and for this example 5.0 oz/yd^2 or 0.035 lb/ft^2 is assumed. The total ballonet weight is therefore

$$W_{Ballonet} = 38,189 \text{ ft}^2 \times 0.035 \text{ lb/ft}^2 = 1337 \text{ lb}$$

These fabric weights will be used in Sec. 9.13 to build a weight summary for the CargoStar.

8.10 Summary

The structural design of an airship requires the engineer to balance not only the usual factors such as weight, risk, cost, and schedule but also to consider other factors such as lifting gas retention and purity, in-service inspection and maintenance access, hangar size limits and construction methods, and environmental effects from sunlight, among others. Referring back to the CargoStar example, "the design team may be asked to extend" the service life of the envelope by guarding against ultraviolet light degradation of the hull material is one approach. This approach may involve a marked increase in overall material weight. On the other hand, a mid-life hull replacement may prove to be a more cost effective and safer plan; ultimately a clear understanding of the total product life cycle is needed to make these fundamental choices.

References

[1] Miller, T., and Mandel, M., *Airship Envelopes: Requirements, Materials and Test Methods*, International Airship Conference, 2000.

[2] Bruhn, E.F., *Analysis & Design of Flight Vehicle Structures*, Jacobs Publishing, Inc., Indianapolis, IN, 1973.

[3] *Certification Specifications for Transport Category Airships*, CS 30T, EASA, 2003.

[4] Khoury, G.A., and Gillett, J.D., *Airship Technology*, Cambridge University Press, UK, 1999.

[5] *Type Certification – Airships*, AC No: 21.17-1A, U.S. Dept of Transportation – Federal Aviation Administration, 1992.

[6] Wakefield, D.S., *Membrane Engineering: Principles and Applications*, IASS Asia Pacific Conference on Shell and Spatial Structures, Beijing, 1996.

[7] Wakefield D.S., *Non-Linear Viscoelastic Analysis and the Design of Super-Pressure Balloons: Stress, Strain and Stability*, AIAA 20th Aerodynamic Decelerator Systems Technology, 18th Lighter-Than-Air Systems Technology and Balloon Systems Conferences, Seattle, WA, AIAA-2009-2813, 2009.

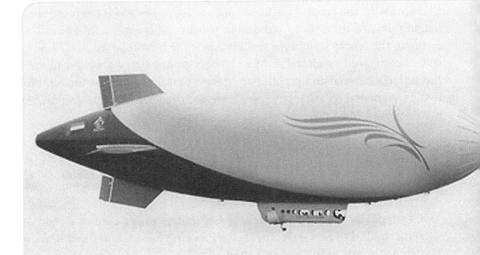

SkyShip

- Weight Estimating—Most Challenging Part of Conceptual Design
- Estimating Weight: Hull Group, Tail Group, Landing Gear/ ACLS, Propulsion System, Avionics/Electrical
- Sample Problem: Solar Hale
- Sample Problem: CargoStar
- Impact of Maximum Speed

Estimating component weights for aircraft and airships is the most challenging part of conceptual design. Much of the weight estimating analysis is based on historical data, which is especially challenging for airships as the historical data base is very sparse compared to aircraft.

Estimating the weight of an aircraft or airship involves as much art as it does science.

Anonymous

9.1 Introduction

T his chapter will discuss the subsystems in an airship/hybrid and offer some WERs (weight estimating relationships) to estimate the weights. The WERs were developed from historical aircraft and airship data. The weight in all cases is in pounds.

Estimating the empty weight, W_E, of an airship/hybrid is the most challenging part of the conceptual design process. It is especially difficult to estimate the empty weight for airships since the historical database is very sparse compared to aircrafts. Most design groups carry a weight margin through conceptual and preliminary design to account for the uncertainty in the weight estimates and the inevitable and dreaded "weights growth." At the Lockheed Martin Skunk Works the margin on W_E is 6%.

9.2 Hull

The hull group is shown in Fig. 9.1 and consists of the following:

Envelope: fabric, seams, adhesives, structure (in the case of a rigid or semi rigid), nose reinforcement (for mooring mast loads), load bearing patches, joints, doublers, and attach fittings

Septums/catenary curtains: fabric, seams, adhesive, joints, and attach fittings

Ballonet: fabric, seams, adhesives, fans, and ducting (discussed in Sec. 9.3)

Tails: structure, fabric/covering, control surfaces, actuators, mounting and bracing wires

Gondola: structure (less payload, subsystem and equipment items)

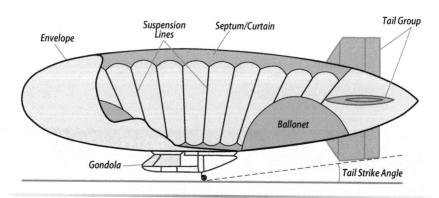

Figure 9.1   The hull group.

The airship/hybrid is assumed to be a non-rigid vehicle, so there would not be any structure in the envelope to hold the hull shape. The non-rigid airship/hybrid maintains its hull shape by its internal gas pressure. The hull structure and material was discussed in Chapter 8.

9.2.1 Envelope Weight Estimating

The fabric weight for the envelope, ballonet, and septum/curtain is determined using the methodology discussed in Sec. 8.4 for the envelope and ballonets and in Sec. 8.7 for the payload/gondola septum or curtain suspension system). This methodology is demonstrated in Sec. 8.9 for the CargoStar sample problem and in Sec. 9.5 for the Solar HALE.

9.2.2 Tail Weight Estimating

The empennage consists of the horizontal tail (the fixed stabilizer and moveable elevator) and the vertical tail (the fixed fin and the moveable rudder). The empennage can have many arrangements as shown on Fig. 3.24. The fixed horizontal stabilizer and vertical fin could be fabric with pressurization (like the envelope except using air) or a light weight structure with a thin film covering. The moveable elevator and rudder control surfaces are usually a light weight structure because of the air loads, gust loads, and weather. The control surfaces are typically 20–25% of the total tail planform area.

The fabric stabilizer and fin looks like a flat, many-lobe air mattress. It would need a pressurization system to maintain an air pressure of approximately 0.2 psi. The analysis for a fabric tail would proceed similar to that of the envelope by determining a maximum skin load (minimum acceptable breaking strength for the fabric), determining a fabric density from Fig. 8.4, and estimating the fabric weight of the fixed stabilizer and fin as follows

$$W_{FSF} = \text{(fabric density)(tail surface area – control surfaces)} \, F_{AF} F_T \quad (9.1)$$

where F_{AF} accounts for the tail external attach fittings and is equal to 1.26 and $F_T = 2.36$ and accounts for the internal structure of the tail (shown in Fig. 9.6a) and includes the cover tape, internal structural tape, spar fabric curtains and spar internal attachments.

The weight of a stabilizer and fin made of light weight structural materials is expressed as

$$W_{SSF} = F_{PSQ} \text{(tail total planform area – control surfaces)} \, F_{AF} \quad (9.2)$$

where F_{PSQ} is the lb/ft^2 to fabricate the tail surface out of lightweight structural materials. The values for F_{PSQ} are (from Fig. 20.1 in [2]).

$$F_{PSQ} = 1.0 \text{ for max } q > 10 \text{ lb/ft}^2$$

(flight at 80 kt and 4000 ft like CargoStar)

$$F_{PSQ} = 0.3 \text{ for max } q < 1 \text{ lb/ft}^2$$

(flight at 26 kt and 65,000 ft like Solar HALE).

The tails are lashed to the envelope (see Fig. 8.6) and braced with cables (see Fig. 1.10). Assume four cables per side and estimate the cable distance from the top of the tail to the hull (60 deg attach angle to hull). Cable bracing weights are discussed in Sec. 9.10.

The weight of the control surfaces is

$$W_{CS} = F_{PSQ} \text{ (control surface planform area)} \tag{9.3}$$

The control surface actuator weights are dependent on the size of the control surface and the air loads (dynamic pressure). The actuator weights are estimated as follows

$$W_{act} = F_{act} \text{ (control surface planform area) } F_{instal} \tag{9.4}$$

where

$$F_{act} = 0.79 \text{ for max } q > 10 \text{ lb/ft}^2$$
$$F_{act} = 0.08 \text{ for max } q < 1 \text{ lb/ft}^2$$
$$F_{instal} = 1.15 \text{ (installation factor)}$$

9.2.3 Gondola Weight Estimating

The gondola is a piece of structure suspended below the airship that houses the crew, payload, and most of the systems. Since most of the airship weight is located in the gondola, the gondola should be located such that the c.g. of the airship is directly beneath the c.b. (typically at 45% of the airship length). The gondola should be as streamlined as possible to reduce the pressure drag and flow separation on the aft end. Figure 9.2 shows the 13-passenger gondola for the Skyship 600 shown on the first page of this chapter.

This gondola is a light weight structure. Since manned airships and hybrids fly at altitudes less than 12,500 ft (FL 12.5), the gondola does not need to withstand an internal pressure. It would be designed for one of three payload categories:

Category		Typical payload
1	Light	ISR and comm equipment (unmanned, less than 1000 lb)
2	Medium	Crew station + systems + passengers or light cargo (less than 4000 lb)
3	Heavy	Payload compartment (vehicles, troops, pallets, TEUs ... more than 5000 lb) + crew station + systems

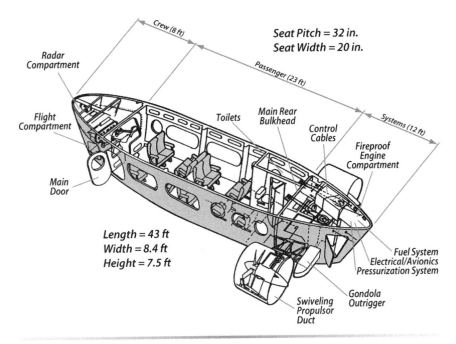

Figure 9.2 The Skyship 600 gondola seats 13 and includes radar and flight deck. It has two swiveling ducted propellers for lift and propulsion.

The gondola weight is estimated knowing the gondola dimensions and a payload category.

Gondola Weight Estimating Relationship:

Category 1: $W_{Gond} = 0.15$ (Payload weight)   (9.5)

Category 2: $W_{Gond} = 353 \, [(\ell/10)^{0.857} \, (w + h)/10 \, (V_{max}/10)^{0.338}]^{1.1}$ (9.6)

Category 3: $W_{Gond} = 1.875 \, S_{Gond} + \text{crew station} + \text{systems}$ (9.7)

where ℓ, w, and h are the gondola dimensions in feet, $V_{max} = $ maximum speed (kt), and $S_{Gond} = $ gondola surface area (4 sides + top/bottom, ft²). System and crew station weight is determined using Eq. (9.6).

If the landing gear is a wheel/strut arrangement it would typically be attached to the gondola. If the airship uses an ACLS, the two or more ACLS pads would be located away from the gondola on the hull lower surface to give a three or four point contact with the ground. Operational considerations and weight estimates for the ACLS system are discussed in Sec. 9.7. Engines are attached to the gondola on an outrigger or to the hull on tripods.

9.3 Pressure Control System

The pressure inside the airship is low, on the order of 0.2 psig, just high enough to hold the shape of the airship envelope but not to require a heavy envelope fabric. A time tested rule of thumb for the internal pressure is 1.2 q, where q is the maximum dynamic pressure that the airship will encounter in its flight envelope. This rule ensures that the nose of the airship will not cave in during maximum q operation.

If the airship volume were completely filled with helium at sea level (SL) the ΔP across the envelope would increase as the airship ascended to the point that either the fabric would burst or the helium would vent to keep the ΔP constant with altitude. The former is not an option and the latter is not affordable since helium is expensive and the envelope would lose its shape as the airship descended. The resolution to this dilemma is the ballonet.

The ballonet is one or more flexible airbags inside the envelope attached to the lower surface. The ballonets are filled or emptied with air to keep the ΔP constant and hold the envelope shape as the airship changes altitude. The ΔP across a ballonet is very small, typically the same as for the envelop. A design issue for the ballonet material is the constant flexing of the fabric as they inflate and deflate. A typical ballonet arrangement is shown in Figs. 1.8 and 9.1.

The size of the ballonets, as a fraction of the envelope volume, depends upon the maximum pressure altitude of the airship (see Fig. 9.3). The ballonet would be completely filled at SL and completely empty at the pressure altitude. Figure 9.3 shows airships with pressure altitudes up to 9000 ft with

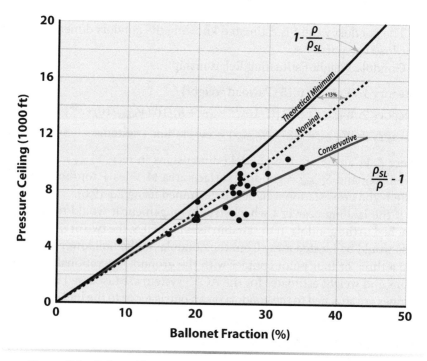

Figure 9.3 Ballonet fraction vs pressure altitude for existing airships.

a few up to 12,000 ft, but none above that except for the stratospheric ISR airships such as ISIS that station keeps at 65,000 ft. The reason is that airships very seldom fly above 9000 ft. Most of the time airships fly at 4000 ft to be above the ATC (air traffic control areas) and out of the general aviation traffic. Airships do a lot of route planning such as flying around high terrain rather than over it.

Taking a closer look at Fig. 9.3 shows that ballonet volume can be calculated several ways compared to the actual data from airship designs. Ballonets would never be sized using the minimum line. However, many designers use the "nominal" line as this is the minimum increased by 13%. The 1.13 factor accounts for the effects of superheating the internal gases, allowance for flying on non-standard pressure days, unusable volume, and the ability to transfer air fore and aft for c.g. control. The third equation is labelled "conservative" but is still a good fit to the data. These data show that the designer has some latitude in choosing ballonet size based on how robust the design needs to be.

The HALE ISR airships (such as the Solar HALE example of Chapter 5) fly at 65,000 ft but do not have ballonets since they stay aloft for months or even years with infrequent ascents and descents. If shape is important at launch then the airship is filled with helium to its normal pressure and launched. During ascent the helium is vented to maintain pressure until it reaches its operational altitude. Since the airship is completely filled with helium at the launch altitude, its buoyant lift has to be modulated by ballast to keep ascent rates within reason. This ballast (usually a water/alcohol mix) is dropped during ascent to keep the rate of climb at a manageable level (usually 1500 f/m). For airships or balloons whose initial shape is not important a measured amount of helium is injected into the envelope that expands as the airship ascends. At operational altitude the measured amount of helium has exactly expanded to provide the targeted envelope volume and pressure (Chapter 13). At the end of the mission the helium would be vented and the limp airship would float to its landing in a remote area.

The flow rate in/out of the ballonet ft^3/m is usually sized by the descent rate. Airship regulation CS30T specifies a maximum rate of descent of 1200 f/m [1]. A notional pressurization system that would service a fore and aft ballonet and an ACLS is shown in Fig. 9.4. It consists of the ballonets, ducting, blower fans, and valves. The scoops (especially the forward facing) should retract when not needed to reduce the airship drag. The system operates as follows:

1. During descent: Inflate forward and aft ballonets B_1 and B_2; open valves V_2, V_3 and V_7; close valves V_4, V_5 and V_6; operate fan F_1.
2. During ascent: Deflate B_1 and B_2, open valves V_1, V_4 and V_5, close valves V_2, V_3, V_6, and V_7; operate fan F_2 to exhaust air from B_1 and B_2.

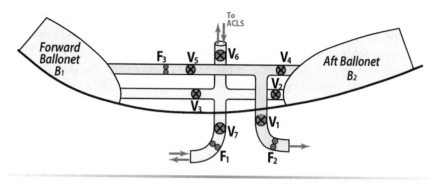

Figure 9.4   Notional ballonet and ACLS pressurization system.

3. c.g. control: Shift air from B_1 and B_2, close valves V_1, V_2, V_3, and V_7; open valves V_4 and V_5, operate fan F_3 to move air between B_1 and B_2.
4. Operate ACLS (airship is at low speed with ballonets full): Close valves V_2, V_3, and V_7; open V_6 and operate fan F_1 to either blow or suck ACLS.

The weight of the ballonets is estimated by assuming a fore and aft ballonet of equal volume each and spherical in shape. Using the area density for ballonets from Chapter 8 the ballonet weight is

$$W_{bal} = (\text{area density})(\text{total surface area of ballonets})(1.13) \quad (9.8)$$

The ballonet ducting is made of the same fabric as the ballonets. The ducting weight is estimated as

$$W_{Duct} = (\text{area density})(\text{duct circumference})(\text{length})(\text{number}) \quad (9.9)$$

The weight of the blower fans in lb is given by the parametric equation

$$W_{BF} = 0.308 \, \dot{w}_{air} \quad (9.10)$$

where $\dot{w}_{air}$ is the required flow rate in ft^3/s for either the ballonet system or the ACLS.

The weight of the valves in lb is estimated as follows:

$$W_{Valve} = (.07)(\text{duct cross-section area in in.}^2)(\text{number}) \quad (9.11)$$

Sample Problem 9.1: Ballonet and Blower Fan Sizing

Assume an airship has a maximum pressure altitude of 10,000 ft ($\sigma = 0.7385$) and weighs 50,000 lb. Using

$$\text{Weight} = \text{Buoyant Lift} = (\text{Volume})(0.065)\sigma \quad (5.44)$$

gives an envelope volume = 1,041,612 ft^3 and ballonet volume = 0 ft^3 at 10,000 ft. At SL (σ = 1) the required envelope volume = 769,231 ft^3 and the ballonet volume is 272,381 ft^3 or 26% of the envelope volume. With helium filling 74% of the volume the buoyant lift at SL is 50,000 lb. As air is emptied from the ballonet, the helium will expand but the buoyancy remains constant. If the cruising altitude for the airship is 4000 ft (σ = 0.8881) the helium volume needs to be 866,153 ft^3 to develop 50,000 lb of buoyant lift. Thus, the ballonet would have to empty to 175,459 ft^3.

If the airship descends to SL from 4000 ft at 1200 f/m (descent time = 200 sec), the blower would have to fill the ballonets with air at the rate of $\dot{w}_{air}$ = (272,381 − 175,459)/200 = 484.6 ft^3/s. Assume that the pressurization system consists of three blowers of 200 ft^3/s capacity each (each blower fan operating at 80% capacity). The weight of the blowers for the ballonets would be (0.308)(484.6)/(0.8) = 186.5 lb.

9.4 Hydro Carbon Fuel Propulsion System

The hydro carbon propulsion system is comprised of

- Engine (piston or turbine)
- Propeller
- Fuel tank
- Starter
- Engine controls
- Engine mount/installation

The engine weight (W_{Eng}) is estimated by determining the horsepower required as discussed in Sec. 4.3. A decision is made as to the number of engines and an off-the-shelf engine selected from Fig. 5.8 or a vendor catalog.

The engine is either mounted to the gondola (on an outrigger, see Fig. 9.2) or attached to the envelope (on a tripod) as shown in Figs. 9.5a and 9.5b. For conventional or hybrid airships weight isn't as important as it is for a high altitude ISR airship. Hence, there are significant differences between the propulsion systems and the structure that attaches the engine/propeller to the side of the envelope. Remember to consider the propeller diameter to make sure the tripod has enough standoff distance from the envelope. The weight of the engine mount and installation would be estimated as follows:

$$W_{EngMt} = F_{EngMt} N_E W_{Eng} \qquad (9.12)$$

Figure 9.5a   Possible engine mount scheme for hull-mounted internal combustion/turboprop engines on conventional or hybrid airship designs.

where

$F_{EngMt} = 0.57$ for a piston or turbine engine on a gondola-mounted outrigger (see Fig. 9.2)

$F_{EngMt} = 0.64$ for a piston or turbine engine on a hull-mounted tripod (see Fig. 9.5a)

$F_{EngMt} = 1.2$ for an electric motor on a hull-mounted tripod (see Fig. 9.5b)

Figure 9.5b   Possible engine mount scheme for hull-mounted electric motors on a high altitude design.

N_E is the number of engines.

The fuel tank(s) is usually located in the gondola and feeds the outrigger-mounted engines and the tripod-mounted engines. The tripod-mounted engines located on the hull would have long fuel lines. The weight of the fuel tanks, pumps, and lines is estimated as

$$W_{FT} = 2.49 \, (\text{Fuel})^{0.6} \, (N_T)^{0.2} \, (N_E)^{0.13} \, [1/(1 + Int)]^{0.3} \qquad (9.13)$$

where Fuel is the total fuel in gallons (a gallon of aviation gas weighs 6.0 lb), N_T is the number of separate fuel tanks, and Int is the percentage of fuel tanks that are integral.

The weight of the engine controls is estimated as

$$W_{EC} = 60.27 \, (\ell_{EC} \, N_E/100)^{0.724} \qquad (9.14)$$

where ℓ_{EC} is the distance (in feet) from the controller to the engine.

The weight in lb of an electric starting system is estimated as

$$W_{Start} = 50.38 \, (NE \, W_{Eng}/1000)^{0.459} \qquad (9.15)$$

The propeller weight in lb is estimated as follows:

$$W_{Prop} = K_P \, N_P \, (N_{BL})^{0.391} \, (d_p \, h_p/1000)^{0.782} \qquad (9.16)$$

where $K_P = 31.92$, N_P = number of propellers, N_{BL} = number of blades per propeller, d_P = propeller diameter in feet and h_p = engine shaft horsepower.

9.5 Solar Energy Propulsion System

The solar energy propulsion system is comprised of

DC electric motor
Propeller
Solar cells
Energy storage device (rechargeable batteries or regenerative fuel cells)

These elements of the solar propulsion system are discussed in Chapter 5 and the component WERs are listed in Table 5.2. The weight build-up for a solar powered airship will be detailed in the following example.

Sample Problem 9.2: Solar HALE

We return to the Solar HALE introduced in Sec. 5.12, discussed in Sec. 7.4, and shown on Fig. 7.2. An initial TOGW = 5000 lb and "24/7" loiter at 65,000 ft was assumed giving a lifting gas volume of 1.04×10^6 ft^3.

The solar power system was sized using solar cells of 32% efficiency and regenerative O_2/H_2 fuel cells (RFC) for the nighttime energy storage device. All that remains is to estimate the component weights using the WERs discussed in Chapters 5, 8, and 9 in order to check the assumed 5000 lb weight.

The hull/envelope weight was discussed in Chapter 8. For the volume = 1.04×10^6 ft^3 and $FR = 3.2$ the surface area of the envelope is estimated [using Eq. (3.19)] to be 58,685 ft^2. The envelope weight is estimated as follows.

The 1.2 $q_{max} = P_{SP}$ rule of thumb is not appropriate for a HALE vehicle since $q_{max} \approx 0.17$ lb/ft^2 is very low. Thus, we assume $P_{SP} = 0.2$ psi. The buoyancy pressure delta on the upper skin is determined from Eq. (8.5) as

$$\Delta P = (0.065)(89.5) = 5.82 \text{ lb/ft}^2 = 0.04 \text{ psi}$$

The skin limit load acting on the envelope is [from Eq. (8.6)]

Envelope limit skin load = $(0.2 + 0.04)(45)(12) = 129.6$ lb/in.

The minimum fabric breaking strength = (hoop skin load)(FS) = (129.6) (4) = 518.4 lb/in. [using Eq. (8.6b)]. From Fig. 8.4 the fabric density for Dyneema is 3 oz/yd^2 = 0.021 lb/ft^2 giving an envelope fabric weight [using Eq. (8.7)] of

$$W_{env} = \text{(fabric density)}(S_{wet})_{env} F_{MA} F_{AF} = (0.021)(58,685)(1.2)(1.26) = 1863 \text{ lb.}$$

Since there is no ballonet, we have $W_{bal} = 0$.

The load suspension curtain fabric area is assumed to be 20% of the solar HALE side area or 3014 ft^2. The curtain load is 1.5 times the envelope maximum skin load = $(1.5)(518.4) = 777.6$ lb/in. From Fig. 8.4 the curtain fabric density using Dyneema is 4 oz/yd^2 = 0.028 lb/ft^2 giving a curtain weight based on Eq. (8.13).

$$W_{curtain} = \text{(fabric density)(curtain fabric area)} F_J = (0.028)(3014)(1.06) = 90 \text{ lb}$$

The curtain suspends the following items (lb):

Installed payload	575
Water ballast and tank	425
Avionics	67
RFC stacks and tanks	364
Radiators (reject extra heat)	65
Equipment bay	140
Total	1636

We will use five 0.125-in. diameter, 7-strand galvanized cables that weigh 0.029 lb/ft for suspension lines, as shown in Fig. 9.1. From Table 9.3 this cable has a breaking strength of 2000 lb. The five cables will give a FS > 4. The weight of the suspension cables is 5(100 ft)(0.029) = 15 lb.

As the Solar HALE ascends to 65,000 ft it will vent helium and carry ballast to keep the ascent rate at a manageable level (i.e. ~1500 f/m). The water ballast will be carried in a cylindrical tank similar to a fuel tank. Assuming the water ballast at 400 lb (62 lb/ft^3) the tank would be 6.5 ft^3 (1 ft diameter × 8 ft long) or approximately 48 gallons. Using Eq. (9.6) gives 25 lb for the water ballast tank.

The fabric stabilizer and fin will be a pressurized fabric resembling a flat, many-lobe air mattress. The tail is shown in Fig. 7.3 with a root chord of 35.3 ft. Assuming a tail t/c = 15% gives a maximum lobe diameter of 3.5 ft. The total exposed planform area for the four tails is 2805 ft^2, which gives a total surface area of approximately 2(1.2)(2805) = 6732 ft^2. The control surface area of the elevators and rudders will be assumed to be 20% of the total planform area.

The fabric structure of the stabilizer/fin is shown in Fig. 9.6a. The tail super pressure will be assumed to be 0.2 psi. Since the tail is pressurized with air the buoyancy force is zero. The hoop skin load (envelope limit skin load) = (0.2) (3.5/2)(12) = 4.2 lb/in. The minimum fabric breaking strength = (hoop skin load)(FS) = (4.2)(4) = 16.8 lb/in. From Fig. 8.4 the fabric density for Dyneema is 1.5 oz/yd^2 = 0.0105 lb/ft^2 giving a tail fabric weight [using Eq. (9.1)] of

$$W_{FSF} = (0.0105)(6732 - 1346)(1.26)(2.36) = 168 \text{ lb}$$

We need to include a pressurization system of 148 lb (assumed) to keep the stabilizer and fin inflated to 0.2 psi.

At this point it is a good idea to check our assumption of a pressurized fabric tail versus a light weight structure tail. Using Eq. (9.2) the light weight structure tail weight would have been (0.3)(0.8)(2805)(1.26) = 848 lb. Thus the fabric pressurized tail at 168 + 148 = 316 lb was a good choice.

The rudders and elevators are light weight structure at 0.3 lb/ft^2 or (0.3)(0.2)(2805) = 168 lb. This light weight structure is shown in Fig. 9.6b and is typical of the structure on the Aerovironment Helios. The actuator weights are estimated from Eq. (9.4) as

$$W_{act} = F_{act} \text{ (control surface planform area)} = (0.08)(0.2)(2805) = 45 \text{ lb}$$

The stabilizer and fin are lashed to the hull of the airship and braced with cables. We will assume that each tail has four bracing cables on each side, each cable is 40 ft long and made from ¼ in. Vectran. From Table 9.4

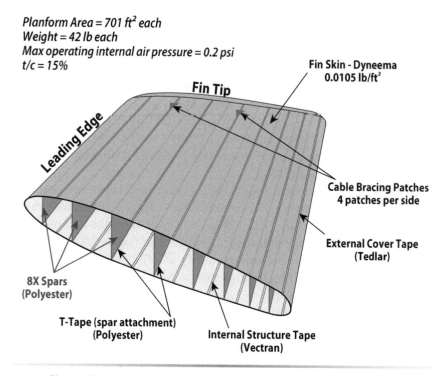

Planform Area = 701 ft² each
Weight = 42 lb each
Max operating internal air pressure = 0.2 psi
t/c = 15%

Fin Skin - Dyneema
0.0105 lb/ft²

Fin Tip

Leading Edge

Cable Bracing Patches
4 patches per side

External Cover Tape
(Tedlar)

8X Spars
(Polyester)

T-Tape (spar attachment)
(Polyester)

Internal Structure Tape
(Vectran)

Figure 9.6a Solar HALE stabilizer/fin pressurized fabric structure.

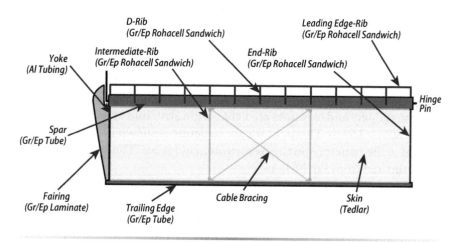

Planform area = 140 ft² each
Weight = 42 lb each

D-Rib
(Gr/Ep Rohacell Sandwich)

Leading Edge-Rib
(Gr/Ep Rohacell Sandwich)

Yoke
(Al Tubing)

Intermediate-Rib
(Gr/Ep Rohacell Sandwich)

End-Rib
(Gr/Ep Rohacell Sandwich)

Hinge
Pin

Spar
(Gr/Ep Tube)

Fairing
(Gr/Ep Laminate)

Trailing Edge
(Gr/Ep Tube)

Cable Bracing

Skin
(Tedlar)

Figure 9.6b Solar HALE elevator/rudder lightweight structure.

the breaking strength is 9400 lb and the weight is 0.022 lb/ft. The weight of the cable bracing is $(4)(8)(40)(0.022) = 28$ lb.

The weight of the control cables from the vehicle management systems (VMS) to the tails is given by Eq. (9.14) where ℓ_{EC} is the control cable length and N_E is the number of engines.

$$W_{EC} = 24.27[(2)(134)/100 + (2)(170)/100]^{0.724} = 90 \text{ lb}$$

The solar power system weights from Chapter 5 are

RFC weight (includes fuel cell stacks and O_2, H_2 and water tanks) = 365 lb.

Installed solar cells (2500 ft^2) = 250 lb

The propulsion system is four electric motors at 1 kW each on hull-mounted tripods (see Fig. 9.5b). From Table 5.2 the specific energy of an electric motor is 0.2 kW/lb (includes motor, controller and propeller). Thus, the weight of the electric motor unit is 1/0.2 = 5 lb each. The weight of the four tripod engine mounts is [from Eq. (9.12)]

$$W_{EngMt} = F_{EngMt} \, N_E \, W_{Eng} = (1.2)(4)(5) = 24 \text{ lb}$$

The weight of the four power cables from the solar cells to the motors and the four control cables from the VMS to the motors is given by Eq. (9.14) (using $N_E = 4 + 4 = 8$).

$$W_{EC} = 24.27 \, [(\ell_{EC} \, N_E)/100]^{0.724} = 24.27[(90 \times 8)/100]^{0.724} = 101 \text{ lb}$$

The fuel system weight (tanks, pumps, lines, etc.) is zero since the motor is electric and the power is solar.

Solar HALE weight summary	
Weight group	**Weight (lb)**
Hull (envelope, tails, bracing cables, and suspension curtains)	2332
Solar power system (solar cells, RFCs, and tanks)	615
Propulsion system (motors, mounts, power cables)	145
Pressurization system (tail, assumed)	148
Water ballast and tank	425
FCS (flight computer, sensors, actuators, control cables)	185
Avionics/communications/DL	67
Installed payload (EO/IR sensor, control unit, 1.15 install)	575
Miscellaneous (assumed equipment bay, radiators, etc.)	205
Margin (6.1%)	303
TOGW	5000

The remaining subsystem weights are

Landing gear = 0
Avionics/communications/DL = 67 lb
Flight control system (flight computer, sensors, etc.) = 50 lb

The Solar HALE meets the assumed TOGW of 5000 lb with a 6.1% margin. The conceptual design stage should carry at least a 6% weight margin because aircraft and airship weights *always* increase from initial design to production.

9.6 Electrical

The electrical system would include the power generation system (such as batteries, fuel cells, APU, or engine-driven generators), the distribution system, and the power conditioning system. Batteries, APU or fuel cells would be vendor items with unique weights. The WER for an electrical system consisting of engine-driven generators and conventional power conditioning and distribution is expressed in terms of the total weight of the fuel system and electronics equipment (the primary users of electrical power on the airship). The weight is estimated as follows:

$$W_{Elect} = K_{elect}(W_{FS} + W_{TRON})^{0.51} \qquad (9.17)$$

where W_{FS} = weight of the fuel system (lb) and W_{TRON} = weight of the installed avionics/electronics (lb). The K_{elect} accounts for the type of mission as follows:

$K_{elect} = 12.57$ for commuter/passenger airships
$K_{elect} = 33.73$ for long range transport airships

9.7 Landing Gear and ACLS (Air Cushion Landing System)

Airship and hybrid landing gears are designed for a 4 f/s sink rate whereas USAF and commercial transports are sized for 10 f/s sink rate and Navy carrier suitable aircraft have a 24 f/s sink rate requirement. All landing gears are sized for a landing weight heaviness $W_{H_L} = (1 - BR)(TOGW) - 50\%$ fuel. During landing the airship/hybrid main landing gear or rear edge of the aft ACLS pads will strike the ground first, and then the nose gear or nose ACLS pad will contact the runway for a three-point run-out. The landing event is the critical condition for tail strike. The geometry should be checked to determine the angle between the fully compressed main gear or rear edge of aft ACLS and the tail or aft body. This angle should be at least 10 deg to prevent damage to the delicate tail structure.

9.7.1 Air Cushion Landing System

Figure 9.7 shows a cross-section of an ACLS pad. It consists of a plenum surrounded by a flexible pressurized fabric called a curtain. The curtain is pressurized to 0.1 psi by a dedicated pressure system which is independent of the main pressurization system as shown in Fig. 9.7. The reason for the separate system is that the curtain needs a positive pressure during suckdown when the main pressure system is supplying a negative pressure to the plenum. The curtain is a very durable, abrasion resistant fabric as there will be some rubbing along the runway during ground operations. During cruise the curtain is evacuated (slight negative pressure) to retract and cover the plenum to reduce the drag of an open cavity.

Figure 9.8 shows a view of the LM P-791 Hybrid Demonstrator looking up into the ACLS plenums. The figure shows the curtains extended. The P-791 was a demonstrator and did not have the capability to retract the curtains and cover the plenum during cruise. The result was a large cruise drag increment due to the open cavities.

Hybrids will land heavy ($W_H > 0$) on a tricycle landing gear or an ACLS. The ACLS is similar to a hovercraft and permits the hybrid airship to operate with unmatched maneuverability on the ground or any other reasonably flat surface including water. The ACLS can reverse the blower fans for a negative 0.1 psi pressurization and suck the vehicle down when it is parked. This grip mode allows the hybrid to remain parked in ~25 kt winds.

The ACLS consists of several ACLS pads on the bottom of the hybrid. The sizing condition for the ACLS plenum is either

1. The landing sink rate at $W_{H_L} = (1 - BR)(TOGW) - 50\%$ fuel compressing the rear edge of the aft pads. This sizing analysis results in a plenum area A_P (ft^2) of *one* ACLS pad aft of the c.g. to be

$$A_P = 0.23 \, W_{H_L} \, V_{SR}/(N_{Pad} \, P_P) \tag{9.18}$$

Where V_{SR} = landing sink rate, f/s (typically 4 f/s), N_{Pad} = number of ACLS pads aft of the c.g. (typically two) and P_P = plenum pressure, lb/ft^2 (same as hull pressure). If there is a single nose ACLS pad its plenum area would be 20% of $A_P \, N_{Pad}$.

2. The total lift of the ACLS system (plenum pressure × the total plenum area) during ground operations should equal the *maximum heaviness* = $(1 - BR)(TOGW)$.

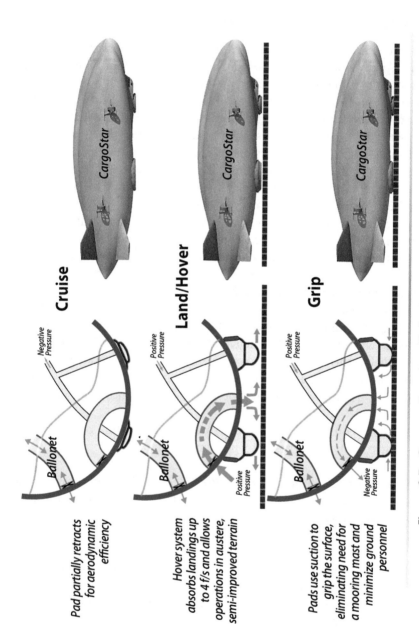

Figure 9.7 Cross-section of an Air Cushion Landing System (ACLS) showing the three modes of operation.

Figure 9.8   Lockheed Martin P-791 hybrid demonstrator showing ACLS pads.

Sample Problem 9.3: ACLS Sizing for the CargoStar

At this point the ACLS plenums for the CargoStar discussed in Sec. 6.5 can be sized. First the two aft pads are sized for a landing sink rate of 4 f/s and a plenum pressure of 0.1 psi. The landing weight heaviness W_{H_L} = $(1 - 0.75)(93,000) - (23,000)/2 = 11,750$ lb. From the sizing analysis above [Eq. (9.18)] the plenum area of one pad is $A_P = (0.23)(11,750)(4)/[(2)(0.1)$ $(144)] = 375$ ft^2 for a plenum pressure of 0.1 psi. Assuming a single nose pad (carrying 20% of the heaviness) gives a total ACLS lift of $[(2)(375) + 172]$ $(14.4) = 13,287$ lb which is less than the maximum heaviness of 25,000 lb at takeoff. Thus, we need to size the ACLS for taxi takeoff.

If we size the ACLS plenum area to generate 25,000 lb at taxi takeoff, the total plenum area is $25,000/14.4 = 1736$ ft^2 or 694 ft^2 for each of the main pads and 347 ft^2 for the nose pad.

9.7.2 Landing Gear and ACLS Weight Estimating

Airships typically have a single landing gear (such as the Skyship 600 and ZP4K) and attach to a docking mast for ground operations. However others such as the Sentinel 1000 have tricycle gears. Airship landing gear weight can be estimated using the following W_{ER}:

$$\text{Single landing gear weight} = 24.2 \, [2(W_{H_0})/1000]^{0.84} \qquad (9.19)$$
$$\text{Tricycle landing gear weight} = 31.2 \, [2(W_{H_0})/1000]^{0.84} \qquad (9.20)$$

where 20% of the gear weight is for the nose gear and 80% for the main gear.

The ACLS weight in lb is given by

$$\text{ACLS weight} = 1.6 \text{ (total plenum area)} \tag{9.21}$$

9.8 Crew and Passenger Accommodations

All weights that follow are in lb.

$$W_{Seat} = K_{Seat} \text{ (number of seats)} \tag{9.22}$$

where

$$
\begin{aligned}
K_{Seat} &= 55 \text{ for flight deck seats} \\
K_{Seat} &= 32 \text{ for reclining passenger seats} \\
K_{Seat} &= 17 \text{ for troop seats} \\
W_{Bunk} &= 28 \text{ (number of bunks)}
\end{aligned} \tag{9.23}
$$

$$\text{Lavatories: } W_{Lav} = K_{Lav} \text{ (number of crew + passengers)}^{1.33} \tag{9.24}$$

where $K_{Lav} = 5.6$ for long range flights and $K_{Lav} = 2.3$ for short range flights.

$$\text{Food/water: } W_{Food} = 5.06 \text{ (number of crew + passengers)(days)} \tag{9.25}$$

where Eq. (9.25) includes galley provisions.

9.9 Avionics and Electronic Equipment

The avionics equipment consists of the communications and navigation gear, VMS, radar, autopilot, instrumentation, and sensors. The weight for the avionics and electronic equipment is best obtained from a careful study of the requirements and vendor catalogs. Table 9.1 contains weights and volume information on common avionics equipment. These weights should be increased by 15% to account for installation.

Avionics equipment must be maintained frequently, so the equipment must be located for easy access by the ground crew. This equipment is usually located in the gondola. It must not be stacked such that a piece of good equipment would have to be removed to get to the faulty item.

9.10 Miscellaneous

Weights and volumes for common avionics equipment can be found in Table 9.1. Fuel weights are given in the following table:

Fuel	Gallon weighs (lb)	Cubic foot weighs (lb)
JP-4	6.5	48.6
Aviation gas	6.0	44.9
#2 diesel	7.15	53.5

Table 9.1 Weights and Volumes for Common Avionics Equipment

Item*	Model designation	Volume (ft³)	Weight (lb)
Intercom system	AIC-25	–	19.2
UHF communications	ARC-109	–	51.0
VHF/UHF	Link 16/SATCOM		44
HF comm	Link 22		130
Ka comm	CDL/SATCOM		18
	ARC-150	0.21	11.0
UHF DF homing	705 CA	–	5.0
Air-to-ground IFF	APX-64	–	53.0
	APX-92	0.11	13.0
TACAN	ARN-52	–	61.0
	ARN-100	1.1	46.0
ILS-VOR	ARN-584	–	27.0
	RCS-A VN- 220	0.05	3.5
Gyro compass	ASN-89	0.21	8.4
Inertial navigation system	AJQ- 20	–	207.0
	LN-30	1.08	44.0
HF radio	ARC-123	–	78.4
Autopilot system	–	–	168.5
Air data computer	AXC-710	0.5	14.0
Radar altimeter	APN-167	–	38.2
Range only radar	SSR-1 (GE)	0.55	25.0
Flight data recorder	–	0.3	15.6
EO/IR target system (raytheon)	AN/AAS-52 target system	18″ diam 0.31 elect units	130 25
EO/IR target system (L3-wescam)	MX-10	10.24 in. diameter by 14 in. height	37 lb

* Abbreviations: UHF, ultrahigh frequency; DF, direction finder; IFF, identification friend or foe; TACAN, tactical air navigation; ILS-VOR, instrument landing system, very-high-frequency omni-directional radio; ECM, electronic countermeasures; EO/IR, electro-optical/infra-red.

Table 9.2 Control and Fuel Line Tubing Weights

OD (in.)	Stainless tubing (lb/f)	Aluminum tubing (lb/f)
0.25	0.067	0.024
0.375	0.105	0.037
0.5	0.142	0.05
0.75	0.218	0.077
1.0	0.293	0.104
1.25	0.369	0.13
1.5	0.444	0.157

Control lines, instrumentation wires, and fuel lines oft times will have to run from the gondola to the tail group or hull-mounted engines. The tail group and hull-mounted engines are usually braced using bracing cables as shown in Fig. 1.10 and Fig. 9.5 and discussed in Chapter 8. Vectran rope is sometimes used for the bracing because it has tremendous strength to weight (see Table 9.4). Tubing, fuel lines, and bracing cables are presented in Tables 9.2 and 9.3.

9.11 Ballast

The airship/hybrid must always have the buoyancy ratio $BR \leq 1$ (otherwise it is an untethered balloon). If the heaviness cannot be controlled with aerodynamic lift or some other feature (see Sec. 4.7.4) then ballast is a last

Table 9.3 7-Strand Galvanized Cable Strength and Weights

Diameter (in.)	Breaking strength (lb)	Weight (lb/f)
3/32	1000	0.0174
1/8	2000	0.029
5/32	2800	0.045
3/16	4200	0.065
7/32	5600	0.086
1/4	7000	0.11
5/16	9800	0.173
3/8	14400	0.243

Table 9.4   12-Strand Vectran Rope Strength and Weights

Diameter (in.)	Breaking strength (lb)	Weight (lb/ft)
1/8	2700	0.006
3/16	5500	0.013
1/4	9400	0.022
5/16	14000	0.036
3/8	19500	0.046
1/2	35000	0.088
5/8	55000	0.14

resort. The important thing to remember about ballast is that at some point it may be dropped from altitude (see the Solar HALE example in Sec. 9.5). Consequently, water is a better ballast than lead shot. The water would be mixed with alcohol to prevent it from freezing.

Sample Problem 9.4: CargoStar Weight Build-Up

We return to the CargoStar example that was introduced in Chapter 6. The initial W_G was estimated at 93,000 lb that gave a W_{OE} of 37,000 lb using Fig. 6.1. The CargoStar specifications are given on Fig. 6.2. This section will use the weight methodologies presented in Chapter 5 (IC/turboprop engine weights and propeller sizing), Chapter 8 (envelope, septum/curtain and ballonet) and Chapter 9 to develop the CargoStar W_E.

In Chapter 8 the CargoStar was used as an example of the methodology for estimating the weights of the envelope, ballonets, and septum. Assuming a maximum speed of 65 kt (q = 14.32 lb/ft^2) these estimated weights are

W_{env} = 4780 lb
W_{Sept} = 1322 lb
W_{Bal} = 1337 lb

The weight of the pressurization system is assumed to be 800 lb using historical data shown on Fig. 9.10.

From Fig. 7.1 the CargoStar tail volume coefficients were estimated to be $C_{VT} = (\ell_{VT})(S_{VT})/[(\ell_B)Vol^{2/3}] = 0.061$ and $C_{HT} = (\ell_{HT})(S_{HT})/[(\ell_B)Vol^{2/3}] = 0.0684$. Using $\ell_{HT} = \ell_{VT} = 0.38 \ell_B = 108.3$ ft and $Vol^{2/3} = 13,463$ ft^2 the total horizontal tail area is 2423 ft^2 and vertical tail area is 2161 ft^2 which gives a total planform tail area of 3247 ft^2 that is canted 42 deg from the horizontal.

Assuming the tails to be constructed from light weight structural material with 20% control surfaces gives a fixed tail weight [using Eq. (9.2) for maximum $q > 10$ lb/ft^2] of 3273 lb. The weight of the control surfaces [from Eq. (9.3)] is 649 lb and the actuators [from Eq. (9.4) plus 15% installation] is 590 lb.

The crew station and system section will be separate from the cargo bay. The cargo bay will contain the fuel tanks and be located directly beneath the c.b. with attention paid to easy roll on/off for payload handling. There will be a tunnel between the crew station and the cargo bay for in-flight access.

The crew station plus system section of the gondola is assumed to have dimensions of length = 25 ft, width = 7 ft and height = 7 ft. Using Eq. (9.6) (category 2) gives a crew station plus system section weight of 2429 lb.

Assuming a cargo bay for the 34,000 lb payload with dimensions length = 55 ft, width = 10 ft and height = 10 ft gives a wetted area of 2400 ft^2. Using Eq. (9.7) (category 3) gives a cargo bay weight of 4500 lb.

The four GTSIOL-550 reciprocating engines weigh 640 lb/engine and are mounted on the hull as shown in Fig. 6.2. The engines would be mounted on a tripod as shown in Fig. 9.5a. The weight of the tripod engine mounts is estimated using Eq. (9.12).

$$W_{EngMt} = (0.64)(4)(640) = 1638$$

The fuel tank will be two tanks mounted outside of the cargo bay. The weight of the fuel tanks, pumps and fuel lines is estimated using Eq. (9.13). For 3667 gallons of fuel, four separate fuel tanks and no integral tanks the weight of the fuel system is 542 lb.

The total weight of the engine controls is 272 lb for an assumed line length from the crew station to each engine of 200 ft [from Eq. (9.14)].

The weight of the electric starter on each engine is 20 lb [from Eq. (9.15)].

The weight of each 14 ft, 3-bladed propeller and controls for the GTSIOL-550 engine is 232 lb [from Eq. (9.16)].

The CargoStar uses an ACLS instead of a conventional tricycle landing gear. Sample Problem 9.3 solved for the plenum area of the 3 point ACLS ... 347 ft^2 for the nose pad and 694 ft^2 for each of the two main pads. The critical case for the sizing is taxi-takeoff. Using Eq. (9.21) the ACLS weight is determined to be 2778 lb.

Assuming a crew of 5 the crew accommodations for a 5000 nm mission (assume 10 days) are:

Flight deck seats	275 lb
Bunks (3)	84
Lavatories	73
Food, water, and provisions	253

The avionics set (from Table 9.1) is as follows (includes 15 % for installation):

Intercom	19 lb
UHF communications	51 lb
INS	207 lb
Auto pilot	168 lb
Air data system	14 lb
Radar altimeter	38 lb
Flight computer	14 lb

The CargoStar weight summary is as follows:

Component	Weight (lb)	Percent of W_E
Envelope	4780	13.9
Septum	1322	4.0
Ballonet	1337	4.0
Tails (light weight structure)	3922	12.0
Gondola/crew station + system section	2429	7.5
Payload bay	4500	13.8
Propulsion	5478	17.0
Engines	2560	
Mounting/tripod	1638	
Engine controls	272	
Electric starters	80	
Propellers	928	
Fuel system	542	1.7
Pressurization system	650	2.0 (Fig. 9.10)
ACLS	2778	8.5
VMS/actuators/avionics (installed)	1178	3.5
Electrical	1170	3.6
Accommodations	685	2.0
Miscellaneous	1700	5.0 (Fig. 9.10)
Empty weight, W_E (no margin)	32,471	
Margin	1940	6.0
Empty weight, W_E	34,411	
Crew + equipment + trapped fuel	1490	
Operating empty weight, W_{OE}	35,901	

The W_E build-up for the CargoStar with a 6% margin is 34,411 lb. Assuming 1250 lb for the 5-man crew plus equipment and a 1% trapped fuel weight gives an $W_{OE} = 34{,}411 + 1490 = 35{,}901$ lb. The W_G = payload + W_{fuel} + W_{OE} = 34,000 + 22,000 + 35,901 = 91,901 lb which is within 1% of the original assumed W_G of 93,000 (see Sec. 6.5). At this point the Cargo-Star conceptual design is considered closed based upon the requirements and assumptions put forth in Sec. 6.5.

Before running off to make a wind tunnel model of the configuration shown on Fig. 6.2, however, the design should be examined to see if it can be made better. For example, one of the reported major attributes of hybrid configurations is the mission flexibility and reduced infrastructure. The CargoStar of Fig. 6.2 will land slightly heavy (about 2600 lb) after burning off its mission fuel. Thus, before the 34,000-lb payload can be removed the vehicle will 1) have to be tied down, 2) fuel tanks filled and ballast added, 3) engine thrust vectored up, or 4) ACLS operated in full suck-down mode . . . all of which are counter to mission flexibility and reduced infrastructure. For the return trip (without payload) approximately 32,000 lb of ballast would have to be loaded onboard the CargoStar. In order to remove these operational inconveniences the take-off BR would have to be about 40%. This is a major design change but one that should be examined before design closure. The redesign for $BR = 0.4$ would be about the same size and volume as Fig. 6.2 but $W_G = 158{,}000$ lb and $S_G = 3624$ ft. For $S_G < 2000$ ft, $BR \approx 0.55$ and $W_G = 118{,}000$ lb. The decision on BR will be a designer judgment call and call into play enlightened performance compromise. The selection of hybrid airship BR will be discussed further in Chapter 12.

9.12 Impact of Maximum Speed

It is useful at this point to examine the impact of maximum speed on the airship W_G. The airship is a very efficient transport vehicle when the mission time is not critical. The power required (and resulting engine weight) increases as the cube of the maximum speed (see Sec. 4.3). The density of the envelope and septum fabric increases as the square of the maximum speed since the fabric breaking strength depends on the internal pressure, which is a function of maximum dynamic pressure (see Sec. 8.9).

The discussion will quantify the impact on CargoStar by increasing the maximum speed for the vehicle from 65 kt to 100 kt. The metric will be the W_E increase. The impact is shown in the table as follows:

Max speed	q at SL	P_R	Engine wt (4)	Envelope wt	Septum wt	Σ wt
(kt)	(lb/ft²)	(hp)	(lb)	(lb)	(lb)	(lb)
65	14.3	2096	2560	4235	1322	8117
80	21.7	3663	4406	5490	1690	11,586
100	33.9	6927	7314	7058	2209	16,581

The weight increase in increasing the maximum speed from 65 kt to 80 kt is 3469 lb or 10.1% of the W_E, which uses up the margin. Requiring an airship to go 80 kt or more is not very efficient and results in a small payload or small fuel fraction. There needs to be a critical and compelling reason for the requirement.

An engine sized for 80 kt has some good news and some bad news. The bad news is that the engines are oversized for cruise (see the propeller design discussion in Chapter 6) and would have to be throttled back outside of the RPM for best $BSFC$.

The good news is that the engines sized for 80 kt result in very large engines with a static thrust of approximately 7300 lb (each engine). With these large engines the take-off ground roll distance is approximately 356 ft compared to 1108 ft for CargoStar with engines sized for 65 kt.

9.13 Summary

Figures 9.9 through 9.11 show the distribution of weight items as a percentage of the empty weight for airships, hybrids, and a long endurance

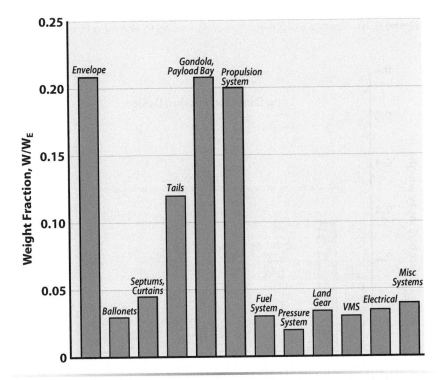

Figure 9.9 Typical weight fraction (referenced to W_E) for a conventional airship.

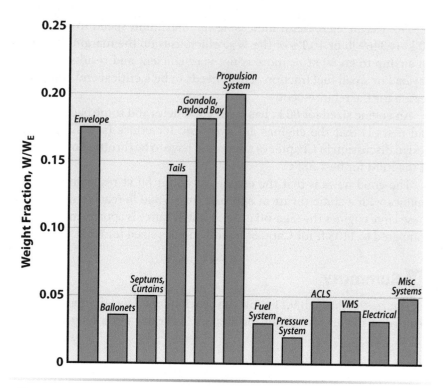

Figure 9.10 Typical weight fraction (referenced to W_E) for a hybrid airship.

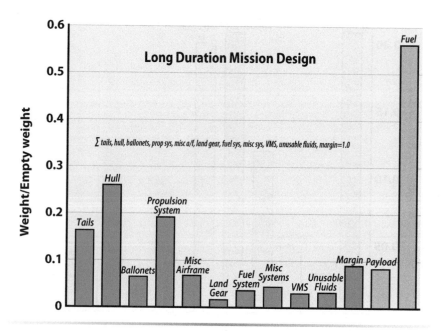

Figure 9.11 Weight fraction for a conventional airship—long duration mission.

airship respectively. This statistical data is useful as a sanity check for estimated weights. For example Fig. 9.10 suggests that the CargoStar envelope and propulsion weights are low and the gondola/cargo bay and ACLS weights are high. However, the reader is reminded of the second paragraph in this chapter that mentions the challenge of determining good weight and is encouraged to continue. Weight data (like aero data) is very proprietary among manufacturers and is difficult to obtain resulting in a small statistical data base available to the airship community. However, the data in Figs. 9.9 through 9.11 should give useful information for making reasonable estimates of weights for the various subsystems.

References

[1] Anon, CS-30T *Certification Specifications for Transport Category Airships*, European Aviation Safety Agency (EASA)", September 2003.
[2] Nicolai, L.M., and Carichner, G.E., *Fundamentals of Aircraft Design: Volume 1— Aircraft Design*, AIAA, Reston, VA, 2010.

Chapter 10 | Stability and Control

Chris Atkinson

- Lateral-Directional Stability and Control
- Longitudinal Stability and Control
- Parametric Tail Sizing
- Roll Stability and Control
- Dynamic Characteristics of Airships
- Airship Control Methods

The USS Akron attempts to moor at Camp Kearny in San Diego, California.

The tails need to be bigger.
Every flight controls engineer

10.1 Introduction

This chapter covers the basics of stability and control of airships, including static stability and dynamic stability mode approximations. The analysis of stability and control for airships has many significant differences from that of conventional airplanes. Table 10.1 lists some differences between airplanes and airships from a stability and control point of view. Most of these differences are either related to the effects of buoyant lift used by airships or to the relatively low numbers of airships produced compared to airplanes.

One of the differences between airships and airplanes listed above is the lack of well-defined military or commercial specifications for airships. Table 10.2 provides a short list of the airship design and certification specifications that are available. The FAA design criteria document for airships, FAA-P-8110-2, is not a Federal

On May 11, 1932, the USS Akron attempted to moor at Camp Kearny in San Diego, CA. The combination of lifting gas heated by the sun and low fuel weight made the airship lighter than neutrally buoyant and almost impossible to control during mooring. Three ground crewmembers were carried into the air clinging to the mooring lines. Two of them, Aviation Carpenter's Mate 3rd Class Robert H. Edsall and Apprentice Seaman Nigel M. Henton fell to their deaths. The third, Apprentice Seaman C. M. "Bud" Cowart was eventually pulled aboard the airship. The Akron was able to successfully moor at Camp Kearny later that day.

Table 10.1 Airplane and Airship Stability and Control Comparison

Airplanes	Airships
Vehicle mass is greater than displaced air mass.	Vehicle mass is approximately equal to displaced air mass.
Vehicle speed is greater than winds.	Vehicle speed is approximately equal to winds.
Important forces are lift, drag, thrust, and weight.	Buoyancy, thrust vector, and added mass also important.
Directionally stable.	Directionally unstable.
Center of gravity position is critical.	Centers of gravity and buoyancy are critical.
Control power increases with airspeed.	Propulsive control power decreases with speed.
Linear aerodynamics over a small range of angles.	Non-linear aerodynamics over a large range of angles.
Well-defined inceptors: stick or yoke, pedals, and throttle.	No commonly accepted inceptor layout.
Well-defined design criteria.	No well-defined military or commercial specs.
Large body of historical data and requirements.	Little historical data and requirements.

Table 10.2 Airship Certification Documents

Document	Description
FAA-P-8110-2	FAA Airship Design Criteria Airworthiness Requirements for the Type Certification of Conventional, Near-Equilibrium, Non–rigid Airships (Not Federal Aviation Regulations)
AC 21.17-1A	FAA Advisory Circular, Type Certification-Airships
FAR Part 21	FAA Federal Aviation Regulations Certification Procedures for Products and Parts
CS-30N	European Aviation Safety Agency Certification Specifications for Normal and/or Commuter Category Airship
CS-30T	European Aviation Safety Agency Certification Specifications for Transport Category Airships

Aviation Regulation (CFAR) like the design criteria for airplanes and helicopters. Instead, FAA-P-8110-2 is a set of requirements intended to provide an equivalent level of safety to that prescribed in FAR 21.17 (b) for special classes of aircraft. The FAA Advisory Circular AC 21.17-1A calls out FAA-P-8110-2 or "Other Airworthiness Criteria" as acceptable criteria for certification of airships under FAR Part 21. More commonly used certification standards for airships are the European Aviation Safety Agency (EASA) standards CS-30N and CS-30T. CS-30T is unique in that it is the only certification specification for a transport category airship. Most existing airships are of the normal commuter category, but recently, a number of new airship and hybrid airships designs have been proposed in the transport category.

Most of the aerodynamic characteristics of the airship used for stability and control analysis are in the form of force and moment coefficients and static and control derivatives as described in Chapter 3. Some additional aerodynamic derivatives used for stability and control analysis are defined in Table 10.3.

Rate derivatives are important for the stability and control of airships, because they have significantly more effect on the overall stability of airships than they do in conventional aircraft. The rates have dimensions of angle per unit time, which means that as the time scale changes, the rate derivatives would change as a function of time dependent parameters such as airspeed. To make the rate derivatives independent of airspeed, they are defined in terms of non-dimensional rate coefficients, $\hat{p}$, $\hat{q}$, and $\hat{r}$.

The acceleration derivatives describe the effect of the force required to accelerate the added mass of air surrounding the airship. Added mass

Table 10.3 Additional Aerodynamic Coefficients and Derivatives for Stability and Control

Non-dimesional rates	Rate derivatives	Acceleration derivatives	
$\hat{p} = p\ Vol^{1/3}/V_\infty$	$C_{L_{\hat{q}}} = \Delta C_L/\Delta\hat{q}$	$C_{L_{\dot{w}}} = \Delta L/\Delta\dot{w}\,\rho\ Vol$	$C_{Y_{\dot{v}}} = \Delta Y/\Delta\dot{v}\,\rho\ Vol$
$\hat{q} = q\ Vol^{1/3}/V_\infty$	$C_{m_{\hat{q}}} = \Delta C_m/\Delta\hat{q}$	$C_{m_{\dot{w}}} = \Delta M/\Delta\dot{w}\,\rho\ Vol^{4/3}$	$C_{n_{\dot{v}}} = \Delta N/\Delta\dot{v}\,\rho\ Vol^{4/3}$
$\hat{r} = r\ Vol^{1/3}/V_\infty$	$C_{n_{\hat{r}}} = \Delta C_n/\Delta\hat{r}$	$C_{L_{\dot{q}}} = \Delta L/\Delta\dot{q}\,\rho\ Vol^{4/3}$	$C_{Y_{\dot{r}}} = \Delta Y/\Delta\dot{r}\,\rho\ Vol^{4/3}$
	$C_{Y_{\hat{r}}} = \Delta C_Y/\Delta\hat{r}$	$C_{m_{\dot{q}}} = \Delta M/\Delta\dot{q}\,\rho\ Vol^{5/3}$	$C_{n_{\dot{r}}} = \Delta N/\Delta\dot{r}\,\rho\ Vol^{5/3}$
	$C_{l_{\hat{p}}} = \Delta C_l/\Delta\hat{p}$	$C_{l_{\dot{p}}} = \Delta l/\Delta\dot{p}\,\rho\ Vol^{5/3}$	

primarily affects the dynamic characteristics of an airship. If added mass is not taken into account the dynamic mode approximations could be off by over 100%. The acceleration derivatives are discussed in more detail in Appendix C on added mass.

Typically aerodynamic data for stability and control analysis comes from wind tunnel tests or non-linear computational fluid dynamics (CFD). Some of the aerodynamic parameters can be estimated based on the geometry of the airship hull and tails. Many of the main aerodynamic parameters are highly non-linear over the range of angles that the airship will see in flight. Thus, linear approximations often will be insufficient to analyze the stability of an airship. Linear CFD codes like potential flow panel code are useful in estimating the dynamic rate and acceleration derivatives.

The Goodyear ZP4K airship is an example of a typical airship that will be used in this chapter for stability and control analysis. The general layout of the ZP4K is shown in Fig. 10.1 along with the some of the major parameters of the airship in Table 10.4 [1].

The initial sizing of the ZP4K tails based on the methods in Chapter 7 result in a horizontal tail volume coefficient of 0.063 and a vertical tail

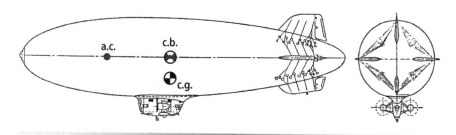

Figure 10.1 Goodyear ZP4K.

Table 10.4 Goodyear ZP4K Airship Parameters

Reference volume	Vol	527,000 ft^3
Length of body	ℓ_B	266.5 ft
Maximum diameter	d	62.1 ft
Mass	m	2317 slug
Roll moment of inertia	I_x	1,127,364 slug ft^2
Pitch moment of inertia	I_y	3,131,566 slug ft^2
Yaw moment of inertia	I_z	3,131,566 slug ft^2
Vertical offset c.g. to c.b.	Δz_{cg}	10 ft
Tail moment arm c.b. to $\bar{c}/4$	ℓ_T	101 ft
Tail moment arm c.b. to hinge line	ℓ_{T_δ}	117.8 ft
Horizontal tail area	S_{HT}	1108 ft^2
Vertical tail area	S_{VT}	888 ft^2
Horizontal tail volume coefficient	C_{HT}	0.063
Vertical tail volume coefficient	C_{VT}	0.0517

volume coefficient of 0.0517. These tail volume coefficients indicate that the ZP4K is well within the typical range tail sizing based on Fig. 7.1, where the ZP4K is shown as point 1.

10.2 Lateral-Directional Stability and Control

The lateral-directional stability of an airship is determined by the interaction between the lateral and directional motion of the airship. Lateral motion is a sideways translation of the airship, and directional motion is rotation around the vertical or yaw axis of the airship.

The lateral motion of an airship is caused by the side force resulting from the sideslip angle of the airship. When an airship translates sideways, the resulting sideslip angle creates a side force in the opposite direction of the translation. This side force creates a stable system where small disturbances in sideslip angle will decrease over time and not cause significant motion of the airship.

The directional motion of an airship is caused by the yaw moment resulting from rotation around the yaw axis. Typically, airships are directionally unstable, because the aerodynamic center is forward of the center of gravity. The yaw moment resulting from a small disturbance in sideslip angle will tend to increase that sideslip angle.

Lateral and directional motions are both influenced by the sideslip angle and the yaw rate of the airship. To determine the overall stability of the lateral-directional system, the contributions of these the two characteristics must be considered together.

The conditions for a steady state turn of an airship are when the yaw rate, r, and sideslip angle, β, are both constant, or when the yaw acceleration, $\dot{r}$, and rate of change of sideslip, $\dot{\beta}$, are both zero as shown in Eqs. (10.1) and (10.2). For a simple yaw condition, the yaw acceleration is simplified as the yaw moment divided by the yaw moment of inertia. The aerodynamic yaw moment is written in terms of the static yaw moment coefficient, which is a function of sideslip angle and the dynamic yaw moment yaw rate derivative.

$$\dot{r} \approx \frac{N}{I_z} = \frac{(C_n + C_{n_{\hat{r}}}\hat{r})\, q_{dyn} Vol^{2/3} l_B}{I_z} = 0 \qquad (10.1)$$

The sideslip angle rate is simplified as the body axis sideways acceleration divided by the true airspeed. The sideways acceleration is equal to the side force divided by mass plus a Coriollis term due to the rotation of the body axis relative to an inertial axis. The side force is written in terms of the static side force coefficient and the dynamic side force yaw rate derivative.

$$\dot{\beta} \approx \frac{\dot{v}}{V_\infty} = \frac{\frac{Y}{m} - V_\infty r}{V_\infty} = \frac{(C_Y + C_{Y_{\hat{r}}}\hat{r})\, q_{dyn} Vol^{2/3}}{m V_\infty} - r = 0 \qquad (10.2)$$

By solving the equations above for the static yaw moment and side force coefficients, we can see in Eqs. (10.3) and (10.4) that for a steady state turn, both coefficients are proportional to yaw rate.

$$C_n = -C_{n_{\hat{r}}}\hat{r} \qquad (10.3)$$

$$C_Y = \frac{m V_\infty r}{q_{dyn} Vol^{2/3}} - C_{Y_{\hat{r}}}\hat{r} \qquad (10.4)$$

By taking the ratio of C_n to C_Y, the yaw rates cancel out producing a constant critical ratio of C_n/C_Y for a steady state turn as shown in Eq. (10.5).

$$\frac{C_n}{C_Y} = \frac{-C_{n_{\hat{r}}}}{\dfrac{2m}{\rho\, Vol} - C_{Y_{\hat{r}}}} \qquad (10.5)$$

Note that for a neutrally buoyant airship, the mass is equal to the mass of displaced air, which is equal to the air density times the displaced volume, so the C_n/C_Y ratio is simplified further in Eq. (10.6).

$$\frac{C_n}{C_Y} = \frac{-C_{n_{\hat{r}}}}{2 - C_{Y_{\hat{r}}}} \qquad (10.6)$$

If we plot this C_n/C_Y ratio on a graph along with the actual static C_n and C_Y coefficients over a range of sideslip angles, the yaw stability characteristics of the airship can be determined. The critical C_n/C_Y line represents a steady state turn at increasing yaw rates. Where the static C_n and C_Y coefficients intersect the critical line the airship will be in a steady state turn. Above the critical line, the airship is unstable, so the sideslip angle and yaw rate will increase until it gets to the critical line. Below the critical line, the airship is stable, so the side-slip angle and yaw rate will decrease until it gets to the critical line.

Figure 10.2 shows the static C_n and C_Y coefficients for the Goodyear ZP4K airship. The sideslip angle points on this graph actually represent negative sideslip angles because the side force is positive.

For sideslip angles from 0 deg to 5 deg, the static yaw moment coefficient is above the critical line, which means that the turn rate and sideslip angle will increase. Past 5-deg sideslip, the static yaw moment is below the critical line, so the turn rate will slow and the airship will go to a lower side-slip angle. The airship will naturally reach equilibrium at the 5-deg point and stay in a steady state turn in those conditions.

Lines have been added to the graph representing the static C_n and C_Y coefficients with the rudder deflected. With a positive 30-deg rudder deflection, the equilibrium point moves out past 10-deg sideslip and crosses the critical

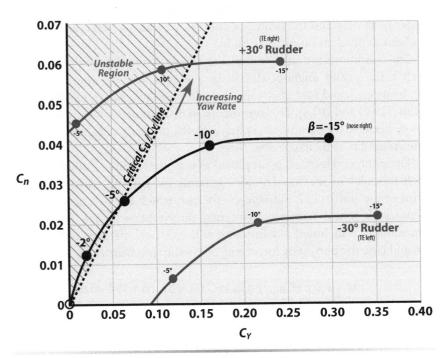

Figure 10.2 Lateral-directional stability graph (ZP4K).

line at a higher yaw rate. This point represents the maximum achievable turn rate with that rudder deflection. With a negative 30-deg rudder deflection, the static C_n vs C_Y line is below the critical line in all cases. This means that the airship has enough control power to return to zero sideslip from all conditions. If the critical line crossed the negative 30-deg rudder line, then there would be some conditions under which the airship would be unable to stop a turn even with 30 deg of rudder deflection in the opposite direction.

The vertical tails and rudder area can be sized based on criteria from this graph: achievable turn rate, ability to stop a turn under all conditions, and sideslip angle for steady state turn. Increasing the size of the tails will have the effect of decreasing the static slope of C_n vs C_Y by moving the aerodynamic center aft. It will also increase the slope of the critical line by increasing the magnitude of the dynamic yaw rate coefficients.

10.3 Longitudinal Stability and Control

The longitudinal stability of an airship is determined by the interaction between the vertical heave and pitch rotation motion of the airship. The longitudinal stability is analyzed using methods similar to those used in yaw but with two additional factors: weight and pendulum stability. Since an airship is not necessarily neutrally buoyant, the static lift is not always zero for an equilibrium condition. Also, the vertical distance between the center of gravity and the center of buoyancy creates a pendulum stability moment dependent on pitch attitude.

The conditions for an instantaneous steady state pitch up are when the pitch rate, q, and angle-of-attack, α, are constant, or when the pitch acceleration, $\dot{q}$, and rate of change of angle-of-attack, $\dot{\alpha}$, are zero as shown in Eqs. (10.7) and (10.8). This represents an instantaneous steady state condition, because with a non-zero pitch rate, the pitch attitude, and thus pendulum stability will change over time.

The pitch acceleration is simplified as the pitch moment divided by the pitch moment of inertia. The aerodynamic pitch moment is written in terms of the static pitch moment coefficient, which is a function of angle-of-attack, and the dynamic pitch moment pitch rate derivative. The pendulum stability is included as a function of the vertical distance between the c.b. and c.g., the buoyancy force, and the airship pitch attitude.

$$\dot{q} \approx \frac{M}{I_y} = \frac{(C_m + C_{m_{\hat{q}}}\hat{q})\, q_{dyn} Vol^{2/3} l_B - \Delta z_{cg}\rho g\, Vol\, \sin\theta}{I_y} = 0 \quad (10.7)$$

The angle-of-attack rate is simplified as the body axis vertical acceleration divided by the true airspeed. The vertical acceleration is equal to the vertical force divided by mass plus a Coriollis term due to the rotation of the body axis relative to an inertial axis. The vertical force is written in terms of the static lift coefficient, the dynamic lift pitch rate derivative, gravitational force, and buoyancy force.

$$\dot{a} \approx \frac{\dot{w}}{V_\infty} = \frac{\dfrac{-L_{aero} + F_g - L_{buoy}}{m} + V_\infty q}{V_\infty}$$

$$= \frac{-(C_L + C_{L_{\hat{q}}}\hat{q})q_{dyn}Vol^{\frac{2}{3}} + (m - \rho\, Vol)g}{mV_\infty} + q = 0 \qquad (10.8)$$

In the pitch case, when we solve the equations above for the static pitch moment and lift coefficients shown in Eqs. (10.9) and (10.10), they are not directly proportional to pitch rate, but both lift and pitch moment have an offset at zero pitch rate due to the weight and pendulum stability.

$$C_m = \frac{\Delta z_{cg}\, \rho\, g\, Vol^{\frac{1}{3}} \sin\theta}{q_{dyn}l_B} - C_{m_{\hat{q}}}\hat{q} \qquad (10.9)$$

$$C_L = \frac{qmV_\infty + (m - \rho\, Vol)g}{q_{dyn}\, Vol^{\frac{2}{3}}} - C_{L_{\hat{q}}}\hat{q} \qquad (10.10)$$

If we plot the critical C_m vs C_L line on a graph along with the actual static C_m and C_L coefficients over a range of angles-of-attack, we can determine the pitch stability and static trim conditions for the airship. Again the intersection between the static C_m vs C_L line and the critical line is an instantaneous steady state pitch rate. For this point to be a trim condition for the airship, it must occur at a weight condition and pitch attitude so that it is at the zero pitch rate point.

Figure 10.3 shows the static C_m and C_L coefficients for the Goodyear ZP4K airship. The weight and pendulum stability effects were calculated assuming the airship is at 30 knots true airspeed, 15-deg pitch attitude, with a 10 ft vertical distance between the c.g. and c.b., and at a buoyancy ratio of 95%. We can see from this graph that the critical line has been shifted so that the zero rate point is at about 0.1 C_L. This is a result of the static lift necessary to overcome the additional net weight of the airship. The zero rate point has also been shifted to about 0.02 C_m by the pendulum moment resulting from the 15-deg pitch attitude.

The critical line intersects the static C_m vs C_L line at the zero pitch rate point and about 7-deg angle-of-attack. Since it is at zero pitch rate, this is the static trim position for the airship with no elevator deflection. The slope of the static C_m vs C_L line is less than the slope of the critical line, so the airship will be statically stable at this point. Note that changing the control surface deflection would change the trim pitch attitude and angle-of-attack.

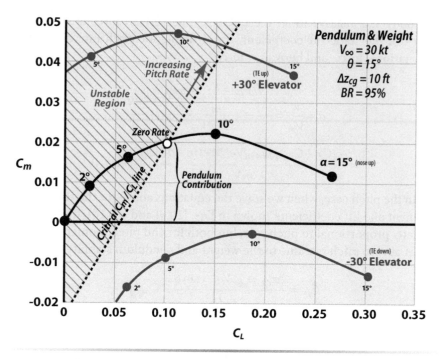

Figure 10.3 Longitudinal stability graph—ZP4K airship.

The horizontal tails and elevator area can be sized based on criteria from this graph: achievable pitch rate, ability to trim pitch over a range of buoyancy ratios, and achievable pitch attitude and climb angle. Increasing the size of the tails will have the effect of decreasing the slope of C_m vs C_L by moving the aerodynamic center aft, and it will increase the slope of the critical line by increasing the magnitude of the dynamic pitch rate coefficients.

10.4 Parametric Tail Sizing

The tail sizing of an airship can be refined by parametrically estimating the effects of the geometry of the tail surface on aerodynamics. The estimations of the tail surface effectiveness are based on approximations from finite wing theory. The parameters used for these approximations are listed in Table 10.5. These parameters are defined for a pair of opposite tail surfaces including the area between the tails to create a full span surface equivalent to the wing of a conventional aircraft.

The lift curve slope of the tail surface is estimated using a simple formula from finite wing theory based on tail aspect ratio and sweep angle [2].

Table 10.5 Parametric Tail Sizing Parameters

Number of tail pairs	N
Exposed tail area	S_T
Total tail area	$S_{T total}$
Tail moment arm c.b. to tail	ℓ_T
Tail moment arm c.b. to hinge line	ℓ_{T_δ}
Tail span	b_T
Tail aspect ratio	AR_T
Tail sweep angle at maximum thickness	Λ
Tail lift curve slope	$(C_{L\alpha})tail$
Tail dihedral angle	Γ
Reference volume	Vol
Length of body	ℓ_B

$$\left(C_{L_\alpha}\right)_{tail} = \frac{2\pi AR_T}{2+\sqrt{4+AR_T^2\,(1+\tan^2\Delta)}} \tag{10.11}$$

$$AR_T = \frac{b_T^2}{S_{T total}}$$

The tail surface effectiveness is then estimated using the equations below for the contribution of the tails to the aerodynamic derivatives [1].

$$(C_{n\beta})_T = N\frac{(C_{L\alpha})tail\,S_{Ttotal}\,\ell_T\eta_M}{Vol^{2/3}\,\ell_B}\sin^2\Gamma \qquad (C_{Y\beta})_T = -N\frac{(C_{L\alpha})tail\,S_{Ttotal}\,\eta_F}{Vol^{2/3}}\sin^2\Gamma$$

$$(C_{n\hat r})_T = -N\frac{(C_{L\alpha})tail\,S_{Ttotal}\,\ell_T^2}{Vol\,\ell_B}\sin^2\Gamma \qquad (C_{Y\hat r})_T = N\frac{(C_{L\alpha})tail\,S_{Ttotal}\,\ell_T}{Vol}\sin^2\Gamma$$

$$(C_{n\delta_r})_T = N\frac{(C_{L\alpha})tail\,\tau S_T\,\ell_{T\delta}\eta_{M\delta}}{Vol^{2/3}\,\ell_B}\sin\Gamma \qquad (C_{Y\delta_r})_T = -N\frac{(C_{L\alpha})tail\,\tau S_T\,\eta_{F\delta}}{Vol^{2/3}}\sin\Gamma$$

$$(C_{m\alpha})_T = -N\frac{(C_{L\alpha})tail\,S_{Ttotal}\,\ell_T\eta_M}{Vol^{2/3}\,\ell_B}\cos^2\Gamma \qquad (C_{L\alpha})_T = N\frac{(C_{L\alpha})tail\,S_{Ttotal}\,\eta_F}{Vol^{2/3}}\cos^2\Gamma$$

$$(C_{m\hat q})_T = -N\frac{(C_{L\alpha})tail\,S_{Ttotal}\,\ell_T^2}{Vol\,\ell_B}\cos^2\Gamma \qquad (C_{L\hat q})_T = N\frac{(C_{L\alpha})tail\,S_{Ttotal}\,\ell_T}{Vol}\cos^2\Gamma$$

$$(C_{m\delta_e})_T = N\frac{(C_{L\alpha})tail\,\tau S_T\,\ell_{T\delta}\,\eta_{M\delta}}{Vol^{2/3}\,\ell_B}\cos\Gamma \qquad (C_{L\delta_e})_T = -N\frac{(C_{L\alpha})tail\,\tau S_T\eta_{F\delta}}{Vol^{2/3}}\cos\Gamma$$

Table 10.6 Parametric Correction Factors and Approximate Values

Tail moment interference factor	η_M	0.4
Tail force interference factor	η_F	0.5
Control surface moment interference factor	$\eta_{M\delta}$	0.9
Control surface force interference factor	$\eta_{F\delta}$	1.0
Control surface effectiveness factor	τ	0.5

Typically, airships have four tail surfaces, or two pairs of tails. The tail dihedral angle, Γ, is used to distinguish between vertical and horizontal tails. With horizontal tails at 0 deg dihedral, the sine terms make the lateral-directional derivatives zero, and with vertical tails at 90-deg dihedral, the cosine terms make the longitudinal derivatives zero. For x-tails with dihedral angles near 45 deg, each tail surface contributes to both lateral-directional and longitudinal stability.

For the angle-of-attack, sideslip, and rate derivatives, the sine and cosine terms are squared, because the dihedral angle effects both the force component from the tail and the component of airflow hitting the tail. The control surface terms sine and cosine terms are to the first power, because only the force component is effected. This may seem to increase the effectiveness of x-tails, because two tails at 45 deg would produce a multiplier of 1.41. This apparent advantage of x-tails is lost when the control surface deflections are allocated between pitch and yaw control power.

The tail surface effectiveness contains a set of factors, η, to correct for tail body interference, and a factor, τ, to estimate the effectiveness of the control surfaces. These factors are determined experimentally in wind tunnel tests. Approximate values for these correction factors are shown in Table 10.6 for typical airship tails. The control surface effectiveness factor assumes tails with a ratio of control surface area to total tail area of approximately 25%.

The tail surface effectiveness estimates can be used to analyze the effects of a changing the size, location, or planform of tails on the stability of an airship. The tail surface contribution must be combined with the aerodynamics hull of the hull based on wind tunnel tests like the data shown in Chapter 3.

10.5 Solar HALE

We return to the example of Solar HALE from Chapters 5 and 7 to revisit the tail sizing based on stability and control requirements.

Sample Problem 10.1: Parametric Tail Sizing

Table 10.7 lists the requirements that will be used to refine the tail sizing. These requirements are typical of what might be required for high altitude maneuvering of a Solar HALE type airship. The Solar HALE

Table 10.7 Solar HALE Stability and Control Requirements

Turn 360 deg in 1 min at 65,000 ft, 20 kt true airspeed using 50% control surface deflections or less.
Return to zero sideslip angle and zero yaw rate from the stable zero control surface deflection point using 50% control surface deflections or less at all speeds.
Achieve 8 deg/s instantaneous pitch rate at zero pitch attitude, 30 kt true airspeed, neutral buoyancy, using 50% control surface deflections or less.
Climb and descend at 100 ft/min at 65,000 ft, 30 kt true airspeed, from −5% to 5% heaviness, using 50% control surface deflections or less.

example has x-tails, so it is assumed that half of the deflection is allocated to pitch control and half to yaw control.

First, we will analyze the initial tail sizing from historical tail volume coefficients to determine where we stand relative to the requirements. The Solar HALE tail parameters from the initial tail sizing are shown in Table 10.8.

The aerodynamic derivatives for the tail contribution to the airship, shown in Table 10.9, are determined using the parametric tail sizing equations. These derivatives are all per radian.

A set of bare hull aerodynamics obtained from a wind tunnel test are shown in Table 10.10 along with approximate rate derivatives for the bare hull. Because the bare hull is axially symmetric, the lateral-directional and longitudinal coefficients are the same with the sign of sideslip angle changed compared to angle-of-attack.

The combined aerodynamics of the airship are determined by adding the bare hull and tail contributions together. The tail angle-of-attack and

Table 10.8 Solar HALE Initial Tail Sizing

Number of tail pairs	N	2
Exposed tail area	S_T	1392 ft^2
Total tail area	S_{Ttotal}	2529 ft^2
Tail moment arm c.b. to tail $\bar{c}/4$	ℓ_T	100.1
Tail moment arm c.b. to hinge line	ℓ_{T_δ}	120.7 ft
Tail span	b_T	85.5 ft
Tail aspect ratio	AR_T	2.87
Tail sweep angle at maximum thickness	Λ	26.5 deg
Tail lift curve slope (per radian)	$(C_{L_\alpha})_{tail}$	3.13
Tail dihedral angle	Γ	50.4 deg
Reference volume	Vol	1,040,000 ft^3
Length of body	ℓ_B	286 ft
Vertical offset c.g. to c.b.	Δz_{cg}	28.7 ft
Air density at 65,000 ft	ρ	0.000176 slug/ft^3

Table 10.9   Solar HALE Initial Tail Aerodynamics

$(C_{n_\beta})_T$	0.128	$(C_{m_\alpha})_T$	−0.088
$(C_{Y_\beta})_T$	−0.458	$(C_{L_\alpha})_T$	0.314
$(C_{n_{\hat{r}}})_T$	−0.317	$(C_{m_{\hat{q}}})_T$	−0.217
$(C_{Y_{\hat{r}}})_T$	0.906	$(C_{L_{\hat{q}}})_T$	0.620
$(C_{n_{\delta_r}})_T$	0.124	$(C_{m_{\delta_e}})_T$	0.103
$(C_{Y_{\delta_r}})_T$	−0.327	$(C_{L_{\delta_e}})_T$	−0.271

sideslip derivatives are multiplied by the angle-of-attack and sideslip angle, respectively, and the elevator and rudder control derivatives are multiplied by the elevator and rudder angles to create an offset to the bare hull aerodynamics. The combined rate derivatives are used to calculate the slope of the critical line for the stability graphs.

$$C_n = (C_n)_H + (C_{n_\beta})_T \beta + (C_{n_{\delta_r}})_T \delta r$$
$$C_{n_{\hat{r}}} = (C_{n_{\hat{r}}})_H + (C_{n_{\hat{r}}})_T$$
$$C_Y = (C_Y)_H + (C_{Y_\beta})_T \beta + (C_{Y_{\delta_r}})_T \delta r$$
$$C_{Y_{\hat{r}}} = (C_{Y_{\hat{r}}})_H + (C_{n_{\hat{r}}})_T$$
$$C_m = (C_m)_H + (C_{m_\alpha})_T \alpha + (C_{m_{\delta_e}})_T \delta e$$
$$C_{m_{\hat{q}}} = (C_{m_{\hat{q}}})_H + (C_{m_{\hat{q}}})_T$$
$$C_L = (C_L)_H + (C_{L_\alpha})_T \alpha + (C_{L_{\delta_e}})_T \delta e$$
$$C_{L_{\hat{q}}} = (C_{L_{\hat{q}}})_H + (C_{L_{\hat{q}}})_T$$

Table 10.10   Solar HALE Bare Hull Aerodynamics

$\alpha, -\beta$	$(C_m)_H, (C_n)_H$	$(C_L)_H, (C_Y)_H$
0 deg	0.000	0.000
2 deg	0.014	0.003
5 deg	0.035	0.012
10 deg	0.070	0.032
15 deg	0.109	0.044
20 deg	0.134	0.118
25 deg	0.142	0.245
30 deg	0.144	0.384
35 deg	0.139	0.554
$(C_{m_{\hat{q}}})_H, (C_{m_{\hat{r}}})_H$		−0.073
$(C_{L_{\hat{q}}})_H, (C_{Y_{\hat{r}}})_H$		0.024

The resulting lateral-directional and longitudinal stability graphs are shown in Figs. 10.4 and 10.5. The control surfaces are assumed to have a maximum deflection of 30 deg. Since the requirements indicate 50% control surface deflection, lines for 15 deg of deflection are also shown.

The turn rate requirement of 360 deg in 1 min, or 6 deg/s, is displayed on the lateral-directional stability graph as the point on the critical line where the yaw rate is 6 deg/s. The C_n and C_Y values at this point are calculated using Eqs. (10.3) and (10.4).

The turn rate requirement point on Fig. 10.4 is outside of the range of control surface deflections between −15 deg and +15 deg. This means that the airship with the initial tail sizing does not meet the turn rate requirement. The airship has excess control power for meeting the second requirement to return to zero sideslip and yaw rate from the stable zero control surface deflection point. This stable point with zero control surface deflection occurs at about −18 deg of sideslip. Only a few degrees of negative control surface defection are required to return the airship to zero sideslip and zero yaw rate. This means that the vertical tail area could be decreased to take advantage of the natural instability of the airship and meet the turn rate requirement while still meeting the requirement to return to straight forward flight.

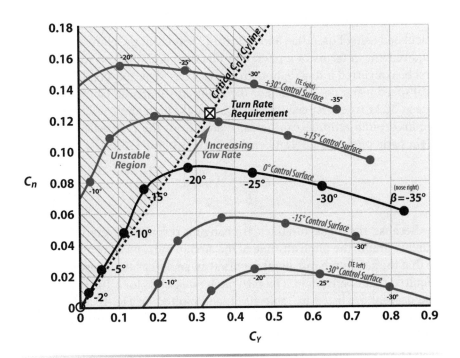

Figure 10.4   Solar HALE initial lateral-directional stability graph.

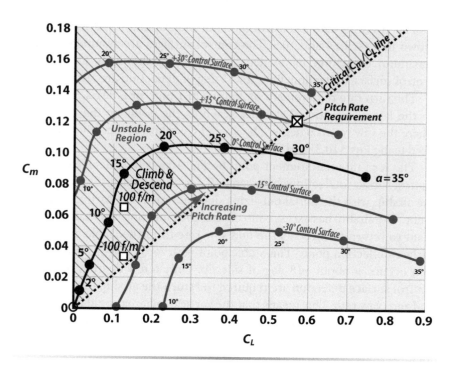

Figure 10.5   Solar HALE initial longitudinal stability graph.

For the longitudinal stability graph, the pitch rate requirement point is calculated using Eqs. (10.9) and (10.10). To represent the climb and descent requirement in terms of a pitch attitude, the climb or descent rates must first be converted into a flight path angle, γ, based on the airspeed and vertical speed using Eq. (10.12). For the required 100 ft/min climb or descent rate at 30 kt true airspeed, the resulting flight path angle is ±1.9 deg. The required pitch attitude can be calculated from the relationship between flight path, pitch attitude, and angle-of-attack in Eq. (10.13).

$$\gamma = \sin^{-1}\left(\dot{h}/V_\infty\right) \tag{10.12}$$
$$\theta = \alpha + \gamma \tag{10.13}$$

Because the angle-of-attack is changing throughout the longitudinal stability graph, the pitch attitude of the airship must be adjusted iteratively so that the zero pitch rate point occurs at an angle-of-attack resulting in the required flight path angle. The zero pitch rate points for a 100 ft/min climb and descent are listed in Table 10.11, and plotted on Fig. 10.5. Because of the symmetry of the airship around neutral buoyancy, the heavy cases with a climb rate are equivalent to light cases with a descent rate, so only the heavy cases are plotted on the graph.

Table 10.11 Solar HALE Initial Climb and Descent Zero Pitch Rate Points

Vertical speed $\dot{h}$	$(F_g - L_{buoy})/L_{buoy}$	γ	α	θ	δe
100 ft/min	5%	1.9°	12.6°	14.5°	−4.1°
−100 ft/min	5%	−1.9°	9.2°	7.3°	−10.1°
100 ft/min	−5%	1.9°	−9.2°	−7.3°	10.1°
−100 ft/min	−5%	−1.9°	−12.6°	−14.5°	4.1°

The zero pitch rate points show that the initial tail sizing meets the climb and descent rate requirement, because at most 10.1 deg of control surface deflection are required to maintain the climb or descent. This means that, like the lateral-directional sizing, the longitudinal tail size could be reduced.

To find a refined tail size that meets the requirements, the exposed tail area is scaled, while maintaining the same span between the tail surfaces. A scale factor of 80% relative to the initial sizing was found to meet all of the requirements. The tail dihedral angle was kept constant at 50.4 deg, however, this parameter could also be adjusted to balance the lateral-directional and longitudinal stability.

A comparison between the geometry of the initial tail size and the new tail size is shown in Fig. 10.6. The updated tail sized parameters are listed in Table 10.12. Since the span between the two tail surfaces has been kept constant, the aspect ratio and lift curve slope of the new tails are greater.

The new lateral-directional stability graph in Fig. 10.7 shows that the turn rate requirement is now met using less than 15 deg of control surface deflection. The critical line is now closer to the −15-deg deflection line at

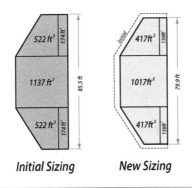

Figure 10.6 Solar HALE reÿned tail size.

Table 10.12 Solar HALE New Tail Sizing Parameters

Exposed tail area	S_T	1113 ft^2
Total tail area	S_{Ttotal}	2130 ft^2
Tail span	b_T	79.9 ft
Tail aspect ratio	AR_T	3.00
Tail lift curve slope	$(C_{L\alpha})_{tail}$	3.19

around −15 deg of sideslip, but it still remains above the line, so the airship can still return to forward flight from the stable zero control surface deflection point.

The new longitudinal stability graph in Fig. 10.8 and climb and descent zero rate points in Table 10.13 show that the climb and descent requirement can be met using 13.9 deg of control surface deflection. The pitch rate requirement is also met using less than 15 deg of control surface deflection.

Then new tail sizing with 80% exposed area relative to the initial tail sizing now meets all of the requirements listed in Table 10.7. Typically, these results obtained from parametric tail sizing would be tested with a

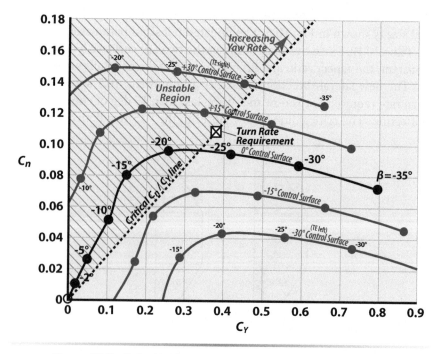

Figure 10.7 Solar HALE new lateral-directional stability graph.

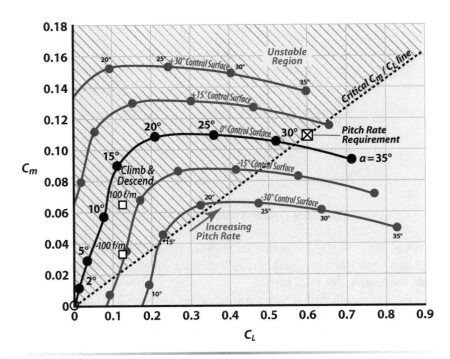

Figure 10.8 Solar HALE new longitudinal stability graph.

wind tunnel model to verify the aerodynamic coefficients before determining a final tail size.

10.6 Roll Stability and Control

The roll stability of an airship is dominated by the roll pendulum effect due to the vertical distance between the center of gravity and the center of buoyancy. Unlike airplanes, the lateral-directional motion is not coupled with the roll axis, so rolling an airship will not make it turn. As illustrated in Fig. 10.9, when an airplane is rolled, a component of the lift generated by the wing is directed to the side and causes the aircraft to turn. For an

Table 10.13 Solar HALE New Climb and Descent Zero Pitch Rate Points

Vertical speed $\dot{h}$	$(F_g - L_{buoy})/L_{buoy}$	γ	α	θ	δe
100 ft/min	5%	1.9°	12.8°	14.7°	−7.3°
−100 ft/min	5%	−1.9°	9.4°	7.5°	−13.9°
100 ft/min	−5%	1.9°	−9.4°	−7.5°	13.9°
−100 ft/min	−5%	−1.9°	−12.8°	−14.7°	7.3°

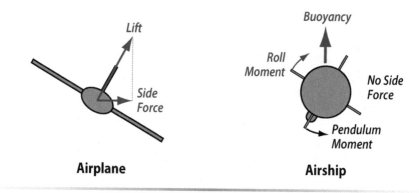

Figure 10.9   Aircraft and airship roll comparison.

airship, most of the lift is due to buoyancy, which will remain pointing upward even when the airship is rolled. Since no side force is created by a roll, the airship does not turn.

To turn an airship, a side force must be generated by applying a yaw moment to the airship to create a sideslip angle as shown in Fig. 10.10. This side force also changes the direction of the net force on the airship. The rolling moment produced by pendulum stability will align the center of gravity in the direction opposite the net force. So, although applying a roll moment to an airship will not cause it to turn, applying a yaw moment to an airship will cause it to roll.

The ratio of the horizontal side force to the gravitational force is the radial load factor, n_r. This ratio determines the turn rate, $\dot{\psi}$, of the airship,

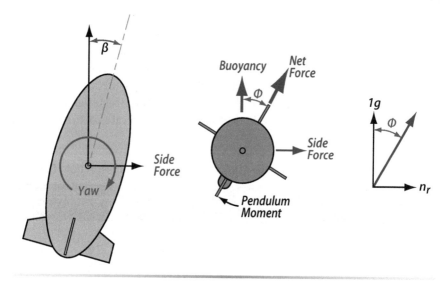

Figure 10.10   Airship turning.

depending on airspeed shown in Eq. (10.14). The radial load factor is equal to the tangent of the bank angle that the airship will achieve due to pendulum stability.

$$n_r = \frac{\dot{\psi}\, V_\infty}{g} = \tan\phi \tag{10.14}$$

Most airships have little or no control power in the roll axis, because it can not be used to turn the airship, and because the pendulum stability keeps the roll axis upright. Typically, no tail or control surface sizing is required for roll axis stability or control power.

10.7 Dynamic Characteristics of Airships

The dynamics of airships can be estimated using simple approximations of the pitch, roll, and yaw modes in the form of state-space equations similar to the dynamic mode approximations for airplanes [3]. State-space equations are differential equations in the linear algebra form show, where x is the vector of states, $\dot{x}$ is the rate of change of the states, u is the input vector, y is the output vector, and A, B, C, and D are matrices. For this analysis, the input and output vectors are not needed since we are primarily interested in the open-loop modes of the system. The input and output equations can be used to define inputs and outputs of the system for analyzing closed-loop control below.

$$\dot{x} = Ax + Bu$$
$$y = Cx + Du$$

The A and B matrices are determined by solving for the translational and rotational accelerations of the airship in terms of the states using equations of motion that include added mass shown in Eq. (10.15). V and ω are the translational and rotational rate vectors, $\dot{V}$ and $\dot{\omega}$ are the translational and rotational acceleration vectors, $\bar{M}$ is the vehicle mass matrix, $\bar{M}_a$ is the air mass matrix, F and M are the forces and moments on the airship, m is the mass, and J is the moment of inertia matrix. For each mode approximation, a subset of the states is expanded in terms of the aerodynamic stability derivatives.

$$\begin{bmatrix} \dot{V} \\ \dot{\omega} \end{bmatrix} = (\bar{M} + \bar{M}_a)^{-1} \begin{bmatrix} F - \omega \times mV \\ M - \omega \times J\omega \end{bmatrix} \tag{10.15}$$

The modes of the state-space system are found using the eigenvalues of the A matrix. The eigenvalues are defined as the values of λ such that the determinant of $A - \lambda I$ is equal to zero. For a simple 2×2 A matrix the eigenvalues can be found by solving the quadratic formula using the individual elements of the A matrix as shown in Eq. (10.16). For larger A matrices, the eigenvalues can best be found using computer algorithms.

$$A = \begin{bmatrix} a & b \\ c & d \end{bmatrix}$$

$$\lambda = \frac{(a+d)}{2} \pm \frac{\sqrt{(a-d)^2 + 4bc}}{2} \tag{10.16}$$

The resulting eigenvalues will be either real numbers or pairs of complex conjugates. The natural frequency, ω_n, and damping ratio, ζ, of the mode represented by the eigenvalues are determined from the real and imaginary components using the formulas below. Eigenvalues with positive real components represent unstable modes, and those with negative real components represent stable modes. An important measure for unstable modes is the time to double, T_2, which is the time it will take for a small disturbance to double in amplitude due to the instability of the airship.

$$\lambda = \eta \pm \omega i$$

$$\omega_n = \sqrt{\eta^2 + \omega^2} \qquad \zeta = -\frac{\eta}{\omega_n} \qquad T_2 = \frac{\ln 2}{\eta}$$

10.7.1 Lateral-Directional Mode Approximation

The lateral-directional mode approximation defines the state vector as the body axis side velocity, v, and the body axis yaw rate r. The input to the system is the rudder deflection angle, δ_r. The A and B matrices for the lateral-directional modes are determined using Eqs. (10.17 a–d).

$$\begin{bmatrix} \dot{v} \\ \dot{r} \end{bmatrix} = A \begin{bmatrix} v \\ r \end{bmatrix} + B \begin{bmatrix} \delta_r \end{bmatrix} \tag{10.17a}$$

$$A = (\bar{M} + \bar{M}_a)^{-1} \begin{bmatrix} C_{Y_\beta} \dfrac{q_{dyn} Vol^{2/3}}{V_\infty} & C_{Y_{\dot{r}}} \dfrac{q_{dyn} Vol}{V_\infty} - m V_\infty \\ C_{n_\beta} \dfrac{q_{dyn} Vol^{2/3} \ell_B}{V_\infty} & C_{n_{\dot{r}}} \dfrac{q_{dyn} Vol \ell_B}{V_\infty} \end{bmatrix} \tag{10.17b}$$

$$B = (\bar{M} + \bar{M}_a)^{-1} \begin{bmatrix} C_{Y_{\delta_r}} q_{dyn} Vol^{2/3} \\ C_{n_{\delta_r}} q_{dyn} Vol^{2/3} \ell_B \end{bmatrix} \tag{10.17c}$$

$$(\bar{M} + \bar{M}_a)^{-1} = \begin{bmatrix} m + C_{Y_v}\rho \, Vol & C_{Y_r}\rho \, Vol^{4/3} \\ C_{n_v}\rho \, Vol^{4/3} & I_z + C_{n_r}\rho \, Vol^{5/3} \end{bmatrix} \tag{10.17d}$$

10.7.2 Roll Mode Approximation

The roll mode approximation defines the state vector as the body axis roll rate, p, and the bank angle, ϕ. The A matrix for the roll modes are determined using Eqs. (10.18 a–b).

$$\begin{bmatrix} \dot{p} \\ \dot{\phi} \end{bmatrix} = A \begin{bmatrix} P \\ \phi \end{bmatrix} \tag{10.18a}$$

$$A = \begin{bmatrix} (I_x + C_{\ell_p}\rho \, Vol^{5/3})^{-1} & \left[C_{\ell_p} \dfrac{q_{dyn} \, Vol \, \ell_B}{V_\infty} & -\rho g \, Vol \Delta z_{cg} \right] \\ 1 & 0 \end{bmatrix} \tag{10.18b}$$

10.7.3 Longitudinal Mode Approximation

The longitudinal mode approximation defines the state vector as the body axis vertical velocity, w, the body axis pitch rate, q, and the pitch attitude, θ. The input to the system is the elevator deflection angle, δ_e. The A and B matrices for the longitudinal modes are determined using Eqs. (10.19 a–d).

$$\begin{bmatrix} \dot{w} \\ \dot{q} \\ \dot{\theta} \end{bmatrix} = A \begin{bmatrix} w \\ q \\ \theta \end{bmatrix} + B[\delta_e] \tag{10.19a}$$

$$A = \begin{bmatrix} (\bar{M} + \bar{M}_a)^{-1} & \begin{bmatrix} -C_{L_\alpha} \dfrac{q_{dyn} \, Vol^{2/3}}{V_\infty} & -C_{L_{\hat{q}}} \dfrac{q_{dyn} \, Vol}{V_\infty} + mV_\infty & 0 \\ C_{m_\alpha} \dfrac{q_{dyn} \, Vol^{2/3} \ell_B}{V_\infty} & C_{m_{\hat{q}}} \dfrac{q_{dyn} \, Vol \, \ell_B}{V_\infty} & -\rho g \, Vol \, \Delta z_{cg} \end{bmatrix} \\ 0 & 1 & 0 \end{bmatrix}$$

$$\tag{10.19b}$$

$$B = (\bar{M} + \bar{M}_a)^{-1} \begin{bmatrix} -C_{L_{\delta_e}} q_{dyn} \, Vol^{2/3} \\ C_{m_{\delta_e}} q_{dyn} \, Vol^{2/3} \ell_B \end{bmatrix} \tag{10.19c}$$

Table 10.14 Example Goodyear ZP4K Flight Conditions

Airspeed	V_∞	30 knots (50.6 f/s)
Air density	ρ	0.0023769 slug/ft^3
Dynamic pressure	q_{dyn}	3.047 lb/ft^2

$$\bar{M} + \bar{M}_a = \begin{bmatrix} m + C_{L_{\dot{w}}}\,\rho\,Vol & C_{L_{\dot{q}}}\,\rho\,Vol^{4/3} \\ C_{m_{\dot{w}}}\,\rho\,Vol^{4/3} & I_y + C_{m_{\dot{q}}}\,\rho\,Vol^{5/3} \end{bmatrix} \qquad (10.19d)$$

Table 10.14 shows the flight conditions for the Goodyear ZP4K airship used in the dynamic stability analysis. Table 10.15 shows approximate values for the stability derivatives of the Goodyear ZP4K around zero angle-of-attack and zero sideslip angle.

The resulting modes from the dynamic approximations using the Goodyear ZP4K example values are listed in Table 10.16 and plotted in Fig. 10.11.

The lateral-directional modes show the coupling between the unstable yaw motion and the stable sideslip motion resulting in two first-order modes, one stable and one unstable. This agrees with the static analysis, which shows that the airship should be slightly unstable directionally around zero sideslip angle. The time to double for this unstable mode is approximately 32 sec. As a rule of thumb, the time to double of an unstable

Table 10.15 Example Goodyear ZP4K Stability Derivatives

Yaw		Roll		Pitch	
C_{n_β}	−0.347	C_{ℓ_β}	−0.114	C_{m_α}	0.254
C_{Y_β}	−0.559	C_{ℓ_p}	0.017	C_{L_α}	0.690
$C_{n_{\dot{r}}}$	−0.394			$C_{m_{\dot{q}}}$	−0.394
$C_{Y_{\dot{r}}}$	1.17			$C_{L_{\dot{q}}}$	1.17
C_{n_β}	0.125			C_{m_q}	0.125
C_{n_β}	0.936			$C_{L_{\dot{w}}}$	0.936
C_{n_β}	−0.027			$C_{m_{\dot{w}}}$	−0.027
C_{n_β}	−0.090			C_{L_q}	−0.090
C_{n_β}	0.00076			$C_{m_{\delta_e}}$	0.00085
C_{n_β}	−0.002			$C_{L_{\delta_e}}$	−0.003

Table 10.16 Goodyear ZP4K Dynamic Modes

Lateral-directional modes		
λ	ω_n	ζ
0.0217	0.0217	−1
−0.664	0.664	1
Roll modes		
λ	ω_n	ζ
−0.304 + 0.403i	0.504	0.602
−0.304 − 0.403i	0.504	0.602
Longitudinal modes		
λ	ω_n	ζ
−0.0549 + 0.0832i	0.0996	0.551
−0.0549 − 0.0832i	0.0996	0.551
−0.452	0.452	1

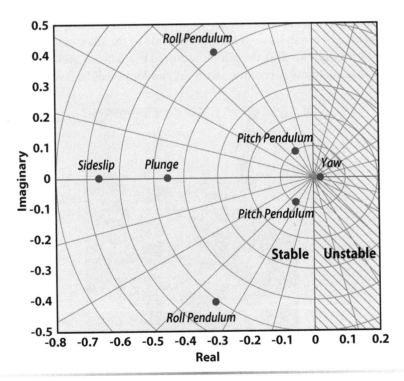

Figure 10.11 Example Goodyear ZP4K dynamic modes.

mode in an aircraft controlled by a pilot should not be less than 12 sec. The roll modes show the effect of the roll pendulum producing a pair of second-order modes. The longitudinal modes show a pair of second-order modes due to the pitch pendulum effect as well as a first-order mode resulting from the stable vertical plunge motion. These modes are typical of the dynamics seen in most airships. These mode approximations can be used as a basis for analyzing the handling qualities of the airship and the stability of closed-loop flight control systems.

10.8 Airship Control Methods

Simple airships like the Goodyear airships are controlled primarily by elevator and rudder control surfaces on the tails and by the engine throttles. The elevator and rudder are controlled manually by the pilot through a system of cables and pulleys with no boost system. The elevator is usually controlled by a large wheel mounted to side of the pilot seat to give the pilot enough mechanical advantage to overcome the aerodynamic hinge moment of the elevator. The rudder is controlled by pedals similar to those on airplanes, because the pilots legs are strong enough to overcome the rudder hinge moments. The throttles are controlled by levers or knobs similar to those in airplanes. Figure 10.12 shows a typical airship cockpit. The elevator control wheel can be seen on the right side of the pilot seat.

To fly a simple airship, the pilot is required to manually stabilize the yaw axis by making many corrections with the rudder pedals to keep the airship

Figure 10.12 Typical airship pilot controls.

pointed at the desired heading. Turns are achieved by applying a rudder input to allow the natural instability of the airship build up a steady sideslip angle and yaw rate. Once the desired heading is achieved, a rudder input is made in the opposite direction to arrest the yaw rate. The pitch axis is controlled by adjusting the elevator wheel to achieve the pitch attitude necessary to control the airships altitude and vertical speed. Due to the buoyant lift, airships can climb and descend at considerably steeper angles than airplanes. Because the buoyant lift is independent of the attitude and airspeed, the airship can point up or down at large pitch angles, provided that it has enough pitch control power to overcome the pendulum stability.

The buoyancy ratio of an airship will change throughout a flight as fuel quantity and other conditions change. The relative temperature between the lifting gas and the outside air, or superheat, determines the pressure of the lifting gas. Airships use ballonets, or air chambers within the envelope, to control the hull pressure. The pressure of the lifting gas determines the weight of air in the ballonets and thus the weight of the airship. If the airship is heavier than air, then a positive angle-of-attack and forward speed are needed to generate lift to overcome the excess weight. If the airship is lighter than air, then a negative angle-of-attack and forward speed are needed to hold the airship down.

The requirement of forward speed makes the process of takeoff and landing more complicated when the airship is heavy or light. Ground operations for most airships require the use of a well-trained ground crew line that is shown in Fig. 10.13. A takeoff under heavy conditions requires a short ground roll to generate enough speed to lift the airship off the ground. A short ground roll is also necessary on landing, because the airship must land with forward velocity. Typically airship do not takeoff lighter than air, because they are ballasted to be neutrally buoyant before takeoff. However, due to fuel burned during a flight and changes in air temperature, airships will often need to land light. This is a complicated process that requires

Figure 10.13 Airship ground operations with ground crew.

coordination between the pilot and ground crew. The ground crew's job is to catch and hold ropes that hang off of the nose of the airship and to hold the airship in place while the pilot uses the engines to force the airship to the ground. Other members of the ground crew will then load ballast onto the airship until it is heavier than air and able to stay on the ground by itself.

Airships are typically kept in hangars or moored by the nose to a mast allowing the airship to swing around and point into the winds while it is on the ground. Because of the buoyancy of airships and because they are greatly affected by winds and atmospheric conditions, an airship cannot be parked like an airplane.

Newer airships such as the Zeppelin NT, shown in Fig. 10.14, take advantage of engine thrust vectoring and fly-by-wire flight control systems to overcome many of the limitations of earlier airships. The main thrusters on the Zeppelin NT are capable of vectoring in the pitch axis to produce a vertical thrust component to overcome the weight of the airship during takeoff and landing. Note that the thrusters and gondola on the Zeppelin NT are placed toward the front of the envelope near the aerodynamic center so that the weight and vertical thrust balance the aerodynamic lift when the airship takes off or lands. A system of tail thrusters, shown in Fig. 10.15, is used to control pitch and yaw at low speeds when the aerodynamic control surfaces are not effective. The Zeppelin NT uses a side stick controller along with throttles and vector levers to command the control surfaces and thrusters through a fly-by-wire flight control system.

Even newer airships, like the experimental Lockheed Martin P-791 Hybrid Airship Demonstrator, are controlled completely through thrust vectoring shown in Fig. 10.16. The P-791 airship has four thrusters that are

Figure 10.14   Zeppelin NT.

Figure 10.15 Zeppelin NT tail thrusters.

capable of vectoring ±90 deg in both pitch and yaw, giving each thruster a full hemisphere of vector angles. The engines are mounted inside a gimbal ring that allows for two axis vectoring, and the vector control is provided by electric actuators. The two forward thrusters are mounted on the side of

Figure 10.16 Lockheed Martin P-791 hybrid airship demonstrator.

the airship, allowing them to thrust forward, backward, up, or down. The aft thrusters are attached to a composite wishbone structure, allowing them to thrust forward, left, right, up, or down. P-791 is controlled by a fly-by-wire flight control system. An open-loop flight control mode allows the pilots to control the thrusters through a mixer by commanding forward, vertical, pitch, roll, and yaw forces and moments. A closed-loop flight control mode actively stabilizes the airship by feeding back sensor signals based on command inputs from the pilot. The P-791 cockpit has a side stick controller, yaw pedals, and forward and vertical thrust levers. The P-791 tail surfaces have no control surfaces that can be moved in flight, so the vectoring thrusters provide both thrust for forward flight and control power.

Another method of controlling an airship is the use of the pressurization system as a means of producing control forces and moments as illustrated in Fig. 10.17. By shifting air between two or more ballonets, the center of gravity of the airship can be moved relative to the center of buoyancy to create a moment on the airship. This can be especially useful for trimming an airship in flight by shifting the c. g. forward to overcome the pitch up moment generated by aerodynamic lift.

Like in airplanes, the stability and handling qualities of airships can be improved using closed-loop flight control systems. Typically these systems use fly-by-wire flight controls so that the computers used to send commands to the controls can also be used to run flight control algorithms. For airships that use a combination of aerodynamic control and thrust vectoring, flight control mixer algorithms are usually needed to allocate the forces to achieve the desired forces and moments. The desired forces and moments are calculated from feedback signals from sensors and pilot control inputs. The simplest feedback loops are rate feedback from rate gyros. A pitch rate command system can be used to improve the handling qualities of the airship in the longitudinal axis. In the yaw axis it is often useful to take advantage of the natural roll characteristics by commanding radial load factor, n_r. Since n_r is proportional to turn rate, this can be done with rate

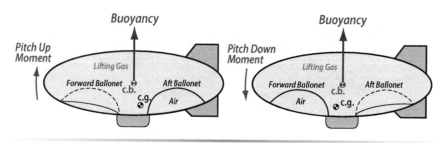

Figure 10.17   Ballonet air shifting to produce pitching moment.

feedback. Since n_r is also related to bank angle, this effectively creates a bank angle control system. If roll moment is available, a roll rate feedback system can be used to add damping to the roll pendulum mode and improve ride quality.

Closed loop flight control systems for airships and conventional aircraft are usually developed using flight simulators to evaluate the handling qualities of the system. Since the main goal of closed-loop flight controls is to improve handling qualities and reduce pilot workload, a real-time pilot-in-the-loop simulation is necessary to get pilot feedback. Typically simulators start out as a simple experimental setups and progress along with the design of the airship cockpit layout and inceptors as a tool for analyzing the effects of design changes on handling qualities. Once a design has reached the flight test phase, the simulation will become an important training tool for pilots and flight test personnel. For an operation vehicle, certified simulations may be created to train pilot and reduce the number of hours a pilot is required to fly the actual vehicle in order to become qualified and maintain proficiency.

References

[1] Ross, S.A. and Liebert, H.R., *LTA Aerodynamics Handbook*, Goodyear Aircraft Corp. Akron, OH, 1954.
[2] Nicolai, L.M., and Carichner, G.E., *Fundamentals of Aircraft and Airship Design, Volume I – Aircraft Design*, AIAA, Reston, VA, 2010.
[3] Nelson, R.C., *Flight Stability and Automatic Control*, McGraw-Hill, 1998.

Chapter 11 / Airship Design

- Review of Conventional Airship Design
- Design Aspects for Aerodynamics, Propulsion, Structures, and Materials
- Airship Performance
- Airship Stability and Control
- Airship Subsystems
- Sample Problem: Design an Axi-Symmetric Airship for a Given Set of Requirements

The A-170 built by American Blimp Company is shown during a remote operation. This workhorse vehicle operates around the world for sightseeing, advertising, and filming sporting events.

Alice: "This is impossible."
Mad Hatter: "Only if you believe it is."
Alice in Wonderland

11.1 Introduction

Although there are other books on airships that might assist a designer, none provide the step-by-step guidance contained in this book. This chapter will discuss how material from previous chapters can be used to design a conventional body-of-revolution airship. We will first discuss general design aspects from each discipline, including aerodynamics, propulsion, structures, and materials. This is followed by a detailed sample problem with a comprehensive summary of the necessary calculations to design an axi-symmetric airship. This example airship design will incorporate traditional aft tails, cabin, crew/passenger car, a nominal gear attached to the gondola, and engines driving propellers. The requirements for the sample problem are based on the known characteristics of the A-170 airship from the American Blimp Company.

Since the mid-1990s there has been a renewed interest in airships due to a new appreciation of their capabilities and the stunning improvements in materials technology as illustrated previously in Fig. 8.2. These new fibers and matrix materials have resulted in new woven and laminated materials that are significantly lighter, more damage tolerant for given loads, and therefore offer a significantly reduced envelope weight.

At the beginning of any new program, requirements and assumptions need to be established. When possible a final design will meet all requirements and be the lightest, or lowest cost, or carry the most payload, among other objectives. However, all requirements cannot be met simultaneously, which makes design compromise necessary. Sometimes the customer prioritizes the requirements, which may also vary over time. Other times the best design is the one that does well on all requirements but is not optimized for any single one. Remember that one of the tasks of the designer is to challenge requirements when they don't make sense or when the best answer becomes highly sensitive to small changes in any requirement.

Most of the tasks associated with a final airship design will be included in the following design process. However, some processes and special systems or components will not be addressed. Specific design issues associated with unique payloads, special materials needs or coatings, actual sensors, etc. will not be individually evaluated either. The resulting airship design at the end of this chapter is an illustration of a typical airship that is in the conceptual phase of its development.

The approach here is to review all design tasks in general terms and then apply them to the actual airship design problem in Sec. 11.12. Design tasks that will be included are definition of speed, cruise and maximum altitudes, envelope sizing, ballonet sizing, envelope pressure, engine cycle, number of engines, propeller size, engine placement, tail sizing, drag estimates, and locations of the a.c., c.b., and c.g. Performance calculations are made using aerodynamic characteristics, propulsion characteristics, and mass properties estimates whose calculations will be explicitly shown.

11.2 Requirements

Initial design efforts should attempt to satisfy all given requirements simultaneously. However, once it becomes obvious that optimizing for one or more requirements is incompatible with optimizing for the others it is incumbent on the engineer to perform sensitivity studies. These will show which of the requirements are most easily relaxed and which are the most sensitive to change. These sensitivity studies compare the variation of one requirement parameter (keeping all other parameters constant) with changes in a fundamental parameter such as range.

Figure 11.1 illustrates a standard way of presenting sensitivity data resulting from trade study results. Usually, one parameter is varied while the others remain constant and the change in the Measure of Merit (MOM), in this case range, is calculated. In this example, the change in range due to varying maximum altitude is slight but changes to propeller efficiency, *FR*, *BSFC*, and drag have modest impacts. However, range is highly sensitive to cruise speed and increased empty weight. Surprisingly, decreases in empty weight do not help since the *BR* is already close to 1.0 and cannot be increased since airships cannot operate when *BR* > 1.0.

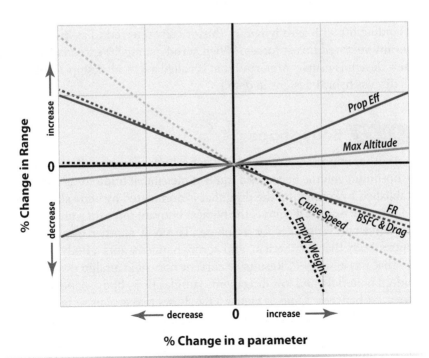

Figure 11.1 Notional sensitivity study results.

11.3 Assumptions and Ground Rules

Assumptions and ground rules should be made with care as they are often the reason for designs having restrictive performance characteristics. Oftentimes large sensitivities can be traced back to unreasonable assumptions or expectations.

11.4 Design Tasks

11.4.1 Body Volume

Envelope or body volume is the single most important design parameter for an airship. It is equivalent to sizing a wing for aircraft. Volume must be closely matched to the basic buoyancy needs but is also affected by how big the ballonets are and the buoyancy ratio, *BR*. For designs with a high altitude requirement the ballonets must be very large, which results in displacing a large amount of lifting gas at sea level where the ballonets are full. Envelope volume would have to be sufficient to create the necessary buoyancy at altitude while also accommodating the lifting gas and fully inflated ballonets at sea level.

Buoyancy ratio is also an important design parameter. The designer selects landing *BR*, which for conventional airships is generally about 95%–98% but generally does not result in a *BR* @ takeoff being less than 85%. This means that the airship is capable of generating at least 2%–5% of its landing lift with aerodynamics (historically referred to as *dynamic lift*) plus any vectored thrust forces. When aerodynamic lift exceeds 10%, then a new design is usually preferred that is called a hybrid airship. Chapter 12 will discuss hybrid designs in detail.

11.4.2 Body Shape

Another important design parameter is body *fineness ratio, FR*. Finding the optimum volume becomes a multi-dimensional trade study since *FR* is established by estimating the drag that is dominated by body skin friction and pressure drags. Of course, the highest buoyant force for a given surface area (envelope weight) is for a sphere. To say it another way, buoyancy wants a body that is spherical, and aerodynamics wants a body of revolution that has a high *FR*. Results of current non-rigid airship designs using modern materials, and low drag (some laminar flow) body shapes generally have a *FR* between 3 and 5. Figure 11.2 shows two groups of body shapes that are made up from ellipsoids and paraboloids, and shows the difference in the aft end shape. How pointed the aft end can be is determined by how much weight from tails or engines is attached there. If the aft end is pointed

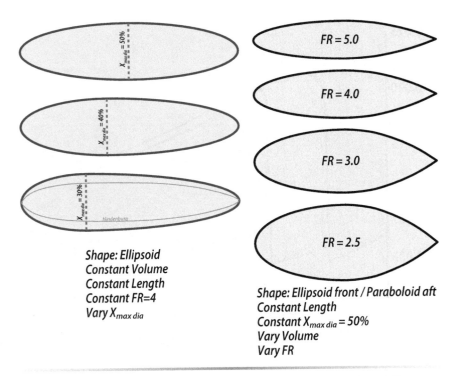

Figure 11.2 Comparison of body geometries—bodies of revolution.

then hard structure must be used to create that shape and/or provide an area for attaching tails and/or engines. Only generously rounded aft ends can be stiffened by pressure alone.

Bodies can also be optimized to create large areas of laminar flow. Figure 11.3 shows five different shapes, all of which have been designed for optimum shapes to maintain laminar flow within different Re regimes. If a body has significant laminar flow then its skin friction drag is significantly lower than if the boundary layer were turbulent everywhere. The first body shape in Fig. 11.3 is designed to have laminar flow up to $Re = 3 \times 10^6$. The second body is designed to have laminar flow for Re up to 10×10^6, and so on. See Chapter 3 and Sec. 11.9 for further discussions.

Body size, shape, and fineness ratio for an airship are determined by buoyancy requirements and not by passenger compartment volume, payload volume, or fuel volume. However, there are optimum contours and fineness ratios that can be defined to minimize zero lift drag (C_{D_0}). The relatively slow speed and reasonable FR of airships results in skin friction creating most of the total drag. This turns into an exercise that attempts to find the largest L/D including buoyancy for a given shape. Remember that the sphere is the minimum skin friction drag shape for a given volume (smallest surface area for a given volume). However, the

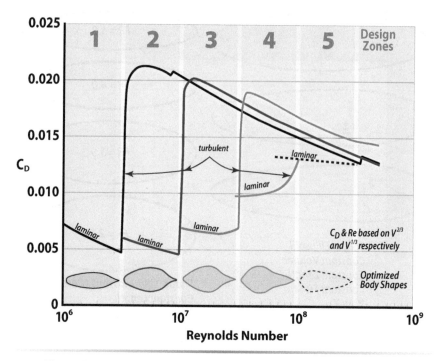

Figure 11.3   Comparison of body contours that are optimized for various *Re* zones.

sphere has significant pressure drag that only increased fineness ratio can reduce. Figure 11.4 shows how the buoyant *L/D* varies with fineness ratio and that there is an optimum near *FR* = 3 over a wide range of airship speeds. It is interesting that the more modern pressure stabilized (non-rigid) airships are close to this optimum fineness ratio and yet the Zeppelins (Hindenburg et al.) in the 1920s–1930s had fineness ratios almost twice as large. Obviously, other design considerations (constant cylindrical sections) resulted in a shape that was not aerodynamically optimal. Early Zeppelins were different because of their rigid structure, which favored structurally efficient cylindrical shapes that also made analyses easier and manufacturing less expensive. This was done at the expense of higher drags.

Instead of repeating equations and figures over again we have created Appendix D, which contains many of the most important equations and figures. Equations for ellipsoid shapes and characteristics are included along with numerous figures with general application throughout the book.

The Bodensee (LZ-120) was the first airship that intentionally designed its body to have lower aerodynamic drag than the previous cylindrical shapes.

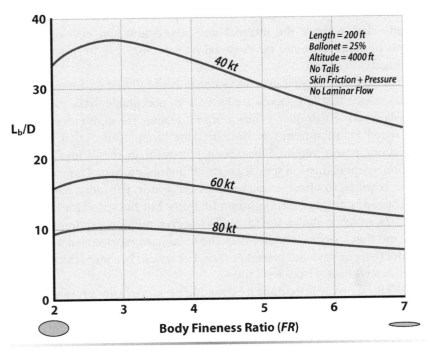

Figure 11.4 Buoyant lift of an ellipsoid body for various *FR* (no tails).

11.4.3 Buoyancy Ratio

Buoyancy ratio is important both at takeoff and landing. At takeoff the *BR* cannot be too low (heaviness too large) or it will be impossible to generate enough aerodynamic lift plus vectored thrust lift for takeoff. There are only two forces that can be generated to overcome heaviness and those are aerodynamic lift, L_{aero}, and vectored thrust. As seen in Chapter 3, bodies of revolution are relatively poor generators of L_{aero} so takeoff heaviness is balanced by small amounts of L_{aero} and vectored thrust. This means that takeoff heaviness, W_{H_0}, is also somewhat small. Typically, takeoff heaviness is no more than 20% of W_G, which means *BR* @ takeoff > 80%.

The *BR* at landing is narrowly restricted as well. When the airship is on the ground or moored to a mast it should have a small amount of heaviness to keep it from being blown around by nominal gusts of wind. Historically, the *BR* for conventional airships at landing is between 95%–98% or 2%–5% heavy.

11.4.4 Speed

Although design effort is always expended to maximize airship speeds, operating at or near maximum speeds always reduces range and endurance. Remember that drag increases with V^2 and power increases with V^3.

This tends to limit an airship's maximum speed to less than 100 kt. Higher speeds also increase the internal design pressure and envelope stress, which requires a heavier envelope fabric (see Sec. 9.13 for a trade study on speed).

The fact that an airship is being considered to meet customer requirements means that low speeds are likely to be acceptable to the user. Lower speeds mean less drag and thus more efficiency. For endurance missions the speed for minimum fuel flow (maximum endurance) that occurs at minimum power required has been discussed in detail in Chapter 4. Station-keeping mission speeds greater than minimum fuel flow are sometimes required to offset strong intermittent winds. For range missions the best speed is not that for minimum fuel flow but the speed for maximum L_{aero}/D, which is shown in Fig. 11.5. However, mission requirements may demand that speeds be higher than the maximum range speed to increase productivity or payload throughput. Higher speeds become desirable when there is a cost associated with time.

While speed is important its dynamic pressure is not an issue for any airship that has low thrust/drag ratios. Nose stiffening is added to handle mooring loads. However, for hybrid airships that have more thrust and generally higher speeds, the dynamic pressure on the nose could require added material thickness or stiffening to prevent local depressions or deformations. The dynamic pressure is mainly responsible for the hull internal pressure. Other speed issues such as flight path upsets resulting in a terminal dive speed do not apply to airships. Historically, battens have been used to stiffen the nose for non-rigid airships with speeds above 50 kt.

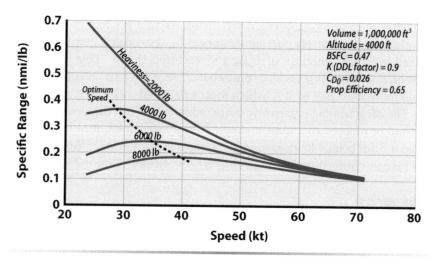

Figure 11.5   Cruise efficiency.

Another method of stiffening the nose is to simply have a higher internal pressure. This, of course, increases the envelope material load, which would require more weight. A recognized design rule is for the internal pressure to be about 20% greater than the maximum expected dynamic pressure, q.

11.4.5 Altitude

There are two flight altitudes that need to be part of the design process. Obviously, the cruise/loiter altitude is where an airship spends a vast majority of its flight time. This altitude is generally low because engine driven propellers are the propulsive units of choice and both propellers and reciprocating engines perform more efficiently at lower altitudes. While efficiency is best near sea level most airships cruise or loiter at 3000–5000 ft density altitude to clear most terrain. Maximum altitude capability is strictly for terrain clearance so flight paths can be direct regardless of enroute mountains.

There are two design philosophies regarding sizing the envelope and its ballonets. Often the maximum altitude determines the size of the ballonet because flights will be flown at this altitude on occasion. However, the envelope can also be sized along with its ballonet using the nominal cruise altitude, which yields a smaller envelope and ballonet. The smaller volume does not generate sufficient buoyant lift for maximum weight but is enough to lift the airship's minimum weight, which is needed to meet FAA regulations. Sometimes maximum altitude is the same as the operational altitude. This occurs whenever high altitudes are needed for Intelligence/Surveillance/Reconnaissance (ISR) missions. Higher altitudes mean larger Fields of Regard (FOR) for onboard sensors.

11.4.6 Ballonet

Earlier, the concept of using a ballonet to maintain constant ΔP across the envelope (which maintains shape for a constant hull volume) was introduced. Ballonets are essential for airships that vary their altitude during flight and do not want to vent their lifting gas (usually helium). For airships and aerostats designed for one altitude to remain at for their entire lives, a ballonet is not necessary and can be eliminated to reduce weight. Based on the data in Table E.1 rigid airships did not have ballonets. Instead of ballonets the hydrogen-filled gas cells were not filled all of the way, allowing for expansion when operating at higher altitudes. Even if there was an occasional overpressure on a gas cell, the loss of some hydrogen was not viewed as critical. Early in the 20th century helium was very expensive compared to hydrogen. In the 21st century helium is 5 to 10 times more expensive than hydrogen.

For non-rigid airships it is the ballonet that is continuously adjusting its volume to the lifting gas density/temperature changes to maintain a constant ΔP across the hull. Modern airships favor the non-rigid airship design that uses a ballonet. A full discussion of ballonet materials and characteristics can be found in Chapter 8.

Airships usually have 2–4 ballonets in a body of revolution airship. This allows ballonets to be spherically (or hemispherically) shaped to minimize their weight. This also allows fore and aft ballonets to be selectively filled for modest control of the c.g. location which can help in reducing longitudinal trim demands and trim drag.

Since non-rigid airships maintain a constant pressure differential (ΔP) across the envelope at all altitudes the ballonets become extremely large (60% to 70% of volume) as operational altitudes become greater than 20,000 ft. See Fig. 9.3 for altitude effect on ballonet sizing. If an airship only ascends once and remains on station for long periods, it is not necessary to have a ballonet. ISIS and Solar HALE are examples of buoyant vehicles without ballonets and whose operational altitudes are above 60,000 ft.

For airships without ballonets launch is more difficult and comes with its own set of problems. During ascent the airship must vent lifting gas so that it does not overpressure the hull as the ambient pressure decreases. Once at altitude all of the extra lifting gas will have vented and only that gas needed for shape and buoyancy remains. However, there is a significant problem with this launch scenario. If the airship has its volume full of lifting gas at sea level it will have a tremendous buoyant force that will quickly accelerate the ascending airship to unsafe or uncontrollable speeds. Something must be added to this launch to make ascent a relatively benign event. This "something" is ballast weight to keep ascent speeds safe. The difficult part is to have the ballast weight deplete as the airship ascends so that ascent speeds are modest and there is no more ballast weight once the airship reaches its operational altitude.

Contrast this with a weather balloon whose simple initial shape easily changes into its final shape as the small amount of lifting gas at launch expands to its final volume at operational altitude. Balloons are discussed in detail in Chapter 13.

11.4.6.1 Ballonet Sizing

One of the important sizing efforts for an airship design is to determine what the maximum volume of the ballonet must be. The fundamental parameter is the maximum operational altitude. It is at this altitude that the lifting gas expansion just fills its envelope volume and the ballonet volume is zero. At sea level conditions the lifting gas has contracted and the ballonet has to exactly fill the evacuated space. This establishes a theoretical minimum ballonet volume (see Fig. 9.3). More precise ballonet sizing should also consider other operational considerations. First, it is not possible to deflate a ballonet completely so there is always some residual air even when deflated. Next, volumes are based on a standard day atmosphere but operations must accommodate flight through pressure regions higher and lower than standard. And lastly, there is the effect of the lifting gas being heated above ambient by the sun (*superheat*).

For estimating purposes Eq. (11.1) gives a theoretical envelope volume based on maximum altitude and the amount of buoyant lift. Equations (11.2a–11.2c) that are shown in Fig. 9.3 give the expressions for estimating the ballonet volume once the envelope volume and maximum altitude are known. Equation (11.2a) is the theoretical minimum. However, further adjustments must be made to reflect actual operations. Other issues include unused ballonet volume (5%), *superheating* the lifting gas by 30°F from the sun ($\Delta Vol\% = [(548.7 \ °R/518.7 \ °R)-1] = 5.8\%$), and flying into a high pressure region ($\Delta Vol = 2\% - 3\%$). Adding up these three increments gives $\Delta Vol\% = 5\% + 5.8\% + 2.5\% \approx 13\%$. So, for sea level standard days a ballonet has to have a volume capability that simultaneously is large enough for the above corrections and is shown as the "nominal" curve on Fig. 9.3 and as Eq. (11.2b). A more conservative approach based on historical design data is presented in Eq. (11.2c). It is recommended that the designer choose a ballonet size between the "nominal" and "conservative" curves on Fig. 9.3.

$$Vol_{env} = L_{buoy} / 0.065 / (\rho_{max \ alt} / \rho_{SL}) \qquad (11.1)$$

$$Vol_{ball} = Vol_{env} \left[1 - (\rho_{max \ alt} / \rho_{SL}) \right] \qquad (minimum) \quad (11.2a)$$

$$Vol_{ball} = Vol_{env} \left[1 - (\rho_{max \ alt} / \rho_{SL}) + 0.13 \right] \qquad (nominal) \quad (11.2b)$$

$$Vol_{ball} = Vol_{env} \left[(\rho_{SL} / \rho_{max \ alt}) - 1 \right] \qquad (conservative) \quad (11.2c)$$

Using these relationships, a specific ballonet material, and a calculated ballonet surface area, the weight of the ballonets can be estimated.

11.4.7 Tail Sizing

Chapters 7 and 10 provide a two-step approach to tail sizing. Results from initial tail sizing in Chapter 7 are good enough for conceptual design efforts. After initial designs are completed, tail sizes can be refined using Chapter 10. Since tails have significant weight (generally in the top 3 with the envelope and propulsion system) and that weight is located aft it is important to minimize tail/fin weight. Controlling the c.g. location is also important for airships just as it is for airplanes.

11.4.8 Propulsion

11.4.8.1 Number of Engines

Determining the number of engines is a very involved trade study and is usually performed with other trade studies during conceptual design phase. Airships offer more opportunity for engine placement than typical aircraft so the engine location trade space tends to be larger. Since engines can be placed most anywhere any number of engines can be accommodated.

Determining the number of engines starts with calculating the total maximum power required based on the largest heaviness and likely maximum flight speeds. Maximum power could also be determined by takeoff requirements. Once the power is known then the number of engines can be established by considering which engines are commercially available off-the-shelf that meet the required performance. Keep in mind that no new engine will be created just for an airship so it is important that a good engine is selected from those already available.

11.4.8.2 Engine Placement

There are three general locations where engines can be placed on an airship. Historically, on rigid airships engines were either placed along the envelope and/or attached to the gondola/cab. Since rigids had structure everywhere engines could easily be mounted in an optimal location. Once rigid designs were replaced by non-rigid designs engines were attached to the cab or hung from the envelope structure above the cab as shown in the bottom pictures in Fig. 1.4. As materials have continued to improve in the last 50 years, it is now possible to efficiently attach engines to the sides of a non-rigid airship as shown in Figs. 8.8 and 9.5.

There is, however, a group of designers who favor mounting engines at the end of airships where the bodies close out. Whether or not this is a good location for an engine requires a very complex engineering analysis and is outside the scope of this text. However, the issues that must be honestly considered include potential aft end flow improvements which can delay separation and lower drag, providing aft end structure to mount an engine (aft engine and structure move the c.g. aft), non-uniform flowfield at the aft propeller face, etc. If an engine is placed at the aft end it must also be of sufficient size to entrain enough flow so that the separated flow region is affected.

Even though the only economically viable airship designs are non-rigid, engines can be installed in most locations. There are many parameters that affect engine placement and they include weight of attachment structure, drag of structure, ease of access for engine service, effect on vehicle drag, availability of inherent structure, weight of fuel lines and/or electrical wiring, impact on airship c.g., propeller diameter, ground clearance during all flight modes, ability to vector thrust, etc. Clearly, there is no one answer for all designs and each design must carefully consider which effects are the most important for meeting requirements.

11.4.8.3 Engine Cycle

Several issues contribute to selecting the proper engine cycle. All engine options for airships are used to drive propellers. As discussed in Chapter 5 turbine engines make no sense for flying at low altitudes although very large airships may require so much power that only a turbine cycle can generate it. Even these large turbines (turboprop) would be integrated with very large propellers. Most airships use internal combustion engines even though turboshaft engines are 2–3 times lower in weight (see Fig. 5.8). The reason for this is that turboshaft engines have poor SFC at low speed and altitude. Future mega-airships whose volumes are more than 20 million ft^3 will likely need lightweight engines with much more power that can only come from a turbine engine cycle.

11.4.8.4 Propeller

There are several design issues associated with integrating propellers onto an airship. For early design efforts it is sufficient to establish the propeller diameter, weight, and the efficiency of the propeller for performance calculations. Other engineering decisions regarding a propeller are whether or not to have a shroud and the influence of propeller wakes on vehicle aerodynamics. These aspects are generally decided later in the design process. If there will be thrust vectoring a decision has to be made on vectoring just the propeller and adding a gearbox or vectoring the entire engine propeller combination. One of the problems is that most off-the-shelf engines don't operate vertically because their oil systems are gravity based. This severely reduces the number of candidate engines.

In order to size a propeller the *speed-power coefficient, C_S,* and *advance ratio, J,* must be known. Using Eq. (5.29) C_S is calculated for the flight conditions along with the power required and the propeller rps, *n*. Using C_S in Fig. 5.9b yields the *advance ratio, J,* which enables the propeller diameter, D_{prop}, to be calculated with Eq. (5.33). These calculations can be made for specific blade angles or assuming an optimum blade angle from a variable pitch propeller.

Speed-Power Coefficient

$$C_S = \left(\frac{\rho V^5}{P n^2} \right)^{1/5} \tag{5.29}$$

Proportional advance ratio

$$J = V / n D_{prop} \tag{5.33}$$

11.4.9 Materials and Structure and Weight

11.4.9.1 Materials

The most significant advances in airship technology have occurred in envelope material fabrics. There are two fundamental types of materials used for airship hulls: laminates and weaves. Both of these materials have several layers with one load bearing layer as shown in Fig. 8.3. The weaves are not just simple alternate over-under styles but are usually complex flat weaves that are proprietary to each material manufacturer. Although most of an airship's fabric strength is provided by the inherent fibers that make up the load bearing layer, the fabric's toughness, resistance to abrasion, and rip-stop ability are a direct result of the weave or laminate arrangement. Some simple weave examples are shown in Figs. 8.3 and 11.6.

Based on discussions in Chapter 8 the designer has to select the material that meets the design requirements for the body. There are two basic types of material available. One is multi-layer with the load bearing layer being a woven material. The other material choice also has a load bearing layer but is a laminate made from layers with fibers laid out in various orientations. Both the woven and laminate layers are then bonded with a helium permeability (usually mylar) layer and an external UV resistant layer (usually Tedlar).

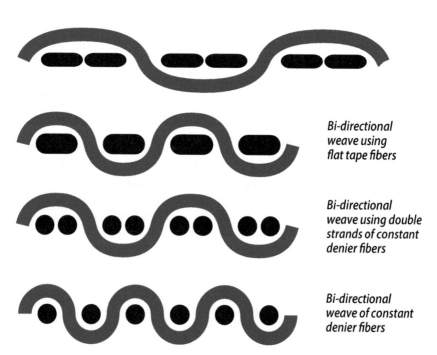

Bi-directional weave using flat tape fibers

Bi-directional weave using double strands of constant denier fibers

Bi-directional weave of constant denier fibers

Figure 11.6 Examples of weaves used for airship envelopes.

The advances of fundamental fiber capabilities over the last 70 years have been impressive compared to metals and composites. Looking at Fig. 8.2 shows the more than 3 × improvement factor fabrics have experienced in *ultimate specific tensile stress* over composites and metals. However, as good as these fabrics are designers are never able to use their full strength capability. The almost synonymous terms that are most often used to express a reduced capability are *margin, knockdown,* and *factor of safety.* This book will use *factor of safety.*

11.4.9.2 Structure

For non-rigid airships, internal pressure must be modulated to maintain a constant pressure differential across the envelope for all operational altitudes. As previously discussed it is the ballonet that is responsible for maintaining this constant pressure differential. Another approach is to make the hull strong enough to withstand the pressure differential throughout the range of operational altitudes. Such a design was constructed in 1929 with the hull of ZMC-2, which was more commonly known as the MetalClad airship. However, in the end it had a ballonet just like all of its contemporary non-rigid airships. Although the ZMC-2 successfully operated for more than 10 years and was lighter than equivalent rigid designs, strangely, it did not change the way airships were designed.

11.4.9.3 Weight Estimation

The envelope is the largest structure on an airship and one of the heaviest along with the gondola/payload bay, propulsion systems, and tails (see Figs. 9.9 and 9.10). This means that material improvements in specific tensile stress will result in lower weight but the overall reduction in total empty weight is modest. Two other envelope material properties that are also important are creep and damage tolerance.

Tails are also a significant portion of the total weight and are constructed in one of two ways. They either use a space frame structure covered with a lightweight fabric or are made of a lightweight fabric that is pressurized similar to an air mattress. Remaining structure such as the cargo bay, crew station, or engine support structure are generally made from current aluminum alloys using efficient truss designs.

Some of the most difficult data to obtain in the aerospace industry are weight relationships for aircraft systems and subsystems. Even harder to get are weights for airships since there are so few new airships and all companies are very protective of their own design data. Nonetheless, some basic data is available. Historical weight data is presented as Fig. 11.7, which comes from the historical database in Appendix E. Both rigid and non-rigid designs are shown even though future designs will be predominantly non-rigid. Fundamental weight estimating relationships are provided in Chapter 9.

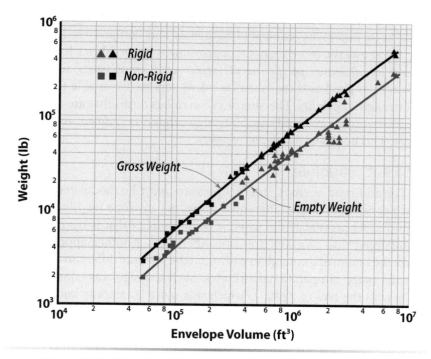

Figure 11.7   Weights of historical airships (bodies of revolution).

11.4.10 Aerodynamics

There are several sources for estimating the aerodynamic characteristics of an airship. Chapter 3 contains numerous figures and equations to assist in these estimations. If there is some wind tunnel data or CFD calculations for the specific design then this is better yet. Miscellaneous drag items such as cables, engine cooling, and interference always have to be added separately.

As discussed in Chapter 3 most of an airship's drag is from skin friction so a significant amount of design time is spent characterizing the actual skin friction coefficients over the entire surface. However, there are two other types of drag that are also a part of total drag and they are pressure drag and drag-due-to-lift. Aft pressure/separation drag on the body should be minimized with careful contouring of the airship envelope downstream of the maximum diameter. Keeping pressure gradients downstream of the maximum diameter as low and smooth as possible will result in an acceptable drag level. The drag-due-to-lift term is generally small since body of revolution airships don't generate large amounts of aerodynamic lift. If these poor lift generating bodies are flown at even modest angles of attack the drag-due-to-lift can become large and unacceptable.

Given that an ellipsoid-like body of revolution has a naturally beneficial pressure gradient over the first 40% or so of surface area, it is proper to account for this effect even in the early stages of conceptual design. Using Fig. D.5 it is easy to estimate the average effective skin friction coefficient, C_{fe} according to how much laminar flow is thought to be likely. Usually, 20% is easy to attain, 30% requires optimized body contouring, and 40% or more requires major CFD and/or experimental efforts. Highly detailed drag reduction efforts that maximize laminar flow regions are usually not part of conceptual design. Often, an initial amount of laminar flow is stated as a goal early on for performance estimates and the aerodynamicist then attempts to create a shape that meets the goal. However, no laminar flow occurs for bodies whose noses are fitted with external structure such as battens.

Total lift is also an important design quantity in that it establishes how much aerodynamic lift is needed. The reason aerodynamic lift is important for an airship is that it establishes how much heaviness can be effectively offset by aerodynamic means. This directly affects how much payload can be off-loaded or how much ballast weight is needed for a return flight.

11.5 Performance: Range, Endurance, Takeoff

Performance includes many parameters that are important to a customer. Some performance is given by the customer as a requirement and other performance measures are derived from requirements and operational scenarios. Aircraft and airship performance measures are similar except for their magnitudes. Most air vehicles are interested in range, payload, speed, endurance, and cost. Transport aircraft are also specifically interested in takeoff performance, passenger comfort, operational costs, and safety. Fighter aircraft are also interested in stealth, high speed cruise, and turning and maneuvering capability. General aviation aircraft are specifically interested in ease of flight, maintenance cost, efficient cruise, stall speed, takeoff and landing distances, and purchase price. Airships, depending on their primary mission, are generally interested in range and endurance, operational cost, controlling or maintaining buoyancy, ballast demands, and ground support.

The standard range-payload curve is as important for cargo transporting airships as it is for transport aircraft. Because of their poor aerodynamic lift abilities normal body-of-revolution airships are unable to carry and/or unload large payloads even though the designer tries to maximize the amount of operational heaviness for any design. Endurance is equally important for other missions and along with range is often plotted against speed to determine what the optimum speed is. Figure 11.8 compares the

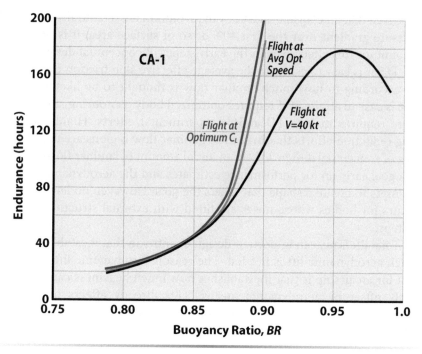

Figure 11.8   Effect of *BR* on endurance for various optimum flight conditions.

endurance for flight at optimum C_L vs flight at the best constant average speed and for a specific constant speed. Flight at a typical constant speed shows the loss of performance as *BR* approaches 1.0. Chapter 4 looks at several other parameters that are plotted vs speed to identify optimum flight conditions.

Performance parameters are usually the MOMs to determine the optimum design for aircraft. These calculations include range, endurance, takeoff, landing, and rates of climb and descent. Although takeoff distances are not very demanding for normal airships they can become more important for hybrid designs. Large rates of climb and/or descent are not performance issues for airships but are safety issues for the ballonet being able to keep up with rapidly changing ambient pressure. It is a safety issue defined by the FAA to make sure the lifting gas can expand so the envelope is not over-pressurized. The FAA requirement is 1200 f/m and is a design requirement for the ballonet.

11.6 Stability and Control

Aircraft acquire their acceptable stability and control from the addition of tails even though there are a few flying wing configurations that do not rely on tails. However, airships also need tails and they need them even more than airplanes because their large bodies of revolution are so unstable in pitch and yaw. Airship pitch instability is reduced by the fact that the c.g. is always below the c.b. (see Fig. 3.12) which results in a pendulum restoring moment for all angles of attack. This lower c.g. position also provides all the roll stability necessary for a body-of-revolution. However, the yaw axis has no moment resisting pendulum contribution from this low c.g. position making it the most unstable of the 3 axes.

Two other characteristics make airships differ from aircraft. First, airships have high moments of inertia relative to their weight, which increases their *time-to-double* amplitude characteristic. Secondly, because of their large volumes the damping terms on the pitch and yaw axes are significantly higher for an airship when compared to any aircraft. It is the damping terms that really distinguish airship handling qualities compared to those of an airplane.

From a stability standpoint, plotting historical vertical tail volume coefficients, C_{VT}, vs envelope volume shows an increasing value for higher envelope volumes (Fig. 7.1). This value can be used to size the vertical tail as described in Chapter 7. However, even though horizontal moments are reduced by the pendulum effect, the horizontal tails are generally larger than the verticals. This increased horizontal area is the result of greater control power demands on the pitch axis. How much control is enough on each axis is a complex issue and will not be discussed here. See Chapter 7 for initial tail sizing and Chapter 10 for a more detailed discussion on tail sizing and other flight control issues.

During design it is necessary to calculate a.c., c.g., and c.b. locations for any airship configuration. The c.b. is located at the centroid of the displaced external gas/fluid, which is easily calculated by any electronic drawing program such as CATIA, Pro Design, or AutoCad. Body of revolution airships typically have their c.b. located somewhere between 40% and 45% of the body length. Remember, only the external shape of the airship can change the c.b. location and redistribution of internal buoyancy, such as moving a ballonet, has no effect on the position of the c.b.

Calculating the a.c. is a little more involved in that a CFD model must be run, a wind tunnel test must be performed, or there is some historical empirical data for a body similar to the proposed design.

There is one other optional term that can be added as well. This term has been referred to historically as *apparent mass*, *virtual mass*, and *added mass*. This book will use the term *added mass* for this effect. Added mass is a complex issue that is usually not included in conceptual designs because it only affects airship performance when the airship is accelerating or decelerating. However, it is easy to use theoretical values for this effect that can improve the answer. Added mass is discussed in detail in Appendix C along with all of its pertinent equations. Plots of theoretical added mass factors are also available in this same Appendix. The added mass term is more important for hybrid designs, which have greater demands for maneuvering and accelerations and decelerations associated with takeoff and landing. Longitudinal accelerations will have their effective forces reduced by 5%–10% because of added mass effects.

11.7 Subsystems

To some it might seem surprising that this is one of the more significant differences between aircraft and airship design. Packaging all of the systems and subsystems into as tight an arrangement as possible requires many thousands of hours by aircraft designers. The internal arrangement must consider access for repair, service, and replacement when size, fit, or form factor require a subsystem or major component to be redesigned. This redesign comes at great cost. This is one of those subtle-on-the-surface and significant-on-the-inside issues that make design, repair, and replacement so much easier for airship systems, subsystems, and components. There is a significant cost difference as well because existing certified components are often easily incorporated without any impact on internal packaging. The airship has an advantage because of its inherently large volumes, internal systems, subsystems, and components. Even if these systems are located inside the gondola or payload bay it is easy to find volumes with good access.

11.8 WT Testing and CFD

Depending on the uniqueness of a design or a greater demand for high fidelity aerodynamic properties, a wind tunnel test may be necessary. There are various technical reports that summarize aerodynamic behavior applicable to airship design such as [2,3,4,5,6]. Small scale wind tunnel testing is inexpensive but data at small scales is generally poor for measuring drag. Larger scale model tests might cost 10 times as much but will give better drag data. Except for drag all other static aerodynamic properties from a small scale test should be measurable within 5%. Drag measurement is difficult for the smaller models as the average model Re is usually close to the boundary layer transition Re, which means the model has an unknown mixture of laminar and turbulent flows that may not accurately represent what is occurring on the full size vehicle.

An alternate to a wind tunnel test is a Navier-Stokes equation solver with shear stresses and turbulence models options turned on. Results from these CFD runs are often poor unless they are performed by an experienced modeller and even then may be hard to explain. However, using CFD to evaluate the effect of "changes" in a configuration are generally good and many forensic options are available to better understand local flow pressures and velocity vectors using these high-end codes. The preferred method is to do both wind tunnel testing and CFD analysis to make minor to modest changes in the design when schedule and funding are available.

11.9 Mooring and Mast Loads

Connecting to a mast is the standard approach to mooring an airship. While it might seem to be a straightforward engineering problem to calculate the mast load, it is not. To calculate this mast reactive load requires good data for the a.c. location, aerodynamic data C_{N_ψ} and C_{Y_ψ} (in ground effect!), and a good estimate of the mass distribution from nose to tail (radius of gyration). It is unusual to have all of this data available during conceptual design. Because the nose load is the result of the difference between two large numbers, answers can vary significantly (>2×) for modest data variations (~±5%). Since these loads do not drive the overall design but only define the loads for which the nose must be properly stiffened it is acceptable that this part of the design be part of the next design phase, the preliminary design.

11.10 Turning

It should not be surprising that an airship does not turn like an airplane nor does it need to. While turning capability can be the difference between life and death for an airplane fighter pilot, it is usually much less important for an airship.

From an aerodynamic perspective airplanes and airships turn in totally different ways. Airplanes combine increased lift, yaw, and roll to perform what is called a *coordinated turn*. Designers build in aerodynamic capabilities and adjust both stability and control characteristics to give an airplane design its proper handling and turning qualities. An airship must also have appropriate turning capabilities but its inherent limitations force turns to be without the use of the roll axis. This type of turn is referred to as *skid-to-turn*, which is similar to how a car turns a corner. The difference is due to the pendulum effect from the c.g. being below the c.b. When an airship tries to roll it is resisted by a very strong pendulum restoring moment. Rather than try to overpower this inherent roll resistance, which would require very large roll control devices, rolling motions are not used.

Historically, airships have been considered to have good turning capability if they can perform a 360-degree turn in less than one minute. Using the turning parameter in Chapter 4 (Figs. 4.22 and 4.23) is a good way to estimate the turning characteristics of any conceptual airship design.

11.11 Historical Airship Database

A historical database of airship characteristics and performance is presented in Appendix E. The database includes weight, volume, year of first flight, maximum speed, horsepower required, maximum operational altitude, and tail sizes. This data is from current literature and there is occasional disagreement among the various sources. When there was inconsistency the value chosen seemed the most likely by this book's authors. Use this data with care. Also note that hull volume and buoyant gas volume are different for rigid designs.

11.12 Sample Design Problem

This design problem will illustrate the steps needed to come up with an acceptable conceptual design of an airship that transports 9 passengers for 725 n mi. Since a full-scale study would require a team of 10–20 engineers, assumptions are made here for some parameters which are normally optimized as part of a much bigger trade study. The requirements, assumptions, and resulting solutions are summarized in Secs. 11.12.1 and 11.12.2.

If the Measure of Merit (MOM) is the smallest design then the goal is to find the smallest volume airship that meets these requirements. If the MOM is cost then the problem can be evaluated from the standpoint of finding the design that burns the minimum amount of fuel (a surrogate for cost) to perform the mission.

11.12.1 Requirements

The list of requirements for the sample problem is actually a short one.

1. Range = 725 n mi
2. Maximum fuel load = 1320 lb
3. Passengers = 9
4. Payload = 9 Pax @ 275 lb = 2475 lb
5. Cruise altitude = 3000 ft Maximum altitude = 10,000 ft
6. Lifting gas is helium
7. Materials are current off-the-shelf polyester fiber based fabrics that are woven and not laminated
8. Cruise speed is 64 f/s
9. Maximum speed = 76 f/s
10. Reserve fuel = 50 lb
11. Crew members = 1

11.12.2 Assumptions

1. Non-rigid body of revolution ($N_L = 2.0$ from Sec. 3.3.1)
2. Buoyancy Ratio (*BR*) at landing = 0.90
3. Body fineness ratio (*FR*) = 4.0
4. Brake Specific Fuel Consumption (BSFC) = 0.45
5. (2) Spark ignition engines capable of vectoring ±90 deg
6. Propeller efficiency, η_P, = 0.75

The following is a step-by-step summary of how an airship is sized. Many of the equations and figures needed here have already been presented in earlier chapters. Some new equations and figures specific to this trade study are presented as needed. In particular, Appendix D will be very helpful.

11.12.3 The Sizing Process

The numbered sections in Sample Problem 11.1 correlate with the step number in the Excel spreadsheet in Table 11.1. A volume is assumed and two separate values of takeoff weight W_G are calculated. The first W_G is calculated based on the amount of fuel necessary to fly the given range of 725 n mi. The other W_G is calculated by estimating the weight using weight relationships given in chapters 8 and 9 and the current chapter.

Requirements:
  Manned vehicle
  Helium lifting gas
  Spark ignition engines & fuel
  Range=725 nmi
  Payload=9 passengers
  Cruise altitude=3000 ft
  Maximum altitude=10,000 ft
  Cruise speed=64 f/s
  Maximum speed=76 f/s
  Crew size = 1

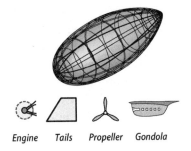

Engine   Tails   Propeller   Gondola

Assumptions:
  Non-rigid ellipsoidal body
  Woven polyester materials
  Engines attached to hull
  Number of tails = 4 & "+" arrangement
  Body laminar flow = 0%
  Body FR = 4
  Buoyancy Ratio (BR) at landing = 0.96
  Prop efficiency=0.75 and BSFC=0.45
  Prop blade angle=optimum
  Propeller rps=20

Solve for:
  Envelope volume
  Envelope length, width, & height
  Propeller diameter
  Buoyant lift
  Heaviness / aero lift
  Component & total drags
  Engine power
  Component weight buildup
  Empty and takeoff weights

Figure 11.9 Airship design problem overview.

Sample Problem 11.1

1. $FR = 4.0$, number of lobes $= 1$ (body of revolution)
2. Assume an initial value for the hull volume $= 1,000,000$ ft^3 (does not have to be very close to the final value). Further iterations will assume new volumes based on the difference between the calculated mission takeoff weight and the takeoff weight from the weight build-up calculations. When these two weights are the same the solution is exact. This is summarized in Table 11.1.
3. Calculate reference value of $Vol^{2/3}$. Given the FR and Volume determines the equivalent diameter, length, width, height, and body aspect ratio. For a given #lobes the lobe diameter can also be calculated. For bodies of revolution $N_{LOBES} = 1$ and the lobe diameter equals the body diameter. Calculate the equivalent body diameter

$$d_e = \text{body diameter} = 78.2 \text{ ft}$$
$$\ell_B = FR \, d_e = 312.6 \text{ ft}$$
$$\text{aspect ratio, } AR = 4 \, d_e/(\pi \, \ell_B) = 0.318$$

4. Using length and diameter calculate the body surface area assuming the shape of a prolate ellipsoid, $S_{wet body}$ using Eq. (D.3) where $w = ht = d_e$

$$\text{Surface Area} = \pi((\ell_B^p \, d_e^p + \ell_B^p \, d_e^p + d_e^p \, d_e^p)/3)^{1/p} \quad \text{where } p = 1.6075$$

$$S_{wet body} \text{ (ellipsoid)} = 61,631 \text{ ft}^2$$

5. Using the assumed envelope volume read horizontal and vertical tail volume coefficients C_{HT} and C_{VT} or calculate C_{VT} from the equation in Fig. 7.1. Be careful with these coefficients as there is no consistent way they are non-dimensionalized. A good approximation for the moment arm, ℓ_{tail}, is 38% ℓ_B. This book uses ℓ_B for the vertical tail reference quantity instead of body width (span).

$$S_{HT} = C_{HT} \, (Vol^{2/3} \times \ell_B)/\ell_{tail} = 0.067 \, (10,000 \times 307.4)/(0.38 \times 307.4) = 1753 \text{ ft}^2$$
$$S_{VT} = C_{VT} \, (Vol^{2/3} \times \ell_B)/\ell_{tail} = 0.059 \, (10,000 \times 307.4)/(0.38 \times 307.4) = 1558 \text{ ft}^2$$

6. The cruise speed for this study is 64 f/s and the cruise altitude is 3000 ft as stated in the requirements which allows the dynamic pressure, q, to be calculated as

$$q = \tfrac{1}{2} \rho V^2 = (0.00218)(64^2)/2 = 4.45 \text{ lb/ft}^2$$

7. Assume no laminar flow (this is conservative). Find the Re for the body. If laminar flow exists over a portion of the body use Fig. D.7 to find an estimate of C_{fe}.

$$Re = \rho V \ell_B/\mu = (.00218)(64)(312.6)/3.66 \, 10^{-7} = 1.18 \, 10^8$$
$$C_{f body} = 0.455/(\log_{10} (Re))^{2.58} = 0.00208$$

The form factor for body drag uses the relationship in Sec. 3.5.3.

$$FF_{3D\,body} = 1 + 1.5/FR^{1.5} + 7/FR^3 = 1.30.$$

Zero lift body drag is now calculated as

$$C_{D_{0_{body}}} = FF_{3Dbody}\, C_f\, S_{wet_{body}}/Vol^{2/3}$$

$$C_{D_{0_{body}}} = (1.30)(0.00208)(61,631)/10,000 = 0.01662$$

8. Drag coefficient for the tails is

$$FF_{tails} = 1 + 1.2\,(t/c) + 100\,(t/c)^4 = 1.23 \text{ for } t/c = 0.15 \quad (\text{Sec. 3.5.3})$$

$$C_{D_{0_{tails}}} = FF_{tails}\, C_f\, S_{wet_{tails}}/Vol^{2/3}$$

Assume $AR_{tail} = 1.0$.

$$\bar{c}_{tail}\,(\text{avg}) = (\bar{c}_{HT} + \bar{c}_{VT})/2 = [(AR_{HT}\,S_{HT}/2)^{1/2} + (AR_{VT}\,S_{VT}/2)^{1/2}]/2$$

$$\bar{c}_{tail}\,(\text{avg}) = [(1 \times 1753/2)^{1/2} + (1 \times 1558/2)^{1/2}]/2 = 57.5 \text{ ft}$$

$$Re = 1.10 \times 10^7 \text{ which leads to } C_{fe} = 0.00296$$

$$S_{wet_{tails}} = 2.2\,(1753 + 1558) = 7284 \text{ ft}^2$$

where 2.2 is the ratio of (wetted area)/(planform area) for the tails

$$C_{D_{0_{tails}}} = (1.23)(0.00296)(7284)/(10,000) = 0.00266$$

9. C_{D_0} of cab/gondola combination, engines, cooling, mounting structure, cables, and landing gear is

$$C_{D_{0_{cab+gond}}} = (0.108\, C_{D_{0_{body}}}\, Vol^{2/3} + 7.7)/Vol^{2/3} \quad [\text{Eq. (3.29b)}]$$

$$= (0.108 \times 0.01662 \times 10,000 + 7.7)/10,000 = 0.00256$$

$$C_{D_{0_{eng\,nac}}} = (\#\text{engines})\,(4.25)/Vol^{2/3} \quad [\text{Eq. (3.31)}]$$

$$= (2)(4.25)/10,000 = 0.00085$$

$$C_{D_{0_{eng\,cooling}}} = (\#\text{engines})\,(2 \times 10^{-6}\, Vol + 4.1)/Vol^{2/3} \quad [\text{Eq. (3.32)}]$$

$$= (2)(2 \times 10^{-6} \times 1,000,000 + 4.1)/10,000 = 0.00122$$

$$C_{D_{0_{eng\,mount}}} = (0.044\, C_{D_{0_{body}}}\, Vol^{2/3} + 0.92)/Vol^{2/3} \quad [\text{Eq. (3.30b)}]$$

$$= ((0.044)(0.01662)(10,000) + 0.92)/10,000 = 0.00082$$

$$C_{D_{0_{cables}}} = (9.7 \times 10^{-6}\, Vol + 10.22)/Vol^{2/3} \quad [\text{Eq. (3.26)}]$$

$$= (9.7 \times 10^{-6}\,1,000,000 + 10.22)/10,000 = 0.00199$$

$$C_{D_{0_{LG}}} = (1.76 \times 10^{-6}\, Vol + 0.92)/Vol^{2/3} = 0.00027$$

10. Interference drag

$$C_{D_{0_{int}}} = (4.78 \times 10^{-6}\, Vol)/Vol^{2/3} \quad [\text{Eq. (3.34)}]$$

$$= (4.78 \times 10^{-6}\,1,000,000)/10,000 = 0.00048$$

11. Total zero-lift-drag

$$C_{D_0} = C_{D_{0_{body}}} + C_{D_{0_{tails}}} + C_{D_{0_{gond}}} + C_{D_{0_{ecm}}} + C_{D_{0_{LG}}} + C_{D_{0_{int}}}$$

$$C_{D_0} = 0.01662 + 0.00266 + 0.00256 + 0.00289 + 0.00187 + 0.00027 + 0.00048$$

$$C_{D_0} = 0.02747$$

12. Drag-due-to-lift, is the other part of total drag that is added to C_{D_0} to obtain a total drag coefficient. The *drag-due-to-lift factor, K,* is obtained from Fig. 3.26 using the body aspect ratio. For a hybrid airship body $AR = $ width$^2/S_{plan}$ and was calculated as $AR = 0.318$ from step {3}. Using the equation in Fig. 3.26 for $AR = 0.318$ which is referenced to S_{plan} gives $K = 1.74$. See [8] for a detailed discussion of Fig. 3.26.

$$K = 1.74/N_L = 0.869 \ (referenced \ to \ Vol^{2/3})$$

where $N_L = 2$ from Sec. 3.3.1

13. The buoyant lift is calculated using 0.0646 lb/ft^3 for helium at sea level conditions. This number can vary slightly for different purities.
Buoyant lift = $0.0646 \times$ Hull volume $\times$ density$_{max \ alt}$/density$_{sea \ level}$

$$L_{buoy} = 0.0646 \ Vol \ (\sigma_{max \ alt})$$
$$= (0.0646) \ (1,000,000) \ (0.9151) = 59,115 \ lb$$

14. Defining weight terms (see Fig. 12.6 for schematic of weight term definitions.

$$W_E + fuel_{unusable} + oil_{unusable} + crew = W_{OE} + payload = W_{ZF} + fuel = W_G$$

Where

$$W_{ZF} = L_{buoy}/BR_{land} - fuel_{res} = 59,115/0.90 - 50 = 65,634 \ lb$$

15. Since

$$W_{ZF} = W_{OE} + PL$$
$$W_{OE} = W_{ZF} - PL$$
$$W_{OE} = 65,434 - 4000 = 61,434 \ lb$$

16. Calculate the *takeoff heaviness* (W_{H_0}) and *landing heaviness* (W_{H_1}) necessary to fly the required range of 725 n mi.
Rewrite Eq. (4.18) as

$$Range = A \left[\tan^{-1} \left(\frac{W_{H_0}}{B} \right) - \tan^{-1} \left(\frac{W_{H_1}}{B} \right) \right]$$

Where $A = \dfrac{326 \ \eta_p}{BSFC \sqrt{K \ C_{D_0}}}$ and $B = q Vol^{2/3} \sqrt{\dfrac{C_{D_0}}{K}}$

Reserve fuel is still on board at landing and is assumed.

$$W_{LAND} = W_{ZF} + fuel_{res} = 65,634 + 50 = 65,684 \ lb$$
$$W_{H_1} = W_{LAND} - L_{buoy} = 65,684 - 59,115 = 6568 \ lb$$

Solving the previous equation for the takeoff heaviness, W_{H_0}, yields-

$$W_{H_0} = B \tan\left[\frac{\text{Range}}{A} + \tan^{-1}\left(\frac{WH1}{B}\right)\right]$$

$A = 326\ \eta_P/[BSFC\ (K\ C_{D_0})^{1/2}] = 326 \times 0.75/[0.45\ (0.869 \times 0.02747)^{1/2}] = 3517$

$B = q\ Vol^{2/3}\ (C_{D_0}\ K)^{1/2} = 4.45 \times 10{,}000\ (0.02747/0.869)^{1/2} = 7921$

$W_{H_0} = (7921) \tan\ [(725/3517) + \tan^{-1}\ (6568/7921)] = 9950\ \text{lb}$

17. Mission fuel (fuel burned) is simply the difference between takeoff heaviness and landing heaviness.

Fuel Burned, $fuel_{burn} = W_{H_0} - W_{H_1} = 9950 - 6568 = 3381\ \text{lb}\ (\textit{mission fuel})$

Total Fuel $= fuel_{burn} + fuel_{res} = 3381 + 50 = 3431\ \text{lb}$

18. Therefore, the gross weight at takeoff to perform the mission is

$$W_G = W_{LAND} + fuel_{burn} = W_{ZF} + fuel_{res} + fuel_{burn}$$

$$W_G = 65{,}684 + 3381 = 69{,}065\ \text{lb}$$

$$BR_{TO} = L_{buoy}/W_G = 59{,}115/69{,}065 = 0.856$$

Although not used in the sizing routine the buoyancy ratio at takeoff, BR_{TO}, is calculated to make sure it is not too low. Any value of BR_{TO} that is less than ~0.8 is an indication that too much aerodynamic lift is needed from the body.

At this point the takeoff gross weight (W_G) to transport 4000 lb of payload along a 725-n mi mission is known. The remaining steps will calculate the W_G that results from a weight buildup which also uses the same assumed volume of 1,000,000 ft^3.

19. In order to estimate maximum engine power and propeller size it is necessary to calculate lift, drag, and power required for the airship. The greatest aerodynamic lift that would be experienced is at start of cruise but at sea level.

$$L_{aero} = W_{H_0} = 9950\ \text{lb}$$

20. Aerodynamic Lift Coefficient (maximum power)

$q_{max} = \text{density at SL} \times V_{max}^2/2 = (0.002377)(76)^2/2 = 6.86\ \text{lb/ft}^2$

$C_{L_{max\ power}} = W_{H_0}/q_{max}/Vol^{2/3} = 9950(6.86)/10{,}000 = 0.145$

21. Total drag at maximum power condition

$$\text{Drag} = (C_{D_0} + K\ C_{L_{aero}}^2)q_{max}\ Vol^{2/3}$$

$$\text{Drag} = (0.02747 + (0.869)(0.145)^2)(6.86)(10{,}000) = 3138\ \text{lb}$$

22. Maximum power per engine @ SL

$$\text{maximum power/engine} = V_{max}\ \text{drag}/\eta_P/NE/550$$

$$\text{maximum power/engine} = (76)3138(0.75/2/550) = 289\ \text{hp}$$

The propulsion section assumes that any size engine is available along with any size propeller. Actually, there are only discrete engines available and one of these engines would have to be chosen that is close to the required power and has the appropriate BSFC.

23. Finding the propeller speed. Although reciprocating engines perform most efficiently (lowest BSFC) for rpm = 2000–2400 they are geared down for the propeller by as much as a factor of 3, for the best propeller speed. The process of finding the best propeller speed is complicated and will not be part of this problem. A value of 20 rps is assumed. This problem also assumes that the engine is sized for maximum speed. In fact, for hybrid airships, where takeoff performance is important, it may be necessary to calculate engine size for a given value of takeoff distance or takeoff climb gradient. Calculations would have to be expanded using Figs. 5.10 and 5.11 to include propeller disk loading to find the engine power for takeoff. It is possible that engines could be sized for takeoff performance rather than for maximum speed. However, for this problem it is assumed that maximum speed determines the maximum power. At this point the characteristics of a specific engine (e.g. Continental 10-360 in Table 5.1) could be used instead of generalized propulsion calculations.

24. Next, the thrust *speed coefficient, Cs,* is calculated and *propeller advance ratio, J,* is found from Fig. 5.9. Assume a variable pitch propeller for this problem. In a comprehensive study of propeller and engine sizing engine rpm and propeller blade angle are varied to get the best performance over a wide range of flight conditions. Variable pitch propellers have the ability to vary blade angle to maximize performance at various flight conditions. However, variable pitch propellers are heavier, more expensive, and require more maintenance.

$$C_S = \left(\frac{\rho V^5}{P n^2}\right)^{1/5} \qquad J = V/n\,D_P$$

$$C_S = [(0.002377)(76)^5/289/550/20^2]^{0.2} = 0.624$$

A curve fit of the J vs C_S data in Fig. 5.9 yields the following equation—

$$J = 0.156\,C_S^2 + 0.241\,C_S + 0.138)$$

For $C_S = 0.624, J = 0.349$

The propeller diameter is now calculated as

$$D_P = (76)/(20)/(0.349) = 9.2 \text{ ft}$$

25. Propeller efficiency

A curve fit of the optimum η_P vs C_S data on Fig. 5.9 is

$$\eta_P = 0.139 C_S^3 - 0.749 C_S^2 + 1.37 C_S + 0.0115 = 0.609$$

Since η_P was assumed to be 0.75 for earlier calculations this is close enough. However, should the assumed value of η_P be too far from the answer it can always be improved by setting the assumed η_P to the actual and recomputing from step {16} (convergence is rapid for the exact solution).

26. Once the internal pressure is calculated the fabric load is easily found for the body diameter.

For polyester weave fabrics the density is obtained from Fig. 8.4 to calculate the weight of the envelope for the assumed material.

$$p_I = 1.2 \, q_{max} + 0.0635 \text{ height}$$

$$p_I = [1.2(6.86) + 0.0635(78.2)]/144 = 0.092 \text{ lb/in}^2$$

The following steps use weight relationships from Chapters 8 and 9.

27. The weight build-up begins with the body and includes factors for manufacturing (1.2) and attachment fittings (1.26). Assume envelope and septum materials have the same areal density and are made from a polyester weave. Assume the factor of safety (*FS*) is 4. See Chapter 9.5.1 for hull weight discussion.

$$\text{hull fabric load (lb/in.)} = \text{internal pressure} \times \text{hull radius}$$
$$\text{hull fabric load} = FS \times 12 \times 0.092 \times 78.2/2 = 173 \text{ lb/in.}$$

It is important to apply the factor of safety (*FS*) to the fabric load and not wait and apply it later to the weight calculation.
Using Fig. 8.4 fabric density

$$(\text{oz/yd}^2) = 0.0453 \times \text{hull load (lb/in.)} + 1.962$$
$$\text{hull fabric density} = 0.0453(173) + 1.962 = 9.75 \text{ oz/yd}^2$$

W_{env} = hull fabric density × (manufacturing) × (attachments) $S_{wet_{body}}$
$$W_{env} = (9.75)(1.2)(1.26)(61,631)/16/9 = 6311 \text{ lb}$$

Assume there is one septum down the middle and it has an area equal to 20% of the sideview area. From Sec. 8.7 septum loads are assumed to be 1.5 × envelope load.

$$\text{septum fabric density} = 0.0453(1.5)(173) + 1.962 = 13.72 \text{ oz/yd}^2$$
W_{sep} = septum fabric density × (0.2) $\pi \, ht \, \ell_B/4)/16/9$
$$W_{sep} = (13.72)(.2)\pi(78.2)(312.6)/4/16/9 = 364 \text{ lb}$$

28. Ballonet weight is calculated for 2 hemispherical ballonets. Design altitude is 3000 ft and hull volume is 1,000,000 ft^3.

$$Vol_{Ball} = Vol_{env} (1/\sigma_{des\ alt} - 1) = 1,000,000 (1/0.9151 - 1) = 92,777 \text{ ft}^3$$
which is based on the conservative empirical data in Fig. 9.3.

surface area of 2 hemi-spherical ballonets = $(4\pi)^{1/3} (3 \, Vol_{Ball})^{2/3} = 9911 \, \text{ft}^2$

$$W_{Ball} = (0.035)(\text{surface area of ballonets})$$
$$W_{Ball} = (0.035)(9911) = 347 \, \text{lb}$$

29. Tail weight assumes a rigid space-frame structural tail concept. Using Eq. (9.2) where $F_{AF} = 1.26$ and $F_{PSQ} = 1.0 \, \text{lb/ft}^2$. For control surfaces that are 20% of the total area the tail weight is the sum of the stabilizer fin weight and the control surface weight. Actuator weights are calculated here but are added to the VMS weights.

$$W_{SSF} = 1.0 \, (S_{HT} + S_{VT}) \, (1.26) \, (80\%) = (1753 + 1558)(1.008) = 3337 \, \text{lb}$$
$$W_{CS} = 1.0 \, (S_{HT} + S_{VT}) \, (20\%)(1) = (1753 + 1558)(0.2) = 662 \, \text{lb}$$
$$W_{tails} = W_{CS} + W_{SSF} = 3337 + 662 = 3999 \, \text{lb}$$
$$W_{act} = (1.15) \, (1753 + 1558) \, (0.79) \, (0.2) = 602 \, \text{lb} \; (part \, of \, VMS)$$

where (1.15) is the installation factor for installing the actuators

30. Crew station/Gondola: Using Eq. (9.6) with gondola dimensions of 26.5 ft/5 ft/9.5 ft yields (assume crew station and gondola are one unit).

$$W_{gond} = 353 \, [(\ell/10)^{0.857} \, (w + h)/10 \, (V_{max}/10)^{0.338}]^{1.1} = 2329 \, \text{lb}$$

31. Weight of engines: (reciprocating) *(assume any size engine is available)* From Fig. 5.8

$$W_{all\text{-}eng} = N_E \, 4.848 \, (\text{power/engine})^{0.7956}$$
$$W_{all\text{-}eng} = (2) \, 4.848 \, (289)^{0.7956} = 880 \, \text{lb} \ldots all \, engines$$

32. Engine mounts, engine controls, and starting:

$$W_{EngMt} = 0.57 \, N_E \, W_{eng} = 0.57 \, (2) \, (880/2) = 502 \, \text{lb}$$
$$W_{EC} = 60.27 \, (\ell_{EC} N_E/100)^{0.724} = (60.27) \, [(50) \, (2)/100]^{0.724} = 60 \, \text{lb}$$
$$W_{Start} = 50.38 \, (N_E \, W_{eng}/1000)^{0.459} = (50.38) \, (2 \times 880/2/1000)^{0.459} = 48 \, \text{lb}$$

33. Propeller weight: Using Eq. (9.16) and using propeller diameter from {32}

$$W_{Prop} = K_P \, N_P \, (N_{BL})^{0.391} \, (d_P \, SHP/1000)^{0.782}$$
$$W_{Prop} = 31.92 \, (2) \, (3)^{0.391} \, (9.2 \times 289/1000)^{0.782} = 210 \, \text{lb}$$

34. Fuel tank weight:

$$W_{FT} = 2.49 \, (\text{Fuel})^{0.6} \, (N_T)^{0.2} \, (N_E)^{0.13} \, [1/(1 + Int)]^{0.3}$$
$$W_{FT} = 2.49 \, ((3381 + 50)/6.0)^{0.6} \, (2)^{0.2} \, (2)^{0.13} \, [1/(1 + 0)]^{0.3} = 141 \, \text{lb}$$

where the fuel is in gallons

35. Pressure system weight: Using Fig. 9.9 estimates the weight as 2% of the empty weight.

$$W_{press} = (0.02)(61,434) = 1229 \text{ lb}$$

36. Landing gear system-tricycle gear:

$$W_{LG} = (31.2)(W_{H_0}/1000)^{0.84} = 215 \text{ lb}$$

37. VMS weight is calculated using Fig. 9.9 as representative of a typical VMS suite for a hybrid airship. The actuator weight is added in here from step {29}.

$$W_{VMS} = 3\% \ W_{OE} + W_{actuators}$$
$$W_{VMS} = (0.03) \ (61,434) + 602 = 2445 \text{ lb}$$

38. Electrical system weight: Using Eq. (9.17) yields

$$W_{Elect} = 12.57 \ (W_{FT} + W_{TRON})^{0.51} = 246 \text{ lb}$$

39. Miscellaneous systems weight: (Fig. 9.9)

$$W_{MSys} = 0.035 \quad W_{OE} = 0.035 \ (61,434) = 2150 \text{ lb}$$

40. Crew and accommodations: Usually Eqs. 9.22–9.25 are used to fill the gondola with seats etc. The (1) crew member is part of the empty weight and the passengers are part of the payload weight.

$$W_{Crew+Acc} = W_{Seats} + W_{Bunks} + W_{Lav} + W_{Food} + W_{Crew}$$
$$W_{Crew+Acc} = (55 \times 1 + 32 \times 9 + 28 \times 0) + 2.3 \ (10)^{1.33} + 5.06 \ (10) + (1)(250) = 443 \text{ lb}$$

where crew members + luggage = 250 lb/person

41. The two final weight items are unusable fluids (gas & oil) and the empty weight margin. Typical values for weight margin on new aircraft designs are usually between 5–10% depending on the complexity of the design. Given that this is a well known design the weights should be more accurate than usual. Use 5% for the margin in this problem. Defining the acceptable margin is often a contentious decision as management wants to make sure the airship is within weight for the customer.

$$W_{uf} = 0.01 \text{ fuel}$$
$$W_{margin} = 0.05 \ W_{OE}$$
$$W_{uf+margin} = 0.01 \times (3 + 50) + 0.05 \times 61,434 = 3106 \text{ lb}$$

42. Adding up the weight items between {27} thru {41} yields the operating empty weight, W_{OE}, based on a system and component weight build up.

43. This final weight is the gross weight, W_G, at takeoff which is based on estimating the weights.

$$W_{G_{wts}} = W_{OE} + fuel + PL = 32,905 \text{ lb}$$

44. This compares the two values of W_G, one based on performing the mission, $W_{G_{perf}}$ {18} and the other based on estimating the total weights, $W_{G_{wts}}$ {43}. A final solution is obtained when $(W_{G_{perf}} - W_{G_{wts}}) = 0$ which yields the weight, fuel, and volume that will perform the mission under the specific assumptions. Table 11.1 shows 4 manual iterations and a final calculation that is exact (where $W_{G_{perf}} - W_{G_{wts}} = 0$) using the "Goal Seek" tool that is available in the Excel Program. The "exact solution" design may not be realistic if the angle of attack to fly at maximum heaviness, W_{H_0} is too large or the ground run distance is too great. It is necessary to calculate the flight angle and ground distance to make sure they are acceptable.

Steps {45} through {48} refer to the values from the "exact" solution case (Vol = 175,866 ft³) in Table 11.1 and not for the case where Vol = 1,000,000 ft³.

45. Using $AR = 0.318$ the data on Fig. 3.8 gives a $C_{L_\alpha} = 0.0088$, which is based on the planform area. Therefore, it is necessary to convert C_{L_α} from one that is referenced to S_{plan} to the traditional one for airships that is referenced to $Vol^{2/3}$. From Eq. (3.1) and its accompanying table find that $N_L = 2.0$.

$$C_{L_\alpha} \text{ (per deg ref to } Vol^{2/3}) = C_{L_\alpha} \text{ (Fig. 3.8 ref to } S_{plan}) N_L$$
$$C_{L_\alpha} = 0.0088 \, (2.0) = 0.0175$$

For the exact solution case in Table 11.1 find the angle of attack at W_{H_0}.

$$C_L \text{ (maximum heaviness)} = W_{H_0}/(q \, Vol^{2/3}) = 2098(4.45 \times 3139) = 0.15$$
$$\text{alpha (maximum heaviness)} = C_L/C_{L_\alpha} = (0.15)/0.0175 = 8.6 \text{ deg}$$

This angle of attack is within 15° which is an angle of attack where lift behavior of low aspect ratio bodies becomes non-linear.

46. It is also necessary to calculate the takeoff ground distance to make sure the design has engines powerful enough plus sufficient aerodynamic lift generated at the maximum scrape angle. Assume a takeoff scrape angle limit of 10°. Calculate the C_L at this angle, find the speed at this angle, calculate the thrust and drag and then compute the takeoff ground distance.

$$C_{L_{to}} = C_{L_\alpha} \, \alpha_{scrape} = (.0175)(10) = 0.175$$

$$\text{Speed at liftoff is } V_{TO} = [2W_{H_0}/(\rho \, C_{L_{to}} \, Vol^{2/3})]^{1/2}$$

$$V_{TO} = 1.1[(2)(1200/(0.002377 \times 0.175 \times 3139)]^{1/2} = 62.4 \text{ f/s}$$

where the 1.1 factor is from FAA regulations on takeoff speeds

47. Thrust during takeoff:
    Calculate the power loading = hp/A = $93/(\pi\,3.8^2) = 2.05$ hp/ft^2. Where hp = 93 is the maximum single engine power during takeoff and A is the propeller disc area. The static thrust T_0 for a variable pitch prop is determined from Fig. 5.11. For a power loading of 2.05, $T_0/\text{hp} = 7.00$ giving a static thrust $T_0 = (7.00)(93) = 651$ lb for each engine.
    Variable pitch prop thrust reduction due to forward speed is determined from Fig. 5.10. Takeoff acceleration is estimated at $0.707 V_{TO} = 44.1$ ft/s. From Fig. 5.10 the $T/T_0 = 0.759$ at 44.1 ft/s gives $T = 494$ lb for each engine. Total average accelerating thrust is $2(494) = 988$ lb.

48. Average drag and acceleration during takeoff:
    The average drag at $V = 0.707 V_{TO}$ is $(0.002377/2)(44.1)^2\,(0.0354)(3139)$ = 258 lb.
    The acceleration is $T - D - \text{gear friction} = T - D - \mu W_{H_0}$.
    Assuming a wheel friction coefficient $\mu = 0.03$ from Table 4.3 gives a gear friction of 63 lb.
    Thus the acceleration force = $988 - 258 - 63 = 666$ lb at 44.1 ft/sec.
    The mass accelerated during takeoff = $W_G/g + (\textit{air mass in ballonet}) +$ (\textit{helium mass})
    $= (12,495/32.17) + (16,316)(0.081)/32.17 + (175,866 - 16,316)(0.0111)/32.17$
    $= 388 + 41 + 55 = 484$ slugs
    The ground roll distance $S_G = \tfrac{1}{2}(\text{mass})\,(V_{TO})^2/(\textit{accel force})_{0.707V_{TO}}$
    $= \tfrac{1}{2}(484)\,(62.4)^2/(666) = 1417$ ft

Figure 11.10 presents a summary of the pertinent design characteristics that were established during the sizing of this hybrid airship. These data come from the right most column in Table 11.1.

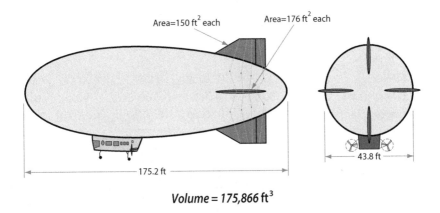

Area=150 ft^2 each Area=176 ft^2 each

43.8 ft

175.2 ft

Volume = 175,866 ft^3

Figure 11.10 Final optimum geometry for sample problem (Table 11.1).

Table 11.1 Summary of sample problem calculations from Excel spreadsheet

Step	Parameter	Manual iterations				Exact
1	$FR =$	4.00	4.00	4.00	4.00	4.00
2	Volume (ft^3) =	1,000,000	500,000	300,000	100,000	175,866
3	Vol$^{2/3}$ =	10,000	6300	4481	2154	3139
	Diameter-equiv (ft) =	78.2	62.0	52.3	36.3	43.8
	Body length (ft) =	312.6	248.1	209.3	145.1	175.2
	Diameter-lobes (ft) =	78.2	62.0	52.3	36.3	43.8
	Body width (ft) =	78.2	62.0	52.3	36.3	43.8
	Body height (ft) =	78.2	62.0	52.3	36.3	43.8
	Aspect ratio =	0.318	0.318	0.318	0.318	0.318
4	Body surface area-ellipsoid (ft^2) =	61,631	38,825	27,619	13,278	19,346
5	Hor tail vol coeff =	0.067	0.062	0.055	0.021	0.043
	Vert tail vol coeff =	0.059	0.054	0.048	0.015	0.036
	Hor tail area (ft^2) =	1753	1020	645	117	353
	Vertical tail area (ft^2) =	1558	900	563	86	299
6	Velocity-cruise (fps) =	64	64	64	64	64
	Qcruise (lb/ft^2) =	4.45	4.45	4.45	4.45	4.45
7	Re-body =	1.18E+08	9.39E+07	7.92E+07	5.49E+07	6.63E+07
	Cfe-body =	0.00208	0.00215	0.00220	0.00232	0.00226
	FF-body-Hoerner =	1.30	1.30	1.30	1.30	1.30
	C_{D_0}-body =	0.01662	0.01716	0.01758	0.01853	0.01803
8	(4) tails Re-tails =	1.09E+07	8.29E+06	6.58E+06	2.69E+06	4.83E+06
	C_{fe}-tails =	0.00296	0.00310	0.00321	0.00374	0.00338
	FF-tails =	1.23	1.23	1.23	1.23	1.23
	CD-tails =	0.00266	0.00255	0.00235	0.00095	0.00190
9	CD-cab/gondola =	0.00256	0.00308	0.00362	0.00557	0.00440
	CD-nac+cool+mount =	0.00289	0.00387	0.00497	0.00918	0.00663
	CD-cables =	0.00199	0.00239	0.00293	0.00519	0.00380
	CD-land gear =	0.00027	0.0003	0.0003	0.0005	0.0004
10	CD-interference =	0.00048	0.00038	0.00032	0.00022	0.00027
11	C_{D_0} =	0.02747	0.02972	0.03209	0.04016	0.03542
12	$K =$	0.869	0.869	0.869	0.869	0.869
13	L-buoy (lb) =	59,115	29,558	17,735	5912	10,396
14	WZF (lb) =	65,634	32,795	19,655	6518	11,502
15	WOE (lb) =	61,434	28,592	15,455	2318	7302
16	Landing heaviness, W_{H_1} (lb) =	6568	3284	1971	657	1155

Step	Parameter	Manual iterations				Exact
	$A =$	3517	3382	3254	2909	2891
	$B =$	7921	5190	3837	2064	2824
	W_{H_0} (lb) =	9950	5120	3214	1286	2098
17	Fuel burned (lb) =	3381	1836	1243	629	943
	Reserve fuel (lb) =	50	50	50	50	50
18	WG-performance (lb) =	69,065	34,677	20,948	7198	12,495
	BR-to =	0.856	0.852	0.847	0.821	0.832
19	Aero Lift@start cr (lb) =	9950	5120	3214	1286	2098
20	C_L@Vmax@SL =	0.145	0.118	0.104	0.087	0.097
21	Drag @ Vmax@SL start cruise (lb) =	3138	1812	1279	691	941
22	Max Power/eng@Vmax@ SL (hp) =	289	167	118	64	93
23	Propeller speed (rps) =	20	20	20	20	20
24	C_S@SL,max power =	0.624	0.697	0.747	0.845	0.783
	J@max power =	0.349	0.381	0.404	0.452	0.422
	Propeller diameter (ft) =	9.2	8.4	7.9	7.1	7.6
25	Prop efficiency-np =	0.609	0.650	0.675	0.718	0.692
26	Internal pressure (psi) =	0.092	0.085	0.080	0.073	0.077
27	Hull fabric density (oz/yd^2) =	9.752	7.665	6.529	4.849	5.605
	Wt-env+lpatch+ seams (lb) =	6311	3125	1893	676	1139
	Wt-septums (lb) =	364	177	105	36	62
28	Volume-ballonet-ft^3 (lb) =	92,777	46,388	27,833	9278	16,316
	Wt-ballonet (lb) =	347	219	155	75	109
29	Wt-tails (lb) =	3999	2319	1460	245	788
30	Wt-Crew Sta+ Gondola (lb) =	2329	2329	2329	2329	2329
31	Wt-engs (lb) =	880	569	431	264	357
32	Wt-eng mounts+ ec+st (lb) =	610	423	340	238	295
33	Wt-props (lb) =	210	128	93	53	75
34	Wt-fuel tanks (lb) =	141	99	79	53	67
35	Wt-Pressure Sys (lb) =	1229	572	309	46	146

(continued)

Step	Parameter	Manual iterations				Exact
36	Wt-landing gear (lb) =	215	123	83	39	58
37	Wt-VMS (lb) =	2445	1207	683	106	338
38	Wt-Elect sys (lb) =	246	230	222	212	217
39	Wt-Misc sys (lb) =	2150	1001	541	81	256
40	Wt-Crew+lav (lb) =	693	693	693	693	693
41	Wt-margin+ ufluids (lb) =	3106	1448	786	123	375
42	Wt-empty, WOE (lb) =	25,274	14,659	10,202	5269	7302
43	WG-weights (lb) =	32,905	20,745	15,695	10,148	12,495
44	(WG-perf) - (WG-wts) (lb) =	36,160	13,933	5253	−2950	0
45	C_L-alpha =	0.0175	0.0175	0.0175	0.0175	0.0175
	Max Flight Alpha (deg) =	12.8	10.5	9.2	7.7	8.6
46	C_L-to @ Alpha–scrape =	0.175	0.175	0.175	0.175	0.175
	TO, Speed@Alpha-sc (fps) =	76.2	68.9	64.7	59.0	62.45
47	Power loading (hp/A) (Fig. 5.9) =	4.37	3.01	2.39	1.62	2.05
	Thrust/thrustSLS (Fig. 5.10) =	0.744	0.749	0.755	0.767	0.759
	ThrustSLS/power (Fig. 5.11) =	5.55	6.22	6.68	7.53	7.00
	Total thrust @ 0.707VTO (lb) =	2389	1556	1189	735	987
48	Drag-ground run (avg-.707V) (lb) =	947	527	357	179	258
	Ground Distance (ft) =	3984	2487	1855	1245	1417

Since the original requirements for this problem were based on the capabilities of the A-170 airship built by American Blimp Company it is instructive to compare the "exact solution" results with the data in Table 11.2. Overall, there is decent agreement. Some differences are to be expected, however. In this case most of the weight difference is the result of using the actual weight of the 360 engine compared to the estimated propulsion weight. Actual propulsion weights are about 600 lb heavier than the estimate, which increases the estimated empty weight to about 8000 lb—close to the actual value of 8366 lb. This also explains the ground run distance difference between actual (~1000 ft) and calculated (1417 ft). Estimated volume is within 3%. The A-170 has an 8% ballonet that gives it a buoyant lift at 3000 ft of 10,080 lb requiring 2331 lb of aerodynamic lift for start of mission.

Table 11.2 Characteristics of A-170 Airship (courtesy of Rudy Bartel, ABC)

Characteristics		Weights	(lb)
Volume (ft^3)	170,297	Gondola assy+pitot sys	3308
Volume $^{2/3}$ (ft^2)	3072	Envelope assy	1495
Ballonet volume (ft^3) (8%)	13,600	Engine pod/pylon/prop	1312
Envelope length (ft)	175.8	Ballonet sys	426
Diameter (ft)	43.0	Ballonet tunnel	26
Hor tail area (2) (ft^2)	410	Ballonet window	3
Vert tail area (2) (ft^2)	410	Horizontal tails (2)	346
Fineness ratio	4.09	Lower vertical tail assy	180
Body aspect ratio	0.311	Upper vertical tail assy	173
Body surface area (ft^2)	19,061	Electrical system	216
Envelope material density (oz/yd^2)	11.3	(2) main landing gear	161
Buoyant lift (lb) @ takeoff	10,123	(1) fwd landing gear	47
Gondola		Nose dish assy (-battens)	160
Length (ft)	26.5	Air system	148
Width (ft)	5.0	Fuel system	92
Height (ft)	9.5	Nose battens assy	82
#passengers	9	Helium plate assy (3)	66
Cruise speed (ft/s)	64.2	Misc fittings, straps, wires	38
Max speed (ft/s)	76.3	Cable assy suspension	31
Cruise alt (ft)	3000	Flight Control Sys	27
Max alt (ft)	10,000	Unusable fuel	17
Range (n mi)	725	Misc cable	12
Payload (9 Pax @ 275 lb)	2475	**Empty Weight**	**8366**
Max fuel (lb)	1320	Fuel	1320
Max endurance@42fps (hr)	23	Crew	250
Crew	1	Payload	2475
Hp/engine	180	TOGW	12,411
BSFC@mid–cruise (lb/hp–hr)	1.768		
Power @ mid–cruise (hp)	16.2		
Engines	(2) 10–360 Lycoming		
Propeller	Constant speed/variable pitch prop		

11.12.4 Trade Studies

This airship design had an $FR = 4$ and an $AR = 0.318$. If the goal is to find the best FR, then this same process can be repeated for many values of FR. The impact of changing the design payload and/or the mission range can also be evaluated. In the end there will be several best solutions that represent the best designs based on lowest weight, smallest volume, least fuel burned, or some other measure of merit. Figure 11.11 shows the carpet plot for the variation of volume for various speeds and *FR*, all of which perform the basic mission. While all of these solutions are real designs they are unconstrained by other limits such as buoyancy ratio, cruise angle of attack, takeoff ground distance, etc. An example of a constraint line is shown overlaid on Fig. 11.11. This line represents the limit to all solutions that have a takeoff ground run ⇐ 2000 ft. In this example only those solutions to the left of the constraint are acceptable. Constraint lines redefine the allowable design space that meets unique customer requirements and/or design realism. The carpet plot in Fig. 11.11 is a standard format for presenting trade study results.

Even though the vehicle from Table 11.1 is a good conceptual starting point, there is still much work to be done. Numerous assumptions were made to facilitate getting a good answer without getting bogged down in complex analyses or costly testing. Therefore, the next step is to start refining these preliminary answers by including specific test data, CFD analyses, structural modeling, more refined weight buildup, an enlarged propulsion study, and include the effects of such phenomena as *added mass* (see Appendix C).

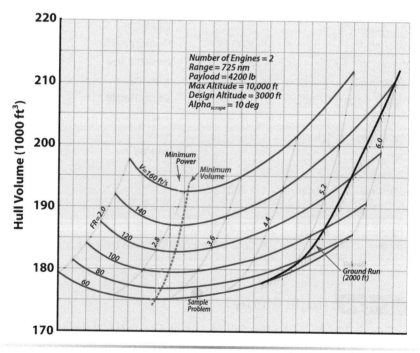

Figure 11.11   Effect of speed and *FR* on airship size.

References

[1] Hoerner, S.F., *Fluid Dynamic Drag: Practical Information of Aerodynamic Drag and Hydrodynamic Resistance*, Chapter 6, Midland Park, New Jersey, 1965.

[2] Lutz, T., Wagner, S., 'Drag Reduction and Shape Optimization of Airship Bodies,' *AIAA Journal of Aircraft*, Vol. 35/No. 3, May 1998.

[3] Abbott, I.H., *Airship Model Tests in the Variable Density Wind Tunnel*, NACA TR-394, 1931.

[4] Freeman, H.B., 'Force Measurements on a 1/40 Scale Model of the U.S. Airship Akron,' NACA TR-432, 1933.

[5] Wright, J.M. & Adams, R.E., 'An Empirical Method for Non-Rigid Airship Preliminary Drag Estimation,' AIAA 91-1277-CP, pp. 59–66, 1991.

[6] Ross, S.A. and Liebert, H.R., *LTA Aerodynamics Handbook*, Goodyear Aircraft Corp., Akron, Ohio, 1954.

[7] *Airship Design Criteria*, FAA2, U.S. Department of Transportation/FAA, Feb 1995.

- Design Tasks
  - Body Volume and Shaping
  - Buoyancy Ratio
  - Ballonet Sizing
  - Tail Sizing
- Performance
- Air Cushion Landing System (ACLS)
- Sample Design Problem
- Defining the Proper Buoyancy Ratio (*BR*)

Before long hybrid airships will play an important military role in both ISR and logistics missions. Supporting the war-fighter is something airships do well. Another important mission for hybrids is carrying cargo to austere natural resource sites year round and returning with a partially processed version of that resource.

We can't solve problems by using the same kind of thinking we used when we created them.

Albert Einstein

12.1 Introduction

I n the 1960s the Aereon Corp designed a 3-hulled rigid airship named the *Aereon III*, which was based on an original design by Solomon Andrews in 1860. Later designs would use a single oblate ellipsoid body with tip vertical tails and a tri-cycle landing gear (see Fig. 12.1). This design became known as the "Deltoid Pumpkin Seed" and was immortalized in a book of the same name [1]. After several modifications the vehicle flew in 1971 but no program emerged.

This hybrid approach would lay dormant for 25 years until Lockheed Martin began studying the feasibility and logistical utility of hybrid airships in 1996. Internally, the program became known as the Aerocraft Program. Although Aerocraft was cancelled by Lockheed Martin in 2000, three design features were studied and verified that would become critical to future hybrid airship designs.

First, the structure must be non-rigid. Since all prior Aereon hybrid airship efforts used rigid structure it was natural to think that Aerocraft should be rigid as well. After two years of study only designs that were non-rigid proved to be feasible.

Second, it was difficult to maintain the shape of pressurized structural designs having oblate ellipsoid shapes. Maintaining an ellipsoidal shape required numerous septums and curtains. After many structural design analyses the idea of lobes was first suggested. Detailed studies showed that merged round lobes enabled the designer to approximate the frontal shape using combinations of numerous circular arc segments.

Figure 12.1   Aereon 26 aka the "Deltoid Pumpkin Seed," 1971.

The third major development that came from the Aerocraft Program was the Air Cushion Landing System (ACLS). First proposed by a young flight controls engineer, this system is a natural fit for any hybrid airship that requires the ability to land at austere sites without the need for landing support personnel. Later ACLS discussions can be found in Sec. 12.7.1.

The concept of a hybrid airship was first introduced in Sec. 1.6. It is recommended that the reader read this section again before continuing with the rest of this chapter.

Now that designing a conventional airship has been discussed with a detailed sample problem at the end of Chapter 11, it is time to take a look at designing a hybrid airship. But first, remember that the term "hybrid airship" is defined in this book simply as one that has the ability to generate aerodynamic lift $\geq$ 10% ($BR \leq 0.9$) of the airship's weight. However, how does the hybrid airship designer establish what the best BR is? A discussion of how to select BR is included at the end of this chapter. This balance of aerodynamic lift vs buoyant lift is a very involved trade study that must include operational considerations as well as the value of speed and time. The greater the amount of aerodynamic lift the less efficient the hybrid flies. But, landing at lower BR enables larger payloads to be freely loaded and unloaded on the ground with a reduced need for ballast. This has significant value to a logistics operator.

Continuing the discussion of the benefits of a hybrid airship, there is a common misconception that the recent popularity of hybrid airships is based on there being more efficient performers than typical axisymmetric designs. Not true. The main benefits of a hybrid are its payload and operational flexibility and reduced dependence on ballast weight and its reduced need for infrastructure (e.g. no mast). This flexibility is the result of being able to generate balancing amounts of aerodynamic lift that are able to offset significant heaviness from large payloads. All of this results in the hybrid having higher productivity (payload $\times$ speed) or lower cost of operations ($/ton-mile). Figure 12.2 illustrates weight as it varies throughout a typical mission where payload is unloaded at the destination and there is no return payload (a very stressing operational scenario for an airship). Having no payload for the return leg is problematic for airships as it creates the need for ballast weight on the return flight. Figure 12.3 continues the discussion of ballast by showing that flying at negative angles of attack and using downward vectored thrust can also significantly reduce the need for ballast. Referring to Fig. 12.3 there are 3 ways of flying a return flight. Case (1) adds ballast that offsets the remaining unbuoyed weight. Case (2) uses flight at negative angle of attack to produce a downward force throughout the flight, thus reducing the amount of ballast needed in Case (1). Case (3) combines flying at a negative angle of

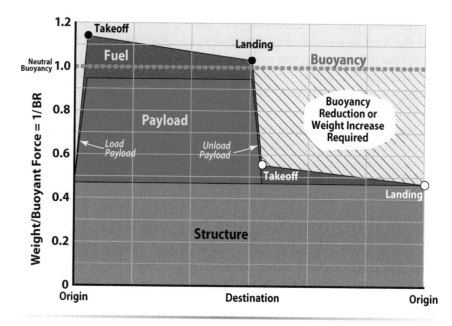

Figure 12.2 Weight variation for a round trip mission with no payload on the teturn leg.

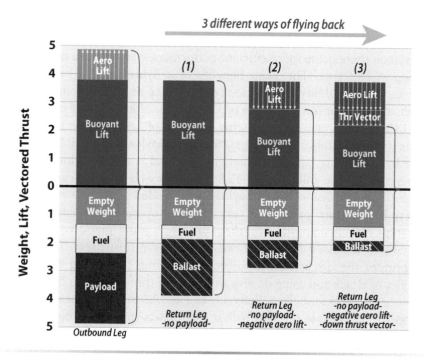

Figure 12.3 Ballast reduction using aerodynamic lift and thrust vectoring.

attack with an added downward thrust to push the airship toward the ground (modest fuel differences for the 3 Cases would modify the ballasts slightly but are ignored for these discussions). Remember that this is a worst case return flight. In general, there is significant returning payload which reduces or eliminates the need for ballast on the return flight.

If a hybrid airship's mission is to transport cargo then it will certainly want to have an Air Cushion Landing System (ACLS) integrated into its design. It is the multi-lobed arrangement of a hybrid airship that so naturally blends the ACLS and hull (see Sec. 9.7.1 and Sec. 12.7.1 for a discussion of ACLS). This is important because an ACLS does not integrate well onto a conventional airship which is a body of revolution. However, the wide stance of a lobed hybrid design enables realistic layouts for an ACLS.

Modern airship design is currently focused more on hybrid airships than on conventional axisymmetric shapes. There has been a renewed interest in all airships because of the now available materials technology that significantly improves their performance and damage tolerance. These new fibers and matrix materials have resulted in new woven and laminated materials that are significantly lighter and tougher overall, which results in reduced envelope weight that also resists damage.

As in Chapter 11, specific design issues associated with unique payloads, special materials needs, actual sensors, etc. will not be individually evaluated. Reductions in envelope weight are available to the hybrid design with its smaller radius lobes. Similar to Chapter 11 all of the fundamental hybrid design tasks are discussed in general terms and then applied to the actual design problem in Sec. 12.11. However, unlike Chapter 11, discussions will concentrate on those design aspects that are different for a hybrid airship vs a conventional airship. Similar design tasks requiring little added discussion include definition of speed, cruise altitudes, envelope sizing, ballonet sizing, envelope pressure, engine cycle, number of engines, propeller size, engine placement, tail sizing, drag estimates, and location of the a.c., c.b., and c.g. In addition to volume and FR examples of design tasks specifically needed for a hybrid airship are: establish takeoff BR (buoyant lift/takeoff weight) and landing BR (buoyant lift/landing weight), find best body aspect ratio (AR) that gives the desired amount of C_{L_α}, size an ACLS system, and calculate takeoff and landing performance.

12.2   Requirements

For any aircraft or airship design every effort should be made to satisfy all given requirements simultaneously. However, once it becomes obvious that one or more requirements are incompatible with the others it is incumbent on the engineer/designer to perform sensitivity studies to show which of the requirements are most easily relaxed. Figure 11.1 illustrates a typical sensitivity plot whose concept is the same for either conventional airship or hybrid airship design. The addition of the ACLS does not change any of the sensitivity studies in that it is captured as an empty weight change. Assumptions and requirements should be made with care as they are often the reason for designs having strange behavior or unexpected performance characteristics. Oftentimes large sensitivities can be traced back to unreasonable assumptions and ground rules. Designers beware!

12.3 Design Tasks

12.3.1 Body Volume

Given that a hybrid airship develops more than 40% of its lift from the lifting gas, envelope volume is an important design parameter. However, planform area and body aspect ratio become significant contributors to generating increased amounts of aerodynamic lift. For a hybrid with a specific mission there is an optimal balance between buoyant lift (*BR*, volume) and body planform area (*AR*) that is only determined by complex trade studies.

One standard design feature of all modern hybrid airships is lobes. Since hybrid airships are relatively new it took time before the concept of lobes was first suggested, fully understood, and then universally adopted. Figure 12.4 shows the geometric differences between a circle, ellipse, and an equivalent area created using lobes. Original hybrid concepts such as the Deltoid Pumpkin Seed [1] and early designs of Lockheed Martin's Aerocraft were rigid designs which used elliptical cross sections without any lobes. It wasn't until 1998 that the idea of using lobes as a surrogate for an ellipse was suggested and quickly incorporated. The reduction in envelope weight for a non-rigid design and the ability to maintain shape under a wide range of internal pressures made lobes an instant design success.

Although the original intent of lobes was to provide a shape that could analytically establish the best septum angle (see Chapter 8.7, and Fig. 8.9), there is another benefit to the structural lobe concept. Since envelope stress is proportional to the envelope radius it stands to reason that the smaller the lobe radii the lighter the envelope. Figure 12.5 presents the straightforward derivation of this simple truth. A lobed arrangement can also provide more width for a given body volume.

Equal Cross-sectional Areas

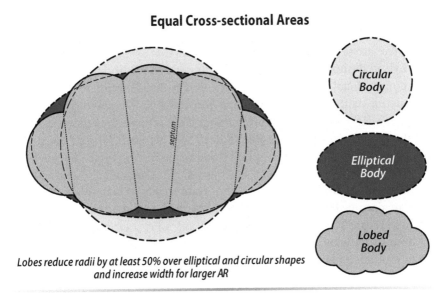

Lobes reduce radii by at least 50% over elliptical and circular shapes and increase width for larger AR

Figure 12.4 Using lobes to create hybrid airship cross sections.

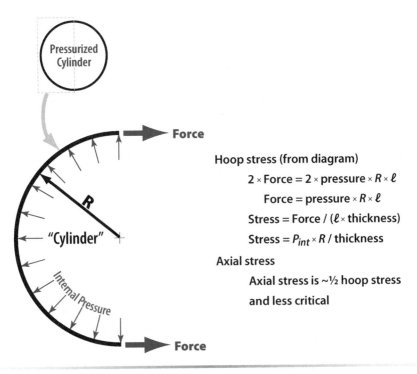

Hoop stress (from diagram)

$$2 \times \text{Force} = 2 \times \text{pressure} \times R \times \ell$$
$$\text{Force} = \text{pressure} \times R \times \ell$$
$$\text{Stress} = \text{Force} / (\ell \times \text{thickness})$$
$$\text{Stress} = P_{int} \times R / \text{thickness}$$

Axial stress

Axial stress is ~½ hoop stress and less critical

Figure 12.5 Hoop stress for a pressurized cylinder.

12.3.2 Body Shape

There are two shaping design parameters that are balanced to optimize any hybrid design: the well known fineness ratio, FR, and aspect ratio, AR. For conventional airships body FR is used to help minimize drag and in association with volume maximizes buoyant lift. Since aerodynamic lift supplies a large portion of hybrid airship lift it is important that this lift be created as efficiently as possible. For aircraft wings it is their AR that is mainly responsible for determining this efficiency. For wings, AR is defined as (span^2/wing planform area). For bodies, AR is defined as (width^2/body planform area). Body AR is just as important to a hybrid airship as wing AR is to an aircraft. Figure 3.26 shows how the drag-due-to lift factor, K, varies with AR regardless of whether it is a bare body of revolution, a hybrid body, or an aircraft wing. This is an important finding [8] that unifies the efficiency of all vehicles that generate aerodynamic lift. Aspect ratio is the primary parameter that determines drag-due-to-lift characteristics regardless of the type of lifting body.

Current non-rigid hybrid designs using modern materials and low drag body shapes generally have a FR between 3 and 5, which is similar to the optimum FR for conventional airships. However, FR is calculated differently for the hybrid since its geometry is more complex than a simple body of revolution. It is standard practice to define the equivalent diameter, d_e, for the circle with the same cross-sectional area as that of the actual airship. This defines $FR = \text{length}/d_e$ for a hybrid airship as shown in Fig. D.3.

When designing hybrid airship bodies consider them to be low-aspect-ratio, large t/c wings much like the lifting bodies of Fig. 3.6. Airfoil design codes can be used to optimize drag, L/D, or a.c. location. The airfoil-like sections can also be optimized to create a substantial laminar boundary layer run. Figure D.5 provides a quick means of estimating the equivalent flat-plate skin friction coefficient for a body with a given amount of laminar flow area. Airships have natural contours that create favorable (proverse) pressure gradients that can result in large areas of laminar flow. With a lot of design work it is possible to have laminar flow exist from the nose to stations just shy of the point of maximum cross section area. A realistic design goal is 30%–40% laminar run.

Body volume, shape, and fineness ratio for an airship are determined primarily by buoyancy requirements and not by passenger compartment volume, payload volume, or fuel volume. However, as discussed earlier, there are optimum contours and fineness ratios that are used to minimize zero lift drag (C_{D_0}). Hybrid airships generate somewhat more lift than bodies of revolution but skin friction is still a large contributor to drag. Design efforts become an exercise that finds the largest total L/D (including buoyancy) for a given shape.

To Lobe or Not to Lobe . . . That is the Question

Prior to 1999 no serious airship design had incorporated the notion of lobes as opposed to a single body of revolution or flattened ellipsoid. During the structural design efforts on the Lockheed Martin Aerocraft Program engineers were trying to establish the optimum intersection angle between the septum and the envelope. Since septums are necessary to maintain non-circular cross-sections under pressure and/or to carry the load of a cab or cargo bay it was important to find the optimum septum angles and arrangement. Designers were certain that there was a best angle for the septum/envelope intersection and began evaluating methods to calculate it. The revelation was supplied by an Aerocraft structural designer. Draw two circles (radii need not be the same) so that they intersect and the angle of the septum is simply the line that passes through the two points where the circles intersect. This unique septum angle results in balanced forces at the septum attachment point, generating a stable shape that does not distort as the internal pressure changes. From that point on hybrid airship designs have used lobes to create the cross-sectional shape, volume, and area equivalent to an ellipsoid. The combination of envelope lobes and septums results in a significantly lower envelope weight when compared to a single ellipsoid with the same volume and fineness ratio. Lobes have been a constant design feature in all Lockheed Martin hybrid airships since then and now all modern non-rigid hybrid airships use lobes in their envelope designs.

12.3.3 Buoyancy Ratio

Although the hybrid airship designer can select BR for either takeoff or landing, often these values are the result of the mission fuel burned and its associated payload. In general the takeoff BR cannot be too low (heaviness too large) or it will be impossible to lift the airship off during takeoff due to a limited tail strike angle of attack. There are only two forces that can be generated to over come heaviness: aerodynamic lift and vectored thrust. Looking at Fig. 3.8 shows that hybrids are about 3×–4× more efficient at generating lift than that of a body of revolution. This greater ability to generate aerodynamic lift during takeoff and landing allows takeoff heaviness to be much higher or conversely the BR can be much lower than for a conventional airship. Theoretically, any BR can be considered ($BR = 0$ is a normal airplane). However, generally speaking BR lower than 40% are costly to the operator and may run into takeoff distance limits.

Configurations below $BR = 40\%$ create more drag-due-to-lift without an offsetting amount of buoyant lift to generate an overall L/D. This was discussed in Chapter 3. The fact that for $BR > 0.4$ hybrid airship configurations become viable is another result of the square-cube law coming into play and working in favor of the buoyant system.

The *BR* at landing is restricted as well. Contrary to conventional airships a hybrid may have an ACLS instead of a landing gear. Operationally speaking *BR* for hybrids are usually 70% or more, or 30% heavy. Other contributions to vertical force in the form of vectored thrust or grip force from the ACLS can become an acceptable landing *BR* for hybrid designs. These ACLS and vectored thrust contributions are significant and can change *BR* at landing by 10%–15%.

12.3.4 Speed and Dynamic Pressure

Although great effort is always expended to maximize airship speeds, operating at or near these high speeds always reduces range and endurance and increases takeoff weight. However, higher speeds will also provide more productivity (payload × speed) and throughput, which is important to the owner/user of the airship. Although lower speeds mean less drag and thus better fuel efficiency it is the cost of delivering a pound of payload over a given distance that matters most to owners and operators. Compared to cargo delivery, mission speed is much less important for a stationkeeping mission that allows the use of low optimum speeds for minimum fuel flow.

From a structural standpoint, hybrid airships have their internal pressure defined by maximum dynamic pressure just as conventional airships do. Chapter 8 discusses this in more detail.

Internal hull pressure is set by the following relationship:

$$\text{Envelope pressure} = 1.2 \times \text{maximum } q$$

12.3.5 Altitude

Cruise and maximum altitudes are established in a similar manner for all airships. Airships spend most of their time at cruise/loiter altitude which is selected to be high enough to safely clear normal terrain but low enough to get the most efficiency from the propellers and engines. Propellers and internal combustion engines perform best at the highest atmospheric pressure and density (i.e. low altitudes). Maximum altitude is strictly for terrain clearance so flight paths can be direct regardless of enroute terrain. Sometimes maximum altitude is the same as the operational altitude. This occurs whenever high altitudes are needed for Intelligence/Surveillance/Reconnaissance (ISR) missions.

12.3.6 Ballonet

Hybrid airships need ballonets for the same reason as any other non-rigid airships. Their size is calculated using the same approach as in

Chapter 11. Because of its flattened shape the hybrid airship will likely have more smaller ballonets than in a conventional airship with the same volume.

12.3.7 Tail Sizing

Refer to Chapters 7 and 10 for discussions on initial and final tail sizing. Results from initial tail sizing in Chapter 7, using the technique of *tail volume coefficients*, are often good enough for conceptual design efforts. Once the size, shape, and volume of the envelope have been established it is appropriate to refine the tail sizes from Chapter 7 with the techniques in Chapter 10. Since tails have significant weight and that weight is located aft it is important to minimize tail/fin weight. Controlling the c.g. location is also important for airships and puts added pressure on the designer to keep the tails as small as possible.

Hybrid airships use modifications of the X-tail arrangement and never use the "+" or "Y" shapes because it reduces the tail strike angle. Because of its takeoff and landing demands the hybrid airship has some added stability and control considerations that are used to refine the tail sizing estimates obtained from Chapter 7.

12.3.8 Propulsion

The layout of a hybrid airship offers added flexibility to the location of engines, which allows for significant thrust vectoring capabilities. As shown in Fig. 12.6 engines or just their propellers can be vectored either up or down depending on flight mode. A certain amount of vectored thrust is necessary for any airship for ground maneuvering when control surfaces become ineffective at low speeds and reverse thrust is necessary for braking. Vectored thrust is also useful during vertical or heavy weight takeoffs and while payloads are being loaded or unloaded. Figure 12.3 Case (3) shows an example of the contribution of vectored thrust during the unloading process.

The technique for identifying the number of engines is similar for hybrid designs with the added burden of providing good takeoff acceleration and landing braking performance. One- and two-engine-out cruise flight may be the determining factor or it may be maximum speed. Each design will be different depending on mission requirements. Other propulsion design efforts such as engine placement, engine cycle, and propeller size are philosophically the same as for conventional airship designs.

12.3.9 Materials

Hybrid airship designers use the same materials technology available to any modern airship. Their added advantage of reduced envelope stress is

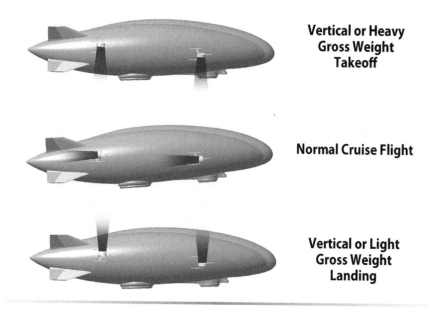

**Vertical or Heavy
Gross Weight
Takeoff**

Normal Cruise Flight

**Vertical or Light
Gross Weight
Landing**

Figure 12.6 Use of thrust vectoring to augment or reduce lift forces.

due to their smaller lobe radii that replace the single radius of a body of revolution. This results in an envelope material that has a lower weight. All of the factors of safety are the same. See Chapter 8 and Sec. 11.4.8.2 for detailed discussions.

12.3.10 Weight Estimation

Weight estimating for hybrid airships has two primary differences compared to conventional airship design. First, the envelope is comprised of lobes that are stabilized by septums. There is a weight reduction associated with the smaller radii of lobes vs a body of revolution but there is an added weight for the stabilizing septums. The overall benefit is still a weight reduction for hybrids with lobes. The other change is the result of replacing a conventional airship's landing gear with an ACLS. The ACLS will be heavier but it is this system that gives the hybrid airship many unique capabilities.

Since there is no public data on weight buildups for hybrid airships, the best guidance is to use Fig. 9.10. Although this data is non-dimensionalized it should enable the designer to estimate weights with sufficient accuracy for a conceptual design. This data will be used in the sample problem at the end of this chapter. A summary of some of the important weight terms is shown in Fig. 12.7.

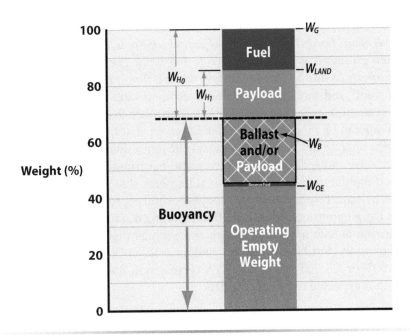

Figure 12.7 Weight definitions.

12.3.11 Aerodynamics

Unfortunately, there is little public data for hybrid designs. Most of the aerodynamic data in Chapter 3 is for axisymmetric bodies with the exception of hybrid data points in Table 3.1. There is no good pitching moment data for hybrids, thus eliminating the possibility of historical validation. This makes the estimation of hybrid performance difficult. However, two critical aerodynamic parameters $(K$ and $C_{L_\alpha})$ can be estimated using Figs. 3.8 and 3.26. This allows the designer to generate a drag polar along with how much lift is generated for an angle of attack. The precision of these estimates is acceptable for conceptual development.

For programs that progress to the preliminary design stage wind tunnel test data and/or CFD estimates are necessary for a successful design. Even having this data is not enough to bring the design risk down to that of a conventional airship. It is hard to be sure that the wind tunnel or CFD is giving the right answer without calibrating these tools to a known hybrid design.

12.4 Performance

Performance includes more parameters for a hybrid airship compared to a conventional one. Both use range, endurance, speed, payload, and cost. Because hybrids carry more payload and are often designed for logistical purposes other design parameters are important. Transportation efficiency such as *pounds of fuel burned to carry a ton of payload one mile,* or *productivity,* which is often expressed as *commercially as ($/ton-mile). The military uses its own productivity measure—(ton-miles/day), which is simply payload × speed.*

Takeoff and landing performance estimates are more necessary for hybrid airships since their takeoff *BR* is often much lower than 1.0 and their behavior can be more like an airplane than an airship. Since conventional airships operate much closer to neutral buoyancy their takeoff and landing profiles are quite different. However, the equations of motion for both takeoff and landing and the FAA requirements that must be met are the same. In many respects the hybrid behaves similarly to a commercial transport. Developing acceptable takeoff and landing distances and speeds must be proven as safe to a certifying agency such as the FAA. Currently, the document for certifying transport airships is titled Certification Specification 30T or CS 30T [6,7]. The FAA certification document for a hybrid airship is expected to release in 2013.

12.5 Weather

Whenever weather and airships are mentioned together it is always in reference to how weather events impact airship operations and performance. However, there is one aspect of global weather patterns that can significantly reduce the flight time and fuel burned for long-range missions. Recognizing that winds are the result of flow within large vortical weather patterns (cyclonics) it is possible to change course to pick up a tail wind regardless of which direction the airship is flying. Figure 12.8 illustrates how this works. Since cyclonics always rotate the same direction (counter-clockwise in the northern hemisphere and clockwise in the southern hemisphere) it is always possible to catch some tailwind benefits.

12.6 Stability and Control

Hybrid airship shapes are unstable on the longitudinal axis just like a body of revolution. A body of revolution is also very unstable on its yaw axis whereas the lower sideview profile of the hybrid design makes for a less unstable body, directionally. Conventional airships have horizontal and vertical tails, which are nearly identical in area. Hybrid designs have similarly sized tails as well but the verticals should end up being a little smaller than the horizontal tails. Another difference is that the pendulum stability contribution for the longitudinal axis is somewhat less for the hybrid simply because the hybrid has less vertical height than the single body of revolution with the same volume.

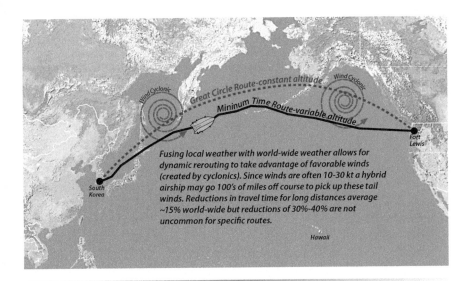

Fusing local weather with world-wide weather allows for dynamic rerouting to take advantage of favorable winds (created by cyclonics). Since winds are often 10-30 kt a hybrid airship may go 100's of miles off course to pick up these tail winds. Reductions in travel time for long distances average ~15% world-wide but reductions of 30%-40% are not uncommon for specific routes.

Figure 12.8 Dynamic route replanning based on weather data.

12.7 Subsystems

Hybrid designs have all the advantages of a conventional airship in the ease of laying out an internal arrangement. This subtle but important benefit is responsible for lowering maintenance costs for most airships by making all systems easily accessible for service or repair. There is also a reduced design cost based on the ease of finalizing the internal arrangement by not having to spend much time efficiently arranging and packaging systems, subsystems, and components.

12.7.1 Air Cushion Landing System

Historically, airships have needed only modest landing gear since their heaviness is generally small. This is not the case for hybrid airship designs. Since heaviness can be 40% or more of the unbuoyed weight, landing gear for hybrid airships need to be more substantial. Many hybrids will operate with takeoff and landing techniques similar to modern commercial aircraft. Integrating an aircraft style landing gear to a hybrid is a difficult task and the design will add significant weight as the structural designer attempts to spread out the point loads associated with landing gear bogeys. With variable winds side loads can be substantial and will quickly overload the gears capability unless it has the ability to caster. However, this doesn't solve all of the problems since the airship has to maintain its position during these winds. Ultimately, operational

Typical commercial transport landing gear. Point loads transmitted into air vehicle structure.

Air Cushion Landing System (ACLS) is very large but distributes the load over a large area. Since the bottom of a hybrid has a large area this system integrates well with ACLS designs.

Figure 12.9   An ACLS uniquely replaces a standard landing gear.

needs will demand that the vehicle be tied down for all ground support activities.

The beauty of an ACLS is that it naturally has a large area, which distributes the loads, allows taxi and ground maneuvers over austere/uneven terrain, and can be reversed to provide suckdown during variable wind conditions. Integrating an ACLS onto a hybrid airship uses existing technology of hovercraft vehicles. Figure 12.9 illustrates substituting an ACLS for a standard landing gear.

Estimating the drag of an ACLS is generally not an issue as modern hybrid designs include retractable ACLS that are faired over in flight. Thus, for cruise flight $\Delta C_{D_{0ACLS}}$ is assumed ≈ 0.0002. If the drag of extended ACLS pads is necessary then see Sec. 3.5.3.

For estimating the weight of an ACLS use the technique from Sec. 9.7.1. The sample problem at the end of this chapter will include an estimate of ACLS weight as well.

12.8   WT Testing and CFD

Wind tunnel testing and CFD play a more important role for the hybrid design than for an axisymmetric body airship. Since there is no data available to the public each company must create its own data that is unique to each design.

12.9 Turning

Requirements for a stable turn are created and discussed in Chapter 10. For the longitudinal (pitch) axis it is the term C_m/C_L that is important and for the directional (yaw) axis the important term is C_n/C_Y. The requirement of turning 360° in 1 min is used as a design goal in Chapter 10 for sizing tails. See Figs. 4.22 and 4.23 for data that estimates turning performance.

12.10 Sample Design Problem

A sample problem that shows the detailed calculation of a single hybrid airship design follows.

12.10.1 Requirements

The list of requirements for the sample problem is actually a short one.

1. Capable of operating out of austere landing sites
2. Range = 1000 nm
3. Payload = 40,000 lb
4. Maximum altitude = 8000 ft
5. Lifting gas is helium
6. Materials are current off-the-shelf Vectran fiber based fabrics
7. Cruise speed is 50 kt or 84.5 f/s at an altitude of 4000 ft
8. Maximum speed = 1.1 × cruise speed @ sea level
9. Reserve fuel = 5% fuel burned

12.10.2 Assumptions

1. Non-rigid, 3-lobed body (N_L = 2.4 from Sec. 3.3.1) with an ACLS
2. Buoyancy Ratio (BR) at landing = 0.70
3. Body fineness ratio (FR) = 3.0
4. Brake Specific Fuel Consumption (BSFC) = 0.48
5. (4) spark ignition engines capable of vectoring ±90 deg
6. Crew members = 3
7. Propeller efficiency, η_P, = 0.65
8. Propeller: 3-bladed variable-pitch propeller with Clark Y airfoil

The following is a step-by-step summary of how an airship is sized. Many of the equations and figures needed here have already been presented in earlier chapters. Some new equations and figures specific to this trade study are presented as needed. In particular, Appendix D will be very helpful.

12.10.3 The Sizing Process

The numbered sections in the Sample Problem correlate with the step number in the Excel spreadsheet in Table 12.1. This problem is solved in the same manner as for the sample problem in chapter 11. In both cases a volume is assumed and two separate values of takeoff weight W_G are calculated. The first value is calculated based on the amount of fuel necessary to fly the given range of 1000 nm. The other takeoff weight is calculated by estimating the weight using weight relationships given in chapter 9 or the current chapter for the given volume.

Sample Problem 12.1: Solution, Step by Step

1. $FR = 3$, number of lobes = 3 ($N_L = 2.4$)
2. Assume an initial value for the hull volume = 2,000,000 ft^3 (does not have to be very close to the final value). Further iterations will assume new volumes based on the difference between the calculated mission takeoff weight and the takeoff weight from the weight build-up calculations. When these two weights are the same the solution is exact.
3. Calculate reference value of $Vol^{2/3}$. Given the FR and volume determines the equivalent diameter, length, width, height, and body aspect ratio. For a given #lobes the lobe diameter can also be calculated. Once the lobe diameter is calculated then the body height and width are easily found.

   Calculate the equivalent body diameter

   $$d_e = (6 \times \text{Volume}/(FR \ \pi))^{1/3} = 108.4 \text{ ft}$$
   $$\ell_B = FR \ d_e = 325.2$$

   diameter of lobes = equiv diameter/(equiv diameter/diameter of lobes)

   $$d_c = d_e/(d_e/d_c)$$
   $$d_e/d_c = -0.0178 \ N_{LOBES}^2 + 0.361 \ N_{LOBES} + 0.575 \quad \text{(Fig. D.3)}$$
   $$d_c = 108.4/1.5 = 72.3 \text{ ft}$$
   $$w = (1 + N_{LOBES}) \ d_c/2 = 144.5 \text{ ft}$$
   $$ht = d_c = 72.3 \text{ ft}$$
   $$\text{aspect ratio, } FR = 4 \ w^2/(\pi \ \ell_B \ w) = 0.566 \text{ ft}$$

4. Using length, width, and height calculate the body surface area assuming the shape of a scalene ellipsoid, $S_{wet_{body}}$ using Eq. (D.3).

$$\text{Surface Area} = \pi \left((\ell_B^p w^p + \ell_B^p ht^p + w^p ht^p)/3 \right)^{1/p} \quad \text{where } p = 1.6075$$

$$S_{wet_{body}} \text{ (ellipsoid)} = 92{,}589 \text{ ft}^2$$

$$S_{wet_{body}} \text{ (lobed)} = (\text{perimeter}_{lobes}/\text{perimeter}_{ellipsoid}) \, S_{wet_{body}} \text{ (ellipsoid)}$$

$$S_{wet_{body}} \text{ (lobed)} = (1.081)(92{,}589) = 100{,}101 \text{ ft}^2 \text{ (Fig. D.2)}$$

5. Using the assumed envelope volume read horizontal and vertical tail volume coefficients C_{HT} and C_{VT} or calculate C_{VT} from the equation in Fig. 7.1. Be careful with these coefficients as there is no consistent way they are non-dimensionalized. A good approximation for the moment arm, ℓ_{tail}, of hybrid airships is 38% ℓ_B. This book uses ℓ_B for the vertical tail reference quantity instead of body width (span).

$$S_{HT} = C_{HT} \, (Vol^{2/3} \times \ell_B)/\ell_{tail} = 0.069 \, (15{,}874 \times 325.2)/(0.38 \times 325.2) = 2889 \text{ ft}^2$$

$$S_{VT} = C_{VT} \, (Vol^{2/3} \times \ell_B)/\ell_{tail} = 0.062 \, (15{,}874 \times 325.2)/(0.38 \times 325.2) = 2575 \text{ ft}^2$$

6. The cruise speed for this study is 50 kt (84.5 f/s) and the cruise altitude is 4000 ft as stated in the requirements which allows the dynamic pressure, q, to be calculated as $q = \frac{1}{2}\rho V^2 = (0.00211)(84.5^2)/2 = 7.53 \text{ lb/ft}^2$

7. Assume no laminar flow (this is conservative).

$$Re = \rho V \ell_B / \mu = (.00211)(84.5)(325.2)/3.66 \; 10^{-7} = 1.59 \; 10^8$$

$$C_f = 0.455/(\log_{10}(Re))^{2.58} = 0.00200$$

The form factor for body drag uses the relationship in Sec. 3.5.3.

$$FF_{3D \, body} = 1 + 1.5/FR^{1.5} + 7/FR^3 = 1.55.$$

Zero lift body drag is now calculated as

$$C_{D0_{body}} = FF_{3D_{body}} \, C_f \, S_{wet_{body}}/Vol^{2/3}$$

$$C_{D0_{body}} = (1.55)(0.00200)(100{,}101)/15{,}874 = 0.01949$$

8. Drag coefficient for the tails is

$$FF_{tails} = 1 + 1.2 \, (t/c) + 100 \, (t/c)^4 = 1.23 \text{ for } t/c = 0.15 \text{ (Sec. 3.5.3)}$$

$$C_{D0_{tails}} = FF_{tails} \, C_f \, S_{wet_{tails}}/Vol^{2/3}$$

Assume $AR_{tail} = 1.0$.

$$\bar{c}_{tail} \text{ (avg)} = (\bar{c}_{HT} + \bar{c}_{VT})/2 = [(AR_{HT} \, S_{HT}/2)^{1/2} + (AR_{VT} \, S_{VT}/2)^{1/2}]/2$$

$$\bar{c}_{tail} \text{ (avg)} = [(1 \times 2889/2)^{1/2} + (1 \times 2575/2)^{1/2}]/2 = 36.9 \text{ ft}$$

$$Re = 1.80 \; 10^7 \text{ which leads to } C_{fe} = 0.00274$$

$$S_{wet_{tails}} = 2.2 \, (2889 + 2575) = 12{,}021 \text{ ft}^2$$

where 2.2 is the ratio of (wetted area)/(planform area) for the tails

$$C_{D0_{tails}} = (1.23)(0.00274)(12{,}021)/(15{,}874) = 0.00255$$

9. C_{D_0} of cab/gondola combination, engines, cooling, mounting structure, cables, and landing gear is

$$C_{D_{0_{cab+gond}}} = (0.108 \, C_{D_{0_{body}}} \, Vol^{\frac{2}{3}} + 7.7)/Vol^{\frac{2}{3}} \qquad \text{[Eq. (3.29b)]}$$
$$= (0.108 \times 0.01949 \times 15{,}874 + 7.7)/15{,}874 = 0.00259$$
$$C_{D_{0_{eng\;nac}}} = (\#engines)(4.25)/Vol^{\frac{2}{3}} \qquad \text{[Eq. (3.31)]}$$
$$= (4)(4.25)/15{,}874 = 0.00107$$
$$C_{D_{0_{eng\;cooling}}} = (\#engines)(2 \times 10^{-6} \, Vol + 4.1)/Vol^{\frac{2}{3}} \qquad \text{[Eq. (3.32)]}$$
$$= (4)(2 \times 10^{-6} \times 2{,}000{,}000 + 4.1)/15{,}874 = 0.00204$$
$$C_{D_{0_{eng\;mount}}} = (0.044 \, C_{D_{0_{body}}} \, Vol^{\frac{2}{3}} + 0.92)/Vol^{\frac{2}{3}} \qquad \text{[Eq. (3.30b)]}$$
$$= ((0.044)(0.01949)(15{,}874) + 0.92)/15{,}874 = 0.00092$$
$$C_{D_{0_{cables}}} = (9.7 \times 10^{-6} \, Vol + 10.22)/Vol^{\frac{2}{3}} \qquad \text{[Eq. (3.26)]}$$
$$= (9.7 \times 10^{-6} \, 2{,}000{,}000 + 10.22)/15{,}874 = 0.00187$$
$$C_{D_{0_{ACLS}}} = 0.0002$$

(ACLS system is faired over during cruise so the drag is very low)

10. Interference drag

$$C_{D_{0_{int}}} = (4.78 \times 10^{-6} \, Vol)/Vol^{\frac{2}{3}} \qquad \text{[Eq. (3.34)]}$$
$$= (4.78 \times 10^{-6} \, 2{,}000{,}000)/15{,}874 = 0.00060$$

11. Total zero-lift-drag

$$C_{D_0} = C_{D_{0_{body}}} + C_{D_{0_{tails}}} + C_{D_{0_{gond}}} + C_{D_{0_{ecm}}} + C_{D_{0_{LG}}} + C_{D_{0_{int}}}$$
$$C_{D_0} = 0.01949 + 0.00255 + 0.00259 + 0.00403 + 0.00187 + 0.0002 + 0.00060$$
$$C_{D_0} = 0.03133$$

12. Drag-due-to-lift is the other part of total drag that is added to C_{D_0} to obtain a total drag coefficient. The *drag-due-to-lift factor*, K, is obtained from Fig. 3.26 using the body aspect ratio. For a hybrid airship body $AR = width^2/S_{plan}$ and was calculated as $AR = 0.566$ from step {3}. Using the equation in Fig. 3.26 for $AR = 0.566$ which is referenced to S_{plan} gives $K = 0.685$. See [9] for a detailed discussion of Fig. 3.26.

$$K = -0.0145\left(\frac{1}{AR}\right)^4 + 0.182\left(\frac{1}{AR}\right)^3 - 0.514\left(\frac{1}{AR}\right)^2 + 0.838\left(\frac{1}{AR}\right) - 0.053$$
$$K = 0.685/N_L = 0.685/2.4 = 0.286 \quad (\textit{referenced to } Vol^{\frac{2}{3}})$$

13. The buoyant lift is calculated using 0.0646 lb/ft³ for helium at sea level conditions. This number can vary slightly for different purities.

Buoyant lift $= 0.0646 \times$ Hull volume $\times$ density$_{\text{max alt}}$/density$_{\text{sea level}}$

$$L_{buoy} = 0.0646 \; Vol \; (\sigma_{max \; alt})$$
$$= (0.0646) \; (2{,}000{,}000) \; (0.7860) = 101{,}551 \; lb$$

14. Defining weight terms (see Fig. 12.7 for schematic of weight term definitions).

$$W_E + \text{fuel}_{unusable} + \text{oil}_{unusable} + \text{crew} = W_{OE} + \text{payload} = W_{ZF} + \text{fuel} = W_G$$

Where $W_{ZF} = L_{buoy}/BR_{land} - fuel_{res} = 101{,}551/0.7 - 1251 = 143{,}822 \; lb$

15. Since $W_{ZF} = W_{OE} + P_L$

$$W_{OE} = W_{ZF} - P_L$$
$$W_{OE} = 143{,}822 - 40{,}000 = 103{,}822 \; lb$$

16. Calculate the *takeoff heaviness* (W_{H_0}) and *landing heaviness* (W_{H_1}) necessary to fly the required range of 1000 nm. Rewrite Eq. (4.18) as

$$\text{Range} = A \left[\tan^{-1} \left(\frac{W_{H_0}}{B} \right) - \tan^{-1} \left(\frac{W_{H_1}}{B} \right) \right]$$

Where $A = \dfrac{326 \, \eta_p}{BSFC \sqrt{K \, C_{D_0}}}$ and $B = q Vol^{2/3} \sqrt{\dfrac{C_{D_0}}{K}}$

Reserve fuel is still on board at landing and since it can be calculated in several different ways it will equal 5% Fuel Burned in this problem.

$$W_{LAND} = W_{ZF} + fuel_{res} = 143{,}822 + 1251 = 145{,}073 \; lb$$
$$W_{H_1} = W_{LAND} - L_{buoy} = 145{,}073 - 101{,}551 = 43{,}522 \; lb$$

Solving the previous equation for the takeoff heaviness, W_{H_0}, yields-

$$W_{H_0} = B \, \tan \left[\frac{\text{Range}}{A} + \tan^{-1} \left(\frac{W_{H_1}}{B} \right) \right]$$

$A = 326 \, \eta_p / [BSFC \, (K \, C_{D_0})^{1/2}] = 326 \times 0.65 \, / \, [0.48 \, (0.286 \times 0.03133)^{1/2}] = 4666$

$B = q \, Vol^{2/3} \, (C_{D_0} \, K)^{1/2} = 7.53 \times 15{,}874 \, (0.03133/0.295)^{1/2}] = 39{,}568$

$W_{H_0} = (39{,}568) \tan [(1000/4666) + \tan^{-1} (43{,}522/39{,}568)] = 68{,}545 \; lb$

17. Mission fuel (fuel burned) is simply the difference between takeoff heaviness and landing heaviness.

$$\text{Fuel Burned} = W_{H_0} - W_{H_1} = 68{,}545 - 43{,}522 = 25{,}023 \; lb \; (\textit{mission fuel})$$
$$\text{Total Fuel} = W_{H_0} - W_{H_1} + fuel_{res} = 25{,}023 + 1251 = 26{,}274 \; lb$$

18. Therefore, the gross weight at takeoff to perform the mission is

$$W_G = W_{LAND} + fuel\ burned = W_{ZF} + fuel_{res} + fuel\ burned$$
$$W_G = 143,822 + 1251 + 25,023 = 170,096\ \text{lb}$$
$$BR_{TO} = L_{buoy}/W_G = 101,551/170,096 = 0.597$$

Although not used in the sizing routine the buoyancy ratio at takeoff, BR_{TO}, is calculated to make sure it is not too low. Any value of BR_{TO} less than 0.6 is an indicator that there needs to be a change in one of the other design parameters such as range, BR at landing, etc.

At this point the takeoff gross weight (W_G) to transport 40,000 lb of payload along a 1000 nm mission is known. The remaining steps will calculate the W_G that results from a weight buildup, which also uses the same assumed volume of 2,000,000 ft³.

19. In order to estimate maximum engine power and propeller size it is necessary to calculate lift, drag, and power required for the airship. The greatest aerodynamic lift is at start of cruise at sea level.

$$L_{aero} = W_{H_0} = 68,545\ \text{lb}$$

20. Aerodynamic Lift Coefficient (maximum power)

$$q_{max} = \text{density at SL} \times V_{max}^2/2 = (0.002377)(1.1 \times 84.5)^2/2 = 10.26\ \text{lb/ft}^2$$
$$C_{L_{max\ power}} = W_{H_0}/q_{max}/Vol^{2/3} = 68,545/(10.26)/15,874 = 0.421$$

21. Total drag at maximum power condition

$$\text{Drag} = \left(C_{D_0} + K C_{L_{aero}}^2\right) q_{max}\ Vol^{2/3}$$
$$\text{Drag} = (0.03133 + (0.295)(0.421)^2)(10.26)(15,874) = 13,346\ \text{lb}$$

22. Maximum power per engine

$$\text{maximum power/engine} = V_{max}\ \text{drag}/\eta_p/NE/550$$
$$\text{maximum power/engine} = (1.1)(84.5)(13,346)/0.65/4/550 = 867\ \text{hp}$$

The propulsion section assumes that any size engine is available along with any size propeller. Actually, there are only discrete engines available and one of these engines would have to be chosen that is close to the required power and has the appropriate *BSFC*.

23. Propeller speed and engine sizing. Although reciprocating engines perform most efficiently (lowest *BSFC*) for rpm = 2000–2400 they are geared down for the propeller by as much as a factor of 3 for the best propeller speed. The process of finding the best propeller speed is complicated but it can be done using Fig. 5.9b. The best speed is usually between 10 rps and 20 rps. A value of 10 rps is assumed for this problem.

For this sample problem it is assumed that the engine is sized for maximum speed. In fact, for hybrid airships where takeoff performance is important it may be necessary to calculate engine size for a given value of takeoff distance or takeoff climb gradient. Calculations would have to be expanded using Figs. 5.10 and 5.11 to include propeller disk loading to find the engine power for takeoff. It is possible that engines could be sized for takeoff performance rather than for maximum speed.

24. Next, the thrust *speed coefficient, Cs,* is calculated and *propeller advance ratio, J,* is found from Fig. 5.9b. Assume a variable pitch propeller for this problem. In a comprehensive study of propeller and engine sizing engine rpm and propeller blade angle are varied to get the best performance over a wide range of flight conditions. Variable pitch propellers have the ability to vary blade angle to maximize performance at various flight conditions. However, variable pitch propellers are heavier, more expensive, and require more maintenance.

$$C_S = \left(\frac{\rho V^5}{P n^2} \right)^{1/5} \qquad J = V/n\, D_P$$

$$C_S = [(0.002377)(84.5)^5/867/550/10^2]^{0.2} = 0.735$$

A curve fit of the J vs C_S data in Fig. 5.9 yields the following equation:

$$J = 0.156\, C_{S^2} + 0.241\, C_S + 0.138$$

For $C_S = 0.73, J = 0.399$.

The propeller diameter is now calculated as

$$D_P = (84.5)/(10)/(0.396) = 21.3 \text{ ft}$$

25. Propeller efficiency
    A curve fit of the optimum η_P vs C_S data on Fig. 5.9 is

$$\eta_P = 0.139\, C_S^3 - 0.749\, C_S^2 + 1.37\, C_S + 0.0115 = 0.667$$

Since η_P was assumed to be 0.65 for earlier calculations this is fairly close. However, if the assumed η_P is not close to the actual η_P the answer can be improved by setting the assumed η_P to the actual and recomputing from step {16} (convergence is rapid).

26. One of the benefits of a hybrid airship is the use of lobes to create the airship's cross-sectional shape. Figures D.2 and D.3 provide the background for determining the lobed geometry. The reduction in lobe diameter compared to the equivalent diameter {3} will give a lower envelope weight. All that needs to be calculated is the internal pressure and then the body weight can be found.

For Vectran laminate fabrics the density is obtained from Fig. 8.4 to calculate the weight of the envelope for the assumed material.

$$p_l = 1.2\ q_{max} + 0.0635\ \text{height}$$
$$p_l = [1.2\ (10.26) + 0.0635\ (72.3)]/144 = 0.117\ \text{lb/in.}^2$$

The following steps use weight relationships from Chapters 8 and 9.

27. The weight build-up begins with the body and includes factors for manufacturing (1.2) and attachment fittings (1.26). Assume envelope and septum materials have the same areal density and are made from a Vectran laminate. Assume the factor of safety (*FS*) is 4. See Sec. 8.9 for hull weight discussion. For a hybrid hull where

$$\text{hull fabric load (lb/in.)} = FS \times \text{internal pressure} \times \text{lobe radius}$$
$$\text{hull fabric load} = 4 \times 12 \times 0.117 \times 72.3/2 = 203.0\ \text{lb/in.}$$

From Fig. 8.4, fabric density (oz/yd^2) = 0.0085 × hull load (lb/in.) + 1.365

$$\text{hull fabric density} = 0.0085\ (203) + 1.365 = 3.095\ \text{oz/yd}^2$$
$$W_{env} = \text{hull fabric density} \times (\text{manufacturing}) \times (\text{attachments})\ S_{wet_{body}}$$
$$W_{env} = (3.095)\ (1.2)(1.26)\ (100,101)/16/9 = 3253\ \text{lb}$$

Assume there are (2) septums each that are 0.75 of the body side area and have a 1.06 factor for seaming. Using the approach in Sec. 8.7 yields:

$$\text{Septum fabric load} = (1.5)(\text{hull fabric load}) = (1.5)(203) = 304.5$$
$$\text{Septum fabric density} = 0.0085\ (304.5) + 1.365 = 3.95\ \text{oz/yd}^2$$
$$W_{sep} = (2)\ (1.06)\ \text{septum fabric density} \times (0.75)\ \pi\ ht\ \ell_B/4)/16/9$$
$$W_{sep} = (2)\ (1.06)\ (3.95)\ (0.75)\ \pi\ (72.3)(325.2)/4)/16/9 = 807\ \text{lb}$$
$$W_{body} = 3253 + 807 = 4060\ \text{lb}$$

28. Ballonet weight is calculated for 2 spherical ballonets in each lobe for a total of 6 ballonets. Maximum altitude is 8000 ft and hull volume is 2,000,000 ft^3.

$$Vol_{Ball} = Vol_{env}\ (1/\sigma_{max\ alt} - 1) = 2,000,000\ (1/\ 0.7860 - 1) = 544,529\ \text{ft}^3$$

is based on empirical data in Fig. 9.3. This ballonet volume differs 27.2% from the theoretical value based on $(1 - \sigma_{max\ alt}, 21.4\%)$ because actual airship designs must also account for unusable ballonet volume, non-standard days, and superheat. This means that the actual ballonet is ~6% larger in this problem than the theoretical minimum size.

surface area of 6 spherical ballonets = $6 \, \pi \, (3 \, Vol_{Ball}/\pi/6)^{2/3}$ = 36,900 ft^2

$$W_{Ball} = (0.035) \text{ (surface area of ballonets)}$$
$$W_{Ball} = (0.035) \, (36,900) = 1292 \text{ lb}$$

29. Tail weight

This problem assumes a rigid space-frame structural tail concept. Use Eq. (9.2) where $F_{AF} = 1.26$ and $F_{PSQ} = 1.0$ lb/ft^2. For control surfaces that are 20% of the total area the tail weight is the sum of the stabilizer and fin weight and the control surface weight. Actuator weights are calculated here as well but are added to the VMS weights.

$$W_{SSF} = 1.0 \, (S_{HT} + S_{VT}) \, (1.26) \, (80\%) = (2889 + 2575)(1.008) = 5508 \text{ lb}$$
$$W_{CS} = 1.0 \, (S_{HT} + S_{VT}) \, (20\%)(1) = (2889 + 2575)(0.2) = 1093 \text{ lb}$$
$$W_{tails} = W_{SSF} + W_{CS} = 5508 + 1093 = 6601$$
$$W_{act} = (1.15) \, (2889 + 2575) \, (0.79) \, (0.2) = 993 \text{ lb } \textit{(part of VMS)}$$

where (1.15) is the installation factor for installing the actuators.

30. Crew gondola and payload bay: Using Eqs. (9.6–9.7) with gondola dimensions of 55 ft / 10 ft / 10 ft and crew station dimensions of 10 ft/10 ft/10 ft yields-

$$W_{crew \, stat} = 353 \, [(\ell_G/10)^{0.857} \, (w + h)/ \, 10 \, (V_{max}/ \, 10)^{0.338}]^{1.1} = 1426 \text{ lb}$$
$$W_{gond} = 1.875 \text{ surface area of gondola}$$
$$W_{gond} = 1.875 \, (2)(55 \times 10 + 55 \times 10 + 10 \times 10) = 4500 \text{ lb}$$

31. Weight of engines: (reciprocating) *(assume any size engine is available)*

From Fig. 5.8 $W_{eng} = N_E \, 4.848 \, (\text{power/engine})^{0.7956}$

$$W_{eng} = (4) \, 4.848 \, (867)^{0.7956} = 4218 \text{ lb}$$

32. Engine mounts, engine controls, and starting: 4050

$$W_{EngMt} = 0.64 \, N_E \, W_{eng} = 0.64 \, (4) \, (4218/4) = 2700 \text{ lb}$$
$$W_{EC} = 60.27 \, (\ell_{EC}N_E/100)^{0.724} = (60.27) \, [(150) \, (4)/100]^{0.724} = 221 \text{ lb}$$
$$W_{Start} = 50.38 \, (N_E \, W_{eng}/1000)^{0.459} = (50.38) \, (4 \times 4218/4/1000)^{0.459} = 98 \text{ lb}$$

33. Propeller weight: Using Eq. (9.16) and using propeller diameter from {32}

$$W_{Prop} = K_P \, N_P \, (N_{BL})^{0.391} \, (d_P \, SHP/1000)^{0.782}$$
$$W_{Prop} = 31.92 \, (4) \, (3)^{0.391} \, (21.2 \times 867/1000)^{0.782} = 1911 \text{ lb}$$

34. Fuel tank weight: Fuel is in gallons and aviation gas weighs 6.0 lb/gal.

$$W_{FT} = 2.49 \, (\text{Fuel})^{0.6} \, (N_T)^{0.2} \, (N_E)^{0.13} \, [1/(1 + Int)]^{0.3}$$
$$W_{FT} = 2.49 \, ((25,023 + 1251)/6.0)^{0.6} \, (2)^{0.2} \, (4)^{0.13} \, [1/(1 + 0)]^{0.3} = 524 \text{ lb}$$

35. Pressure system weight: Use Fig. 9.9

$$W_{PressSys} = (2\%)(W_{OE}) = 0.02\ (103{,}786) = 2076\ \text{lb}$$

36. ACLS landing system: The size of the ACLS pads can be sized by either a landing at a given sink rate or taxiing at takeoff weight. A quick calculation compares the two results. See Chapter 9 for discussion. Where plenum pressure = internal envelope pressure of 0.117 lb/in.^{2} At landing condition:

$$\text{Area of ACLS pads} = (0.23)\ W_{H_1}\ V_{SR}/(N_{PAD}\ P_{PAD})$$
$$\text{Area of main ACLS pad} = (0.23)(43{,}522)(4)/(2)/(0.117 \times 144) = 1188\ \text{ft}^2$$
$$\text{Area of nose ACLS pad} = (0.117)(2)(695) = 163\ \text{ft}^2$$
$$\text{Total ACLS pad area based on landing} = (2)(1188) + 163 = 2539\ \text{ft}^2$$

At takeoff the total pad area = takeoff heaviness/plenum pressure
$$\text{Area of ACLS pads based on takeoff} = 68{,}545/(0.1173 \times 144) = 4057\ \text{ft}^2$$

The weight of the ACLS system is based on takeoff since it has the greatest pad area, the weight is calculated by 1.6 × ACLS pad area [Eq. (9.21)] so,

$$W_{ACLS} = (1.6)(4057) = 6491\ \text{lb}$$

37. VMS weight can be calculated two ways. Figure 9.10 could be used as representative of a typical VMS suite for a hybrid airship or the pieces can be added in separately. Assume $W_{computer} + W_{avionics} = 500$ lb and add in the tail actuator weight from step {29}

$$W_{VMS} = W_{computer} + W_{avionics} + W_{act} = 500 + 993 = 1493\ \text{lb}$$

38. Electrical system weight: Using Eq. (9.17) and assuming the electronic/avionics/VMS equipment weighs 500 lb results in the following –

$$W_{Elect} = 33.73\ [(W_{FS} + W_{TRON})]^{0.51}$$
$$W_{Elect} = 33.73\ [(470 + 500)]^{0.51} = 1125\ \text{lb}$$

39. Miscellaneous systems weight: (Fig. 9.10)

$$W_{MSys} = 0.05\ W_{OE} = 0.05\ (103{,}822) = 5191\ \text{lb}$$

40. Crew and accommodations: Using Eqs. (9.22–9.25) with 2 crew seats, 5 passenger seats, and 3 bunks plus 2 crew members and 3 passengers (not part of payload) yields -

$$W_{Crew+Acc} = W_{Seats} + W_{Bunks} + W_{Lav} + W_{Food} + W_{Crew+Pax}$$
$$W_{Crew+Acc} = (55 \times 2 + 32 \times 5 + 28 \times 3) + 5.6\ (3)^{1.33} + 5.68(3)^{1.12} + (3)(250) = 1148\ \text{lb}$$

where crew members/passengers + luggage = 250 lb/person

41. The 2 final weight items are unusable fluids (gas & oil) and the empty weight margin. Typical values for weight margin on new aircraft designs are usually between 5–10% depending on the complexity of the design. Since fabric weights are less predictable than metallic weights 6% is used in this example.

$$W_{uf} = 0.01 \text{ fuel}$$

$$W_{margin} = 0.06 \, W_{OE}$$

$$W_{uf+margin} = 0.01 \times (25{,}023 + 1251) + 0.06 \times 103{,}822 = 6492 \text{ lb}$$

42. Adding up the weight items between {27} thru {41} yields the operating empty weight, W_{OE}, based on a system and component weight build up.

43. This final weight is the gross weight, W_G, at takeoff which is based on estimating the weights.

$$W_{G_{wts}} = W_{OE} + fuel + PL = 117{,}870 \text{ lb}$$

44. This compares the two values of W_G, one based on performing the mission, $W_{G_{perf}}$ {18} and the other based on estimating the total weights, $W_{G_{wts}}$ {43}. A final solution is obtained when $(W_{G_{perf}} - W_{G_{wts}}) = 0$ which yields the weight, fuel, and volume that will perform the mission under the specific assumptions. Table 12.1 shows 4 manual iterations and a final calculation that is exact (where $W_{G_{perf}} - W_{G_{wts}} = 0$) using the "Goal Seek" tool that is available in the Excel Program.

The "exact solution" design may not be realistic if the angle of attack to fly at maximum heaviness, W_{H_0} is too large or the ground run distance is too great. It is necessary to calculate the flight angle and ground distance to make sure they are acceptable.

Steps {45} through {59} refer to the values from the "exact" solution case (Vol = 934,983 ft^3) in Table 12.1 and not for the case where Vol = 2,000,000 ft^3.

45. Using

$$AR = 0.566 \text{ the data on Fig. 3.8 gives a } C_{L\alpha} = 0.0186$$

which is based on the planform area. Therefore, it is necessary to convert $C_{L\alpha}$ from one that is referenced to S_{plan} to the traditional one for airships which is referenced to $Vol^{2/3}$. From Eq. (3.1) and its accompanying table find that $N_L = 2.4$.

$$C_{L\alpha} \text{ (per deg ref to } Vol^{2/3}) = C_{L\alpha} \text{ (Fig. 3.8 ref to } S_{plan}) N_L$$

$$C_{L\alpha} = 0.0186 \, (2.4) = 0.045$$

For the exact solution case in Table 12.1 find the angle of attack at W_{H_0}.

$$C_L \text{ (max heaviness)} = W_{H_0} / (q \, Vol^{\frac{2}{3}}) = 30{,}960/(7.53 \times 9685) = 0.425$$
$$\text{alpha (max heaviness)} = C_L/C_{L_\alpha} = (0.425) / 0.045 = 9.4 \text{ deg}$$

This angle of attack is within 15 deg which is an angle of attack where lift behavior of low aspect ratio bodies becomes non-linear.

46. It is also necessary to calclate the takeoff ground distance to make sure the design has engines powerful enough plus sufficient aerodynamic lift generated at the maximum scrape angle.

   Assume a takeoff scrape angle limit of 10 deg. Calculate the C_L at this angle, find the speed at this angle, calculate the thrust and drag and then compute the takeoff ground distance. Liftoff speed,

$$V_{TO} = 1.1 \times Speed \, @ \, \alpha_{scrape}$$
$$C_{L_{to}} = C_{L_\alpha} \, \alpha_{scrape} = (0.045)(10) = 0.45$$
$$\text{Speed at liftoff is } V_{to} = \left[2 W_{H_0} / (\rho \, C_{L_{to}} \, Vol^{\frac{2}{3}}) \right]^{\frac{1}{2}}$$
$$V_{TO} = (1.1)[(2)(30{,}960/(0.002377 \times 0.45 \times 9685)]^{\frac{1}{2}} = 85.1 \text{ f/s}$$

47. Thrust during ground run: Calculate the

$$\text{power loading} = hp/A = 349/(\pi \, 18.0^2/4) = 1.38$$

where $hp = 349$ is maximum single engine power during takeoff and A is propeller disc area. Static thrust T_0 for a variable pitch prop is determined from Fig. 5.11. For a power loading of 1.38, $T_0/hp = 7.91$ giving a static thrust $T_0 = (7.91)(349) = 2761$ lb for each engine.

   Variable pitch prop thrust reduction due to forward speed is determined from Fig. 5.10. Takeoff acceleration is estimated at 0.707 V_{TO} = 60.2 ft/s. From Fig. 5.10 the $T/T_0 = 0.669$ at 60.3 ft/s gives $T = 1844$ lb for each engine. Total average accelerating thrust is 4(1844) = 7375 lb.

48. Average drag and acceleration during takeoff:
   The average drag at $V = 0.707 \, V_{TO}$ is $(0.002377/2)(60.2)^2 (0.03366)$ (9562) = 1385 lb.
   The acceleration is $T - D - gear \, friction = T - D$ because *gear friction* is zero due to the ACLS.
   Thus the acceleration = 7375 − 1385 = 5990 lb at 60.2 ft/s.
   The mass accelerated during takeoff = W_G/g + (air mass in ballonet) + (helium mass)

$$= (77{,}834/32.17) + (254{,}563)(0.081)/32.17 + (934{,}983 - 254{,}563)(0.0111)/32.17$$
$$= 2419 + 641 + 235 = 3295 \text{ slugs}$$

The ground roll distance $S_G = \frac{1}{2} \, (mass)(V_{TO})^2/(T - D)_{0.707 \, V_{TO}}$
$$= \frac{1}{2} \, (3295)(85.1)^2/(5990) = 1991 \text{ ft.}$$

This completes the sample problem so some observations are appropriate. Even though the initial value for volume was far from exact, convergence is rapid and intuitive. It is highly recommended that an Excel spreadsheet model be created using "Goal Seek" to find the exact solution. The two checks, one for flight angle of attack and another for takeoff ground distance are important. For this problem the 9.5 deg of angle at initial cruise is acceptable but must be monitored throughout the design cycle. Takeoff ground distances of about 2000 ft are generally acceptable.

49. Ballast needed for neutral weight

This last parameter is not necessarily a design parameter but is important to the operations of a hybrid airship. Since outbound payloads rarely equal inbound payloads some ballasting may need to take place. Although the hybrid airship is specifically designed to reduce or eliminate ballast for large payload missions, sometimes there is a large payload mismatch. This value represents the amount of weight that has to be added back to the airship either in the form of return payload or ballast weight.

$$\text{Ballast, payload offset} = \text{Outbound payload} - \text{Landing Heaviness}$$
$$\text{Ballast, payload offset} = 40,000 - 20,346 = 19,654 \text{ lb}$$

Figure 12.10 presents a summary of the pertinent design characteristics that have been established during the sizing of this hybrid airship. These data come from the right most column (exact solution) in Table 12.1.

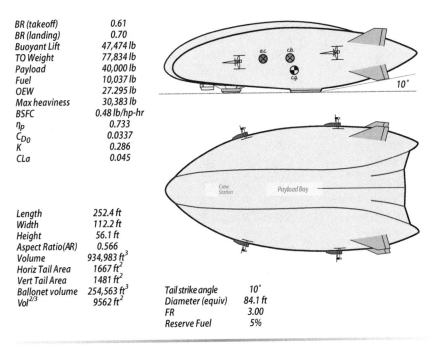

BR (takeoff)	0.61
BR (landing)	0.70
Buoyant Lift	47,474 lb
TO Weight	77,834 lb
Payload	40,000 lb
Fuel	10,037 lb
OEW	27.295 lb
Max heaviness	30,383 lb
BSFC	0.48 lb/hp-hr
η_p	0.733
C_{Do}	0.0337
K	0.286
C_{La}	0.045

Length	252.4 ft
Width	112.2 ft
Height	56.1 ft
Aspect Ratio(AR)	0.566
Volume	934,983 ft^3
Horiz Tail Area	1667 ft^2
Vert Tail Area	1481 ft^2
Ballonet volume	254,563 ft^3
Vol$^{2/3}$	9562 ft^2

Tail strike angle	10°
Diameter (equiv)	84.1 ft
FR	3.00
Reserve Fuel	5%

Figure 12.10 Sample problem "exact solution" characteristics for a hybrid airship.

Table 12.1 Summary of Sample Problem Calculations from Excel Spreadsheet

Step	Parameter	Manual iterations				Exact
1	$FR =$	3.00	3.00	3.00	3.00	3.00
2	Volume (ft³) =	2,000,000	1,500,000	1,300,000	1,000,000	934,983
3	Vol$^{(2/3)}$ =	15,874	13,104	11,911	10,000	9562
	Diameter-equiv (ft) =	108.4	98.5	93.9	86.0	84.1
	Body length (ft) =	325.2	295.4	281.7	258.1	252.4
	Diameter-lobes (ft) =	72.3	65.6	62.6	57.4	56.1
	Body width (ft) =	144.5	131.3	125.2	114.7	112.2
	Body height (ft) =	72.3	65.6	62.6	57.4	56.1
	Aspect ratio =	0.566	0.566	0.566	0.566	0.566
4	Body surface area-ellipsoid (ft²) =	92,589	76,431	69,476	58,327	55,771
	Body surface area-lobed (ft²) =	100,101	82,632	75,113	63,060	60,296
5	Hor tail vol coeff =	0.069	0.068	0.068	0.067	0.066
	Vert tail vol coeff =	0.062	0.061	0.060	0.059	0.059
	Hor tail area (ft²) =	2889	2355	2125	1753	1667
	Vertical tail area (ft²) =	2575	2098	1891	1558	1481
6	Velocity-cruise (fps) =	84.45	84.45	84.45	84.45	84.45
	Qcruise (lb/ft²) =	7.53	7.53	7.53	7.53	7.53
7	Re-body =	1.59E+08	1.44E+08	1.37E+08	1.26E+08	1.23E+08
	Cfe-body =	0.00200	0.00202	0.00204	0.00206	0.00207
	FF-body-Hoerner =	1.55	1.55	1.55	1.55	1.55
	C_{D_0}-body =	0.01949	0.01975	0.01988	0.02012	0.02018
8	(4 tails) Re-tails =	1.80E+07	1.63E+07	1.54E+07	1.40E+07	1.37E+07
	Cfe-tails =	0.00274	0.00278	0.00280	0.00285	0.00286
	FF-tails =	1.23	1.23	1.23	1.23	1.23
	CD-tails =	0.00255	0.00256	0.00256	0.00255	0.00255
9	CD-cab/gondola =	0.00259	0.00272	0.00279	0.00294	0.00299
	CD–nac+cool+mount =	0.00403	0.00440	0.00463	0.00512	0.00526
	CD-cables =	0.00187	0.00189	0.00192	0.00199	0.00202
	CD-ACLS =	0.0002	0.0002	0.0002	0.0002	0.0002
10	CD-interference =	0.00060	0.00055	0.00052	0.00048	0.00047
11	C_{D_0} =	0.03133	0.03207	0.03250	0.03340	0.03366
12	$K =$	0.286	0.286	0.286	0.286	0.286
13	L-buoy (lb) =	101,551	76,163	66,008	50,776	47,474
14	WZF (lb) =	143,822	107,862	93,485	71,537	67,295

Step	Parameter	Manual iterations				Exact
15	WOE (lb) =	103,822	67,862	53,485	31,537	27,295
16	Landing heaviness, W_{H_1} (lb) =	43,522	32,641	28,289	21,761	20,346
	A =	4666	4612	4581	5214	5021
	B =	39,568	33,047	30,240	25,738	24,705
	W_{H_0} (lb) =	68,545	51,025	44,161	32,015	30,383
17	Fuel burned (lb) =	25,023	18,383	15,872	10,254	10,037
	Reserve fuel (lb) =	1251	919	794	1000	502
18	WG-performance (lb) =	170,096	127,164	110,150	82,790	77,834
	BR-to =	0.597	0.599	0.599	0.613	0.610
19	Aero Lift@start cr (lb) =	68,545	51,001	44,141	32,015	30,360
20	CL@Vmax@SL =	0.421	0.379	0.361	0.312	0.310
21	Drag @ Vmax@SL start cruise (lb) =	13,346	9840	8527	6281	5987
22	Max Power/eng@ Vmax@SL (hp) =	867	639	554	354	349
23	Propeller speed (rps) =	10	10	10	10	10
24	Cs@SL,max power =	0.735	0.781	0.804	0.879	0.882
	J@max power =	0.399	0.421	0.432	0.470	0.471
	Propeller diameter (ft) =	21.2	20.1	19.6	18.0	17.9
25	Prop efficiency-np =	0.669	0.691	0.701	0.732	0.733
26	Internal pressure (psi) =	0.117	0.114	0.113	0.111	0.110
27	Hull fabric density (oz/yd^2) =	3.095	2.897	2.809	2.661	2.626
	Wt-env+lpatch+ seams (lb) =	3253	2514	2215	1762	1662
	Wt-septums (lb) =	807	616	540	425	400
28	Volume- ballonet-ft^3 (lb) =	544,529	408,397	353,944	272,265	254,563
	Wt-ballonet (lb) =	1292	1067	969	814	778
29	Wt-tails (lb) =	6601	5379	4851	3999	3803
30	Wt-Crew Sta+ Gondola (lb) =	5926	5926	5926	5926	5926
31	Wt-engs (lb) =	4218	3310	2953	2067	2043
32	Wt-eng mounts+ ec+st (lb) =	3018	2426	2194	1613	1598
33	Wt-props (lb) =	1911	1444	1265	834	823
34	Wt-fuel tanks (lb) =	524	436	399	315	303

(continued)

Step	Parameter	Manual iterations				Exact
35	Wt-Pressure Sys (lb) =	2076	1357	1070	631	546
36	Wt-ACLS (lb) =	6491	4955	4340	3212	3064
37	Wt-VMS (lb) =	1493	1309	1230	1102	1072
38	Wt-Elect sys (lb) =	1157	1105	1082	1030	1022
39	Wt-Misc sys (lb) =	5191	3393	2674	1577	1365
40	Wt-Crew+lav (lb) =	1148	1148	1148	1148	1148
41	Wt-margin+ ufluids (lb) =	6492	4265	3376	2005	1743
42	Wt-empty, WOE (lb) =	51,597	40,649	36,231	28,457	27,295
43	WG-weights (lb) =	117,870	99,951	92,896	79,711	77,834
44	(WG-perf) – (WG-wts) (lb) =	52,226	27,213	17,253	3079	0
45	CL-alpha =	0.0447	0.0447	0.0447	0.0447	0.0447
	Max Flight Alpha (deg) =	12.8	11.6	11.0	9.5	9.4
46	CL-to @ Alpha-scrape =	0.447	0.447	0.447	0.447	0.447
	TO Speed@ Alpha-sc (fps) =	99.2	94.2	91.9	85.4	85.1
47	Power loading (hp/A) (Fig 5.9) =	2.46	2.02	1.84	1.39	1.38
	Thrust/thrustSLS (Fig 5.10) =	0.639	0.648	0.653	0.668	0.669
	ThrustSLS/power (Fig 5.11) =	6.62	7.04	7.24	7.89	7.91
	Thrust @ .707VTO (lb) =	14,685	11,658	10,465	7449	7375
48	Drag-ground run (avg) (lb) =	2908	2216	1943	1448	1385
	Ground distance (ft) =	2313	2120	2035	2075	1991

12.11 Trade Studies

At this point further analyses would include trade studies to find the smallest/lightest hybrid design to better meet the relevant Measures of Merit (MoM). The smallest/lightest design will be the least development and acquisition cost.

Design Trades—Vary volume, *FR*, number of lobes, tail geometry, etc. to find the smallest/lightest vehicle.

Mission Trades—Vary cruise speed, cruise strategy, maximum speed, *BR*, etc. to find the lightest/smallest vehicle.

The MoM depends on the user and mission. For example:

1. Commercial cargo/passenger transport—Typical MoMs are $/ton-mile, ton-mile/day, $/passenger-mile, and passenger-mile/day where the $ is total operating cost (TOC: people, fuel, insurance, facility user fees, etc.).
2. Takeoff ground roll distance S_G—Users would typically like to keep $S_G < 2000$ ft.
3. Minimum capital costs—Minimum Infrastructure such as unimproved area, no mooring mast, no ballast provisions, minimal tie downs, minimal vehicle shelter, minimum personnel provisions, and no fuel. The hybrid design would most likely have an ACLS for landing and takeoff from an unimproved field. For the commercial or military user this would be operation to and from a remote area. Typical commercial missions would be to remote sites where the cost of building a road is prohibitive (such as logging, resupply, and pocket mining). A typical military mission would be a clandestine infiltration/exfiltration of special forces and equipment where there should be no evidence that a remote site landing had ever occurred.

It is interesting to consider a mission trade on takeoff *BR* holding everything else constant. As the takeoff *BR* varies from 0.8 to lower values, the volume, *TOGW*, *TO* ground roll distance S_G, dimensions and empty weight varies as shown on Fig. 12.11. From Fig. 12.11 we observe the following percent changes in going from a $BR = 0.8$ to 0.5 ($BR = 0.8$ is the baseline):

Item	Change (from $BR = 0.8$)	% Change
Volume	$-358{,}968$ ft^3	-28
Body length	-30 ft	-11
TOGW	$+7804$ lb	$+10$
Fuel burned	$+5354$ lb	$+61$
Empty weight	$+2808$ lb	$+10$
Total propulsion wt	$+2256$ lb	$+36$
ACLS weight	$+1985$ lb	$+170$
Total hull weight	-382 lb	-11
Tail weight	-1047 lb	-22

The gross features of the hybrid (volume, dimensions, $TOGW$) do not change much over a broad BR range. The selection of the TO BR for the design of the hybrid airship should be based upon the dominant MoM above. For example if the MoM is #1 (commercial cost) the BR should be around 0.8 to keep the fuel burned to a minimum. If the MoM is to keep the $S_G < 2000$ ft the BR should be greater than 0.65.

If MoM #3 is dominant, the takeoff BR should reflect the situation that the flight out would have no payload and the return flight would have full payload or vice versa. This means that the buoyant lift ≈ payload + ½ fuel and the vehicle is heavy at the remote site.

Selecting the TO BR fixes the buoyant lift. It is observed that a hybrid designed for operation at a TO $BR = 0.8$ could be used occasionally for lower BR operation provided the fuel tanks were enlarged (to carry the extra fuel), the ACLS was sized for the increased heaviness and the engines oversized. The $BR = 0.8$ design would have to add extra payload and fuel to get to $BR = 0.5$. On the other hand, the hybrid designed for a TO $BR = 0.5$ would have oversized fuel tanks, ACLS and engines but an undersized buoyant lift and would have to reduce payload and/or fuel to operate at higher BR such as $BR = 0.8$.

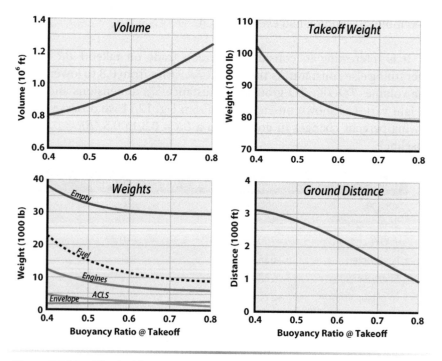

Figure 12.11 Effect of Buoyancy Ratio on various performance parameters.

Even though the vehicle from Table 12.1 is a good conceptual starting point, there is still much work to be done. Numerous assumptions were made to facilitate getting a good answer without getting bogged down in complex analyses or costly testing. Therefore, the next step is to start refining these preliminary answers by including specific test data, CFD analyses, structural modeling, more refined weight buildup, an enlarged propulsion study, and include the effects of such phenomena as *added mass* (see Appendix C).

References

[1] McPhee, J. *The Deltoid Pumpkin Seed*, Ballantine Books, 1979.

[2] Khoury, G.A. and Gillett, J.D., *Airship Technology*, Cambridge Aerospace Series 10, New York.

[3] Blakemore, T.L., *Pressure Airships-Nonrigid Airships*, Ronald Press, New York, 1927.

[4] Fink, R.D., *USAF Stability and Control DATCOM*; DTIC, 1978.

[5] Hoerner, S.F., *Fluid Dynamic Drag*; Hoerner Publishing, 1993.

[6] *Airship Design Criteria*, FAA-P-8110-2, U.S. Department of Transportation/FAA, Feb 1995.

[7] *Certification Specifications for Transport Category Airships, CS-30T*, EASA (European Aviation Safety Agency), September 2003.

[8] Nicolai, L.M. and Carichner, G.E., *Airplanes and Airships . . . Evolutionary Cousins*, AIAA Paper 2012-1178, 60th AIAA Aeronautical Sciences Meeting, Nashville, TN, Jan 2012.

[9] Nicolai, L.M. and Carichner, G.E., *Fundamentals of Aircraft and Airship Design, Volume I—Aircraft Design*, AIAA Education Series, Reston, VA, 2010.

Chapter 13 / Balloon Design

Rodger E. Farley

- Balloon First Principals
- Shape Fundamentals
- Flight Environments
- Membrane Structures
- Preliminary Design Methodology

2008 launch of the record-setting NASA super-pressure pumpkin balloon in Antarctica.

The simplest solution is often the hardest to see.

Anonymous

13.1 Balloon Design, Introduction

B alloons are one of the earliest aeronautical devices and yet, understanding their deceptively simple behavior can be a challenge. Fortunately, their mysterious ways can be described by a first principles approach. The balloon technology addressed in this chapter deals mainly with large plastic high-altitude scientific gas balloons that fly in the stratosphere. These giant balloons have a launch, ascent, float, and termination phase. The stratospheric scientific balloons commonly used range in size from 11 to 40 million cubic feet (MCF), with typical altitudes of 120,000 ft (36.5 km) and gross inflations (buoyant lift) up to 14,500 lb. Serious research to develop these types of stratospheric balloons have their origins with the U.S. Air Force (for cosmic ray effects on pilots), and for cold war high-altitude spying platforms. Many UFO sightings can be traced back to these large balloons. Today these high-altitude vehicles are used by the science community, conducted through the Columbia Scientific Balloon Facility supported by NASA.

What makes a lighter-than-air vehicle possible is the principle of buoyancy or Archimedes' principle where the buoyancy force is equal to the weight of the displaced fluid, such as a boat displacing water. A balloon is slightly different in that the buoyancy is equal to the difference between the weight of the displaced air and the weight of the lift gas. The buoyancy force for the whole volume is known as gross inflation (balloons) or buoyant lift (airships), which changes with temperature and pressure. When a balloon and its payload float in equilibrium, the system density is exactly equal to the density of the ambient air.

A lifting gas such as helium is used because it provides reacting pressure such that thin-film membrane structures can be used in the construction. If one wanted to reinvent the so-called "vacuum balloon" then the additional weight of a lift gas will seem trivial compared to the huge weight penalty from a structure that has to take the full compressive stress of atmospheric pressure (and not buckle). See Fig. 2.6 for further discussion.

There are many types of balloons. Latex weather balloons have a skin that keeps stretching until they burst (usually at a calibrated diameter) while having no fixed float altitude. Hot air balloons have constant pressure and volume but variable temperature differentials between inside and outside the gas envelope. Gas balloons can have variable or fixed volume with an uncontrolled temperature differential. They can be pressurized

The balloon is a 7M ft³ super pressure pumpkin with 200 gores designed by the author and built by Aerostar. An engineering test flight proved that the design/analysis methodology can deliver a meridionally reinforced membrane pressurized structure with no global shape instabilities. Antarctica is where this untested design can fly safely, and the balloon holds the record of 54 days of flight over Antarctica for like craft. The flight provided valuable performance data that also help to fine tune the balloon flight simulation software for stratospheric flight.

as a constant volume/variable pressure device or unpressurized as a constant pressure/variable volume device. Using volume, pressure, and temperature one can come up with combinations to operate at any fixed altitude.

Balloons can be made to have many different shapes, but the most common shape is the "natural shape", whereby the skin stress in the circumferential (transverse) direction is zero. Meridionally lobed membrane structures such as circular parachutes and balloons, where cords or tapes run in a vertical plane, concentrate the loads in the "longitudinal direction", or meridional as it is called. The stresses in the transverse direction (circumferential, hoop) are very small compared to the stresses in the meridional direction which concentrate in the load tapes. This leads to the natural shape, and depending on the internal pressurization and skin weight, will lead to the variation of natural shapes illustrated in Fig. 13.1.

The super pressure shape on the far left is called a pumpkin balloon, and once this shape has been established more pressure will not change its basic shape (with bulges it is a pumpkin, without bulges it is an isotensoid). On the far right is the under-pressure shape, so named because the cone angle of the skin at the bottom is vertical. Reducing its pressure further will shrink the balloon but it will still have the same scale shape. There is an infinite spectrum of intermediate shapes in between these two with one special case in the middle, the zero-pressure shape. An actual balloon only has the intended amount of material at its design point. Intermediate shapes are more complex than the simplified shapes shown in Fig. 13.1.

When we speak of pressure in a balloon, we are more accurately describing the "differential pressure", the difference between internal gas pressure and the external atmosphere. The lift gas will have a measurable gradient of differential pressure from bottom to top (base to apex, in balloon jargon). The zero-pressure shape is notable because at the very bottom, the differential pressure is zero. In other words, at the base point, the lift gas pressure is equal to ambient atmospheric pressure. As one moves upward through the balloon, the differential pressure increases to a maximum at

Figure 13.1   Spectrum of axi-symmetric natural balloon shapes with horizontal tops.

the apex. This is a manifestation of the lifting force created by a less dense lift gas. A hot-air balloon is an example of a zero-pressure balloon since its bottom is open and the differential pressure at the base is zero.

To convey an intuitive sense for high-altitude balloon flight, it helps to look at the evolution of a typical flight. Filling operations consists of inflating a bubble with helium. The volume of this bubble typically is less than 1% of the float volume since it will expand as it rises. After launch (Fig. 13.2) most of the clear plastic polyethylene skin is stretched out vertically encased

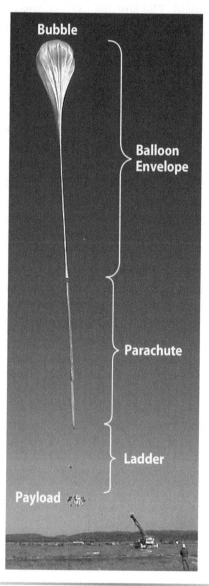

Figure 13.2 Launch in Australia. (CSBF)

in a clear plastic sleeve. With all the lift concentrated in the small bubble, there is a good deal of film stress which is why these balloons have a cap, which is an extra 1, 2, or 3 layers of film in the top region.

As the balloon rises with more buoyancy than it weighs, the balloon accelerates to a speed of approximately 4–5 m/s where the aerodynamic drag force settles into equilibrium with the "excess" buoyancy known as free lift. As the balloon rises the atmospheric pressure decreases with altitude which expands the helium gas. The un-deployed portion of balloon skin begins to slowly fill out, which exposes more balloon skin to the radiant environment and atmosphere. The helium gas cools down in temperature due to the expansion, but the sun and earth are warming the plastic skin, which in turn warms up the helium via internal convection. With the thinner air the sun is stronger and helps the ascent through the coldest portion of the atmosphere, the tropopause. Contrary winds at the tropopause change the flight path. In addition to the direct solar energy being absorbed by the skin, from below the balloon skin absorbs the reflected sunlight known as albedo, and the warm earth bathes the balloon with radiant infrared energy. Above the tropopause into the stratosphere, the ascent slows down considerably where the air temperature begins to get warmer instead of colder due to the concentration of ozone that absorbs the ultra violet component of sunlight. As a result the warmer air rapidly decreases in density, slowing the ascent at about 15 km altitude. The air is so thin at this point that the temperature of the plastic skin is dominated by radiant energy balance. That is, the temperature is mostly the result from the balance of direct sun, albedo, and infrared energy absorbed vs infrared energy emitted from the plastic skin. During ascent the gas temperature lags behind the balloon film temperature, but for the most part the helium is at nearly the same temperature as the skin at float conditions.

At the design float altitude, generally with only 1% of the atmosphere left overhead, the skin is expanded completely to its full volume. The extra helium gas that was used as free lift is now venting through "pony-tail" ducts to the atmosphere whose lower openings are at the same level as the bottom of the balloon. The ducts are attached about 1/3 of the way up from the bottom (see Fig. 13.3) so as to prevent siphoning of air into the balloon. The sun and earth warm up the chilled helium until an equilibrium temperature is achieved. Depending whether the flight is in Antarctica or at mid latitude, the gas may be slightly colder or warmer than the surrounding air. As a typical high-altitude zero-pressure balloon the bottom pressure is equal to atmospheric pressure and the top has a slightly higher pressure than ambient which manifests as buoyant lift. Since the lift gas is less dense than air, the height of helium gas weighs less than the equivalent height of air, and so there is more helium pressure remaining at the top that was not burdened with the job of reacting gas weight. That extra pressure is

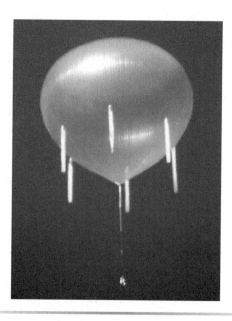

Figure 13.3 40 MCF zero-pressure balloon at float. (Mike Smith, Aerostar)

usefully employed lifting the top of the skin which in turn is able to lift the payload below. The sun is ~1.5 times as strong at high altitude and the earth's infrared is about 2/3 of its maximum strength down at the ground since now it has to pass through the atmosphere where some is absorbed in a complex exchange with clouds and sky. The sun gets higher and the ground gets warmer increasing the up-welling infrared, which in turn heats up the helium via the skin. The helium gas expands venting some gas out of the bottom ducts. At solar noon the gas becomes the warmest it will get, venting the gas to a minimum mass condition which equates to the maximum gas available for the next day.

The sun starts to set, which cools down the skin and the lift gas. The gas contracts (displacing less air) and the balloon slowly sinks. At night there is approximately half the radiant energy available in the form of the up-welling infrared energy. With a much greater amount of contraction the sinking accelerates unless ballast is dropped to equalize the weight with the available lift.

The morning sun warms up the balloon again, but as it is lighter than before it ascends to an even higher altitude. Which means more gas is vented at solar noon, repeating the previous day's cycle of events. Several more diurnal cycles like the previously described and the balloon will run out of ballast.

To terminate the flight a destruct command is given; the payload is pyro-separated and begins to fall. As it falls the destruct line yanks out a banana-peel slice out from one of the gores, ripping the gore the entire length. The balloon bursts and the payload parachutes away. In the meantime the balloon carcass randomly tangles up and impacts the ground in a safe area. The payload settles down in a safe area as well, its impact lessened by the cardboard crush pads on the bottom of the gondola's legs.

13.2 Modern Zero-Pressure Balloons

There are many kinds of balloons, but one that is frequently spoken of is the "natural shape". Natural shape refers to designing the balloon such that the skin has zero or near-zero circumferential (transverse) stress, leaving all the major stress in the meridional (vertical plane) direction as in a parachute. Natural shapes with different loading ratios (suspended weight/gross inflation) and skin density ratios are known as "Sigma shapes" from Justin Smalley's reports [1]. Different Sigma numbers will result in different shapes for a given areal weight density of the membrane film. A heavier film will of course result in a larger required volume for the same payload. Justin Smalley in the 1960s developed the famous Sigma tables by which one could design natural-shaped zero-pressure balloons with zero circumferential stress [1]. The tables would give the shape, film tension to payload ratio T/L, and balloon weight to payload ratio W/L parameterized according to the Smalley Sigma number Σ. Figure 13.4 shows the pertinent characteristics of a zero-pressure balloon.

If the differential pressure between the ambient air and the lifting gas at the base position (bottom of the balloon) is zero, then it is a "zero-pressure" design and has the familiar upside-down onion shape. Hot-air balloons are zero-pressure designs as the bottom is not sealed, and so must be in equilibrium with the ambient pressure at the bottom opening. In their practical construction these balloons are made with vertical "banana-peel slices" known as gores. The polyethylene gores are heat sealed together along their long edges and reinforced at the seams with embedded cords known as load tapes. With internal pressure the gores can bulge out between the loaded tendons to give the lobed appearance although some designs are smooth. Hot air balloons are bulged to gain added volume. A zero-pressure balloon's skin is in stress equilibrium with the buoyancy pressure using the meridional radius of curvature resulting in a smooth skin. Zero-pressure designs have relatively low skin and tendon (or load tape) loads. Being unsealed a helium zero-pressure balloon will vent overboard helium that gets heated if the expansion is above the volume capacity of the balloon. But at night it will drop in altitude due to the cooling of the gas. This is why zero-pressure balloons only last several diurnal cycles, as ballast quickly runs out in an attempt to maintain altitude.

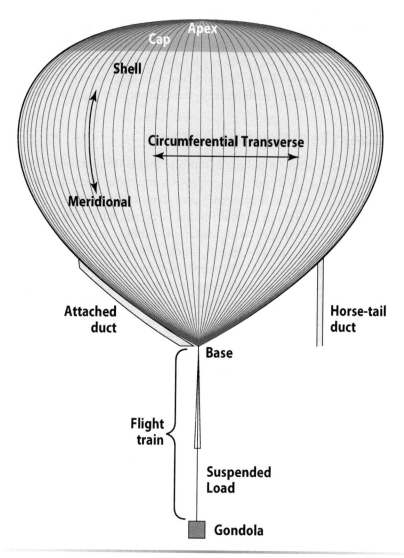

Figure 13.4 Zero-pressure balloon definitions.

These high-altitude balloons are constructed with polyethylene film in the 0.3 to 0.8 mil range (0.0003 in.–0.0008 in.). This special film is co-extruded from several layers that melt into one layer. The purpose is to minimize the chance of ever having pin-holes that line up leading to undesirable porosity above and beyond the normal amount of helium permeability. Although strictly speaking helium permeability is a function of film temperature, film thickness, and differential pressure, the differential pressures would have to be orders of magnitude higher than what is seen in large balloons to make differential pressure of any concern (temperature and film thickness having the greatest effect).

Balloons can bob at float altitudes due to their inherent mass-on-a-spring behavior. Adiabatic expansion/compression is complicit in that if there is a perturbation in altitude, say upwards, there is an expansion and cooling of the gas which in turn contracts to reduce buoyancy. The balloon sinks below the equilibrium point and then compresses with an increase in temperature, increasing buoyancy. The atmosphere as a body behaves similarly with so-called Vaisala-Brunt gravity waves, and when the wind blows over mountains the gravity waves can set up the perturbations necessary to disturb the balloon in the stratosphere. When the super temperatures are just right (near zero), vertical bobbing resonances can occur.

Zero-pressure balloons need ducts to prevent over-pressurization of the envelope which would result in a structural failure. These ducts can hang down or be pulled up to the shell. This is done for some payloads which might have a telescope that can be occluded by a hanging duct.

There are numerous ways to launch large balloons, but the one preferred at CSBF (Columbia Scientific Balloon Facility) is the "dynamic" launch method shown in Fig. 13.5. Launch dates are set according to the upper level winds, such as during turn-around at the equinoxes, but launch times are planned for when surface winds are minimal. Flights are meticulously planned using reliability probabilities leading to a casualty expectation (CE) analysis. If CEs are above a given threshold then there is no flying that day.

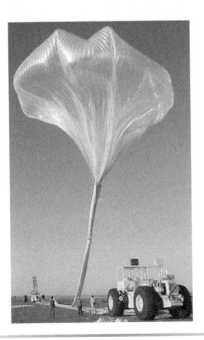

Figure 13.5 Dynamic launch. (CSBF)

13.3 Modern Super Pressure Balloons

If the envelope is sealed and pressurized substantially above a zero-differential base pressure, then these designs are classified as super-pressurized to survive diurnal temperature effects without losing altitude. Super pressure designs generally come in two categories, spherical and pumpkin shaped. In a spherical design all the stress is carried by the skin, whereas a pumpkin design, such as that in Fig. 13.6, mostly separates the roles of structural and gas containment. A super-pressurized balloon in effect has reserve pressure, such that a drop in temperature should not drop the pressure below the point where a large change in shape volume occurs. Since the mass and volume of the gas is a constant, the average density is also constant, which preserves the float altitude. While zero-pressure balloons can have substantial altitude variation as the gas envelope expands and contracts over the course of a day, a super pressure balloon will have a nearly constant altitude profile so long as pressure is maintained in the envelope. That is where the magic lies, in determining the expected temperature variation for the flight environment and seeing that the structure can take the stress for the required super pressure. Using the ideal gas law and taking the derivative dP/dT, it will equal the gas density times the gas constant. This implies that the lower the density, the smaller the ΔP will be for a given ΔT, which implies that the higher it flies then the lower the ΔP that one has to deal with over a diurnal cycle. Higher is easier for a super pressure balloon (>33.5 km, 110,000 ft).

Figure 13.6 Julian Nott's super pressure pumpkin balloon at float.

Figure 13.7 1960s polyester superpressure sphere. (GHOST)

Super pressure spheres (Fig. 13.7) are optimal for small balloons, but for large balloons the skin weight becomes too much. As a pressure vessel the skin stress is proportional to 1/(radius of curvature), so a large sphere will have huge skin stresses while a pumpkin reduces the hoop stress by having a small bulge radius on each lobed gore (small compared to the global radius of the balloon). This is the same principle that reduces skin weight on lobed airships vs bodies of revolution with the same volume.

The cords that look like parachute lines in Fig. 13.8 are called tendons and are made with a very stiff fiber material called PBO. PBO however has degradation issues from ultraviolet and moisture. Without protection measures one must assume that at least 50% of the strength will diminish over the course of the flight. In any case the failure mode is a creep-rupture type which further de-rates the tensile ability. Still, the stiffness and lightness of this material is phenomenal.

The major problem with super pressure pumpkins is an inherent shape instability that can occur when there is an excess of circumferential material, forming what is known as an s-cleft. Two views of the s-cleft are shown in Figs. 13.9 and 13.10. Analyzing for such instability in the intended shape is extremely complex and has occupied years of effort by mathematicians and other researchers. Fortunately, their efforts as well as empirical scaling

Figure 13.8 Super pressure pumpkin lobed bulges.

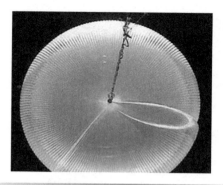

Figure 13.9 Flight 555 Sweden, viewing an s-cleft. (NASA, CSBF)

from flight and laboratory observations have resulted in a much greater understanding of this phenomenon. For instance, we now know that this is indeed a low potential energy state and not an artifact of friction. In fact, modeling shows that the volume slightly increases from the design shape to an s-cleft shape. The majority of the potential energy is in compressing the gas, so having a slightly larger volume option for the balloon (to lower the pressure) becomes irresistible when the conditions are ripe for an s-cleft. Empirical rules of thumb known as cleft factors (CF) are now considered when designing super pressure pumpkins. A cleft-free design is shown in Fig. 13.11. A 7-million ft^3 pumpkin with 200 gores (Flight 591) now holds the endurance record for a balloon of that size with a 54-day flight circling Antarctica 3 times (2008, NASA and CSBF).

With no circumferential stress and no lobing this shape would be an isotensoid which is consistent with Euler's Elastica formulation. A sphere is famous for its maximum [volume/area$^{3/2}$] ratio, but a fascinating fact is that the pumpkin shape has a [volume/gore length^3] ratio greater than a sphere, which is why filament-wound spacecraft propulsion tanks use this shape.

Figure 13.10 27-m pumpkin designed to show an s-cleft simulating flight 555.

Figure 13.11   27-m diameter test pumpkin designed to be cleft free.

13.3.1   Recommended Super Pressure at Design Float Altitudes

This is the super pressure required for any pressurized balloon to stay afloat:

$$\Delta P_{Required} = \frac{P_{air}}{T_{air}} \, \Delta T_D + \Delta P_{Reserve}$$

$$\Delta P_{Required} \approx \left(475 \, e^{-0.155 \, (Alt_{km} - 1)} \right) \Delta T_D + \Delta P_{Reserve} \qquad (13.1)$$

$\Delta P_{Reserve}$ is the minimum differential pressure you wish to have in reserve, in Pascals.

Alt_{km} is the altitude in kilometers

ΔT_D = (maximum gas day temperature − min gas night temperature) in °C

Note: the lower the material α/ε ratio, the lower the required differential pressure.

Desert regions can have up to $\Delta T_D \sim 70°C$, whereas $\Delta T_D \sim 30°C$ to 50°C for most other places when at high altitudes for film $\alpha/\varepsilon \sim 0.2$.

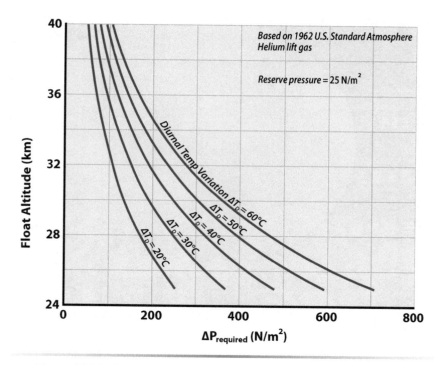

Figure 13.12 Required super pressure as a function of float altitude.

As can be seen in Fig. 13.12, the higher the balloon flies the much easier it gets as far as the super pressure requirement.

13.4 Balloon Physics Fundamentals

As was stated in the introduction, balloons are easily investigated using a first principles approach, which are summarized below:

- Ideal gas law
- Aerostatics
- First law of thermodynamics
- Heat transfer by radiation
- Heat transfer by convection
- Newton's laws

13.4.1 Ideal Gas Law

A given number of gas molecules at a given absolute temperature T in a given volume V will produce the same pressure P no matter the molecular species, that is, for an "ideal gas" Eq. (13.2a).

$$P = n\,RT/V \quad (\text{N}/\text{m}^2 = \text{Pa}) \qquad (13.2\text{a})$$

Where R is the ideal gas constant of 0.83141 Joules/mole/°K, n is the number of moles of molecules (6.022×10^{23} molecules per mole). A customized version of the ideal gas law using the mass density and a customized ideal gas constant R_{gas} has the form shown in Eq. (13.2b).

$$P = \rho\,R_{gas}\,T \quad (\text{Pa}) \qquad (13.2\text{b})$$

where ρ is the mass density (kg/m^3) of the gas.

To customize the equation for any particular gas:

$$R_{gas} = 8314.1/\text{molecular weight} \qquad (\text{m}^2/\text{s}^2/°\text{K})$$

The ideal gas constant for helium $\qquad\qquad\qquad\qquad R_{he} = 2077.2$

The ideal gas constant for diatomic hydrogen $\qquad\qquad R_{h2} = 4148.7$

The ideal gas constant for air (20% oxygen and 80% nitrogen) $R_{air} = 287.1$

These numbers are slightly off from the direct calculation due to the fact that these gases are not ideal, but close. Non-ideal gases have additional degrees of freedom (rotation and internal vibration) and sometimes other sources for creating a potential field such as weak electrical forces. These elements combine to add inaccuracies to the simple statistical bouncing-particle based ideal gas law. This non-linear behavior only becomes noticeable under large pressures as happens when compressing/decompressing from storage tanks.

13.4.2 Buoyancy and Aerostatics

A difference in densities between the ambient air and the lifting gas produce an aerostatic pressure differential within the envelope volume that produces buoyant lift force. In terms of the Archimedes principle, the difference in weight of the displaced heavier fluid and the weight of the lighter fluid is equal to the buoyant force, or gross lift regardless of the shape of the object.

Looking at these two cases (Fig. 13.13) we can set up two equilibrium equations for a cylinder of air of short length L, and one which has a lighter gas substituting the air:

$$P_{base} \times Area = W_{air} + P_{apex} \times Area \qquad (13.3\text{a})$$

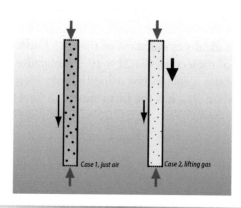

Figure 13.13 Air or helium in a column.

$$P_{base} \times Area = W_{gas} + P_{apex} \times Area + buoyancy_{reaction} \qquad (13.3b)$$

where W_{air} and W_{gas} are the weights of air and gas respectively in the cylinders.

The buoyancy reaction force is holding the case for equilibrium. If we divide by the volume ($Area \times L$), and diminish the length to a differential dz (z going up), we arrive at the aerostatic principle such that ($P_{apex} = P_{base} + dP$):

$$\frac{dP_{air}}{dz} = -g \times \text{density}_{air} \qquad (13.4a)$$

$$\frac{dp_{gas}}{dz} = -g \times \text{density}_{gas} \qquad (13.4b)$$

If we subtract the two equilibrium Eqs. (13.3a and 13.3b) and substitute buoyancy $= -$ buoyancy reaction, we arrive at Archimedes principle:

$$\text{buoyancy} = W_{air} - W_{gas} \qquad (13.5a)$$

If we further divide this by the volume, it leads us to the specific buoyancy parameter b:

$$b = g\,(\text{density}_{air} - \text{density}_{gas}) \qquad (13.5b)$$

The units are Newtons of lift per cubic meter of lift gas.

Figure 13.14 illustrates the pressure distribution in a spherical envelope with the pressures being equal and opposite only at the bottom. The resultant pressure differential is characteristic of a "zero-pressure" design by

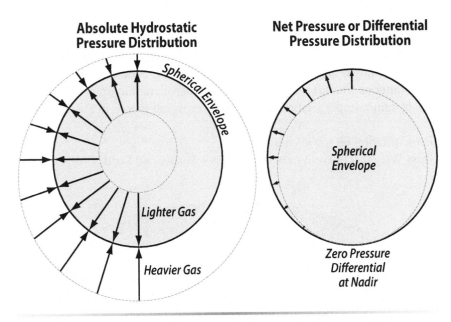

Figure 13.14 Nature of buoyancy in a zero-pressure condition.

subtracting the absolute gas pressure from the absolute air pressure. The pressure differential ΔP distribution from apex to base is simply:

$$\Delta P = \Delta P_{apex} - bZ \qquad (13.6)$$

where ΔP_{apex} is the maximum differential pressure at the apex and Z is the vertical coordinate starting at the apex and pointing in the base direction. The total buoyant lifting force is used to define the gross inflation.

$$GI = b \times \text{Volume} \qquad (13.7)$$

where Volume is the volume of the gas envelope. Under equilibrium conditions the gross inflation = gross weight (which equals suspended weight + balloon envelope structure weight).

Combining the aerostatic differential Eq. (13.4a) with the ideal gas law Eq. (13.2b) as applied to air, integrating vertically upwards in z with a linear temperature profile (known as the temperature lapse rate) results in Eq. (13.8), which is useful for modeling atmospheric pressure at altitude.

$$P_2 = P_1 \left[\frac{T_2}{T_1} \right]^{\left[\frac{-g}{LR_{air}} \right]} \qquad (13.8)$$

where

$$L = \left[\frac{T_2 - T_1}{z_2 - z_1} \right]$$

13.4.3 Force Balance and Free Lift

Any additional lift above and beyond what it takes to lift the gross weight is known as the free lift force, F. During equilibrium ascent the instantaneous free lift will be balanced by aerodynamic drag. Let's take a look at some simple relationships whose terms are shown in Fig. 13.15.

Gross Inflation $GI = b \times$ Volume
Gross Weight = Balloon material weight + Suspended Load

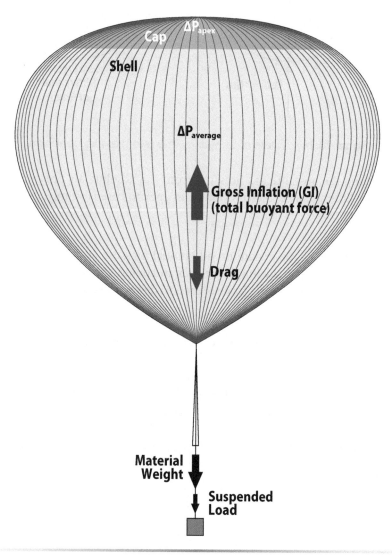

Figure 13.15   Balloon force balance.

Gross Inflation = Gross Weight + Freelift
Gross Inflation = Gross Weight + Drag @ equilibrium ascent
Drag = Freelift @ equilibrium ascent
GI   Gross Inflation, N
G   Gross weight = $W + L$
W   Balloon carcass weight (do not include lift gas)
L   Suspended load
F   Free lift (excess buoyancy)   Free lift ratio = $1 + F/G = GI/G$

For the isothermal and zero-pressure case, the specific buoyancy b with g = acceleration of gravity = 9.807 m/s^2 and air density in kg/m^3:

$$b = g \times \text{density}_{air} \left[1 - \frac{R_{air}}{R_{gas}} \right] \text{N/m}^3 \tag{13.9}$$

Here we define the bulk super temperature and super pressure:

$$\Delta T = T_{gas} - T_{air} \tag{13.10a}$$

$$\Delta P = P_{gas} - P_{air} \tag{13.10b}$$

The next equation combines buoyancy and the ideal gas law to produce the free lift ratio (FreeLift$_{ratio}$ = 1 + F/G) at any condition where T and P are the ambient air temperature and pressure, ΔT and ΔP are the super temperature and pressure of the lifting gas.

$$FreeLift_{ratio} = \frac{M_{gas}}{M_{gross}} \left[\frac{\left(1 + \dfrac{\Delta T}{T_{air}} \right)}{\left(1 + \dfrac{\Delta P}{P_{air}} \right)} \frac{R_{gas}}{R_{air}} - 1 \right] \tag{13.11}$$

M_{gas} is the gas mass, and M_{gross} is the gross mass (G/g). Easily rearrange the equation to determine the required gas mass if the gross mass is known and a launch free lift ratio established (it usually is). Equilibrium is when the free lift ratio = 1.0. Isothermal freelift is when $\Delta T = 0$. For zero-pressure balloons $\Delta P = 0$. Normal day launch free lift ratios are in the 10% range (=1.1), while launching at night could use free lift ratios in the 25% range (=1.25).

Montgolfier Hot Air Balloon

With zero-pressure balloons, the super pressure = 0. Assuming perfect equilibrium (free lift ratio = 1.0), then for a Montgolfier type of balloon it simplifies as follows.

Gross mass (suspended mass + envelope mass, kg) for a zero-pressure Montgolfier hot-air balloon with super temperature ΔT °C as shown in Eq. (13.12).

$$GrossMass = Volume \frac{P_{air}}{R_{air}} \left[\frac{1}{T_{air}} - \frac{1}{(T_{air} + \Delta T)} \right] \qquad (13.12)$$

where

Volume is the gas envelope volume in cubic meters

P_{air} is the atmospheric pressure in pascals (pascal = millibars × 100 = N/m²)

T_{air} is the atmospheric temperature in °K (°K = °C + 273)

Or, in terms of required super temperature (FreeLiftForce in newtons) the super temperature required for the given volume, gross mass, and free lift force is shown in Eq. (13.13) (°C).

$$\Delta T = \left[\frac{1}{T_{air}} - \frac{GrossMass + \dfrac{FreeLiftForce}{g}}{Volume} \frac{R_{air}}{P_{air}} \right]^{-1} - T_{air} \qquad (13.13)$$

13.4.4 Balloon Thermal Fundamentals

Temperatures play a key part in the performance of a balloon. The optical properties of the skin and the energy flux of the environment (the watts per unit area of radiant energy) will determine this temperature. High-altitude balloons fly in an environment dominated by radiant thermal energy such as direct solar, reflected solar (albedo) from the ground, clouds, and sky, and infrared energy (up-welling and down-welling). In the case of an ascending balloon the gas is expanding and the energy flux impinging the skin is constantly changing. To determine the temperature of the balloon skin in this case one must employ the first law of thermodynamics which is a statement of conservation of energy, also known as the energy balance equation (leaving out the macroscopic kinetic and potential energies).

$$\Delta NetHeatEnergy = \Delta InternalEnergy + \Delta Work\ done$$

Or in classical form: $dQ = dU + PdV$.

Internal convection is rather important as the transport mechanism of heat energy from the skin to the lift gas. Conduction however does not play much of a role in such thin film membrane structures.

For both static and dynamic conditions, heat transfer by radiation is an important element to understand. Starting with optical properties, this refers to a surface's solar absorptivity α, transmissivity τ, reflectivity r, and infrared emissivity ε, and transmissivity τ_{IR}. The parameters α, ε, τ, and r are expressed as fractions in that:

$$1 = r + \alpha + \tau$$
$$1 = r_{IR} + \alpha_{IR} + \tau_{IR}$$

where the first equation implies solar wavelengths and the second equation is for IR wavelengths.

From Kirchhoff's law of radiation heat transfer, at any specific wavelength the absorptivity is equal to the emissivity while in thermal equilibrium. This is important because absorptivity is generally given for solar short-wave lengths, but emissivity is given for IR wavelengths. So if we are interested in the absorption of IR radiant energy, we use the given IR emissivity as equal to the IR absorptivity ($\alpha_{IR} = \varepsilon$).

A property of any surface with an absolute temperature T; it emits radiant energy according to the Stefan-Boltzmann relationship:

$$q = \varepsilon \, \sigma \, T^4 \quad IR\ flux,\ \text{Watts/m}^2$$

where σ = Stefan-Boltzmann constant (5.67×10^{-8} W/m²K⁴), ε is the surface emissivity, and T is in °K.

The energy absorbed Q on a surface from a radiant heat source with flux q depends on the absorptivity α of the receiving surface and how much of the receiving surface has a view of the source:

$$Q = \alpha \times \text{SurfaceArea} \times \textit{ViewFactor} \times q \quad \text{Watts}$$

Where ViewFactor is the factor to take into account geometric viewing effects, q is the source flux (W/m²), and SurfaceArea is the exposed surface area of the receiving object.

ViewFactor for a balloon skin surface is the diffuse-radiant viewfactor of the earth, also called Fbe. It is the ratio of the balloon surface area that "sees" the earth divided by the total exposed balloon surface area. In the viewfactor equation below for a sphere, Z is the balloon altitude in meters.

$$R_{earth} = 6,371,000 \text{ m}$$

$$HalfCone_{angle} = \sin^{-1}\left(\frac{R_{earth}}{R_{earth} + Z} \right) \tag{13.14}$$

$$\textit{ViewFactor} = \frac{1 - \cos\left(HalfCone_{angle} \right)}{2} \tag{13.15}$$

At high altitudes the view factor is approximately 0.45. As one ventures higher in altitude, the Earth's horizon "dips" away to be below horizontal due to curvature effects. $Dip = \pi/2 - HalfCone_{angle}$, and the view radius on the surface is simply $= 6371 \times Dip$ (km). The ground view radius is plotted in Fig. 13.16.

Convection effects can be expressed with heat transfer factors that can be calculated for a sphere. Internally there is a natural free convection that is probably much more complicated than what has been researched for spheres, but it is useful enough for balloons. Internal convection heat transfer from film to gas happens faster than one might suppose, taking only 15–20 min to get the gas temperature nearly equal to the film temperature for large balloons at float (from simulations matching flights). Here is the general form of the heat transfer from convection:

$$QConvectionInternal = HCint \times (Tgas - Tfilm) \times Areference$$

$$QConvectionExternal = HCext \times (Tair - Tfilm) \times Areference$$

where *Areference* is the reference surface area and *HC* is the heat transfer coefficient (watts/m^2 per deg K). Equation (13.16) is an approximate HC for external natural convection as a function of altitude and super

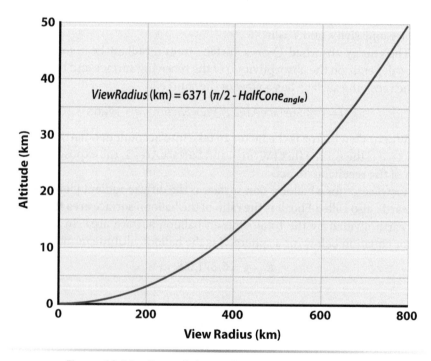

Figure 13.16 Ground view radius of balloon at altitude.

temperature for a sphere which is good enough for preliminary design (Alt is altitude in meters, ΔT in °K).

$$HC_{sphere}\left(Alt,\Delta T\right)=e^{-0.00007\ Alt}\left[0.38+0.7\left(1-e^{-0.06|\Delta T|}\right)\right] \quad (13.16)$$

There are reflections of solar and infrared energy that reflect inside the balloon giving more opportunities for energy absorption. Looking at the diagram one can add up the effective energy absorbed by looking at the grey dots in Fig. 13.17 and seeing that the total energy absorbed is proportional to:

$$\alpha+\alpha\,\tau\left(1+r+r^2+r^3+r^4+r^5+\ldots\right)$$

where we can define an effective reflectivity:

$$r_{effective}=r+r^2+r^3+r^4+r^5=r/(1-r)$$

13.4.5 Balloon Shape Fundamentals

The natural shape equations are a specialization of the membrane-shell equations. So let's first examine the membrane equations for a differential area element using the parameters defined in Fig. 13.18.

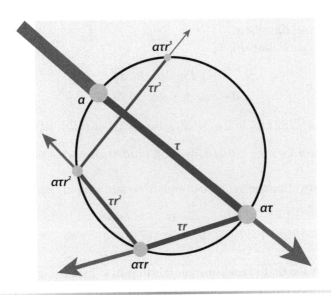

Figure 13.17 Internal reflections.

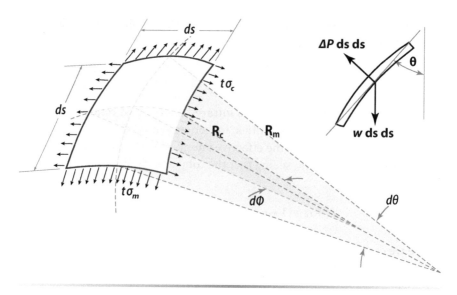

Figure 13.18   Membrane differential element in equilibrium.

This differential element that is ds × ds has a compound curvature with radius of curvature R_m and R_c. It has differential pressure ΔP acting on it as well as weight per unit area w acting in the direction of g. Film stresses σ_m and σ_c multiplied by the thickness t produce a line load N/m along the orthogonal edges. For simplicity there are differential changes not shown on this area element so as to derive the basic equilibrium equation in the surface normal direction.

Geometric relationships:

$$ds = Rm\ d\theta = Rc\ d\phi$$
$$dA = ds\ ds = Rm\ d\theta\ Rc\ d\phi$$

Setting up equilibrium in the ΔP direction with $\sin(d\theta) \sim d\theta$:

$$\Delta P\ dA = 2\sigma_m t\ (d\theta/2)ds + 2\sigma_c t\ (d\phi/2)ds + w\ dA\ \sin(\theta)$$

Finally substituting the geometric equations and simplifying:

$$\frac{\Delta P}{t} = \frac{\sigma_m}{R_m} + \frac{\sigma_c}{R_c} + \frac{w\ \sin(\theta)}{t} \tag{13.16}$$

One can see that membrane equilibrium has "pressure vessel" components from both the meridional and circumferential directions. It is in this manner that a zero-pressure balloon can have vertical wrinkles indicating no circumferential stress even while having buoyancy differential pressure inside it. The meridional curvature is supplying all the stress needed to satisfy equilibrium.

13.4.6 Non-Dimensionalization of Balloon Parameters

Balloon researchers in the 1950s and 1960s were able to discover the commonality of natural shapes by viewing them in non-dimensional form. The purpose of non-dimensionalizing a system is to remove scale effects, such that large or small, a system can be seen as a ratio of forces and geometries. If the ratios between different balloons are the same, then we can expect similar shape regardless of relative size. The first job is to identify the relevant dimensional parameters, which for a simple balloon has the following:

Parameter	Variable name	Dimension	Note
Volume	V	m^3	volume of the gas
Specific buoyancy	b	N/m^3	lift ability of the gas
Gross inflation	GI	N	total lift capacity
Free lift	F	N	excess lift
Gross weight	$G = GI-F$	N	equilibrium, $G = GI$
Suspended load	L	N	
Differential pressure	ΔP	N/m^2	
Total meridional tension	T	N	
Skin areal weight density	w	N/m^2	
Radial coordinate—gore	r or x	m	
Vertical coordinate—gore	z	m	
Gore length	s	m	

The load becomes nondimensionalized as $L_{bar} = L/GI$.
A "natural length" is defined as $\lambda = (L/b)^{1/3}$ which is still dimensional (m).

Parameter	Nondimensional variable	Equation
Suspended load ratio	$Lbar$	L/GI
Skin areal weight Sigma	Σ	$w/(b \lambda \kappa)$
Differential pressure ratio	$Abar$	$\Delta P/(b \lambda)$
Tension ratio	$Tbar$	$T/(b \lambda^3)$

Where $k = (1/2\pi)^{1/3} = 0.541926$ (vestige of early research on spheres).

The early researchers used the suspended load L to non-dimensionalize forces which has some advantages. One can also use gross inflation to non-dimensionalize forces which leads to a very interesting relationship; load ratio $Lbar$ and $Sigma$ become linear, making calculations simple. With this new definition:

The new "natural length" is defined as $\lambda_F = (G/b)^{1/3} = Vol^{1/3}$ (13.17a)

Skin areal weight $\Sigma_F = \dfrac{w}{b\,\lambda_F\,K} = \Sigma\,Lbar^{1/3}$ (13.17b)

Generally speaking the relationship between load ratio *Lbar* and Σ_F for any shape is:

$$Lbar = 1 - \left(K \times SurfaceArea / Volume^{2/3}\right) \times \Sigma_F$$

Every shape has a characteristic (SurfaceArea/Volume$^{2/3}$) ratio that is a constant for rigid shapes regardless of absolute size, and even though natural zero-pressure shapes change with *Sigma number*, the relationship is still surprisingly linear.

For a sphere:

$$Lbar = 1 - 2.62074 \times \Sigma_F$$

For a zero-pressure natural-shape balloon:

$$Lbar = 1 - 2.682\,\Sigma_F \qquad (13.17c)$$

It is interesting to note that in Fig. 13.19 when the load ratio is zero (it can only pick itself up, no payload), the shape becomes the same as a super pressure pumpkin shape (next section), and the classic *Sigma number* goes to infinity. In the classic definition of *Sigma*, there are no *Sigma* numbers that won't produce a viable balloon. Using the *SigmaF* definition allows balloons to be designed that cannot fly but can be used as ground test models for study (if *SigmaF* is greater than 0.37). Most modern scientific zero-pressure balloons (ZPB) have Sigma numbers between 0.1 and 0.2.

13.4.7 Balloon Film Fundamentals

Linear low density polyethylene co-extruded film is used for both zero-pressure and super pressure balloons. Smaller balloons have used Mylar (polyester), but the large scientific balloons are manufactured with polyethylene. The film structural properties are a function of temperature, applied stress, and the length of time the stress is applied. Quite unlike designing with a fabric or a metal, there is no one-size-fits-all approach when dealing with this material. The glass transition temperature is of critical importance as these balloons pass through the tropopause (coldest portion of the atmosphere). The expected film temperatures must remain above this threshold of approximately $-98°C$ ($175°K$). It becomes critical for a super pressure design to know the limits for yield point and tertiary

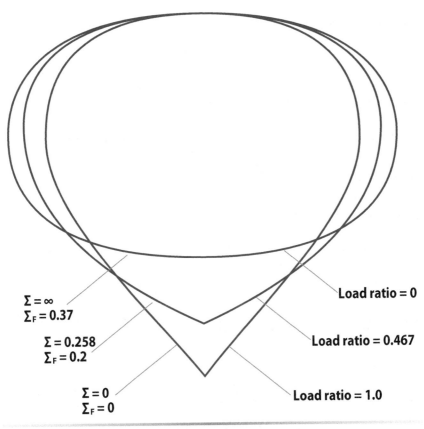

$\Sigma = \infty$
$\Sigma_F = 0.37$

$\Sigma = 0.258$
$\Sigma_F = 0.2$

$\Sigma = 0$
$\Sigma_F = 0$

Load ratio = 0

Load ratio = 0.467

Load ratio = 1.0

Figure 13.19 Natural zero-pressure shapes with different load ratios.

creep which are changing with the environment. Generally speaking, it is a good idea to keep the non-thermal portion of strain to below 2–3%. Strains are divided up into the initial elastic strain, the thermal strain, and the creep strain. Complex visco-elastic constitutive materials models have been developed [2] that guide the usage for super pressure designs. What has been theorized and experimentally verified is that these polyethylene films behave best when in a biaxial stress state. There is a term called *effective stress* where it is minimized when a nearly 1:1 biaxial stress state exists.

In a super pressure pumpkin, additional effects occur with the film material that benefits the stress state over time. As the lobed-bulge creeps in the hoop direction, the radius of curvature reduces, thus reducing the hoop stress. As the material creeps in the meridional direction, more shared load is passed to the tendons reducing the meridional stress. This general beneficial effect has been called "strain arrest".

13.5 Balloon Environments

The high-altitude polyethylene scientific balloon is a thermal vehicle. It lives in a mostly radiant thermal balance from the many heat sources that influence it (Fig. 13.20). When warm, the gas expands and either pressurizes the gas envelope or is vented. If vented, then nighttime cooling makes ballast drops necessary.

These are the basic heat sources that influence balloon flight:

a) Direct sunshine on the skin membrane
b) Reflected diffuse sunshine in the form of ground albedo
c) Reflected diffuse sunshine in the form of cloud albedo
d) Diffuse infrared from the ground/atmosphere (up-welling)
e) Diffuse infrared from the clouds
f) Atmospheric convection
g) Diffuse infrared from the sky (down-welling)

At 33.5 kilometers altitude on an average sunny day in Ft. Sumner, the fraction of heat loads absorbed in the skin of a large scientific balloon (10 to 40 million cubic foot volume) is approximately:

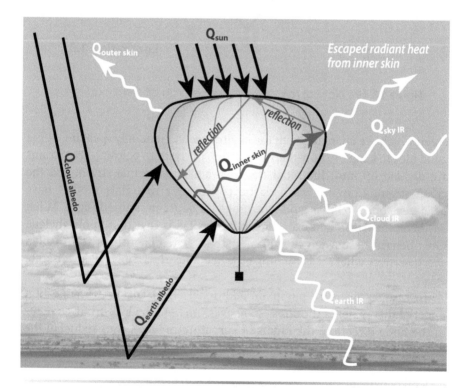

Figure 13.20 Radiant environment for a balloon.

Direct solar	37%
Indirect solar (albedo)	14%
Earth IR	49%
Atmospheric convection	−1%

At night convection has more effect, but only slightly.

It should be noted that direct sun impingement is a collimated source and affects a projected area of the balloon, while the other environments have a diffuse source nature to them requiring the use of the ViewFactor term applied to the full surface area.

Figure 13.21 is a simplification of data from the ERBS spacecraft that mapped IR flux and albedo at the top of the atmosphere (TOA). The max/min albedo and max/min IR flux is used to set temperature bounds on the balloon film. This affects performance and film stress.

As a design guide use these average upwelling IR values (W/m^2) below:

Alice Springs, Australia, April	IR = 267, albedo = 0.24
Circumglobal S. Hemisp., summer	IR = 257, albedo = 0.23
Ft Sumner NM, Aug	IR = 265, albedo = 0.25
Circumpolar Antarctica, Dec, Jan	IR = 192, albedo = 0.63
Kiruna Sweden to Canada, June	IR = 226, albedo = 0.45

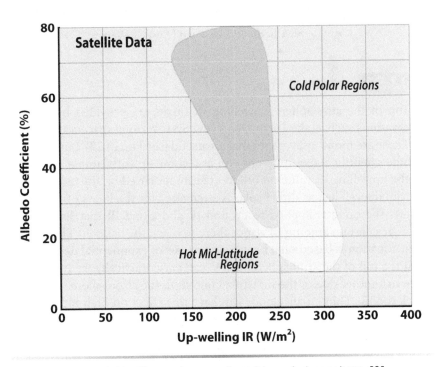

Figure 13.21 Thermal energy flux at top of atmosphere. [3]

13.5.1 Solar Flux

The transmissivity τ_{atm} of optical frequency irradiance thru the atmosphere follows a Beer-Lambert Law format of exponential decay [4], but we will only concern ourselves with values at the top of the atmosphere (TOA). Solar irradiance flux at the top of the atmosphere has a +/− 45 W/m² variation over the course of a year. It can be expressed by the day number (1 thru 365, 1 starting on January 1):

$$I_{Sun} = 1358 + 45\cos(2\pi \ (Day_{number} - 10)/365) \quad W/m^2$$

Solar irradiance flux at the balloon altitude Z is then simply:

$$q_{Sun} = I_{Sun}\,\tau_{atm}$$

where $\tau_{atm} \sim 0.996$ for high altitudes with solar elevation above 22 deg.

13.5.2 Albedo Flux

There is a simple model relating surface albedo and albedo flux. The assumption here is that the albedo number is the total specular + diffuse solar reflection, but treated as all diffuse for simplicity. The albedo flux is proportional to the solar irradiance at the top of the atmosphere and sine of the solar elevation angle above the horizontal:

$$q_{albedo} = \text{Albedo} \times Isun \times \sin(Elv) \quad W/m^2$$

13.5.3 Infrared Flux

Top of the atmosphere upwelling IR fluxes are provided by satellite observations such as the Earth Observing System Aqua spacecraft.

These are monthly averages that are useful (see Fig. 13.22), but for high-altitude scientific ballooning one can also come up with simple models to get the upwelling IR values at different altitudes based on the temperature at ground level. Just as direct solar is attenuated for thickness of the atmosphere, the same is applied to ground IR and cloud IR that has to pass through a certain amount of atmosphere to reach the balloon. The attenuation equation is based on a Beer-Lambert law of exponential decay. Note: IR at the top of the atmosphere (TOA) can vary drastically from one local zone to another due to the moisture content in the atmosphere which will block the IR. Thin cirrus clouds can have this effect on high-altitude balloons. Deserts will have the greatest day/night temperature swings due to little water acting as heat capacity storage. Somewhere on the order of 25°C swing is not unreasonable. Oceans should have the least day/night swing in surface temperature. BEWARE! Most IR data on Web sites is highly

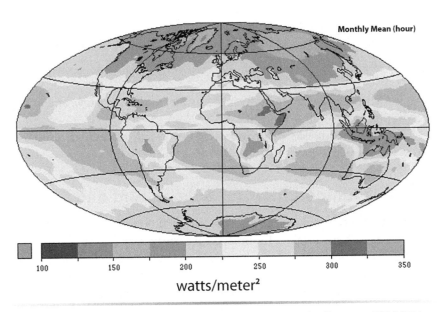

Monthly Mean (hour)

watts/meter²

Figure 13.22 EOS satellite up-welling infra-red data for January 2006, TOA.

processed to reflect ground surface temperatures, not top-of-atmosphere (TOA) temperatures! A simple model [5, 7] for upwelling IR at altitude based on the ground temperature and a single attenuation factor follows.

Upwelling Infrared Environment

IR diffuse radiation at ground level with ground emissivity ε

$$q_{IRground} = \varepsilon_{ground} \; \sigma \, T^4_{ground} \quad \left(W/m^2 \right)$$

Transmission factor of ground IR to account for atmospheric absorption below the balloon is approximated by Eq. (13.18).

$$Transmission_{IR} = A_{IR} \left[\frac{P_{air}}{P_{sea\,level}} - 1 \right] + 1 \qquad (13.18)$$

Maximum attenuation factors:

$A_{IR} = 0.45$ for temperate air masses
$A_{IR} = 0.35$ for dry air masses
$A_{IR} = 0.30$ for very dry air masses (Antactica)

Ground IR diffuse radiation at balloon altitude, W/m^2

$$q_{IRgroundZ} = q_{IRground} \; Transmission_{IR}$$

$\sigma = 5.67 \times 10^{-8}$ Stefan-Boltzman constant W/m² °K⁴

Here is a suggested list of ground emissivities:

Desert = 0.85 (data suggests variation within a region, as low, as 0.8)
Average ground = 0.95
Snow = 0.98

It is fortuitous that in the hottest places the emissivity is low, and in the coldest places the emissivity is high.

The down-welling sky IR environment is non-existent at high altitudes, and at the surface is about 250-300 W/m².

13.6 Bulk Temperatures for Balloon Film and Gas

The high-altitude polyethylene scientific balloon is a thermal vehicle. The measured optical values for the typical zero-pressure linear low density polyethylene StratoFilm 372 are scattered as one might expect with clear plastic. What is shown in the table are a mixture of Edward's mode transflectance measurements [6] and modifications due to flight experience. What is most important is the absorptivity to emissivity ratio, α/ε. At night there is no solar irradiance to absorb, so what determines balloon skin temperature in this case is the magnitude of the infrared environment, the viewfactor, and some atmospheric convection.

Zero pressure balloon (ZPB) Material SF372, 0.0008" thick

	1 layer	2 layers	3 layers	4 layers	LT 400	LT 150	LT 600
					polyester load tapes 25 mm		
α	0.024	0.042	0.057	0.087	0.13	0.069	0.128
ε	0.134	0.234	0.314	0.475	0.793	0.618	0.743
τ	0.916	0.847	0.788	0.667	0.383	0.624	0.479
τ_{IR}	0.866	0.766	0.686	0.525	0.07	0.336	0.189
α/ε	0.176	0.180	0.182	0.184	0.16	0.11	0.17
0.8 mil film							

13.6.1 Steady State Translucent Spherical Balloon Temperature

If we model the balloon as a translucent sphere with a viewfactor of the earth equal to Fbe, the steady state equation boils down to this relationship for skin temperature [7]:

Film-air temperature differential, °K

$$\Delta T_{fa} = T_{film} - T_{air}$$

$$T_{film} =$$

$$\left[\frac{\dfrac{\alpha}{\varepsilon} q_{sun} \left[1+\tau(1+r)\right]\left(\dfrac{1}{4}+\text{Albedo}\sin\left(Elv\right)F_{be}\right)+\left[q_{IR}F_{be}+q_{sky}\left(1-F_{be}\right)\right]\left[1+\tau_{IR}\left(1+r_{IR}\right)\right]-\dfrac{hc_{ext}\Delta T_{fa}}{\varepsilon}}{\sigma\left[1+(1+r_{IR})\,\tau_{IR}\right]} \right]^{\frac{1}{4}}$$

(13.19)

Where

α = solar absorptivity
ε = infrared emissivity (and IR absorptivity)
τ = solar transmissivity
τ_{IR} = infrared transmissivity
Albedo = albedo coefficient of the ground and sky combination
Elv = solar elevation angle above a horizontal plane
F_{be} = view factor that the balloon has of the earth, typically ~ 0.45
q_{sun} = solar flux, W/m^2
q_{IR} = up-welling infrared flux, W/m^2
q_{sky} = down-welling IR, W/m^2
σ = Stefan – Boltzmann constant (5.67×10^{-8} W/m^2 °K^4)

hc_{ext} is the air/film convection heat transfer coefficient, W/m^2/°K

For preliminary design cases the external convection factor hc_{ext} can be ignored for establishing maximum balloon film temperatures. It should be noted that load tapes on the gores run hotter due to more absorption of radiant energy. Their effects can be approximated by modifying the skin optical properties with an area ratio approach of the load tape optical properties vs skin area. Thus "global" properties can be used to get accurate bulk temperatures. More complex models modeling the balloon as a six-sided rectangular box are used for roughing out the skin temperature gradients, or one can turn to software such as Thermal Desktop. Such gradients can run as high as 44°C from coldest to hottest surface for typical balloon polyethylene film.

Taking the translucent sphere equation and graphing it for various α/ε ratios, albedo coefficients, and local solar elevation angles, one can see the expected pattern as a function of upwelling IR in Fig. 13.23.

Entering the x-axis with a known IR flux and moving up to the proper solar elevation angle curve one can quickly determine a bulk film temperature, which for most purposes will also equal the lift gas temperature at float.

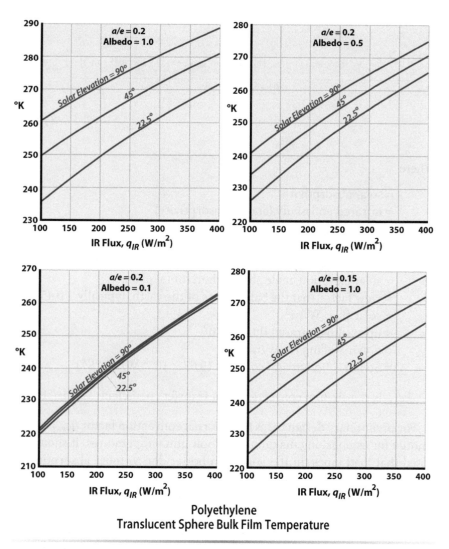

Figure 13.23   Film temperature as a function of IR flux–various albedos.

Three albedo coefficients were plotted here (0.1, 0.5, 1.0) for two α/ε ratios (0.15, 0.20) which are typical for polyethylene film. These results reflect high-altitude conditions (33.5 km) of some small amount of external convection. The graph shows that nighttime film temperatures are independent of α/ε and albedo, as no sun is shining. Film temperatures at night are completely dependent on the upwelling IR environment (view factor included), not the film optical properties.

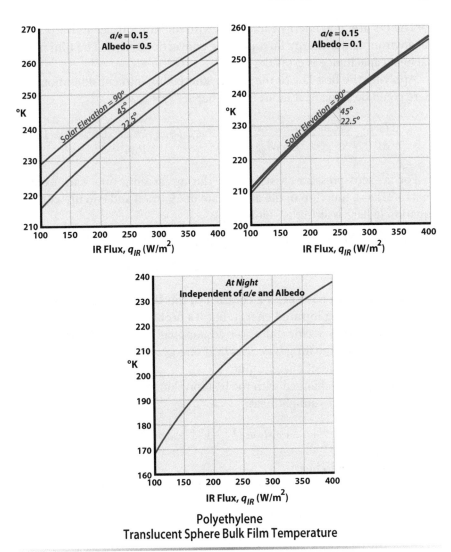

Polyethylene
Translucent Sphere Bulk Film Temperature

Figure 13.23 (continued) Film temperature as a function of IR flux.

13.7 Balloon Transient Thermal Behavior

Applying the first law to the balloon skin (plastic film) leads to these relationships [5, 7]:

$$\underbrace{Q_{Sun} + Q_{Albedo} + Q_{IRplanet} + Q_{IRsky} + Q_{IRfilm} + Q_{ConvExt}}_{\text{Energy IN}} = \underbrace{Q_{ConvectionInternal} + Q_{IRout}}_{\text{Energy OUT}} + \underbrace{c_f + M_{film}\frac{dT_{film}}{dt}}_{\text{Energy Stored}}$$

(13.20)

The Q's are total energy exchanged with the film surface in watts from the environmental energy flux, plus the convection terms. The film material heat capacitance is c_f (W/kg/°K).

Applying the first law to the lift gas and taking into consideration that there could be some heat input from a burner ($\gamma = c_p/c_v$):

$$\frac{dT_{gas}}{dt} = \frac{(Q_{ConvectionInternal} + Q_{burner})}{C_v \, M_{gas}} + (\gamma - 1) \frac{T_{gas}}{\rho_{gas}} \frac{d\rho_{gas}}{dt} \qquad (13.21)$$

For a zero-pressure assumption during ascent, one can express Eq. (13.21) as a function of the ascent rate dz/dt (m/s) and end up with this equation (z here is altitude, + going up):

$$\frac{dT_{gas}}{dt} = \left[\frac{(Q_{ConvectionInternal} + Q_{burner})}{M_{gas}} - g \frac{T_{gas}}{T_{air}} \frac{R_{gas}}{R_{air}} \frac{dz}{dt} \right] \frac{1}{c_p} \qquad (13.22)$$

There seems to have been a concern in the 1970s that the helium gas itself was absorbing infrared energy, but it's probable that those measurements were due to accidental water contamination of the lift gas. These equations assume the lift gas gets its heat energy supplied only by skin/gas convection or by direct heater input.

The energies exchanged with the balloon skin are listed below [5, 7].

Absorbed direct sunlight heat:

$$Q_{sun} = \alpha \, A_{projected} \, q_{sun} \left[1 + \tau \left(1 + r_{effective} \right) \right] \quad \text{(Watts)}$$

Absorbed albedo heat:

$$Q_{Albedo} = \alpha \, A_{surf} \, q_{Albedo} \, ViewFactor \left[1 + \tau \left(1 + r_{effective} \right) \right] \quad \text{(Watts)}$$

Absorbed upwelling IR heat from the planet surface:

$$Q_{IRplanet} = \alpha_{IR} \, A_{surf} \, q_{IRplanet} \, ViewFactor \left[1 + \tau_{IR} \left(1 + r_{effectiveIR} \right) \right] \quad \text{(Watts)}$$

Absorbed IR from the sky:

$$Q_{IRsky} = \alpha_{IR} \, A_{surf} \, q_{IRsky} \, (1 - ViewFactor) \left[1 + \tau_{IR} \left(1 + r_{effectiveIR} \right) \right] \quad \text{(Watts)}$$

Net emitted IR energy from both interior and exterior of the balloon skin:

$$Q_{IRout} = \alpha \, \varepsilon \, A_{surf} \left[1 + \tau_{IR} \left(1 + r_{effectiveIR} \right) \right] T_{film}^4 \quad \text{(Watts)}$$

A_{surf} refers to the external exposed surface area on the bubble, and $A_{projected}$ is the projected area of the bubble illuminated by direct solar exposure. The illuminated projected area of a natural-shape balloon varies with solar elevation angle *ELV*, and uses the top projected area as the reference. Here are some approximation formulas:

For a zero-pressure shape:
$$Area_{projected} = Area_{top} [0.9125 + 0.0875 \times \cos(\pi - 2ELV)]$$

For a pumpkin shape:
$$Area_{projected} = Area_{top} [0.8219 + 0.1781 \times \cos(\pi - 2ELV)]$$

13.8 Meridionally Lobed Membrane Structures

One can modify the classic membrane equations to incorporate weight from load tapes by making the film areal weight density a function of the geometry. One can do the same for a pumpkin balloon with bulging material, as well as adding transverse stress to alter the "natural" shape. The meridional tension *T* is a combination of the load tape and the film, each sharing a portion of this tension according to their stiffness, thermal contractions, and slack strain in the load tapes. The membrane equations consider it as a uniform structure, while structural equations can later separate the tension into the various components. The formulations that follow are generalized for a smooth shape, but with clever manipulations can also be used for shapes with lobed gores. The idea is that there is an equivalent smooth balloon underneath every bulged/lobed balloon.

Most references orient the *Z* axis up and integrate the differential equations from bottom to top. The opposite is done here as the boundary condition at the top is easy to calculate a priori. See the illustrations for definitions and the differential length element d*s*:

Smooth Axi-Symmetric Membrane Equations Adopted for Balloons

t_c = circumferential film line load, N/m = $stress_c \times$ thickness
t_m = meridional film line load, N/m = $stress_m \times$ thickness
w = film weight per unit area, N/m^2 (can vary with location to account for load tape weight)
ΔP = differential pressure = $\Delta P_{apex} - bZ$, N/m^2 (pascals)
T = total meridional tension = $2\pi X t_m$, (N)
b = specific buoyancy, N/m^3
X = radial coordinate, meters
Z = vertical coordinate, starting from the top and pointing down
The change in angle theta, θ:

$$\frac{d\theta}{ds} = 2\pi \frac{-t_c \sin(\theta) - X w \cos(\theta) + X \Delta P}{T} \qquad (13.22)$$

The change in total meridional tension:

$$\frac{dT}{ds} = 2\pi \left[t_c \cos(\theta) - X w \sin(\theta) \right] \tag{13.23}$$

Auxiliary geometric relationships:

$$\frac{dZ}{ds} = \sin(\theta) \quad \text{and} \quad \frac{dX}{ds} = \cos(\theta) \tag{13.24}$$

The circumferential line load t_c can be a prescribed function to alter the shape, or it can be made zero to produce a natural shape. If $t_m = t_c =$ constant, and $w = 0$, then a sphere will result using a constant ΔP. Use Fig.13.24 to identify parameters in the previous equations.

Integrate these equations from the top down. For a zero-pressure design ΔP at the bottom $= 0$. The initial values of ΔP_{APEX} and T_{APEX} need to be adjusted until the solution constraints are satisfied. For a weightless apex plate the angle theta at the top $= 0$.

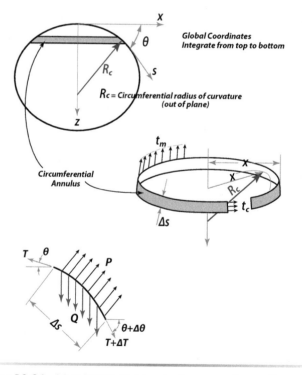

Figure 13.24   Membrane free-body diagram of smooth annulus.

13.9 Preliminary Design and Performance Methodology

Designing a super pressure pumpkin for a particular set of require-ments at high altitude is a daunting task. Instead, we will show the prelimi-nary design steps useful for a natural-shaped zero-pressure balloon, which is quite informative. It helps to have some approximation equations and some historical information to help in quick design studies of this sort. Much use is made from non-dimensional relationships.

Assuming a standard atmosphere, one can generate an approximation formula for helium specific buoyancy b as a function of altitude for zero super pressure. Normal super temperatures change b by ~0.5%, so for this phase of study it is fine. Specific buoyancy b is N/m^3 and Alt is altitude in meters:

$$b(Alt) = 0.003 + 28\,e^{-0.000161\,(Alt + 2600)} \quad 11{,}000 < Alt < 40{,}000$$
$$b(Alt) = -1.2 + 19\,e^{-0.00009\,(Alt + 5500)} \quad 0 < Alt < 11{,}000$$

Next is the film + load tape average areal density w from historic data:

$$w(Alt) = 0.77 - 1.25 \times 10^{-5}\,Alt \text{ N/m}^2 \text{ for "light" designs}$$
$$w(Alt) = 1.43 - 2.7 \times 10^{-5}\,Alt \text{ N/m}^2 \text{ for "heavy" designs}$$

13.9.1 How Big, How Heavy?

We come into this problem with a payload mass and a target float alti-tude. With the altitude we calculate the specific buoyancy b and the film areal density w. As a first guess we will assume that the ballast to suspended mass ratio is 25% ($Ballast_{ratio} = 0.25$). That is load L is 25% ballast weight. If we know our payload mass (kg) which includes the gondola, then the suspended mass is simply:

$$Mass_{Suspended} = \frac{Mass_{payload}}{(1 - Ballast_{ratio})} \tag{13.25}$$

The suspended load $L = g\,Mass_{Suspended}$ where $g = 9.807$ m/s^2.

Now we use a design formula Eq. (13.26) to easily determine the load ratio $Lbar$. This is easily solved with a few iterations starting with $Lbar = 0.6$. Remember $\kappa = 0.541926$, and the 2.682 factor is from Sec. 13.4.6 balloon shape fundamentals Eq. (13.17c).

$$Lbar = 1 - \frac{2.682}{\kappa}\frac{w}{b}\left[\frac{Lbar}{L}b\right]^{1/3} \tag{13.26}$$

Gross Weight = $L/Lbar$ (Newtons)
Equilibrium gross weight = gross inflation GI
Volume = GI/b (m³)

(Balloon mat'l mass, kg) $Mass_{Balloon} = \dfrac{Gross\,Weight}{g} - Mass_{Suspended}$ (13.27)

Turning to non-dimensional relationships mapped out in the following approximation equations for a zero-pressure shape, one can determine the gore length, diameter, height, surface area, number of gores, and load tape tension.

Determine the *SigmaF* number from Eq. (13.17c).

$$\Sigma_F = \frac{(1 - Lbar)}{2.682} \tag{13.28}$$

Determine the natural length λ_F from Eq. (13.17a).

$$\lambda_F = Volume^{1/3} \quad (m)$$

The gore length

$$Sbar = 1.994 - 0.336\,\Sigma_F - 0.049\,\Sigma_F^2$$

Gore length = $Sbar\,\lambda_F$ (m)
The diameter

$$Dbar = 1.305 + 0.164\,\Sigma_F + 0.479\,\Sigma_F^2 - 1.478\,\Sigma_F^3 + 3.667\,\Sigma_F^4$$

Diameter = $Dbar\,\lambda_F$ (m)
The balloon height and differential pressure at the apex

$$Hbar = 1.275 - 0.445\,\Sigma_F - 1.496\,\Sigma_F^2$$

Height = $Hbar\,\lambda_F$ (m)

$$\Delta P_{apex} = Height\,b\,\,N/m^2 \quad (pascals)$$

Surface area

$$Areabar = 4.913 + 2.598\,(\Sigma_F - 0.05)^3$$

Area = $Areabar\,\lambda_F^2$ (m²)

The number of gores can be determined by your manufacturing constraint, in that there is usually a maximum width per gore that can be handled, call that $Width_{max}$

$$N_{gores} = \frac{\pi\,Diameter}{Width_{max}}$$

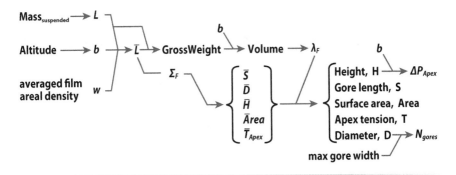

Figure 13.25 Zero-pressure balloon design process—first pass.

The maximum load tape tension is usually at the apex, not the base:

$$Tbar = 1.556 - 0.66\,\Sigma_F - 1.59\,\Sigma_F^2$$

Total Tension at the apex = $Tbar\,\lambda_F^3$ (N)
Tension per load tape at the apex = Total Tension/$Ngores$

Sample Problem 13.1

Figure 13.25 diagrams the zero-pressure solution sequence.

With the basic size established, one can now work it directly with the apex tension, surface area, and gore length to give better structural mass estimates. If the balloon mass is on target, then fine; if not then the parameter w needs to be adjusted until convergence.

Let's say our suspended mass is 2500 kg (5512 lb), our design float altitude is 37.2 km (122,042 ft), and our averaged combined areal density $w = 0.3705\ \mathrm{N/m^2}$ (0.00774 lb/ft²). At that altitude the specific buoyancy $b = 0.04916\ \mathrm{N/m^3}$ (0.000313 lb/ft³). The suspended load L is therefore 24,518 N (5512 lb). Iterating the $Lbar$ equation gives us a load ratio of $Lbar = 0.6027$; that is 60.27% of the gross weight will be suspended weight. Next the gross weight $= L/Lbar = 40,679$ N (9145 lb), and volume $=$ gross weight$/b = 827,454\ \mathrm{m^3}$ (29,220,000 ft³). The characteristic length $LambdaF = Vol^{1/3} = 93.9$ m. Moving into the non-dimensional parameters, $SigmaF = 0.1481$ which allows the calculations of the nondimensional gore length $= 1.943$, diameter $= 1.337$, height $= 1.176$, surface area $= 4.915$, and meridional tension $= 1.423$. Combining with the characteristic length $LambdaF$, the actual sizes are thus: gore length $= 182.427$ m (598.54 ft), diameter $= 125.498$ m (411.76 ft), height $= 110.429$ m (363.32 ft), surface area $= 43323.8\ \mathrm{m^2}$ (466,166 ft²), total apex tension $= 57901$ N (13017 lb). If for manufacturing the maximum gore width is limited to 2.478 m (97.5 in.), then the number of gores is 159. The apex differential pressure is 5.429 Pa.

13.9.2 How Much Ballast?

To answer this question we must first discover the film and gas temperatures at solar noon for the hottest daytime temperature and the coldest gas temperature at night. First approximate the solar declination angle for the date (by day number, Daynumber), then the solar elevation angle at local noon:

$$Declination = 23.452 \sin \left[2\pi \; \frac{284 + Day_{number}}{365} \right]$$

$$Elevation_{noon} = Declination + 90 - Latitude$$

These are in degrees with north latitudes (+), and south latitudes (−).

In Sec. 13.5.1 use the equation to determine the solar flux, q_{sun}. Select from your environment sources what the maximum daytime upwelling IR will be, and the albedo coefficient. Use the table in Sec. 13.6 to select the bulk optical properties for your balloon. Use the equation for a translucent sphere and determine the bulk film temperature at noon (assume = gas temperature). Make the $q_{sun} = 0$ for nighttime and determine the night gas temperature. Select a nighttime parking altitude acceptable to the mission and flight safety; maybe 15,000 ft (4.6 km) lower than daytime altitudes. From your atmosphere model determine the air temperatures for both day altitude and night altitude. With these temperatures one can calculate the super temperatures at noon at the daytime altitude and super temperature at the nighttime parking altitude.

$$\Delta T_{day} = T_{film_{day}} - T_{air_{day}}$$

$$\Delta T_{night} = T_{film_{night}} - T_{air_{night}}$$

Then, the nominal gas mass with zero super temperature is:

$$M_{gas_{nominal}} = \frac{M_{gross}}{\dfrac{R_{gas}}{R_{air}} - 1}$$

For launch multiply by the free lift ratio, usually 1.1, to get the launch gas mass required.

Reconfiguring the free lift ratio equation (Sec. 13.4.3) for a zero-pressure balloon at equilibrium to get the daytime gas mass (remember the super temperature vents out some of the gas):

$$M_{gas_{day}} = \frac{M_{gross}}{\left(1 + \dfrac{\Delta T_{day}}{T_{air_{day}}} \right) \dfrac{R_{gas}}{R_{air}} - 1}$$

The gross mass capacity at night using the daytime gas mass:

$$M_{gross_{night}} = M_{gas_{day}} \left[\left(1 + \frac{\Delta T_{night}}{T_{air_{night}}} \right) \frac{R_{gas}}{R_{air}} - 1 \right]$$

Finally, the ballast that needs to drop to maintain the selected parking altitude:

$$M_{ballast} = M_{gross} - M_{gross_{night}}$$

If the ballast ratio ($M_{ballast}/M_{suspend}$) is less than the originally assumed amount, then the parking altitude and other design parameters are viable. Generally one will find that ~10% of the gross weight needs to be dropped every night, and over a cold storm the upwelling IR can be mostly blocked as to rapidly bring down the balloon where there is just not enough ballast aboard. If one can accept a lower parking altitude at night, then the ballast requirement will go down, 5%–7% of the gross weight being a good target.

13.9.3 Suspended Mass Capability

With your new design (or old one) a graph of the maximum suspended mass capacity of any balloon can be made if you have the balloon envelope material mass (everything above the hook point on the base plate), call it $M_{balloon}$. With your atmosphere model one can make air and lift gas density a function of altitude *Alt*.

$$M_{suspended} (Alt) = Volume \left[\rho_{air} (Alt) - \rho_{gas} (Alt) \right] - M_{balloon}$$

Or, with the definition of specific buoyancy *b*:

$$M_{suspended} (Alt) = Volume \left[b(Alt)/g \right] - M_{balloon}$$

Acknowledgments

I would like to thank Henry Cathey, Gabriel Garde, and Tracy Bohaboj of PSL, Danny Ball of CSBF, Mike Smith of Aerostar, Jim Rand of Winzen, David Wakefield of Tensys, Sergio Pellegrino of Cal Tech, and last but not least Debora Fairbrother of NASA for their years of guidance and inspiration.

About the Author

Rodger E. Farley has worked as an aerospace engineer for 31 years, mostly at the Mechanical Systems Branch at NASA/GSFC and currently serves as a mechanical systems engineer. His experiences encompass aircraft, rotorcraft, spacecraft, and most recently balloon craft. He is the chief designer for the NASA super pressure balloon development, working with a talented team to bring this large vehicle to a working status.

Education:

M.S. Aerospace Engineering, University of Maryland, 1986
B.S. Aerospace Engineering, University of Maryland, 1980
Current Position:
Mechanical Systems Engineer, Code 543
NASA/Goddard Space Flight Center, Greenbelt, Maryland

References

[1] Smalley, J.H., *Determination of the Shape of a Free Balloon—Summary*, Litton Systems Inc., BT-1530, 1965.
[2] Rand, J.L., *A Constitutive Model of StratoFilm 420*, Winzen International, Inc, Report for NASA's Balloon Program Office, 2007.
[3] Cathey, H. M. and Lindsey, B., *Top of Atmosphere Upwelling IR*, Physical Science Laboratory, memo, 2010.
[4] Kreith, F. and Kreider, J.F., *Principles of Solar Engineering*, McGraw-Hill Book Co., 1978: p. 43.
[5] Farley, R. E., *BalloonAscent: 3D Simulation Tool for the Ascent and Float of High Altitude Balloons*, AIAA-2005-7412, 2005.
[6] Bohaboj, T.A., *Radiative Properties of Balloon Materials—Third Set of Materials*, Physical Science Laboratory, memo, 2005.
[7] Farley, R.E., *BalloonAscent: Balloon Flight Simulation Tool Theoretical Manual*, NASA/GSFC internal document, 2004–2011.

INTRODUCTION TO THE CASE STUDIES

The second half of this book will address case studies of air vehicles that have influenced the art and science of design. A case study is supposed to capture the significant decisions and issues in the development of a product, in this case an air vehicle. The value of a case study to a student of design is to experience the mental and physical activity associated with resolving a requirements conflict, a design problem, a marketing question, or a management issue. The reader should "walk the walk" with the program manager or chief engineer from concept definition to delivery of the product. The case study should highlight the lessons learned during the conduct of the program. The reader should takeaway the successes and mistakes of the past. Often it is the mistakes of the past that are of the most value to the designer because "he who forgets the mistakes of the past is destined to repeat them."

The case studies were selected deliberately to embrace a broad cross section of air vehicles—military to commercial to private sector, incompressible to hypersonic, and hydrocarbon-powered to man-powered to no power. The following summaries introduce the nine case studies and highlight the notions that should be taken away by the readers. The authors were carefully chosen for their first-hand experience or intimate knowledge of each subject.

Case Study 1: Lockheed Blackbirds (A-12, YF-12, M-21, and SR-71) by John R. Whittenbury

The family of aircraft nicknamed the Blackbirds (A-12, YF-12A, M-21, and SR-71) produced by Lockheed's Advanced Development Projects (ADP) "Skunk Works" under the leadership of Clarence L. "Kelly" Johnson remain among the greatest aeronautical achievements of the 20th century. They not only achieved operational performance capabilities that would be unequalled even at the beginning of the 21st century, but pioneered a multitude of unproven technologies that had to be matured in parallel with design and manufacturing development. This concurrent approach, which would be viewed as extremely risky in today's environment, was made possible by a unique combination of circumstances. These included an urgent national need during the height of the Cold War; visionary leaders in both government and industry that were willing to take measured risks and accept failures while making progress; secrecy that minimized the number of people involved and associated bureaucracy; and the Skunk Works operating rules crafted and enforced by Kelly Johnson.

Not having the desktop computing tools of today during the Blackbird design phase of the late 1950s/early 1960s, engineers used simplifying assumptions and judgment to solve complex design and analysis problems, combined with testing for confirmation. In the case of the A-12 developed for the CIA, the result was an aircraft that remarkably achieved Mach 3 flight less than four years after project go-ahead, and completed an operational overseas deployment and 29 combat missions less than 10 years after go-ahead. The follow-on SR-71 for the U.S. Air Force began flying operational overseas combat missions less than four years after first flight and remained viable throughout its entire operational lifespan. The hallmark of a great design is one that can evolve without major redesign, and the Blackbird family demonstrated this repeatedly as new missions, sensors, avionics, and capabilities were added until final retirement in 1999.

Case Study 2: X-35 Concept Demonstration Aircraft by James Eshleman

The key takeaway for this case study is that you need a good idea and then good people to execute the idea to win a competition. The Shaft Driven Lift Fan (SDLF) gave the Lockheed Martin team a tremendous advantage in designing an aircraft for the U.S. Marine Corps. The SDLF not only gave an augmentation to the vertical thrust but also a "cool" forward jet (minimizing the thrust sag due to hot gas ingestion) and a long inlet. The long inlet gave Line of Sight (LOS) blockage of the compressor face with a modest turning of the flow in the diffuser. The SDLF was the good idea but it took good people with strong wills and the commitment to work together to get the idea to work.

The Boeing X-32 team undoubtedly had good people also but their selection of the direct lift/vectored thrust (DL/VT) concept, although lower risk than the SDLF, had a much reduced design space for the Marine Corps mission. The Boeing effort was likely doomed from the start, thus illustrating how important the early decisions are in a major development program.

Case Study 3: Boeing 777 by Leland Nicolai

The Boeing 777 "Triple 7" was designed to fit between the B-767 and B-747. It was a much different program than the other aircraft in the Boeing 7XX stable. For the first time the users (All Nippon Airways, American Airlines, British Airways, Cathay Pacific, Delta Airlines, Japan Airlines

Qantas and United airlines) wrote the requirements and had a major role in the development of the airliner. This was a different operating environment for Boeing and required a change in culture and management style. These eight airlines were called the "gang of eight" and the management style for the Triple 7 development was called "Working Together." The decision to spend five billion dollars and five years developing the 777 represented a substantial commitment and risk for the company. But the gamble paid off as the Triple 7 has become the most profitable commercial jet produced by the Boeing Commercial Airplane Company.

Case Study 4: HondaJet by Michimasa Fujino

The entrance of the global giant Honda Motor Company into the business jet community is a story in progress. The entrance of the Honda Company into a very crowded and competitive market was not a surprise based upon Mr. Honda's commitment to expanding human mobility. The story started almost three decades ago with the author's passion for airplanes. This case study is a personal account of the love affair between the author and the HondaJet. The major takeaway for this case study is the importance of designer passion in the pursuit of design excellence.

Case Study 5: Hybrid Aircraft Technology and the Development Journey by Robert R. Boyd

The Hybrid Aircraft offers game-changing air cargo capability by combining low-risk technologies in a revolutionary new class of air vehicle. Lockheed Martin quietly developed and tested the technology over two decades, culminating in a successful P-791 demonstration in early 2006. Six years later there are still no operating Hybrid Aircraft despite substantial technical validation and extensive business planning. Why? This case study explores the complex and nonlinear world of revolutionary technology development through the eyes of the Hybrid Aircraft leadership, offering the aspiring engineer unique perspective, tips, and hazards to watch out for on the long and difficult journey from idea conception to operational acceptance.

Case Study 6: Daedalus by Harold Youngren

On April 23, 1988 the Daedalus Project, a three-year program in education and research at MIT, culminated with a successful 74-mile flight by the human-powered aircraft Daedalus 88 setting two world records that remain unbroken to this day. This case study revisits the design and development and the flights of the three Daedalus aircraft and outlines some of the most significant results obtained in the areas of physiology, meteorology, aerodynamics, structure, performance, stability, and control. The study documents the design characteristics and the performance of the Daedalus airframe as well as the human and financial resources devoted to the project.

Case Study 7: Skyhawk: A Cessna Legend by Conrad F. Newberry

The Cessna C-172 Skyhawk aircraft was designed in 1955 for a single mission, basic cross-country flight, a continuation of the C-170 "Family Car of the Air" concept. The design was revisionist in character. Cessna wanted an inexpensive (first cost and operational), safe, reliable, easily maintained aircraft to compete with similar products of other general aviation manufacturers. The Skyhawk was the result of excellent engineering, production, and marketing collaboration. Some features, such as the wing and fuselage design, were the result of Cessna's long-time corporate experience. Other features, such as the tricycle landing gear, were the result of product improvement and competitive product pressure. Cessna understood customer demands and desires and was determined to meet those demands and desires.

The result was an inexpensive, robust, and dependable aircraft as demonstrated by the C-172 endurance records of Heth/Burkhart set September, 1958 and the current record set by Timm/Cook in February, 1959. Proof of concept is that some 43,000 Skyhawks (more than any other aircraft model ever produced by any company) have been produced over the past 60+ years and the aircraft is still in production. Aircraft companies make their money by building airplanes, not by conducting research. *Cessna obviously got something very, very right!*

Case Study 8: T-46A and Fairchild Republic Company by Leland Nicolai

This case study is about how an excellent aircraft design can be destroyed by the people committed to its creation. The reader should takeaway an understanding of the importance of ethical behavior in the execution of an aircraft development program. A disregard for the fundamental tenets of good program management (such as Kelly's 14 Rules) can terminate an aircraft program and destroy a legendary airplane company. This regrettable end is documented here.

Case Study 9: Foot-Launched Glider Design and Performance by Paul Dees

Who among us has not wanted to fly like a bird with the ability to take off and land easily on our feet? Not only is this possible but it is now quite common through flying foot-launched gliders. They are one of the most prolific types of aircraft, though not commonly discussed or understood in engineering circles. This may be due to a bad reputation from many fatal accidents during the surge of popularity of hang gliding during the early 1970s. The popularity was due to the emergence of the Rogallo wing and the ingenuity of an Australian named John Dickenson. Unfortunately, in the early days many pilots died due to unsafe characteristics at low angle of attack or due to lack of safe training standards.

Thankfully, since then safety by many measures has improved to match that of private aircraft. Foot-launched gliders exist in three primary forms: flex-wing hang gliders, rigid-wing hang gliders, and paragliders. The structural and aerodynamic designs of flex-wings and paragliders differ significantly from traditional aircraft by having designed flexibility purposely in their wings to enable good handling qualities. All three forms are most commonly seen as tailless designs, although tails in rigid wings are now more common. All are lightweight and can be easily transported via automobile or sport utility vehicle (SUV). They make up a numerous and fascinating part of the aircraft design spectrum.

Case Study 1

Lockheed Blackbirds (A-12, YF-12, M-21, and SR-71)

John R. Whittenbury

Clarence L. "Kelly" Johnson with the SR-71 at Beale AFB, CA. (Lockheed Martin)

- The Need for a U-2 Successor
- The Competition for a U-2 Successor: Project GUSTO
- Skunk Works Approach and Tools
- A-12 Technology, Design, and Manufacturing Development
- Fighter and Reconnaissance-Strike Variants
- Flight Test Challenges
- Using the A-12 as a Launch Platform
- Operational and Capability Improvements

> *I can truly say everything on the aircraft, from rivets and fluids, including materials and power plants had to be invented from scratch.*
>
> Clarence L. "Kelly" Johnson

CS1.1 Introduction

The family of triple-sonic aircraft developed by Lockheed's Advanced Development Projects (ADP) "Skunk Works" organization under the leadership of Clarence L. "Kelly" Johnson included several mission-specific variants. The A-12 developed for the Central Intelligence Agency (CIA) to perform overflight reconnaissance established the basic vehicle configuration and blazed the trail of technology innovation that enabled successive variants. These included the YF-12A, a prototype air defense interceptor developed for the United States Air Force (USAF) that demonstrated successful interception of airborne targets from long range at high speed and altitude; the M-21, a variant of the A-12 used as a launch platform for the D-21 unmanned reconnaissance vehicle; and the SR-71, developed for the USAF to perform wide area, multi-sensor synoptic reconnaissance. Though never officially named beyond their respective program codenames, the popular term "Blackbirds" will be used when describing common technologies and features.

This case study provides only an overview of the engineering challenges and operational accomplishments of these aircraft, and the reader is encouraged to review the references cited for greater details of their respective histories and the people that created and operated them.

CS1.2 The Need for a U-2 Successor

The motivation for the initial member of the Blackbird family, the A-12, can be traced to the initial operational use of its subsonic predecessor, the U-2. Also developed by Lockheed ADP during the height of the Cold War under an aggressive schedule that achieved first flight only nine months after go-ahead, the U-2 was designed to perform overflight reconnaissance of denied territory in peacetime, operated by the CIA. (Today's U-2S version, an upgrade of the U-2R built in the late 1960s and U-2R/TR-1 built in the 1980s but with state of the art avionics and sensors, continues to perform global intelligence, surveillance, and reconnaissance missions operated by the USAF.) It was believed that the aircraft's high altitude operations above 70,000 ft would deny detection by early warning radars and engagement by airborne interceptors; even so, operational life expectancy was estimated to be 18–24 months before defenses were expected to catch up. Unexpectedly, the initial series of overflights of then-East Germany, Poland, and Russia in June-July of 1956 were detected and tracked, leading President Dwight D. Eisenhower to strongly consider a suspension of further overflights until the radar detection problem could be addressed. A program codenamed RAINBOW was immediately initiated to reduce the radar detectability of the U-2 [1]. Several techniques developed by the

Massachusetts Institute of Technology (MIT) Lincoln Laboratory and Lockheed ADP were tested but all were parasitic in nature, adding weight and drag through radar-absorbent coatings and external dipole antenna arrays [2]. These treatments were employed on nine operational missions through May 1958, but proved ineffective [3]. An important lesson learned was that design features for reduced radar detectability must be incorporated in the vehicle configuration from inception.

Accordingly, the CIA under the leadership of U-2 program manager Richard M. Bissell, Jr. began studying new design alternatives for a successor to the U-2. One such alternative, ironically already under study by Lockheed, was a supersonic, high altitude, liquid hydrogen (LH$_2$) fuelled reconnaissance aircraft initiated by the USAF in late 1955 under Project SUNTAN. Compared with conventional jet petroleum (JP) fuel, LH$_2$ with its high energy content per unit mass promised a dramatic reduction in thrust specific fuel consumption. But the fuel's low density and cryogenic temperatures (with a boiling point of −423°F) required large insulated fuel tanks that could only be accommodated in the fuselage and not the thin wings. Lockheed's model CL-400-10 SUNTAN design of 1956 (see Fig. CS1.1) resembled a scaled-up F-104 with a crew of two, LH$_2$ fuel tanks in the fuselage together with bicycle landing gear, T-tail with folding ventral fin (patented by Kelly Johnson himself), and Pratt & Whitney Model 304-2 engines mounted in pods on the tips of the trapezoidal wing. Each engine heated and expanded LH$_2$ through a turbine that drove a ducted fan via a reduction gearbox; additional thrust was provided by an afterburner [4].

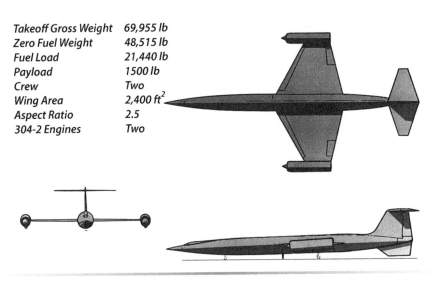

Takeoff Gross Weight	*69,955 lb*
Zero Fuel Weight	*48,515 lb*
Fuel Load	*21,440 lb*
Payload	*1500 lb*
Crew	*Two*
Wing Area	*2,400 ft^2*
Aspect Ratio	*2.5*
304-2 Engines	*Two*

Figure CS1.1 General arrangement of the Lockheed CL-400-10. (Lockheed Martin)

Despite a projected cruise Mach number of 2.5 at an altitude of 100,000 ft, the low supersonic lift-to-drag ratio of the configuration, low fuel mass fraction, and less than anticipated propulsion performance resulted in a less than desired mission radius of 1440 nautical miles. Global cryogenic fuel storage facilities and aerial refueling were required, both of which were considered impractical for LH_2. Given these concerns, Kelly Johnson recommended cancelling the project in March 1957, even after having ordered long-lead material in anticipation of a production contract and demonstrating that LH_2 could be safely produced and stored in large quantities [5]. Though follow-on studies also involving Boeing, Convair, and North American Aviation were conducted into 1958, similar conclusions were drawn and the quest for an LH_2 fuelled reconnaissance aircraft was abandoned. It was clear to Bissell that a clean-sheet approach was needed.

CS1.3 The Competition for a U-2 Successor: Project GUSTO

The first step in defining a U-2 successor was to establish requirements. A reconnaissance payload of at least 500 lb and a mission radius capability of 2000 nautical miles were desired, with availability to perform operational missions 18–24 months from go-ahead [6]. What was less clear was the tradeoff between speed, altitude, and radar cross-section (RCS) on detection and survivability (see Volume 1, Chapter 12 for a description of RCS). On the recommendation of Edwin Land, founder of the Polaroid Corporation who chaired one of President Eisenhower's technological capabilities panels (later nicknamed the "Land Panel") and who would play a major role in subsequent airborne and spaceborne reconnaissance programs, Bissell commissioned the Scientific Engineering Institute (SEI), which had also supported Project RAINBOW, to perform an operations analysis study [7]. One approach was to fly subsonically at high altitude with reduced RCS. The other approach, discovered by Dr. Franklin Rodgers of SEI and later nicknamed the "Rodgers Effect," was to fly supersonically at higher altitudes with moderately reduced RCS [8]. He determined that an aircraft possessing a certain RCS and flying at Mach 3 above 90,000 ft would be very difficult to detect and track by a radar operator observing a pulse-position indicator due to the faint number of radar "blips" per scan [9].

> Radar cross section (RCS) is a measure of radar reflectivity, equal to the cross sectional area of a perfectly conducting sphere having the same reflectivity as the target object. The RCS is expressed in square meters or in decibels relative to a square meter (dBsm).

With initial requirements in hand and two different approaches to the radar detection problem, the CIA-initiated Project GUSTO in the spring of 1958 to develop design concepts. One unusual military concept was a ramjet-powered aircraft carried to altitude by a balloon, but it was shown impractical and not seriously considered. Despite Lockheed's proven capability as demonstrated by the U-2, Bissell needed competition and invited the Convair Division of the General Dynamics Corporation to participate as well, led by its head of advanced development, Robert Widmer [10]. Convair previously studied for the USAF a Mach 4 reconnaissance aircraft dubbed Super Hustler to be launched from a B-58 Hustler at supersonic

> Kelly Johnson performed an independent study of the ramjet-powered aircraft concept, carried to launch altitude by balloon. He calculated that the balloon would have to be a mile in diameter, quipping ". . . and that's a lot of hot air."

speed, and had also performed pioneering studies in threat assessment, operations analysis, and RCS reduction that agreed well with Rodgers' findings. Again using the B-58 as a launch platform, Convair developed a new parasitic aircraft dubbed First Invisible Super Hustler (FISH) that integrated special shaping and treatments for RCS reduction, Marquardt ramjet engines fed by a ventral inlet for cruise at Mach 4 above 90,000 ft, and turbojet propulsion (initially one Pratt & Whitney JT12 and ultimately two pop-out General Electric J85s) for recovery and landing. Convair's studies had shown that disc-shaped planforms provided a reduced radar return across all azimuth angles [11], and so the FISH wing leading and trailing edges were formed from circular arc segments incorporating radar absorbent structures.

Meanwhile, Lockheed studied both subsonic and supersonic designs. The initial GUSTO Model 1 attempted to apply reduced RCS shaping and treatments to a conventional tailed configuration, whereas GUSTO 2 featured a tailless flying wing configuration with two J57 turbojets buried in the saucer-shaped fuselage; both designs employed extensive use of plastic structures with radar absorbing materials [12]. Neither was promising, and Kelly Johnson began a series of supersonic designs dubbed Archangel, so-named because they would fly higher and faster than the Angel, the nickname for the U-2. The initial Archangel 1 was conceived to be the simplest, lowest risk approach to achieving the design objectives, but it did not address RCS reduction. It relied on a new powerplant, the Pratt & Whitney J58, and titanium alloy materials (technologies described in Sec. CS1.5) to meet the high temperature Mach 3 environment. However, Archangel 1 did not meet the altitude-over-target objective of 100,000 ft. After initially adding wing tip ramjets burning ethyldecaborane (a toxic fuel dubbed High Energy Fuel-3 or HEF-3 that promised 35% higher energy content than JP-150 jet petroleum), wing area was enlarged and the J58 engines moved outboard for wing bending relief to create the Archangel 2. This design increased cruise Mach number to 3.2 in order to achieve 100,000 ft, but still did not address RCS reduction, and was deemed too large by the Land Panel [13].

Kelly Johnson decided to try a different approach: downsize the aircraft and use a combination of smaller turbojets and ramjets. The resulting A-3

(Archangel now being abbreviated) employed two Pratt & Whitney JT12 turbojets for takeoff, acceleration, and landing, and two 40-in. diameter Marquardt ramjets burning HEF-3 for cruise. But at only 25% of the takeoff gross weight of Archangel 2 and with half the payload, the A-3 required extremely lightweight structure, and despite the smaller size did not reduce the RCS appreciably. To address this shortfall, the A-4 through A-6 design studies attempted to combine small size with RCS reduction shaping features including a blended wing/fuselage and vertical tail surfaces above the wing. These designs used combinations of turbojet, ramjet, and in the case of the A-5, rocket propulsion (for takeoff assistance) but fell well short of the mission radius objective. Johnson concluded that maximum performance and minimum RCS were mutually exclusive, and pursued a series of more conventional designs focused on maximum performance consistent with an 18–24 month development schedule and with no concessions to RCS reduction. These designs, A-7 through A-9, employed a single J58 turbojet and two wing tip mounted ramjets, but still did not meet the mission radius and altitude objectives [14].

In order to meet the performance objectives, a new design direction was required. The A-1 through A-9 all suffered from low cruise lift to drag ratio (L/D) and poor mission fuel fraction, both of which hampered mission radius. The L/D and fuel fraction were improved by combining a long, slender fuselage for fineness ratio (to reduce supersonic wave drag) and large fuel volume with a delta wing planform that provided high supersonic L/D and excellent low speed characteristics. Together with a single vertical tail and twin turbojet engines mounted underneath the wing, the new configuration met the radius objective with acceptable altitude performance in a lightweight, elegant design. The initial configuration of this type, the A-10, used General Electric J93-3 turbojets then already under development for the North American Aviation B-70 Valkyrie Mach 3 bomber; concerns with J58 development schedules prompted consideration of the J93 as an alternative. The next configuration, formally proposed in March 1959, was the J58-powered A-11 (together with a J93-3 powered derivative, the A-11A) which added aerial refueling capability and the option for a mixed fuel load of JP-150 and HEF-3. Kelly Johnson remarked that the design had come full circle, in that the A-11 was similar in overall size and weight to the original Archangel 1 [15]. Table CS1.1 shows the evolution of the Archangel design series.

When Lockheed's A-11 and Convair's FISH were proposed in June 1959, the CIA and Land Panel were dissatisfied. The A-11 offered no RCS reduction, whereas the final, enlarged FISH configuration (Fig. CS1.2) relied upon building a

The competition nearly ended in November 1958 when the Land Panel recommended selection of the FISH instead of the A-3. But despite giving Convair a development contract in December 1958, the Government prudently continued the competition as a risk reduction measure.

Table CS1.1 Archangel Design Evolution

Model	Dimensions	Weights	Cruise performance
A-1	Length = 116.67 ft Span = 49.6 ft Height = 23.58 ft	Zero fuel = 41,000 lb Fuel = 61,000 lb Takeoff GW = 102,000 lb	Mach no. = 3.0 Altitude = 83–93 kft Radius = 2000 nm
A-2	Length = 129.17 ft Span = 76.68 ft Height = 27.92 ft	Zero fuel = 54,000 lb Fuel = 81,000 lb Takeoff GW = 135,000 lb	Mach no. = 3.2 Altitude = 94–105 kft Radius = 2000 nm
A-3	Length = 62.3 ft Span = 33.8 ft Height = 14.6 ft	Zero fuel = 12,000 lb Fuel = 22,600 lb Takeoff GW = 34,600 lb	Mach no. = 3.2 Altitude = 95 kft Radius = 2000 nm
A-4	Length = 58.33 ft Span = 35.0 ft Height = 17.21 ft	Zero fuel = 24,600 lb Fuel = 33,300 lb Takeoff GW = 57,900 lb	Mach no. = 3.2 Altitude = 92 kft Radius = 1370 nm
A-5	Length = 46.0 ft Span = 32.5 ft Height = 16.92 ft	Zero fuel = 18,500 lb Fuel = 31,820 lb Takeoff GW = 50,320 lb	Mach no. = 3.2 Altitude = 90 kft Radius = 1557 nm
A-6-5	Length = 64.0 ft Span = 47.2 ft Height = 22.85 ft	Zero fuel = 29,200 lb Fuel = 33,750 lb Takeoff GW = 62,950 lb	Mach no. = 3.2 Altitude = 90 kft Radius = 1287 nm
A-7/8/9 Series	Length = 93.75 ft Span = 47.5 ft Height = 22.85 ft	Zero fuel = 27,200 lb Fuel = 43,700 lb Takeoff GW = 70,900 lb	Mach no. = 3.2 Altitude = 91.5 kft Radius = 1637 nm
A-10	Length = 109.5 ft Span = 46.0 ft Height = 19.25 ft	Zero fuel = 33,300 lb Fuel = 52,700 lb Takeoff GW = 86,000 lb	Mach no. = 3.2 Altitude = 90.5 kft Radius = 2000 nm
A-11	Length = 116.67 ft Span = 56.67 ft Height = 21.03 ft	Zero fuel = 36,800 lb Fuel = 55,330 lb Takeoff GW = 92,130 lb	Mach no. = 3.2 Altitude = 93.5 kft Radius = 2000 nm

new, lengthened version of the B-58 for launch due to increased weight and drag in the mated configuration. Accordingly, Lockheed was directed to redesign the A-11 to incorporate RCS reduction features at the expense of cruise altitude, and Convair was directed to redesign FISH for self-sufficient operations using only turbojet propulsion (no ramjets) and obviate the need for a new "B-58B" launch platform.

Convair's redesign, dubbed KINGFISH and developed in approximately two months, began as a scaled-up FISH but rapidly evolved into a new configuration. Twin J58 engines were fed by two-dimensional, upper surface mounted mixed compression inlets and exhausted through two-dimensional single expansion ramp nozzles (SERNs) (Fig. CS1.3). Like FISH, KINGFISH employed brazed stainless steel honeycomb sandwich structure to withstand Mach 3+ cruise temperatures, pyroceram materials in the edges together with inlet treatments to reduce RCS, and a crew capsule escape system (similar to the B-58). Convair quickly modified their full-scale FISH RCS pole model, used to measure radar signatures, into a 7/10 scale facsimile of KINGFISH and collected data to substantiate their proposal [17].

> RCS "pole models" are sub-scale or full-scale test articles exposed to radar energy in order to measure their RCS. They are so named because they are mounted on poles or pylons to isolate them from radar energy reflected from the ground surface.

Lockheed's redesign of the A-11 saw the transformation of the aircraft into a configuration that would still appear ahead of its time fifty years later (see Fig. CS1.4). The basic layout was

Cruise Mach No.	4.0
Cruise Altitude	90,000 ft
Range	3,900 nm
Span	37.0 ft
Length	48.5 ft
Height	9.8 ft

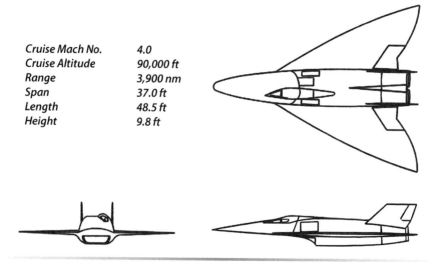

Figure CS1.2 Final Convair FISH configuration as proposed in June 1959. [16]

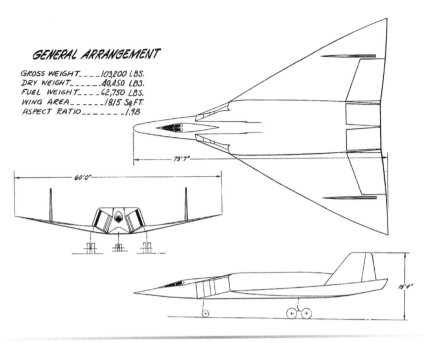

GENERAL ARRANGEMENT

GROSS WEIGHT____103,200 LBS.
DRY WEIGHT_____40,450 LBS.
FUEL WEIGHT____62,750 LBS.
WING AREA_____1815 Sq.FT.
ASPECT RATIO_____1.98

Figure CS1.3 Convair KINGFISH general arrangement.
(Eric Hehs/Lockheed Martin)

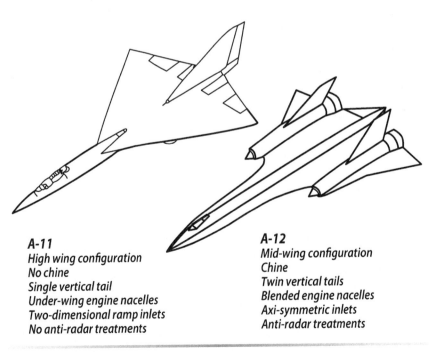

A-11
High wing configuration
No chine
Single vertical tail
Under-wing engine nacelles
Two-dimensional ramp inlets
No anti-radar treatments

A-12
Mid-wing configuration
Chine
Twin vertical tails
Blended engine nacelles
Axi-symmetric inlets
Anti-radar treatments

Figure CS1.4 Configuration transformation: A-11 to A-12.

Table CS1.2 Comparison of Lockheed A-12 & KINGFISH Characteristics [20]

Characteristic	Lockheed A-12	Convair KINGFISH
Speed	Mach 3.2	Mach 3.2
Total range	4120 nm	4000 nm
Range at altitude	3800 nm	3400 nm
Cruise altitude		
Start	84,500 ft	85,000 ft
Mid-range	91,000 ft	88,000 ft
End	97,600 ft	94,000 ft
Cost (12 aircraft w/o engines)	$96.5 million	$121.6 million

developed by aerodynamics lead Richard Fuller and configurator Ed Baldwin, working with electromagnetics lead L. D. McDonald and Edward Lovick, Jr. who offered suggestions on shaping and treatments [18]. Reduced side profile, shallow side slopes, and smooth curvilinear planform elements were key attributes of the new configuration. The wing was repositioned from the upper to mid-fuselage, and the engines moved from under-wing pods into blended mid-wing nacelles with axisymmetric, free-stream inlets replacing the A-11's two-dimensional ramp inlets. The single, large vertical tail was replaced with two smaller surfaces, each mounted atop the engine nacelle and canted inboard 15 deg to scatter incident radar energy up and away from its source. To further reduce fuselage sidewall slopes and control scattering of radar energy, chines of ogival planform were added and blended into the wing leading edge along with fillets and fairings along the fuselage and nacelles, and radar absorbing materials and structures were incorporated into the perimeter edges and control surfaces [19]. Lockheed submitted their new design, the A-12, in August 1959, together with Convair and their KINGFISH (see Table CS1.2 for a comparison of the competing designs).

The two designs were evaluated in the categories of Analysis and Design (the ability to meet the performance and RCS requirements), Models and Components (pole model and component-level test data to validate the RCS signature), Materials Research (the maturity of high-temperature materials), and Subsystems (fuel, hydraulic, environmental control, etc.) [21]. Although some members of the joint DOD/CIA/USAF selection panel favored KINGFISH because of its lower RCS, the A-12 was lower cost and deemed to be lower risk, and Richard Bissell notified Kelly Johnson on August 28, 1959 that Lockheed had

The A-12 performance requirements were ambitious for early 1960. The design cruise Mach number was 60% greater than the dash Mach number of the F-104 Starfighter, and the cruise altitude was 70% higher than that of the B-58 supersonic bomber.

won the competition. Project GUSTO was terminated and Project OXCART was begun—an ironic codename for what would become the world's fastest aircraft. However, Lockheed first had to prove that the A-12 "anti radar" approach was viable before the CIA would commit to a full go-ahead (Convair continued certain KINGFISH risk reduction activities during the same period as a backup should the A-12 fail). After quickly building a full scale pole model and elevation post and collecting RCS data during the fall of 1959, Lockheed provided evidence that the A-12 was worthy of further development. But the additional weight of the RCS reduction treatments resulted in a significant decrease in cruise altitude to maintain range, and this greatly alarmed Richard Bissell, who had promised President Eisenhower higher penetration altitudes into denied airspace. Kelly Johnson quickly reduced the zero fuel weight by 1000 lb, added 2000 lb of fuel, reduced holding fuel to 2700 lb, and reduced the reconnaissance payload to 600 lb. The weight reductions and additional fuel allowed the A-12 to achieve an altitude over the target of 91,000 ft, which satisfied Bissell but worried Johnson because no more performance margin remained. With the RCS and performance objectives met, Lockheed was awarded a full go-ahead for 12 aircraft on January 30, 1960. But in order to meet the scheduled first flight date of May 1961, Johnson and his Skunk Works would have to move out rapidly into unchartered territory.

CS1.4 Skunk Works Approach and Tools

The OXCART program began with the understanding that Lockheed ADP would conduct the effort in the same manner as the U-2: quickly, quietly, and on budget. With the Skunk Works operating rules (summarized in Volume 1) as the foundation, Kelly Johnson formed a relatively small engineering team to execute the greatest challenge ever faced in aircraft design; peak engineering headcount on the A-12 was only 150. Secrecy meant limiting the number of people accessed to the program, and each engineer was responsible and accountable for their design from "cradle to grave." For example, an engineer working on a particular component was responsible for performing requirements analysis and design trade studies; interfacing with other design disciplines to ensure integration compatibility; developing initial layouts followed by production drawings; coordination with the materiel organization to select and order parts; working together with shop personnel as the component was produced, often creating sketches or making "red mark" drawing corrections on the spot; overseeing installation and test of the component; and ensuring that it operated as intended. The present-day term "concurrent engineering" captures very well what the Skunk Works was practicing 50 years ago.

In the days before computer-aided design tools were available, engineers relied on descriptive geometry to develop drawings on vellum or mylar. Per Skunk Works practice, each drawing contained all information needed to produce the part, and had only four approval signatures in the title block: engineer, stress, material, and project engineer (the latter often signed off by Kelly Johnson himself). Although an IBM mainframe computer was available to perform more complex calculations such as those associated with the method of characteristics for supersonic inlet design, engineers normally used sliderules and mechanical calculators. Because tools such as the sliderule required the operator to expect the magnitude of the answer, engineers had to develop an intuitive understanding of the reasonableness of their calculations.

One of the Skunk Works operating rules was to minimize paperwork, but record important work thoroughly. For example, the A-12 design specification was only 41 pages in length (versus present day specifications that can run in the hundreds of pages), but it captured all necessary data. Military standards were used as guidelines, and deviations where necessary were noted in the design specification. By capturing the intent of the military standards, engineering documentation was considerably streamlined without compromising safety or capability.

The Skunk Works organization (Fig. CS1.5) minimized bureaucracy. Kelly Johnson served as ADP Vice President and Chief Engineer, but strong

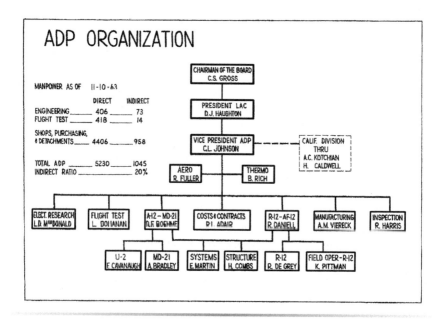

Figure CS1.5 Lockheed ADP organization in November 1963. [22]

lead engineers including Richard Boehme (A-12 project engineer), Henry Combs (structures), Ben Rich (thermodynamics/propulsion), Richard Fuller (aerodynamics), David Campbell (propulsion and inlet design), Ed Martin (systems and payloads), L. D. McDonald (electromagnetics) and John McMaster (flight controls) played pivotal roles in the day-to-day execution of the program, and continued to be involved as the A-12 developed and evolved into other variants. Later the organization expanded to include AF-12 project engineer Richard DeGrey and R-12 project engineer J. R. "Russ" Daniell (who both later served as SR-71 program manager) with Art Bradley leading the MD-21 project (discussed in Sec. CS1.8.2).

CS1.5 A-12 Technology, Design, and Manufacturing Development

The A-12 schedule called for first flight in May 1961, less than 18 months from full go-ahead, and to meet this aggressive schedule technology development, design, and manufacturing would have to occur in parallel. No airbreathing aircraft had ever sustained flight at Mach 3, let alone for over an hour. An entire suite of new technologies, design features, and manufacturing techniques were required to accommodate the wide range in flight conditions between high speed/high altitude cruise and takeoff, aerial refueling, and landing. At the Mach 3.2 design point, aerodynamic heating due to skin friction created leading edge stagnation temperatures in excess of 800°F and average surface temperatures above 550°F as shown in Fig. CS1.6; even higher temperatures, above 1100°F, occurred in the engine nacelle and exhaust areas. However, the aircraft also had to accommodate the transition to sub-zero ambient temperatures during subsonic aerial refueling at 25,000–30,000 ft. Accommodating these wide-ranging conditions affected every aspect of the design.

CS1.5.1 A-12 Configuration Design and Aerodynamics

The A-12 configuration was driven by the need to combine shaping for RCS reduction with a low-drag basic design that minimized trim drag at cruise. The fuselage chines added during the A-11 to A-12 transformation were key. Besides RCS reduction, the chines created a lifting surface that, being forward of the center of gravity, was de-stabilizing in the pitch axis and helped mitigate the rearward shift of the aerodynamic center (and accompanying nose-down pitching moment) during transition to supersonic speed. Together with active center of gravity (CG) control via fore/aft fuel transfer and a stability augmentation system (SAS), low

During cruise, aerodynamic heating and thermal expansion caused the aircraft to grow by about 2.5 in. in length.

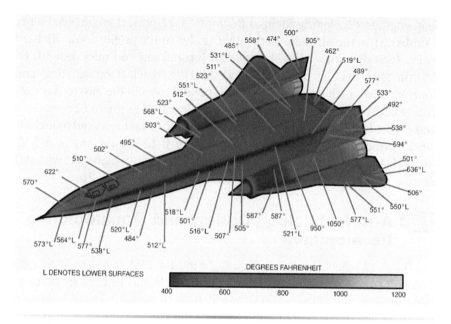

Figure CS1.6 Surface temperatures at M = 3.2 (SR-71 shown). [23]

stability margins could be maintained with correspondingly small trim deflections and associated trim drag. While de-stabilizing longitudinally, the chines improved lateral/directional handling by streamlining the fuselage and reducing side force and yawing moment due to sideslip. They also provided a convenient place to route subsystems (and later integrate additional bays for mission equipment) before smoothly blending into the 2.5% thick bi-convex airfoil delta wing. As shown in Fig. CS1.7, the overall trimmed maximum lift-to-drag ratio benefitted from the addition of chines to the fuselage [24].

The aerodynamic design was developed through numerous wind tunnel tests conducted in the Lockheed 4 foot by 4 foot and NASA Ames Unitary Plan facilities. Throughout these tests, the design underwent several refinements. Smaller, all-moving rudders sized for engine-out control (made challenging by the wide separation of the engines) replaced conventionally-hinged rudders, and the rudder trailing edges were swept forward to improve structural stiffness and reduce flexibility effects on control power. The elevons that provided pitch and roll control were extended in chord length for improved effectiveness. Following testing that revealed excessive bending and nacelle carry-through loads in the outboard wing, and after it was too late to change the angle of incidence without incurring cost and schedule impact, conical camber was introduced in the outboard leading edge to reduce the effective angle of attack in the presence of the nacelle up-wash airflow (Fig. CS1.8). The conical camber improved the span-wise

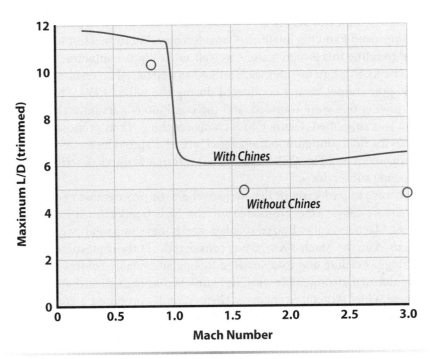

Figure CS1.7 Addition of the chine improved trimmed maximum L/D. [25]

Figure CS1.8 Outboard wing leading edge conical camber (SR-71 shown). [NASA]

lift distribution and also helped reduce the rolling moment due to sideslip for improved handling qualities. Considerable wind tunnel test hours were used to refine this design feature, as well as the chine configuration [26].

The RCS design features were refined as well based on results of the full scale pole model testing conducted during the fall of 1959. The original sharp wing tips were rounded, and the continuous curvature chine planform was simplified. Figure CS1.9 compares the A-12 as proposed in 1959 versus the final configuration, whereas Fig. CS1.10 provides the A-12 general arrangement and Fig. CS1.11 points out internal features described in the following subsections.

It was realized early in the Archangel design process that conventional aluminum alloys were unsuitable for the high temperatures experienced during Mach 3+ cruise. Brazed stainless steel honeycomb sandwich construction used on the Mach 3 XB-70 was considered, but the sophistication of the brazing procedure and its associated tooling and quality control challenges were deemed incompatible with the Skunk Works manufacturing approach. Instead, chief structures engineer Henry Combs proposed a conventional stiffened structure, but employing an unconventional material: titanium [29]. Titanium maintained strength at elevated temperatures and offered nearly twice the strength to density ratio of stainless steel, however it had never been used for the majority of an airframe. Three titanium alloys were used: Ti-5Al-2.5Sn, referred to as A-110; Ti-13V-11Cr-3Al, referred to as B-120; and Ti-6Al-4V, referred to as C-120. Higher temperature areas of the airframe required more exotic "super alloy" materials, including Hastelloy X for the airframe-mounted exhaust ejector flaps, and René 41 for the nozzle flap

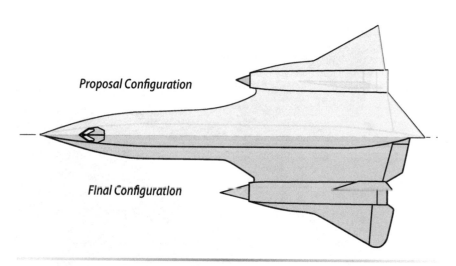

Figure CS1.9   A-12 proposal configuration vs final configuration.

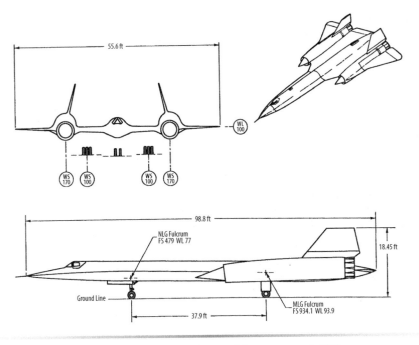

Figure CS1.10 A-12 General Arrangement. [27]

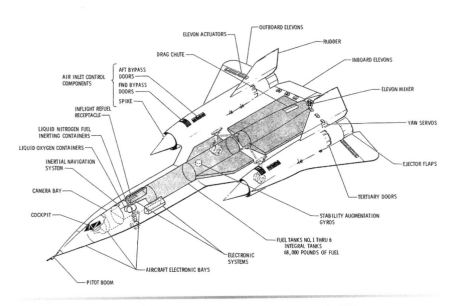

Figure CS1.11 A-12 Internal Arrangement. [28]

support ring. Stainless steel was also used for concentrated load areas including the nacelle fittings and rudder posts, as well as for rivets and bolts for usage above 550°F. Table CS1.3 lists the materials used and their properties.

The selection of titanium alloys created many manufacturing challenges. Before new drill bits, profiler cutting heads, cutting fluids, and optimized machining speeds/feeds were developed, the rate of metal removal was only 5% of that possible when machining aluminum alloys [31]. Heat treatment processes at the Skunk Works had to be revised to mirror that of the supplier, Titanium Metals Corporation, to prevent embrittlement of titanium parts that would otherwise shatter if dropped on the floor. Early failure of wing panels spot-welded and washed with water in the summer was traced to the heavy use of chlorine in the Burbank water system during those months. Cadmium-plated tools could not be used due to their incompatibility with titanium. A new process called "hot sizing" was developed that used a hot press (custom-built by Bliss) and matched tooling to form sheet metal titanium parts at a temperature of 1450°F. For complex parts, this process had to be repeated in stages at considerable cost.

Elaborate quality control procedures were instituted. For every batch of ten parts, three test coupons were produced. The first coupon was evaluated for tensile strength, the second for notch sensitivity (using a "notch bend" test in which the sample was bent through a small radius around a quarter-inch cut), and the third for a reheat test if required. Careful records were maintained that traced every part to its original stock material.

The airframe comprised a forward fuselage and an aft fuselage/wing assembly joined at fuselage station (FS) 715. This manufacturing joint,

Table CS1.3 The Airframe Employed Titanium, Steel, and Nickel Alloys [30]

Material	Alloy	Structure	Design temperature
Titanium	13V-11Cr-3Al (B-120)	Wing and Fuselage; Rivets	500°F
	5Al-2.5Sn (A-110)	Engine Inlet/Nacelle; Wing Panels	900°F
	6Al-4V (C-120)	Wing Beams/Fittings; Bolts	500°F
Steel	A-286	Nacelle Fittings; Bolts	Up to 1200°F
	4340 Nickel Plated	Rudder Posts	500°F
Nickel Alloys	Rene′ 41	Special Pins, Exhaust Ejector	Up to 1600°F
	Hastelloy X	Exhaust Ejector	Up to 1600°F

which facilitated airframe sub-assembly fabrication and transport to the final assembly/flight test location, provided versatility by allowing the forward fuselage to be reconfigured for alternate variants without necessitating redesign of the entire airframe. (Although in actuality there would be many differences forward and aft of FS715 in future Blackbird versions.) The fuselage used light gage B-120 annealed sheet metal skins riveted or spot-welded to ring stiffeners formed of B-120 aged sheet. In some areas, chemical milling was used to reduce skin thickness from 0.020 in. down to 0.016 in. in order to reduce weight. Four discrete longerons of extruded C-120 aged material carried fuselage bending loads; these longerons were inherently not fail-safe and made the FS715 joint the most critical structural area on the aircraft [32]. The fuselage was fabricated in upper and lower halves that were mated to the continuous carry-through wing structure. Removable, non-structural fillets blended the cylindrical fuselage into the upper and lower wing surfaces, and provided access for subsystem routing.

The wing structure comprised an inboard wing/nacelle and an outboard wing/nacelle, the latter of which was hinged to provide engine access. Wing bending loads were carried across the nacelles by frames and into multiple, continuous wing beams (spars) that provided fail-safe redundancy. The inboard wing was built as left and right sub-assemblies joined at the centerline. Chord-wise ribs provided support for the hat-section stiffened wing skins that employed chord-wise corrugations (facing inward from the outer mold line contour) to accommodate thermal expansion relative to the cooler wing beam caps by carrying shear but not bending loads (see Fig. CS1.12). These distinctive corrugations were added following a test in which the 4-foot by 6-foot wing box test article warped when heated to cruise temperatures in an oven [33]. The inboard wing nacelles each provided a stub fin for mounting a steel post for the all-moving rudder.

The fuselage and nacelle chines were considered fairings and not primary structure, and used silicone-asbestos composite skins attached to annealed B-120 frames. These composite materials, developed by Lockheed ADP to provide radar absorbent structure with operating temperatures up to 600°F, were also used in the wing leading edges, elevon trailing edges, rudders, and in portions of the inlet spikes (see Fig. CS1.13). A band of composite material in the nose provided isolation of the pitot mast, which acted as the high frequency (HF) radio antenna exciter, from the rest of the structure that acted as the HF antenna. The alternating triangular pieces of composite material in the wing perimeter edge were used to attenuate incoming radar energy much like the pyramid structures used in the walls of an anechoic chamber [34]. The rudders used

Kelly Johnson joked that the corrugated wing skins led him to being accused of "trying to make a Ford Tri Motor go Mach 3."

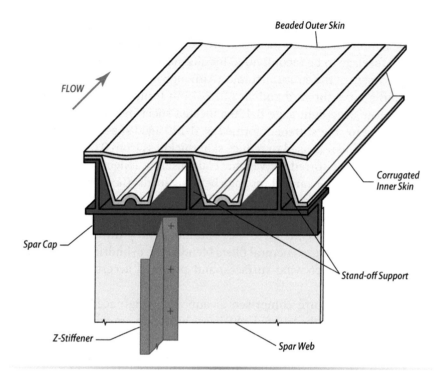

Figure CS1.12   Wing box construction.

titanium for the basic frame, but the composite portions were difficult to qualify. During a test flight, large areas of the composite material came apart due to expansion of trapped air inside; the solution was to drill a 1/8-in. diameter hole at the base of the rudder to permit the pressure to equalize [35]. Early A-12s used all-titanium rudders, as did the later YF-12A, M-21, and SR-71 (though composite rudders were also used interchangeably on the SR-71).

The initial A-12s were assembled in Building 82 of Lockheed's Burbank plant, where engineering offices were co-located and assembly of the first 20 U-2s had also taken place, but later A-12s and follow-on Blackbirds were fabricated in Building 309/310, originally built for Lockheed's double-decker R6V Constitution transport assembly line (and later used in the 1980s for the F-117A Stealth Fighter production line). It was impractical to test fly the A-12 from Burbank airport due to security and runway length restrictions, so the aircraft were disassembled and transported to a remote flight test location using custom-built shipping containers; Fig. CS1.14 shows the A-12 production line and one of the shipping containers for the fuselage and inboard wing assembly. A total of twelve single-seat A-12s were built, plus one two-seat trainer designated the A-12T (discussed in Sec. CS1.5.11).

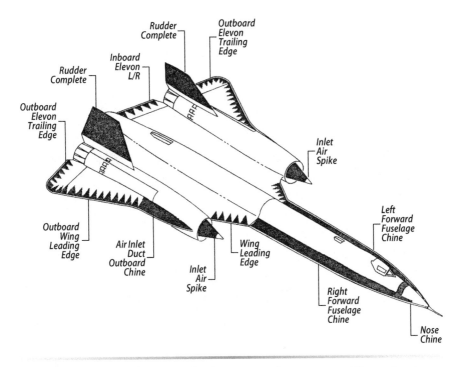

Figure CS1.13 Silicone-asbestos composites were used throughout the airframe. [36]

One of the challenges in designing the structure was the characterization of material properties and design allowables at elevated temperatures to inform the "Basis for Structural Design" document used by engineers. This required fabrication of material coupons and full-size structural articles that were tested within a large oven at the Skunk Works. In addition to the wing box unit described previously that led to the corrugated wing skins, a fuselage section that included the cockpit, windshield/canopy, nose equipment bay, and payload equipment bay was tested to characterize the behavior of highly curved thin gage skins and cockpit/equipment bay cooling and insulation. Some of these tests included a person in the cockpit, protected by a full pressure suit, to provide real-time feedback.

The extreme thermal environment and its variation over the flight profile, coupled with the limitations of the analytical tools, made determination of external and internal loads very problematic. Simplifying assumptions and approximations were imposed with acknowledgement that thermal loads would be conservative. On external surfaces, steady-state temperature estimates assumed a skin emissivity of 0.5 for unpainted areas and 0.9 for black surfaces with a constant radiation heat sink temperature of −60°F (outer space value) for upper surfaces and +60°F (earth surface value) for lower surfaces, and radiative effects between external surfaces were not included in

Figure CS1.14   A-12 production line; shipping container in foreground. (Lockheed Martin)

structural sizing calculations [37]. With the exception of the engine nacelle areas, internal heat transfer effects on skin temperatures were not included and the skin assumed to be perfectly insulated from internal heat transfer. Internal joints assumed uniform temperatures and joint efficiency factors were not adjusted for temperature effects [38].

To characterize the effects of loads on the airframe, a full-scale static test article was fabricated that included the fuselage, inboard wing, and left hand nacelle/outboard wing. Loads were applied using hydraulically actuated pads bonded with RTV adhesive to the test article, and internal strains were measured for comparison to predicted values using strain gages installed throughout the structure. Because the structure was most critically loaded in the transonic region, testing at room temperature was adequate. The static test article was tested to limit load, 130% of limit load, and ultimate load (150% limit load); testing was completed by August 1962 (after first flight). However, structural integrity of the rudders (both all-titanium and all-plastic versions) remained challenging and required further testing.

Airframe weight had to be minimized in order to maximize onboard fuel for long-range performance, but with sufficient strength to accommodate aerodynamic and inertia loads. Because the mission profile was dominated by cruise, the normal load factor limits were established at +2.5 g and −1.0 g, similar to transport category aircraft, with reductions in permissible positive load factor at increased weights. At supersonic speed, the increased lifting effectiveness of the fuselage caused the FS715 joint to be critical in up-bending for a maneuver load factor of +2.5 g, whereas at subsonic speed the joint was critical in down-bending. The design airspeed envelope ranged from a minimum equivalent airspeed (KEAS) of 135 KEAS to a design limit speed (V_L) of 500 KEAS and a design high speed (V_H) of 450 KEAS [39]. These equivalent airspeed limits created a narrow flight envelope as shown in Fig. CS1.15.

> The maximum speed of the aircraft was limited by a maximum permissible compressor inlet temperature of 427° C, corresponding to approximately Mach 3.38 on a standard day (−56° C at altitude). Normally, Mach number was limited to 3.2, but could be increased to 3.3 if warranted by the tactical situation.

An important design consideration for any airframe is access for inspection, both during construction and in service. By building the fuselage in upper and lower halves, open access was provided. Small removable panels provided access to the fuselage fuel tanks for tank inspection and sealing. The removable upper wing skins provided access to fuel tank areas for equipment replacement and for tank sealing as well. The ability to seal (and re-seal) the fuel tanks became an ongoing challenge throughout the life of the A-12 and follow-on Blackbirds due to the thermal expansion and contraction of the airframe.

Also considered part of the airframe structure, the landing gear used a forward–retracting two-wheel nose gear and main gear struts that retracted inward between the wing beams and fuel tanks. The struts were made from B-120 titanium forgings, the largest of their kind at that time. Because the main wheel assemblies had to fit within the 64-in. diameter fuselage and between the fuselage fuel tanks that acted as a heat sink for main wheel bay cooling, an unusual three-wheel arrangement on a single axle was used to distribute the load. Any two of the main wheels could carry limit load in the event of a tire blowout [41]. The tires themselves (25×6.75, 16 ply Type VII for the nose and $27.5 \times 7.5 \times 16$, 22 ply Type VIII for the mains) were pressurized with dry nitrogen to maintain uniform tire pressure throughout the flight envelope and prevent tire explosions, and the tires were impregnated with aluminum oxide to reduce temperatures via radiative cooling. Nevertheless, main gear tire explosion cans were added later to further insulate the tires from high temperatures and contain tire explosion shrapnel from damaging fuel and hydraulic lines running through the main wheel bay. To supplement the wheel brakes during normal landings or takeoff aborts, the A-12 carried a 45-foot diameter drag chute in an upper aft fuselage compartment that was kept relatively cool (80°F to 200°F) by surrounding fuel.

Note: Above 50,000 ft Minimum airspeed is 300 KEAS
Maximum altitude restriction-
with derichment - 85,000 ft
without automatic inlet operation - 80,000 ft

Normal Operating Speed

250 KEAS
300 KEAS
350 KEAS
400 KEAS
450 KEAS
500 KEAS

Minimum Demonstrated Airspeed

Design High Speed (V$_{jl}$)

3.2 Mach
Design Mach No.
V$_{jl}$ and V$_{L}$

3.0 Mach
Max Mach Without
Automatic Inlet Operation

Design Limit Airspeed (V$_{L}$)

Normal Bank Angle 30°
While Above 2.5 Mach

Minimum Airspeed Restriction - 135 KEAS

Altitude (1000 ft)

Mach Number

Figure CS1.15   A-12 Flight Envelope. [40]

CS1.5.2 Powerplant (JT11D-20/J58)

Key to achieving the A-12's high speed, high altitude cruise performance objectives was the Pratt & Whitney JT11D-20 powerplant, developed under the leadership of program manager Jack McDermott and project engineer William Brown. Originally a scaled-down version of the J91 turbojet built as a competitor to the General Electric J93 for the B-70 bomber, the engine was initially developed under the even-numbered Navy designation J58-P-2 for a never-built Mach 3 dash version of the Vought F8U-3 Crusader III. For the A-12 application that called for continuous operation at Mach 3+, the engine was completely redesigned under the Pratt & Whitney model designation JT11D-20, but was still often referred to as the J58 [42]. The JT11D-20 design requirements were a leap beyond those of the contemporary Pratt & Whitney J57 and J75 turbojets then in production (see Table CS1.4). These requirements, and in particular the extreme continuous operating temperatures, required a new set of materials (see Table CS1.5).

The single rotor engine used a nine-stage, 8:1 pressure ratio compressor and a two-stage turbine. Early in design while assessing high speed operability, engineer Robert Abernethy determined that the compressor was deep in stall and the turbine choked, with corresponding temperatures that threatened to melt the afterburner due to lack of cooling flow. He developed and patented a bleed-bypass design (perhaps not coincidentally named Recover Bleed Air after his initials) that bled air from the 4th compressor stage via 24 circumferential doors, bypassed it around the remaining compressor stages and combustor via 6 external tubes, and reintroduced

Table CS1.4 JT11D-20 Design Requirement Versus J57/J75 [43]

	J57 AND J75	**JT11D-20**
Mach number	2.0 for 15 min (J75 only)	3.2 Continuous
Corrected airflow turndown ratio (cruise/maximum)	90%	60%
Altitude	55,000 ft	100,000 ft
Compressor inlet temp	−40°F to 250°F (J75 only)	−40°F to 800°F
Combustor exit temp	1750°F (Takeoff) 1550°F (Continuous)	2000°F (Continuous)
Max. fuel inlet temp	110°F to 130°F	350°F
Max. lubricant inlet temp.	250°F	550°F
Thrust/weight ratio	4.0	5.2
Military power operation	30-min Time Limit	Continuous
Afterburner operation	Intermittent	Continuous

Table CS1.5   The JT11D-20 Required Advanced Materials [44]

Material	Description	Temperature capability	Application
Ti-8-1-1	Titanium alloy; good creep resistance	850°F	1st stage compressor blades
Ti-5-2.5	Titanium alloy; weldable	850°F	Compressor inlet case
Waspalloy	Nickel-base alloy; strong and oxidation-resistant at temperature; difficult to weld	1400°F	Most components
Inconel 718	Nickel-base alloy; easy to weld	1300°F	Diffuser case
Hastelloy X	Non-hardenable nickel-base alloy; good for low stress applications; high oxidation resistance	2000°F	Burner components
Astroloy	Nickel-base alloy; creep and tensile strength superior to Waspalloy	1500°F	Turbine discs
MAR-M-200DS	Directionally solidified nickel-base superalloy	1500°F	1st stage turbine vanes
IN-100	Cast nickel-based superalloy	1500°F	1st and 2nd stage turbine blades; 2nd stage turbine vanes; afterburner nozzle flaps

it in the afterburner, resulting in improved compressor stall margin, increased afterburner cooling flow, and additional thrust [45]. The bleed bypass opened as a function of compressor inlet temperature and engine speed to hold corrected airflow constant. Without this innovation, the A-12 would never have achieved its performance requirements.

The initial JT11D-20 engines had fixed-geometry inlet guide vanes that were cambered to provide best performance at cruise Mach number, but at the expense of airflow and thrust at transonic conditions. The later, higher thrust JT11D-20B version incorporated variable inlet guide vanes that could operate in either the fully open "axial" position for takeoff and acceleration to supersonic speed, or the partially closed "cambered" position for most efficient cruise.

The fuel/air mixture was combusted in a can-annular chamber with eight cylindrical combustion liners (cans). Due to the high flashpoint of the fuel (described in Sec. CS1.5.5), a standard high-energy ignition system was incapable of initiating combustion. After investigating a hydrogen-based ignition system, the solution was to incorporate a pyrophoric (combusting on contact with oxygen) chemical ignition system using tri-ethylborane, abbreviated TEB. Located on top of the engine, a nitrogen pressurized, 600-cubic centimeter container provided 16 shots of TEB; each shot could be used for either starting the engine or initiating after-burner operation via throttle movement. Because TEB was so critical to engine and afterburner operation, mechanical TEB counters (advanced by throttle lever movement) on the pilot's throttle quadrant provided an indi-cation of TEB shots remaining.

Engine start was accomplished using either a custom-built start cart equipped with two coupled 400-hp Buick (later 465-hp Chevrolet) V8 racing engines or an air turbine starter connected to a direct-drive gearbox on the bottom of the engine. During early operations, it was found that the engine would not start because of depressed inlet airflows that were not adequately characterized from wind tunnel tests. As a temporary fix, an inlet access panel was removed for ground starts, and later "suck-in" doors were added in the nacelle. Finally, twelve "start bleed" doors were added to the engine to aerodynamically un-load the compressor during starting, and this solved the problem (see Fig. CS1.16) [46].

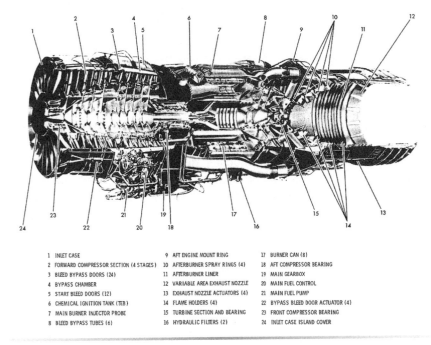

1 INLET CASE	9 AFT ENGINE MOUNT RING	17 BURNER CAN (8)
2 FORWARD COMPRESSOR SECTION (4 STAGES)	10 AFTERBURNER SPRAY RINGS (4)	18 AFT COMPRESSOR BEARING
3 BLEED BYPASS DOORS (24)	11 AFTERBURNER LINER	19 MAIN GEARBOX
4 BYPASS CHAMBER	12 VARIABLE AREA EXHAUST NOZZLE	20 MAIN FUEL CONTROL
5 START BLEED DOORS (12)	13 EXHAUST NOZZLE ACTUATORS (4)	21 MAIN FUEL PUMP
6 CHEMICAL IGNITION TANK (TEB)	14 FLAME HOLDERS (4)	22 BYPASS BLEED DOOR ACTUATOR (4)
7 MAIN BURNER INJECTOR PROBE	15 TURBINE SECTION AND BEARING	23 FRONT COMPRESSOR BEARING
8 BLEED BYPASS TUBES (6)	16 HYDRAULIC FILTERS (2)	24 INLET CASE ISLAND COVER

Figure CS1.16 Pratt & Whitney JT11D-20 engine. (Lockheed Martin)

Due to the thermal environment and the large "turndown" ratio between maximum and minimum fuel flows, one of the most challenging components on the engine was the main fuel control. Besides metering fuel to the combustion chambers, the control generated signals to operate the compressor bleeds, fuel shutoff, fuel manifold dump, variable geometry afterburner exit nozzle, and permit afterburner operation. After much trouble, the main fuel control was eventually replaced with a Bendix unit from the XB-70's General Electric J93 engine [47].

One of the unusual features of the engine was that the main burner fuel flow could be "trimmed" to maintain exhaust gas temperature (EGT) within limits. This required the pilot to monitor EGT during flight and increase or decrease EGT as required using manual switches (later, an automatic engine trim system for the SR-71 reduced pilot workload). To prevent EGT from rapidly increasing in the event of an inlet unstart (described in Sec. CS1.5.3) with its associated reduced airflows, a fuel de-richment system reduced the fuel/air mixture in the burner cans for EGT values at or above 860°C. Besides combustion, fuel served as a hydraulic and cooling fluid for the engine. The compressor start and bypass bleed valves were actuated with fuel, as were the afterburner nozzle exit flaps. After being heated, hot fuel was immediately burned.

The engine required new lubricants capable of high temperature operation. Normal grease tended to thicken with increasing temperature, and a new formulation was developed that was thinned with a solvent and capable of operating up to 550°F. The engine oil developed under Pratt & Whitney Aircraft Specification 524B was solid at 0°F, and would not pour until 40°F was reached. The oil was preheated using a ground cart before being loaded in the engine oil system, and had a maximum operating temperature of 500°F.

Each engine was connected via a power takeoff shaft to an Accessory Drive System (ADS) within each nacelle, remotely located to isolate the ADS and its accessories from the high temperatures of the engine bay. A double universal joint accommodated displacements between the engine and ADS due to thermal expansion. Each ADS drove two hydraulic system pumps, a fuel circulation pump for the heat sink system, and a variable frequency AC generator (the later YF-12A and SR-71 used a constant-speed drive to energize the generator). A self-contained ADS lubrication system was cooled by a fluid-to-fuel heat exchanger.

The unusual mid-wing engine nacelle arrangement complicated engine installation and removal. To gain access to the forward engine mount located at top centerline and the two outboard rear engine mounts, the entire outboard wing and outer nacelle was hinged and raised as shown in Fig. CS1.17; the inboard rear mount was accessed via a door on the inboard

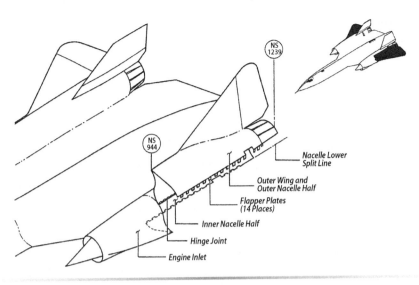

Figure CS1.17 The outer wing and nacelle folded for engine access. [48]

side of the nacelle. A special engine installation and removal dolly accommodated the translations necessary to clear the surrounding airframe structure.

After building 11 experimental and 10 development engines, Pratt & Whitney built 51 prototype engines designated YJT11D-20A that powered the A-12 and later the AF-12 (YF-12A) aircraft. These so-called "YJ" engines produced 20,500 lb of thrust at military power and 31,500 lb in maximum afterburner. An improved "J" afterburner was introduced in 1965 that increased thrust to 22,900 lb and 32,500 lb for military power and maximum afterburner, respectively; these engines were nicknamed "JJ" engines. A total of 99 production engines were later built for the SR-71 program, initially with "J" afterburners but then upgraded to model "K" configuration with two-position inlet guide vanes and nozzle/afterburner improvements. This definitive version, designated JT11D-20B, provided 24,500 lb of thrust in military power and 34,000 lb in maximum afterburner.

The JT11D-20 scored many firsts, including first engine rated to operate continuously in military power or in afterburner; first engine to use fuel as a hydraulic fluid; first engine to use directionally solidified turbine airfoils; first production dual-cycle engine; and first application of a variable orifice afterburner spray bar that provided more efficient metering and distribution of fuel across the flight envelope [49]. Today, over 50 years since it's development begun, the JT11D-20 remains a technological marvel and achievement that has not been surpassed.

CS1.5.3 A-12 Inlet (and Automatic Inlet Control System)

The blended wing/nacelle configuration led to selection of axially symmetric, free-stream inlets (as opposed to under-wing two-dimensional inlets, originally selected for the A-11, that could benefit from precompression in the wing flow field). In order to slow down the Mach 3.2 airflow to subsonic speeds suitable for engine ingestion without incurring a large total pressure loss, a mixed (external and internal) compression inlet was selected. In addition to establishing the series of oblique shockwaves for external compression, the prominent inlet centerbodies or "spikes" controlled the ratio of inlet capture area to throat area. Each spike was full forward from takeoff up to Mach 1.6. Beyond Mach 1.6, the spike retracted aft to move the terminal normal shock into the inlet, a process known as "starting" the inlet (much in the same fashion as starting a supersonic wind tunnel). The spike position was automatically controlled as a function of Mach number, angle of attack, and angle of sideslip, but could be manually controlled as well. Using an IBM 3-60 computer and a method of characteristics code developed in parallel with the inlet design, a spike half-cone angle of 13 deg was selected so that the oblique "shock-on-lip" condition occurred at the design cruise Mach number of 3.2 for maximum pressure recovery with the spike fully retracted 26 in. At this condition, the inlet capture area increased 112%, while the throat area decreased 54% as compared to the spike full-forward position [50]. Each spike was canted inward and downward to improve total pressure recovery. Porous slots around the circumference of the spike at its maximum diameter bled off the spike boundary layer, which was routed through the centerbody and its support struts then overboard through louvers. Boundary layer air from the cowl surrounding the spike was bled through 32 "shock trap" tubes positioned around the cowl that used one-way valves to help stabilize the normal shock.

A set of bypass doors (later designated the "forward" bypass doors) matched the inlet airflow to engine demand and controlled the position of the normal shock by varying the backpressure. To minimize drag due to the momentum loss of bypassed air, the inlet was sized at the Mach 3.2 design condition such that the forward bypass doors were nearly closed. However, the more closed the forward bypass doors were, the less stable the normal shock position was. The optimum, but least stable position for the normal shock was at the throat; opening the forward bypass to place the normal shock slightly aft of the throat was less efficient but provided more stability.

Nestled between the forward bypass door slots around the inner circumference of the inlet cowl were 16 individual fairings nicknamed "mice." The mice were added in November 1963 during the A-12 flight test program in an effort to cure a "duct rumble" condition at and above Mach 2.4 suspected to be caused by flow separation from the aft surface of the inlet spike. The cross sectional area of the mice removed cross sectional flow area from the inlet duct, which in turn reduced flow expansion and created a more favorable pressure gradient that prevented flow separation.

Another set of bypass doors was added immediately in front of the compressor inlet face to correct unacceptable engine restart characteristics resulting from lower engine idle/windmill airflows as well as use of foreign object elimination screens in the inlet. Originally nicknamed the "onion slicer" because of their resemblance to the kitchen utensil, the "aft" bypass doors improved transonic inlet/engine airflow matching and allowed the forward bypass doors to close down further, reducing drag. Aft bypass airflow joined shock trap airflow to provide engine bay cooling and additional mass flow for the exhaust ejector described in Sec. CS1.5.4.

The overall inlet system (Fig. CS1.18) contributed a large percentage of the overall thrust by achieving a 40:1 compression ratio at the Mach 3.2 cruise condition. At cruise speed, the inlet produced 54% of the total thrust, with 17.6% by the engine and 28.4% by the ejector (described subsequently), compared with Mach 2.2 values of 13%, 73%, and 14%, respectively [51]. Figure CS1.19 shows the inlet, engine, and ejector thrust contributions as a function of speed.

Just as the inlet could be started, it could be "unstarted" as well if the normal shock moved forward of the throat and was expelled. The immediate loss of thrust created large yawing and rolling moments that were extremely disorienting and required the pilot to manually extend

> Kelly Johnson considered the air inlets and their control system to be the greatest challenge he faced on the program.

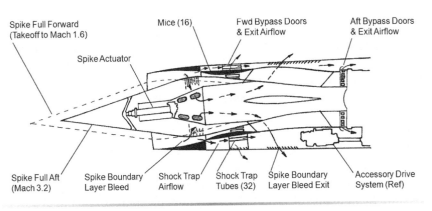

Figure CS1.18 Inlet configuration and airflows.

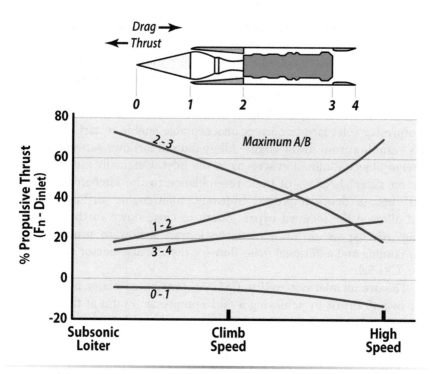

Figure CS1.19   Inlet, engine, and ejector thrust contribution vs speed. [52]

the spike and open the forward bypass doors for the affected inlet to recapture the normal shock and "restart" the inlet. Unstarts were frequent during initial operations with the original inlet control system that used pneumatic controls. Via a series of orifices, bellows, cams, and linkages, local pitot and static pressures were converted into electromechanical signals to the hydraulic actuators for the spike and forward bypass doors. However, because of the close mechanical tolerances and thermal expansion/contraction within the system, it was impossible to get repeatable and reliable performance. The pneumatic system was eventually replaced by an electronic system developed jointly by Lockheed ADP and AiResearch, and the frequency of unstarts dropped considerably, saving the A-12 program from possible cancellation. Later inlet control system features for the SR-71 included a g-bias to reduce the likelihood of unstarts in turns, a shock expulsion sensor for automatic restart, and a cross-tie that overrode the automatic spike and bypass door schedule for both inlets.

The inlet spike and forward bypass were normally positioned automatically (the aft bypass doors were manually selected), but the pilot could operate them manually as well. Initially, the only indication of inlet operation was a compressor inlet manifold pressure gauge, and controls were limited to switches for moving the spikes forward and opening or closing the forward bypass doors. Later, inlet spike and forward bypass door position indicators were added, as well as an improved compressor inlet pressure gauge with a "barber pole" indicator that displayed the optimum compressor inlet pressure corresponding to the flight condition if above 250 KEAS and Mach 1.8. These indicators, together with rotary inlet spike (with Mach schedule) and forward bypass door control knobs provided much better pilot control of manual inlet operation [53].

CS1.5.4 A-12 Exhaust Ejector

Each engine's exhaust, together with engine bay airflow originating from the shock trap and aft bypass, was expelled through an airframe-mounted ejector nozzle, one of the first examples of its kind. The ejector employed a series of annular tertiary air doors that were spring-loaded in the open position, and a set of free-floating exit flaps made from Hastelloy X nickel alloy attached to a support ring of Rene' 41. During low speed flight up to Mach 1.1, the tertiary doors provided additional mass flow to fill the ejector, and the exit flaps closed to form a convergent nozzle. Above Mach 1.1, the shock trap/aft bypass airflow provided enough additional mass flow to fill the nozzle, pushing the tertiary doors closed, whereas the exit flaps opened between Mach 0.9 and 2.5 to form a convergent-divergent nozzle that expanded the exhaust gases to supersonic speed (see Fig. CS1.20).

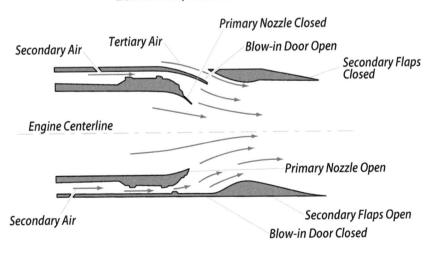

Figure CS1.20 Exhaust ejector configuration at low (top) and high (bottom) speeds. [54]

Originally the ejector was to be part of the engine, but Lockheed ADP and Pratt & Whitney jointly agreed that an airframe-mounted ejector was lighter weight. Pratt & Whitney remained responsible for ejector performance and wind tunnel testing, but testing of the ejector in isolation from the wing caused under-prediction of drag because the wing interference effects were not present.

CS1.5.5 Fuel System

To accommodate the wide range of operating pressures and temperatures, a low vapor pressure, high flashpoint fuel, initially designated Special Kerosene 1 (SK-1), PWA 523E (with PSJ-67A lubricity additive) and later JP-7, was developed by the Shell Oil Company in association with Ashland, Monsanto, and Pratt & Whitney. The A-12's six integral fuel tanks contained a total of 10,590 gallons (68,300 lb at an average density of 6.45 lb/gal) of fuel as shown in Fig. CS1.21, and were interconnected

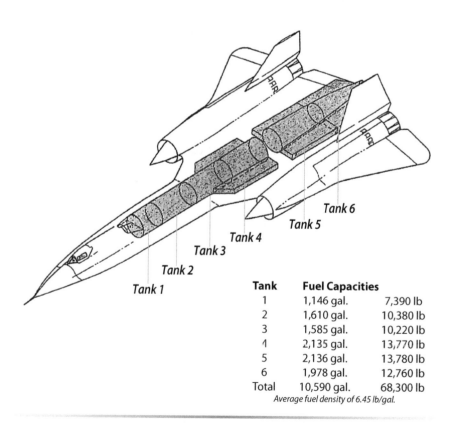

Tank	Fuel Capacities	
1	1,146 gal.	7,390 lb
2	1,610 gal.	10,380 lb
3	1,585 gal.	10,220 lb
4	2,135 gal.	13,770 lb
5	2,136 gal.	13,780 lb
6	1,978 gal.	12,760 lb
Total	10,590 gal.	68,300 lb

Average fuel density of 6.45 lb/gal.

Figure CS1.21 A-12 fuel tanks and quantities. [56]

by left and right fuel manifolds and a single vent line. The tanks were pressurized and inerted with gaseous nitrogen contained in liquid form by two dewars (one 75 liters, the other 106 liters) in the nose landing gear bay. Because the nitrogen in the ullage space was dumped during aerial refueling, the amount of remaining liquid nitrogen determined the number of remaining aerial refueling opportunities and thus aircraft maximum range on a single

> The A-12 (and later Blackbirds) all leaked fuel on the hangar floor when the airframe was cool. Normally this would present a safety hazard, but the flashpoint of JP-7 was so high that a lit match could not ignite the fuel.

mission. Boost pumps in each fuel tank transferred fuel forward (via the right manifold) or aft (via the left manifold) for c.g. control, commanded by float valves in each tank or manually selected by the pilot. Fuel dump valves in each manifold were opened to reduce aircraft weight in an emergency. Fuel tank sealing remained a challenge for the A-12 and follow-on Blackbirds; ultimately an elastomeric polyester material developed by Dow Chemical and 3M was selected that could withstand the thermal expansion and contraction of the airframe [55].

Fuel was also used as a heat sink to cool the cockpit and equipment bay air, engine oil, TEB tank, accessory drive system (ADS) oil, and hydraulic fluid. A "smart valve" in the fuel heat sink system sent hot fuel to the engine for combustion if below 295°F, or if above back to fuel tank 4 for reuse (unless tank 4 was full).

The capability to aerial refuel from KC-135Q tankers, carrying JP-7 fuel, was key to global reconnaissance operations (see Fig. CS1.22). An upper fuselage mounted, single point receptacle was used for both aerial refueling and ground refueling using a special adapter (later, the SR-71 would have a dedicated ground fueling receptacle as well). Aerial refueling was typically performed at 32,000 ft, with initial contact made in non-afterburner. As fuel was transferred and weight increased, minimum afterburner was selected for one engine, with the other modulated in non-afterburner to maintain sufficient thrust.

Mockups and test rigs were used throughout the design and development of the fuel system. The major fuel system test rig consisted of a full-size working model of the fuel tanks, fuel transfer system, aerial refueling system, engine fuel control units, afterburner fuel control units, fuel gauging system, and fuel dump system. The entire assembly was mounted on a fulcrum in the "Fort Robertson" area of the Lockheed Burbank plant (which had earlier been used to perform liquid hydrogen testing for the CL-400 program), and was raised or lowered to evaluate fuel system operation for simulated climb and descent angles.

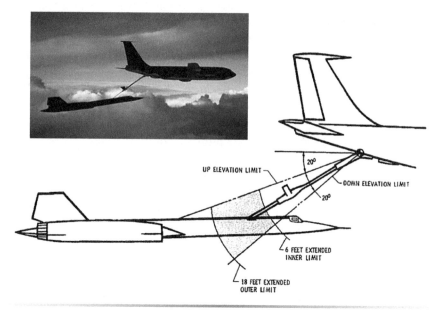

Figure CS1.22 Aerial refueling via KC-135Q. [57]

CS1.5.6 A-12 Air Conditioning and Pressurization System

To keep the cockpit and equipment bays at suitable pressure and temperature throughout the extremes of the flight envelope, the A-12 was equipped with two air conditioning systems, left and right, that were functionally similar. The left air system serviced the cockpit, nose compartment, pilot's pressure suit, inverters, and inertial navigation system (INS), whereas the right system serviced the electronics and equipment bays. In the event of left air system failure, a crossover permitted the right air system to supply the cockpit and associated equipment. Each system took high pressure bleed air from the compressor ninth stage at a temperature above 1250°F and ducted it through a ram air heat exchanger in each inlet that cooled the air to 850°F, primary and secondary fuel/air heat exchangers that further cooled the air to approximately 160°F, a bleed air filter (nicknamed the "Waspatrap" because it trapped Waspalloy particles scraped from the engine compressor casing by rubbing blade tips), and finally to an air cycle refrigeration unit where it was cooled to −30°F before entering the cockpit. The refrigeration unit was housed in an air conditioning bay; this practice,

> The environmental control system capacity was sufficient to cool approximately forty houses of 1500 ft^2 each.

used on other Skunk Works aircraft, facilitated development of the refrigeration unit as a single package, versus a fractionated system installed in multiple locations. The integration of the fuel system with the air conditioning system represented one of the first implementations of a fuel thermal management system, now used routinely in many aircraft. The cockpit and bays were insulated with three in. of layered fiberglass/aluminum foil to maintain an approximate 70°F temperature at cruise conditions where the external boundary layer temperature was approximately 710°F and the skin temperature approximately 550°F [57]. Application of high emissivity black paint also helped reduce surface temperatures by 35 deg through radiative heat transfer. High pressure air was also provided to the canopy and equipment bay hatch inflatable seals, as well as the windshield defog system. The cockpit and nose were normally pressurized to an altitude of 26,000 ft, whereas the equipment bay was pressurized to an altitude of 28,000 ft. The overall environmental control system represented a highly successful collaboration between AiResearch and Skunk Works engineers [58].

> When Kelly Johnson ordered a sample of what was advertised to be a high temperature hydraulic fluid, he received a canvas bag containing the "fluid" in powdered form with instructions to heat into a liquid. Deciding that thawing out the hydraulic system with a blowtorch was not a good idea, he pursued other options.

CS1.5.7 A-12 Hydraulic System

The A-12 had four separate 3350-psi hydraulic systems designated A, B, L, and R. Systems A and B were redundant and powered the flight controls only; either system could assume full load. A reserve oil tank could be switched into either the A or B system in the event of a hydraulic leak. System L powered the left inlet spike and bypass doors, landing gear, normal nose wheel steering, normal wheel brakes/anti-skid, the retractable antenna for the ultra-high frequency (UHF) radio, and normal operation of the aerial refueling receptacle, whereas System R powered the right inlet spike and bypass doors, alternate nose wheel steering and wheel brakes, emergency landing gear retraction, and alternate aerial refueling receptacle operation. A variable-volume piston pump driven off the respective accessory drive gearbox pressurized each system. The high temperature environment required a special hydraulic fluid developed by the Pennsylvania State University Petroleum Refinement Laboratory to Lockheed specification SP-302 and military specification MIL-H-27601A, with an operating temperature range of −65°F to 600°F. To prevent fluid leaks, induction or furnace brazing was used at all possible joints instead of threaded fasteners [59].

CS1.5.8 A-12 Flight Controls

The flight control system included both a manual flight control system and, to accommodate the low levels of static stability necessary for low trim drag during cruise and to reduce pilot workload, an automatic flight control system with a full-time three-axis stability augmentation system (SAS). The overall system was designed to allow the pilot to safely land in the event of any two electronic failures; any single mechanical failure combined with any single electronic failure; any single hydraulic failure, and any single engine failure. Although a fly-by-wire system was initially considered, it was considered immature, and Lockheed pursued a simple, reliable system employing analog computers [60].

Inboard and outboard elevons provided pitch and roll control, and twin all-moving rudders provided better yaw control for engine-out operation than hinged rudders. The outboard elevons, which were rigged 3° higher in the neutral position than the inboards to reduce wing root bending moments, provided approximately 85% of the roll control power; both inboard and outboard elevons had approximately the same authority in pitch. Pilot inputs via a conventional control stick and rudder pedals were transmitted by dual-redundant cables and pushrods to dual-redundant irreversible hydraulic servo valves that actuated the surfaces. Because such a system did not transmit control surface air loads back to the stick and rudder pedals, springs were used to provide artificial feel in proportion to the commanded deflection. The elevons could deflect up to 10° trailing edge-down and 24° trailing edge-up in pitch, and up to 12° trailing edge up and down in roll. When pitch and roll were commanded simultaneously (as in a coordinated turn), elevon deflection range was increased to 20° trailing edge-down and 35° trailing edge up [61]. The rudders could deflect +/–20°. Above Mach 0.5, surface limiters were manually engaged to reduce elevon deflection range to +/–7° and rudder deflection range to +/–10°. To accommodate thermal expansion and contraction, the control cables were made of Elgiloy, a low coefficient of thermal expansion material used in watch springs, and tension regulators absorbed slack in the system. A mixer assembly in the tail cone took pilot pitch and roll commands and converted them into appropriate mechanical commands to the inboard elevon control valves; pushrods and torque tubes that ran across the engine nacelles slaved the outboard elevons to the inboards. The mixer assembly was easily the most complex mechanism in the aircraft, and built with the precision of a Swiss watch. The mixer assembly also included electrical actuators for pitch and roll trim, located downstream of the artificial feel springs so that control stick position was not affected by trim position. Rudder trim actuators were combined with the artificial feel springs in the stub fins, and thus rudder trim was reflected in rudder pedal position. In the event of a runaway trim actuator, the pilot could use a trim power switch to quickly and simultaneously cut power to all trim actuators before isolating the fault.

The SAS operated in all three axes. The pitch axis could command a maximum of elevon 2.5° trailing edge-up or 6.5° trailing edge-down, the roll axis a maximum elevon differential of 4° between left and right, and the yaw axis 8° left or right [62]. The pitch and yaw axes both used two independent channels, A and B, and a monitor/voting channel, M, whereas the roll axis, being less critical, had no monitor channel. Using logic circuits, the M channel compared the functioning of the A and B channels in the pitch and yaw axes and automatically disconnected a failed channel, whereas for the roll axis the pilot could select a single channel. In the event of two failures, the affected channel was completely disconnected. If both pitch channels became inoperative due to electronics failures or overheating of the pitch/yaw rate gyro package (that was buried in the number 3 fuel tank for cooling and to minimize flight control system coupling with fuselage bending modes), a back-up pitch damper provided pitch rate signals to either channel to facilitate aerial refueling and landing.

The autopilot operated via the SAS in the pitch and roll axes. Pitch modes included attitude hold, knots equivalent airspeed (KEAS) hold, and Mach hold. It was never possible to obtain truly satisfactory Mach hold operation due to low damping of the supersonic phugoid, airframe flexibility effects, and outside air temperature gradients that caused the local speed of sound to change quickly, causing the airplane to hunt in the pitch axis to maintain constant Mach number [63]. Roll modes included attitude hold, heading hold, and automatic navigation via steering commands from the inertial navigation system (INS). A Mach trim system associated with the autopilot fed a pitch elevon signal as a function of Mach number to improve speed stability during manual flight between Mach 0.2 and 1.5. Because neither the autopilot nor Mach trim system were required for safety of flight, they were not redundant.

To qualify the flight control system in the high temperature environment, an "iron bird" was constructed that included primary flight control system components installed in an oven at the Skunk Works. Control surface mass simulators were servo-driven by commands from a simulated cockpit and SAS/autopilot, which received signals from an analog computer programmed with the aircraft equations of motion. The A-12's flying qualities were developed through piloted simulations at Lockheed's Rye Canyon Research Laboratory and the NASA Ames Research Center's NE-2 two-axis full motion simulator; a variable stability JF-100C aircraft operated by NASA was also used [64]. Though the SAS normally operated full-time and the autopilot was normally used during cruise, the aircraft could be flown without augmentation or autopilot operation, but with difficulty.

CS1.5.9 A-12 Electrical System

Each remote accessory drive energized a 30 kVA generator that produced 115/220 VAC, three-phase power supplied to an AC bus and a 200-ampere transformer-rectifier unit. The 28 VDC power output from each transformer-rectifier was provided to "monitored" and "essential" DC busses. Two 25 ampere-hour silver-zinc emergency batteries powered the DC bus if either the generators and/or transformer-rectifiers failed. The monitored bus powered the equipment bay payload and INS, whereas the essential bus serviced all other functions as well as four 600 volt-ampere inverters that powered the SAS, inlet control system, avionics, and cockpit instruments. To accommodate maximum operating temperatures ranging from 800°F to 1000°F, new wire insulation, connectors, terminal strips, and potting compounds were required.

CS1.5.10 A-12 Communications, Navigation, and Identification Systems

The A-12 was equipped with a high frequency (HF) radio for long range voice and status communications, and an ultra high frequency (UHF) radio for voice communications and air-to-air ranging for aerial refueling tanker rendezvous. Radio navigation aids included tactical air navigation (TACAN), automatic direction finding (ADF), and an instrument landing system (ILS) receiver, but the primary means of navigation was an inertial navigation system (INS) produced by Minneapolis-Honeywell. A flight reference system (FRS) provided a secondary source of attitude and heading reference signals. A unique piece of equipment was the integrated periscope and sun-compass, which provided a display at the top of the instrument panel. The periscope provided a variable magnification view beneath the aircraft for navigation and reconnaissance target identification purposes, whereas the sun compass was used as a cross-check or in emergencies to determine true heading by measuring the bearing of the sun relative to a pre-computed value based on latitude and longitude. A film strip projector integrated with the periscope displayed a moving map of the mission route or pilot checklists, and a destruct system could be manually activated by the pilot prior to ejection or automatically during ejection to destroy the film strip along with water soluble maps to prevent them from falling into an adversary's hands.

CS1.5.11 A-12 Crew Accommodations

An early design decision was that the pilot would be protected from the high speed, high altitude environment in the event of cockpit depressurization or emergency escape by a full pressure suit and ejection seat combination, rather than an escape capsule as used in the B-58 and XB-70 that permitted a "shirt sleeve" environment. The pressure suit/ejection seat occupied less volume than the escape capsule, and was deemed more reliable [65]. The A-12, AF-12, and early SR-71s used a modification of the F-104's Lockheed model C-2 ejection seat, whereas later SR-71s (and the YF-12A during NASA service) used a Lockheed SR-1 ejection seat, later used in the U-2R/S and the Space Shuttle Orbiter *Columbia* during its first four missions.

The V-shaped windshield and curved canopy panels used laminated quartz glass. Pressurized air from the air conditioning system provided defog capability. The canopy was manually raised and lowered, with support provided by a nitrogen boost counterbalance system. Special sunshades provided a means to shield the instrument panel from glare, which could be extreme at high altitude. The instrument panel was conventional, with air conditioning and landing gear controls on the left, flight instruments and inlet controls in the center, and engine, fuel, and electrical system controls on the right. Throttles, radio, and mission payload controls were located on the left console, and SAS/autopilot, INS, and radio navigation system controls were on the right. A caution/warning light panel on the center stand alerted the pilot when necessary, and multiple circuit breaker panels provided the means to isolate system faults.

Kelly Johnson evaluated the crew accommodations firsthand when he made a flight with test pilot Louis W. "Lou" Schalk in the A-12T trainer on August 27, 1963 (Fig. CS1.23). This aircraft, Article 124, replaced the equipment bay behind the cockpit with a raised second cockpit for the student or instructor pilot, or in this case Kelly Johnson. He believed that the designer and builder had a responsibility to test their own product, both as a means of inspection and to maintain competency to design future aircraft [66]. He noted room for improvement in the areas of seat comfort and forward visibility through the V-shaped windshield.

CS1.5.12 Mission Equipment

The A-12 was equipped with a pressurized equipment bay (or "Q-bay") aft of the cockpit that accommodated one of four camera packages: Type I (Perkin-Elmer), a panoramic system with two cameras that provided a 63 nmile swath with 30% stereo overlap and 12-in. resolution at nadir; Type II (Eastman Kodak), a mechanically simpler set of two panoramic cameras, built as a lower-risk backup to the Type I, that provided a swath width of 60 nm with 30% stereo overlap and 17-in. resolution at nadir; Type III, a modification of the U-2's "B" camera that too was developed as a back-up to the

Type I; and Type IV (Hycon), an 8-in. resolution spotting camera similar in design and concept to the "B" camera, but incorporating improvements in optics, film transport, and vibration control [67]. Interchangeable lower Q-bay hatches matched window configuration to the camera, a design feature leveraged from the U-2. When carrying the Type I, the Q-bay could be purged with helium to reduce optical degradation resulting from convective turbulence. Despite Kelly Johnson's concerns about the complexity of the Type I, which combined advanced features such as twin reflecting cube scanners imaging onto a single piece of film and concentric film supply/take-up reels to minimize center of gravity shift, Dr. Roderic M. Scott's design performed extremely well and was the only camera used operationally by the A-12 [68]. In addition to the cameras, the A-12 also carried systems to record electronic signals and to provide additional protection (besides speed, altitude, and low RCS

Well before flying in the A-12, Kelly Johnson flew as a flight test engineer in the original Lockheed Model 10 Electra. Earlier, while a graduate student at the University of Michigan in 1933, Johnson performed wind tunnel tests of the new design and solved a directional stability issue by replacing the Electra's small single vertical tail with twin vertical tails tip-mounted on the horizontal stabilizer. He joined Lockheed that same year, initially as a tool designer but rapidly moving into the engineering department.

Figure CS1.23   Kelly flew the A-12T with Lou Schalk on 27Aug1963. (Lockheed Martin)

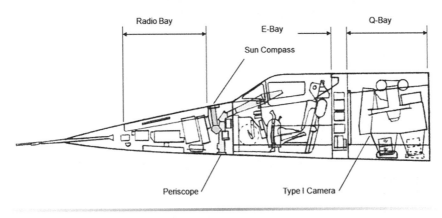

Figure CS1.24 A-12 forward fuselage internal arrangement.

features) against surface to air missile threats [69]. An electrical "E"-bay was located immediately forward of, and accessed through, the Q-bay and contained the SAS, autopilot, air data computer, air data transducer, FRS, TACAN, ADF, and IFF equipment.

Figure CS1.24 shows the internal arrangement of the A-12 forward fuselage including the nose radio bay, cockpit, E-bay, and Q-bay (shown with the Type I camera).

CS1.5.13 A-12 Mission Profile and Performance

A typical A-12 mission profile began with takeoff in maximum afterburner, accelerating to an initial climb speed of 400 KEAS and then continuing the climb at Mach 0.9. Unless an aerial refueling was planned (to top off the tanks if takeoff was performed at less than full fuel to improve safety in the event of a takeoff abort), the climb continued at Mach 0.9–0.95. At 38,000 ft, the pilot disconnected the autopilot (if engaged) and initiated a 6000–8000 ft/min rate of descent to accelerate through Mach 1, losing nearly 10,000 ft, and then resumed the climb at Mach 1.15, maintaining 450 KEAS. At Mach 1.5, fuel flow was reduced 6000–8000 lb/hr per engine, and at Mach 1.7 the pilot selected position "B" (50% open) for the aft bypass doors as the inlet spikes began to retract. The climb continued up to Mach 2.6 and an altitude of approximately 60,800 ft, at which point maximum afterburner was selected and airspeed reduced by 10 KEAS for every increase in 0.1 Mach, either manually or using the autopilot KEAS hold feature. The aft bypass doors were set to position "A" (15% open) at Mach 2.7, and finally closed at Mach 3.0. At approximately 76,000 ft and Mach 3.2, the cruise climb segment began, where altitude increased as fuel was burned to maximize range. Descent was initiated by

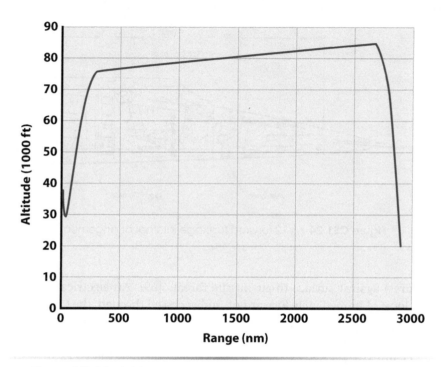

Figure CS1.25   A-12 mission profile—end of aerial refueling to 7500 lb fuel remaining.

retarding the throttles to military power to establish a 300 KEAS descent speed, and the aft bypass doors were selected in reverse order (position A at Mach 3.0, position B at Mach 2.7, and closed at Mach 1.7). Final approach to landing was flown at 165 KIAS with 5000 lb of fuel remaining. Figure CS1.25 shows the altitude versus range profile from the end of refueling to end of descent with 7500 lb of fuel remaining.

CS1.6 Fighter and Reconnaissance-Strike Versions

The A-12 design was optimized to perform overflight reconnaissance, but Kelly Johnson recognized that the basic design offered the potential to perform alternate missions as well. Only four months after full go-ahead for the A-12, he developed concepts for an air defense fighter version carrying two trapeze-launched MB-1 Genie air-to-air missiles, and a bomber version carrying a single 2000-lb weapon in a central rotary bay. These concepts would lay the foundation for two other Blackbird variants: the AF-12 (later renamed YF-12A), and R-12 (later renamed SR-71) described herein.

CS1.6.1 Air Defense Fighter: AF-12 (YF-12A)

Following cancellation in late 1959 of the F-108 Rapier Mach 3 interceptor then under co-development by North American Aviation with the B-70 Valkyrie Mach 3 strategic bomber, General Howell Estes of the USAF Air Defense Command (ADC) asked Kelly Johnson whether the F-108's advanced Hughes Aircraft Company AN/ASG-18 radar and GAR-9 missiles could be integrated into an air defense interceptor version of the A-12 to demonstrate interception of airborne targets (representing incoming bombers) at long range from a high speed, high altitude platform [70]. A feasibility study showed that it was possible, and under Project KEDLOCK three A-12 production slots were reprogrammed for what was to be called the AF-12 (and later redesignated YF-12A in 1964).

> The YF-12A's AN/ASG-18 radar/AIM-47A missile later evolved into the AN/AWG-9 radar/AIM-54 Phoenix missile system for the U.S. Navy's Grumman F-14A Tomcat.

The project required close collaboration between the Skunk Works and Hughes Aircraft Company, the latter led by L. A. "Pat" Hyland with AN/ASG-18 project engineer Clare Carlson. The forward fuselage was completely redesigned to accommodate the air defense mission requirements, taking advantage of the FS715 manufacturing joint. The nose chine was trimmed aft to accommodate the AN/ASG-18 radar with its 40-in. diameter antenna, along with infrared search and track (IRST) sensors (using lead selenide detectors cooled with liquid nitrogen) integrated into the leading edge of the shortened chine. The resulting increase in fuselage diameter and reduction in chine length, together with the addition of a second cockpit for a Fire Control Officer (FCO), reduced directional stability and required addition of three ventral fins (left and right fixed located underneath each nacelle, and one centerline that folded for takeoff and landing to provide clearance; see Fig. CS1.26). Four fuselage bays were added, three of which each accommodated an 820-lb GAR-9 (later redesignated AIM-47A) missile and the fourth dedicated to fire control system avionics (Fig. CS1.27). A Freon liquid cooling system was added to cool the radar transmitter, radar receiver, antenna drive, high voltage power supply, and the IRST sensor heads, whereas an ethylene glycol-Freon system cooled the missiles [71]. To accommodate the increased electrical power demanded by the fire control system and inertial reference platform (approximately 49 kW), the A-12's 30 kVA generators were replaced with 60 kVA units energized by constant speed drives. Camera pods could be installed underneath each nacelle and on the left fuselage underside to photograph missile launches, and were cooled using self-contained liquid nitrogen systems [72]. The airframe design used titanium in place of the

Figure CS1.26   The AF-12 (YF-12A) used three ventral fins. (Lockheed Martin)

silicone asbestos composites, as "anti-radar" treatments were not required for the air defense interception mission.

The YF-12A was limited to three prototypes, but Kelly Johnson pursued a production contract throughout the 1960s. In July 1961, General Curtis LeMay inquired whether Lockheed could produce ten per month [73], but a production contract did not materialize. In September 1962, Johnson developed a concept for an "AFR-12" that could be configured as a fighter or as a reconnaissance aircraft, using an airframe common with the R-12

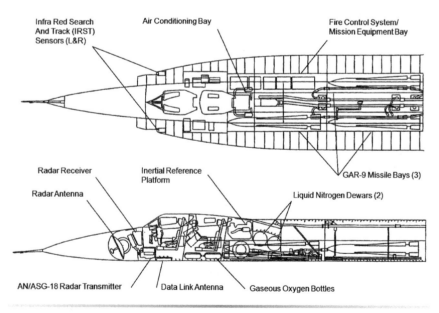

Figure CS1.27   AF-12 (YF-12A) fuselage internal arrangement.

(discussed in Sec. CS1.6.2), folding fin AIM-47B missiles, and, because of production readiness and integration concerns with the AN/ASG-18, a new radar. This design was later referred to as the AFR-112, AF-112D, and finally the F-12B. A revised nose radome of streamlined cross section enabled elimination of the ventral fins. Lockheed ADP, together with Hughes Aircraft Company and Pratt & Whitney, proposed a 94 aircraft production program in 1967, and later in the spring of 1969 a 71 aircraft program was offered to no avail. But Secretary of Defense Robert S. McNamara had already decided that a modified version of the F-106 (ironically never built) would be a less costly alternative to the F-12B, and the ability to produce any further Blackbirds was permanently decided with the order in February 1970 to destroy large jig production tooling; detailed tooling was retained for spares [74].

CS1.6.2 Reconnaissance-Strike: RS-12, R-12, and SR-71

In January 1962, Kelly Johnson delivered an unsolicited proposal for a bomber version of the A-12, nicknamed "B-12" and employing conventional free-fall weapons dropped from a centerline rotary weapons bay, to Air Force Under Secretary Dr. Joseph Charyk, USAF Colonel Leo Geary, and finance officer Lew Meyer. Johnson was given a verbal go-ahead to do a six-month engineering study, and began construction of a new engineering building to accommodate the anticipated engineering staff [75]. Johnson's proposal was forwarded to Wright Patterson Air Force Base for evaluation, in particular to the Weapon System 110A Project Office that was responsible for overall management of the B-70 Valkyrie program. There, a small team led by Major Ken Hurley took Johnson's concept a step further by recommending that the overhead bombing approach be replaced by a more sophisticated offset method using standoff weapons targeted by a side-looking radar (SLR) (similar to the approach being proposed for a reconnaissance-strike version of the Valkyrie designated RS-70) and using an astro-inertial navigation system from the cancelled GAM-87 Skybolt air-launched intercontinental ballistic missile for precision navigation and targeting [76]. Johnson initially resisted the concept because of its increased complexity and cost, but ultimately adopted the approach because of the improved survivability it would provide during the attack phase [77]. In addition to providing weapon targeting and navigation updates, the SLR could be used for post-nuclear strike reconnaissance. With this expansion in mission capability, the B-12 became the Reconnaissance-Strike or RS-12. Goodyear Aerospace and Westinghouse competed for the nose-mounted SLR, while Minneapolis-Honeywell and Hughes Aircraft Company competed for the aircraft and weapon guidance system. The weapon, a

Figure CS1.28 RS-12 mockup showing nose SLR cavity
and extended missile. (Lockheed Martin)

trapeze-launched missile based on the GAR-9 with a nuclear warhead, was integrated into fuselage chine bays and a second cockpit added in place of the A-12's Q-bay for a Reconnaissance-Strike Officer (RSO). The RSO's cockpit included a Recorder Correlator Display (RCD) that processed and displayed SLR imagery in near real-time. All of these systems were modeled in a full-scale mockup of the RS-12 fuselage (Fig. CS1.28) that was used to evaluate the placement of equipment and cockpit controls.

Throughout the spring of 1962, the RS-12 secretly competed with the RS-70 for the reconnaissance-strike mission. With its incorporation of A-12 RCS reduction treatments, it became clear that the RS-12 would be much more survivable than the RS-70, as well as less expensive because of its smaller size. In fairness, the RS-70 (Fig. CS1.29) was based on an aircraft originally sized to carry a 10,000 lb weapon payload with an unrefueled range of 6000 nm that resulted in a takeoff gross weight in excess of 530,000 lb. However, the closely held RS-12 may have been seen as a threat to the higher profile RS-70 program and was curtailed; the RS-70 was later cancelled as well. But two XB-70 prototypes were completed to demonstrate technologies for long range, high altitude flight at Mach 3, making their first flights in May 1964 and July 1965—ironically after the A-12 had achieved Mach 3.2 in secret. The first XB-70 was later operated by NASA and provided data in support of the United States Supersonic Transport (SST) program before being retired in February 1969.

Kelly Johnson then pursued a reconnaissance optimized version, the R-12, that carried multiple sensors covering radar, infrared, optical, and radio frequency bands, thus providing a wide area synoptic reconnaissance capability. The R-12 retained the second cockpit, now for a Reconnaissance *Systems* Officer (RSO), but no weapons provisions. When President Lyndon

RS-70 GENERAL ARRANGEMENT

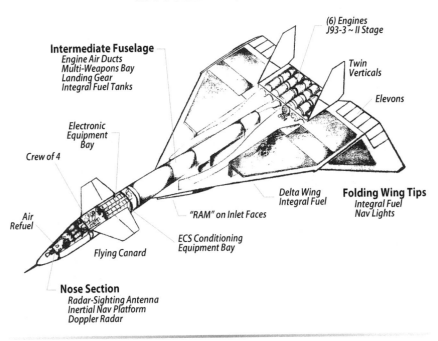

Figure CS1.29 The RS-70 was the competitor to the Lockheed RS-12. [78]

Johnson revealed the existence of the aircraft on July 24, 1964, it was under a new designation, SR-71, ostensibly for "Strategic Reconnaissance." However, the aircraft continued to be referred to within the Skunk Works as the R-12 or SR-12 (as Kelly Johnson's personal log was titled) and there is some evidence [79] that the term RS-71 was also used, possibly as a follow-on to the RS-70 designation series. Nevertheless, SR-71 would become the enduring designation for the ultimate development of the Blackbird series. Three variants were produced: the SR-71A of Fig. CS1.30 (29 examples), the SR-71B trainer (two examples, in which the aft cockpit was elevated and equipped with a second set of flight controls for an instructor pilot), and the single SR-71C (built as a replacement trainer after the loss of SR-71B 61-7957 using the aft section of YF-12A 60-6934 and the forward fuselage of the SR-71 static test article). The SR-71B and SR-71C (Fig. CS1.31) both used fixed ventral fins identical in configuration to those of the YF-12A in order to compensate for the decrease in directional stability created by the elevated instructor's cockpit.

The SR-71 incorporated numerous design refinements compared with the A-12 (see Fig. CS1.32). Single-piece forgings replaced built-up

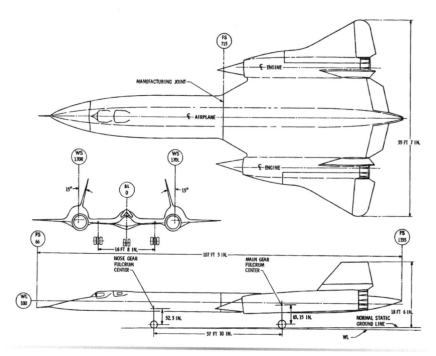

Figure CS1.30 SR-71A General Arrangement. (USAF)

**Trainer
Versions**

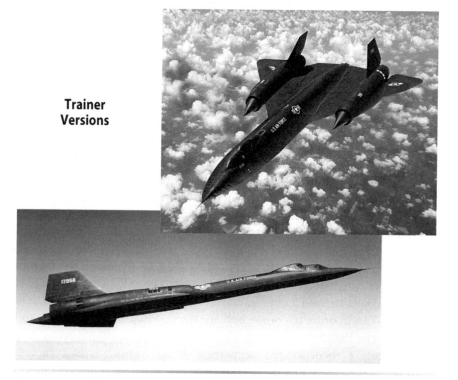

Figure CS1.31 SR-71B (bottom) and SR-71C (note shorter tail cone). (USAF)

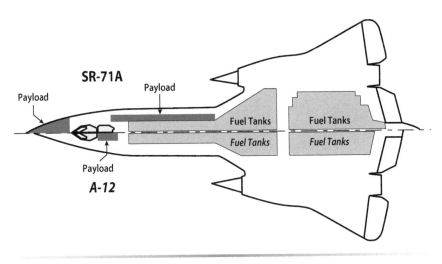

Figure CS1.32 Payload and fuel accommodations: A-12 vs SR-71.

components such as the nacelle carry-through rings, which reduced parts count (though the first six SR-71s used built-up nacelle rings). The chine planform was broadened for increased lift and a removable and interchangeable nose added. A lengthened fuselage tailcone increased the fineness ratio to reduce supersonic wave drag and provided additional fuel volume; extending the wing fuel tanks outboard of wing station 72 also increased fuel volume. But the addition of the RSO cockpit eliminated the Q-bay and the ability to carry the A-12 camera packages, and a new set of cameras had to be designed to fit within chine compartments distributed along the length of the fuselage. Twin 60 kVA generators, similar to those in the YF-12A, provided additional power for the SR-71's expanded complement of mission systems, shown in Figs. CS1.33 and CS1.34.

The removable and interchangeable nose forward of the cockpit pressure bulkhead carried the Side-Looking Radar (SLR) (initially the Goodyear GA-531 and GA-531A Product Improvement Program (PIP) and later the higher resolution GA-531B CAPability REconnaissance (CAPRE) system) receiver, transmitter, and antenna, with the radome integrated into the nose lower surface. Later, an alternate nose carried an Itek Optical Bar Camera (OBC) that provided high-resolution panoramic images with stereo capability (the camera is still in use today on the U-2).

Letter-designated bays located abreast of the nose landing gear bay and in the chines carried other sensors and avionics equipment. The small C-bay directly aft of the RSO cockpit carried a Fairchild F489 Terrain Objective Camera (TROC) to record the aircraft ground track and for mapping purposes. The D-bay to the right of the RSO cockpit was added to house a portion of the aircraft defensive electronics (nicknamed "DEF")

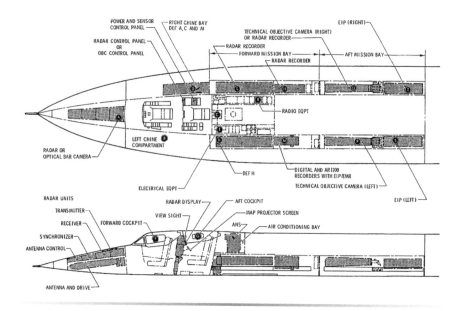

Figure CS1.33 Multiple bays accommodated SR-71 avionics and mission payloads. [80]

systems. The electrical load center was housed in the electrical (E) bay, whereas radio equipment was housed in the radio (R) bay. The K and L bays originally housed an electromagnetic reconnaissance system (EMR) to record electronic intelligence (ELINT), and later the DEF H system and a radar recorder, respectively. The EMR system was later replaced by an ELINT Improvement Program (EIP) system located in the S and T bays which each originally housed an Itek Model 9085 Operational Objective Camera (OOC) for wide-area panoramic stereoscopic coverage at 2-ft resolution of "operational objective" targets (these cameras proved disappointing and were withdrawn from service after the OBC became available). The P and Q bays each housed a Hycon HR-308 Technical Objective Camera (TEOC) that could be pointed either automatically by the flight plan tape or manually by the RSO to image "technical objective" targets at high resolution for technical exploitation. The M-bay originally housed an infrared (IR) camera produced by HRB-Singer [81] that was later removed and replaced with the DEF E system and ultimately additional recorders for the DEF and EMR/EIP systems. A mission recorder system (MRS) monitored and recorded mission and aircraft system parameters for later play-back.

As described earlier, the SR-71 relied upon an astro-inertial navigation system (ANS) adapted from a system originally developed by Nortronics for the GAM-87 Skybolt air-launched intercontinental ballistic missile

Figure CS1.34 SR-71 mission payloads were modular and reconfigurable. (Lockheed Martin)

(ICBM). Designated NAS-14 for the SR-71 application, the ANS used a telescopic star tracker mounted to the outer gimbal of the inertial platform. Looking through a quartz window and compensating for the refraction caused by external shockwaves across the window surface, the star tracker searched until it acquired the navigation stars from its catalog. By comparing the star azimuth and elevation angles against known values for a particular location, the system corrected drift of the inertial platform and updated present position. A digital computer with only 16K of random access memory (RAM) (expanded from 4K for the Skybolt application) performed the calculations, using a portable chronometer located in the RSO cockpit for time correlation [82]. The NAS-14 was installed in the center of the air conditioning bay aft of the RSO cockpit.

Other navigational systems included a General Electric SR-3 flight reference system (FRS) that provided a standby attitude and heading reference (later replaced by a Litton SKN-2417 inertial navigation system (INS)); a viewsight that provided an optical display of the terrain and a means to update the ANS; and projected map displays in each cockpit. Like the RS-12, the RSO cockpit included a radar Recorder Correlator Display (RCD) that displayed SLR imagery in near real time for reconnaissance target identification and navigation updates using radar fix points.

While the SR-71 was optimized for strategic reconnaissance, Kelly Johnson continued pursuit of a bomber version. The B-71 study of early 1965 replaced the sensor bays with four weapon bays, each carrying a Short Range Attack Missile (SRAM) with modified fins to accommodate launch at Mach 3.2. The final bomber concept, nicknamed "Bx" and shown in Fig. CS1.35, carried four SRAMs or penetrator missiles targeted by a multi-mode radar; Kelly Johnson was intrigued by the penetration potential against hardened targets of a weapon launched at Mach 3. On both the B-71 and Bx, the chine span was shortened and the chine radius enlarged to provide better weapon bay integration, and to compensate for the reduction in lifting surface, the entire fuselage forward of FS715 was tilted up by 1°48', which also reduced trim drag. The Bx also added a lengthened tail cone for additional sensors and defensive systems. This so-called "Big Tail" was later built and tested on SR-71A 61-7959 in 1975-1976. But neither the B-71 nor Bx were pursued, and the SR-71 remained the ultimate version of the Blackbird family.

CS1.7 Flight Test Challenges

CS1.7.1 A-12 Flight Test Program

Flight test of the A-12 began nearly a year later than planned, and not with the intended engine. Because of continuing development difficulties with the initial YJT11D-20 engines, Kelly Johnson decided in September 1961 to use Pratt & Whitney J75-P-19W afterburning turbojets (with water injection to boost thrust to 26,500 lb in afterburner) for initial test

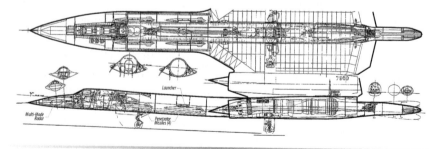

Figure CS1.35 The "Bx" was the ultimate bomber concept of the SR-71. (Lockheed Martin)

flights [83]. The engine inlets and nacelles required extensive modifications to accommodate the engine's smaller diameter and different mounting scheme (Fig. CS1.36). With these lower-thrust engines, maximum Mach was limited to 1.6 and maximum altitude to 50,000 ft, and the inlet spikes were locked forward. These engines powered other A-12s, including the A-12T trainer throughout its career, until arrival of the YJT11-D20.

In early March 1962 after arrival and re-assembly at the flight test location, the first A-12 (Article 121; the term "article" was used instead of "aircraft" for security purposes) was filled with fuel. Unfortunately, 68 leaks developed due to lack of adhesion between the original Viton fuel tank sealant and titanium, and the sealant had to be completely stripped and replaced [84]. To enable engine runs to proceed, two 316-gallon P-38 vintage drop tanks, one above each inboard wing, were temporarily used to provide a source of fuel. On the first high speed taxi test on April 25, 1962, chief test pilot Louis W. "Lou" Schalk elected to briefly lift off the runway for the traditional "hop," but misalignment of the rudder pedals and nose-wheel steering resulted in large rudder deflections during the ground roll that, once airborne, created oscillations that were exacerbated because the SAS was disengaged for what was intended to be a taxi test only. The first intended flight the following day (with SAS engaged) saw non-structural fillet panels shed during takeoff due to a failure of the forward fillet mounting bracket. The team had to work around the clock over the weekend to re-design and repair the fillets in time for the "official first flight" on Monday, April 30 (see Fig. CS1.37) that was witnessed by Government dignitaries and senior Lockheed management. Airspeed was limited to 250 knots and altitude to 30,000 ft. A low altitude flyby and maneuver at the conclusion of the 59-min flight reportedly left Kelly Johnson speechless.

The YJT11D-20A engine was initially flown on October 5, 1962 in Article 121's left nacelle only, with a J75-P-19W in the right. Initial challenges included lower than expected thrust, inconsistent fuel consumption,

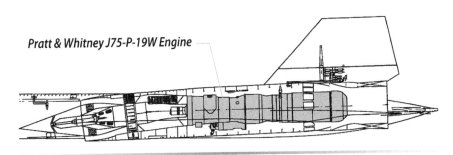

Figure CS1.36 J75-P-19W engine installation used for initial flight tests. (Lockheed Martin)

Figure CS1.37   First official flight of the A-12 on April 30, 1962.
(Lockheed Martin)

thrust "jumps" at different throttle settings, and problems with afterburner lighting [85]. On January 15, 1963, the second A-12, Article 122, flew with two YJT11D-20A engines for the first time, and Article 121 followed on March 9, 1963 with two fully instrumented engines necessary to begin the flight envelope expansion program towards the design cruise speed and altitude [86]. Progress was slowed by a spate of foreign object damage (FOD) incidents between April and May 1963 that required 18 engine removals and addition of FOD screens in the inlets, and loss of Article 123 on May 24, 1963 due to pitch-up resulting from erroneous airspeed indications caused by pitot tube icing; pilot Ken Collins ejected safely. Mach 3.06 at an altitude of 72,000 ft was finally achieved on July 20, 1963 in spite of continued propulsion system troubles. These included engine main fuel control anomalies that eventually required replacement of the original fuel control with a Bendix unit from the General Electric J93 engine used on the XB-70; a "duct rumble" condition experienced between Mach 2.0 and 2.4 that was suspected to be caused by flow separation from the aft expansion surface of the inlet spike; unacceptable engine in-flight restart capability; and problems with the inlet control system leading to inlet instability and unstarts. As described in Sec. CS1.5.4, the duct rumble condition was corrected by the addition of "mice" to tailor the inlet duct cross sectional area distribution to reduce flow expansion and separation. The aft bypass doors were added to provide additional bypass area to facilitate inflight engine re-starts, and also improved inlet/airflow matching during normal operation.

But inlet operation remained problematic until the Lockheed/AiResearch electronic inlet control finally replaced the original pneumatic control system. With this change, made at the direction of CIA Deputy Director of Science and Technology Dr. Albert D. Wheelon [87], inlet operation became more reliable and repeatable, and inlet unstarts were reduced in frequency. The A-12 was now able to achieve the design Mach number of 3.2, reached on November 26, 1963 by Lockheed test pilot William C. "Bill" Park. On February 3, 1964, Lockheed test pilot James D. "Jim" Eastham, who was also the project pilot for the AF-12, made the first sustained flight at design speed, with 15 min above Mach 3.2 and a maximum Mach number of 3.33 up to an altitude of 84,000 ft. On July 9, 1964, thermal gradients during descent led to seizure of Article 133's outboard elevon servo valve, causing the aircraft to roll uncontrollably to the left on short final; pilot Bill Park ejected successfully. Another accident involving Article 126 on December 28, 1965 was caused by the pitch and yaw rate gyro harnesses being hooked up to the stability augmentation system in reverse, and led to adoption of unique electrical connectors; operational pilot Mele Vojvodich was able to eject safely [88].

With the ability to now attain and sustain high speed, high altitude cruise conditions, another issue arose: range performance. The aircraft burned more fuel during transonic acceleration than expected, which led to adoption of a diving maneuver to accelerate to supersonic speed before continuing the climb (suggested by Jim Eastham, who used a similar technique in the YB-58 he flew for Hughes Aircraft Company while testing the radar and missile systems destined for the AF-12) as well as a faster equivalent airspeed climb schedule [89]. The aircraft also suffered more drag than predicted during cruise. The higher drag was eventually traced to a combination of causes. Because exhaust ejector wind tunnel tests were performed in isolation and not in proximity to the wing and its interference effects, ejector drag was under-predicted. Fuselage boattail drag was under-predicted by the wind tunnel model support sting corrections. The "dimples" in front of the ejector tertiary air inlet doors added to the drag as well [90]. But one of the largest contributors was drag due to nacelle leakage and the momentum loss of air dumped overboard through the forward bypass doors during cruise. Improved sealing of the nacelles and the addition of the aft bypass doors to allow the forward doors to close down during cruise helped mitigate this contributor, but nevertheless range performance fell approximately 20% short of expectations. Another factor was that the original specification range of 4068 nautical miles from takeoff to initial aerial refueling was based on a reserve fuel load of 2700 lb, which was raised to 5000–7500 lb for greater safety margin.

It was difficult to obtain specific range data for long range cruise segments because 180-deg turns were required to remain within the confines of the special use airspace; longer, straight line cruise segments were required. A mission plan named SILVER JAVELIN took the A-12 on a round-robin trip from the test location to multiple waypoints across the United States. On December 21, 1966, Bill Park flew the SILVER JAVELIN route in Article 129, which was specially instrumented with additional inlet duct instrumentation. In the matter of 6 hr, he flew a total of 8700 nautical miles with two refuelings, and on the third leg demonstrated a range in excess of 3000 nautical miles from completion of refueling to landing [91]. Other tests proved systems reliability and mission payload functionality in the high speed, high temperature environment. The A-12 had now proven its operational capability and was ready to leave the nest.

CS1.7.2 AF-12/YF-12A Flight Test Program

Following three months of integration at Burbank and one month of reassembly at the test location, the AF-12 flight test program began on August 7, 1963, when Jim Eastham took serial 1001 (USAF serial 60-6934) on its trouble-free maiden flight (see Fig. CS1.38). This aircraft remained a Lockheed test aircraft, whereas its sisters 1002 (USAF serial 60-6935) and 1003 (USAF serial 60-6936) were assigned to the Air Force.

Whereas Lockheed conducted the A-12 flight test program for the CIA, a Combined Test Force (CTF) comprised of Lockheed and Air Force crews conducted the AF-12 (and later SR-71) test program. Lockheed was responsible for Category I flight tests, which demonstrated basic airworthiness. The Air Force was responsible for Category II flight tests, which verified performance capability and systems operation.

The AF-12 became the first version of the Blackbird to be revealed to the public on February 29, 1964 by President Lyndon B. Johnson under the designation "A-11" (deliberately chosen by Kelly Johnson to mask the existence of the covert A-12 program). After the announcement, which

Figure CS1.38 The AF-12 makes its first landing on August 7, 1963. (Lockheed Martin)

indicated that two "A-11 type aircraft" were based at Edwards AFB, two AF-12s were hurriedly flown from the test location to Edwards, where the exhaust gases from one of the aircraft promptly set off the fire extinguishing system in the receiving hangar [92].

The first unpowered AIM-47A missile drop occurred on April 16, 1964. It was discovered that more missile nose-down pitching moment was needed to prevent the missile from firing through the cockpit once ignited, and the second launch on May 28, 1964 was safely conducted [93].

On September 30, 1964 the now re-designated YF-12A was officially unveiled to the press at Edwards AFB. After CTF director Colonel Robert L. "Fox" Stephens and his FCO, Lieutenant Colonel Daniel Andre were forced to return to the ramp in 60-6936 prior to takeoff due to oxygen system depletion, pilot Jim Eastham and flight test engineer John Archer hurriedly suited up and flew 60-6934 for the demonstration [94]. With the YF-12A now acknowledged, there was eagerness to demonstrate some of the potential of the aircraft, as well as tempt the Soviets to demonstrate the capabilities of their secretive MiG-25 in response. On May 1, 1965 the YF-12A flown by USAF crews set nine official world absolute and class records for speed and altitude, both sustained and over a closed course, after having been previously developed and test flown in secret by Jim Eastham and USAF Colonel Sam Ursini. But the records did not reveal the true performance capability of the aircraft.

Envelope expansion proceeded slowly due to unavailability of higher thrust engines with "J" afterburners, trouble with constant speed drives, and the higher priority of the A-12 program. Finally, in December 1964 the YF-12A reached Mach 3.0, and on January 9, 1965 Jim Eastham took 60-6934 to Mach 3.23, with five minutes above Mach 3.2. On March 18, 1965 at the Pacific Missile Range off Point Mugu California, a AIM-47A missile was fired at a target for the first time, which it hit 36.2 mi away. Another launch on September 28, 1965 made at Mach 3.2 and 75,000 ft missed the target flying 36 mi away by only six ft [95]. After one more west coast launch on March 22, 1966, 60-6934 and 60-6936 deployed to Eglin AFB, Florida, where two additional launches were made. The most successful launch was against a JQB-47 target flying at low altitude that was hit in the horizontal stabilizer (yet managed to land and later serve as a target once again for the YF-12A at the White Sand Missile Range) [96]. After a final launch on September 21, 1966, the evaluation was concluded and the YF-12A fleet placed in storage.

CS1.7.3 The SR-71 Flight Test Program

With Lockheed test pilot Robert J. "Bob" Gilliland at the controls, the first SR-71A (serial 2001, USAF serial 61-7950) made its maiden flight from

Palmdale Airport, Air Force Plant 42 on December 22, 1964, beginning the Category I test program (see Fig. CS1.39). Like the YF-12A, the SR-71 was tested by a combined test force (CTF) led by Colonel "Fox" Stephens. The first SR-71A was joined by 61-7951 and 61-7952 that served as contractor test aircraft, whereas 61-7953, 61-7954, and 61-7955 were operated by the USAF for Category II testing.

Like the A-12, the SR-71 suffered degradation in range performance due to higher than predicted drag. In an effort to reduce the trim drag contributor, test flights were performed that explored scheduling the fuel burn sequence to place the center of gravity (CG) closer to the aft limit. While reducing trim drag, the aft CG reduced the longitudinal stability of the aircraft, and created the potential for an unrecoverable condition in the event of an upset. Unfortunately, this occurred during a test flight of 61-7952 on January 25, 1966. After entering a 35-deg bank turn while in the autopilot automatic navigation mode at Mach 3.17 and an altitude between 77,000 and 78,000 ft, the right inlet unstarted, which immediately increased the bank angle to 60 deg. The resulting disturbance in the pitch axis could not be arrested with control inputs, and a pitch-up developed that immediately resulted in breakup of the aircraft. It was suspected that the structural deflections led to actuation of the external emergency canopy release, which jettisoned both canopies. Lockheed test pilot Bill Weaver was ripped from the aircraft (his ejection seat was found in the wreckage), whereas flight test engineer Jim Zwayer ejected. While Weaver survived, Zwayer

Figure CS1.39 The first SR-71A on an early test flight. (Lockheed Martin)

was killed, despite nominal operation of his ejection seat and parachute. Several improvements were made to the SR-71 following the accident. While the pilot could already control fuel transfer by manually selecting fuel tank boost pumps, manual forward and aft fuel transfer switches were added that provided additional control. Direct read-out CG position indicators were added to both cockpits; the only prior indirect indication of CG position was to monitor the pitch trim indicator for excessive elevon trim deflections. Finally, the inclination of the removable nose, normally six deg nose-down relative to the fuselage waterline, was reduced by two deg in the nose-up direction to increase nose lift and reduce trim drag (see Fig. CS1.40) [97].

While these modifications were made, the Category II performance testing was suspended, then resumed in October 1966. With the change in configuration and installation of higher thrust JT11D-20B engines, all previous performance data was invalidated and had to be re-run. Unfortunately, more aircraft losses were suffered later on that led to further design improvements. These included an automatic pitch warning (APW) system that used an eccentric motor to shake the control stick if angle of attack and pitch rate became excessive (there was no natural warning) and a stick pusher to provide pitch-down recovery; an improved instrument panel layout that placed the standby attitude indicator directly above the primary attitude director indicator for easier correlation of attitude references; and more prominent emergency bailout advisory lights and relocated pilot's control switch. After a series of takeoff accidents in which tire blowouts caused the magnesium main wheels to catch fire and ignite fuel tank fires with their shrapnel, the main wheel materials were replaced with aluminum.

Category II performance testing concluded that the SR-71 was capable of unrefueled missions of approximately 3000 nautical miles, with over

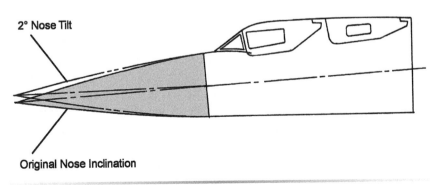

Figure CS1.40 Two-deg nose tilt reduced trim drag.

one hour at Mach 3.2 above 80,000 ft. The range performance fell short of the specification by 5–10% when consistent assumptions for fuel allowances were used. The original engine start/taxi/takeoff/acceleration to climb speed fuel allowance of 3000 lb was increased to 6500 lb based on experience, and the allowance for fuel remaining after descent to 25,000 ft was increased from 6000 lb to 10,000 lb [98].

Flight test data was captured using high speed recorders and cameras that filmed panels of instruments (designated Automatic Observer (AO) panels) located in the chine compartments; engineers had to record instrument indications frame by frame to reconstruct the time history of data. Later, time correlated data from the ANS was used that greatly increased data reduction efficiency [99].

Although test aircraft 61-7950, 61-7952, 61-7953, and 61-7954 did not survive, 61-7955 enjoyed a long career testing systems (described in Sec. CS1.9) that included the Digital Automatic Flight and Inlet Control System (DAFICS); Advanced Synthetic Aperture Radar System (ASARS-1); improved DEF systems; improved ANS navigation control/display unit; pilot's peripheral vision display; and improved cockpit lighting. In January 1985, the aircraft was retired and replaced with 61-7972, which was more representative of the operational fleet.

CS1.8 Using the A-12 as a Launch Platform

The performance capability of the A-12 and its basic configuration provided intriguing possibilities for use as a high speed, high altitude platform to launch other vehicles. Some of these possibilities remained design studies, whereas others resulted in flight hardware.

CS1.8.1 AP-12 Satellite Launcher Design Study

In 1962, Lockheed ADP and Lockheed Missiles and Space Company explored together the feasibility of launching a 900–1000 lb reconnaissance satellite into a 80 nm orbit using a modified Polaris A3 missile carried and launched by a modified A-12, designated AP-12. Air launch over the Pacific Ocean provided flexibility in launch azimuth and orbital inclination. The missile and tailcone fairing were mounted on a rail system underneath the AP-12 (a top side arrangement was initially studied but found impractical). For takeoff, the missile was positioned between the nose and main landing gear; a gap between the main landing gear and missile tailcone fairing provided clearance for the main landing gear to retract (Fig. CS1.41). After takeoff, the missile slid aft against the tailcone to close the gap, and the entire assembly slid further aft to achieve the proper center of gravity location for cruise and missile launch. After cruising out at Mach

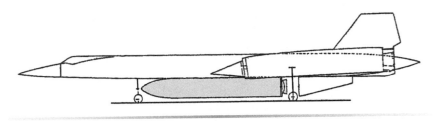

Figure CS1.41 AP-12 with modified Polaris A3 missile (takeoff position).

3.2 to the launch area (with two subsonic aerial refuelings in between), the AP-12 initiated the launch maneuver, which called for a 2-g pull-up until a flight path angle of 15 deg was attained (the higher the flight path angle, the more payload could be placed in orbit). A Launch Control Officer located in a second cockpit in place of the Q-bay monitored launch conditions and commanded missile release. Upon completion of the study, Kelly Johnson concluded that the concept was marginally feasible, and it was not pursued further [100].

CS1.8.2 The MD-21 Reconnaissance System

Following the interception of Francis Gary Powers in his U-2C over Sverdlovsk on May 1, 1960 that brought an end to manned overflights of the Soviet Union, discussions were held on the feasibility of making an unmanned drone version of the A-12, nicknamed Q-12. Kelly Johnson maintained that the A-12 was too large and complicated to make such a concept practical, and instead advocated using the A-12 as a launch platform for a smaller unmanned vehicle [101]. In July 1962, authorization was received to perform a 6-month design study, and by December a full-scale Q-12 mockup was completed (Fig. CS1.42). Of titanium construction with silicone asbestos edges and planform shaping for reduced RCS, the Q-12 was to be launched at Mach 3.2 at altitudes between 85,000 and 90,000 ft. Using a Marquardt XRJ43-MA20S-4 ramjet engine similar to that used in the Boeing BOMARC missile and a fixed-geometry inlet optimized for supersonic cruise at Mach 3.3, the Q-12 would travel 3000 nm and reach a maximum altitude of 95,000 ft [102]. An auxiliary power unit driven by inlet bleed air provided electrical power to onboard systems and hydraulic pressure to the actuators for the left/right elevons and single centerline rudder. At the end of the cruise segment, planned to occur over international waters, a hatch containing the Hycon HR-335 reconnaissance camera/film and high-value avionics was ejected and recovered via parachute, snagged during descent by a JC-130H or recovered in the water, while the disposable airframe was destructed using an on-board explosive charge.

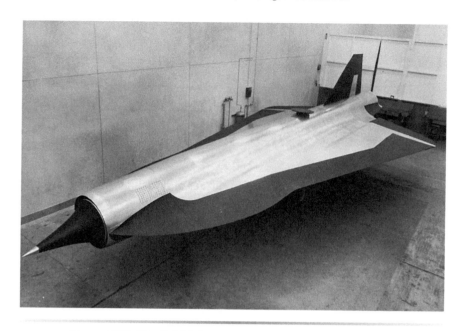

Figure CS1.42 Q-12 full scale mockup. (Lockheed Martin)

In December 1962, Lockheed ADP was given authorization to design and fabricate 20 drones (under Project TAGBOARD) and two A-12 based launch platforms (article numbers 134 and 135 under Project WEDLOCK) to the launch vehicle configuration [103]. To avoid confusion with the A-12 program, the numerical designation was transposed to create the M-21 (Mother) launch platform and D-21 (Daughter) drone, the latter re-designated from Q-12. When mated, the pair became the MD-21 reconnaissance system (Fig. CS1.43). Modifications to achieve the M-21 configuration included replacement of the Q-bay with a cockpit for the Launch Control Officer (LCO) enclosed by a modified SR-71 RSO cockpit canopy faired into the outer mold line contour; LCO controls for in-flight checkout, performance monitoring, and launch of the D-21; an aft-viewing periscope for the LCO to observe and film launch events; a centerline pylon on the upper surface for mounting the D-21; relocation of the drag chute compartment aft of the D-21 mounting pylon; a stellar-monitored inertial navigation system; fuel system interfaces to enable in-flight refuel of the D-21 and transfer of D-21 heat

In 1999, the author served as Skunk Works project engineer for a NASA-sponsored study that investigated use of four remaining D-21 vehicles in their possession as testbeds for rocket-based combined cycle engines and other high-speed aircraft technologies. The extensive modification program was not pursued; two of the vehicles were given to museums (Blackbird Airpark and March Field Air Museum) and the other two were returned to the U.S. Air Force.

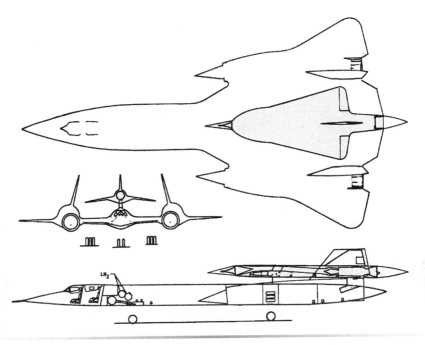

Figure CS1.43 MD-21 General Arrangement.

sink fuel to the M-21; environmental control system interfaces to provide partially-cooled air to the D-21 auxiliary power unit; and a transmitter to destruct the D-21 after launch if necessary [104].

One of the most challenging aspects of the program was the launch maneuver. Lockheed ADP performed extensive wind tunnel testing and simulation to characterize the ability of the D-21 to transit the M-21's supersonic flowfield. The initial launch technique was to perform a 0.9-g push-over (by monitoring a sensitive accelerometer) and rely on D-21 aerodynamic lift for separation, holding roll angle and rudder deflection to zero to provide maximum clearance for the D-21 between the M-21's rudders (which was only 22 in. on each side with rudders neutral).

The first MD-21 mated flight (Fig. CS1.44) occurred on the afternoon of December 22, 1964 with Lockheed test pilot Bill Park at the controls – the same day as the maiden flight of the SR-71 which occurred earlier that morning. But first launch of the D-21 would not occur until March 5, 1966 due to a multitude of challenges including D-21 elevon flutter; poor transonic acceleration of the MD-21 (leading to installation of higher thrust JT11D-20B "K" engines of 34,000 lb thrust each); safe separation without M-21 tail strikes of the D-21 nose fairing (ultimately eliminated) that covered the ramjet inlet before launch; deployment of the

Figure CS1.44   MD-21 maiden flight on December 22, 1964.
(Lockheed Martin)

parachute recovery system; and determination of separation forces on the D-21. The first launch was successful, but Kelly Johnson prophetically described the launch technique as ". . . the most dangerous maneuver we have ever been involved in [for] any airplane I have worked on" [105]. Two more successful launches occurred, but the fourth on July 30, 1966 ended in disaster when the D-21 collided with the second M-21 (Article 135) shortly after launch. Unlike the first three launches, this launch was made in level flight at one-g (there was concern that the 0.9-g pushover maneuver would add unnecessary pilot workload during operational missions). The M-21 rapidly pitched up and broke apart over the Pacific Missile Range, filmed by flight test engineer Keith Beswick who was flying with Art Peterson in Article 134 as chase. Both pilot Bill Park and Lockheed flight test engineer/LCO Ray Torick successfully ejected, but Torick tragically drowned following his water landing. Kelly Johnson was devastated, and the entire program was re-evaluated by the National Reconnaissance Office (NRO), which had taken over the effort from the CIA in 1963 [106]. Ironically, prior to the accident Johnson had already proposed an alternate launch method using a B-52H carrier aircraft and solid rocket motor mounted to the D-21 as a means of boosting it to ramjet ignition speed and altitude, and the program was re-structured to pursue this approach. The surviving M-21, Article 134, was placed in storage and later on display at the Museum of Flight in Seattle Washington,

I was working in flight test on the MD-21 Program known internally as the "Mother-Daughter" Program at the time of the crash. As a young engineer I experienced the devastation of losing a crew member who I had worked with at the Nevada test site. It was an early reminder of how things can go wrong when you are pushing the technology envelope.
Grant Carichner

together with D-21 serial number 510 mounted in its captive carry position.

CS1.9 Operational Service and Capability Improvements

CS1.9.1 A-12

Even though the CIA received its first A-12 in September 1963, a limited operational capability was not achieved until November 1964 in response to threats by Soviet Premier Nikita Khrushchev to shoot down U-2 aircraft if they overflew Cuba following the U.S. elections. Under Operation SKYLARK, A-12 articles 125, 127, 128, 130, and 132 were equipped with higher thrust YJT11-D20A engines with "J" afterburners, automatic forward bypass doors, and aft bypass doors to provide a Mach 2.9, 76,000 ft penetration altitude, 1700 nmile tanker-to-tanker range capability for operations over Cuba if so ordered. Phase II of SKYLARK, achieved in March 1965, improved operational Mach number to 3.05. A "Major Minimum" program begun on May 1, 1965 incorporated defensive electronics, the Lockheed/AiResearch inlet control, increased liquid nitrogen and liquid oxygen to facilitate three in-flight refuelings and longer missions, and a beefed-up FS715 joint to accommodate the higher bending moments resulting from the added equipment weight [107].

Despite readiness to perform operational missions, it would not be until May 1967 that three A-12s (Articles 127, 129, and 131) from the 1129th Special Activities Squadron (nicknamed "The Roadrunners") deployed to Kadena Air Base, Okinawa under Operation BLACK SHIELD. Over the next year, the aircraft (nicknamed "Cygnus" after the constellation) would fly a total of 29 operational missions, including 24 over North Vietnam, two over Cambodia and Laos, and three over North Korea [108]. Despite several attempts, no aircraft were shot down, but tragically pilot Jack Weeks and Article 129 were lost at sea on June 4, 1968 during a routine functional check flight following an engine change. In March 1968, the SR-71 joined the A-12 at Kadena, and began to take over the mission following a Government comparative study of the two platforms that decided in December 1966 to retire the A-12 fleet for cost savings and to avoid a perceived duplication of reconnaissance capability. Ironically, an improved A-12B version had also been initiated in 1966. Only Article 122 was modified to this configuration, which included an upgraded cockpit/instrument panel; upgraded defensive systems; provisions for a side-looking radar; higher capacity environmental control system, 60 kVA generators, and improved electrical system, all leveraged from the SR-71; and JT11D-20B engines.

This aircraft was stored at Palmdale in September 1967, and later joined by the rest of the remaining A-12 fleet where they were squeezed into a corner of Lockheed's Air Force Plant 42, Site 2 facility out of sight. In June 1968, the OXCART program was closed down, but the accomplishments of the 1129th Special Activities Squadron were recognized in a special ceremony on June 26, 1968 where the pilots were awarded the CIA Intelligence Star for valor [109]. The existence of the A-12 remained classified until 1982, and its history did not begin to come to light until nearly a decade later. In 2007, more A-12 documents were declassified in conjunction with the 40th anniversary of A-12 operational missions, and Article 128 was mounted on display at CIA Headquarters as a fitting tribute to the OXCART program participants and their accomplishments.

CS1.9.2 YF-12A

The three YF-12A demonstrators were never intended to be operational aircraft, but nevertheless performed more duties following completion of the original test program in November 1966. After being removed from storage, the first aircraft (USAF serial 60-6934) and its fuselage/wing assembly aft of FS715 were used to build the SR-71C replacement trainer. Meanwhile, NASA and the USAF entered into a joint research program using the remaining aircraft. Under Phase I, the Air Force Systems Command (AFSC) used the third aircraft (USAF serial 60-6936) to help develop advanced radars and interception tactics, and also served as a high speed, high altitude threat simulator. A ruptured fuel line and fire on landing approach to Edwards AFB resulted in loss of the aircraft in June 1971 but former A-12 pilot Lt Colonel Ronald "Jack" Layton and FCO Major Billy Curtis escaped safely. Under Phase II, the second aircraft (USAF serial 60-6935) was provided to NASA in 1969 after removal of the AN/ASG-18 radar fire control system and addition of upgrades that included the electronic inlet control, JT11D-20B engines, and SR-1 ejection seats. Up until its retirement in November 1979 to the National Museum of the United States Air Force, serial 60-6935 performed a multitude of experiments ranging from landing dynamics evaluations to high speed boundary layer and heat transfer studies [110]. NASA also operated a "YF-12C" that was actually the second SR-71A (USAF serial 61-7951) re-identified using USAF serial number 60-6937 (the next serial number after the third YF-12A, but ironically already

During a subsonic test flight with the Coldwall experiment, the YF-12A suffered loss of its center ventral fin, but the stability degradation was deemed acceptable, and the fin was left off. Later, a replacement ventral fin was tested, made from a beryllium/aluminum alloy developed by Lockheed called "Lockalloy."

Figure CS1.45 NASA YF-12A with "Coldwall" experiment and "YF-12C" in formation. (NASA)

belonging to an A-12) so as not to reveal that NASA had an SR-71 in its possession [111]. This aircraft was used for propulsion system testing as well as a high-speed chase for the YF-12A; Fig. CS1.45 shows one such flight during which the YF-12A tested a "Coldwall" boundary layer and heat transfer experiment. A Cooperative Airframe/Propulsion Control System (CAPCS) tested in the YF-12C provided the foundation and impetus for a digital automatic flight and inlet control system retrofit for the SR-71 fleet.

Lockheed ADP explored modifying the YF-12C to achieve Mach numbers of 3.5 or 4.0 for high-speed research and "zoom climb" capability up to approximately 95,000 ft (see Fig. CS1.46). Both options called for addition of SR-71B style ventral fins for enhanced directional stability, higher temperature capability materials in the wing leading edges and rudders, new windshield and canopy glass panels, and an inlet water injection cooling system supplied by fuel tank number 2 (converted to carry water instead of JP-7). The Mach 3.5 inlet used a new, longer inlet spike with 40 in. of travel (versus 26) and enlarged inlet capture area, whereas the Mach 4.0 option used an even longer spike with 50 in. of travel and an all-new inlet up to nacelle station (NS) 944. The radical modification was not pursued, and the YF-12C was retired in 1978.

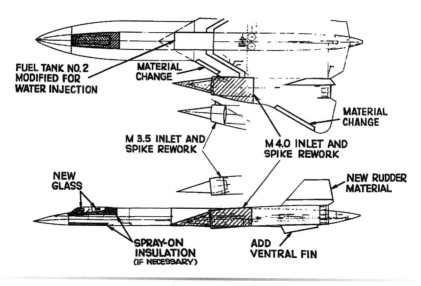

Figure CS1.46 YF-12C modifications to achieve Mach 3.5 or Mach 4.0.

CS1.9.3 SR-71

On January 7, 1966, the Strategic Air Command's 4200th Strategic Reconnaissance Wing at Beale AFB, California received its first SR-71 aircraft, an SR-71B trainer delivered by Wing Commander Colonel Douglas T. Nelson and instructor pilot Colonel Raymond Haupt, followed by an SR-71A on April 4 of that year. In June 1966, the 4200th Wing was redesignated the 9th Strategic Reconnaissance Wing, with the 1st and 99th Reconnaissance Squadrons both operating the SR-71. In parallel with the Category II flight test program, Category III operational testing was conducted with operational crews.

To help train Air Force pilots and RSOs, a SR-71 mission simulator was developed by Link in only 15 months and delivered to Beale AFB in September 1965. The simulator (see Fig. CS1.47) comprised a hydraulically actuated motion-base pilot's cockpit (with integral instructor's station) and fixed-based RSO cockpit connected to instructor control stations and a bank of digital computers that could replicate systems operation and flight dynamics of the aircraft, as well as all manner of emergencies as determined by the instructors. Neither cockpit had outside visual displays, using instead opaque windows that could be illuminated to

After the final USAF/NASA SR-71 program shutdown, the simulator was transported from Edwards AFB to the Frontiers of Flight Museum at Dallas Love Field. It is hoped to make the simulator operational once again to give visitors an appreciation of the challenges in flying the SR-71.

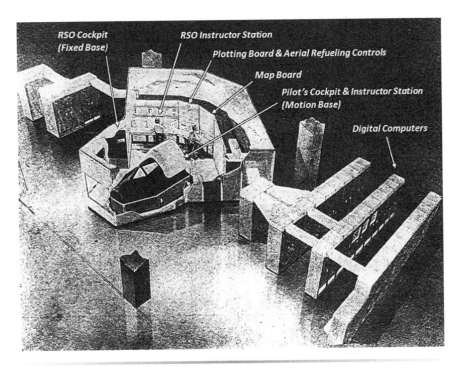

RSO Cockpit (Fixed Base)

RSO Instructor Station

Plotting Board & Aerial Refueling Controls

Map Board

Pilot's Cockpit & Instructor Station (Motion Base)

Digital Computers

Figure CS1.47   SR-71 mission simulator as installed at Beale AFB.

represent instrument flying conditions in daylight, but the RSO could observe simulated terrain looking through the viewsight. The pilot and RSO could train together or individually. Besides training operational crews, the simulator served as an engineering tool for re-creating abnormal situations, developing new procedures, and to evaluate new operational capabilities.

In December 1966 the Government decided that the SR-71 would take over the role of the A-12, starting with operations in Southeast Asia. During 1967, preparations were made, including installation and test flights of defensive systems, and in March 1968 the SR-71 flew its first wartime operational mission from Operating Location 8 (OL-8, later re-designated Detachment 1) of the 9th SRW at Kadena Air Base, Okinawa. The SR-71 continued to operate from Okinawa over the next 21 years (see Fig. CS1.48) as part of a three airplane detachment, gaining the nickname "Habu" after being compared to the dark pit viper indigenous to the island, as well as from a two airplane detachment established at RAF Mildenhall, England. The SR-71 was supported by a fleet of KC-135Q aerial tankers that made global operations possible. A full description of the SR-71's operational

Figure CS1.48   SR-71A landing at Kadena AB, Okinawa; Det 1 hangars in background. (USAF)

missions is not within the scope of this case study, but the reader is referred to [112] for a definitive account.

New sensors introduced in the early 1970s included the optical bar camera (OBC) and CAPRE radars described in Sec. CS1.6.2. The aircraft's defensive systems also went through several upgrades to address current and anticipated threats. In 1975 and 1976, a "Big Tail" modification was evaluated in SR-71 61-7959 that accommodated 864 lb of payload, including an aft-facing defensive system and a modular bay that could house an OBC. This particular aircraft was also modified with deeper chine bays to accommodate left and right OBC cameras in lieu of the OOC cameras. The Big Tail pivoted about fuselage station 1300, articulating up 8.5° to provide ground clearance for takeoff rotation and landing flare, and down 8.5° to provide drag chute clearance (Fig. CS1.49, [113]). "Big Tail" was not pursued, but demonstrated the versatility of the basic SR-71 airframe.

The early to mid 1970s also saw a number of world records set by the SR-71 and her crews

> The maintenance tasks required by the SR-71 were more time-consuming and demanding than any other Air Force aircraft due to the extreme thermal environment and its effect on the aircraft structure and subsystems. Approximately 400 maintenance man-hours per flight hour (measured in the 1981 timeframe) were required and involved a dedicated and highly skilled team of Air Force maintenance personnel and civilian contractor technical representatives.

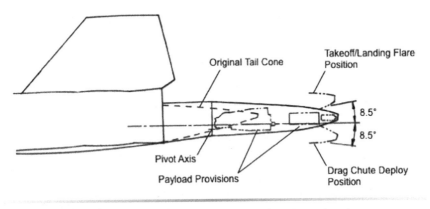

Figure CS1.49   The Big Tail articulated up and down for clearance.

from the 9th SRW. After a non-stop flight of 10.5 hr and 15,000 nm in 61-7968 with multiple aerial refuelings, pilot Lt Colonel Thomas B. Estes and RSO Lt Colonel Dewain C. Vick received the Mackay Trophy in 1971 for the "most meritorious flight of the year" and the Harmon Trophy in 1972 for "the most outstanding international achievement in the art and/or science of aeronautics in the preceding year." On September 1, 1974, pilot Major James V. Sullivan and RSO Major Noel F. Widdifield flew 61-7972 to the Farnborough Airshow in England, setting a New-York to London speed record of 3461.53 statute miles in 1 hr 54 min 56.4 sec for an average speed of 1806.95 mph. On the return trip to Beale AFB on September 13, 1974, Captain Harold B. "Buck" Adams and RSO Major William C. Machorek set a corresponding London to Los Angeles speed record of 5446.87 statute miles in 3 hr 47 min 39 sec for an average speed of 1435.59 mph. The year 1976 also saw a series of world speed and altitude records set during July 27/28. These included a world speed record of 2193.167 mph, set by pilot Captain Eldon W. "Al" Joersz and RSO Major George T. Morgan in 61-7958; a world altitude in horizontal flight record of 85,068.997 ft set by pilot Captain Robert C. Helt and RSO Major Larry A. Elliott, also in 61-7958; and a world speed over a closed course record of 2092.294 mph set by pilot Major Adolphus H. "Pat" Bledsoe and RSO Major John T. Fuller in 61-7962.

By the late 1970s, the analog automatic flight control system (AFCS) and automatic inlet control system (AICS) had become less reliable and maintainable, compounded by "vanishing vendors" for replacement parts. Encouraged by the results of the Cooperative Airframe/Propulsion Control System (CAPCS) tested in the NASA YF-12C in 1978, the Digital Automatic Flight and Inlet Control System (DAFICS) program was initiated to replace the analog computers and achieve enhanced redundancy, reliability, maintainability, and supportability. A trade study determined that a three-computer configuration combining the AFCS and AICS functions in each computer was superior to making dual-redundant digital versions of the separate AFCS and AICS computers. The new system, developed by Honeywell, combined the SAS, autopilot, automatic pitch warning, Mach trim, and air data functions. A new pressure transducer assembly located near the cockpit front pressure bulkhead/nose disconnect (FS 248) permitted the nose boom pitot-static, angle-of-attack, and angle-of-sideslip sensor pressure lines to be shortened, which reduced pressure lags in the system. The digital air data computations were of higher accuracy and improved overall system operation. Fuel savings were realized from more precise control of the inlets, and system mean time between failures increased by a factor of eight. The system was retrofitted to the entire SR-71 fleet by 1983 [114, 115, 116].

The structural life of the aircraft also saw improvement. The original design life was 3000 hr, but by conducting an Aircraft Structural Integrity Program (ASIP), Lockheed ADP and the USAF were able to extend the service life to 4500 hr. It was estimated that the SR-71 fleet had sufficient remaining airframe life to last well into the 21st century.

Another major improvement in reconnaissance capability came in 1985 with the operational fielding of the Advanced Synthetic Aperture Radar System-1 (ASARS-1). Using a specially contoured nose with a one-piece high temperature radome, the new radar offered several advantages over the CAPRE, including higher resolution and the ability to image in turns. Like the CAPRE, ASARS-1 could be used to update the aircraft present position by imaging fix points with known geographic coordinates. Radar imagery was stored in an Ampex Digital Cassette Recorder system incremental (DCRsi™) located in the L bay, along with the ASARS-1 computer and data handler [117]. Because the ASARS-1 Integrated Processor Display (IPD) in the RSO cockpit required greater depth ahead of the instrument panel than the CAPRE RCD, there was no longer room for the periscopic viewsight assembly. Instead, an electro-optical viewsight was installed that used separate electro-optical camera and cathode ray tube (CRT) display units, but provided less resolution than the periscopic unit.

To help improve situational awareness and reduce the potential for disorientation during night flying, a peripheral vision display (PVD) was added to the pilot's cockpit during the mid-1980s. This device, which was mounted inside the cockpit canopy on the right-hand side, projected a horizontal line of laser light matching the local horizon across the instrument panel to provide a peripheral vision cue.

Despite these system enhancements that promised to keep the SR-71 viable well into the 21st century, the SR-71 program faced challenges in the late 1980s that threatened its continuation. The inability to transmit sensor data near real-time was cited as a shortcoming, even though a line of sight data link to ground stations had been test flown and a beyond line of sight satellite communications data link had been defined but not funded [118]. The lack of persistence over a surveillance objective was also given as a drawback. But the capability offered by the SR-71 was unique and unmatched by any other operational platform: high speed, survivable, global, wide area synoptic reconnaissance with an unpredictable flight path/time of arrival to defeat cover-up of reconnaissance objectives by adversaries. Nevertheless, advocates for the SR-71 were unsuccessful in reversing the phase-out of the program. On October 1, 1989, operations were suspended except for minimum proficiency flying, and on November 22, 1989 the SR-71 program was terminated. A retirement ceremony for the SR-71 was held at Beale AFB on January 26, 1990, and the SR-71 fleet was dispersed to museums, except for four SR-71A models (61-7962, 61-7967, 61-7971, and 61-7980) kept in flyable storage and the SR-71B trainer (61-7956) that was undergoing periodic depot maintenance (PDM) at Lockheed's Air Force Plant 42, Site 2 facility at the time of program termination. On its final flight March 6, 1990 to Washington Dulles International Airport, where it would remain in storage until put on display at the National Air and Space Museum Stephan F. Udvar-Hazy Center, 61-7972 once again set several world speed records between coasts and cities. Crewed by pilot Lt Colonel Ed Yeilding and flight test engineer Lt Colonel J.T. Vida, who led the SR-71 Flight Test Force at Palmdale, 61-7972 flew coast to coast in only 1 hr 7 min 53.69 sec, for an average speed of 2124.51 mph. Table CS1.6 provides a summary of SR-71 sorties and flight hours as of January 1990.

Table CS1.6   SR-71 Operational Missions Summary and Flight Hours as of January 1990

Operational Sorties Flown	3551
Total Sorties Flown	17,300
Flight Hours (Operational Sorties)	11,008
Flight Hours (Total)	53,490
Mach 3+ Total Flight Time (Operational Sorties)	2752
Mach 3+ Time (Total)	11,675

Having already operated the Blackbird in the 1970s, NASA knew the value of the aircraft as a high speed, high altitude research platform, and arrangements were made to transfer 61-7971, 61-7980, and 61-7956 to the NASA Dryden Flight Research Center (61-7962 remained at Palmdale and was later transported to England for display in the American Air Museum at Duxford). The SR-71B PDM was completed to bring the aircraft back into service to train flight crews, and the Link SR-71 mission simulator, which was undergoing refurbishment and upgrade in Binghamton, New York, was delivered to NASA Dryden for training as well. Over the next nine years, the NASA SR-71s provided a unique test capability. A large upper surface "canoe" was installed on 61-7980 (also designated NASA 844) and used to test the Linear Aerospike SR-71 Experiment (LASRE), a 12% scale, semi-span model of the Lockheed Martin X-33 equipped with a linear aerospike engine to gather in-flight thrust data. Safety concerns prevented the LASRE from being "hot fired," but cold flow tests were conducted [119].

Meanwhile, Congress decided to restore a limited USAF SR-71 operational capability in Fiscal Year 1995 to address perceived shortfalls in airborne reconnaissance capability. Aircraft 61-7967 and 61-7971 were refurbished at Lockheed's Plant 10 facility, with 61-7968 remaining in standby storage. The Air Force agreed to share with NASA the SR-71B and simulator for pilot training, and Hangar 1864 (not far from NASA Dryden) was selected as the operational location for what would become Detachment 2 of the 9th SRW. Former SR-71 flight crews volunteered for reassignment to the program and two new crewmembers were selected and trained.

The ASARS-1, OBC, and TEOC sensors were brought out of retirement, together with upgraded defensive systems, but new capabilities were added as well. A 300 nmile line-of-sight Common Data Link (CDL) capability for ASARS-1 radar imagery was provided via a steerable antenna enclosed within a high temperature radome that protruded beneath the lower fuselage ahead of the nose landing gear bay. Radar data could be transmitted to the ground at 274 megabits per second or stored onboard via the DCRsi™. The TEOC was fitted with new lenses for enhanced resolution, and a 25-megapixel electro-optical backplane was tested that enabled real-time transmission of digital photography to the ground via the CDL. To process ASARS-1 imagery, a podium-size Clip-In Kit (CIK) was developed that replaced three van-sized processing systems to reduce the deployment footprint [120].

Despite these improvements and being ready to perform operational missions from Edwards AFB or from forward deployed locations on short notice, Detachment 2 was never called upon. A presidential line item veto on October 15, 1997 zeroed funding for the program, but questions on the constitutionality of the veto allowed Detachment 2 to continue operations until September 1998. NASA's operations were coming to a close as well, and at the Edwards AFB air show on October 9, 1999, NASA 844 made what turned out to be the final flight of an SR-71; a planned flight the next day was cancelled.

CS1.10 Summary

It is amazing to realize that the entire A-12 program—from initial go ahead to termination after successfully completing 29 operational missions—occurred within the span of a decade, all while breaking technological barriers in every discipline. The program provided the basis for the YF-12 that successfully demonstrated interception of high and low-flying targets at long range from a high speed, high altitude platform, and the SR-71 that provided 24 years of service to the USAF Strategic Air Command, followed by a successful reactivation for Air Combat Command five years later. The hallmark of a great design is one that is able to incorporate technology improvements without a fundamental re-design, and the SR-71 did so in many areas, including sensors, avionics, and flight/inlet control systems.

The A-12, YF-12, M-21, and SR-71 programs were made possible by a combination of factors. They were fostered by an urgent national need that demanded speed of action and secrecy. They were led by visionary individuals, both in Government and industry, who were willing to take risks and accept failures on the road to success. They were designed by engineers who were not afraid of stretching beyond the state of the art, who accepted responsibility for their product throughout the entire design life cycle, and who did not accept that it "could not be done" according to conventional wisdom. They were built by skilled workers that learned how to fabricate a completely new set of materials under an extremely aggressive timeline. They were flight tested and operated by aircrews, engineers, and maintainers that turned an experimental aircraft into a operational system that routinely flew hours at Mach 3.2 above 80,000 ft for long duration, worldwide reconnaissance missions. The result was a family of aircraft (summarized in Table CS1.7) that remain among the greatest aeronautical achievements of the 20th century, and a testament to Kelly Johnson and his Skunk Works.

Upon its retirement on March 6, 1990, the SR-71 was flown from its birthplace in Palmdale, CA to the Smithsonian Steven F. Udvar-Hazy Air and Space Museum at Dulles Airport, VA in 64 min, 20 s at an average speed of 2148 mph, with one 311 mi leg at 2190 mph (both still stand as World speed records for an operational aircraft). The Blackbird family remains among the greatest aeronautical achievements of the 20th century.

Table CS1.7   Blackbird Model Characteristics (Not Including A-12T and M-21)

Model	A-12	YF-12A	SR-71A	SR-71B	SR-71C
Crew	1	2	2	2	2
Wing Area (ft^2)	1795.0	1795.0	1795.0	1795.0	1795.0
Wing Span (ft)	55.62	55.62	55.62	55.62	55.62
Length (ft)	101.6	101.7	107.417	107.417	103.67
Fuselage Diam. (in)	64.0	64.0	64.0	64.0	64.0
Height (ft)	18.45	18.45	18.5	18.5	18.5
Zero Fuel Weight (lb)	52,700	59,150	60,000	59,000	58,714
Payload Weight (lb)	600	3 x 820	1500–4000	1500	1500
Fuel Weight (lb)	69,800	64,850	80,000	80,000	69,300
TO Gross Weight (lb)	123,600	124,000	140,000	139,000	128,014
Engine	YJT11D-20A	YJT11D-20A	JT11D-20B	JT11D-20B	JT11D-20B
Number Produced	12	3	29	2	1
Number Surviving	6	1	18	1	1

About the Author

John R. Whittenbury received his BS degree in mechanical engineering from the University of the Pacific in Stockton, CA and his MS degree in aeronautics and astronautics from Stanford University. After working at the NASA Ames Research Center while a student, he joined the Lockheed Martin Aeronautics Company (Skunk Works) in 1992 as a design engineer on the NASA ER-2 high altitude earth resources aircraft. Over a nine-year period he served as an engineer and project manager on a number of advanced development activities in the Skunk Works, and continued a life-long fascination with the Blackbirds. He is currently a senior engineering manager for a major aerospace corporation and is a Senior Member of the American Institute of Aeronautics and Astronautics.

References

[1] Pedlow, Gregory W. and Welzenbach, Donald E., *The CIA and the U-2 Program, 1954–1974*, Center for the Study of Intelligence, Central Intelligence Agency, Langley, VA, 1998.

[2] Ibid.

[3] Ibid.

[4] Mulready, Dick, *Advanced Engine Development at Pratt & Whitney: The Inside Story of Eight Special Projects 1946–1971*, Society of Automotive Engineers, Inc., Warrendale, PA, 2001.

[5] Sloop, John L., *Liquid Hydrogen as a Propulsion Fuel, 1945–1959*, NASA SP-4404, National Aeronautics and Space Administration, Washington D.C., 1978.

[6] Whittenbury, John R., "From Archangel to OXCART: Design Evolution of the A-12, First of the Blackbirds," Lockheed Martin Skunk Works, Palmdale, CA, 2001; public release authorization AER200305006.

[7] Suhler, Paul A., *From RAINBOW to GUSTO: Stealth and the Design of the Blackbird*, American Institute of Aeronautics and Astronautics, Inc., Reston, VA, 2009.

[8] Ibid.

[9] Ibid.

[10] Bissell, Richard M., with Lewis, Jonathan E. and Pudlo, Frances T., *Reflections of a Cold Warrior: From Yalta to the Bay of Pigs*, Yale University Press, New Haven, CT, 1996.

[11] Suhler, *From RAINBOW to GUSTO: Stealth and the Design of the Blackbird.*

[12] Ibid.

[13] Whittenbury, "From Archangel to OXCART: Design Evolution of the A-12, First of the Blackbirds."

[14] Ibid.

[15] Johnson, Clarence L., "Archangel Log," Lockheed Corp., Burbank, CA, released in abridged form, April 1993.

[16] Whittenbury, "From Archangel to OXCART: Design Evolution of the A-12, First of the Blackbirds."

[17] Pedlow, Gregory W. and Welzenbach, Donald E., *The Central Intelligence Agency and Overhead Reconnaissance: The U-2 and OXCART Programs, 1954–1974*, Center for the Study of Intelligence, Central Intelligence Agency, Langley, VA, 1998; declassified October 2004.

[18] Whittenbury, "From Archangel to OXCART: Design Evolution of the A-12, First of the Blackbirds."

[19] Suhler, *From RAINBOW to GUSTO: Stealth and the Design of the Blackbird.*

[20] McIninch, Thomas P., "The OXCART Story," Studies in Intelligence, Volume 15, No. 1, Winter, 1971.

[21] Hehs, Eric, "Super Hustler, FISH, Kingfish, and Beyond (Part 3: Kingfish)." Code One Online. Web. 30 September 2011.

[22] Johnson, Clarence L., "History of the OXCART Program," SP-1362, Advanced Development Projects, Lockheed California Company, Burbank, CA, 1968.

[23] Law, Peter V., "SR-71 Thermal Environment," part of "Case Studies in Engineering: The SR-71 Blackbird," Course Ae107, presented at the Graduate Aeronautical Laboratory, California Institute of Technology, Pasadena, CA, 1990.

[24] Meyer, Jerry E., "SR-71 Configuration Development," part of "Case Studies in Engineering: The SR-71 Blackbird," Course Ae107, presented at the Graduate Aeronautical Laboratory, California Institute of Technology, Pasadena, CA, 1990.

[25] Ibid.

[26] Johnson, "Archangel Log."

[27] A-12 Technical Manual A12-2-2 Airframe, Advanced Development Projects, Lockheed California Company, Burbank CA, 1966.

[28] Whittenbury, "From Archangel to OXCART: Design Evolution of the A-12, First of the Blackbirds."

[29] Altizer, John, "SR-71 Structures and Materials," part of "Case Studies in Engineering: The SR-71 Blackbird," Course Ae107, presented at the Graduate Aeronautical Laboratory, California Institute of Technology, Pasadena, CA 1990.

[30] Ibid.

[31] Johnson, C.L., "Development of the Lockheed SR-71 Blackbird," Lockheed Horizons, Issue 9, Winter 1981/82.

[32] Altizer, John, "SR-71 Structures and Materials."

[33] Johnson, C.L., "Development of the Lockheed SR-71 Blackbird."

[34] Merlin, Peter W., *From Archangel to SENIOR CROWN: Design and Development of the Blackbird*, American Institute of Aeronautics and Astronautics, Reston, VA, 2009.

[35] Suhler, *From RAINBOW to GUSTO: Stealth and the Design of the Blackbird.*

[36] Altizer, John, "SR-71 Structures and Materials."

[37] Ibid.

[38] Zuchowski, Brian, AFRL-RB-WP-TR-2010-3069, Final Report, Task Order 0015, Air Vehicles Integration and Technology Research (AVIATR), Lockheed Martin Aeronautics Company, August 2010; approved for public release, distribution is unlimited.

[39] A-12 Utility Flight Manual, 15 September 1965; Changed 15 June 1968; released August 2007.

[40] Ibid.

[41] Altizer, John, "SR-71 Structures and Materials."

[42] Connors, Jack, *The Engines of Pratt & Whitney: A Technical History, American Institute of Aeronautics and Astronautics*, Reston, VA, 2010.

[43] Brown, William H., Webb, William L., and Dees, W. Stanley, "JT11D-20 Engine— A Successful Leap into the Technical Unknown," part of "Case Studies in Engineering: The SR-71 Blackbird," Course Ae107, presented at the Graduate Aeronautical Laboratory, California Institute of Technology, Pasadena, CA 1990.

[44] Ibid.

[45] Abernethy, Dr. Robert B., "More Never Before Told Tales of Pratt & Whitney," presented to the Roadrunners and the J58 Eagles Reunion, 2004.

[46] Ibid.

[47] Brown, Webb, and Dees, "JT11D-20 Engine—A Successful Leap into the Technical Unknown."

[48] A-12 Technical Manual A12-2-4 Powerplant, Advanced Development Projects, Lockheed California Company, Burbank, CA, 1966.

[49] J58/JT11D-20 Axial Flow Gas Turbine/Single-Spool Turbojet Fact Sheet, Pratt & Whitney, courtesy Roadrunners Internationale, accessed 2010.

[50] Campbell, David H., "F-12 Series Aircraft Propulsion System Performance and Development," American Institute of Aeronautics and Astronautics Paper 73-821, presented at the 5th Aircraft Design, Flight Test, and Operations Meeting, St. Louis, MO, 1973.

[51] Ibid.

[52] Anderson, Tom, "SR-71 Propulsion Integration, Engine Inlet and Exhaust," part of "Case Studies in Engineering: The SR-71 Blackbird," Course Ae107, presented at the Graduate Aeronautical Laboratory, California Institute of Technology, Pasadena, CA 1990.

[53] Campbell, "F-12 Series Aircraft Propulsion System Performance and Development."

[54] Anderson, Tom, "SR-71 Propulsion Integration, Engine Inlet and Exhaust."

[55] George, M. F. Jr., and Burton, R. V. Jr., "Development and Flight Test Evaluation of Fuel Tank Sealants for Mach 3+ Aircraft," American Institute of Aeronautics and Astronautics Paper 79-24124, 1979.

[56] A-12 Utility Flight Manual.

[57] Ibid.

[58] Law, "SR-71 Thermal Environment."

[59] DeGrey, Richard P., "Hot, High-Speed Hydraulics," Hydraulics & Pneumatics, December 1982.

[60] Loschke, Robert, "SR-71 Flight Controls," part of "Case Studies in Engineering: The SR-71 Blackbird," Course Ae107, presented at the Graduate Aeronautical Laboratory, California Institute of Technology, Pasadena, CA 1990.

[61] A-12 Utility Flight Manual.

[62] Ibid.

[63] Loschke, "SR-71 Flight Controls."

[64] McMaster, John R., and Schenk, Frederick L., "The Development of the F-12 Series Aircraft Manual and Automatic Flight Control System," American Institute of Aeronautics and Astronautics Paper 73-37474, 1973.

[65] Johnson, "Development of the Lockheed SR-71 Blackbird."

[66] Johnson, Clarence L., and Smith, Maggie, *Kelly: More Than My Share of It All*, Smithsonian Institution Press, Washington, D.C., 1985.

[67] "Oxcart Sensors," Central Intelligence Agency, 1972; approved for release August 2007.

[68] Pedlow, Gregory W. and Welzenbach, Donald E., *The Central Intelligence Agency and Overhead Reconnaissance: The U-2 and OXCART Programs*, 1954–1974.

[69] Memorandum for Committee on Imagery Requirements and Exploitation, "Comparison of SR-71 and A-12 Aircraft," 26 September 1967; approved for release August 2007.

[70] Johnson, Clarence L., "AF-12 Log," Lockheed Corp., Burbank, CA, released in abridged form, April 1993.

[71] Eastham, James D., "YF-12A Pilot's Information and Questionnaire," SP-564, Advanced Development Projects, Lockheed California Company, Burbank, CA, 15 July 1963.

[72] Ibid.

[73] Johnson, "AF-12 Log."

[74] Ibid.

[75] Johnson, Clarence L., "SR-12 Log," Lockheed Corporation, Burbank, CA, released in abridged form, April 1993.

[76] Byrnes, Donn A. and Hurley, Kenneth D., *Blackbird Rising: Birth of An Aviation Legend*, Sage Mesa Publications, Los Lunas, NM, 1999.

[77] Johnson, "SR-12 Log."

[78] NASA CR-115703, B-70 Aircraft Study, Final Report Volume II, April 1972.

[79] Memorandum from John A. McCone, "Discussion At Lockheed," 26 September 1967; approved for release August 2007.

[80] SR-71A Utility Flight Manual (U), Issue E: 31 October 1986, Change 2: 31 July 1989, Lockheed Advanced Development Company, Burbank, CA; declassified 1991.

[81] Byrnes and Hurley, *Blackbird Rising: Birth of an Aviation Legend*.

[82] Ibid.

[83] Johnson, "Archangel Log."

[84] Ibid.

[85] Ibid.

[86] Memorandum for the Record, "Factors Affecting A-12 Flight Test and Mach Number Extension," 21 July 1963; approved for release August 2007.

[87] Wheelon, Dr. Albert D., "DDST Support of OXCART," Presented at the Roadrunners Internationale Meeting in Las Vegas, Nevada, 2 October 2003.

[88] Johnson, "Archangel Log."

[89] Inteview with James D. Eastham, Rancho Palos Verdes, CA 29 November 2003.

[90] Anderson, "SR-71 Propulsion Integration, Engine Inlet and Exhaust."

[91] Johnson, "Archangel Log."

[92] Jenkins, Dennis R. and Landis, Tony R., *Lockheed Blackbirds*, Warbird Tech Series Volume 10, Specialty Press, North Branch, MN, 2004.

[93] Johnson, "AF-12 Log."

[94] Inteview with James D. Eastham, Rancho Palos Verdes CA, 12 August 2006.

[95] Jenkins and Landis, "Lockheed Blackbirds."

[96] Byrnes and Hurley, *Blackbird Rising: Birth of an Aviation Legend*.

[97] Abrams, Richard, Lusby, William A. Jr., Evenson, Mervin L., and Skliar, William L., SR-71A Category II Performance Tests, 69-AFFTC-39459, Air Force Flight Test Center, Edwards, CA, 1970.

[98] Ibid.

[99] Ibid.

[100] Johnson, Clarence L., "Feasibility Report—Modification of A-12 Vehicle for Air Launched Orbital Reconnaissance System," SP-404, Advanced Development Projects, Lockheed California Company, Burbank, CA, 7 September 1962; approved for release 25 June 2002.

[101] Johnson, Clarence L., "Q-12 Drone Log," Lockheed Corporation, Burbank CA, released in abridged form, June 1993.

[102] Merlin, *From Archangel to SENIOR CROWN: Design and Development of the Blackbird.*

[103] Memorandum for Deputy Director of Science and Technology from Deputy Director of Special Activities, "TAGBOARD Program," 17 August 1967; approved for release August 2007.

[104] "M-21 Model Specification," SP-662, Advanced Development Projects, Lockheed California Company, Burbank, CA, 11 May 1966.

[105] Johnson, "Q-12 Drone Log."

[106] Memorandum for Deputy Director of Science and Technology from Deputy Director of Special Activities, "TAGBOARD Program."

[107] Memorandum for Director of Central Intelligence, "OXCART Program Status Briefing Notes," 26 February 1965; approved for release 1996.

[108] Robarge, David, *Archangel: CIA's Supersonic A-12 Reconnaissance Aircraft*, Central Intelligence Agency, Washington D.C., 2007.

[109] McIninch, "The OXCART Story."

[110] Merlin, Peter W., Mach 3+: NASA/USAF YF-12 Flight Research 1969–1979, SP-2001-4525, National Aeronautics and Space Administration, 2002.

[111] Ibid.

[112] Crickmore, Paul F., *Lockheed Blackbird: Beyond the Secret Missions*, Osprey Publishing, Oxford, England, 2004.

[113] Jenkins and Landis, *Lockheed Blackbirds.*

[114] Loschke, "SR-71 Flight Controls."

[115] DeGrey, Richard P., Nelson, Roger L., and Meyer, Jerry E., "SR-71 Digital Automatic Flight and Inlet Control System," Society of Automotive Engineers, Inc., Technical Paper 851977, 1985.

[116] Tilden, Lieutenant Colonel Thomas V., "Digital Flight and Inlet Control in the SR-71," Society of Experimental Test Pilots, 1987.

[117] The SR-71 Reconnaissance System Executive Handbook, Lockheed Martin Skunk Works, Palmdale, CA, 1995.

[118] Graham, Colonel Richard (USAF Retired), *SR-71 Revealed: The Inside Story*, Motorbooks International Publishers & Wholesalers, Osceola, WI, 1996.

[119] Merlin, *From Archangel to SENIOR CROWN: Design and Development of the Blackbird.*

[120] The SR-71 Reconnaissance System Executive Handbook.

X-35 Concept Demonstration Aircraft

James Eshleman

- Review of the Historic X-35B Flight and Early Evolution of the X-35
- Overview of the Joint Strike Fighter Program
- Fundamental Characteristics of Flow Fields Associated with V/STOL Aircraft
- Concept Assessments for V/STOL Propulsion
- The SSF, ASTOVL, and JAST Programs
- Flight Test Phase of the JSF/CDA Program

This was the end of the "Ha Trick" flight that epitomize how the F-35 would have capabilities not available o any other aircraft. The mis was to take off vertically, accelerate to and cruise at supersonic speeds, and ret to the origin with a vertica landing. This historic miss is described on the next pa

If it looks good . . . it flies good.

Anonymous

CS2.1 "Hat Trick" Aviation Milestone

On the morning of July 20, 2001, the X-35B, call sign "Hat Trick," taxied to runway 22 at Edwards Air Force Base (EAFB) with Marine Corp Major Art "Turbo" Tomassetti at the controls. Turbo was a graduate of the U.S. Navy (USN) Test Pilot School and a veteran Harrier pilot. He had also completed several previous flights in the X-35B including the aircraft's second hover. But this morning would be different. Not only would this be a first for Turbo, it would be a first for aviation. On that July day, the X-35B would make a short takeoff (STO), climb to altitude, perform a level acceleration to supersonic speeds, and then return to EAFB for a vertical landing (VL). While each of these tasks had been completed individually, this was the first time they would be performed in a single flight. Known as "Mission-X," this was the graduation exercise for the program and the Lockheed Martin team. It also represented the final step in a competition that would result in the award of the largest military contract in history. That competition could trace its beginning back more than a decade and would eventually involve every major aerospace company in the United States and many others around the world.

> The airplane won't amount to a damn, until they get a machine that will act like a hummingbird—go straight up, go forward, go backward, come straight down and alight like a hummingbird. It isn't easy . . . somebody is going to do it.
> —*Thomas Edison (1847–1931)*

Clearly Thomas Edison underestimated the utility of the airplane and its impact on society. Arguably, helicopters provided the utility Edison referred to. However, the helicopter is speed limited and aircraft designers have worked steadily to develop configurations with helicopter utility and fixed wing speed.

CS2.2 Joint Strike Fighter (JSF) Program Overview

Research on Vertical and Short Take Off and Landing (V/STOL) aircraft has gone on for decades. Several configurations have made it to flight hardware. However, only a very few produced operational systems. From the period after World War II and into the Vietnam era, the concept of "Vertical Envelopment" came into being. This doctrine allowed large numbers of troops and equipment to be dispersed without roads or air bases. At the time, vertical envelopment was achieved only through the use of helicopters. During the Cold War, fixed wing V/STOL would get another boost when runway denial tactics were believed to threaten the effectiveness of conventional air forces.

With the advent of the Harrier, and later, the AV-8B, the U.S. Marine Corps (USMC) envisioned the use of fixed wing V/STOL aircraft operating in conjunction with their helicopter forces to fulfill the vertical envelopment doctrine. Fixed wing V/STOL would provide rapid response Close Air Support (CAS) and fill a role similar to heavy artillery of other forces. The ability to provide basing flexibility and rapid response, led the USMC to commit to a full V/STOL force in the late 1970s.

Through the next two decades, this requirement would grow into a replacement for the USMC AV-8B's, Royal Navy's Sea Harrier, as well as the F-16, F-18C, D, and the A-6. These various programs would eventually become the three variants of the Joint Strike Fighter with the potential of producing 3100 aircraft for the USMC, USAF, USN, and eight other allied partners.

CS2.3   Early Efforts

The X-35 evolved from as clean a sheet as possible, taking into consideration the lessons learned from more than 40 years of V/STOL research and development. In the late 1980s, a small design study was undertaken within Lockheed by Advanced Development Projects (ADP), better known as the "Skunk Works." Off an empty hallway, in a small room that could hold no more than two desks, a couple of file cabinets, and a whiteboard, a small group of engineers with various specialties argued about what made a good V/STOL jet. This group was small and, except for only a few grey beards (Grant Carichner, Paul Bevilaqua, and Paul Shumpert), young. In fact, most were under 30, and a few (Brian Quayle, Pat Tait, and myself) would stay with the project through completion. Shumpert and Bevilaqua would eventually become joint patent holders for the Shaft Driven Lift Fan (SDLF) propulsion system that would power JSF. All were well versed in the history of V/STOL development and the pitfalls that had befell previous attempts (Fig. CS2.1). Paul Bevilaqua had worked on the ill-fated Rockwell XFV-12A program and Shumpert had worked the Lockheed XV-4A/B Hummingbird aircraft. Brian Quayle had just finished the Lockheed U.S./ U.K. V/STOL study, I came to Lockheed after four years of V/STOL research and large scale testing at NASA Ames, and Pat Tait had worked Integrated Flight and Propulsion Controls (IFPC) on the YF-22 program for several years.

The group's early efforts were high level discussions that didn't involve technical drawings, formal specifications, or analysis. The main product of this was a prioritized list of salient features such a system should possess; without preconceived notions of what it would look like or, in some cases, hardly a clue as how to achieve them. Features such as thrust augmentation during STOVL operation, compact, and efficient transfer of energy forward of the center of gravity, STOL overload capability, benign ground environment, controllability in STOVL flight, "uncompromised up-and-away

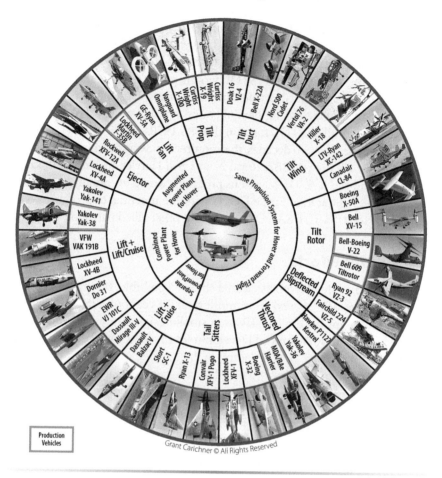

Figure CS2.1 V/STOL aircraft summary (2008). [1]

performance, affordability, and others." These characteristics will be discussed in more detail later in the chapter.

Setting the stage on the technical side, modern fighter aircraft were operating with thrust-to-weight ratios approaching, and in some cases exceeding 1.0. The YF-22 was demonstrating the utility of integrated flight-propulsion controls. The same advances in high speed computing that enabled integrated flight-propulsion controls, also fueled an earlier, two decades long, push in analytical tool development leading to several other engineering advancements. These included analysis of transonic aerodynamics, high angle-of-attack separated flows, advanced structures, and combustion modelling leading to ever higher temperature engines.

On the political side, the Cold War was still brewing. Satellite photos of the Yak-141 supersonic V/STOL fighter, combined with the nearly decade-long Soviet war in Afghanistan, provided a backdrop of urgency. The U.K.

experience in the Falklands, with sea-based V/STOL attack jets (Harriers), provided real-world examples of current limitations as well as what could be achieved. Combine that with concern regarding availability of overseas basing, vulnerability to runway denial tactics, and a ground swell of interest began to form regarding the suitability of V/STOL aircraft to respond in these situations. As a result, vertical takeoff (VTO) capability was becoming less important and STO became the primary basing and mission driver along with VL capability at end of mission weights. STOVL became the watch word to emphasize this basing and operational concept.

The United States and United Kingdom had been working cooperatively on "white world" (acknowledged) STOVL programs for several years. These conceptual efforts were aimed at a follow-on to the AV-8B/Harrier family of aircraft with supersonic capability. In the mid to late 1980s these efforts were winding down with significant interesting work but no real configurations. And because of the international nature of the program, observables technologies were kept out of the studies.

In the meantime, NASA was working several low level efforts with the Lockheed Skunk Works to explore integration of observables technologies on STOVL configurations. While these efforts were low level compartmentalized efforts with limited funding, they did serve to clarify a number of issues and focus later efforts.

Both the U.S./U.K. and NASA efforts helped illuminate the issues as well as the potential benefits associated with STOVL aircraft and in early 1988 the landscape began to clarify itself. The USMC was absolutely committed to becoming an all V/STOL force. They wanted an aircraft that could replace both the AV-8B and the F-18(C/D) they were operating. While the Marines were more concerned Infra-Red, IR, and acoustic signature, they knew that for significant funding, procurement numbers, and political backing to form any such aircraft would need to consider USAF demands for low Radar Cross Section (RCS) as well.

No one was interested in another subsonic STOVL fighter. The only question was how supersonic such aircraft would need to be. The F-16 and F-18 both demonstrated that agility and acceleration (quickness) were now more important than maximum speed to success in the air battle. Advances in missile technology only added to this argument.

With momentum building, NASA Ames established a supersonic STOVL project office with Press Nelms as its chief. Sam Wilson, who would later become a critical central figure to the effort, was his deputy. A parallel classified effort was led by Roy Presley.

CS2.3.1 STOVL Strike Fighter (SSF)

In 1988, with the USMC as its primary service advocate and the Defense Advanced Research Projects Agency (DARPA) providing funding and programmatic support, what would evolve into the X-35 began to take shape. Initial funding was $100K for feasibility studies followed by two conceptual design contracts of about $2M each. The early Lockheed program, first led by Jerry Rising and later by Grant Carichner, operated out of a small corner of building 311, ADP headquarters at the Burbank airport. This effort was quite small, totaling around 20 people including management, security, and administrative staff.

The role of the USMC and DARPA as advocates, and the tireless efforts of two individuals in particular, proved critical to the early success and evolution of the program. First was retired Colonel, AV-8A test pilot, and PhD (as he was fond of pointing out) Bill Scheuren. Bill was a true warrior and Marine Corps to the core. He only celebrated November 10, the founding of the Marine Corps, as "his" birthday, and would tell everyone, "I'm 219 years old today," or however many years it happened to be. He would also remind everyone that the USMC was older than the United States. His dedication was absolute, his attitude was unrelenting, and more than any other individual, Bill Scheuren was the driving force behind the program at that time.

Sam Wilson became Bill's chief engineer. I had known Sam while working for NASA at Ames Research Center (ARC). We had a common background, sharing experiences with many old guard NASA V/STOL mentors, including Dave Hickey, Dave Koenig, Pres Nelms, Wally Deckert, and Dick Kuhn. Sam rotated through a management cycle at NASA headquarters in Washington D.C. where he became associated with DARPA and elected to stay on there. He was a V/STOL advocate with a keen sense of how to integrate the technical and political aspects of a program. He was steeped in the history of V/STOL technology and as committed as anyone to seeing it become a reality.

Both Sam and Bill worked closely with Marine Corps Aviation headquarters on various early capabilities trade studies. Their efforts resulted in a summary of "Desired Operational Capabilities for Advanced STOVL Aircraft," which outlined a wide variety of operational capabilities, most of which had never before been incorporated in any V/STOVL aircraft. Ultimately, this served as the top level requirements spec for these early efforts.

We were always aware that DARPA was also funding our competitors to work their own concepts against the emerging requirements. At that time, those included McDonnell Douglas and General Dynamics, each with their unique approach towards V/STOL propulsion and the aircraft design process. That competition was healthy, fierce, and, I think, critical to our overall success.

The major ADP programs at this time were for customers such as the USAF and CIA. Our primary experience was with aircraft such as the SR-71 (still operating at that time), F-117A, and U-2/TR-1. Lockheed's last practical ventures into the world of fixed wing V/STOL had been the XFV-1 "Pogo" in the early 1950s and XV-4 "Hummingbird" in the early 1960s. We fully recognized the limitations of our experience and set out to strengthen our operational knowledge. Along these lines, we enlisted the services of retired USMC Colonel Russ Stromberg. Russ was an AV-8B test pilot and squadron commander who had led development of the "Marine Corps/ DARPA Desired Operational Capabilities for Advanced STOVL Aircraft" document. He brought considerable technical, practical, and programmatic experience. This was invaluable from the standpoint of understanding what current aircraft brought to the fight, what pilots, ground crews, and commanders found lacking in existing systems, and how Marine aviation worked in general.

Through Russ, we came to understand and appreciate the high level of integration that Marine Air operations had achieved and depended upon. Basing flexibility, both ashore and at sea, is critical for Marine Corps aviation to effectively support their combat units. As a truly organic element of the Corps, the close relationship between Marine aviation and their ground counterparts is constantly reinforced. The Marines were looking for a single, advanced STOVL aircraft that not only combined but exceeded the close air support capabilities of the AV-8B and the air-to-air capabilities of the F/A-18.

At this point, there was no requirement for operating off aircraft carriers (CVs). The primary emphasis was operation from austere land bases and air capable amphibious warfare ships, primarily LHAs and LHDs, in roles similar to the AV-8B. However, the program did have extremely aggressive observability goals. At that time the F-117A was still a true black program and industry, as well as the military, used every opportunity to push application of observables technology to find out what could be practically achieved. The concept of balanced observables was only beginning to catch on (an implication of this mind set is discussed later).

Now the hard work started. How to determine what such an airplane looks like and how to achieve all of these "desires" in a single package? To our advantage, there had been a long history of V/STOL development; a few successes, and a great many failures. This is illustrated by a diagram referred to as "The V/STOL Wheel of Misfortune," as seen in Fig. CS2.1. Never in my life had the value of history been so blatantly apparent. We had the advantage of the post mortems from many failed V/ STOL aircraft concepts and programs. Drawing from those failures sharpened our sense of what was important and what was not. We could all agree that fresh ideas, tempered by historical pragmatism were needed, as none of us wanted to create another slot on the wheel of misfortune.

All Marine pilots are also trained as infantrymen and can be distinguished from other services by their camouflage flight helmet covers.

CS2.4 V/STOL Flow Fields

Before going further, it is important to understand some fundamental characteristics of flow fields associated with V/STOL aircraft. In general, V/STOL flow fields are driven by the jet exhaust and are affected by the presence of the ground (ground-effect) and relative forward velocity as seen in Fig. CS2.2a. Reference [3] provides a general survey of V/STOL related ground-effect flow physics.

A vehicle in a steady state hover, as seen in Fig. CS2.2a, is subjected to the effects of ambient air being entrained into the exhaust jets or "lift posts." In this case the vehicle is hovering well away from the ground and is said to be out of ground-effect (OGE). As entrained ambient air accelerates into the exhaust jets its static pressure is reduced. This creates a lift loss phenomenon referred to as "suck-down," which typically ranges from 2% to 5% of the jet thrust.

A vehicle hovering near the ground is said to be in ground-effect (IGE) and typically exhibits a more complex set of characteristics as shown in Fig. CS2.2b. Near an impingement surface such as the ground, a jet flows outward to form a wall jet. The increased surface area associated with wall jets results in significantly higher flow entrainment and correspondingly higher levels of suck-down as the ground is approached.

A multi-jet vehicle forms wall jets about each lift post that collide and form upwardly flowing regions referred to as "fountains." The fountain flows eventually impinge on the vehicles lower surface as it descends to land. Fountain impingement may be a source of upward force that acts to offset the lift loss resulting from suck-down. A visual example of the fountain formed by impinging jets is seen in Fig. CS2.2c. In this figure, smoke

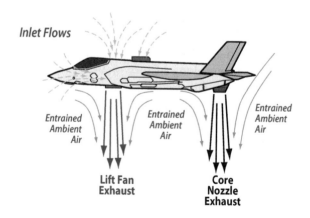

Figure CS2.2a   Vehicle hovering in ambient air (Out of ground-effect, OGE).

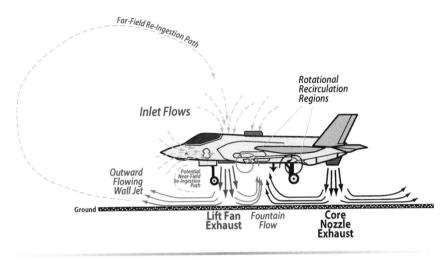

Figure CS2.2b   Vehicle hovering in ground-effect (IGE).

generated on the inboard side of the core flow on the Grumman 698 full scale twin tilt nacelle model flows inboard and makes the fountain flow visible.

An important characteristic of fountain flows is that they can form areas of high velocity rotational flow as they impinge on the underside of the vehicle. If strong enough, these rotational areas may produce large

Figure CS2.2c   Smoke flow visualization of the fountain on the Grumman 698 full scale tilt nacelle model.

negative pressures that can completely negate any lifting effect of the fountain. For some multi-jet configurations this results in higher levels of suckdown than an equivalent single jet.

An area that often receives little attention but needs to be considered during design is the moments and trim changes that result from these induced forces. While typically not a designing factor, these induced moments are generally destabilizing and must be taken into account when sizing control systems. Similarly, the forces and moments resulting from inlet flow momentum or "ram-drag" also need to be taken into account. We will see later that the destabilizing effect of the inlet in forward flight was an important factor in early SSF configuration development.

As exhaust flows mix with ambient air some hot exhaust gases can be drawn into the inlet resulting in thrust losses, and if serious enough, engine stall. Any configuration producing hot exhaust gasses will experience some level of hot gas ingestion (HGI). This is caused by general mixing of hot exhaust flows with ambient air resulting in slightly elevated far-field temperatures as seen in Fig. CS2.2b. While far field ingestion may result in thrust loss (we will discuss later why this is not stated as a definite effect), a greater concern are near-field flow paths that allow exhaust gasses to be pulled into the inlet before significant mixing occurs. While there may be several potential causes for near-field HGI, the main culprit is usually hot fountain flows that move along the surface of the vehicle and into the vicinity of the inlet where they can be sucked in. This can result in significant inlet temperature rise in localized regions that can cause engine surge, stall, and possibly flame-out. In general the hotter the front lift posts, the greater the chances of encountering hot gas ingestion (HGI) issues.

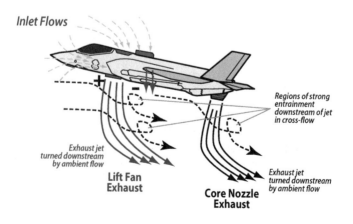

Figure CS2.2d Vehicle transitioning (out of ground-effect, OGE).

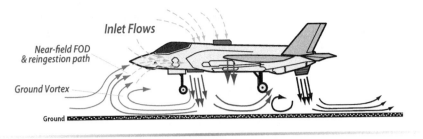

Figure CS2.2e Vehicle transitioning (in ground-effect, IGE).

As the vehicle moves through the air as in transition to forward flight, or simply by hovering over a stationary point during a head wind, additional flow field complexities become apparent as seen in Fig. CS2.2d. The ambient flow turns the exhaust jets and sets up a kidney-shaped vortex pattern in the downstream jet. This increases the jet's entrainment and results in generally negative pressures in their wake. In some cases, pressures slightly above ambient can be encountered just ahead of the jet caused by the blockage effect. The net result is generally increased levels of lift-loss and positive (nose up) pitching moments as forward speed increases (unless the jet is located very near the trailing edge of a planform such as a jet flap).

As an aircraft with downward deflected jets moves along or near the ground (as in a STO or rolling VL), the forward flowing wall jet interacts with the free stream to form a horse-shoe shaped ground vortex as seen in Fig. CS2.2e. The ground vortex causes a general up swell and rotation in the local flow field that can result in not only HGI but also water and FOD from the operating surface being drawn into the inlets. The geometry of the ground vortex is a complex function of the vehicles speed, jet deflection and thrust, configuration of the forward lift post(s), and jet exit characteristics.

A discussion of V/STOL flow field physics is not complete without acknowledging the effect of these flows on the acoustic and thermal environments that aircraft structures are subjected to. These turbulent, high speed flows impinge on the aircraft surfaces and create extreme acoustic environments in which overall sound pressure levels (OASPL) can exceed 130db locally. In addition to the acoustics, fountain impingement results in surface temperatures that can be a significant percentage of the jet exhaust. Repeated exposure to the combination of these effects will certainly drive material, design, and manufacturing choices.

Going into what would eventually become the Joint Strike Fighter program we were fortunate to have an excellent understanding of these effects as they had been well documented by over 40 years of outstanding V/STOL research by folks like Kuhn, Foley, Wyatt, Koenig, Dudley, Margason, and many others. Even with that, the process of educating engineers, managers, and operators who had worked exclusively CTOL programs on the characteristics and issues associated with V/STOL configurations was an ongoing process.

CS2.5 V/STOL Concept Assessments

As a result of the increased interest in the mid 1980s, Lockheed had provided a small amount of internal funding to very quietly study V/STOL propulsion concepts. The United States and United Kingdom had just spent several years and a significant amount of money studying some interesting ideas that we realized would not pan out as viable systems. Headed up by Paul Bevilaqua and Paul Shumpert the Skunk Works studies looked at the problem of STOVL propulsion from a fresh perspective. As mentioned earlier, this effort started off as simple brainstorming sessions in which we stepped back from the details of the design and looked at what was needed for an idealized STOVL propulsion system. We leaned heavily on the lessons from previous successes and failures in order to develop a set of key traits. A partial list of the major lessons learned follow.

CS2.5.1 Thrust Augmentation during STOVL Operation

Hovering vehicles need a T/W of about 1.12 to 1.20 to be practical. This high level of thrust provides good STO and in-flight acceleration (the AV-8B is one of the quickest airplanes in existence from 0 to 400 knots). However, these engines are oversized for cruise and have poor fuel specifics. The correspondingly large inlets generally force V/STOL configurations into an early, steep, and significant transonic drag rise that tends to limit the range and top speed of current generation V/STOL aircraft. The ideal STOVL aircraft would employ a propulsion system sized for up-away performance with a means of augmenting thrust during STOVL operation. Many such systems had been proposed, however all had other significant drawbacks that will be discussed shortly.

CS2.5.2 Compact and Efficient Transfer of Energy Forward of the c.g.

Many aircraft were operating with T/W approaching, and for some weight conditions exceeding, 1.0. Unfortunately, all of that thrust was typically exhausted out the back of the airplane in a direction nominally along the flight path. While Vertical Attitude Takeoff and Landing (VATOL) configurations had been tried several times, none were successful for a number of reasons (we kept VATOL in the early conceptual mix despite its many drawbacks). The ideal system would employ a propulsion system capable of shifting an appropriate amount of thrust ahead of the c.g. to allow it to land in a horizontal attitude. It would also allow that transfer to occur efficiently and not require large amounts of volume and/or weight to do it.

CS2.5.3 STO Overload Capability

Conventional Takeoff and Landing (CTOL) aircraft enjoy the ability to trade payload for takeoff and landing distance. All aircraft are mission limited by how much weight they can get off the ground. Even with air-to-air refueling, mission flexibility is always increased with higher takeoff weights. The Harrier family had demonstrated the utility of STO overload by allowing the aircraft to take off at weights above what it could hover at. This requires the aircraft to have good low-speed handling characteristics, a wide transition envelope, and the ability to vector thrust effectively. In this case, the ideal STOVL system would allow the operator to fully exploit the physical capability of the system to trade between takeoff distance and payload. The system would be capable of VTOL at one end of the spectrum and CTOL at the other.

> My 'Aha' moment regarding the shaft driven concept came when Bevilaqua and Shumpert explained to me that the temperature of the deflected engine exhaust was significantly lower because of the work extracted by the shaft. This gave us augmented thrust, an ability to move thrust around for STOVL modes, and a vertical exhaust that was hundreds of degrees Fahrenheit lower than basic engine exhaust temperatures. From that time on I was convinced that the shaft driven concept was the best solution.
>
> —*Grant Carichner*

CS2.5.4 Benign Ground Environment

Ground environment encompasses many things on a STOVL aircraft. Of primary importance, impingement temperatures must be low to allow the aircraft to operate repeatedly from common surfaces. Asphalt softens (or melts), concrete spalls, metals expand and buckle, and surface treatments are ablated away. Operational experience with the AV-8B revealed that the wing tip reaction control jets (roll) would erode the edges of asphalt roads forcing the aircraft to constantly relocate after a few passes. Exhaust jet pressure, velocity, and profile are the next considerations as these parameters determine the ability of the aircraft to operate in proximity to people and other equipment. Since the entire premise behind STOVL aircraft is to reduce the amount of real estate required to operate, the size of keep out zones must also be included in the aircraft's basing requirements. The last major ground environment consideration is acoustics as sound pressure level (SPL) affects the performance and health of personnel and equipment. The noise environment associated with operation of modern jet aircraft already requires exposure limits and protection requirements for personnel. Equipment and structures in the proximity of an operating STOVL aircraft need to be designed and qualified to withstand the associated SPL. Since SPL roughly scales with the eighth power of jet exit velocity, acoustics can get out of hand quickly. Here, the ideal STOVL system would have cool exhaust jets that derive thrust from increased mass flow, rather than velocity and temperature.

CS2.5.5 Controllability in STOVL Flight

Nearly every V/STOL concept developed to date had met with control issues to some degree. Controllability in STOVL flight can be a fundamental limitation as the aircraft is operated at airspeeds below that required for wing borne flight. STOVL vehicles generate additional lift by moving large masses of air, which results not only in ram drag, but also moments about all three axes as a result of turning large airflows. Consequently, these forces must be balanced as additional forces and moments used for control must be generated by the propulsion system itself. This increases requirements for fineness of control, rates, and reliability of the propulsion system as it now becomes an integral part of the flight control system. Additionally, any performance degradation resulting from control application needs to be accurately reflected in the aircrafts performance modelling. From a controllability standpoint, the ideal STOVL system would be able to rapidly and precisely generate an abundance of control power about each axis with minimal cross coupling and performance degradation.

CS2.5.6 Uncompromised Up-and-Away Performance

Any fighter/attack aircraft must have outstanding air-to-air combat capabilities as well as generally good speed, range, payload, and survivability characteristics. The aircraft must be able to win the fight as well as serve as a stable weapons and sensor platform. It must have the speed and range to attack targets well inside enemy territory and be able to loiter on station as well. Such aircraft need speed to engage threats at a distance and to escape when necessary. They also need to carry a wide mix of weapons and sensors all while incorporating sufficient levels of cross spectrum observables technology to allow them to survive against advanced and projected threats. In short, the ideal STOVL aircraft needs to be able to do everything its

CTOL counterpart can, and also allow force commanders to say, "Oh, and by the way, it lands vertically."

CS2.5.7 Affordability

Affordability is extremely critical in the development of STOVL systems because they usually:

1. Require new technologies that drive up development cost.
2. Contain significant components, subsystems, and features that are not common with other aircraft increasing initial fly-away and life cycle cost (LCC).
3. Increase deployment cost due to the need for unique logistics elements.
4. Require additional training and certification of both flight and ground personnel.
5. Impact existing infrastructure requiring modification of already fielded systems and/or development of new operating procedures.

To overcome these affordability concerns, a STOVL aircraft must offer compelling new capabilities. Ideally, a new STOVL aircraft would maximize usage of existing technologies; incorporate common or only modified versions of existing components, systems, and software; be straightforward to operate and maintain; and utilize existing infrastructure, logistics, and procedures.

CS2.5.8 Scoring the V/STOL Concepts

At this point we knew what we wanted our STOVL propulsion system to do, but it wasn't even a concept yet. We had little more than this list of desirable traits on a whiteboard and a rationale as to why they were important.

At the heart of any good aircraft is a balanced design. A V/STOL aircraft is, by necessity, one of the most integrated systems devised by man. A key point is that it can't tolerate much compromise, but it must be very carefully balanced. As history has taught, designs compromised to provide V/STOL capabilities were doomed from the start. To be successful, a V/STOL design must be balanced. While this may seem like semantics, it is indicative of a general mindset and approach to the task of synthesizing an airplane. Considerable engineering design and engine cycle analysis went on at the conceptual and preliminary levels to sort through a myriad of potential options [2].

We found the simple process of concept scoring to be surprisingly useful. In this process, a list of various potential options is generated and scored against a list of fundamental characteristics. While the numerical scores provide some insight, the real value is in the arguments that take place. For this to be

> "We sure spend a lot of money on affordability."
> —*Jim Eshleman*

	Thrust Augmentation during STOVL Operation	Compact and efficient transfer of energy forward of the c.g.	STO overload capability	Benign ground environment	Controllability in STOVL flight	Uncompromised up & away performance	Affordability	TOTAL SCORE	Example Platform
Lift + Lift /Cruise	5	4	3	2	3	5	1	23	Yak-38,
RALS	5	3	3	1	3	3	2	20	Hummingbird (XV-4B)
PCB	5	3	4	1	3	2	2	20	
FIW	4	3	3	5	2	1	3	21	XV-5A
DL/VT	2	3	4	2	3	2	4	20	Harrier Family, AV-8A/B
Fuselage Ejector	3	1	2	3	2	1	4	16	Hummingbird (XV-4A)
FIF (gas coupled)	4	4	4	4	4	3	3	26	
FIF (mechanically coupled)	4	5	4	4	5	4	3	29	

Figure CS2.3   STOVL Propulsion Concept Scoring System.

effective two things are needed. First, each concept needs an advocate with a strong personality who has invested some time and effort making it work. Second, the process needs a respected moderator who can intercede when necessary. The junior folks made great advocates and we would take turns arguing one concept or another. Paul Shumpert, who we affectionately referred to as "The Grey Fox," made a great moderator. He had tremendous patience and knowledge, and was well respected by everyone.

Typically we used a scoring system of 1 to 5, with 5 being best. This made a 4 only 20% below a 5, but 1 was half as good as a 2 and represented a fundamental flaw or weakness. It is important to note these rankings are determined against our particular programs requirements. A configuration being developed for another role might arrive at a completely different conclusion. To see how the SDLF system stacked up against others, consider Fig. CS2.3 (note that the early SDLF concept was simply called the Fan-In-Fuselage or FIF).

It is worth discussing how these fundamental discriminators vary between the configurations considered in our early SSF efforts. This provides an appreciation of the range of variables considered in those early design efforts and allows us to understand why and how the SDLF propulsion system emerged as the strongest candidate coming out of that effort.

CS2.5.8.1   Lift + Lift/Cruise (L + L/C)

This configuration is one of the simplest approaches to V/STOL propulsion. A lift engine is located (typically oriented vertical or nearly so) in the forward portion of the aircraft to provide vertical thrust forward of the c.g. A vectoring nozzle is generally provided for the cruise engine(s) to allow it to vector downward during V/STOL operation. This concept has

Figure CS2.4 Lift + Lift/Cruise Aircraft in Flight, XV-4B Hummingbird (top) and YAK-38 (bottom).

been employed with some success on the YAK-38, YAK-141, and Lockheed Hummingbird (XV-4B) aircraft as seen in Fig. CS2.4.

CS2.5.8.2 Remote Augmented Lift System (RALS)

The RALS propulsion concept ducts fan bypass air to a downward oriented combustion chamber and nozzle typically located ahead of the c.g. Fuel is routed to the combustion chamber and during V/STOL operations the RALS acts much like an afterburner providing a substantial increase in thrust along with afterburner level jet exit temperatures.

CS2.5.8.3 Plenum Chamber Burning (PCB)

PCB is a propulsion concept explored by the United Kingdom on the Pegasus family of V/STOL engines in which fuel is ducted to the forward nozzles and burned in the relatively cool fan air. This approach results in high exhaust temperatures similar to the RALS. The PCB concept was explored in full-scale tests in which a PCB equipped Harrier airframe

Figure CS2.5 PCB test rig for Harrier aircraft.

was plunged towards the ground to simulate the landing dynamics as seen in Fig. CS2.5. These tests showed high levels of inlet temperature rise due to exhaust gas reingestion and resulted in some spectacular engine stalls.

CS2.5.8.4 Fan in Wing (FIW)

This concept had been successfully developed and flown in the Ryan XV-5A aircraft as seen in Fig. CS2.6. In this approach, large fans are embedded in each wing (an additional smaller fan was located in the nose of the XV-5A) and exhaust gas is ducted from cruise engines to power these fans, typically through a set of tip turbines. This concept augments vertical thrust by increasing the mass of air being used for propulsion and has relatively low exhaust temperatures and velocities.

CS2.5.8.5 Direct Lift Vectored Thrust (DL/VT)

Direct lift vectored thrust systems have been the most successful fixed wing V/STOL propulsion systems to date. This concept is best known for its usage in the Harrier family of aircraft in which thrust from a single cruise engine is vectored downward to provide direct vertical lift during hover as seen in Fig. CS2.7. This approach has been successfully employed

Figure CS2.6 XV-5A hovering at NASA Ames Research Center.

Figure CS2.7 AV-8B in hover showing the thrust vectoring nozzles directed nearly vertically downward.

by the AV-8A and B and would become the approach selected by Boeing for the X-32. The Hybrid Fan Vectored Thrust (HFVT) concept was a variant of this approach employed on a promising study configuration during the U.S./U.K. program.

CS2.5.8.6 Fuselage Ejector (FE)

The FE concept is one in which engine air is ducted to a series of ejector nozzles that direct flow downward. The entrainment of this flow acts to pull in ambient air and increase the overall thrust of the system while also reducing exhaust temperatures. This concept was employed with limited success on the Lockheed Hummingbird (XV-4A). Other aircraft configurations using this concept were the Rockwell XV-12A and the General Dynamics E-7 concept. A precursor to the E-7 undergoing full scale testing the NASA ARC 40 × 80 foot wind tunnel is seen in Fig. CS2.8.

CS2.5.8.7 Fan in Fuselage (FIF)

For the purposes of this discussion, FIF is broken into two approaches in Fig. CS2.3 to allow the fan coupling approach to be evaluated. Gas coupled fans would duct flow (either fan air, compressor bleed air, or core flow) forward to power either a single or series of tip driven lift fans located

Figure CS2.8 Early fuselage ejector concept in the Ames 40 × 80 wind tunnel.

forward of the c.g. in the fuselage. This concept looks similar to the L+L/C concept with the lift engine replaced by the lift fan and energy being provided by flow from the cruise engine rather than fuel burned separately in the lift engine.

The mechanically coupled FIF concept is similar to the gas driven concept except that motive force for the lift fan is provided by some mechanical means from the cruise engine. This is the concept that ultimately resulted in the Shaft Driven Lift Fan (SDLF) propulsion system and can trace functional similarity from earlier Tandem Fan V/STOL and SST concepts. While not widely known, years earlier GE had successfully modified an engine to drive a shaft which provided useable torque.

CS2.5.9 Thrust Balance Considerations

Consider the free-body diagram of a generic hovering vehicle (Fig. CS2.9) with lift posts located fore and aft of the c.g. at XF and XA respectively. For the vehicle to operate in a stable, steady-state hover, the simple equation:

$$Xc.g. = (F_A * X_A + F_F * X_F)/(F_A + F_F) \qquad (1)$$

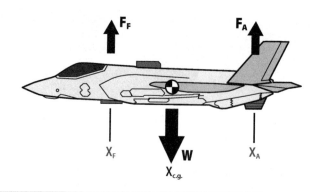

Figure CS2.9 Simple hovering vehicle with fore and aft lift posts.

describes the relationship between the fore and aft jets and their locations relative to the c.g.

This simple hover balance equation, along with some basic assumptions based on the propulsion systems ability to shift thrust fore and aft, are used to generate the vertical thrust curves seen in Fig. CS2.10. The values shown in this figure, while representative of an X-35 class of vehicle, are purely generic. It should be noted that this description works equally well when considering the lateral axis with thrust transfer between left and right sides of the airplane for roll control.

For an ideal hovering propulsion system, thrust could be moved freely between the fore and aft lift posts without loss as required to compensate for c.g. location or provide pitching moment for longitudinal control. The resulting hover thrust relationship results in the ideal curve of Fig. CS2.10, in which total vertical thrust is constant across the c.g. range.

Next, consider a lift engine system such as the L+L/C concept. With this system no thrust can be transferred between the two lift posts and the only ability to compensate for longitudinal c.g. location or generate pitching moment is to throttle one of the engines down from its maximum thrust level. The hover thrust relationship for these concepts is seen in Fig. CS2.10 as the curve labelled "thrust spoiling systems." For this system, maximum vertical thrust is generated at only one c.g. location in which both the lift and lift/cruise engines are operating at maximum thrust. In the example system illustrated in Fig. CS2.10, this occurs when the c.g. is located at FS 354 and a total vertical thrust of 36,000 lb is generated. As the c.g. moves forward, the lift engine thrust is maintained at maximum and the lift/cruise engine thrust is reduced and the total vertical thrust capability of the system is reduced accordingly.

The SDLF system allows thrust to be effectively transferred from the rear lift post to the lift fan through the drive shaft and behaves nearly like the ideal system. Thrust transfer is accomplished by modulating the aft nozzle exit area and consequently the amount of work taken out at the low pressure turbine. This work, in the form of shaft horsepower, is transferred to the lift-fan and generates thrust at the forward lift post. The process can be accomplished at high rates because the rotational speed of the system stays constant as the amount of the thrust transferred changes. Aft nozzle area can be modulated at high rates and the changing work extracted from the low-pressure turbine results in torque changes through the drive train. Lift-fan thrust is modulated by inlet guide vanes moving in concert with aft nozzle area. This allows thrust to rapidly shift between the fore and aft post with minimal loss. In fact, as more thrust is transferred forward, lift-fan airflow increases and the propulsive efficiency of the system increases slightly until the airflow limit of the lift-fan is reached. This characteristic is seen on the curve labeled "SDLF" in Fig. CS2.10. The SDLF system is located within the vehicle such that its operation encompasses the necessary c.g. range with margin for additional pitch control without approaching the airflow limits of the lift-fan. The detailed internal workings of the SDLF system have been described by Paul Bevilaqua in a 2009 Wright Brothers Lecture [2].

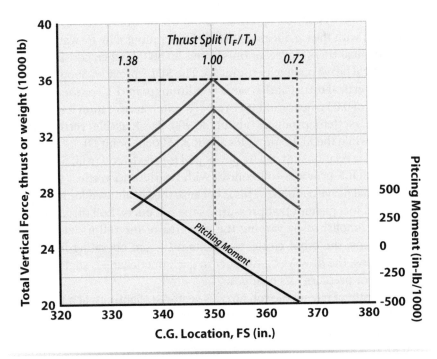

Figure CS2.10   Typical hover thrust balance (front/aft).

Lockheed Martin along with Pratt and Whitney performed critical risk reduction testing and large-scale technology demonstrations in the early 1990s using a highly modified F100-220/229 engine to prove the claims made previously and demonstrate the operating characteristics of the SDLF propulsion system. These results are discussed in Sec. CS2.5.10.

Any V/STOL propulsion system that does not efficiently allow thrust to be transferred between lift posts will suffer usable hover thrust degradation similar to the L+L/C case used in this example. The variation will only be in degree. Systems such as RALS, PCB, and DL/TV suffer significant losses in usable hover thrust as well, just not to the extent of the L+L/C concept.

CS2.5.10 Control Margin and Usable Hover Thrust

A fundamental advantage of the SDLF as a V/STOL propulsion system is its natural ability to provide all axis hover control without additional reaction control systems (RCS). As a general guideline, a hovering vehicle needs adequate thrust margin in hover to allow it to accelerate upward at about 0.1 g out of ground-effect (OGE), and also provide a combined maneuver control margin of about 0.12 rad/sec^2 pitch acceleration while maintaining 0.05 g vertical acceleration [4]. The required pitching moment (PM) to generate this acceleration can easily be determined and that relationship is also shown by the lower PM curve on Fig. CS2.10. While the sample L+L/C system can produce about 35,000 lb of vertical thrust for operations with the c.g. located at FS 354, operations 4 in. forward result in usable vertical thrust levels of only about 30,600 lb when considering the 0.1 g vertical heave maneuver (point A). The SDLF provides about 32,000 lb of usable vertical thrust at this same condition (point B). For the combined maneuver of 0.05 g heave and 0.12 rad/sec^2 pitch acceleration it is seen that the L+L/C system produces about 30,900 lb of usable vertical thrust (point C) while the SDLF provides about 33,500 lb (point D).

A similar relationship can be developed for the lateral (roll) axis. In this case, the SDLF provides roll control with nearly zero vertical thrust loss and minimal heave coupling. This is because the fan air used for roll control is gathered in a plenum that pressurizes the roll ducts. Roll thrust modulation is accomplished by varying the exit area of the roll nozzles. As one nozzle opens, the other closes to maintain the fan operating line. When this happens, the flow Mach number in each duct changes slightly, resulting in minor pressure loss changes.

As a result of having the roll nozzles vectored slightly aft at a fixed angle to improve STO and transition performance they produce a slight roll-yaw coupling. This was a compromise made to simplify the demonstrator system. After considering vectoring roll nozzles on the F-35B, the production program elected to fix the roll nozzles as well.

An additional advantage of the SDLF system is the large amount of relatively cool air from the forward lift fan. This effectively shields the inlets from the hot exhaust gases of the aft jet during hover IGE operations as seen in Fig. CS2.2b. As energy is extracted from the core flow to power the lift fan, temperature and velocity of the aft exhaust jet is reduced. Fan bypass air was diverted to the roll jets allowing the core flow to expand to the full nozzle area and increasing the amount of power that can be extracted by the low pressure spool to drive the lift fan. Slowly but surely, details of the SDLF system were worked out and it ultimately became a compact and efficient V/STOL propulsion system that addressed a number of serious short comings associated with previous V/STOL concepts.

CS2.5.11 Relative Ranking of Each System

In the early stages of the program these concepts were considered from the standpoint of first order physical principles and not tied to a specific aircraft configuration beyond what the propulsion concept would fundamentally dictate. This allowed the pros and cons of each to be evaluated on the basis of its own fundamental merits and ranked relative to the others. While the scoring certainly carries some level of subjectivity, the mechanically coupled fan-in-fuselage approach emerges as the most promising concept with the gas coupled system just behind it (Fig. CS2.3). With that, a reasonable point spread accuracy of this ranking is probably about +/- 3 points. It is insightful to discuss the reasons for scoring certain elements either a 1 (deal killer) or 5 (near optimum).

It is critical to understand that these scores are derived against high level requirements and concept-of-operations (CONOPs) for the SSF. An aircraft being designed for a different role and against a different set of operational requirements would be scored differently with perhaps a different concept emerging as the most suitable.

CS2.5.11.1 Thrust Augmentation during STOVL Operation

Considering the ability of each of these approaches to provide thrust augmentation during STOVL operation (Fig. CS2.3, column A) the L+L/C, RALS, and PCB configurations (rows 1–3) all rate the high score of 5 because each provides significantly increased levels of STOVL thrust.

The FIW system receives a score of 4 in this regard because it provides high levels of thrust augmentation by pulling energy away from the cruise engine (similar to the FIF concepts). However, because the FIW concepts are packaged in thin areas of the airplane they will likely be limited in what can be done within geometric constraints.

The DL/VT approach is given a score of 2 in this regard because it does not provide STOVL mode thrust augmentation above what the engine is basically sized for in up-and-away flight. This results in either low levels of STOVL thrust or a cruise engine that is oversized for conventional operation. While a score of 2 may not be a deal killer by itself, it points to a fundamental weakness of the concept that must be compensated for in other ways.

Fuselage ejectors appear appealing from the standpoint of STOVL thrust augmentation. However, in practice they are sensitive to installation factors that reduce the thrust augmentation realized in full scale applications. Both the Lockheed Hummingbird (XV-4A) and Rockwell XFV-12A suffered from these effects and as a result, the ejector systems given a lower relative score of 3.

The gas coupled FIF concept provides significant thrust augmentation due the larger mass of total air being used for vertical lift. All the fan concepts effectively increase the bypass ratio of the cruise engine resulting in higher levels of basic propulsive efficiency. The gas coupled fan suffers from duct pressure losses and limits on the amount of power/flow that can be extracted without adversely affecting the operation of the cruise engine. For these reasons the concept is given a relative score of 4 in this area.

The mechanically coupled FIF concept also provides STOVL thrust augmentation levels similar to the gas coupled system, however it does not suffer from the duct losses and flow concerns. The simple drive shaft is one of the most efficient means to transfer mechanical energy. As a result the mechanically coupled FIF concept could arguably be scored above its gas coupled counterparts; however, within the levels of uncertainty here it is given a relative score of 4 as well.

CS2.5.11.2 Compact and Efficient Transfer of Energy Forward of the c.g.

As discussed earlier, the ability of any V/STOL system to effectively move energy (or thrust) as required is critical to its overall performance and operation. The L+L/C configuration transfers energy forward of the c.g. in the form of a fuel line which on the surface may seem very compact and efficient. However the L+L/C system actually operates on the principle of "thrust spoiling" (Fig. CS2.10) and only produces its highest level of thrust at only one c.g. location. For any operation requiring thrust split variation from nominal, the thrust level of the system must be decreased to maintain longitudinal (pitch) balance. As a propulsion system, L+L/C does not transfer thrust forward of the c.g. but it is a compact and relatively nonintrusive element to incorporate into the airplane. It is generously given a relative score of 4 in this category.

The RALS system also generates high levels of thrust forward of the c.g., however it requires ducting to bring the fan bypass air forward. The profile of these ducts is external to the existing engine and inlet ducts. This results in volume growth at or near the aircrafts maximum cross sectional area.

This system also acts on the thrust spoiling principle similar to the L+L/C configuration. As a result of the increased cross sectional area resulting from ducting, the RALS system is given a relative score of 3.

PCB is given the same score as the RALS because this system suffers from the same thrust spoiling limitations and operates on similar principles. The location of the forward posts is fixed very near the fan as seen in Fig. CS2.11. Even though PCB does not require internal ducting, the geometry constraints and relatively large size of the burning nozzles generates similar increases in maximum cross sectional area. The geometric inflexibility makes the concept increasingly sensitive to longitudinal c.g. movement and likely requires an additional pitch control mechanism such as reaction control system (RCS) used on the Harrier family to be incorporated. As a result, the PCB system is given a score of 3 in this ranking.

FIW is also given a relative score of 3 in this category because this system typically does not transfer energy forward of the c.g. By definition, FIW transfers thrust laterally out the wings and as a consequence, the fans remain near the c.g. This provides no means to affect pitch control unless another system is incorporated such as a forward fuselage fan similar to the XV-5A. While differential thrust of the wing fans can provide good roll control, this axis is generally much less demanding than pitch.

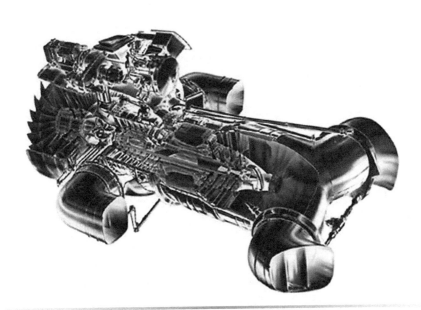

Figure CS2.11   Simplified cut-away of Pegasus Plenum Chamber Burning (PCB) Engine.

DL/VT is also given a relative score of 3 in this area for reasons similar to PCB. The forward lift posts can effectively be moved forward by shifting the engine forward while lengthening the aft ducting to maintain positioning of the rear post, this results in a bulky and volumetrically inefficient propulsion system installation that also results in reduced inlet length.

The fuselage ejector concept (all ejector concepts for that matter) is an extremely inefficient use of critical volume. Ejectors require large amounts of open volume as particular care must be exercised in forming their interior geometry and inlets. The XV-4A is a good example of this as nearly the entire fuselage volume of this subsonic aircraft was taken up by propulsion and ejector systems. Combine this with concern over installed performance and this concept was given a relative score of 1, indicative of a fundamental weakness of the system.

The gas coupled FIF concept provides true transfer of energy between the front and rear lift posts and provides an expanded STOVL operating envelope over many of the other concepts under consideration. Drawbacks of the system in this area are few and result mainly from the volume and pressure losses associated with ducting to the lift fan. The volume required for transfer ducting is likely acceptable on subsonic configurations, however it can be problematic for aircraft with supersonic requirements. For this reason it is given a relative score of 4.

The mechanically coupled FIF concept has the same advantages of the gas coupled system. Additionally, the mechanically coupled system requires less volume to transfer energy forward and has minimal transmission losses. Installation of a mechanically coupled lift fan are similar to those of the lift engine, however the concept is only packaging a fan system within the volume, not an engine complete with combustor and turbine. However, it does suffer from geometric constraints imposed by the level of complexity willing to be considered in its gear box. The mechanically coupled FIF system is a near ideal system in that there are minimal losses associated with transfer of energy between forward and aft lift posts. Consequently, this system is given the high score of 5 in this area.

CS2.5.11.3   STO Overload Capability

STO overload capability is dependent upon thrust and the ability to vector. Thrust available determines acceleration and can be used to contribute direct lift during takeoff and once airborne. This allows the aircraft to achieve flight at speeds lower than would otherwise be possible. The ability to vector allows thrust to be used as required through the STO process. Since thrust augmentation in STOVL operation is already a ranking category, STO overload capability is primarily determined by a configurations ability to use that thrust effectively through vectoring.

For thrust vectoring to reach its full potential, it must be rapid, trimmable, and usable across a wide range of vector angles (including forward for airborne braking stops). Most of the configurations considered in Fig. CS2.3 were scored with an average value of 3 in this category because of their inability to vector over a wide range (RALS, L+L/C, and FIW). The Fuselage Ejector concept is given a below average value of 2 because it typically presents very large ram drag components due to the amount of flow being turned by the system. Ram drag on the fuselage ejector concept comes right at the point in transition when thrust performance is needed most.

The remaining configurations are all given above average values in this category because they are able to effectively vector their thrust across a wide range of conditions. None are ideal because each has slight limitations in either their ability to vector completely aft or to use thrust split (Fig. CS2.10) as a means of pitch control.

CS2.5.11.4 Ground Environment

Factors affecting the ground environment of STOVL aircraft are primarily temperature, jet velocity, and acoustics. The afterburner level jet exit temperatures of the RALS and PCB concepts drive their scores to 1 in this area. This represents a serious if not fatal drawback to their basing flexibility. The L+L/C and DL/VT concepts are not much better and receive scores of 2 because they present full engine core temperature to the landing surface at critical low heights. On the other hand, FIW presents a very benign ground footprint with low temperatures, jet velocities, and acoustics. In fact, at the end of its flight test program, the Ryan XV-5A underwent a series of tests to determine its utility as a search and rescue vehicle. The FIW concept is a near ideal STOVL system in this regard and receives a score of 5.

One might think fuselage ejector systems should have a ground environment similar to the FIW; however the worse single jet determines ground environment. Because of layout considerations most fuselage ejector concepts use fan air to power the ejectors leaving unmixed hot core flow to be vectored down at the back of the airplane. This provides an impingement post as severe as L+LC or DL/VT. Core flow can be used to power the ejectors; however it requires ducting hot exhaust flow forward through the aircraft.

Both the gas and mechanically coupled FIF concept receive above average scores of 4 in ground environment because energy is taken out of the core streams to power the fan systems. This significantly reduces the exit temperatures, velocities, and corresponding acoustical energy of the aft nozzle system. Except for FIW, none of the other systems actively remove energy from the worst case lift post to reduce its adverse effects on the landing surface and surrounding environment.

CS2.5.11.5 Controllability in STOVL Flight

Controllability in STOVL flight is critical to the success of any concept. Numerous V/STOL concepts have either failed or had severe operating restrictions based on limitations of their control systems. In this context "control" means the ability to generate desired (either upsetting or restoring) forces and moments with adequate rate and without unduly compromising performance.

Four configurations (L+L/C, RALS, PCB, and DL/VT) are given average scores of 3 because they are able to provide controlling forces and moments either through nozzle vectoring, thrust spoiling, or with implementation of an additional reaction control system (RCS) similar to the Harrier.

On the other hand the FIW and fuselage ejector systems receive below average scores of 2 because of the adverse effects of ram drag induced moments. Both vehicles experience increasing nose up pitching moment as they accelerate through transition. Implementation of these STOVL lift systems typically results in serious degradation of the vehicle basic aerodynamic performance as well. Cross winds will also induce rolling moment upsets. Both concepts require additional systems to be incorporated into the design to provide pitch and roll control.

Both FIF configurations receive high scores in this area for the reasons discussed earlier in Sec. CS2.5.8.2. The gas coupled system receives a score of 4 because this system has greater lag characteristics than the mechanically coupled system. From a control standpoint, the mechanically coupled FIF or SDLF concept is a near ideal system from this standpoint and receives a high score of 5. Forward and aft lift posts can be modulated to provide nearly uncoupled pitch, roll control is provided by differential thrust from the roll jets. Yaw control (typically the least critical axis) is provided by side-to-side vectoring of the aft nozzle which couples yaw with side-force slightly. These control moments are provided quickly and with minimal impact on performance.

CS2.5.11.6 Uncompromised Up and Away Performance

In order to fight and win, any STOVL jet must be able to compete and excel in the air. Its effectiveness as a weapon system cannot be adversely compromised to make it STOVL. Simply put, it has to be a good jet.

Here the L+L/C configuration stands out against the rest. This configuration represents the least impact of STOVL systems on what is otherwise a CTOL airframe. Include the volume for the lift engine and nozzle, install an upper surface inlet, put a vectoring nozzle on the cruise engine, and add a reaction control system. With that, a CTOL airplane is converted to a

STOVL jet with the major configuration requirement being that it has a bifurcated inlet. At a given weight, the STOVL version can do everything its CTOL counterpart can save for the loss in fuel capacity. For these reasons the L+LC configuration is given the high score of 5 in this area.

The FIW configuration does not fare as well in this area. To install fans in the wing, either the fan system must conform to the geometric constraints of the wing or the wing geometry (including airfoil shape) must be altered to accommodate the fans. Mechanisms are required to provide inlets and vectoring of the fan flow during STOVL operation and closing/sealing up the wing surfaces when not. Add to this the reduction in volume for wing spars and efficient load paths and the result is a complex and highly compromised wing.

The fuselage ejector configuration suffers from similar constraints to the FIW. Ejectors must either be packaged within the fuselage or formed by deploying panels to form an ejector box. In either case, critical fuselage volume near the c.g. is consumed for the ejector right where modern LO configurations would opt to locate internal weapons bays. Both the FIW and fuselage ejector configuration become limited in space near the c.g. for external carriage. Since these constraints are so severe and fundamental to the configurations, both are given low scores of 1.

The PCB and DL/VT concepts are given below average scores of 2 because of the volume required to package their propulsion systems near the point of maximum cross sectional area. This degrades transonic acceleration and maximum supersonic speeds for both configurations.

The RALS and gas-coupled FIF systems suffer from similar volume constrains, only less severely. Both configurations require ducting of engine fan (or perhaps core) air forward to either burn in a RALS device or power the gas-coupled lift fan. As a result, these configurations are given an average score of 3.

The RALS and gas-coupled FIF systems suffer from similar volume constrains, only less severely. Both configurations require ducting of engine fan (or perhaps core) air forward to either burn in a RALS device or power the gas-coupled lift fan.

The mechanically coupled FIF configuration provides energy transfer forward through a drive shaft contained within the volume of the inlet and does not significantly increase the maximum cross section of the aircraft.

CS2.5.11.7 Affordability

From an affordability standpoint the L+L/C concept is scored lowest because no modern lift engine exists within western inventories. Consequently a complete engine development program would have to be undertaken. While the RALS and PCB configurations do not require an engine development program, they will likely require significant infrastructure investment to accommodate the severity of their ground environments. The DL/VT configuration is likely the most affordable since the concept could be based on an evolution of Harrier systems to a larger core engine (as it turned out, this was harder than first thought for the thrust levels and corresponding core temperatures required). The fuselage ejector system would also be relatively affordable to develop and field as it had only a few high temperature moving parts and could make use of existing engines. Each of the fan concepts are scored average on affordability because while they do not require development of an engine or significant infrastructure investment, they do require development of the fan system and associated power transfer devices.

CS2.5.11.8   Overall Ranking

This process was doubly important at the time as it allowed us to evaluate our competitor as well. As we came to understand the technical requirements and performance capabilities of a mechanically driven lift fan, it became apparent it was the best option.

Our competitors at the time were working their concepts, and Lockheed needed to ensure the SDLF concept developing from this would remain with us. Up to this point our propulsion studies had been conducted on company money, which made it easier to apply for an exclusive patent on the concept. While it took several years; on May 11th, 1993, Lockheed Corporation, and in particular, Paul Bevilaqua and Paul Shumpert were awarded U.S. Patent No. 5,209,428 for the "Propulsion System for a Short Takeoff and Landing Aircraft."

Meanwhile, our closest competitors were also working on their concepts (a gas coupled lift fan at McDonnell Douglas and a fan-in-wing concept at General Dynamics), which were unknown to us at the time. Northrop was exploring L+L/C and Boeing had not yet entered the field.

A wave of aerospace consolidation was also sweeping the industry. Lockheed became Lockheed Martin after merging with Martin Marietta, Northrop became Northrop Grumman after acquiring the Grumman Aircraft Company, and Rockwell was bought by Boeing.

CS2.6   The Dust Cover

We knew we had a superior STOVL propulsion concept. Now we needed an airplane to put it in. During this period the program's funding was small and our ability to conduct developmental testing was limited to a single crude low speed wind tunnel model and a very simple small scale (5%) cold flow hover model shown in Fig. CS2.12.

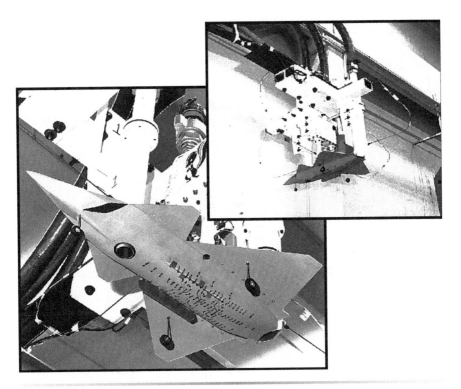

Figure CS2.12 Early STOVL Strike Fighter 5% scale hover model.

CS2.6.1 Planform Considerations

Incorporating low observability (LO) features into any aircraft configuration requires carefully balancing aerodynamic performance against radar cross section (RCS) and survivability. As Carichner and Nicolai [1] discuss in Sec. 12.4.2, the leading edge (LE) and trailing edge (TE) sweep angles define the major characteristics of the configurations RCS pattern. Since it is typically desired to keep the forward sector as clean as possible minimum sweep angle criteria are generated. Radar return spikes from the LE and TE tend to merge and form a broad lobe if the sweep angles are near the same value. If the LE and TE sweep angles are identical, the spikes overlay each other and result in fewer total spikes in the overall RCS pattern. This approach was taken with the B-2 and YF-23 planforms. If the LE and TE angles cannot be aligned, then they must be sufficiently far apart so as to produce two thin spikes and not a single broad lobe as was done with the YF-22 and F-117A (Figs. 12.12 and 12.13).

By considering the general wing planform geometry shown in Fig. CS2.13 the relationship between LE and TE sweep angles, aspect ratio (AR), and

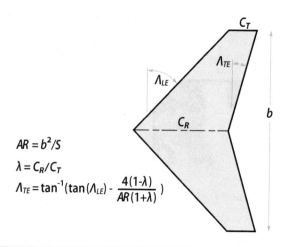

$$AR = b^2/S$$

$$\lambda = C_R/C_T$$

$$\Lambda_{TE} = \tan^{-1}(\tan(\Lambda_{LE}) - \frac{4(1-\lambda)}{AR(1+\lambda)})$$

Figure CS2.13   General wing planform geometric relationships.

taper ratio (λ) can be derived. That relationship is presented graphically in Fig. CS2.14 for $\lambda = 0.2$.

By establishing LE and TE sweep criteria, the available design space becomes apparent. A general set of those constraints could be:

1. LE sweep angle to be greater than 30 deg.
2. TE sweep angle to be greater than 15 deg.
3. Difference between LE and TE sweep angles to be more than 15 deg.

With those simple constraints applied to the design space of Fig. CS2.14 the available planform geometries become apparent. For the STOVL Strike Fighter (SSF) a trapezoidal wing did not offer suitable volume for fuel, structure, or the necessary roll posts, especially after volume is subtracted for leading and trailing edge control surfaces. They also present integration issues with LO configurations that benefit from a defined chine line and shallow body side angles. Trapezoidal wings in the available design space are also subject to pitch up at high angles-of-attack which would increase the rate and control power requirements in pitch. Consequently the SSF planform became the diamond delta that offered the highest aspect ratio within acceptable geometric parameters.

The early SSF centered around a canard configuration that placed the wing well aft on the body. This provided more room for integration of the lift fan as the wing carry through structure was well aft of the lift fan cut out. This also allowed the c.g. to be further aft reducing the amount of forward thrust and consequently power extraction required. The volume

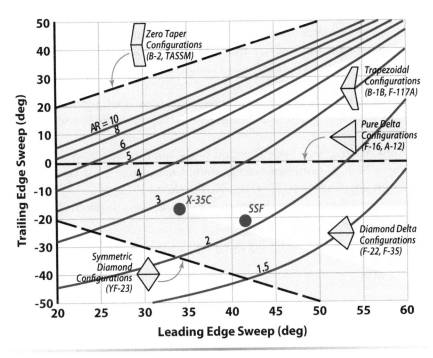

Figure CS2.14 Geometric planform constraints.

of the canard was used to fill in the area distribution between the lift fan bulge and the wing. It could be argued that a properly configured canard would provide higher CLmax and consequently better instantaneous turn rate. The longitudinal characteristics of the design were closing nicely.

CS2.6.2 Competitive "What Ifs"—In Flight Thrust Vectoring

A potential drawback of the SDLF that we wrestled with early on was its inability to employ thrust vectoring in up-and-away flight like the Harrier. To vector thrust for direct-lift (vectoring thrust down to augment the lift vector) as the Harrier could, the SDLF would have to go through the fan engagement process at high speeds and that would have been a deal killer. To employ thrust vectoring (TV) of the aft nozzle only (for pitch augmentation) would have meant additional complexity on an already complex, risky, expensive, and heavy nozzle. Perhaps not a deal killer—but close.

The reality was that in-flight thrust vectoring was something the Harriers had available to them, so combat pilots learned how and when to use it in a fight. As it turns out, those applications were very limited. We were able to leverage our YF-22 experience with pitch thrust vectoring to show that this capability was really not that useful for a single-engine light weight fighter going against advanced (third- and fourth-generation) fighters. The simpler the argument, the more effective it was at getting this point across and Clive Whitmore from the YF-22 helped considerably in this area.

From a direct lift standpoint, thrust vectoring in up-and-away flight has always been a loser, despite claims from some Washington D.C. think tanks. Fortunately, the technical argument was pretty simple and went like this:

Consider the Free-Body Diagram of the aircraft shown in Fig. CS2.15 pulling a high g, sustained (steady state) turn using thrust vectoring for direct lift augmentation. First, resolve the thrust (T) into stability (L & D) axis and sum the forces in the lift and drag directions respectively:

For lift

$$L + T \sin(\alpha + \delta) = n\,W \qquad\qquad (CS2.1)$$

For drag

$$D - T \cos(\alpha + \delta) = 0 \qquad\qquad (CS2.2)$$

Realizing that $D = L/(L/D)$ and using a little algebra, the following relationship is obtained:

$$n = (T/W)[(L/D) \cos(\alpha + \delta) + \sin(\alpha + \delta)] \qquad\qquad (CS2.3)$$

Note that throughout this development, nozzle deflection (δ) only appears linearly combined with angle-of-attack (α) in the term ($\alpha + \delta$). For the purpose of sustained load factor, thrust vectoring serves the same purpose as angle-of-attack, to incline the thrust vector relative to the flight path. With the goal of maximizing sustained load factor, the function in (3) is maximized at values of ($\alpha + \delta$) given by

$$\tan(\alpha + \delta) = 1/(L/D) \qquad\qquad (CS2.4)$$

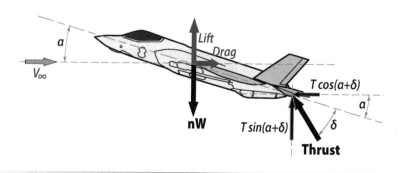

Figure CS2.15 4-force free body diagram of an airplane in a steady state turn.

Consider a sample aircraft with the longitudinal aerodynamic characteristics given in Fig. CS2.16 (L/D_{max} of 7.19 at 5.9-deg angle-of-attack). The optimum ($\alpha + \delta$) is 7.92 deg and the corresponding sustained load factor $n_{z_{opt}}$ is 7.259 g. This is a thrust vector angle of only +2.02 deg (down). If the airplane had no thrust vectoring capability ($\delta = 0$), n_z would be

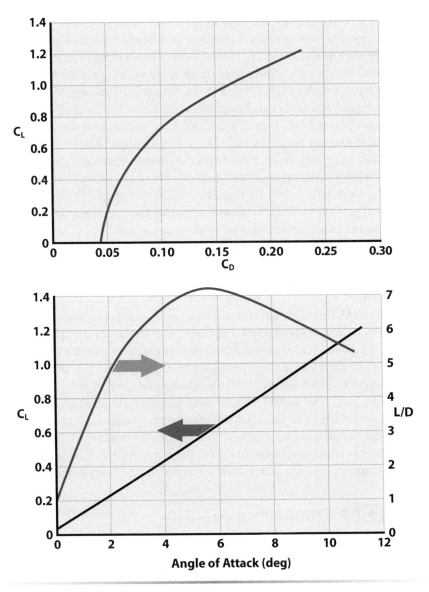

Figure CS2.16   Longitudinal aerodynamic characteristics for a sample fighter airplane.

7.255 g, or only 0.004 g less than with optimum thrust vectoring. For an airplane with a combat weight of 36,000 lb, the increased load factor would amount to only 144 lb of force normal to the flight path. Ignoring the increased complexity, cost, reliability, and potential impact to combat readiness, if the thrust vectoring system only added 144 lb to the weight of the airplane, it would result in a net loss in performance. Not a good trade.

Instantaneous load factor presents a more complex argument because the aircraft is allowed to decelerate during the maneuver, which trades energy for a momentary increase in turn rate. In reality, the AV-8 could only exploit this momentary advantage in a limited portion of its flight envelope (below about 250 kt). Here the AV-8 had a clear P_S advantage and could recover the lost energy faster than its early competitors (third generation fighter such as the F-4 or F-14) could. A little recognized factor in these early engagements that Russ Stromberg is fond of pointing out is that 3rd generation fighters typically did not have robust anti-departure features. Both the F-4 and F-14 had spin modes that could be difficult if not impossible to recover from, while the AV-8 on the other hand could readily recover from a departure and was nearly impossible to spin. Consequently, an early AV-8 pilot could fly fearlessly below about 250 kt while others flying 3rd generation fighters became tentative and much less aggressive at these speeds. This advantage was neutralized against 4th generation aircraft such as the F-16, F-15, F/A-18, Su-27, and Mirage because they had much greater Ps capability in the low speed end of the flight envelope and advanced flight controls allowed their pilots to fly with confidence at these low speeds.

Along with this came advances in weapons technologies such as helmet-mounted sights, off-bore-sight weapons, and the general application of stealth. Emphasis had changed from close in visual air combat maneuvers (ACM) to beyond visual range (BVR) with first-detection, kill-at-a-distance capabilities. Incorporation of in-flight TV into any design presents significant risk, cost, and weight. The weight increase alone would adversely impact performance in all areas. Advancement of weapons and threat systems had progressed to a point that effectively countered the limited increase in performance brought by TV. After much time, analysis, and discussion, the proponents of thrust vectoring in up-and-away flight fell silent.

CS2.6.3 Competitive "What Ifs"—Vertical Tails

During the late '80s through the early '90s, the Northrop B-2 was being unveiled and the A-12 Avenger (General Dynamics and McDonnell Douglas) was also becoming public knowledge. Initial observability goals for the SSF required configurations without vertical tails as well, despite the fact that the SSF was to be a fighter capable of supersonic flight.

Historic trends in fighter aircraft design was to progressively larger vertical tail volumes, now the challenge was to design a modern fighter with none. Configurations capable of supersonic flight want to be long and thin (high fineness ratio) with great care given to the development of their longitudinal area distribution for wave drag. Without vertical tails, such configurations tend to be highly directionally unstable.

All manner of deployable aerodynamic devices such as yaw vanes, split ailerons, and forebody vortex flow control systems were considered. None of these proved suitable and yaw thrust vectoring was ultimately considered as the design baseline. While considerable effort was spent trying to overcome the inherent stability-and-control deficiencies, it was a simple back-of-the-envelope calculation that best illustrated the dilemma inherent with yaw thrust vectoring relative to tailless fighter designs.

Using a simple analytical model and some conservative assumptions, it can be shown that yaw thrust vectoring simply cannot provide adequate yaw control power. Since a tremendous marketing effort had been sunk into yaw thrust vectoring and tail-less fighter configurations by a number of organizations through the industry, these results were not well received.

Using a simple 1-DOF approach along with some basic assumptions a straightforward relationship can be developed to evaluate the amount of yaw control power needed to overcome a lateral gust

$$N = I_{ZZ} \ddot{\psi} \qquad \text{(CS2.5)}$$

Here, N is the yawing moment, I_{ZZ} is the moment of inertia about the yaw axis, and $\ddot{\psi}$ is the corresponding angular acceleration. Consider a gust that produces a sideslip angle β where N_U represents the upsetting moment and N_C is the controlling moment; this equation can be rewritten as either

$$N_U + N_C = I_{ZZ} \ddot{\beta} \qquad \text{(CS2.6)}$$

or

$$N_C = I_{ZZ} \ddot{\beta} - N_U = I_{ZZ} \ddot{\beta} - q S b C_{N_\beta} \ddot{\beta} \qquad \text{(CS2.7)}$$

Here q, S, and b are the usual dynamic pressure, reference area and span while C_{N_β} is the configuration's total directional stability (including inlet effects). Allowing the gust recovery response to be of the form (Fig. CS2.17)

$$\beta = \beta_0 e^{-at} \qquad \text{(CS2.8)}$$

so that

$$\ddot{\beta} = a^2 \beta_0 e^{-at} \qquad \text{(CS2.9)}$$

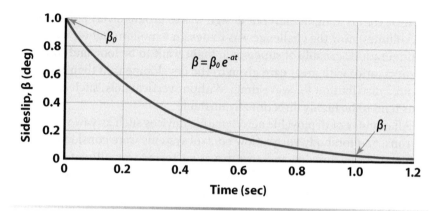

Figure CS2.17   Prescribed analytical gust recovery response.

allows Eq. (CS2.7) to be rewritten as

$$N_C = \beta_0 e^{-at} \left(I_{ZZ} a^2 - q S b C_{N_\beta} \right) \qquad\qquad \text{(CS2.10)}$$

or

$$C_{N_C} = \beta_0 e^{-at} \left(I_{ZZ} a^2 / q S b - C_{N_\beta} \right) \qquad\qquad \text{(CS2.11)}$$

and

$$N_{C_0} = \beta_0 \left(I_{ZZ} a^2 - q S b C_{N_\beta} \right) \qquad\qquad \text{(CS2.12)}$$

$$\beta_1 / \beta = \beta_0 e^{-at} / \beta_0 = e^{-at} \qquad\qquad \text{(CS2.13)}$$

and

$$a = -\ell_n \left(\beta_1 / \beta_0 \right) \qquad\qquad \text{(CS2.14)}$$

Choosing a value for a from the table in Fig. CS2.18 prescribes how quickly the system recovers from an initial gust upset (β_0).

The relationship between gust velocity, true airspeed, and side slip is seen in Fig. CS2.19.

For a fictitious aircraft with characteristics similar to the SSF configuration of the day, the general yaw control power needed to recover from a side gust is shown in Fig. CS2.17. Those results are summarized in Fig. CS2.20

β_1/β_0	0.01	0.02	0.03	0.05	0.08
a	4.6	3.9	3.5	3.0	2.5

Figure C2.18   Time constant values for gust recovery after 1 sec.

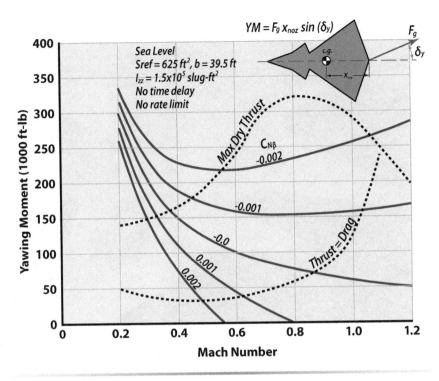

Figure C2.19   Gust induced side-slip at sea level.

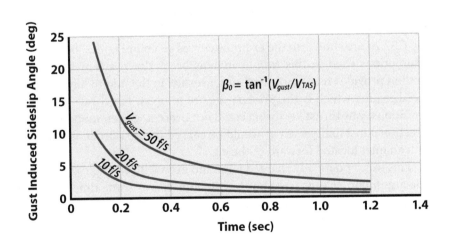

Figure CS2.20   Yaw moment required to overcome a Beta Gust with thrust vectoring.

Figure CS2.21 STOVL Strike Fighter (SSF) configuration, WT Model and artist rendering.

for various levels of thrust and directional stability and thrust for 20 deg thrust vector angle. Note that the yawing moment available curves of Fig. CS2.20 are similar to the example curves given in Fig. 3.23b. While the assumptions used in this simple analysis break down at low speeds, the method provides reasonable first-order results in the mid to high subsonic speed regime which is the primary area of interest. For thrust equal drag conditions, yaw thrust vectoring could not overcome even modest levels of directional instability such as would be caused by the ram drag associated with an inlet located forward of the c.g.

This was a politically charged issue and even with a simple and undeniable argument, change was slow to catch on. It was some time before the SSF configuration was finally drawn with even modest vertical tails as seen in the artist's rendering of Fig. CS2.21.

CS2.7 The Long Pause, SSF to ASTOVL

After completion of the DARPA funded SSF contract the program languished from a technical standpoint. The concepts backbone was the SDLF, and while this new propulsion concept looked promising, it was still a paper concept. While history could point to a few hardware programs that had performed some isolated elements of the effort such as clutches, fans, super-critical drive shafts, and power extraction, nothing approached the level of integration required by the SDLF. Promising or not, it was still high risk and expensive. For nearly two years DARPA and the services struggled with how to proceed with the effort. It was decided to insert a risk-reduction plug into the effort that became the Advanced Short Take Off—Vertical Landing (ASTOVL) program.

In the period between SSF and ASTOVL, many of us worked the A/F-X program, which was to be a conventional carrier based attack aircraft to fill the hole created by the demise of the A-12 program. The A/F-X program consisted of about five different industry teams and many of us were lucky enough to be assigned to the LM-Boeing-Grumman team. As it turned out this was a very good experience as we were able to expand our associations and learn new skills all while maintaining a close core team. The LM portion of our A/F-X team was run by Rudy Burch who would later become the LM ASTOVL program manager. Rudy was a no-nonsense guy who managed and didn't get in the way of his engineers.

Ultimately, the A/F-X program was cancelled as part of the Bottom-Up-Review (BUR) conducted by DoD in the mid 1990s. The BUR also called for the formation of the Joint-Advanced-Strike-Technology (JAST) program which would eventually become the Joint Strike Fighter (JSF). Some may note this phase of development was called the "Common Affordable Light-weight Fighter" or CALF program. The name never really stuck, primarily because calves mature into something other than fighter jets. Amidst this shuffle, DARPA established the ASTOVL program as an independent entity.

It was here that Rick Rezabek entered the program as chief engineer of the LM ASTOVL program. Rick was a rising star at the Skunk Works and would go on to become chief engineer of the Concept Demonstration Aircraft (CDA) portion of the JSF effort. When Rudy and Rick came aboard to take over the ASTOVL program, it was a case of the right guys, at the right time. Both were committed to doing the right things and letting common sense prevail. For many of us, this was the model of how to get things done, and it worked.

The request for proposal (RFP) asked for large scale demonstrations of two competing lift fan concepts. The two concepts were to be a gas driven and a mechanically driven system. Each competitor could submit two responses (one shaft driven and one gas driven) but only one contract would be awarded for each concept and no company could be awarded both. At this point Lockheed Martin (LM) had been granted patent rights to the SDLF and our response was designed to ensure that only LM would be able to pursue it. We submitted a single proposal for the SDLF concept, effectively forcing the gas driven version on our competitor.

CS2.8 The Advanced Short Takeoff and Vertical Landing (ASTVOL) Program

In March 1994 two contracts were awarded, one to LM for the SDLF ($33M) and one to McDonnell Douglas for the Gas Coupled Lift Fan (GCLF, $28M). General Dynamics (GD/FW) was out of the competition with their Fan-in Wing concept and shortly after the ASTOVL award LM bought the GD Fort Worth Division. Almost immediately we began consolidating the expertise from our former competitor.

At this point Boeing had been considered fourth in a three man race and very few were taking them seriously. However, Boeing understood the coming effort could evolve into the last manned fighter the United States would produce for some time. In a bold corporate move, Boeing offered to perform the ASTOVL work using their own money. In March of 1994 they received a DARPA contract for $6M that would be matched by Boeing to study designs based upon a DL/VT concept. The following year they received an additional $10M for the DL/VT concept which was again matched by Boeing.

All three ASTOVL contractors were required to design both demonstrator and operational aircraft, and to perform large-scale powered-model (LSPM) demonstrations to reduce risk. These test were intended to validate the propulsion concepts, to show that hot-gas ingestion would not be a problem, and to demonstrate that there was sufficient control power for transition from hover to cruise.

Our ASTOVL team was rounded out by tremendous support from the engine companies. We had selected Pratt & Whitney (PW) as the prime propulsion system integrator led by Charlie Price and John Sprague with Rolls Royce (RR) providing the aft nozzle, roll ducts, and roll nozzles, and Allison Advanced Development Company (AADC) providing the lift-fan, gear box, and drive shaft. AADC eventually had the term "Lift-Fan" trademarked, so its usage in reference to the JSF Lift-Fan^{TM} is hyphenated with the trademark symbol. Pratt & Whitney would also provide excellent support to the LSPM effort through Mike Wade and Robert Marshall of their West Palm Beach test organization.

The program was about risk reduction and technology demonstration. The competition centered on large scale demonstrations in key technical areas including the basic SDLF propulsion system, airframe integration, transition performance, control system conceptual development, and some areas of manufacturing technologies.

The centerpiece of the effort was a series of Large Scale Powered Model (LSPM) demonstrations of the SDLF system integrated into a wind tunnel model airframe. The LSPM was tested at the NASA Ames Research Centers National Full Scale Aerodynamics Complex (NFAC). This allowed the integrated SDLF propulsion system to demonstrate its operating characteristics at large scale as well as provide valuable scale effects data and detail design information to further refine the concept.

A big part of demonstrating our understanding of the combined aerodynamic and propulsive induced effects came down to establishing an integrated force and moment (F&M) accounting system. The key to this was developing the accounting system in an integrated fashion with the wind tunnel testing and analytical efforts such that all the puzzle pieces fit together. Conceptually, this is no different than any other airplane program. However, the ASTOVL concepts would have considerably more propulsion induced effects to take into account than other aircraft. Fortunately, there is a wealth of lessons learned from 50 years of V/STOL testing to draw from along with the diligent survey work of folks like Dick Kuhn and Dave Koenig [3, 5].

Our approach was to take the inlet and exhaust problems separately. Individual small scale inlet and jet effects models were tested in the early phase of the program and their results combined according to the F&M accounting system to predict the large scale results. The LSPM would be the check to validate how well that approach worked. The effort was as much a validation of the F&M accounting system as anything. The ASTOVL wind tunnel program centered about the following:

1. Small scale (12%) low speed wind tunnel model (unpowered with flow through inlets) for configuration validation and development (Fig. CS2.22).
2. Inlet compatibility model (~11%) for evaluation of the STOVL inlet system performance and determination of inlet induced F&M (Fig. CS2.23).
3. Small scale (12%) jet effects and interaction (JEI) model for evaluation of hover and transition characteristics (Fig. CS2.24).
4. Large Scale (91%) Powered Model (LSPM) to determine large scale characteristics with real jet effects (temperature, swirl, exhaust profiles, etc., Fig. CS2.25) on the complete configuration.

We engaged in a 4-month period of configuration refinement that allowed us to perform detailed configuration trade studies and incorporate the results into the configuration. This was lead by Brian Quayle and produced a set of very comprehensive configuration level trade studies that resulted in our ASTOVL baseline. This configuration, known as Configuration 140, is seen in Fig. CS2.22 during a low speed wind tunnel test in San Diego.

Figure CS2.22 Early ASTOVL low speed WT Model (12% Scale) of Configuration 140.

A key to our ASTOVL test efforts was the adoption of a model reference configuration (MRC) based on the 140 design. Once the configuration's aerodynamic characteristics were validated, we could freeze wind tunnel and CFD model lines on a single consistent MRC while the aircraft lines continued to evolve. Because ASTOVL testing concentrated on low speed characteristics, the MRC lines were tailored towards low speed testing requirements (no supersonic area ruling for example). To gain every advantage possible in getting the LSPM built on time, the MRC was lofted using flat panels and simple planar curves to allow fabrication of simple roll formed skins. Even wing contours were lofted this way. The only complex surfaces allowed were associated with the inlets and forebody. This is probably one of the only times that a set of wind tunnel models was laid out with producibility as a major concern. The challenge was hard enough, so there was no place for complexity that did not directly contribute to the quality of the end data product. The LSPM design effort was lead by Pete Taylor who would succumb to cancer before the program concluded. He and Mike

Figure CS2.23   ASTOVL inlet compatibility model.

Wade of P&W (who would survive his bout with cancer during the program) had their names placed on either side of the LSPM canopy.

One of the first tests of the program was the low speed ASTOVL inlet compatibility model (Fig. CS2.23). This was needed to validate inlet

Figure CS2.24   ASTOVL Jet Effects Interactions (JEI) model.

distortion levels and performance for the LSPM that would come later. It also provided a thorough ASTOVL inlet performance data base that would have been cost prohibitive with the LSPM. With our experience on the YF/F-22 Caret inlets, up-and-away inlet performance was not considered high risk. The ASTOVL inlet compatibility model provided data in three primary areas:

1. Inlet pressure recovery and distortion at the engine face.
2. Inlet pressure recovery and distortion at the lift-fan face.
3. Forces and moments induced by turning of the inlet flows. The data accumulated in this area during the ASTOVL program was used throughout the ASTOVL, JAST, and later JSF efforts.

While the inlet compatibility model investigated the effects of air flow feeding the engine and Lift-Fan^{TM}, the Jet-Effects-Interactions (JEI) model in Fig. CS2.24, investigated effects of the various exhaust plumes on the vehicles aerodynamics. This model was a 12% representation of the MRC with an outer metric (force measuring) shell and inner non-metric flow system that simulated the various jet streams. This approach allowed direct measurement of thrust by removing aerodynamics and induced jet effects. It also eliminated the need for complex thrust calibrations. Careful attention was paid to matching LSPM features. Surface pressure instrumentation layouts enabled direct comparison of small and large scale results. This would later prove to be critical when LSPM comparison data became available. Later in the JSF program even greater attention to modelling detail would be undertaken as will be discussed in Sec. CS2.13.

The centerpiece of the ASTOVL effort was the Large Scale Powered Model (LSPM) demonstrations. These centered around producing a working SDLF propulsion system based on the Pratt & Whitney F100-220/229 engine packaged in a large scale wind tunnel model. The LSPM would be tested both in hover and transition at the NASA ARC National Full-Scale Aerodynamics Complex (NFAC). This propulsion seystem was a unique build using the F100-220 as the basic engine but with a -229 turbine to allow greater horsepower extraction to drive the Lift-Fan^{TM}.

The LSPM started out as an 86% scale model representation of the MRC. Based on a compromise between thrust and jet exit area (geometric), this scale provided the best overall match of propulsion characteristics using the F100-220/229 based SDLF to emulate the F119-611 engine. As the ASTOVL program became part of the Joint Advanced Strike Technology (JAST) program, the configuration changed to an aft tail, and the LSPM scale was re-indexed to 91%. This evolution is discussed later.

Realizing that outdoor hover testing of the LSPM would be subject to San Francisco bay area weather, we immediately committed to an aggressive schedule that delivered the LSPM to NASA ARC in April of 1995. This allowed us to lockdown the schedule for the Outdoor Aerodynamic Research Facility (OARF) over the spring and summer months and push our competitors to the fall and winter.

The task of designing, fabricating, and checking out the LSPM from scratch, along with the various other small scale design, test, analysis, efforts necessary to support became the focus of our lives for the next 24 months. Almost immediately we set up engineering and design areas adjacent to the shop fabricating the LSPM. This allowed shop personnel to immediately ask questions and occasionally harangue the engineering staff. We generally operated with the support of Lockheed Martin management as an independent collocated team. This allowed a high level of concurrency. We were literally still designing the back of the model while the front was being assembled.

Every feature had to earn its way on. The LSPM was fabricated from water jet cut mild steel plate welded together as seen in Fig. CS2.25. No more sophisticated than building a bridge, this was an example of Kelly Johnson's KISS (Keep It Simple Stupid) principle at its best. Our competitive approach was equally simple; set a pace that no one else could keep up with; and then maintain it.

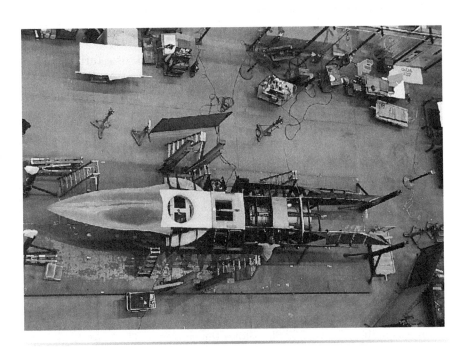

Figure CS2.25   Lockheed Martin LSPM ASTOVL model under construction in Palmdale.

Figure CS2.26 ASTOVL Large Scale Powered Model (LSPM)—Hover Test at NASA Ames.

Meanwhile the folks at NASA Ames had the task of refurbishing the OARF facility with a model support system that would meet the ASTOVL programs needs. That effort was headed up by Paul Askins and Pete Zell, with senior oversight from Rich Margason. The result was the very clean and non-intrusive system seen in Fig. CS2.26 that allowed us to test the 52,000 lb model with engine and lift-fan running.

For 11 months the LSPM was run through any number of conditions including thrust calibrations, hover in ground-effect, and transition at forward speeds up to about 100 kt in the 80 × 120 wind tunnel. We could not have met this challenge without an outstanding test support crew made up of Mark Buchholz, Mark Post, Jim Winner, Paul Siegmund, Tom Dragoo, and Bob Tems from LM, and Robert Marshall, Vijae Kapoor, Mike Wright, and Bill Paris from P&W. Again, much credit goes to the entire NASA ARC support team. Various images of the LSPM undergoing testing at the NASA ARC NFAC are seen in Figs. CS2.27a thru CS2.27d. Without their support and use of the national asset NFAC facilities, the ASTOVL program would not have been executable. Figure CS2.28a shows the Boeing X-32 LSPM on the test stand and Fig. CS2.28b shows the first flight of the X-32B demonstrator.

As the LSPM returned from P&W, "last minute" completion efforts were performed by a small team of fitters and welders led by Bob Mahafe

Figure CS2.27a   The LSPM undergoing hover testing at the NASA ARC OARF.

Figure CS2.27b   Surface flow visualization of the LSPM at near gear height.

Figure CS2.27c   Smoke flow visualization of the LSPM—Hover Test at the NASA ARC OARF.

working 12 hr shifts, 7 days per week, for about 3 weeks straight. The welders spent most of that time in welding leathers during the mid summer months of Palmdale with temperatures reaching close to 110F.

The program owes each one of these guys a huge debt of gratitude as their skills and efforts have gone largely unnoticed and unrecognized.

I am still impressed by the sight of one welder (Eloy Martinez) standing on a ladder with his head and welding helmet shoved into a compartment working the back side of a skin panel that he could not see. With a welding stinger in one hand, he looked through a mirror that he held in the other and welded up the panel.

Eloy's reward was an all expenses paid trip to NASA Ames to help us fit up a new set of horizontal tails over a single weekend as the configuration transitioned to meet new JAST requirements.

With the combined effects of a light tail wind, maximum roll thrust offset, at a near gear height condition with thrust split approaching maximum nose down (thrust decreasing on the lift fan) we did encounter an HGI induced stall during LSPM testing. While part of the run schedule, this condition was well outside of any operational envelope. The stall-dump protection system operated as it should, and automatically throttled the

Figure CS2.27d   LSPM installation (l) and ready to test (r)—NASA ARC NFAC 80 ft by 120 ft wind tunnel.

Figure CS2.28a   Boeing's large scale powered model (LSPM) for demonstrating the DL/VT concept on the Boeing ARC OARF test stand.

Figure CS2.28b   Boeing JSF X-32B demonstrator lifts off on its maiden flight (March 29, 2001). Notice the direct lift jets located at the aircraft c.g.

engine to idle. We performed a post run inspection, conferenced, and were back up and running within 45 min with no ill effects.

Unfortunately, the stall occurred while a group of JAST visitors from the LM Fort Worth Division were standing at the perimeter fence. By the end of the next run my phone was ringing with upper management wanting explanations of what we thought we were doing. I explained that it was a planned run (with the exception of the tail wind which was within established test

For me, it was an awakening as to how risk adverse the aerospace industry was becoming. Just 12 years earlier I was called into my NASA bosses office (Dave Hickey) to explain why I stopped a run at this very same facility. My explanation to him did not include the fact that at less than one year out of college, I had no intention of damaging a $4M dollar piece of test equipment in order to get one more data point (although I did dent it in a couple of places). Dave very simply explained to me that "the very purpose of this kind of testing was to find limits and boundaries, and we can't do that without crossing them once in awhile." I think both decisions were the right ones. In one case pulling back, and in the other pushing ahead.

limits), that we expected the possibility of HGI, and had specifically monitored the inlet temperatures as we progressed through the run.

Rudy Burch and Rick Rezabek earned their pay that day and endeared themselves to me as both of them supported our actions and didn't allow the test to be micro-managed from afar.

There is simply no substitute for the experiences gained from being able to break things in an environment forgiving of mistakes.

CS2.9 What We Learned

The first significant piece of validation data was the installed thrust augmentation of the SDLF system above the basic F100-220/229 engine. We spent about three months calibrating the SDLF system and measuring its thrust in the various operating conditions. When all was complete we had achieved an augmentation ratio of about 1.35, or 35% more vertical thrust than the basic installed thrust of the F100-220/229 engine.

The next critical piece of information came from quantifying the hover flowfield well enough to understand the HGI mechanisms and demonstrate that this risk was mitigated. The configuration relied on the use of the weapons bay doors to redirect fountain flow away from the aircraft. This served to increase hover performance by providing a lift cushion as well as force the hot exhaust flows away from the inlets. Without the weapons bay doors deployed, fountain flow goes vertically up the sides of the fuselage towards the main and lift-fan inlets. With the weapons bay doors opened, the fountain flow was redirected completely and no temperature rise was measured at the lift-fan. At close ground heights and high thrust splits (nose down) a slight inlet temperature rise was measured at the engine face.

Full scale inlet pressure recovery levels in hover and transition were better than predicted. This allowed us to justify higher performance levels in hover and transition, further reducing the configuration's risk in this critical area. In addition, the pressure data provided an understanding of the flow physics in the inlet system that allowed us to isolate the limited HGI coming through the main inlets. As flow entered the auxiliary inlets it pushed the main inlet flow down and separated the boundary layer flow from the upper surface of the main ducts. This formed a distinctive horizontal shear layer as shown on the engine face in Fig. CS2.29. Consequently, the areas of elevated temperatures could be positively identified as coming from the lower portion of the main inlet.

Pressure recovery levels in the auxiliary inlet flow were noticeably higher than the main inlet flow that travelled the length of the duct. A small separation bubble was also noticed on the top of the inlet duct just downstream of the auxiliary inlet and ahead of the engine face. With this data we

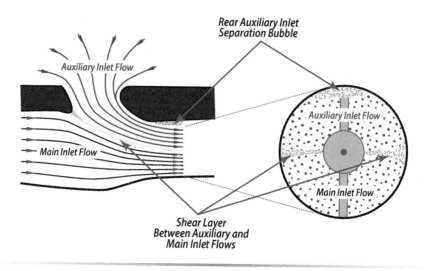

Figure CS2.29 LSPM provided high quality inlet pressure recovery data and flow details.

were able to validate our inlet CFD models and provide confidence in our understanding of the airflows through the inlet system during hover and transition operation.

Rapid thrust split control was another critical demonstration performed by P&W during the SDLF installation and checkout tests in West Palm Beach. In a simple sense, this involved rapidly changing the nozzle exit area in concert with the lift-fan inlet guide vane (IGV) angle. As the nozzle exit area increased, more power could be extracted from the core stream. As this occurred the IGVs would also open to allow the lift-fan to absorb the power by changing pressure ratio across the fan at constant RPM. A change in thrust split from 0.65 to 0.95 (near maximum nose down to near maximum nose up) was accomplished in 0.6 sec as seen in Fig. CS2.30. Because the LSPM propulsion system had a simple feed forward control scheme, an overshoot occurred in either direction that the system would slowly correct from. This was an anomaly of the model control system that would be corrected when the propulsion system was developed for flight. This was our first experimental validation of the SDLF's theoretical characteristics described earlier in Sec. CS2.5.9 and seen in Fig. CS2.10.

An interesting flow feature appeared near the LSPM roll nozzles that would require correction on the flight article. An area of lower pressure was measured on the sidewall of the fuselage just inboard of the roll nozzle as seen in Fig. CS2.31. This suction area was strong enough to pull oil streaks up the sidewalls of the LSPM near the roll jets. Venting through the small isolation gap around the roll nozzles on the smaller JEI model

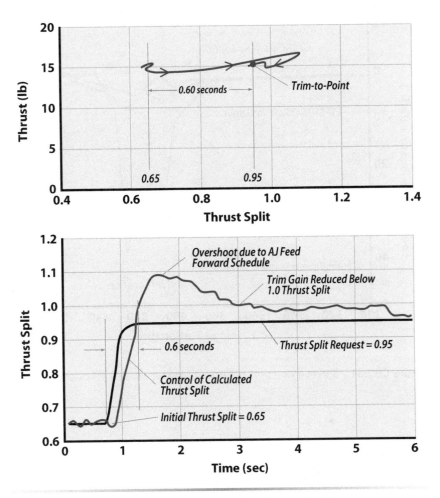

Figure CS2.30   Demonstration of rapid thrust split response.

prevented this phenomenon from forming there. This represented the largest anomaly discovered between the large and small scale models and was the result of this small modelling detail.

The roll nozzles also produced a disproportionately large lift-loss during transition due to their location relative to the wing. Low pressure downstream of the roll jets acted on nearly the entire length of the wing root causing a significant lift loss as illustrated in Fig. CS2.32. This would also be corrected on the flight article as a result of shifting the wing forward when horizontal tails were incorporated and moving the roll posts further outboard away from the fuselage sidewalls (the additional moment arm also increased the amount of rolling moment available through the roll jets).

In general, there was a scale benefit when comparing data between the LSPM and JEI models. This allowed us to take credit for a little increased

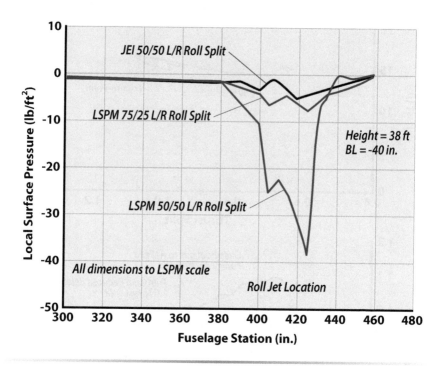

Figure CS2.31 Surface pressure differences between 12% JEI and 86% LSPM.

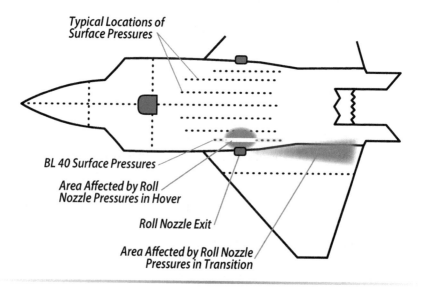

Figure CS2.32 Location of surface pressure differences between the JEI and LSPM.

performance during hover. These comparisons would also serve to point out the need for extreme care in modelling of this type in the JSF program that was to come.

In all, we put over 200 hr on the F100-220/229 based SDLF in environments ranging from hover to transition at high combined angles of attack and sideslip. During this testing we also demonstrated full scale gear meshing loads in 238 total hours of lift-fan operation (including rig testing at RR-Allison). But most important of all, we did what we said we were going to do.

CS2.10 The ASTOVL Competition

Ultimately the ASTOVL competition came down to a three way race between Lockheed-Martin with their SDLF system, McDonnell Douglas with a gas-coupled lift fan system, and Boeing with their $1 contract working a direct lift/vectored thrust concept on their own money. As time went on, DARPA provided discrete statement of work tasking and additional funds to Boeing. Lockheed-Martin had committed to a test that allowed us to lock up the prime test windows at NASA ARC that left our competitors dealing with the prospect of either 1) testing outdoors in the fall and winter, 2) slipping a year, 3) finding another test site, or 4) failing to deliver.

About halfway through the ASTOVL effort, McDonnell Douglas acknowledged they would not be able to fulfill their obligations and backed out of the competition. Boeing selected option 3 and with their $1 they had a great deal of flexibility. On land leased from an Indian tribe in Washington, Boeing built their own outdoor test site as seen in Fig. CS2.28a. Boeing developed their DL/VT concept around a simple delta wing planform. The strength of the Boeing design was its perceived simplicity. The airframe was straight forward and employed innovative manufacturing techniques Boeing could leverage from their commercial experience. The propulsion system was also thought to be low risk with the DL/VT concept being used flown on the Harrier. In reality there were several critical shortcomings to the design, including:

1. No augmentation of STOVL thrust.
2. High probability of hot gas ingestion.
3. A fundamentally short inlet duct that did not lend itself well to an LO aircraft. In reality, the configuration would have required some form of front frame device that would have negated the pressure recovery advantage of the short duct.

Despite all of this, Boeing would eventually deliver an X-32 aircraft capable of meeting the basic program objectives. This was a testament to an innovative and determined competitor.

CS2.11 The Transition to JAST, Enter the Navy, Consolidation, and a Farewell to Canards

As mentioned earlier, the Joint Advanced Strike Technology (JAST) program was started in late 1993. Initially this program had a broad focus on general strike technologies, but as time went on the focus came to be a new strike aircraft. The overlap with programs such as the CALF and ASTOVL programs became progressively more apparent and eventually these programs were rolled into a single JAST program (later to become the Joint Strike Fighter or JSF).

With JAST came a new customer, the Navy with the additional requirement for a version of the airplane capable of cat/trap operations from aircraft carriers (CVs).

Carrier operations require lower approach speeds that drive configurations towards high lift devices as angle-of-attack becomes limited by visibility and handling quality concerns. Aerodynamic balance of a canard configuration shifts the wing further aft than a conventional layout. The consequence of this is that high lift devices, such as large trailing edge flaps, produce a significant nose down pitching moment that must be trimmed by the canard. To handle this, the canard grows and requires large nose up deflections to balance the vehicle. Downwash aft of the canard unloads the inboard portion of the wing and partially negates the benefit of the high lift devices. It becomes immediately apparent that a canard configuration is incompatible with slow approach speeds required of CV based aircraft. As a result the configuration was redrawn with an aft tail and configuration numbering was rolled from 100 series (canard/ASTOVL configurations) to 200 series (aft tail/JAST configurations).

For many, this was hard to accept; however the logic was simple and undeniable. With the Navy as a customer and CV operations as a requirement, an aft tail configuration was the answer. Unfortunately, this change occurred as the LSPM was being delivered to NASA ARC, so our design team set out to provide a modification kit that would allow the model to be converted between aft tail and canard configurations while under test. Since the small scale test program that provided correlation data had already been completed we tested the LSPM in both configurations. The LSPM in the aft tail configuration is seen on left side of Fig. CS2.27a while the canard configuration can be seen on the right. An overlay of the LSPM fitted with aft tails and the early JAST 220B configuration is seen in Fig. CS2.33.

Two other significant configuration changes came about as the program transitioned from ASTOVL to JAST. First, the vectoring flap aft nozzle would be changed out for a 3-Bearing Swivel Duct/Nozzle (3BSD/N) and the F-22 style caret inlets would be changed to a Diverterless Supersonic Inlet (DSI) configuration that had been under development. While both design changes came with challenges, they would also save critical weight.

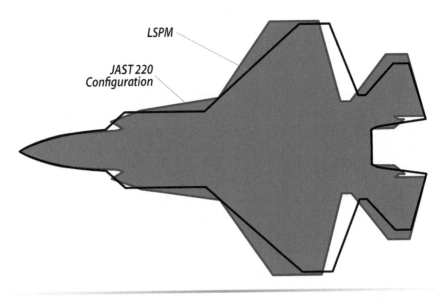

Figure CS2.33 Overlay of LSPM with aft horizontal tails and the JAST 220B configuration.

The 3BSD/N drew its lineage from work done at Convair that had been put on the shelf years before. Several trips were made to the former Soviet Union in attempts to develop collaborative agreements regarding similar technology they had developed for the Yak-141. Unfortunately, nothing of significance was obtained from them.

CS2.12 No Pause This Time, JAST to JSF

While the Lockheed-Martin ASTOVL program was busily executing to plan, the rest of the Lockheed-Martin enterprise was working details on the new 200 series configurations and developing the myriad of proposal issues needed to support such an effort. Staff was growing by leaps and bounds as program offices and engineering spaces were established at the LM Fort Worth facility (LMTAS) to handle the Preferred Weapons System Concept (PWSC) while the Skunk Works (LMSW) pursued details for a set of Concept Demonstrator Aircraft (CDA) or "X" airplanes.

As LSPM testing completed in early 1996 there was a dash to incorporate the results into our JSF configuration as rapidly as possible as the final RFP was due out in March. The summer of 1996 was spent developing the planning for the upcoming CDA program and providing a winning proposal response. We expected a tough competition against every remaining major U.S. aerospace contractor including Boeing, McDonnell Douglas, and Northrop Grumman.

As the program transitioned, Rick Baker took over as Program Manager of the CDA effort as Rudy Burch prepared to retire after completing the

ASTOVL program. Rick was a practical leader who listened to his engineers and was well respected by all who worked for him. He wisely allowed our flight sciences team to begin long lead wind tunnel work at risk and several months in advance of program award.

Another critical addition to the LMSW CDA team was our new Deputy Chief Engineer, George Law, who transferred out from LMTAS. George was a brilliant engineer with a background in structures that nicely complemented Rick Rezabek's background in flight sciences. George fit in immediately with the Skunk Works culture and helped to foster relations across the program. George was instrumental in handling the hardware side of the CDA effort while Rick spent a lot of his time working programmatic details. While George had a PhD, he never used the title, and earned respect across all aspects of the program through genuine brilliance combined with an outstanding work ethic. George was a true leader and critical to the programs eventual success.

In November of 1996, concept demonstration contracts were awarded to Lockheed-Martin and Boeing. Soon after, Boeing announced the purchase of McDonnell Douglas, again demonstrating how serious they were about this competition and their role in the future of military aerospace. The LM team was now facing a competitor comprised of its two arch rivals in the aerospace industry. This would be countered through teaming with both Northrop-Grumman and British Aerospace (BAe). These teammates brought a wealth of knowledge, experience, talent, and stability to the newly formed team. The Brits in particular, were a tremendous help in many critical aspects and are worthy of discussion later in Sec. CS2.18.

For me, the fall of 1996 was spent establishing wind tunnel test windows and requirements at a number of facilities across the U.S., U.K., and the Netherlands. We also started building a 12% scale loads model and Jet Effects model for early testing. Once the contract was awarded, we were immediately in the wind tunnels and the pace did not let up.

Our early Jet Effects Interactions (JEI) testing started within one week of contract award in the NASA Langley V/STOL tunnel. The JSF JEI model had more detail than our earlier ASTOVL work as seen in Fig. CS2.34. Eventually we would model the full 3BSN duct as well as install flow turbulence screens and swirl vanes to ensure that flow entrainment effects due to jet exit characteristics (turbulence, velocity profile, and swirl) were accounted for at small scale.

In February 1997, our first high speed test of the new configuration at the Calspan wind tunnel in Buffalo, New York unveiled a directional stability issue at transonic Mach numbers and elevated angles-of-attack. We felt the issue might be caused by flow spillage over the top of the new inlet configuration and put together a quick fix alternative using caret inlets. While the caret inlet configuration cured the problem, Russ Killingsworth and his S&C

Figure CS2.34 Details of the JSF JEI model.

team were able to determine the source of the instability as an asymmetric vortex burst caused by flow spillage from the DSI inlet as we thought. Through a series of loft changes, Russ was able to delay the burst to the point it was no longer an issue. However, it would result in a late loft and wing incidence change after our early start loads testing had been completed.

Immediately upon completion of the Calspan test we met in Fort Worth to discuss the results and make a final configuration decision. Still dressed in my testing clothes from Buffalo, I came out in favor of the loft change, which drew some strange looks from the LMSW side of the management chain. They were familiar with my typical mindset, which was to resist just about any configuration change unless it was proven absolutely necessary. So Rick Baker called me off to the side to explain. There were two major reasons behind my decision. First, I felt it was better to bite the bullet on this issue now rather than push a configuration with a known shortcoming into flight. Unless we were willing to put caret inlets back on the airplane, I felt we might need every bit of help we could get with integration of the new DSI configuration. Second, we felt we had a critical short fall in STO performance and the increased wing incidence would help that out. After Rick understood the reasoning, he was good with the change as well.

One of the last major configuration variables to freeze was location of the clutch between the lift-fan and the main engine. Our earlier ASTOVL configurations had it as part of the lift-fan and when disengaged, the drive shaft was left to spin within its casing. A major last minute effort was launched to try to pull the clutch back to the engine face and allow the drive shaft to stop turning when the lift-fan was not engaged. While this may seem like a first order common sense good thing to do, like most design decisions, the devil is in the details.

Two things killed the decision to pull the clutch back to the engine face. First, are the structural requirements to hang on to a clutch as it engages several thousand (even at reduced throttle settings) shaft horsepower. All of that structure must pass through the main inlet duct and present minimal pressure loss to the flow entering the engine. Second, the blockage caused by the clutch itself which is now located only inches ahead of the engine face. Ultimately, Pratt & Whitney stood up and said that this was not the way to go due to structural dynamics issues of increasing the overhanging mass at the front of the engine. The clutch remained at the Lift-Fan^{TM}.

CS2.13 The Government Team

Parallel to the LM effort was a dedicated government team made up of various specialists from the Navy, Air Force, Marine Corp, NASA, and DARPA. We enjoyed a very open relationship with our government counterparts. One of the smartest aspects of the program was to allow the government team to send rotating representatives to operate on site for extended periods of time.

Along with that access, came an increased level of responsibility and judgment on the part of our on-site government counterparts. It's not productive for the simple day-to-day squabbles that inevitably arise in such an enterprise to be blown out of proportion and reported back to the program office. However, real news does need to be reported in an accurate, timely, and fair manner. To their credit, our government counterparts rarely missed the mark and, to my knowledge, never in any significant way. Their sensitivity to these issues provided an environment of overall trust and shared responsibility. The net result was that the government had unprecedented access to our daily workings throughout the program.

Government leadership flowed from the top down, and our military services provided some great examples. The early JAST program was organized and directed by USAF General George Muellner who came from the F-22 program. No stranger to cutting edge aircraft technology, General Muellner orchestrated the formation of the JAST program office and integrated the ASTOVL and CALF efforts. His leadership from 1993 to 1995 took the effort from an amorphous wish list of technologies to a focused strike aircraft program with a clear direction.

Rear Admiral Craig Steidel took over from General Muellner and led the transition from JAST to JSF. Admiral Steidel oversaw the early competition and Concept Demonstration Phase (CDP) contract award. He also took the program through its early development including the Initial Design Review (IDR). Between 1995 and 1997, I had several opportunities to brief Admiral Steidel on my parts of the program and found him to be one of the fairest and sharpest individuals I have ever dealt with. I was constantly impressed by his ability to genuinely absorb highly technical information and almost instantly incorporate it into his big picture view. Admiral Steidel could bring out the very best in those around him.

Major General Leslie Kenne (USAF) took over from Adm. Steidel in 1997 and led the program from design into production and early hardware testing. General Kenne also oversaw the early engine development tests conducted by Pratt and Whitney. Always polite and soft spoken, the petite General Kenne commanded respect and could easily get the attention of any captain of industry when she desired it.

General Mike Hough (USMC) took over leadership of the program in 1999 from General Kenne. General Hough led the program through some of its greatest and most difficult times. It was early in General Hough's tenure that LM reported higher than budgeted manufacturing costs that took us deep into the red. General Hough and his staff helped us work through the debacle and got the program back on track. He personally addressed the program and referred to what we had been through as a "life-changing moment"; and it was. General Hough also led the program through flight testing and the final JSF down select in 2001. He was someone you just wanted to work for, and it didn't matter if you were on the government or contractor side of the effort.

Supporting each of these program directors were equally dedicated officers at the class desk level. USMC Col. Mike Nyalko filled this function the longest, working the program back in the days of SSF. Mike was a veteran test pilot who made some of the early AV-8B flight tests including the first shipboard trials in 1984. He was an excellent facilitator and a technically astute engineer. To this day, the program and the country owe him a great debt of gratitude.

As the program transitioned to JSF, a dedicated class desk position was formed. The first class desk officer was USAF Lt. Col. Dave Chafee. Col. Chafee worked out the various systems engineering processes required to manage the program through the early design stages. He worked tirelessly to build and maintain coordination and trust between the government and contractor teams. This would serve as a model for those who filled the role later in the program.

Next was USMC Col. Eugene "Gator" Fraser. Gator was a dynamic individual who was easy to like and to work for. He led the program through many of the early propulsion trials at P&W. An F-18 pilot first and foremost, he would later become commander of the NAS Pax River flight test wing. In fact, Gator was in that role when the X-35C made its way to Pax River for CV suitability testing. It was a tremendous benefit to have someone with a deep familiarity with the program in command at that time the aircraft were being tested at Pax.

Following Col. Fraser was USN Captain Steve Burris who was also a test pilot and distinguished naval aviator. Captain Burris (commander at the time) went by the call sign "Torch," but was soft spoken, technically brilliant, and a tremendous leader. Torch would lead the LM side of the government effort through flight testing and down select. Since the JSF/CDA program, he has moved on to various other command assignments within the Navy where he served with distinction and honor before retiring.

Lisa Phanneschlag (later Lisa Nyalko) was my primary government counterpart and we were particularly open and frank with one another. If she didn't like something, she let me know and I worked hard to make sure she was never surprised. Other members of her government team included Mark Thompson, Paul Sing (Singshinsuk), Kevin McCarthey, Tim Naumowicz (NASA), and Mark Fellows (USAF).

These folks had complete access to our daily activities and were team mates in every respect. They also made it easy to work hard. That level of cooperation and trust was one of the keys to our success and can't be overstated. A program that cannot build and maintain credibility and trust among its customer community is doomed.

CS2.14 Our British Teammates

The JSF program brought with it international interest and support. This was a tremendously stabilizing influence and added credibility to the program. British Aerospace (BAe) provided excellent technical support in the form of Bob Burton (flight test leadership), Dave Allen (flight controls), Paul Bloxham (flight test), Bill Ellison (flight sciences), and of course Simon Hargreaves (STOVL test pilot). Each of these folks possessed a wealth of knowledge and experience in STOVL aircraft design, test, and operations. They fit seamlessly into the engineering team and made immediate and profound impacts.

Bob Burton provided senior leadership as the deputy flight test lead and helped us navigate the nuances of STOVL flight testing. Being familiar with Harrier testing, Bob helped the team understand the particular issues associated with the hover and transition flight regimes and was instrumental in formulating our flight test plans in this area. Paul Bloxham was Bob's right hand man and would serve as the flight test director on the initial STOVL flights including the first hover. Together, they put together and helped execute a STOVL flight test plan that allowed the X-35B to perform in 2 months, what took the Kestrel several years to accomplish.

Bill Ellison helped our flight sciences group considerably as I was able to hand him the task of providing an air data system for the airplane without a second thought. Through Bill's diligence and expertise, I never lost a moment of sleep over air data.

Dave Allen helped our flight controls organization and ultimately became the deputy lead for STOVL flight controls. Technically brilliant and well versed in the issues of STOVL flight, Dave was usually one step ahead of the rest of us in this area.

There simply are no words to express the impact that Simon Hargreaves made on the X-35 program. His calm, unassuming demeanor masked the confidence and professionalism of a highly competitive, combat tested, fighter pilot who was driven to succeed.

CS2.15 Weight Control

Weight control and weight margin will always be a major topic during the development of any V/STOL airplane. Because of the narrow margins involved, the ability to project final weights is critical to program success. For this reason, any new V/STOL design should have significant weight and performance margin in its early development.

Through this process we developed a set of performance based metrics to capture meaningful elements of the aircraft's development. One of these was hover bring-back weight, or the maximum amount of fuel and payload the aircraft could perform a VL with. This metric allowed us to evaluate design changes that might affect multiple disciplines such and weight, thrust, and control law implementation.

For these reasons our NAVAIR customer prescribed weight margin based on the level of program maturity. They used historical trends to project the airplane's weight growth and final weight margins. This was a constant source of (usually) good natured banter between the contractors (us) and our government counterparts (them). We would make the argument that the historical trends should not be applied to a V/STOL airplane because weight had to be maintained at an even more critical level for an aircraft that was to take off and land vertically. NAVAIR would argue that they would remove the trend lines just as soon as we proved to them that they did not apply (basically build the airplane).

Since Flight Sciences was responsible for all aspects of aircraft performance including STOVL, the weight statements had special meaning to us. Every two weeks a weight statement would be issued and I would go down the list and challenge almost every addition. On one occasion a designer came in to tell me that George Law was considering using button head fasteners on a set of skin panels over the nozzle between the vertical tails. Being an aero guy, the designer thought he would find a sympathetic ear and that I would argue to have the button heads replaced with flush rivets if not from a drag standpoint, at least for aesthetic reasons. But I was aware of George's reasoning; the button heads meant the skins could be milled thinner in the area of the fasteners saving weight in the rear of the airplane. In fact, the button heads saved about 7 lb and a chemical milling process, so I explained to the designer that "they look beautiful to me."

Afterwards I went to Rick Rezabek and George to advise them that there would be no resistance to this change from the aero group and Rick approved the change to button head fasteners. I quipped that apparently they had "the wrong aero guy." To which Rick responded, "No, I think we've got the right one."

This is just one example of our ability to get together and make system level trades quickly and it was one of our team's greatest strengths. It also drove many in management crazy who were more familiar and comfortable with formalized and drawn out aerospace decision making processes. The Skunk Works approach of driving decision making responsibility to lowest possible levels ran counter to many who clung to power through the decision making process with nothing more than a veto.

It was common practice to weigh each part as it came in and allocate the lightest to ship 301, which was designated as the STOVL jet. This practice worked well up to the point the program discovered we were running over budget. That required us to pull out all stops on getting the first airplane assembled. At that point the forebody for ship 300 was further along with the heavier parts so it was mated with the aft body of ship 301. This resulted in losing the few pounds of incremental weight savings on the STOVL jet.

We continued to build the airplanes and track our weight progress against NAVAIR's historical trends. As one of those who argued against the historical trends, I was happy that we were tracking either flat or with a very shallow increase well below the historical trends. This continued up to about 70% through completion of the first airplane when we were hit with higher than projected heat rejection numbers from the lift-fan and hydraulic systems. A significant redesign effort was undertaken to add additional heat exchangers to the airplanes and we ended up right on NAVAIRs historical weight growth trends.

History would not be so easy to beat, and I no longer argue against NAVAIRs historical weight growth curves. While we did everything we legitimately could to save weight, we only managed to match the curves. Even so, I don't look at this as a failure as we were developing a significantly more complex design with greater levels of risk than had been attempted previously.

CS2.16 Another Design Challenge—Lift Fan Integration and the Inlet Ducts

General packaging around the lift-fan was one engineering challenge we anticipated from the LSPM back in ASTOVL. Now we needed to make a flight-weight structure accommodate the four foot lift-fan hole as well as all the transport elements from the cockpit. Then throw in a set of operating lift-fan inlet and nozzle doors to round things out.

Koi Marcucelli and Dave Fallabella, two very talented engineers had been working this issue since ASTOVL. The solution was a single piece (per side), composite, structural inlet duct along with some very well balanced engineering integration. Allison Advanced Development accommodated tighter lines along the sides of the lift fan nozzle than they would have liked, as did Don Santman and the rest of propulsion folks as the inlet contours were wrapped tightly around the base of the lift fan. To get as much volume as possible around the base of the lift fan for inlet duct structure, it was pushed up as much as the loft limits on the upper surface would allow. The challenging contours lead to lift-fan inlet flow distortion issues later in the program. This was dealt with through additional lift-fan inlet flow distortion testing and transition envelope flight restrictions. These issues would continue to challenge F-35 designers.

Early on there was considerable debate as to the configuration of the inlet duct and particularly, whether the two halves should joint prior to the engine face. Don Santman, our propulsion lead, was the calm voice of reason through this process. Don had worked similar configurations on the F-104 Starfighter and U-2. A fact not particularly well known is that the U-2 made use of a number of legacy F-104 elements including much of the fuselage and aft portions of the inlet system. Both aircraft had a set of side inlets with serpentine ducts that fed into a single engine. While not as serpentine as the X-35 inlet, the resemblance was inescapable, especially near the engine face. The answer was to keep the two inlet ducts isolated up to about 1" forward of the engine face. This kept the flow from each side isolated and the two ducts could not "talk" to each other in asymmetric flight conditions. The inlet ducts were separated by a septum that formed the passage for the drive shaft as well as housing the hinge mechanism for the auxiliary inlet doors.

Inlet design is a critical aspect of any vehicle with signature constraints. The inlet ducts are typically one of the largest contributors to Radar Cross Section (RCS) in the forward sector. As a result, they are usually highly contoured to hide the engine face and require special consideration at any joints or mating surfaces. In order to minimize joints and provide a sound structural unit, the X-35 would have a single piece structural duct.

The task that fell to Koi and his design team was not only to design the duct itself, but how to build and qualify it. With the help of Alliant Tech Systems, they came up with an approach of filament winding the inlet duct over a collapsible internal mandrel. It turned out that design of the mandrel was as much a challenge as designing the duct as it had to be designed in multiple pieces that formed a rigid mandrel while being able to be disassembled after the duct was wrapped and cured around it.

CS2.17 Engine Dynamics and STOVL Control

Engine control dynamics are always a concern on carrier based aircraft due to the need to rapidly increase thrust for glide path control while on approach to the carrier. When approaching the carrier, aircraft routinely experience sink as a result of "flying through the burble" created by the wake of the superstructure and movement of the carrier itself through the air mass just ahead of the approaching aircraft.

Since engines with large diameters, such as a single engine jet versus a twin (with two smaller engines) tend to have higher rotational inertias, they typically also have longer spool up times. The program breathed a collective sigh of relief early in the JAST program when the Navy announced they were committed to a single engine carrier based aircraft.

This was only the beginning of our engine dynamics and control issues as STOVL brings an entirely new level of requirements to the propulsion system. With a STOVL aircraft, the engine is not just a thrust producer, it is a primary flight control in hover and low speed transition. Consequently, it must produce three force and three moment components all with the necessary rates, precision, and control power.

This drove slew rates on each of the nozzles as well as failure recognition and accommodation. For example, in the event of an aft nozzle runaway condition, a set of locks would engage to freeze the nozzle in place. The aircraft now had to be designed to be flyable with the aft nozzle failed in any given position. This is an example of just one of the many problems that the Integrated Flight and Propulsion Controls (IFPC) group had to grapple with as similar failure accommodation had to be incorporated for the other SDLF functions as well.

Eventually it was determined that if an aft nozzle runaway occurred within a very small envelope of time during the rotation portion of a STO, the aircraft would be unrecoverable and the pilot would likely not have sufficient time to eject safely. This topic received considerable scrutiny and required buy-in by the Flight Test and especially pilot communities.

Eventually the issue became validating that the propulsion actually produced the desired characteristics. This task fell to the test group at Pratt & Whitney and represented a potentially thorny issue as the program began its final push to close these qualification issues out. The engine test stands in Florida had all been designed for static testing of engines or in the worst case, testing throttle-up dynamics of a conventional system producing only axial thrust. When the SDLF produced a significant control input such as a thrust split transient, the test stand would oscillate at its various natural frequencies and modes. Unfortunately, many of the stands natural frequencies were right in the range of interest for SDLF validation testing.

A concentrated effort to understand the dynamics of the test stand and its responses to test inputs was undertaken jointly by P&W and LM. One approach was to model the propulsion system in detail using high dynamic response pressure instrumentation in an attempt to determine SDLF dynamics without altering the test stand. Another approach was to selectively stiffen (or in some cases soften) the test stand to push its natural frequencies sufficiently either above or below the areas of interest. Both approaches were employed and collectively the SDLF characteristics were sufficiently characterized.

The task of STOVL propulsion integration was the responsibility of a team lead by Scott Winship and Kathy Zapka. It had many talented individuals such as Steve Wurth, Jim Winner, Mark Post, and Mark Smith.

CS2.18 Tail Hook and Speed Brake

The program dealt with several late arriving design challenges. With any fast paced technology demonstration program, decisions have to be made early and are based on experience and the best information available at the time. There are invariably changes and late emerging requirements that require a nimble technical and management approach. Two of these late arriving challenges were the integration of a tail hook and speed brakes.

Tail hook: Since the X-35C would only perform Field Carrier Landing Practice (FCLPs) maneuvers as part of its CV suitability evaluation, the aircraft was not originally envisioned with a tail hook. An FCLP is a carrier approach to a land based runway without arresting wires. The FCLP results in runway contact and a go-around much like a bolter maneuver. Ship 300 (X-35C) can been seen performing these maneuvers in Fig. CS2.35.

As the requirements to operate a technology demonstrator at Pax River matured, it became apparent that a tail hook was required. EAFB has longer runways and miles of lake bed for run out in the event of braking failure. Pax River did not have that luxury and the base is much more tightly

Figure CS2.35 Ship 300 (CV) performing FCLPs at PAX River.

integrated and closer to the local community. As a result, ship 300 was equipped with a USAF style tail hook (from the F-117A) that mounted externally to the airframe.

Led by Koi Marcucelli, the challenge was developing a mounting scheme that would take hook loads into the existing structure in the event an arrestment was required. The solution employed a set of double pinned joints that closely controlled the hook loads into a bridge structure that fit up under the engine that beamed the loads out into the keels. Installation of the bridge work required Electron Beam (EB) welding a set of clevis brackets for attachment of the hook fittings on to one of the existing titanium bulkheads. This allowed the tail hook to be mounted externally the airframe. While this solution was simple, straightforward, and effective; it was also heavy and a bit clunky. But most importantly, it worked.

Speed brakes: Another late arriving design change implemented on both X-35 aircraft was the speed brakes. The flight controls folks had been complaining for some time about handling deficiencies during maneuvers requiring precise speed control at low power settings such as in-flight refueling. This went on until one day I received a phone call from Tom Morgenfeld, our chief test pilot, inviting me to come down and fly the simulator. This was an opportunity I jumped at even knowing full well I was being set up in the process.

After flying the simulator around the virtual upper desert for a few minutes, Ed Burnett, who had been running the simulation lab, asked if I would like to try forming up with a tanker for a little in-flight refueling exercise. At throttle settings required for maintaining formation with the tanker, thrust response of the P&W F119 was very non-linear and sensitive. In fact, it was all I could do just to keep the simulated X-35 under the shadow of the tanker. After a few frustrating minutes of this, Tom Morgenfeld commented over my shoulder, "a little sensitive isn't it, why don't you let me give it a try." He climbed in the seat and the simulation was reset with the airplane approaching the tanker. Then Tommy proceeded to fly a clean rendezvous with the tanker as if nothing at all was amiss. From this I learned two important lessons.

First, the X-35 had some significant handling qualities deficiencies in this regime. Second, I realized how extraordinary a pilot Tom Morgenfeld was. He had developed a feel, and managed to adapt to flying an airplane that I could barely keep in the sky under the same conditions.

The airplane needed more drag to get the engine up on the thrust curve, so we set out to develop a set of speed brakes that could be retrofitted onto both airplanes. They were quickly lofted up and added to the high speed wind tunnel model that was about to undergo final validation testing. This data was used to determine the aerodynamic impacts as well as generate design loads. The speed brakes were also added to the STOVL Jet Effects model to evaluate the potential to use them as Lift Improvement Devices (LIDs) during hover. As it turned out, they helped capture some fountain lift and provided a small improvement in our VTOL performance as well.

The speed brakes were mounted to the airplane with simple bolt on brackets. Since the majority of the underlying structure had already been fabricated, we had to limit the load the speed brakes could impart into the structure. This was accomplished by sizing the actuators and limiting their mechanical advantage such that the local structure could not be overloaded. For this reason, the speed brakes are often seen in the slightly open position (Fig. CS2.36) because aero loads at high speed were sufficient to pull them open slightly. As with the tail hook and so many other design issues, this was an example of perfect being the enemy of "good enough." The speed brakes worked, and we moved on. Ship 301, the X-35A, is seen in Fig. CS2.37 ready to begin flight testing.

It is important to note that any technology demonstration program will face late emerging requirements changes and design challenges. The examples given here are only two from a list that also included modifications to the Lift-Fan inlet door seals and the addition of heat exchangers in all airplane variants. These changes were dealt with through necessary and prudent engineering design decisions within programmatic constraints.

This is also why demonstration programs need to design for robustness with plenty of margin.

Figure CS2.36   Ship 301 (STOVL) with F-16 Test Chase in the airspace of EAFB.

Figure CS2.37   Ship 301 (CTOL) ready to begin test work ups.

CS2.19 Flight Test

The flight test phase of the JSF/CDA program is arguably one of the most successful by an X airplane. First and foremost, it led to the winning one of the largest aerospace production contracts of all time. But its success was much more than that. That success can be measured by the numbers and accomplishments.

CS2.19.1 First Flight and the X-35A

On October 24, 2000 with Tom Morgenfeld at the controls, the X-35A (ship 301) lifted off from Palmdale runway 7 at about 0906. The short flight took the aircraft up to 10,000 ft and about 250 kt. While a few handling quality checks were made, the real objective was simply to move the airplane and test team to Edwards Air Force Base (EAFB) where the flight test program would begin in earnest.

Performance of the X-35A and Dick Burton's flight test team was exemplary. In a period of one month, the X-35A would complete 27 flights, accumulate 27.4 hr of flight time and complete 100% of its planned test points. A total of 6 pilots, including company test pilots from LM and BAe, as well as the U.K. Ministry of Defense and U.S. Government would fly and evaluate the airplane.

Two key factors led to the aircraft's initial success. First, the aircraft was highly reliable which allowed it to fly on an almost daily basis. A large part of this reliability is owed to the level of simulation and checkout that had been performed prior to the airplane ever moving under its own power. In fact, no software updates were required over the entire flight test program. The closest thing to an anomaly was encountered when Tom Morgenfeld took the aircraft supersonic for the first time on November 21, and a small unplanned pitch excursion was encountered. It was determined this was caused by the shockwave transiting the air data system and the excursion immediately smoothed out as the aircraft became supersonic.

The second key factor to the initial success of the X-35A was the ability to perform in-flight refueling. The first aerial refueling flight occurred on November 7, only 2 weeks and 9 flights after the aircraft made its first takeoff from Palmdale. Aerial refueling allowed the airplane to stay up longer and complete more test points per flight. Several photos of the X-35A undergoing flight test can be seen in Figs. CS2.38a–CS2.38c.

CS2.19.2 CV Testing

The X-35C took off from Palmdale on December 16, 2000 with LM test pilot Joe Sweeney at the controls. This flight was similar to the first flight of

Figure CS2.38a X-35A in straight and level flight.

Figure CS2.38b View of the tanker and ship 301 from chase.

Figure CS2.38c X-35A on approach to EAFB.

the X-35A in that it served to relocate the aircraft to EAFB and provided initial systems and handling quality checks on the CV variant. The aircraft would remain at EAFB and perform initial airworthiness workups before becoming the first X airplane to ferry across the United States. In early February of 2001, Joe Sweeney would pilot the airplane on its first ferry leg from EAFB to the LM plant in Fort Worth Texas. USMC Major Art Tomassetti would then take it on to PAX River to complete CV suitability evaluations.

During this time, the X-35C would complete 73 flights, accumulate 58 hr of flight time, and perform 250 field carrier landing practice approaches (FCLPs). The X-35C performed as well as planned or better, demonstrating precise glide path control and responsiveness. Several intentional approach misalignments were performed and the aircraft responded well to each of them. Figure CS2.36 shows the X-35C completing an FCLP at PAX River with Chief Engineer George Law in the foreground.

Images of the X-35C during various flight test activities at PAX River can be seen in Figs. CS2.39 and CS2.40. In all, the X-35C was flown by 8 pilots and expanded the flight envelope to 1.2 Mach number and 20 deg AoA.

Figure CS2.39   The X-35C returning to PAX River with an F/A-18 as chase.

Figure CS2.40   The X-35C performing low speed handling quality evaluations at altitude.

CS2.19.3 STOVL Flight Testing

Upon completion of flight testing at EAFB, the X-35A was returned to Palmdale to undergo retrofit to the X-35B (STOVL version) as seen in Fig. CS2.41. That work included not only removing the conventional engine and replacing it with the SDLF version, but also reconfiguring a set of heat exchangers that had been located in the roll duct bays on the X-35A. With that complete, the aircraft began a series of ground tests and STOVL work-ups on a specially designed hover pit facility constructed at the LM Palmdale plant.

The hover pit was designed to allow the X-35B to take off and land vertically while ducting the exhaust gases away from the aircraft to reduce the risks of HGI and other ground-effects described in Sec. CS2.4. The hover pit facility can be seen in Fig. CS2.42 with the X-35A and X-35C in the foreground.

This facility proved instrumental in allowing the X-35B to progress quickly into the STOVL flight regime. There are two approaches to initial STOVL flight, one is to start from conventional wing borne flight and progressively slow the airplane down eventually transitioning to a hover. The second approach is to perform a VTO and press the airplane up to a hover. The press-up approach was chosen by LM as we felt the risks associated with hover could be mitigated with the hover pit. The transitional (or build down) approach requires the aircraft to fly through the low speed region

Figure CS2.41   The X-35A undergoing reconfiguration to the X-35B.

Figure CS2.42 The Palmdale hover pit facility.

dominated by jet induced effects on the very first flight while close to the ground and does not lend itself well to an incremental approach. Lessons learned from aviation history (Fig. CS2.1) show the majority of V/STOL aircraft have some degree of trouble in the transition regime.

The press-up approach off the hover pit allowed the X-35B to establish a stable hover condition with minimal risk. Then, during transition, the pilots would have an established point to take the aircraft at each end of the speed range. Early in the flight test planning our British team mates, led by Bob Burton, came out in favor of the press-up approach and helped facilitate planning for the maneuver.

The aircraft had a minimal fuel requirement at shutdown because fuel was used as a heat sink for on board systems such as hydraulics. Performing the first press-up at light weights would maximize performance margin, but also meant operating near minimal fuel conditions with limited hover time. Operating at heavier weights would reduce performance margin but gave the pilots plenty of time in hover and lots of fuel to reject heat into.

One suggested approach was to intentionally overload the aircraft insuring the airplane would not be able to hover. Then, as fuel burned off, the airplane would just barely come unstuck from the surface. This approach would require the pilot to maintain task focus for a potentially extended period as the airplane burned off enough fuel to reach its maximum unstuck weight.

This could be compared to a batter in baseball having to step into the box and wait for a pitcher to throw a pitch without the ability to call time out. The pitcher on the other hand could wait as long as he wanted to throw the pitch, and the batter would get only one pitch to hit. As it would turn out, both Rick Baker and Gary Ervin had played baseball. They immediately understood the analogy and realized it was not a good task to give the pilot.

The task then became how to determine how much fuel (weight) we could safely hover with after an initial light weight (minimum fuel) press-up had been accomplished. The flight sciences and propulsion teams developed an ingenious approach to solving that problem. With weight and fuel burn very well known, the mass of the vehicle could be accurately determined at any point in time. Flight test data provided Power Lever Angle (PLA or throttle position) which could be entered into the installed engine deck along with the actual atmospheric conditions and several other measured variables. This allowed the installed vertical thrust to be determined fairly accurately. The engine deck could also be used to determine the amount of vertical thrust remaining. Data from on board accelerometers provided the acceleration of the vehicle and completed the $F = ma$ equation. With this, it became a complex yet straight forward task of post processing the flight data from an initial light weight hop to determine how much additional weight the airplane could safely carry into hover.

Morning came early on June 23rd, 2001 as we prepared for the first hover press up of the X-35B. As the moment approached, Simon began his throttle up and the nose wheel lifted first as was usual. The main landing gear struts began to extend as the airplane became progressively lighter on the wheels. He held the aircraft steady just prior to lift off while a last set of checks were made. During this period a typical slight roll oscillation began as a result of the force feedback through the main landing gear. The "go" call was made and Simon continued his throttle up and the airplane steadily rose off the hover pit. What was planned to be a short 5 ft hop, turned into a 15 or 20 ft press up that lasted several seconds. The airplane had a considerable amount of vertical thrust and broke ground with a little more vertical acceleration than was anticipated. This took a little explaining by the flight test team to ensure the government that Simon was not showboating. The first press up with Simon Hargreaves at the controls is seen in Fig. CS2.43.

The next day Simon again performed a light weight press up with the express objective to gather performance data and allow us to determine hover margin. After that press up was completed we waited for the performance numbers to be calculated to determine how much fuel could be

Figure CS2.43 First press up of ship 301 (STOVL) off the hover pit in Palmdale.

taken on board for the next press up. After a few minutes a call came over the radio from test director Paul Bloxham (call sign Blossom). With a typically measured British accent the call came in: "Hat Trick One, we'd like you to taxi to the fuel pits and prepare to take on . . . ah . . . get this . . . 3500 lb of fuel."

Vertical performance of the X-35B was being demonstrated at Palmdale which is a little over 2500 ft MSL. As it turned out, our competitor had also achieved hover that same day. However, they hovered at PAX River near sea level and the X-32B had been stripped down to reduce weight. It was at that point we knew we had won the competition, as long as we could just complete the remaining contractual requirements.

The additional 3500 lb of fuel turned out a bit over estimated as the second hover attempt of the day could not quite break loose from the ground.

The X-35B was relocated to EAFB where the remainder of the STOVL flight test program was conducted. As described in the introduction; On July 20th, 2001, USMC Major Art "Turbo" Tomassetti performed the first ever flight consisting of a short takeoff followed by a level acceleration to supersonic and a vertical landing. The X-35B in conventional flight mode is seen in Fig. CS2.44 with shock diamonds trailing from the exhaust. Other images of the X-35B performing various flight test maneuvers can be seen in Fig. CS2.45a–d.

Figure CS2.44 The X-35B in conventional flight.

On August 6, the X-35B made its final test flight (Fig. CS2.46). Flown by Tom Morgenfeld, the flight lasting some 3.7 hr and included 6 in flight refuelings (Fig. CS2.47). While all planned test points had already been completed, this last flight satisfied the minimum flight time requirements

Figure CS2.45a The X-35B performing a short takeoff.

Figure CS2.45b The X-35B in conventional flight with a Harrier as chase.

Figure CS2.45c The X-35B performing a vertical landing at the VTOL pad at Edwards AFB.

Figure CS2.45d   The X-35B coming into the hover.

set in the contract and relocated the aircraft to Palmdale. In total, the X-35B made 17 VTOs, 27 VLs, 14 STOs, and 6 short landings in a total of 39 flights.

A summary of flight test milestones is seen in Table CS2.1 and important statistics are found in Table CS2.2.

CS2.20 And the Winner Is...

On Oct 26, 2001 the Boeing X-32 team and the Lockheed Martin X-35 team were assembled in Seattle and Ft. Worth, respectively, to hear the government decision on the winner of the 22 aircraft, $24B JSF EMD contract. AF Sec James Roche took the podium to announce the winner. "Both proposals were very good," he said. "Both demo programs were very good. But on the basis of strengths, weaknesses, and degrees of risk to the program, it is our conclusion... that the Lockheed Martin team is the winner of the JSF program on a best-value basis" [6].

One of Lockheed Martin's strengths was the SDLF concept, which provided augmentation of the F119 thrust resulting in 37,300 lb of hover lift compared with the Boeing DL/VT concept which provided only 33,000 lb. This hover lift margin was represented as a lower risk system at meeting the USMC requirements. Most government evaluators agreed that the

Figure CS2.46 Ship 301 (STOVL) completing the flight test program at Edwards AFB.

X-35 was a pretty airplane and as Kelly Johnson often said, "If it looks good . . . it flies good."

The cited weaknesses of the X-32 were the hot gas ingestion during vertical landing causing the hover thrust to sag and the engine to overheat, the blocker vanes in the inlet front frame reducing the total pressure recovery (and thrust), the reduced hover lift (the X-32 did not do a vertical take-off during flight test), and the real concern that Boeing was not demonstrating their EMD configuration.

In retrospect it can be argued that the program winner was decided 15 years earlier when the decision was made to go forward with the SDLF concept. The adage that "90% of a program is defined in the first 10% of the decisions made", was never more true than here. The long term pragmatic approach that looked at history and put together a V/STOL propulsion system that was fundamentally sound, proved to the deciding factor. Boeing did an outstanding job of getting performance and capability out of their platform and the competitive environment ensured that neither team left anything on the table.

For their efforts, the team of Lockheed Martin, Rolls-Royce, Pratt & Whitney, BAE Systems, Northrop Grumman, and the Joint Strike Fighter Program Office received the 2001 Collier Trophy. The aircraft remain on

Figure CS2.47   The X-35B performing an in-flight refuel from a KC-135 tanker.

display in the Smithsonian Institution (Ship 301) and the United States Association for Naval Aviation (Ship 300).

CS2.21 Epilogue

Certainly much of this has not been written in the analytical style of a typical design case study. It is a story of the evolution of the JSF program through the Concept Demonstration Phase and flight of the X-35s as told from the perspective of one person, me. Unquestionably there are other valuable stories and meaningful perspectives to learn from. I have intentionally given much attention to individuals and their contributions to the team over a long period and ask for understanding from the many, many people I have left out. The program was a success because of contributions by everyone from one end of the organizational structure to the other.

This story is also told mainly from the perspective of what came before the CDA phase of the JSF program. Much has been written and reported since the award of the JSF CDP contracts in November of 1996, but very little was told about ASTOVL, and almost nothing has been told about what came before it.

Finally, the competition was not won by one team solving equations or riveting bulkheads together better than the other. The most important

Table CS2.1   X-35 Flight Test Milestones

Accomplishment	Aircraft S/N	Flight no.	Date
Initial Ramp Taxi (301)	301	–	10/13/00
Initial Ramp Taxi (300)	300	–	12/9/00
First Flight of Acft. 301 (X-35A)	301	1	10/24/00
First Flight of Acft. 300 (X-35C)	300	1	12/16/00
First Flight of Acft. 301 (X-35B)	301	28	6/23/01
First Supersonic Flight	301	25	11/21/00
First USAF Piloted Flight	301	5	11/03/00
First USN Piloted Flight	300	4	12/22/00
First USMC Piloted Flight	301	13	11/10/00
First RAF Piloted Flight	301	22	11/18/00
First Aerial Refueling	301	10	11/7/00
First Hot Pit Refueling	301	20	11/18/00
Maximum AoA Demonstration (20°)	301	18	11/15/00
Maximum Positive g (5.0)	301	16	11/12/00
Maximum Mach Number (1.20)	300	70	3/10/01
Maximum Speed (535 KCAS)	300	70	3/10/01

lesson to take away is that competitions are won by people. These are strong-willed individuals committed to working together, respecting one another despite their many differences, willing to accept responsibility, and getting the job done right.

Table CS2.2   Important X-35 Flight Test Statistics

Event	X-35A (301)	X-35C (300)	X-35B (301)	Total
First Flight	10/24/00	12/16/00	6/23/01	–
Total Flights	27	73	39	139
Total Flight Hours	27.4	58.0	21.5	106.9
Total Contractor Flights	17	41	22	80
Total Government Flights	10	32	17	59
USAF	7	3	–	10
USN	–	22	–	22
USMC	1	6	9	16
RAF	2	1	8	11
Most Flights Per Week	8	17	10	
Most Flights Per Day	4	6	3	
Average Flights Per Day	6.3	5.9	7.0	
Air Refueled Flights	8	4	5	17
Total Pilots Checked Out	8 (2 LM, 1 BAE Sys, 1 USAF, 1 USMC, 1 RAF, 2 USN)			

About the Author

Jim Eshleman was the Flight Sciences lead engineer on the Joint Strike Fighter (JSF) Concept Demonstration Aircraft (CDA/X-35). He worked at the Lockheed Martin Skunk works from 1986 through 2000. Before joining Lockheed Martin, he was a V/STOL and powered lift research engineer at the NASA Ames Research Center. Since leaving Lockheed Martin, Jim has worked as a consultant and project engineer on several aircraft and defense systems including Northrop Grumman's Unmanned Combat Air System Demonstration (X-47B/ UCAS-D). He lives in Southern California with his wife Suzanne and has three sons.

References

[1] Nicolai, L.M., Carichner, G.E., *Fundamentals of Aircraft and Airship Design*: Volume I—Aircraft Design, AIAA Education Series, Reston, VA, 2010.

[2] Bevilaqua, P., "Genesis of the F-35 Joint Strike Fighter," *Journal of Aircraft*, Vol. 46, No. 6, November–December 2009.

[3] Kuhn, R. E. and Eshleman, J. E., "Ground-Effects on V/STOL and STOL Aircraft – A Survey," AIAA-85-4033, October 1985.

[4] Franklin, James A., "Criteria for Design of Integrated Flight/Propulsion Control Systems for STOVL Fighter Aircraft," NASA TP 3356, April 1993.

[5] Koenig, David G., "V/STOL Wind Tunnel Testing," NASA TM 85936, May 1984.

[6] Fallow, J., "Uncle Sam Buys an Airplane," *The Atlantic Monthly*, pp. 62–74, June 2002.

- Boeing's Gamble
- The Customer is Always Right
- A Paperless Airplane for the 21st Century
- The Boeing 777 Takes Off
- The Boeing 777 Gamble Pays Off
- The Airliner War

The quote was a new one for Boeing. Prior to the 777, Boeing would set the requirements and build the airplane. Customers would "take it or leave it." At the start of the Triple 7 program Boeing assembled eight customers (called the gang of eight) who set the requirements and oversaw the airplane's development.

Rule #1: The customer is always right.
Rule #2: When the customer is wrong see Rule #1.

CS3.1 Boeing's Gamble

In 1986 the airlines were telling Boeing that they wanted an airplane that was bigger than the 767 and smaller than the 747. It took Boeing several years before they put aside their derivative 767 sales brochures and decided that the airlines really wanted a new airplane. According to the Boeing numbering system the new airplane would be the 777 . . . the "Triple Seven." It would fit in between the 767 and the 747 as shown on Fig. CS3.1.

> The aircraft is the Boeing 777-200LR, which holds the world's record for the longest nonstop flight by a commercial jet airliner. It flew 11,644 nm from Hong Kong to London.

The B-777 would be a twin engine, 21st century airplane with all new features and it would compete head on with the Airbus A340 and the McDonnell Douglas MD-11/12. The vision was that the Boeing 777 family would be at least four different airplanes. The first airplane, the A-market plane, would be designed for regional and domestic service. It would handle 300 to 440 passengers over the medium range routes of the North Atlantic, across the United States and throughout Asia where an airline would lose money on a 747 but a 767 would be too small.

The B market would be the long, thin routes ranging from 5550 nm up to 7000 nm. The "thin" referred to the passenger volume that would not be profitable for a 747. And then there would be stretch versions of the A-market and B-market airplanes. Eventually the stretch B-market would carry 300 passengers over 8500 nm routes . . . which is a 16 to 17 hr flight.

In early 1989 the 777 design team presented their A-market design shown on Fig. CS3.2 to the corporate leaders with the specifications summarized in Table CS3.1 [1]. The response was positive and the Boeing Commercial Airplane Company launched a major new program at a time when the aerospace market was in a downturn. In the early 1980s Boeing ramped up too fast and was overstaffed in 1988. There would be layoffs from 1988 to 1992. The decision to spend $5B and five years developing the Boeing 777 represented a substantial commitment and risk for the company. The 777 program had to succeed or Boeing risked losing its leadership position in the commercial transport market.

The Boeing 777 program leader was Phil Condit, an experienced and well-respected manager. Condit left the program in 1992 to become president of the Boeing Company, and Alan Mulally took over as program leader. Mulally was a youthful, charismatic leader with broad experience, having worked on nearly every Boeing "7" airplane from the 727 through the 767 [1].

CS3.2 The Customers Are Always Right

The Boeing 777 design phase was different from previous Boeing jetliners. For the first time, eight major airlines (All Nippon Airways, American,

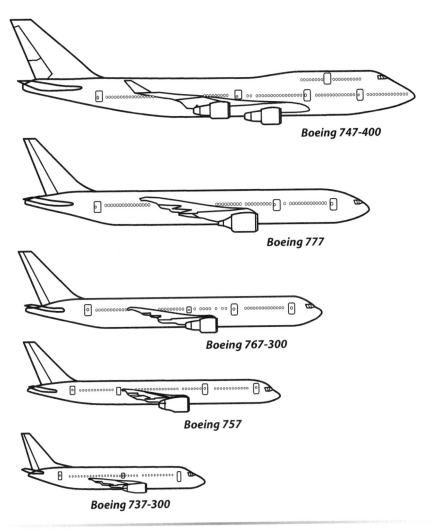

Figure CS3.1 Boeing family of commercial transports.

British Airways, Cathay Pacific, Delta, Japan Airlines, Qantas and United) had a role in the development of the airplane. This was a departure from industry practice where the airplane builder typically conducted the design process with little input from the airlines. The eight airlines that contributed to the 777 design became known as the "gang of eight" and the management style was called "Working Together." The consensus of the "gang of eight" was that they wanted a cabin cross-section close to the 747's capacity of at least 325 passengers, fly-by-wire controls, a glass cockpit, flexible interior, and 10% better seat-mile costs than the A330/340 and MD-11/12 [2].

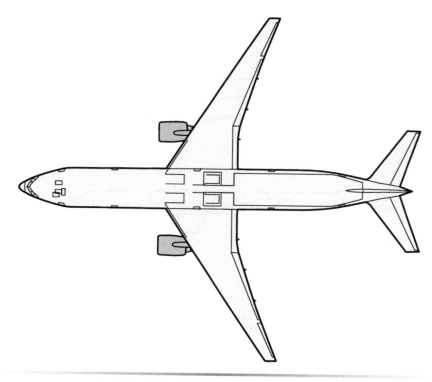

Figure CS3.2 Boeing 777 planview.

Table CS3.1 Boeing 777 Specifications

Seating capacity	305 (3 class) to 400 (2 class)
Length (ft)	209
Wing span (ft)	200
Wing sweep (degrees)	31.6
Cabin width (ft)	19.25
Fuselage width (ft)	20.33
Empty weight (lb)	307,000
Max takeoff weight (lb)	545,000
Cruise Mach	0.84 at 35,000 ft
Max cruise Mach	0.89 at 35,000 ft
Max range (nm)	5200
Engines (2 @ 77K lb SLS)	PW 4077 RR 877 GE 90-77B

In October 1990, United Airlines became the Boeing 777's launch customer when they placed an order for 34 aircraft with an option for 34 more. United selected the PW 4074 turbofan to power its replacement for the aging DC-10s. United wanted the new transport to handle three different routes: Chicago to Hawaii, Chicago to Europe, and non-stop from Denver to Hawaii. Given the over water portions of United's Hawaii routes an ETOPS (Extended Twin-engine Operations) certification was essential . . . and the more time the better.

The FAA (and JAA) ETOPS rule is that the twin engine airplane is limited to operating over routes that contain an adequate airport that can be reached within X hr with one engine inoperative. The CONUS to Hawaii routes required X to be at least 2 hr with 3 hr preferred. United wanted 3 hr and Boeing made the decision to have the B-777 certified for 3 hr ETOPS at delivery of the first aircraft in 1995, which had never happened before. ETOPS certification typically didn't happen until the transport had been in service for a period of time.

CS3.3 A Paperless Airplane for the 21st Century

The Boeing 777 was the first commercial transport to be designed entirely by computer. The program used two linked computer systems: CATIA (Computer-graphics Aided Three-dimensional Interactive Application sourced from Dassault Systems and IBM) and EPIC (Electronic Preassembly in CATIA). All design drawings were created using CATIA so there was no need for paper drawings. EPIC allowed a virtual B-777 to be assembled, in simulation, to check not only for parts interference but also to verify proper fit of the many thousands of parts. Essentially EPIC could eliminate the use of physical mock-ups. The Boeing 777 management was initially not convinced of the program's abilities and built a physical mock-up of the nose section to test the results. It was so successful that all further mock-ups were cancelled at considerable cost saving.

Boeing introduced a number of advanced technologies on the B-777 design. A fully digital fly-by-wire (FBW) active flight control system was installed replacing the conventional mechanical system of pulleys and cables. Boeing was several years behind Airbus with this 21st century flight control system. The FBW system offered the airlines a software-configurable flight control system and reduced maintenance by eliminating the re-rigging of the control system as a mechanical pulley-cable system would stretch and wear.

The B-777 wing employs a supercritical airfoil design that is swept back at 31.6 deg and optimized for cruising at Mach 0.84. The supercritical airfoil permitted a slightly thicker wing section for increased fuel volume.

The Triple 7 aspect ratio of 8.7 is the highest of all the Boeing 7 series family (except for the 737-200 at $AR = 8.8$) giving it improved range, landing and take-off, and a higher cruising altitude and C_L.

The high lift system for the B-777 is a double slotted Fowler flap inboard, a single slotted flap outboard of the engine and a full span leading edge flap (see Fig. CS3.3). This system gives the airliner a maximum C_L of 2.8. The aircraft features the largest landing gear and the biggest tires ever used on a commercial airliner [5]. Each tire of a B-777-300ER six-wheel main landing gear can carry a load of 59,500 lb, heavier than other wide-bodies such as the B747-400. The aircraft has triple redundant hydraulic systems with only one system required for landing. A ram air turbine is fitted in each wing root fairing.

Boeing was especially sensitive to customer desires. Several design decisions bear this out. Boeing had customer feedback that the toilet seat and lid in the lavatory of the 727 through 767 aircraft were prone to fall down with a loud bang, disturbing the passengers in the vicinity of the rest rooms. Boeing engineers designed a restraint system for the seat and lid that would let them descend slowly and quietly [1]. Not a big deal but a nice touch.

The next design decision example had a more substantial impact on the program. ALCOA was marketing an alloy of aluminum and lithium that

Figure CS3.3 Extended flaps and landing gear of an American Airlines 777-200ER.

offered good fracture toughness and was 20% lighter than the conventional aluminum alloys. It was more expensive but with judicious application it could "buy its way onto the airplane" by saving structural weight. Airbus had 30,000 lb of the material and had saved up to 2200 lb of structural weight by using it on one of their airplanes. The B-777 structural designers wanted to use it for the in-spar ribs and the stanchions that hold the floor beams. This application would save 400 lb over the conventional aluminum. When the aluminum-lithium parts were being fabricated and assembled the material showed cracks, particularly where edges had been machined or holes drilled. Testing of the parts and assemblies confirmed that the cracks would not grow under loading and would not be a structural problem. The cracks were more of a cosmetic issue. But psychologically that could be a lifetime problem for the airlines as they tried to convince their maintenance people and passengers that there was no damage. So well into the design phase the decision was made to not use aluminum-lithium on the aircraft. In addition to losing the 400-lb weight saving, there was the extra work associated with a drawing change and Boeing had to pay ALCOA a $2M cancellation fee.

CS3.4 The Triple "7" Takes Off

By the start of the B-777 production in January 1993 Boeing had 118 firm orders with options for 95 more from ten airlines. The production plan had an unprecedented amount of foreign subcontracting. The international suppliers included Mitsubishi Heavy Industries and Kawasaki Heavy Industries in Japan making fuselage panels, Fuji Heavy Industries in Japan making the center wing section, Hawker DeHavilland in Great Britain making the elevator and ASTA in Australia building the rudder [3 and 4]. Japan Aircraft Development Corporation (JADC), representing the Japanese aerospace contractors, made an agreement with Boeing to be a cost-sharing partner for 20% of the entire B-777 program. The total investment in the B-777 program was estimated at $5B from Boeing with an additional $2B from their suppliers.

On April, 9 1994, the first Triple 7 was rolled out in Renton, WA with ceremonies around the world. The first flight took place three months later on 14 June 1994 . . . 43 months after program start (Table CS3.2). This started an 11-month flight test program with nine aircraft. Five of the aircraft were powered with Pratt & Whitney PW 4000 engines, two with General Electric GE 90 engines and two with Rolls Royce Trent 800 engines. To satisfy the 3-hr ETOPS requirement for United Airlines, eight 180-min single engine diversion test flights were performed. At the successful conclusion of the flight testing, the B-777 was awarded simultaneous airworthiness certificates by the FAA and JAA on 19 April 1995.

Table CS3.2   Boeing 777 Milestones

Event	Date	Elapsed-months
Program launch (UA signed for 34 aircraft)	Oct/1990	
Rollout	Apr/1994	40
Full-scale static tests begin	Jun/1994	43
Wing completes design limit load	Jun/1994	43
First flight	Jun/1994	43
Wing completes ultimate load (destroyed)	Jan/1995	51
FAA/JAA aircraft certification	May/1995	55
First delivery (United Airlines)	May/1995	55
Fatigue tests complete two lifetimes	Mar/1997	83

On 15 May 1995 Boeing delivered the first PW 4084 powered B-777-200 to United Airlines (see Fig. CS3.4). Fifteen days later the FAA certified the aircraft for 180 min ETOPS operation making it the first twin engine aircraft to have this certification at its entry into service. Longer ETOPS clearance of 207 min was approved later.

In November 1995, Boeing delivered the first B-777 with General Electric GE-77B engines to launch customer British Airways. The aircraft was

Figure CS3.4   United Airlines B-777-200; UA was the launch customer for the B-777. [5]

Table CS3.3 Boeing 777 Family

Variant	777-200	777-200ER	777-200LR	777-300	777-300ER
Seating capacity (3 Classes)	305 400 (2 class) 440 (max)	301 400 (2 class) 440 (max)	301 400 (2 class) 440 (max)	368 451 (2 class) 550 (max)	365 451 (2 class) 550 (max)
Length (ft)	209	209	209	242.3	242.3
Wing span (ft)	200	200	212.6	200	212.6
Wing sweep (degrees)	31.64	31.64	31.64	31.64	31.64
Cabin width (ft)	19.25	19.25	19.25	19.25	19.25
Fuselage width (ft)	20.33	20.33	20.33	20.33	20.33
Empty weight (lb)	307,000	315,000	326,000	353,600	366,940
Max fuel weight (lb)	201,500	293,900	347,850	293,900	311,300
Max takeoff weight (lb)	545,000	656,000	766,000	660,000	775,000
Cruise Mach @ 35 Kft	0.84	0.84	0.84	0.84	0.84
Max cruise Mach @ 35 Kft	0.89	0.89	0.89	0.89	0.89
Max range (nm)	5235	7700	9380	6015	7930
Engines (2)	PW 4077 RR 877 GE 90-77B	PW 4090 RR 895 GE 90-94B	GE90-110B	PW 4098 RR 892 GE 90-94B	GE 90-115B
Thrust (×2) (lb)	PW 77,000 RR 77,000 GE 77,000	PW 90,000 RR 95,000 GE 94,000	GE 110,000 GE 115,000	PW 98,000 RR 95,000 GE 94,000	GE 115,000
2009 unit cost ($M)	NA	205–231	237–263	NA	257–286
Delivered a/c (Jul 09)	88	412	36	60	211

pressed into service several weeks later. The first Rolls Royce Trent 877 powered B-777 was delivered to Thai Airways in March 1966, completing the entry of all three engines into service. All three engine-airframe combinations had secured ETOPS-180 certification from the point of entry into service.

In November 2005, a B-777-200LR ("LR" for longer range) flew 11,664 nm on a special flight from Hong Kong to London, setting a World record for the longest commercial jetliner flight. In regular service, the B-777-200LR (the "C" market variant) is capable of flying 9380 nm in 18 hr. The -200LR (shown on the title page) features an increased maximum TOGW and three optional auxiliary fuel tanks in the rear cargo hold. It also features raked wingtips, redesigned main landing gear, and additional structural strengthening.

As of the writing of this book the Triple 7 has been in commercial service for 18 years with more than 56 customers. The public reception of the aircraft has been great and it is the flagship of the Boeing fleet of transports. As of July 2012 Boeing had orders for 1379 "Triple 7" aircraft of all variants, with 1030 delivered (Table CS3.3). The most common variant used Worldwide is the B-777-300ER, with 612 aircraft delivered, and the Emirates operates the largest B-777 fleet with 78 aircraft [5].

CS3.5 The Triple "7" Gamble Paid Off

The competition between Boeing and Airbus (referred to as the "Airliner War") is a result of the two companies' domination of the large jet airliner market since the 1990s, which is itself a consequence of numerous corporate failures and mergers within the global aerospace industry over the years. Airbus began its life as a consortium, whereas Boeing took over its arch rival, McDonnell Douglas, in 1997. Other airplane companies, such as Lockheed and Convair in the United States and Dornier and Fokker in Europe, pulled out of the civil airliner market well before 1990 after disappointing sales figures and economical problems. The collapse of the Eastern Bloc and its trade organization Comecon around 1990 put the former Soviet aircraft industry a distant third to Boeing and Airbus [5].

In the decade between 1999 and 2008 Airbus received 6378 orders (all transport models), while Boeing received 6140. However in the "head-to-head" market fight between the Boeing 777 and the Airbus A 340, Boeing is the clear winner. As of August 2011 Boeing had delivered 956 "Triple 7s" and Airbus had delivered 375 A340s (even though Airbus had a two year head start into the market place) [5]. The Triple 7 has replaced the Boeing 747 as the most profitable commercial jet in the Boeing stable.

In December 2011 the FAA certified the Boeing 777 airliner for ETOPs flights up to 5.5 hr away from the nearest available airport. Not only can the airplane now access remote regions, but can fly more direct routes, reducing both fuel and travel time. The 5.5 hr ETOP is the longest for any current two-engine transport.

References

[1] Sabbagh, Karl, *21st Century Jet—The Making of the Boeing 777*, Macmillan General Books, London, UK, 1995.

[2] Birtles, Philip, *Boeing 777: Jetliner for a New Century*, MBI Publishing Co., Galtier Plaza, Suite 200, 380 Jackson St, St. Paul, Mn, 55101, 1998.

[3] Yenne, Bill, *Inside Boeing: Building the 777*, MBI Publishing Co., Galtier Plaza, Suite 200, 380 Jackson St, St. Paul, Mn, 55101, 2002.

[4] Sharma, K.J. and Bowonder, B, *The Making of the Boeing 777: A Case Study In Concurrent Engineering*, International Journal of Manufacturing Technology and Management, Vol. 6, No. ¾, pages 254–264, 2004, Publisher: Interscience, Geneva, Switzerland.

[5] Anon, *Boeing 777*, Wikipedia, Retrieved 30 Oct 2012.

Case Study 4   *HondaJet*

Michimasa Fujino

- Honda's Aviation Challenge
- New Concept for the Light Jet
- Development of HondaJet Key Technologies
- HondaJet World Debut
- Decision to Commercialize the HondaJet
- Reflection on Mr. Honda

The HondaJet design team was challenged to deliver 21st century performance with elegance. The result is the business jet shown above. This case study is a personal account of the love affair between the author and the HondaJet design. The story starts three decades ago with the author's passion for airplanes.

I finally realized that all of my life experiences have worked in concert to bring me here.

Michimasa Fujino

CS4.1 Honda's Aviation Challenge

It may be somewhat unique within the very competitive arena of today's global business environment that Honda has always been focused on the future and that the company's top management regularly has its eyes set twenty or more years ahead. From its very beginnings, Honda has remained steadfastly committed to becoming a world leader in providing meaningful forms of transportation for the benefit of people across the globe. As a result, Honda's advancement into aviation can easily be understood as a natural progression of the company's core philosophy to offer products of the highest performance, quality, and value—from motorcycles . . . to automobiles . . . to advanced future technologies—that help to further enhance and expand human mobility.

Honda's aerospace endeavors began in 1986, when top management decided to establish a new research center, named the Honda R&D Fundamental Research Center, and form teams to prepare new products for the 21st century. Several new projects were started, including humanoid robotics, gas turbine engines, ultra-light automobiles, and airplanes. Without any advanced notice, I was assigned to the airplane project. From that point in time, my long journey began—a journey I could never have imagined would find me devoted to the airplane business for more than 25 years.

In 1986, I was working in Honda's automobile division after having graduated from Tokyo University two years earlier. I decided to become an automobile engineer, even though I had earned a degree in aeronautical engineering. I was fascinated by the challenging and dynamic opportunities available within the Japanese automobile industry from both the technical and business perspectives. The very competitive and independent spirit of the Japanese automobile industry fosters a unique environment that encourages the exploration of products encompassing new concepts and technologies. Honda, in particular, has always demonstrated a challenging spirit and independence to penetrate into new markets at its own risk and to boldly expand its boundaries. As I had been independent from my earliest childhood, I joined Honda without hesitation. I was enjoying automobile design and development as a young Honda engineer when I was told by my boss that I was to be transferred to the airplane project. Frankly speaking, I was quite surprised by the announcement.

It is well known that airplane development and certification are not easy tasks. These endeavors require large up-front investments in technology, people, and infrastructure. In addition, the very complex certification process can take a great deal of time to complete. However, if you consider this from a different viewpoint, these challenges are exactly why airplane development—especially development from a clean sheet design—can be such an exciting endeavor and one that not many people have the opportunity to undertake during their lifetime. So, I decided to accept this much larger challenge and join the airplane project team.

Our team was told by top management that we needed to develop from scratch a Honda-original airplane. Complicating our task, the research themes given to us included a forward-swept wing design, Advanced Turbo Prop (ATP) propulsion, all-composite structure, and so forth. Every one of these research themes challenged the limits of available technologies and directly challenged our team as well, because we did not have any experience in aviation technologies.

The first airplane design effort was named MH01. The project was to modify the wing and empennage of an existing single engine turboprop airplane to composite structure in order to demonstrate the advantage of composite application to primary structures. Following MH01, we developed our next airplane for the first time from a clean sheet design. This second airplane, named MH02, was a pure experimental aircraft that consisted of various new technologies [1]. There are pros and cons to applying state-of-the-art technologies to a new airplane design, and some of our goals, such as the ATP, were not realized. However, many things were learned from the challenges we faced. By the end of the MH02 project, we were ready to design an airplane from a clean sheet of paper. Our experience seemed sufficient to fabricate and test an airplane as well. Essentially, we understood the fundamentals of airplane design. We needed computer programs to analyze aircraft performance, aerodynamic characteristics, stability and control characteristics, loads, and vibration and flutter. Many drawings were generated and numerous stress analyses were performed. Actually, our team conducted structural tests, system tests, and ground vibration tests, and I even had to sand tooling molds by hand. Over time we learned and came to understand the theory needed to design an airplane, but also gained valuable practical experience. Through this project, we created the in-house capability to develop aircraft, and all of these experiences would become very important in the pursuit of future airplane designs.

CS4.2 New Concept for the Light Jet

Although we had developed many airplane design capabilities, the MH02 project ended. Some board members thought we should not continue the airplane project. The automobile industry was very competitive globally, and board members thought we should concentrate the company's resources on automobile research and development. This opinion was driven in part by the fact that the airplane business is very specialized and by the resulting perception that it is very difficult for a new entrant to get into this business. Furthermore, compared to the automobile business, it can take much longer to recoup the investment in aviation. But, most importantly, without unique Honda technologies and concepts that would bring new value to the industry and customers, there would be little meaning for Honda to get into the aviation business.

During the MH02 project, I had numerous opportunities to use commercial airlines, general aviation and business jets. Based on this experience, I envisioned great potential for high performance light jets. I thought if an airplane could be designed having both high fuel efficiency and high speed, without sacrificing cabin volume and luggage space, there would be a potential demand in the business jet market. So, the new design targets for a Honda advanced light jet became:

1. Maximum efficiency—improved fuel efficiency by 20% compared to conventional jets. The target was to realize a direct operating cost per seat mile close to that of a first-class ticket on a commercial airliner under full seating capacity. This direct operating cost should be the lowest among light jets.
2. Maximum cabin space—increase cabin space by 20%. Even though it may be categorized as a light jet, the cabin space should be plush and passengers' feet should never overlap in a club seating arrangement. There also should be a roomy and comfortable lavatory, which passengers do not hesitate to use. Cabin space should be the best in the class.

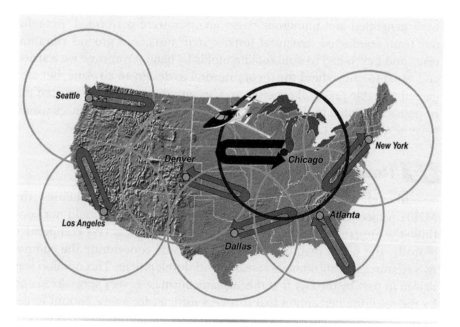

Figure CS4.1 Round trip range without refueling.

3. Maximum luggage space—maximized luggage space capable of storing six golf bags. Many light jet users would use their jets to go play golf, so this is an important requirement.
4. Maximum speed—cruise speed that should exceed 400 kt, even though it may be classified as a light jet, it should be the fastest in its class.
5. Ownership experience—not only high performance and high efficiency, but also highly attractive to satisfy customers' ownership experiences from quality and aesthetic points of view. The airplane should look beautiful.
6. Optimum range—capable of flying from New York to Miami nonstop and operating out of most major U.S. hub cities via round trips without refueling as shown in Fig. CS4.1.

To achieve these design goals, it is necessary to reduce the overall airplane size as much as possible in order to reduce drag and operating costs, but we also need to maximize space for the cabin and luggage. Achieving these two opposing requirements is a true technical challenge that required the development of various new technologies.

CS4.3 Development of HondaJet Key Technologies

CS4.3.1 Over-The-Wing Engine Mount (OTWEM) Configuration

Current business jet designs have engines mounted on the rear fuselage. If we could mount the engines on the wings, the carry-through structure required to mount the engines on the rear fuselage would be eliminated. This would allow the fuselage internal space to be maximized without increasing the size of the fuselage. Since light business jet designs are very close to the ground, it is impossible to install the engine under the wing. Where else could the engines be located? Nearly fifteen years ago the idea of an OTWEM arrangement was sketched on the backside of a calendar as shown in Fig. CS4.2.

It was immediately recognized as a technical challenge, both from aerodynamic and aeroelastic standpoints, to employ an OTWEM configuration for a high-speed jet aircraft. In general, locating the engine nacelle over the wing causes unfavorable aerodynamic interference and induces a strong shock wave that results in a lower drag divergence Mach number. Also, mounting the engine on the wing significantly changes the vibration characteristics of the original clean wing and, as a result, influences aeroelastic characteristics as well. So, initially, it was thought to be a poor idea to mount the engine over the wing from both aerodynamic and aeroelastic standpoints. It was critical to the design objectives that the OTWEM

Figure CS4.2 Original sketches compared to final design.

configuration be part of the design, so the initial research was focused on "how to design an OTWEM configuration that minimizes the wave drag increase at high speeds (high Mach number) and achieves higher cruise efficiency and minimized aeroelastic disadvantage."

Wave Drag Reduction

Aerodynamic design was initially focused on how to minimize aerodynamic interference from the nacelles and pylons by optimally contouring the nacelle, pylon, and fairing. After several design iterations I glanced at my bookshelf and happened to see an old textbook on aerodynamics authored by Ludwig Prandtl. This book reminded me that when there were no computers for numerical simulation or computational fluid dynamics, aerodynamic flow was analyzed by using analytical methods with complex functions $(x + iy)$ to represent the flow around a body. For example, superposition of basic flow functions, such as source-sink, circulation, uniform parallel flow, etc., are used to construct the resulting flow around a body. These theoretical methods renewed the thought process that led to a more evolved design.

The new approach focused on how to "create flow and pressure distribution" that would reduce wave drag rise by superimposing flow fields around bodies, instead of attempting to minimize the interference between two bodies. In other words, an optimum pressure distribution and flow was sought by superimposing two flow fields. I refer to this as "favorable interference." The HondaJet design needed a Natural Laminar Flow (NLF) wing airfoil section to greatly reduce drag. However, an NLF airfoil exhibits high wave drag at high Mach number because of its pressure distribution required for laminar flow. So, it was critical to reduce the upper surface shock wave strength to increase the drag divergence Mach number when we used an NLF wing for high speed aircraft.

I attempted to use a flow field in front of the nacelle to decelerate flow where local wing flow velocity is reaching the speed of sound and a shock wave is formed when an aircraft flies at high Mach number. I thought both (1) that this effect of nacelle flow field to the wing pressure distribution would be smaller at lower Mach number and a favorable pressure gradient could be maintained at most climb and cruise speeds, and (2) that the effect would be larger and could delay shock wave at high Mach number. As a result, we would be able to achieve laminar flow under most flight conditions and, at the same time, reduce the shock wave strength at high Mach number, achieving a higher drag divergence Mach number.

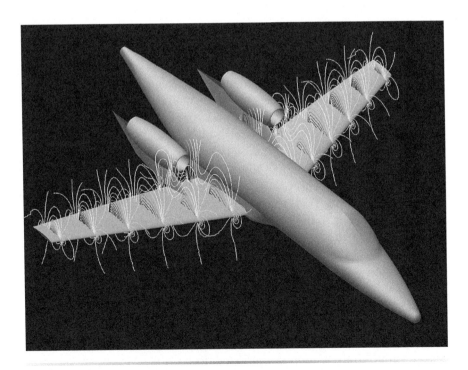

Figure CS4.3 Off-body pressure contours of the OTWEM configuration.

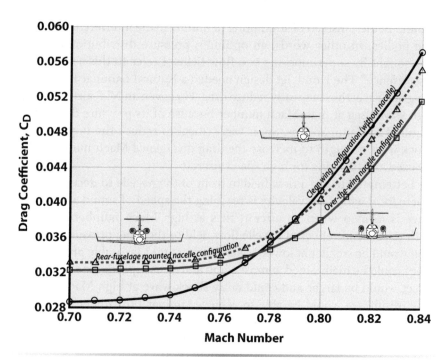

Figure CS4.4 Comparison of drag divergence for various nacelle configurations.

Based on this approach, hundreds of computer simulations were performed for different OTWEM configurations having different nacelle locations relative to the wing. These runs confirmed that there was an effect of chord wise and vertical nacelle locations on wave drag. The simulation results demonstrated that the strength of the shock can be reduced and that drag divergence occurs at a higher Mach number than that for the clean-wing configuration when the nacelle is located at optimum position relative to the wing (Fig. CS4.3). A transonic wind tunnel test was conducted in the Boeing Transonic Wind Tunnel (BTWT) to validate the simulation results. The optimum over-the-wing nacelle configuration exhibits lower drag than the conventional rear-fuselage engine-mount configuration [2] (Fig. CS4.4). The concept of positive interference of the OTWEM configuration was also confirmed at the NASA NTF (National Transonic Facility) at high Mach number and high Reynolds number testing as well (Fig. CS4.5). Ultimately, shock wave strength was reduced and drag divergent Mach number was increased for the optimum OTWEM configuration while maintaining laminar flow under most flight conditions. The final aerodynamic configuration of the HondaJet's OTWEM is based on these results.

Figure CS4.5   NASA NTF wind tunnel test.

Stall Characteristics

Another critical technical evaluation of the OTWEM configuration is to determine its stall characteristics. Docile stall behavior is very important for an airplane. Analytical studies were performed to design wing twist distributions and pylon shapes. Then, wind tunnel tests were conducted to measure the design's stall characteristics. The upper wing surface stall pattern of the OTWEM configuration from a 1/6-scale, low-speed wind-tunnel test is shown in Fig. CS4.6. The wing stall pattern begins at 55% semi-span and separation propagates inboard. At the stall angle of attack, the region of the wing between the fuselage and the nacelle is not stalled, resulting in good stall characteristics. Additionally, there is adequate stall margin over the outboard portion of the wing. The lift curves obtained from the 1/6-scale tests with and without nacelles are shown in Fig. CS4.7. The zero-lift angle and the maximum lift coefficient of the OTWEM configuration are higher than that of the clean-wing configuration. Consequently, there is no disadvantage with respect to the lift characteristics for this OTWEM configuration [3].

Inlet Distortion Evaluation

I believed it was important to investigate the inlet-flow distortion for the OTWEM configuration at high angles of attack because the separated

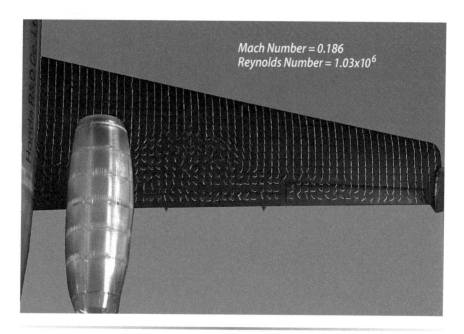

Figure CS4.6 Stall pattern.

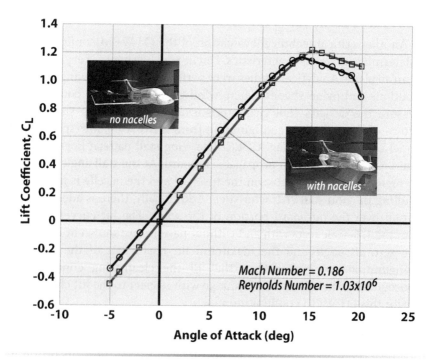

Figure CS4.7 Comparison of lift curve with and without nacelles.

flow from the wing at high angles of attack could enter into the engine inlet. The inlet shape and length were carefully designed by taking these conditions into account. To evaluate these characteristics, a 1/6-scale, powered-model test using DC motor engine simulators was conducted in the Honda Low-Speed Wind Tunnel. The inlet and duct shapes were scaled from the full-scale nacelles, and forty total-pressure probes were mounted at the fan face (Fig. CS4.8). An investigation was conducted to determine if the measured total-pressure distortion exceeded the limits for high and low mass-flow conditions at various angles of attack and sideslip angles. Examples of the distortion pressure patterns are shown in Fig. CS4.9. The inlet total-pressure distortion is less than 0.1% up to the stall angle of attack and less than 2% at post stall angles of attack. Similar tendencies were obtained from tests with low and high mass-flow ratios. The results demonstrate that the distortion does not exceed the limits specified by engine requirements within the flight envelope.

The OTWEM configuration has a higher cruise efficiency than that of a conventional rear-fuselage engine-nacelle configuration and, at the same time, the cabin volume is maximized. Figure CS4.10 shows the final configuration of the HondaJet OTWEM configuration [3].

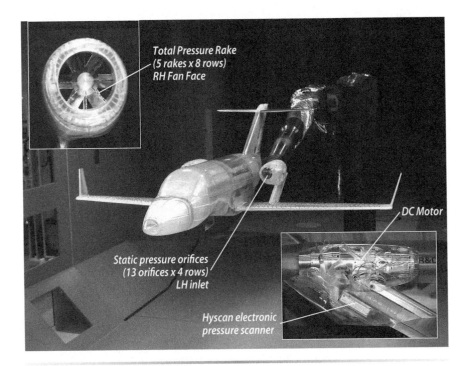

Figure CS4.8   1/6 scale engine simulator.

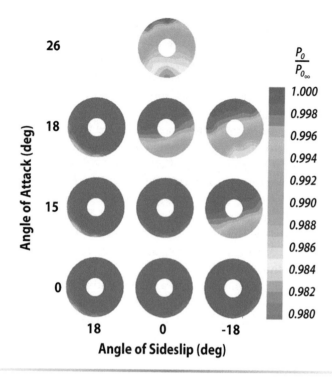

Figure CS4.9 Pressure distribution pattern obtained from powered model test.

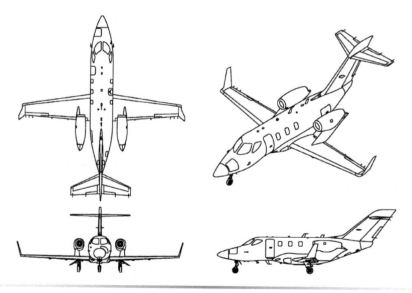

Figure CS4.10 General arrangement.

Aeroelastic Characteristic

Mounting the engine on the wing, however, significantly changes the vibration characteristics of the original wing and, as a result, influences the aeroelastic characteristics. In addition, the nacelle aerodynamic load and interference will affect the flutter characteristics. Positioning the engine ahead of the elastic axis of the wing to increase the flutter speed is a well-known design rule, which has a marked effect on the configuration of modern transport aircraft. For the present OTWEM configuration, however, the engine is positioned aft of the elastic axis of the wing, and the aeroelastic characteristics must be carefully calculated. Also, the aerodynamic effect on the flutter characteristics induced by having the engine nacelle positioned over the wing must also be included, especially in the transonic flight regime. It is necessary to validate flutter characteristics for the over-the-wing engine nacelle configuration.

The flutter characteristics of the OTWEM configuration were determined using extensive theoretical studies and numerous wind-tunnel tests. The location of the engine mass and the stiffness of the pylon relative to that of the wing are also important parameters for wing-flutter characteristics. Theoretical analyses were performed using the ERIN code and NASTRAN, and substantiated with low-speed and transonic wind-tunnel flutter tests at the National Aeronautical Laboratory Transonic Flutter Wind Tunnel. The study shows that the symmetric flutter mode is more critical than the anti-symmetric mode for the OTWEM configuration [4].

The effects of the aerodynamic load and the interference due to the engine-nacelle installation over the wing were studied as well. The result showed that the effects are small for this OTWEM configuration. Furthermore, the engine-pylon vibration characteristics influence the flutter characteristics. The study showed that the flutter speed is highest when the engine-pylon side-bending frequency is close to the uncoupled first wing-torsion frequency (about 0.9 to 1.0 times the uncouple first wing-torsion frequency). The flutter speed is also lowest when the engine-pylon pitching frequency is about 1.25 times the uncoupled first wing-bending frequency (Fig. CS4.11). Based on these results, spanwise location of engine, wing stiffness and mass distributions were designed to satisfy the flutter-clearance requirements.

CS4.3.2 Natural-Laminar-Flow Airfoil (SHM-1)

To maximize the performance of the HondaJet, a new natural-laminar-flow airfoil, the SHM-1, was designed using a conformal-mapping method [5]. The pressure gradient on the upper surface is favorable to about 42% chord,

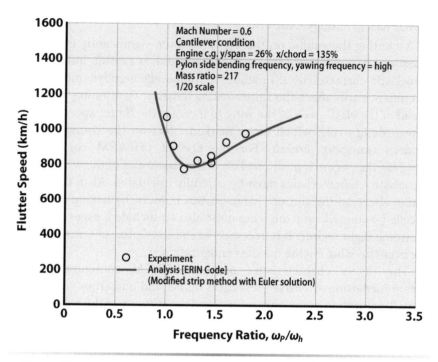

Figure CS4.11 Correlation of the flutter analysis with experimental data.

followed by a concave pressure recovery, which represents a compromise between maximum lift, pitching moment, and drag divergence. The pressure gradient along the lower surface is favorable to about 63% chord. Wing leading-edge geometry was designed to cause transition near the leading edge at high angles of attack to minimize the loss in maximum lift coefficient due to roughness. The upper-surface trailing-edge geometry was designed to produce a steep pressure gradient and, thereby, induce a small separation. By incorporating this new trailing-edge design, the magnitude of the pitching moment at high speeds is greatly reduced. The SHM-1 airfoil and its associated pressure distribution are shown in Fig. CS4.12. The airfoil has been tested in both low-speed and transonic wind tunnels, as well as full-scale flight testing using a gloved T-33 aircraft (Fig. CS4.13, Full-scale flight testing validated the performance of the airfoil at full-scale Reynolds number and Mach number. The laminar-to-turbulent boundary-layer transitions were visualized in real time using an infrared (IR) camera during the T-33 flight tests (Fig. CS4.14). As designed, the airfoil exhibits a high maximum lift coefficient, and yet, has docile stall characteristics and a low profile-drag coefficient.

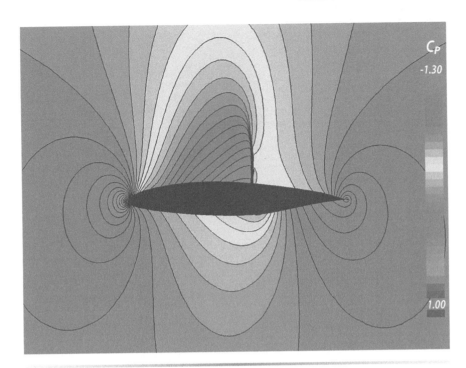

Figure CS4.12   Airfoil pressure fields showing location of upper surface shock.

Figure CS4.13   T-33 flying testbed.

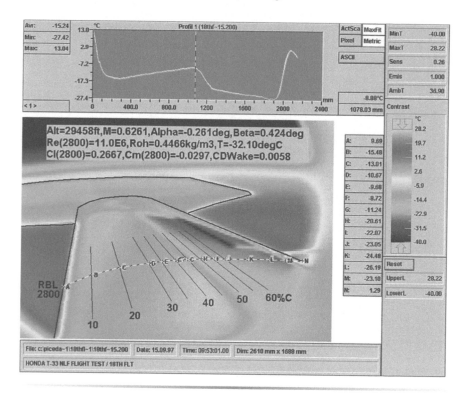

Figure CS4.14   Boundary layer transition.

CS4.3.3 Composite Fuselage

Composite material is now widely used in the aviation industry to reduce structural weight by taking advantage of its superior mechanical properties such as specific strength. However, careful evaluation is needed especially for composite material application for light jets because of its cost and the relative size of the aircraft. The weight benefit is often limited by the necessary minimum gauge of the structure and other design constraints. As a result, it is not always easy to take advantage of the characteristics of composite material for aviation applications. In addition, unique characteristics of composite material, including strength "knock down factor" for hot wet conditions, compression after impact (CAI), and inter-laminar shear strength, etc., are design constraints that must be considered, which have negative impacts on actual weight reduction. For the HondaJet structure, composite material is applied mainly to the fuselage taking into account all of the design aspects and constraints described above.

The HondaJet's fuselage is constructed entirely of graphite composites. The material is a 350-deg-F cure epoxy pre-preg reinforced by carbon fiber. The matrix is Cytec 5276-1 high-damage-tolerance, epoxy resin, while the reinforcement is TOHO G30-500 high-strength, intermediate-modulus fiber. As shown in Fig. CS4.15, the cockpit and tail sections are a honeycomb sandwich construction to maintain the compound curves, which are especially important for the laminar-flow nose. The sandwich structure also has the advantage of reduced cost due to the ease with which it can be fabricated into complex, three-dimensional contours. An integrally stiffened panel structure is employed for the constant cross-section portion of the cabin. The stiffened panel structure reduces weight because of its high efficiency structure and also maximizes the cabin volume. The general frames and stringers have identical dimensions in the constant cabin section, so the numbers of molds for the frames and stringers are minimized. The constant fuselage section can be easily extended to satisfy future fuselage stretching. A feature of the fuselage fabrication is that the sandwich panel and the stiffened panel are co-cured integrally in an autoclave to reduce weight and cost. It was a technical challenge to cure the honeycomb sandwich structure under the pressure (85.3 psi) required for the stiffened panel, but a new method called the "picture-frame stabilizing method" prevents core crushing.

Another feature of the HondaJet composite fuselage is the buckling tolerance design that has been adopted to the stiffened panel. Shear buckling is allowed under limit load. The skin thickness and ply orientation of each structural skin bay were designed for optimum stress level and contribute to weight reduction as well (Fig. CS4.16).

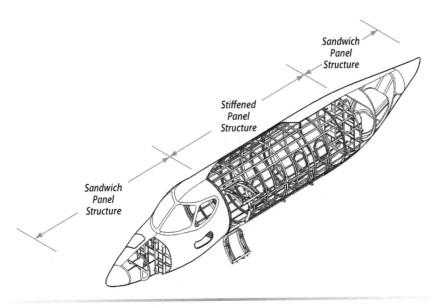

Figure CS4.15 Fuselage structure.

Figure CS4.16   Fuselage skin.

Through the application of composite material for the HondaJet's fuselage, we have achieved lower weight along with both an affordable fabrication cost and the best contour for aerodynamics (Fig. CS4.17).

CS4.3.4 Advanced Cockpit Design

Within the automobile industry, design cycles are very short and the number of models is very high compared to the aviation industry. As a result, the automobile design process from concept to 3D definition is not only very sophisticated, but also very short and efficient. For the HondaJet, the primary goal for the cockpit design was to achieve a high degree of flight safety while incorporating automobile interior and cockpit design processes. This approach achieves a high degree of integration of cockpit functionality, human factors consideration and interior aesthetics. In order to realize this objective, we conducted several systematic design studies for cockpit layout, machine human interface to define basic dimensions for seating, flight controls and switches, and visibility pattern assessment (Fig. CS4.18). Then, several concept sketches were drawn, and, based on the sketches and defined geometry, very high fidelity 3D surfacing data was created. At this point virtual design studies for aesthetics, including

Figure CS4.17 Composite fuselage.

evaluation of textures were undertaken. These studies utilized high-performance workstations and software that projected images on what is called a "powerwall," which offers four-times higher resolution display than a standard high-definition display. In addition, more detailed simulation analyses, such as ray tracing analysis of the instrument panel display's reflection on the windshield, etc., were conducted to define and verify glare shield geometry.

When the original concept of the HondaJet was started in the 1990s, cockpit and avionics systems on light business jets were not very well integrated and still relied heavily on conventional instruments. The goal was to incorporate a highly integrated system with large Primary Flight Displays (PFDs) and a Multi-Function Display (MFD) presenting all flight information. The HondaJet flight deck features a Honda-customized Garmin all-glass avionics system incorporating an advanced layout with three 14-in. landscape-format displays and dual touch-screen controllers for overall avionics control and flight plan entries. The cockpit is shown in Fig. CS4.19. All information from flight and engine instrumentation to navigation, communication, terrain, traffic data, and the like is uniquely integrated and digitally presented on the large-format, high-resolution, dual PFDs and single MFD. The PFDs have optional Synthetic Vision capability and the

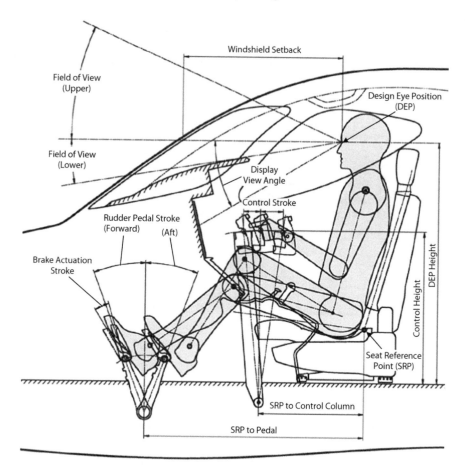

Figure CS4.18 Pilot geometry.

MFD features split-screen capability with satellite weather, graphical synoptics, etc. Intuitive touch-screen multi-function controllers provide a low-workload user interface that is ideally suited to our high-performance light jet aircraft. The HondaJet Avionics Suite, integrated into our human-centric cockpit design, represents a significant enhancement in both capability and user experience.

CS4.3.5 Innovative Interior Design Concept

A challenge for the interior design of light business jets is that the cabin volume is limited by the fuselage dimension, and it is very difficult to achieve adequate space to satisfy passengers' comfort requirements. The volume desired for each passenger depends on the trip duration, and an example is shown in Fig. CS4.20. Based on the HondaJet aircraft's designed

Figure CS4.19 Actual HondaJet cockpit.

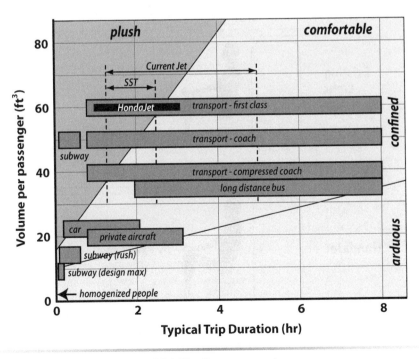

Figure CS4.20 Passenger volume.

trip duration passengers were allocated 60 ft^3 [6]. This volume is somewhere between that available in a commercial airline's business and first class cabin area. These are plush accommodations for light jets. By applying the OTWEM configuration, it is possible to maximize the HondaJet cabin volume without increasing the outside dimensions of the aircraft and, consequently, airplane drag and weight. The HondaJet cabin volume is approximately 20% larger than similarly-sized business jets. By virtue of the OTWEM design, cabin length can be much greater and hence, seat-pitch increases significantly compared to conventional light jets. As a result, extra leg space has been realized. Within the HondaJet cabin, passengers' feet do not overlap each other—a common discomfort in many light jets and a definite advantage for the HondaJet (Fig. CS4.21).

The private lavatory was one of the more important interior design attributes for the HondaJet. Research indicated there was a true hesitation for many light jet passengers to use a lavatory in flight. Therefore, designing a lavatory that passengers would not hesitate to use would further enhance the passenger's overall comfort. The result is a lavatory that is very spacious. Taking this one step further, two sky light windows were installed on the ceiling to provide a source of outside light into the lavatory to enhance an already spacious environment (Fig. CS4.22). By offering this

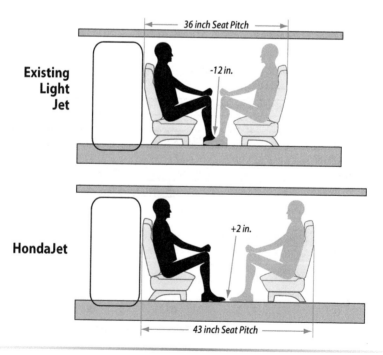

Figure CS4.21   Seat foot and leg room.

Figure CS4.22 Lavatory.

roomy and bright lavatory, the emotional hesitation to use the onboard lavatory is eliminated and the overall level of passenger comfort and relaxation is increased substantially. Considering that the purchase decision and the desired use of light jets are often based on emotional reactions to, and interactions with, the aircraft, I believe the lavatory design is a definite selling point that results in a positive HondaJet ownership and user experience.

Important improvements in the HondaJet interior design in the areas of aesthetics and human factors have been realized by utilizing Honda's automobile interior design expertise. It likely can be said that the HondaJet is the first business aircraft to fully utilize automobile interior design processes. My general impression is that the typical current business jet interior design philosophies and its design processes are still not as refined as those in today's automobile industry. The character of most current business jet interiors is expressed by the use of high-end materials, such as walnut wood or gold plating. However, compared to modern luxury automobile interior designs, there would seem to be many potential areas for improvement in business jet interior aesthetics and ergonomics.

By creating high fidelity 3D surfacing computer models, we have employed highly realistic computer graphics to make virtual material selections for the HondaJet interior. In addition, a precise, full-scale interior mock-up—both cockpit and cabin—was fabricated, and many design parameters were assessed and finalized using production design parts. Integration of aesthetics, human interface, and ergonomic considerations gives the HondaJet interior a more modern, yet timeless, image. The original interior concept sketch and final production interior are shown in Fig. CS4.23 respectively.

A very important design consideration is the ease of installation of each interior part during the completion phase of aircraft final assembly. Automobile designers use sophisticated techniques to design each interior part to have exceptional fit and finish in final assembly by using each part's overlap or gap to achieve high tolerance management. This cost-effective design technique takes into account slight variations that may occur with each installation and results in a consistent, high-quality look and feel. Furthermore, as the installation time for aircraft interior parts generally is much longer than for automobiles, such business aircraft parts historically

Figure CS4.23 Interior layout.

may not be well designed for a mass production concept. To improve parts installation time—as well as the quality of fit and finish—automobile interior design experience and expertise has played a crucial role in HondaJet interior design.

CS4.3.6 Paint Scheme

Paint is as important as other technologies for automobile manufacturers. The paint scheme of the HondaJet is significant because this is my final "art work" contribution. Generally speaking, aeronautical engineers are not usually concerned with colors and paint schemes, but to me they are just as important as they are for cars. I wanted to indicate physical phenomena and forms through colors. We came up with more than 150 proposals and finally decided to use the paint scheme shown in Fig. CS4.24. I wanted to reflect the movement of air flowing over the fuselage—dark blue on a white background with silver edges making for vivid contrasts. The coloring is a bit eccentric for this kind of business jet but matches very well its shape. I wanted to use a blue that seems to swallow you up. Many airplane designers may not pay much attention to paint, but I put a lot of thought into choosing a design to express both my passion and effort that I have put into this aircraft.

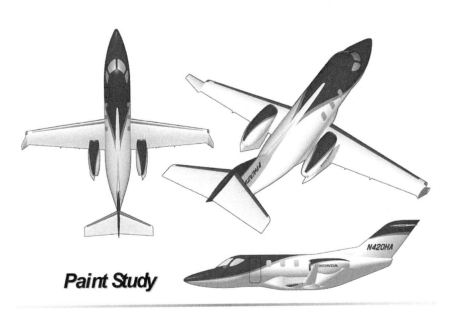

Figure CS4.24 Paint scheme.

CS4.4 HondaJet World Debut

CS4.4.1 First Flight

It took almost six years from the original proposal to completion of design and fabrication of the HondaJet. There were many discussions and arguments about the HondaJet concept—and even skepticism about the project—but we finally reached the point of being able to prove the aircraft's concept was valid. Our resources were limited, but everyone on our very small team worked hard to make the HondaJet a reality [7]. After completing the final assembly and all ground tests, the HondaJet was pulled out of the hangar and appeared on the ramp under the sunlight. It was the first time for the HondaJet to appear under the sky. During its steering test, the HondaJet ran on the ramp as if a figure skater were making "figure-eights" on the ice. I must admit I became rather emotional when I saw the HondaJet taxiing under its own power. It looked like it was moving under its own will. At last, I felt as if my "daughter" would become independent of me as she tried her wings for the first time. Following steering tests and low- and high-speed taxi tests, the HondaJet made a successful maiden flight on December 3, 2003 (Fig. CS4.25).

Figure CS4.25 First flight.

CS4.4.2 HondaJet World Debut at Oshkosh AirVenture

Although we could prove from the flight test results that the HondaJet met performance goals and showed potential [8], it was the opinion of some of Honda's top management that it was difficult to get into the airplane business. In order to gain a better understanding from the board of directors of the commercial potential, a HondaJet would be displayed publicly to get reactions from industry people and potential customers. The airplane was still experimental and there was no plan to commercialize the HondaJet at that time. This being the case, the perfect venue for displaying the HondaJet was AirVenture Oshkosh, which is primarily an experimental aircraft show.

On July 28, 2005, in the clear morning air and under the strong summer sunlight, the HondaJet landed at the Oshkosh airport. During taxiing to the ramp, thousands of airplane enthusiasts stopped to stare at the HondaJet. In an instant, the HondaJet was surrounded by people in AeroShell Square, and everybody was very excited by this dynamic new airplane design. I still cannot forget the pride and excitement I felt seeing the beautiful HondaJet painted in deep blue and bright white under the clear sky in Oshkosh. It was the first time the HondaJet had appeared in public (Fig. CS4.26). Although I had been working on the airplane project within Honda for

Figure CS4.26   Oshkosh 2005.

almost twenty years, I really could not have imagined that this moment would be realized.

Many airplane fans at Oshkosh talked to me and said, "I have never seen such a beautiful airplane." It was the most wonderful compliment, and one that I had never experienced before. In my heart, however, my emotions were more complex at that time, as there was still much debate within Honda's top management as to whether the airplane program should be terminated or proceed. Some in Honda's top management still felt it was difficult to commercialize the HondaJet. I worried that this showing of the HondaJet at Oshkosh might be both the plane's public debut and the closing ceremony on the project. However, I was determined not to give up. I was hoping that I could commercialize the HondaJet and deliver it to customers by changing the general opinion of the board of directors.

At Oshkosh, the HondaJet surprisingly attracted many people and drew tremendous interest. I was asked many questions about Honda's plan for commercialization of the HondaJet. Even though we did not have any plan at that time to manufacture the HondaJet, someone sent a $50,000 deposit check to purchase a HondaJet. Many media wrote positive comments about the design and performance of the airplane, and the reaction was overwhelming and much more positive than even I had hoped. Based on these overwhelming reactions, Honda's top management gradually changed its opinion and realized the great potential for the HondaJet in the market. I could feel that both the internal and external atmosphere in support of the HondaJet project was growing positively after the Oshkosh debut.

CS4.5 Decision to Commercialize the HondaJet

Following Oshkosh and armed with the tremendously positive public and industry feedback to the HondaJet, I approached Honda's board of directors with a proposal to commercialize the airplane. After several discussions with the board about the HondaJet and the proposed business plan, the decision to commercialize the airplane was finally made in March 2006. The president of Honda Motor Company at that time, Mr. Fukui, made his final decision after a few minutes of silence. He spoke as if he were convincing himself, "Honda is a mobility company. We should pursue the future through the HondaJet." For a moment I could not even believe what I had just heard. Using a Japanese expression, it was the moment the "mountain moves." Based on this decision, we finally could make a formal announcement of our commitment to commercialize the HondaJet at Oshkosh AirVenture in July 2006. It was exactly twenty years after I joined the airplane research project (Fig. CS4.27).

In October 2006, Honda formally started to take orders for the Honda-Jet at the National Business Aviation Association (NBAA) annual convention

Figure CS4.27   HondaJet news conference 2006, Oshkosh, WI.

in Orlando, Florida. I prepared as best as I could, but still I could not sleep the night before the event. On October 17 at NBAA, in front of the actual HondaJet being displayed on a slowly rotating turntable in the center of our exhibit, we held a press conference and formally announced sales of the HondaJet. Nearly 1000 people attended the press conference, and the exhibit was filled to capacity with customers eager to purchase the airplane (Fig. CS4.28). With so many guests, our exhibit certainly stood out within the convention hall. On the first day of NBAA, more than 100 HondaJets were sold. There were many customers who actually signed a contract on the spot after seeing the HondaJet displayed at our exhibit and experiencing the cabin mock up, which highlighted Honda's design innovations for light business jet. Customers were literally waiting in long lines to put down a deposit to purchase the HondaJet, and it was an absolutely unbelievable scene. Many NBAA attendees said that the HondaJet aircraft were "selling like hotcakes" and that such a scene had not been witnessed before in the history of NBAA. Salesmen and Honda associates, including me, were very excited and running around to take care of customers.

It was 1986 when I attended NBAA for the first time. It was held that year in Anaheim, California. I was a young engineer and had just joined Honda's airplane project. In Japan, there is no comparable event to NBAA. I vividly remember the moment when I entered the NBAA exhibition hall. I was struck by the gorgeous exhibits and was in awe of the stunning business

Figure CS4.28 Press conference 2006, NBAA.

jets. Ever since that first NBAA, it has been my dream to bring a jet I designed to this magnificent show. However, I did not really ever imagine that such a moment would be realized in that wonderful way in 2006.

CS4.6 Reflection on Mr. Honda

In the summer of 2009, I had the opportunity to go to the home of and speak with Mrs. Sachi Honda, the wife of Honda Motor Company founder Mr. Soichiro Honda. When I presented a HondaJet model to her and reported the commercialization of HondaJet, she mentioned to me that, "If Mr. Honda were still alive, how happy he would be to see the HondaJet launched." Mr. Honda always had a special passion for airplanes, and it was his dream to build an airplane from his youngest age. After he graduated from school, he joined a small automobile repair shop as a technician. He was an exceptional individual with a passion for engineering, and he eventually established Honda Motor Company, which became one of the world's most pre-eminent engineering companies.

I have one memory of Mr. Honda that I will never forget. I had an opportunity to see Mr. Honda only once when I was in my late twenties. I was a young engineer working at Honda R&D Japan. At the office, all Honda associates wear white uniforms, but one morning when I went into the restroom, I came across a gentleman with thin hair in a Hawaiian Aloha shirt. He looked very different, so I immediately recognized him as Mr. Honda. Although Mr. Honda had already retired from the company, he made occasional visits to various parts of the company see what was happening. At that time, I was working for the airplane project, which was very confidential . . . even internally . . . and it was not allowed for me to discuss anything about my job with anyone. My boss told me that the airplane project was so confidential that company management did not even let Mr. Honda know that Honda R&D was undertaking the airplane project. According to my boss, if Mr. Honda knew that Honda R&D was working on the airplane project, he could not resist returning to the company because of his strong passion for airplanes. So, we had to be very secretive with him as well.

When I looked at Mr. Honda, he looked at me as well, and our eyes met each other for a moment. I was about to say, "I am now working on the Honda airplane project!" But I was so young, and I did not have the courage to speak to him. I just bowed slightly, and he bowed slightly as well, and we passed each other without a word. After a few years, he passed away, and there was no opportunity to see him again. When I am facing tough challenges, I always think that, if I had spoken to him at that time, what would he have said to me? Would he have encouraged me to continue with the airplane project? Was he really looking forward to seeing the HondaJet fly? Unfortunately, I will never know. I am sure that he had a passion to achieve his dream to create an airplane. However, he probably did not imagine that his intention would be realized with the establishment of a company in the United States rather than in Japan. And, he probably did not expect his dream would start with the most advanced light business jet instead of a single-engine propeller aircraft.

I will never forget Mr. Honda's clear and bright eyes. I want to keep alive for the future his vision and challenging spirit, and I hope we will see many HondaJets flying all over the world.

About the Author

Michimasa Fujino (CS4.29) is the founding president and CEO of Honda Aircraft Company, the Honda subsidiary responsible for Honda's overall airframe business strategy, and the further development, marketing, sales, and production of the innovative HondaJet advanced light business jet. Prior to leading the formation of Honda Aircraft Company, Fujino rose through the engineering ranks to become a vice president with Honda R&D Americas Inc. and the large project leader for the HondaJet. In this capacity, he led all engineering tasks from design through experimental verification, fabrication, and flight testing of the HondaJet. Fujino created the unique Over-The-Wing engine mount (OTWEM) configuration for

Figure CS4.29 Author.

the light business jet. Fujino also developed a new natural laminar flow (NLF) airfoil and fuselage nose for the HondaJet. Not only did he design, build, and sell his concept, HondaJet, he also drove the formation and development of a new airplane manufacturing company, Honda Aircraft Company. Fujino, who joined Honda R&D Co. Ltd. in Japan in 1984, graduated from Tokyo University with a degree in aeronautical engineering. He now resides with his wife and three children in Greensboro, North Carolina.

References

[1] Fujino, M., "Aerodynamic and Aeroelastic Design of Experimental Aircraft MH02, DOT/FAA/CT-94/63," *Proceedings of the 1994 AIAA/FAA Joint Symposium on General Aviation Systems*, May 24–25, 1994, pp. 435–459.
[2] Fujino, M. and Kawamura, Y., "Wave-Drag Characteristics of an Over-the-Wing Nacelle Business-Jet Configuration," *Journal of Aircraft*, Vol. 40, No. 6, November–December 2003, pp. 1177–1184.

[3] Fujino, M., "Design and Development of the HondaJet," *Journal of Aircraft*, Vol. 42, No. 3, May–June 2005, pp. 755–764.

[4] Fujino, M.; et al., "Flutter Characteristics of an Over-the-Wing Engine Mount Business-Jet Configuration," AIAA Paper 2003-1942, Apr. 2003.

[5] Fujino, M., et al., "Natural-Laminar-Airfoil Development for a Lightweight Business Jet," *Journal of Aircraft*, Vol. 40, No. 4, July–August 2003, pp. 609–615.

[6] Torenbeek, E, *Synthesis of Subsonic Airplane Design*, Delft University Press and Kluwer Academic Publishers, The Netherlands, 1982, pp. 68–69.

[7] Murman, E, "Lean Aerospace Engineering," AIAA Paper 2008-4, 46th AIAA Aerospace Sciences Meeting and Exhibit, 7–10 January 2008, pp. 16–17.

[8] Fujino, M.; et al., "Flight Test of the HondaJet," ICAS 2004-4.10.1, 24th Congress of the International Council of the Aeronautical Sciences, August 29–September 3, 2004.

Case Study 5 — Hybrid Aircraft Technology and the Development Journey

Robert R. Boyd

- Technology Development
- Forming a Concept for a Value Proposition
- Finding Opportunity Space
- The Capability Gap
- Proof of Concept
- System Efficiency

In January 2006, the P-791 Hybrid Aircraft demonstrator lifted off in Palmdale California. The validation of this revolutionary technology occurred over the next 10 months, yet five years later no Hybrids are operational. The story serves as a poignant example of the very human process of technological change.

The more things change, the more they stay the same.
French Proverb

I t was a cold, dry dawn in the Palmdale desert as the first rays of sunshine illuminated the wispy clouds drifting by on the morning of January 31, 2006. The hangar doors of building 602 stood open; flurries of activity ebbed and flowed around the large P-791 airship standing just inside (Fig. CS5.1). A small group of engineers huddled against the cold awaiting movement from the strange multi-lobed airship looming over them. For them, the leadership of the P-791 team, the previous year had sailed by in a torrent of design and fabrication work on the vehicle now humming with life just yards away. Today was judgment day, first flight, the day when your designs either fly or they don't. Unbeknownst to the rest of the world, in a few minutes the product of thousands of hours of their labor would take to the sky. For the project manager, the author, there was strangely no sense of worry that failure was nearby, that the two people strapped into the odd craft were in any danger. The simulations had been thorough enough, the ground checks comfortably passed, and the inherent safety of the partially buoyant vehicle gave all at hand a level of comfort not normal for first flights. As the ship came to life, balancing on the four-legged hover pads, and began to slowly move from the huge hangar bay, the engineers turned to face the dawn, walking slowly toward the waiting runway. Only one thought kept echoing through the manager's mind: would this be the only manned hybrid airship ever flown?

CS5.1 Introduction

On the surface, development of new products and services appears to revolve around new technology, scientific discovery, or singular inspiration. Scientists, engineers, and technicians around the world work diligently toward these ends, deeply entrenched in the belief that the world around us is entirely deterministic. Two plus two is four, force is always equal to mass times acceleration, and new applications of technology merely need to demonstrate quantitative value to be widely accepted and implemented. Unfortunately for the aspiring engineer, humans are far from deterministic. Just beneath the surface in the world of development lurks the most challenging requirement: wide acceptance of a new product or service requires *change* from the current pattern of behavior exhibited by existing customers. Those customers are human. In other words, to have a successful new business, some cultural change must occur. This challenge should not be underestimated, for it typically dwarfs the technical challenges of developing the new product or service.

This article explores the world of technology development from the initial inspirational technology concepts through acceptance and implementation. The reference frame for this exploration is an emerging technology, Hybrid Aircraft, viewed through the eyes of the author from inception to current day. Perhaps because of its relative simplicity and/or

Figure CS5.1 Lockheed Martin P-791 aircraft exits hangar for first flight.

because of the large cost risk hurdles endemic in any aircraft development, Hybrid Aircraft technology serves as an ideal example to expose the subtle, but ever present human challenges in the development world. Hybrids have relatively low technical risk, perform much needed tasks, and are supported by robust business plans, yet they still struggle through an uphill battle to gain wide acceptance. There are a number of references available on the subject of effecting change and mountains of material on the human psyche; this work is not intended to capture the breadth and depth of human understanding but to offer the technical developer a glimpse into what may lie ahead. Some practical advice for navigating the challenging waters is extended along the way. Hopefully you will become intrigued in the development cycle and maybe learn something new about aircraft in general. Perhaps you are already involved in technology development, or aspire to move into this dynamic world; whatever your motivation, realize your path ahead will be a long, winding, and often frustrating journey.

CS5.2 The Road Less Traveled—Technology Revolution

Physics has the remarkable ability to not change. The only thing changing is our understanding of the world around us and our application of that knowledge to improve existing products/processes or respond to new challenges. In the sphere of products and services, these changes are often divided into two categories: evolutionary and revolutionary. Evolutionary change is common in the industrial world, arising from the incremental modification of a product or service from the existing state to something slightly better. Examples would be upgrading avionics or re-engining an aircraft. To some extent the discussions here apply to evolutionary change, but the primary focus will be on the road less traveled: technology revolution. There are only a handful of people collected in just a few places around the world who follow this road less traveled. I have been fortunate enough to work many years in one of those, the Lockheed Martin Skunk Works in Palmdale California. Having collaborated with many other innovative development groups, I have noted one commonality: a relentless desire for change. Note that desire for change is not a common human trait. In fact for roughly one-third of the population it is a major fear (reference 'temperaments' in Meyers-Briggs psychology). Another third will likely initially believe change to be more trouble than it is worth. Not to mention the sizeable contingent of people who are currently profiting from the way things are done now, making the proposed change not just a fear, but a threat. No matter the technology revolution proposed, there will be plenty of opposition. For this reason it is key to have environments where ideas can evolve safely in their early stages. Successful technology revolution organizations always have these places and seek to attract people who have a passion for change. Production aircraft development requires a wide range of skills and sizeable resources—not something easily done in one's garage. If this is your passion, find one of the few places left where aircraft innovation occurs and go there.

So, to find this road less traveled, where should one start? Different schools of thought have emerged on this subject. Some favor the Ah Ha! category, where inspiration just hits you, like the apple falling for Newton. In popular lore this often occurs just after a major defeat. There are a number of famous anecdotes about products and services appearing inspirationally from failures, like Post-It notes from a weak glue formulation or the ESPN network from lousy Boston Whalers coverage. While these "phoenixes" are interesting, compared to the scale of all technology development, they are rare. A more common class of innovation fits in the "Rubik's cube" category, where challenges are mulled over for an extended time, and different patterns or approaches tried, until one finally works. From practical evidence, this seems to yield success much more often. This category of revolutionary change comes from either applying a new technology approach to an existing need or combining mostly existing technologies to fulfill an underserved need. Some like to theorize more and test less, while others use a more action-oriented development approach. Both approaches seem to work; best to stick with a method that keeps your interest high for extended durations. As an example, the Internet revolutionized information transfer once affordable routers and high speed data links were developed and widely implemented: revolution arising from the application of new technology to existing needs (transferring information from one repository to another). The breadth and depth of this change has gone far beyond simple information transfer. Few could have imagined the Internet enabling business models that challenge deeply embedded business constructs like the postal system, newspapers, the telephone system, or even now the entire media broadcast system. Given that inspiration remains difficult to schedule, the best bet to define a revolutionary product or service will likely be to follow a Rubik's Cube approach.

For the Hybrid Aircraft, there was no 'Ah Ha!' moment. The current product spectrum revolves around a straightforward task: moving stuff from here to there. The twenty year path to define the current product spectrum followed a series of twists, turns, dead ends, successes, disappointments and frustrations along the way. As with all aircraft today, it was not a singular genius, but an effective team that yielded the result. Careful thought and modeling of the underlying physics was combined with verifiable tests to validate concepts. Surprisingly, the business models were at least as challenging, maybe more so, than the technical models. Often termed 'revolutionary' or 'disruptive', the Hybrid Aircraft products offer a new capability, affordable transport without expensive infrastructure, which could have a major impact on world economics. As we approach the first wave of products to emerge from this technology, it is instructive to understand the steps that got us here and why they are so important to effect the change required for a technology revolution.

CS5.2.1 First Step on the Journey: The Value Proposition Well Spring

Unfortunately for the dreamers of the world, an idea by itself isn't worth much. To be worth something, an idea needs to have compelling value to potential customers . . . and only customers can define value. So what is a technology developer to do? You can't define the value proposition, but you can help find it and help shape it. Understanding the value proposition of the new product or service is pivotal to the success of the technology development path, and absolutely life or death for the revolutionary technology (see blue box on next page). Finding the value proposition for a revolutionary technology may seem simple, but it isn't, mostly because all customers have one simple thing in common: they are humans. So, think of yourself as a sculptor staring at a lump of clay. How much clay do I have? What does the buyer want to do with this sculpture? Why are they buying it? If you

Revolutionary Product Value Proposition— A Life or Death Issue

Consider optical disc players. What we now call a DVD player is actually the second generation of a product known as an optical disc player. The first was the Laser Disc, a product such an utter failure that the second generation required a completely new name. Why did the first version fail while the second was wildly successful? Draw your own conclusions, but I contend that it was a value proposition problem. In the late 1970s and early 1980s when the Laser Disc first became affordable, there was another new and exciting video technology: videotape. Up until that time consumers had never been able to control the content on their televisions, only watch what was broadcast. Laser Disc offered the value of watching movies when you wanted to rather than when the networks decided to show them. It was upstaged by videotape which offered not only this capability, but also the ability to record programs from network television or other videotapes. Laser Disc proponents tried to counter this point by emphasizing the quality of the picture and the durability of the media, but to no avail. It wasn't until the music industry convinced consumers that optical music media (CDs) were a significant quality improvement over vinyl records (clarity, smaller size, durability) that the idea of quality in optical media began to take hold. With that cultural change and the frustration with videotape quality decrease over time, the consumer was ready for the 'new' DVD technology that then supplanted the video market almost entirely.

cannot reasonably write down in a few sentences, or preferably a few words, what the value proposition is for your revolutionary technology, how do you expect your customer to do so? If they do not understand the value, you have little hope of selling them anything no matter how "gee whiz" your idea may be.

Finding the value proposition is not one of those tasks best done sitting alone at your desk. It requires significant interaction with others, preferably customers, to really get to the core principles. In my mind, the opportunity (or value) space comprises the intersection of what is needed (defined by customers) with what is possible (usually defined by developers). Boeing, in developing the 777, brought forward significant improvements in composites, aerodynamics, electronics, and propulsive efficiency. A committee of Boeing developers and airline customers was formed to help define which of these technologies had the most value to the fare-paying passengers and the airlines themselves. By effective modeling of the benefits of the new capabilities, a solid performance model was developed and numerous advance orders were garnered (reducing investment risk). As the developer, you must understand the heart of the technology: how it works, what is easy and what is difficult, what drives costs in the short term and the long term, what hurdles are left to cross and how long it will take to get there, what are the sensible limits of capability, etc. (Fig. CS5.2). If you do not, nobody else will. More importantly, the customers you interface with and the resource providers who enable the next steps will quickly discover your lack of substance and either dismiss the concept altogether or go find someone else who actually does understand the technology. In either case, this important audience will not take you seriously and your path may grow cold. Initially, you may not even know who your real customers are. Only by understanding their business, what works and what doesn't, the value they deliver to their customers and how their business is usually done, can you bound the value space. This part is challenging, especially for the engineer. Very little in your training will prepare you for the seemingly random and outrageous desires and demands of your customers or the highly nonlinear ways the status quo evolves. The best advice is listen to (not just hear) what they are saying. Very often customers themselves do not know what they really want. You often get the "I will know it when I see it" response; possibly true, but not terribly helpful. Another favorite is "we've always done it this way," to which I always ask the simple question, "why?" My question usually puts them on the defensive but if you are with a

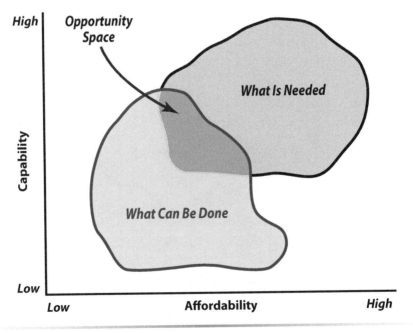

Figure CS5.2 Finding opportunity space.

particularly enlightened customer may yield a very interesting line of discussion. Do not make the mistake of talking to only one customer and thinking you've got it. One customer does not make for a revolution. It will take many to push through the adversity ahead, so better to build a cadre of supporters before the tension rises than to wait and try to find advocates in the later stages of product development. As you map out your value space, existing capabilities and gaps will start to become evident. Zero in on the places where there are lower barriers to entry, and where there is a match between need and technical possibility. These places are the fertile ground to find strong value propositions. An example map of transport cost versus forward speed is shown in Fig. CS5.3, revealing a potential gap between ground and air modes of transportation. Hybrids are positioned specifically to address this gap in the transportation value space.

Once a concept for a value proposition is formed, it needs to be shaped into a solid and defensible statement. Just like the lump of clay, it needs to be worked over time, carving away the parts that do not need to be there, adding bits as needed, and shaping the final result. At first, just try to write it down. You will find this to be oddly difficult. It may seem obvious to you, but the words may not fit right, or there are too many, or the story is circular. If you can't write it down, you can't explain it. If you can't explain it, nobody will understand it. If nobody understands it, there will be no acceptance, nor any resources to develop it. Many a technologist falters at this

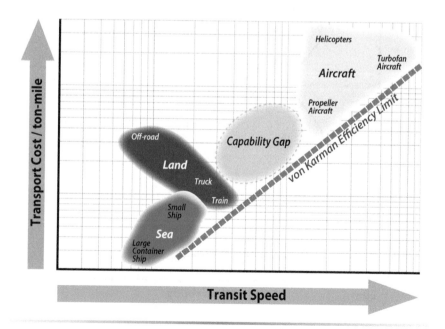

Figure CS5.3 The capability gap in transportation cost/speed space.

point, choosing to retreat back into the science (more exploring of what is possible) rather than find the value in what they do that will transition into reality, or worse, face a painful truth: there may not be a sensible value proposition. Shaping your value proposition requires some work. This is the point where non-technical factors such as cost, schedule, competition, current practice, cultural bias, and many others creep into the discussion. Just because it can work doesn't mean anyone will pay for it. A top level business plan should be sketched out to support the claims. Unless your customer is a technologist, it is best to be brutally honest with yourself here. Imagining massive cost reductions with scale or 'just one miracle' to make it work or assuming 100% market share are common traps that can generate an unrealistic value proposition. These kinds of assumptions make for great bumper sticker headlines and can get you past the casual observer. The serious development resource providers will not be fooled. Given that any new aircraft will require serious resources (tens to hundreds of millions of dollars) to develop, it is best to be honest up front. Even if you get started up the "proof of concept" mountain, without a strong value proposition the path will not go very far. Once you have a starting point, share it with your customers, refine it, shorten it, find the key points that need to be verified for it to work and get their buy in. This process never really ends, just dissolves into product evolution once your revolutionary concept becomes widely accepted.

The current Hybrid Aircraft value proposition represents a long evolution of thought and dialog, blending the possible with the needed and sharpened by what makes financial sense. Initially, the idea of using buoyant lift technology (airships) to augment current transport modes seemed to make sense. Assumedly it would be more cost effective and there seemed to be few major technology hurdles. In concert with our potential shipping customers, we worked to define the 'middle market', the gap in transportation that is faster than a ship yet cheaper than an airplane (a wide gap). Fortunately, the shipping industry is fairly straightforward: cargo moves either in standard containers or in bulk, markets are point to point, and the key driving variables are delivery time (block speed) and cost per ton of cargo per mile traveled ($/ton-mile). As with most things, the devil was in the details. Airships are large, making them difficult to fit into existing spaces or for that matter even land on existing runways. Airships are also particularly susceptible to shifting winds on the ground, making loading and unloading cargo challenging as the airship weather vanes (or 'kites') around the mooring mast. One lingering technical challenge is the simple fact that unlike aerodynamic or direct (thrust based) lift, buoyant lift cannot easily be turned off. This is a major problem when you want to unload half the aircraft's gross takeoff weight yet stay within a narrow range of allowable weight states (more like 10% of the gross weight) lest the airship float away. Significant time was spent working solutions to these problems, which eventually led to the current concept of the Hybrid Aircraft, blending buoyant lift with aerodynamic lift. One unexpected item was the natural problem of blending airship envelopes

> **Hybrid Aircraft Value Proposition (Simplest Form)**
>
> "Affordably move stuff to and from places with lousy infrastructure."

with regular landing gear, a problem that grows with size. Our investigation into air cushioned systems led to both a better technical concept and an interesting market space: remote cargo delivery. Although not directly considered in the initial concept, it became clearer with time that the best opportunity, and thus the strongest value proposition, arose not from areas with well-developed airports, rail, and road systems, but from places that had lousy transportation capability. For a long time the Hybrid value proposition read "no infrastructure" but after enough long-winded push back from numerous sources that "no" was too extreme to be realistic, it was softened to the more appropriate word "lousy" resulting in the simplest form (see box). Note that up front and prominent is the word "affordable," which speaks to the fundamental fiscal conservatism of the transport industry, the primary customer base. The value proposition also helps define what the concept is not. In the case of the Hybrid, we are not seeking to alter the way cargo is packed for transport, nor impose new infrastructure burdens, or compose a grand plan for realignment of the fundamental nature of the shipping industry. For the Hybrid Aircraft, keeping perceived change to a minimum was critical given the very traditional customer base not accustomed to taking on significant risks. This value proposition has guided the approach to technical and business validation over the past decade of development.

Once a strong value proposition begins to evolve, key driving points emerge and pressure builds to verify these points. For the Hybrid Aircraft the key points were ability to operate with little infrastructure and cost reduction below existing transport modes. The next step on the journey is often the most exciting for the technology developer, providing evidence to back the claim, often described as proof-of-concept.

CS5.3 The Mountain of Proof

As the technology journey continues past the initial steps, the terrain becomes much more challenging. Especially with aircraft development, the resources required to validate both technical and business claims are sizeable. Yet a first article, or demonstrator, is normally produced before larger resources for production will be released. There is a sensible reason for this tied to the balance between financial risk (capital outlays) and reward (returns); more risk requires more reward. Demonstration reduces risk, thereby reducing higher return requirements for the larger production investment. A smaller investment in a demonstrator limits financial exposure in case it fails to meet performance, costs more, or takes longer than expected (virtually every aircraft produced falls prey to this, Fig. CS5.4). Financial managers typically seek to balance portfolio risk levels across a wide range of industries and projects so that no one project failure can destroy the entire portfolio. At these early stages of a project, risk will always be rated higher, so the returns must be strong (typically >30% Internal Rate of Return, or IRR). To make matters worse, the fraction of a financial portfolio devoted to higher risk projects is usually quite small and no matter the source of funds, there will always be other projects vying for the same resources. That is why resources for demonstration are often called "the hardest money to find." Despite this, with a strong value proposition established and a sensible plan for technical and business validation proposed, one has a reasonable chance to garner the required resources. With resources in hand, the project can begin the long climb up the mountain of proof.

The first element of proof that comes to mind for every engineer is the technical side. Performance validation sets the stage for the ramp up to production by proving that all technical claims are reasonable (hopefully with margin). As you devise a plan for proving technical viability, think

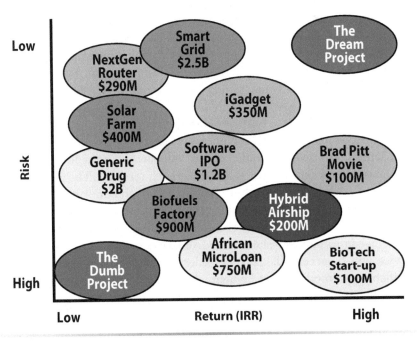

Figure CS5.4 The investor's view.

ahead to the next stage of the journey. What will the next, less risk tolerant, resource sponsor want to see? Will the plan cover all the bases? Will they understand why the elements proven in test (probably at reduced scale) will translate to reliable day-to-day operations? Especially when planning demonstrations for a revolutionary technology, framing the planned activities in an operational context familiar to the customer is critically important. Remember, the demonstration activity is not really executed for the current resource sponsor, but with an eye toward the next level sponsor. The demonstration level sponsor will only generate returns if the project moves forward into production, thus sponsors are keenly interested in what the next step will require. This predisposition is somewhat relieved if you are working for government sources since direct financial return is not their primary motivator. More and more, however, as federal budgets tighten, stronger emphasis is placed on technology transition to the field so even these sources are compelled to appreciate the end user's concerns. Despite what you may think, new product or service development projects are not science projects. They are focused on developing a functional capability that does not exist (revolutionary) or improving a capability that does exist (evolutionary). Sometimes developers lose sight of this objective and development efforts stagnate under the weight of cost and schedule challenges due to over specification of accuracy, excessive test points, virtuoso detail design, tweaking without end, or other common engineering ailments. A solid technical plan for validation is a requirement, but also expected as an ante, for development professionals. Early failures can spell disaster, so a 'walk then run' approach to development is key to success. The technical plan can lose an opportunity by being incomplete or poorly connected to objectives, but it rarely discriminates one developer from another. Technical discrimination is typically gained by what the product does in the end, not how you got there.

Similarly, a defensible business plan is a required element of proof to move the technology development forward to the next stage. The business plan defines how revenues and expenses will pan out in the future production phases; thus defining when and how much return will be garnered for the risk of investing. As much as engineers hate to admit it, this piece is more important than the technical proof. Without a solid business plan there will be no development regardless of the gee-whiz factor offered. There needs to be clear statements from customers that the product or service will be accepted (advance orders in commercial aircraft terms). Verifiable cost models must be defined for both non-recurring and recurring activity. It takes a team of business professionals to accomplish this task and should not be trivialized. Remember, others are fighting for the same resources, and the business plan can discriminate vividly if done well. In a revolutionary concept, the business plan presents a particularly difficult challenge for the revenue side. Given that the revenues will be derived from doing something differently, there will by definition be no past performance data available to use for revenue projections. This will be perpetually unsettling to the resource sponsors as it is a financial risk that cannot be retired in any other way than by just doing it. No amount of study or independent assessment will close the risk; possibly just make them feel a little better about it. For this reason an uncommon amount of understanding and insight about the future market will be required on the side of the investor. As a developer, there is only so much you can do to mitigate this barrier, but understanding that it is there can improve your interactions with resource sponsors. Aircraft development poses a particular challenge since the costs of technical proof are high and production return periods are long; neither point helps a financier agree to a commitment.

For the Hybrid Aircraft, technology validation was a decade-long climb. Every aspect of the technical and business construct was challenged, argued, redefined, re-argued, and then re-defined again. What began as a simple technical point of using efficient buoyant lift to move cargo more cost effectively became a complex morass of overlapping assumptions, emotions, numbers, desires, and objectives. Despite extensive planning and re-planning, it was not a linear, predictable program execution. To achieve the final piece of proof, the P-791 demonstrator, a small core group of people painfully, yet patiently, stepped through successes and failures over many, many years. It began with a phone call.

Fifteen years prior to the flight of P-791, a visionary group of transportation leaders approached the Skunk Works with a challenge to develop a new air transport that would have two to three times less operating cost than current systems, yet retain a two-day or less transit time across the Pacific Ocean. The primary drivers for air transport cost are the fixed airframe costs (principal, interest, and insurance) and the major direct costs (fuel and maintenance). Looking at the existing transports, most are low-purchase-cost, relatively old, converted passenger carriers. Clearly, there would need to be a major change to realistically compete. Evolving transport airframe technology looked to only provide 10–20% gains at best; not good enough to make the difference required. What was needed was a dramatic increase in efficiency to achieve the cost reductions sought. The answer, surprisingly, was not found in the lab, but in the history books: airship technology. Airships rely on buoyant lift, the third and least used form of lift behind aerodynamic and direct lift. Buoyant lift has the advantage of very high efficiency, but only at lower airspeeds. The graph illustrates this point by using lift to drag ratio (L/D), a common measure of aircraft performance. Aerodynamic lift and drag vary with the square of velocity, so we are used to velocity canceling out in the L/D equation for fixed wing aircraft. Given that payload throughput increases linearly with speed, fixed wing aircraft are driven to increase speed (more productivity) until C_L/C_D changes (transonic drag rise), with best transport efficiencies in the 15–20 range for L/D. Unlike aerodynamic lift, buoyant lift does not increase or decrease with speed (see Fig. CS5.5). This has an interesting effect; L/D is not just a function of lift and drag coefficients anymore, but a function of many variables. Most notably, L/D declines with velocity squared. Faster is always worse for a buoyant system. Efficiency does increase with the unusual term (Vol/S), which is the ratio of the buoyant volume divided by the drag reference area (mostly the surface area). This ratio increases as the vehicle becomes larger. So, ideally, best efficiency is gained by flying a very large airship slowly. Realistically, cross-oceanic pay-loads would be in the range of 200–1000 tons. At those weights and travel-ing a modest 50 kt, you can easily show L/D ratios above 200, more than ten times higher than fixed wing aircraft. This simple and incontrovertible bit of physics sent us down the buoyant lift path; a path that turned out to be more revolutionary than one would think for an old technology.

Airships have other benefits beyond just efficiency that help the trans-port mission. Airships of hundred-tons cargo capacity are rather large, so the common problem of filling a fixed wing aircraft fuselage cargo bay

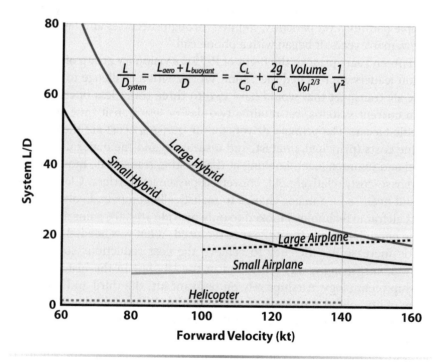

Figure CS5.5 System efficiency is different for Hybrid Aircraft.

volume before reaching maximum weight capacity, cubing out, can be alleviated. Airships like to fly low (air is more dense), so the cabin can be designed as unpressurized which reduces structural weight. Most dramatically, since the time of the great airships in the 1920–30s, structural hull fabrics have increased their tensile strength to weight ratios more than twenty-five times (Vectran/Kevlar/carbon fiber versus rubberized cotton). This makes the modern non-rigid airship hull (Fig. CS5.6) much lighter than the rigid hulls of yesteryear. These factors combined make the ratio of useful weight (payload) over gross weight greater than 50% (70% or more for short hauls) as opposed to 30–40% for existing fixed wing transports. This high useful weight fraction has the effect of reducing the aircraft fixed cost per ton-mile, thus adding improvement to the already substantially lower fuel cost. Early in the design process it was obvious we had stumbled on to something interesting (higher useful weight fractions), but if it was this easy, why had it not been done before?

Of course, it was not easy. Once design detail moved past the conceptual level, issues began to arise that were not immediately obvious. The initial design was not in fact a hybrid, but a "pure" airship, single hull, relying wholly on buoyant force for lift. One of the drawbacks to buoyant lift, we came to understand, is the simple fact that you cannot turn the lift

Figure CS5.6 American Blimp A-170 on the mast.

off. We take this benefit for granted with the other forms of lift (aero and direct). This annoying fact means the airship has to remain near neutral buoyancy always: day or night, cold or hot, empty or full of cargo. The last one is the hardest. Recall that 50–70% of the weight of the aircraft at maximum gross weight is payload! That means no matter how you design the buoyant lift, the lift balance will be way off at one payload extreme or the other. In airship terminology, "buoyancy ratio" indicates the ratio between net force (difference between buoyant lift and weight) and total weight (no buoyant lift), expressed as a percentage. More weight than lift is "heavy" and more lift than weight is "light." A pure airship must remain very close (~3%–5%) to neutral buoyancy (weight = lift). If you can change the net weight 50% just by loading and unloading the aircraft, 3%–5% is not going to cut it. Another problem of near neutrality revolves around how you moor the airship to a point on the ground. If you have seen an airship on the mast, this will be an easy concept, but if not, imagine a tall pole (mast) with the nose of the airship tied to the top of it. This is the traditional way of mooring an airship. The pole allows the airship to rotate around it as the wind blows (known as kiting). When winds become gusty, it also allows the airship to bob up and down (even to the extreme of going vertical).

Practically, this means there are no concentrated load points (tie downs) on the airship hull except at the nose, making the overall structure lighter. This works fine when you are flying advertising missions and just have to load a few people and some fuel, but if your task is to move cargo, it doesn't work so well. Imagine trying to load or unload 500 tons of containers from a ship that is constantly moving around. How long would that take? Building hangars at every location to obviate the wind problem was considered, but that sizeable fixed cost significantly reduces the business returns. The intractable challenges of near neutral buoyancy led us to consider a different way forward, one where aerodynamic lift is blended into the system in a substantive way (Fig. CS5.7). The Hybrid Aircraft was born . . . again.

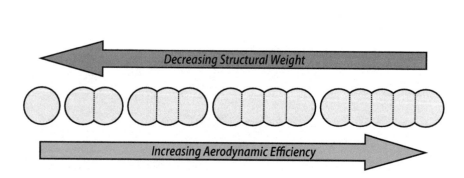

Figure CS5.7   Airship envelope tradeoff between efficiency and weight.

Initial Hybrid Aircraft designs did not look like they do today. At first, it was thought that a highly aerodynamic shape would offer the best performance. Such a shape would be able to carry a large percentage of aerodynamic lift, possibly even allowing heaviness to approach 50%. If this were possible, then the intractable problem of no "off" switch for the buoyant lift could be handled at the required cargo mass fractions. Significant effort was put into aerodynamic analysis to optimize shapes of this type. The best shapes ended up being multi-lobed structures, not the traditional single lobe airship design. Modest camber was put into the envelope to carry aerodynamic lift load (a lifting body) and provide a relatively straight under body for cargo bay integration. To efficiently carry load up to half the gross takeoff weight, the number of lobes, or blended circular sections, was approaching six or seven to achieve reasonable performance. However, in a multi-lobed lifting body, aerodynamic efficiency is counterbalanced by structural mass, which increases with the number of lobes (more septums and more surface area).

Both rigid and non-rigid structures were designed in great detail during a five-year period under the program moniker Aerocraft, an internally funded Lockheed Martin project. Six and seven lobe structures are heavy, so heavy that the gains in aerodynamic performance (resulting in fuel savings) were not outweighing the structural mass increase. Additionally, the rigid structures bore weight increases of 2–3 times their non-rigid counterparts. As the effort slowed following the tech bubble burst around the year 2000, it was becoming obvious that a shape in the 2–3 lobe range of non-rigid construction (carrying 20%–30% heaviness) would have best blend of aerodynamic and structural efficiency. For reasons entirely unrelated to aerodynamic performance, this also helped solve another persistent challenge, that of landing.

Landing represents the most dramatic event in the life of an aircraft; dramatic in the sense that if not done correctly, it can result in severe damage to either the aircraft or structures on the ground and severe trauma to people in the vicinity. Landing is also dramatic in the sense that the loads and stresses seen in the aircraft structure are the most concentrated and dynamic when the wheels touch the ground. The issue is the arresting of downward momentum for a very large object by a relatively small structure (the landing gear). While landing gear also serves the purpose of supporting the aircraft in takeoff, taxi, and parking, the loads imparted at landing are much higher and therefore size the gear. So what's the big deal? Aircraft gear have been sized for decades without issue, why is a hybrid different? The tricky part lies in two not so obvious points: 1) for a buoyant vehicle mass and weight are not different by g (gravitational constant) and 2) hybrid landing gear must serve another function in parking apart from holding the aircraft up off the ground. The first point seems to be a violation of natural

Figure CS5.8 Lockheed Martin ACLS test rig.

physics, but it isn't. Buoyant vehicles have a natural net weight (heaviness) which is the difference between the buoyant lift vector and the 'weight' of the components of the aircraft. Net weight is the measurement you would get if you put the aircraft on a scale. For a typical aircraft, this number is the same as the aircraft mass multiplied by the gravitational constant (g). For a buoyant aircraft, the mass is much larger than the net weight divided by g. This fact is important in only one situation, when you are trying to significantly change the momentum of the buoyant aircraft. During taxi or sitting on the ground, the landing gear only has to counter the net weight of the system, but at the moment of landing (touchdown), the gear has to arrest the downward momentum of the system. For a typical aircraft, this is a trivial difference, as downward momentum (mass times velocity) is sensibly related to the overall net weight. Discovery of the painful fact that for a buoyant system, the effective momentum is tied to mass, not net weight, threw the entire landing gear plan into disarray. Initially, the Aerocraft had wheels (lots of them) arranged in large trucks much like a 747 or other large transport, sized for the net weight of the aircraft like the FAA certification documents require. Oops! Correct calculation by mass, not by net weight, showed that the sizeable number of wheels baselined were too low by nearly a factor of four. One sketch of the 'full momentum' wheel arrangement showed the concept to be so ridiculous that further study simply made no sense. Something different had to be done. One good rule of thumb in dealing with buoyant structures is to reduce stress concentrations as much as possible. Conventional landing gear are the exact opposite, funneling a tremendous amount of stress though the landing strut(s). But what other options were there? Our thinking on the subject eventually led us to a dramatically different approach: the air cushion landing system (ACLS). In practice, ACLS looks a lot like a hovercraft, generating upward force by a trapped cushion of air inside what looks to be a large inner tube (Fig. CS5.8). By spreading the landing momentum over a large inflated surface area, the huge amount of load was not reacted by a single load member, but distributed over a large area of the airship hull. It looks odd, but it works. In fact, it works incredibly well! To prove that it worked we built a large apparatus that spun a heavy arm in a circle with a hover pad on the end and could be dropped to simulate landing loads. Couple that with some fairly interesting sensing equipment and we could not only prove that it worked, but also refine the spring and damping response of the pads to assure a comfortable and safe touchdown. This rig also verified the question of how quickly the pad seals on descent (very quickly). The ACLS plan also solved the second, and more insidious problem, which was keeping the aircraft from moving while parked on the ground. This is the subtle second task that standard aircraft gear performs in the parked condition. They hold up the aircraft and also keep it from moving around when the wind blows, a much more problematic issue with the large helium envelope. Recall from the discussion of airships on the mast, that loading and unloading operations are severely impacted if the aircraft keeps moving around. The second function of the ACLS is to act as a suction cup when parked, to grip the ground with enough lateral force capacity to keep the aircraft from moving around. Many are confused by this, thinking the ACLS keeps the aircraft from lifting off (not true); the function is to keep it from moving side to side. So, with all the elements in place, by about 2004 it was time to finally put them all together and prove it.

I cannot say that convincing the resource sponsors to build a Hybrid technology demonstrator was easy, but in the end it turned out to be the only way to definitively answer key operational questions. Would the aircraft handle well in real weather? Does this ACLS grip function actually scale up? Were takeoff and landing operations safe at 20% heaviness? Was this aircraft difficult to build? Would pilots actually fly in it? All these questions were answered with a 120-ft long demonstrator designed and built in an intense thirteen-month period from early 2005 to early 2006. This aircraft (Fig. CS5.9) was known as P-791 (a randomly picked number from the LM naming system) and thanks to some visionary leadership at Skunk Works, resources were committed in late 2004 to launch the project on one

Figure CS5.9 Lockheed Martin P-791 Demonstrator.

condition: it needed to fly quickly, preferably in twelve months. So the small group of design engineers that had been working on elegant shape optimization and detailed component design turned overnight into a bunch of panicked airship builders already behind schedule before even starting. Given resource constraints, the smallest size possible was chosen based upon the critical objectives required to prove. Most critical was takeoff, landing, and ground handling performance. For this reason, the aircraft only needed to fly near the ground and near the operating field. It was important to demonstrate that a human pilot, or even an unmanned approach, could control the aircraft and verify that such a craft could be operated safely. Detailed performance issues such as cruise drag would be difficult to simulate at small scale, so drag optimization was not attempted. A power to weight (really mass or inertia) ratio was selected to be very low, such that a more powerful aircraft (with higher top speed) would handle better; simply put, if you could handle P-791, anything larger would be easy. The twelve months of design and construction flew by quickly for the design team. We had the benefit of few design reviews or financial exams. Surprisingly, the assembly process went much more smoothly than expected, with the components attached to the envelope in just a few weeks after arrival in the hangar. At least those that were there. In the end, we would have made the goal delivery date except for a bizarre shipping incident where the large ACLS pads became lost in transit. I can say it is quite shocking to send an expediter down to customs in Los Angeles to pick up the cleared ACLS shipment only to find there are no containers. The large containers were simply not there despite being checked through several destinations and clearing customs. After nearly a month of searching, the containers were found in an alternate warehouse at Heathrow, and finally were delivered to Palmdale. Only a few weeks behind our breakneck goal, we took the vehicle out for taxi test runs in early January 2006. Confident that the safety issues had been addressed, early on the morning on January 31, we took the vehicle out for the first flight. It is impossible to describe the thrill that goes with seeing the proof of concept successfully validated for a project that consumes so much of your life—for us more than a decade.

For engineers, successful completion of a demonstrator program can be a source of great satisfaction and often represents a major milestone in life and career. For the more science-oriented, this may be the end of the journey, since the rest of the trek revolves more around functional detail than technical discovery. Coming down from the mountain of proof it may seem that journey's end is near, but it is not. There is still a long trek across a hot, dry, punishing valley to go.

CS5.4 The Valley of Obfuscation

The title of this section may seem odd, but it represents an intentional blend of a common phrase with the author's observations from many years wandering this valley. The common phrase is "Valley of Death," as in the place where projects go to die. The reference comes from the development industry where an inordinately large number of projects successfully complete their demonstration and validation activities yet never move on to real products or services. From my perspective, projects terminate in this phase due to a lack of understanding more than any preordained fate or lack of detail. This lack of understanding has little to do with the project team or even the technical evaluators; the apparent lack of understanding seems to be the natural reaction of the culture at large to the idea of change. Recall that a significant portion of the population has a natural fear of change, and more than half see change as more trouble than it is worth. This portion of the population has not been paying attention to your effort before now. Passing the project deftly through the gatekeepers on the mountain of proof, though necessary, has done little to sway the larger portion of the world that is now being asked to adopt your change. In fact, you have only scratched the surface. The word obfuscation was specifically chosen to paint an image of life in this valley. At every turn the environment itself and the people you meet in it will confuse, twist, blur, ignore, manipulate, and irrationally disagree with facts and evidence presented. Some are motivated by genuine opposition since your change may represent impact on a current revenue stream they profit from. Many will only consider your project for a few moments and may render decision without actually considering any facts. Most will just take a wait-and-see attitude, letting others step up to the risks of executing a cultural change. This group may be the most energy draining of all. Executing a change requires a decision to be made and this 'wait and see' group does not really oppose your idea; they just are not willing to step up and stand behind it. Witness the vast number of Americans who know they need to lose weight but the relatively few who are willing to change anything in order for that to happen. To find your way out of this valley, momentum from a significant group must be created and sustained, preferably from stakeholders in the market your changes address. These 'early adopters' can point you in the direction of success and if your change can be made worth their while, may guide you there. Often it takes a visionary leader, such as a Howard Hughes or a Steve Jobs, to rally enough momentum to effect large scale change. In the aircraft industry, the major players are the airlines or the flight operations portions of the defense agencies. Since the markets are large, the number of players involved in the aircraft world is very large. Swaying a group this large can be a lifelong task. Consider the blended wing body (BWB) passenger transport conceived in the 1980s. It would seem that tested and validated fuel economy, lower cost functions and superior interior space and environment would make adoption straightforward, yet it lingers. Why? Arguments against range from uncertain construction costs to not enough windows for passengers or unsafe egress to airports not prepared to interface with a different airplane class. If such arguments seem trivial to you, maybe you do have the heart of a developer; if they seem sensible and real, then you fit in with the operating majority. At the end of the day, the BWB just doesn't look like a passenger airplane, so it just doesn't feel right to many key decision makers. At face value, the mountain seemed your biggest obstacle, but in reality you will find the Valley of Obfuscation to be significantly more challenging.

For the Hybrid Aircraft, the trek through this valley was particularly difficult because implementing the technology leads to a fundamental change in the way cargo networks are structured. The peculiarity of no 'off' switch for the buoyant lift permeates the operational planning; so much so that some operators have fallen into denial about their own operations instead of considering change. The simplest way to set up a cargo network with a buoyant vehicle is to make sure there is a reasonable balance of cargo mass moving to and from each node. With a balanced network, there is never a problem with the airship getting too light (Fig. CS5.10). Not surprisingly, since moving around transports of any type (trucks, trains, ships, planes) without revenue cargo (known as 'deadheading') hurts the balance sheets of the operator, most networks in existence are relatively balanced. Operators discount weak demand legs until someone puts on cargo to at least partially cover the cost of moving the transport. Also, multi-point routes are devised to connect alternate node points rather than directly move back along a weak route. After researching these facts, it was shocking

Figure CS5.10 Hybrid airships can accommodate large cargo bays.

to find the number of operators whose resistance to change was so strong that they went into complete denial that they were already operating balanced networks. The discussion would dwell on the 'ballast requirement' for legs with little cargo; one operator when presented with the data showing that he had no routes in that category actually said, "well, if we had a route like that in the future, then it would be a problem, so I'm not interested." There are technical approaches to avoid the ballasting issue, but these do come at a cost, a cost that commercial cargo operators seem unwilling to pay once they understand how much the cost rises.

We also got a wide range of concern over the availability of helium around the globe; not many know it is used in welding, so it is actually rather commonly found. I even had a tense discussion with a military officer who was absolutely sure that helium was highly flammable and that I was lying about it being an inert noble gas (which side of the periodic table was that?). Other concerns range from where will I park it? to what about the weather? or how do I load and unload it? All reasonable thoughts, but what never ceases to amaze me is when presented with sensible approaches to these concerns, the most common response is to look unhappy and fixate on the risk far out of proportion to its likelihood. Cue the Hindenburg reference. One of the all time favorites is the "pop like a balloon" panic. It is unfortunate that we are all familiar with small party balloons that rupture with a dramatic flair at the slightest prick. In our minds, this experience obviously translates directly to a buoyant airship (a big balloon). But in reality, it doesn't. The pressure across an airship envelope is measured in inches of water column (tenths of a psi), so small holes really have very little effect. More like seepage in a ship really, not very dramatic. But since most people have never seen an airship pricked (they actually get shot at regularly without significant effect), there is no personal reference to verify that it is not an issue. This was a perpetual stalling point until I acquired a video of an aerostat envelope that had become untethered in Iraq. The video is shot from a helicopter as the troops are emptying a mini-gun into the envelope to try and bring it down. The stream of tracers clearly goes in one side, then out the other, with no effect on the envelope. Sometimes people make me play it several times over, looking for the computer animation trick (there is none). For the few who are willing to look past these "significant weaknesses," there is one key attribute to the Hybrid Aircraft that continues to generate tremendous operator pull: independence from infrastructure.

Costly fixed infrastructure (ports, rails, roads, airports, etc.) has been the bane of transportation systems for centuries. Improved infrastructure generates higher throughput, and therefore revenues, but at staggering upfront investment costs. Often these upfront costs are borne by governments instead of commercial enterprises, but commerce pays back those governments in the form of taxes and/or usage fees. By an interesting coincidence, the Hybrid Aircraft ACLS system required for sensible technical reasons also has a powerful operational advantage: ACLS allows ground operations on unimproved fields (Fig. CS5.11). This means that operators can access markets currently served by ineffective modes (trucks on dirt roads, small boats over shorelines, ice roads, etc.) without the upfront infrastructure investment. This seemingly small advantage represents a major revolution for oil and gas exploration, mining, logging, and other operations where infrastructure is not available at reasonable cost. In recent conflicts, operations in areas with lousy infrastructure have also become the norm for the military. The wide range of possible landing locations (including on water) changes the way operators look at cargo networks when Hybrids are involved. Cargo nodes that existed only due to the peculiarities of the transport mode (like ports for ships) can be skipped entirely, allowing the cargo to move directly to the next distribution point. Skip the port and go directly to the trucking depot, where large shipments are broken down and dispersed to many separate final destinations. This saves both money and time. With continued understanding, this concept, or some other combination of advantages, will hopefully inspire visionary leaders to commit resources to move the Hybrid into production.

The most difficult aspect of wandering this desolate valley is wondering when, or even if, a way out can be found. Like Moses searching for the Promised Land, the revolutionary technology developer may be trapped here for years. Maybe, like the Laser Disc developers, you just need to wait

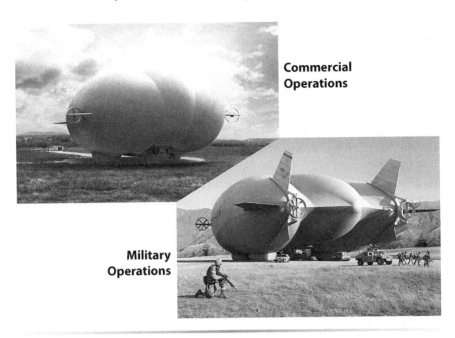

Figure CS5.11 Unimproved field operations are a strong pull for Hybrid Aircraft concepts.

for the DVD. Maybe it will be time to move on to another product or service, leaving the wandering to others. Maybe it will be time to stop. In many ways it becomes a test of faith: do you believe in your product enough to endure against the resistance, or not? From my observation of technology revolutions, when the resistance finally fades it does not fade slowly, surprisingly, it collapses like an avalanche.

CS5.5 The Wild River of Acceptance

With a lot of hard work and a little luck you may reach this final stage of the revolutionary technology journey. After many battles with the naysayers and the protectionists, you have finally moved into a production phase. Maybe it was a visionary customer or a fortunate stroke of luck. Only in hindsight will you be able to understand what factors actually allowed you to move forward, but that will have to wait because now you are riding the wild river. It is a fast and unpredictable, exciting and frightening, cold yet exhilarating ride. There is no telling where it leads, but while you are on it there is no time for rest. History shows a long list of revolutionary concepts that once they finally broke down the resistance, just took off. Consider the recent introduction of the iPad, which was not the first attempt at a tablet computer, or large PDA, but it is by far the most successful. Even at the introduction, many critics believed it was going to be a flop. Who wants an oversized iPhone? Yet the consumer saw something different. The consumer saw an easier to read iPhone; a quicker, more portable internet surfer, an electronic book that could also entertain their kids. Millions of units later, the customer still defines the value proposition. Consider the cellular phone. Once nationwide coverage was strung together and monthly packages became the norm, usage skyrocketed. It skyrocketed so fast that the splinter mobile division of AT&T ended up buying the mother company less than a decade later. Facebook went from a concept to more than 100 million users in less than four years, and two years later was at 500 million users. Going back a little farther, consider the last major revolution in commercial aircraft: the jet engine. From the debut of the first successful jet transports in the late 1950s, it was less than a decade before all the large piston driven transports ended production. When the revolution dawns, it either succeeds or fails in short order.

For the Hybrid Aircraft at the time of this writing, the concept has not yet achieved wide acceptance (Fig. CS5.12). It seems that in a few places there are visionaries who see where it can make a difference in the world, but so far not enough have gathered together to move it forward. One day, hopefully, we will look back and discover what finally pushed the product out of the Valley, but as of now one can only speculate. Only time will tell. Regardless, it has been a fascinating journey, leading to places, people, and knowledge I could never have imagined. If revolutionary development is your choice, hopefully it will be as fascinating for you.

Figure CS5.12   Access to remote areas drives the economics of the business case.

The ongoing Hybrid Aircraft development process serves as a poignant example of how difficult it is to effect revolutionary cultural change. After years of struggle through the revolutionary technology development journey, one old, slightly modified, adage rings most true:

The more things change, the more people remain the same.

Acknowledgments

The author would like to acknowledge the contributions of the Lockheed Martin Aeronautics Advanced Development Programs organization and the American Blimp Company that significantly enhance the depth and clarity of this article.

Robert R. Boyd, PhD

Dr. Robert (Bob) Boyd is the Hybrid Lift Portfolio Senior Program Manager at Lockheed Martin Aeronautics, Advanced Development Programs, or "Skunk Works" in Palmdale California. In this position, he manages multiple programs including both internal development efforts (IRAD) and externally funded efforts related to heavy lift and ISR Hybrid Aircraft systems. He has been responsible for all Hybrid Aircraft development at Lockheed Martin over the past ten years, including several variant of lift vehicles. He managed the LM Walrus team for the 2006 DARPA effort. Dr. Boyd received the LM NOVA award for his leadership as Program Manager of the P-791 Hybrid Aircraft Demonstration program, the results of which you have likely seen, but not heard much about. Such is the way of the Skunk Works. . .

During his tenure at Lockheed Martin, he has worked a wide variety of programs including low and high altitude airships, high speed concepts including missiles, aircraft and space launch systems, in-space architectures such as tethers, advanced unmanned aircraft and heavy transports. His leadership responsibilities represent more than $150M in contracted development work.

Dr. Boyd holds a PhD and BS in Aerospace Engineering from The Ohio State University, has authored papers, has six patents granted and others pending, and is a graduate of the DAU Advanced Program Managers' Course. Outside of Lockheed Martin, he has served as Assistant Dean for Academics at the Ohio State University College of Engineering, Technical Fellow at NASA Glenn Research Laboratory, co-founded two independent businesses.

Dr. Boyd has been happily married for more than twenty years and has three teenage sons. He volunteers his time in youth sports, coaching and officiating multiple sports. For his most impassioned sport, soccer, he has been a volunteer referee for more than thirty years. He also enjoys disc golf and is an avid home remodeler.

Case Study 6 — *Daedalus*

Harold Youngren

- Origins, Goals, and Overview of the Daedalus Project
- Phase 1—Feasibility Study; Preliminary Design Development
- Phase 2—Prototype Design, Construction, and Testing (Prototype Aircraft)
- Phase 3—Daedalus Design, Construction, and Testing
- Costs
- Mission and Flight Planning

After a long flight the Daedalus human-powered vehicle is magically captur[ed] by National Geographic. Although this project was completed over 25 years a[go] and many papers and magazine articles have be[en] written about it, this is the only historical account fro[m] the point of view of the ch[ief] engineer.

Engineering is done with numbers.
Engineering without numbers is only an opinion.
Anonymous

CS6.1 Project Summary

CS6.1.1 Introduction

The story of Daedalus, aviation's oldest and most enduring myth of flight, was brought to world-wide attention on April 23, 1988, when a 30-year-old Greek cyclist named Kanellos Kanellopoulos took off from Heraklion, Crete in Daedalus 88, a 31 kg human-powered aircraft built by a team of engineers and students at the Massachusetts Institute of Technology (MIT). When Kanellopoulos landed in the surf at the island of Santorini 4 hr and 72 mi later, he had established two FAI World records and symbolically re-created the mythical flight of Daedalus. The Daedalus records remain unbroken to the present day, a testament to the difficulty as well as the planning and preparation that led to that flight.

The Daedalus project was extensively covered in the popular press and several non-technical accounts of the flight are available. The details of the project were documented in a book [1] as well as in a NOVA documentary [2]. In addition, the Daedalus Project also spawned a host of technical papers covering the research and design aspects of the program.

This study revisits the design, development, and the flights of the three Daedalus aircraft and outlines some of the most significant results obtained in the areas of physiology, meteorology, aerodynamics, structures, performance, and stability and control. The study documents the design characteristics and performance of the Daedalus airframes and its engines, as well as the human and financial resources devoted to the project.

CS6.1.2 Genesis

The Daedalus Project had its beginnings in earlier Human-Powered Aircraft (HPA) projects at MIT, whose Aeronautics and Astronautics program has harboured, tolerated, and even encouraged a succession of remarkable student projects for HPA's stretching back from the BURD I in the late 1960s through the BURD II, the much-flown Chrysalis (built during Paul MacCready's Albatross project, in 1979), to the Monarch B, a vehicle that won the Kremer Speed Prize in 1984. In the mid-1980s the veterans of some of these projects were hungry for a new challenge with less definite goals than those formalized in the Kremer Prize competitions.

The Kremer Prizes, which were sanctioned by the Royal Aeronautical Society, had since 1959 framed and motivated much of the early development of HPA's. The initial Kremer Prize, for a 1 mi flight around a figure-eight course, was won finally in 1977 by a group led by Paul MacCready with the Gossamer Condor. Shortly thereafter, a new Kremer Prize was announced for a flight across the English Channel (a boggling jump to 22.6 mi). This new prize was won by MacCready's team in 1979 with the same pilot, Bryan Allen, flying a lighter and faster aircraft of similar design, the Gossamer Albatross. Allen's successful flight, a 2 hr 49 min test of will and determination, was at the very limits of his endurance [7].

After the English Channel flight the focus of the Kremer Prize competitions shifted from distance and duration tasks to speed and sidestepped the compelling question: What were the true limits for human-powered flight? In 1985 an MIT-based team set out on three-year program to push the limits of low-speed flight and human physiology to break the Gossamer Albatross distance and duration records in a modern recreation of the mythical flight of Daedalus.

CS6.1.3 Myths and Motivation

History describes many poetic, romantic, and mythical references to flight. Perhaps none is more famous than the Greek myth about Daedalus, a master craftsman (and prototypical engineer) who flew to freedom from imprisonment on the island of Crete using wings he fashioned himself. Until the mid 1980s, such a voyage remained purely in the realm of imagination: from Crete to a major land mass is a distance of more than one hundred kilometers, almost three times the distance achieved by a human-powered aircraft.

There were several reasons for undertaking a venture such as the Daedalus Project. The first and most important was education. Human-powered vehicles have supplemented the academic curriculum at MIT since 1968, providing a hands-on design experience for several generations of undergraduate and graduate students. Approximately three dozen students were ultimately involved in Daedalus. The project drew together students and faculty in aeronautical and mechanical engineering and from such diverse fields as classical literature, archaeology and anthropology, meteorology, and the medical sciences. Through a partnership with the Smithsonian's National Air and Space Museum, the project promoted awareness and enthusiasm for engineering, science, the arts, and athletics.

The project also fostered new research in areas related to low-speed flight. A large number of papers were published in the technical literature documenting the project research in low-speed aerodynamics, composite structures, low-power avionics, long-duration exercise physiology, and structural dynamics. Some of the design methodology used for the Daedalus project found its way into computer tools (such as the XFOIL airfoil code) that have been made freely available and are now used throughout the aeronautics industry. The Daedalus legacy provides a valuable resource for modern projects in aircraft design, human-powered vehicles and even high altitude long-endurance (HALE) aircraft.

CS6.1.4 Overview of the Daedalus Project

The project began informally in 1984 with preliminary studies exploring the potential for a very long-range HPA and a record flight that would significantly increase both distance and duration. This initial work lead to a more formal framework for a project organized loosely into three phases to reduce risk, spread over three years. This included a Feasibility Study, a prototype vehicle design and testing phase and a Final phase to build two Daedalus flight vehicles and make a record flight in Greece (see Fig. CS6.1). Each of these phases also involved finding and recruiting new project sponsors. The three phases will be summarized and then expanded further in the following material.

CS6.1.4.1 Phase I Feasibility Study

In the spring of 1985, a team of engineers and students from the MIT Department of Aeronautics undertook a feasibility study with sponsorship from the MIT School of Engineering and the Smithsonian's National Air and Space Museum. During the ensuing year this working group met seven times and jointly undertook a program of research in physiology, meteorology, aerodynamics, structures and flight controls. The weather research showed that a flight with low winds (less than 3 kt) and cool temperatures was feasible in the spring and fall. Physiological testing established the baseline power levels that could be obtained from athlete-pilots over a long record flight.

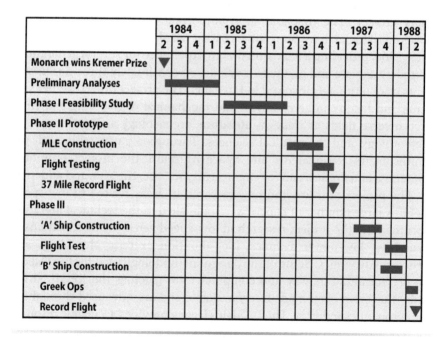

Figure CS6.1 Timeline for the Daedalus Project (from [17]).

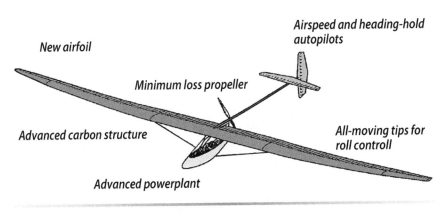

Figure CS6.2   April 1986. Preliminary design developed during Phase I for a 30 m (98.4 ft) span composite airframe with an empty weight of 30 kg (66 lb) (from [3]).

Based on these results, the team developed a preliminary design (Fig. CS6.2) for an aircraft optimized for the proposed Daedalus flight. In addition the group held discussions with organizations in Greece, with potential financial sponsors in the United States, and with scholars in the U.S. and Great Britain. The conclusions to the study [3], announced in April 1986 at the Smithsonian, were that the proposed flight was feasible and would be justified by its educational and research benefits, and that the state of development of some crucial technology justified the intermediate development of a prototype aircraft before undertaking the expedition to Greece.

CS6.1.4.2   Phase II—Prototype Design, Construction & Test

With the financial support of the Anheuser-Busch Company, development of the prototype Michelob Light Eagle began at MIT in June 1986. The aircraft made its first flight in October 1986 at Hanscom Field, MA. Following modifications to increase its span from 102 ft to 114 ft in an effort to reduce induced drag to compensate for a weight overrun, the Light Eagle was shipped to NASA's Dryden Flight Research Facility in January 1987. Figure CS6.3 shows the Light Eagle in flight on the dry lake with original project pilot Lois McCallin. During an intensive 28-day test period at Dryden, the Light Eagle set four world records, including a new absolute distance record of 58.7 km (36.5 mi) set by Glenn Tremml on January 22, 1987. An extensive series of flight experiments, including heart rate measurements, glide tests, flow visualization, and evaluation of an experimental fly-by-light control system were performed to provide quantitative data for design of the follow-on Daedalus aircraft. The most significant results from these tests were (1) verification of the correlation between anaerobic threshold (AT) and long-duration capability, and (2) the discovery that, while most of the

Figure CS6.3 The Daedalus prototype, Michelob Light Eagle with pilot Lois McCallin, in flight over Rogers Dry Lake, California, in January 1987. The final version of Light Eagle had a span of 114 ft and an empty weight of 92 lb. NASA photo by Mike Smith.

wing had laminar flow as designed, the outermost wing was nearly fully turbulent. Reducing the weight and correcting the wing transition problems became key design goals in the Phase III Daedalus development program.

CS6.1.4.3 Phase III—Daedalus Design, Construction, & Testing

With the financial sponsorship of the United Technologies Corporation, construction of two refined Daedalus airframes for the record attempts began at MIT in June of 1987. The first aircraft was completed in October 1987 and delivered to NASA-Dryden for testing in November. Due to poor weather, the test program progressed slowly and the aircraft had accumulated only about an hour of testing when on February 7, 1988 Daedalus 87 was seriously damaged in a crash on Rogers Dry Lake. The accident was due to loss of control in thermal induced gusts caused by a combination of insufficient dihedral and reduced rudder authority. The backup aircraft was completed and shipped to California in late February. While the damaged aircraft was repaired in Massachusetts, the new airframe completed qualification testing in California, and both were transported to Greece on March 26, 1988.

CS6.1.4.4 The Flight From Crete

The Daedalus crew was in Heraklion, Crete and flight ready on March 30, 1988, and was on daily alert for the next several weeks, waiting for weather. After several near-launch attempts in what turned out to be marginal conditions, favorable weather arrived on Saturday April 23. Kanellos Kanellopoulos, one of five pilots who had been training for the mission, took off at dawn. Kanellopoulos's flight went smoothly, sped along by a mild 3 mph tailwind. Figure CS6.4 shows the Daedalus in mid-flight, staying high above the waves. After almost four hr aloft, as he approached the shore at Santorini for landing, mid-day turbulence from the hot beach raised local wind speeds to near flight velocity. Kanellopoulos found it difficult to maneuver in the high turbulence as he tried to land on the beach. At 10:58 am, just yards from the sand, a gust snapped the tail boom at the vertical stabilizer. The aircraft pitched up, the main wing spar failed and Daedalus 88 fell into the water. The pilot escaped and was greeted on shore by the crowd that had gathered there. The flight was subsequently certified for two official FAI World records, including the absolute world distance and duration records for human-powered aircraft (72 mi, 3 hr 54 min).

Figure CS6.4 April 1988. The Daedalus enroute to Santorini with pilot Kanellos Kanellopoulos. Span of the final version was 112 ft and the empty weight was 68.5 lb. (Courtesy of Steven Finberg)

CS6.1.4.5 Program Costs

The direct cost of the Daedalus Project was $1.2 million (in 1989 $) [17]. Some 70,000 person-hours were recorded but an estimate of 100,000 hr seems more in line with actual work habits of the team members. At rates typical for the aerospace industry, these 50 man-years would have cost approximately $5 million. In addition to direct expenses roughly another $700,000 in materials and services were provided to the project through a network of 42 domestic and 25 foreign corporate sponsors.

CS6.2 Phase I Feasibility Study

CS6.2.1 Phase I Overview

The Daedalus Project officially began in May 1985 when the MIT Department of Aeronautics and Astronautics teamed with the Smithsonian Institution's National Air and Space Museum (NASM) to conduct a feasibility study of the proposed flight. The one-year study (see the project timeline in Fig. CS6.1) examined the myth of Daedalus and the geography of Crete to establish a flight path for the recreation of the myth, vital for selection of flight routes to define the "mission" that would design the aircraft. From a technical standpoint the study focused on four major research areas, including: meteorological conditions in the Aegean region; the physiology of long-duration exercise to characterize the engine; low Reynolds number wing design; and the design of large all-carbon primary structures. The feasibility study included aircraft sizing work that established design parameters for the Daedalus airframes. A preliminary design of a prototype aircraft based on this initial study was proposed. The major conclusions of the feasibility study are summarized as follows [3].

CS6.2.2 Geography

The story of Daedalus is very old. Historical accounts of the myth do not define the specific flight path chosen by Daedalus (and his son Icarus, in some versions of the myth), but it is believed to have started on Crete. The geographical setting for the story, encompassing Crete and the southern mainland of Greece, is shown in Fig. CS6.5.

The shortest distance from Crete to Greece is about 60 mi from Akra Spatha, a mountainous peninsula that is about 1000 ft high, to the southern tip of Greece in Neapolis. The nearest town and airport to this remote mountain take-off site is Maleme at the southeastern base of Akra Spatha. If used as an alternative launch site this would add more than 11 air mi to the journey. Alternative interpretations of the myth include an escape by the mythical aviators from the famed Labyrinth, which is thought to have

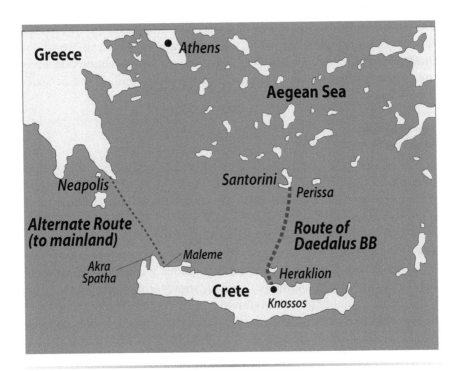

Figure CS6.5   Map of the Aegean Sea showing the geography of Crete, Southern Greece, and Daedalus flight routes (from [17]).

been located at the palace of Knossos near Heraklion, farther east on the north coast of Crete. A flight from Heraklion to Santorini (a remnant of a volcanic island that erupted in Minoan times) would entail a journey of 74 air mi (or 72 mi straight line distance). All versions of the myth leave quite a lot of room for interpretation. Whether it is 60 mi or 72 mi, it is a long way over the sea.

CS6.2.3   Meteorology

The historical accounts of the myth do not define the specific flight path chosen by Daedalus but at 60 mi or more it would be a gruelling test of human endurance compounded by the changeable weather. The weather for a flight had to be favorable in terms of surface winds (magnitude and direction) and temperature. By using data from the U.S. Weather Service, the Hellenic National Meteorological Service, and data from automatic, locally deployed, weather stations donated by Aanderaa Instruments Company, the study concluded that extended periods of calm weather lasting up to 6 hr occur most frequently during the summer months, but that temperatures during this period were likely to be high requiring significant pilot cooling. When wind and temperature constraints were combined, the highest probability of acceptable weather conditions occurred during the spring and fall in March, April, and September and occurred most frequently at night.

CS6.2.4 The Human Engine

The powerplant (our human engine), like the weather, has not changed fundamentally in the thousands of years since the myth. The challenge of human-powered flight derives from the limited power the pilot can supply to the airframe as well as the way in which this power is generated. Like an engine, the human body combines fuel and oxygen to produce energy and waste byproducts. There are two pathways for human-power production: (1) aerobic processes that depend on continuous transport of oxygen and fuel (such as glycogen and lipids), and (2) short term anaerobic processes that activate at higher power levels and use the muscle tissue as stored fuel directly without oxygen. At power levels above the anaerobic threshold (AT), a build-up of lactic acid in the blood inhibits further high power output. For an effort that must be sustained for many hours, the power must be below the pilot's AT level. As a result, the power a human can produce depends very strongly on the duration of the effort and at higher power levels, when the AT is crossed, the sustainable duration falls quickly from hours to minutes.

It turns out that power alone is not the appropriate metric for HPA powerplants. This is because the pilot is also the payload, representing the largest single mass element in the Daedalus vehicles, with approximately two-thirds of the gross weight. Since the power required for flight varies as (gross weight)$^{3/2}$, the power required from the pilot may be simplified to a single parameter: power per unit pilot weight, or specific power, which is constant for a fairly wide range (±15%) of pilot weights.

Although the production of mechanical power by humans has been the subject of many investigations in the past and the physiology of long-duration exercise is reasonably well understood, it was surprising to find that few rigorous scientific measurements (correcting for body mass, environmental conditions and training) existed for durations beyond 1 hr. It soon became clear that the "engine manual" for the Daedalus powerplants was still a work in progress.

The limits of endurance and power for the human engine were investigated with the assistance of Ethan Nadel, a physiologist at Yale University. In Phase I, a Boston area pilot/triathlete, Lois McCallin, was recruited and tested as a human engine for the Daedalus flights. It was determined that short duration ergometer tests measuring the maximum rate at which an athlete could take up oxygen (VO2max, a measure of metabolic power) could also predict the specific power that an athlete could sustain for an extended period. These experiments demonstrated that elite (top 0.5% of the general population) endurance-trained cyclists are capable of producing continuous power at approximately 70% of their maximum oxygen uptake for a period of 4 hr. Based on tests with Lois and other athletes, using their ergometer results as a guide, a nominal "engine" specific power of 3.0 W/kg was chosen for sizing the Daedalus aircraft.

The relationship between oxygen uptake and mechanical power production is shown in Fig. CS6.6 for the semi-recumbent cycling position (measured for Glenn Tremml, a second pilot/triathlete recruited later as a more powerful "engine" for Phase II flying). This relationship is linear since it is due to aerobic oxidation of fuel (glycogen or lipid) in the muscle, though cycling technique may influence the biomechanical efficiency of power production and alter the slope. This maximum oxygen uptake (67 ml/min/kg, which was the VO2 limit in this test) is typical of endurance-trained athletes in this cycling position. At 70% of that maximum, this pilot-athlete produces approximately 3.3 watt/kg, the specific flight power that was actually required for the Light Eagle.

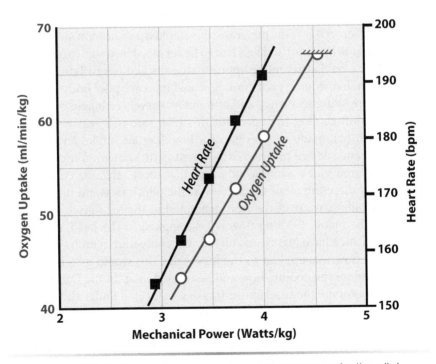

Figure CS6.6 Heart rate and oxygen uptake vs power for the pilot Glenn Tremml (from [8]).

Oxygen uptake is the best metric for mechanical power but is impractical for in-flight measurements since subjects must breathe through a mouthpiece to a flow meter. The alternative, more practical metric is heart rate. Oxygen and fuel are delivered to the muscles by circulating blood at a rate proportional to the cardiac output, which in turn is related to the heart rate (cardiac output = heart rate × stroke volume). The result is that heart rate is also linearly proportional to mechanical power production as shown in Fig. CS6.6. Heart rate can therefore be used to estimate the mechanical power delivered to the pedals. Heart rate is not a perfect metric due primarily to the autonomic response of the pilot to mental stress. At higher power levels this autonomic response is suppressed and heart rate can be a reasonable indicator of power production.

Countermeasures such as appropriate training, preloading the body with glycogen, pilot cooling, and adequate water supply to replace fluid loss were tested as aides to help the pilot avoid reaching his physiological limits during the flight.

CS6.2.4.1   Pilot Cooling

Since the human engine operates with a thermodynamic efficiency of approximately 25%, for a typical flight power of 225 W some 675 W of heat had to be removed from the pilot to maintain a steady body core temperature. Roughly 600 W of this had to dissipated by evaporation. With a heat of vaporization of 0.7 W-hr per gram, this implied a minimum sweat rate of roughly 900 ml/hr. This fluid loss had to be replaced by water (or preferably a drink that would also replace appropriate salts and electrolytes) to avoid dehydration. Inadequate pilot cooling would increase pilot body core temperatures, lowering efficiency and ultimately severely reducing duration.

Convection cooling to cope with this high rate of sweat evaporation (almost 2 lb/hr), made the design of a low-drag air intake and efficient cooling system a design priority. Experiments with simulated cockpit flows and an infrared video camera indicated that most effective cooling was obtained by directing all of the airflow at the pilot's face and upper body. During pedalling, most of the heat generated in the leg muscles is transported to the blood. Cooling flow is best applied to the head, neck, and shoulders, which have highly vasculated skin compared to limbs and torso. No cooling flow was applied to the legs, since this would give little extra heat removal for the cooling flow and associated drag. In the Daedalus aircraft and prototype the fuselage air intake was located under the wing (for higher pressure recovery) and was designed as a low-diffusion, low loss duct to direct cooling flow over the pilot's head and shoulders. The cooling flow exit was located at the trailing edge of the fuselage pod near the tailboom to minimize drag.

Several radical concepts were examined during the project including supercooling the drinking fluid for use as a heat sink for actively cooled undergarments, similar to those for space suits. The heat loads encountered by the Light Eagle and Daedalus pilots turned out to be much more severe than those for space suit design.

CS6.2.5 Mission and Flight Planning

In approaching the Daedalus flight the project team was not constrained by any predetermined competition rules. It did, however, attempt to stay within the definitions set by the Federation Aeronautique Internationale (FAI) for human-powered aircraft. Despite somewhat hazy flight route information at this early stage, the project team considered four major flight profiles (summarized in Table CS6.1).

These included

A. Self-launching sailplane, a small, fast aircraft with a high load factor that would be launched, climb in thermal or slope lift, and use human-power in an extended glide to the mainland. This option was rejected because it would not qualify under the FAI definition of a human-powered aircraft.

B. Medium altitude cruiser, that would attempt to use tailwinds, flying below the altitude where convective lift is well developed. This option was rejected when the weather data showed that tailwinds are almost never present on the projected route.

C. Sea skimmer, that would have a size and power level comparable to Gossamer Albatross but use advances in technology to fly 50% faster. This was the option eventually selected.

D. Low and slow, which would extend the Albatross concept by using extensive wire bracing to produce a wing with over 110 foot span. This option was rejected as being very inflexible in operation and not an exciting technological challenge.

Table CS6.1 Summary of Flight/Design Options Considered for Daedalus (data from [3])

Flight/design option	A sailplane	B med alt	C sea skim	D low&slow
Speed (mph)	29	20–23	15–17	11–12
Cruise altitude (ft)	1000	500	15	15
Load factor (g)	6	4	3	2

Two further flight profiles were considered in the selection of the load factor for the prototype [3]. One involved taking off from the cliffs of Akra Spatha, some 1,000 ft above the northern shore of Crete, using the potential energy to assist the pedalling of the pilot in a powered glide to the Greek mainland. This would be the shortest distance from Crete to the mainland (approximately 60 mi), and despite a structural weight penalty due to the higher load factor required for safety, initial calculations suggested the potential energy of a high launch would offset the weight increase.

The second alternative flight profile was to fly relatively close to the sea with as fast and light a plane as possible. Although the route distance would increase to over 71 mi (as there are no acceptable take-off sites at the base of the cliffs) the airplane could take advantage of ground effect. Since both options appeared feasible, it was decided to include the capability of doing the high altitude flight. In addition, the aircraft was designed to be capable of operations at night or in haze, when weather data indicated the longest periods of calm wind occur.

At the end of Phase I, the overall design goals for the Daedalus prototype were specified as (1) a range of at least 60 mi, sufficient to cover the shortest distance between Crete and the mainland; (2) a speed of 15–17 mph, to minimize exposure to weather changes; (3) a specific power of 3.0 W/kg of pilot weight, equal to 204 W (0.27 hp) for a 68 kg (150 lb) pilot; (4) a flight control system to allow operation at night or in light fog; (5) a structure capable of sustaining 3 g, sufficient for safe launch from the cliffs at Akra Spatha.

CS6.2.6 Conceptual Design and Configuration Sizing

The basic scope and dimensions of an HPA are set by the physics of flight at the low power levels imposed by human physiology. As an example, for a pilot weight of 65 kg (and assuming the plane weighs 32.5 kg, ½ the pilot weight) the 3.0 W/kg specific power implies an output power of 195 W (0.26 hp). Using an efficiency of 85% for propeller and drive train leaves 166 W (0.22 hp) for airframe power. At a speed of 15 mph (6.7 m/s) this gives a drag of 24.7 N (4.9 lb), implying an airframe L/D of 37. This is a very high glide ratio for an aircraft that flies at these low Reynolds numbers, implying that these aircraft have to be very "clean" aerodynamically. Using realistic estimates for drag, to fly at 15 mph at that L/D, a wing with at least 30 m span is required with an aspect ratio of 35 or higher. The demands on aerodynamic cleanliness can be relaxed somewhat by reducing the flight speed. For example, at a speed of 10 mph (4.47 m/s), the same 166 W airframe power only requires an L/D of 25.7. However, the wing area must more than double, so that the resulting large and fragile machine would be poorly suited for long-distance flights in all but perfect conditions.

Given the mission and the set of design goals as well as the meteorology data and "engine specifications" it was possible to begin the conceptual design of an aircraft. The following sections detail the Phase I preliminary design work in aerodynamics, materials, structures and weights used in sizing the aircraft for the Daedalus mission. Utilizing this data, a computer program for HPA sizing was used to explore the design space to select and size an aircraft configuration that would meet the design goals.

CS6.2.6.1 Aerodynamics

The design goal for the Daedalus aircraft was to minimize power at the required airspeed while providing a stable, controllable and safe platform for the pilot. This favored a very high aspect ratio wing operating at a high cruise lift coefficient with "minimal" fuselage pod and boom supporting small stabilizing and control surfaces for low parasite drag. As a result the wing profile drag accounted for a high fraction (roughly 40%) of the total parasite drag. This placed high demands on the performance of the wing airfoils. These airfoils were designed for extensive laminar flow for low drag and had to be compatible with both structural and manufacturing constraints. The wing also had to be reasonably efficient across the narrow, power-limited HPA speed range (11–16 kt) and have moderate pitching moment to minimize wing torsion loads.

Ground effects, which lower induced drag very close to the ground, were potentially significant for the large wingspans and low flight altitudes of typical HPA configurations. Depending on the altitude selected for cruise, ground effect can have a drastic effect, reducing the optimum span and aspect ratio in a sizing optimization. The design team felt than an aircraft optimized in ground effect would produce a design that would be excessively penalized at higher altitudes. Since the high altitude flight option also precluded ground effects and due to concern about ground effects over the sea (over waves) they were not included in any of the wing optimization studies. By not including ground effects in the design, a performance margin was built into the Daedalus aircraft, giving the pilots an extra degree of freedom in the vertical dimension that they would not have had with a design optimized for low altitude. This may have been fortunate decision as some of the flight testing in Phase III showed that flight near the surface may incur additional power losses that are not included in classical ground effect models.

The aerodynamic control surfaces were sized using historical tail volume coefficients from previous HPA designs. The tail areas were assumed to be a fixed fraction of the wing area and also fixed aspect ratio. The fuselage was assumed to be of fixed size with fully turbulent flow. Bracing wire drag was calculated as a function of the size of the aircraft and the particular bracing scheme considered.

CS6.2.6.2 Airfoil Design

The aerodynamic requirements for the wing airfoils were (1) high lift coefficient (CL > 1.0), (2) low drag and (3) low pitching moment. The wing Reynolds numbers were relatively low, roughly 500,000 at the center wing while the tips operated at 200,000, making transition and laminar separation bubbles a factor in the design. Customized airfoils for the wing were designed by aerodynamicist Mark Drela using methodology described in [5]. The design strategy was to minimize losses in the transitional separation bubbles without resorting to mechanical turbulation. The airfoils were also designed with structural and manufacturing considerations for the foam leading edge (LE) shell and acrylic foam/Kevlar trailing edge (TE).

The airfoils were designed with Drela's ISES CFD code, a streamline-based 2D Euler airfoil design method with a unique fully-coupled integral boundary layer method to capture viscous effects. ISES was capable of predicting airfoil performance with laminar separation bubbles that are characteristic of operation in this flow regime. ISES was used to design airfoil shapes to match specified pressure distributions and was used to evaluate 2D viscous performance. The methodology used for this work later found its way into the XFOIL program that was released as open source subsequent to the Daedalus project and has now become an industry standard for airfoil design.

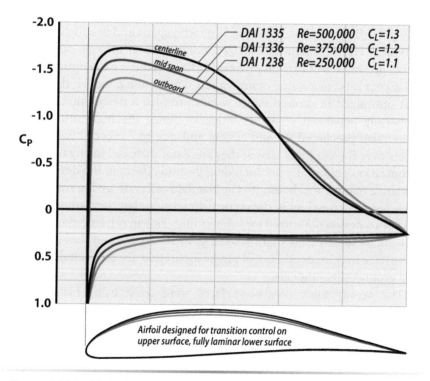

Figure CS6.7 High lift, low Reynolds Number laminar flow airfoils designed for the Light Eagle wing (from [5]).

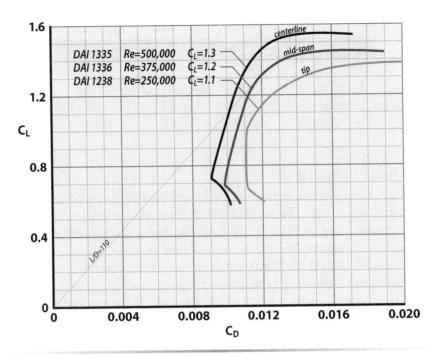

Figure CS6.8 Lift-drag polars for Light Eagle wing airfoils (from [5]).

The DAI1335, DAI1336, and DAI1238 airfoils, used on the wing at the center, middle, and tip respectively, are shown in Fig. CS6.7 with their design point pressure distributions. The corresponding drag polars are shown in Fig. CS6.8. These airfoils use fully laminar lower surfaces, and extensive upper surface laminar flow (50–70% chord).

At the time the Daedalus airfoils were designed, the ISES code and this design approach had not been validated with later experiments at NASA and elsewhere so this design of new, custom and untested low Reynolds number wing airfoils represented a serious risk factor for the program and was one of the reasons a prototype aircraft was built and tested.

CS6.2.6.3 Materials Selection

Carbon fiber-epoxy materials were used throughout the structure due to the requirements for low weight. Since much of the structure was designed to meet stiffness requirements, special attention was paid to the use of newer high modulus fibers, in particular the pitch-based products made by Union Carbide (now Amoco Performance Products) Types P-75 and T-40 as shown in Table CS6.2. The P-75 material was selected for the Daedalus prototype structure based on its high modulus and nominal availability.

Table CS6.2 Comparison of Carbon Fiber Material Properties (data from [3])

Characteristic	T-300	P-75	T-40	T-50
Strength (ksi)	201	135	470	205
Modulus (msi)	20.5	35.0	25.0	35.0
Specific gravity	1.52	1.81	1.62	1.62
Cost (1987 $/lb)	65	300	100	125

CS6.2.6.4 Structures

Since weight was a critical factor in terms of flight power, the wing structure, which was the heaviest component of the airframe, received considerable attention. The design goals for the wing structure were high strength and stiffness in bending and torsion and low weight. Extensive external wire bracing has been used in previous HPAs to reduce structural weight. At low flight speed the reduced structural weight offsets the drag penalty of bracing wires in terms of flight power. As flight speed increases the optimum extent of external bracing decreases. Three wing structural configurations were examined: an externally braced multi-wire design, a fully cantilevered design and a one-wire design. The one-wire aircraft was roughly analogous to a strut-braced wing, with a single tension wire replacing the strut.

Structural designs were developed for the three configurations. Figure CS6.9 shows the spar weight trends for an ultimate load factor of 2 g. One of the unique aspects of the Daedalus wing design is that the tubular spars are smaller than the wing thickness as a result of buckling stability constraints in the spar tube laminates, which act to reduce the optimal spar diameter. This effectively removes wing thickness from the optimization for weight and performance and makes wing spar weight a function of span and loading, not chord or aspect ratio.

Several different spar concepts for the single-wire braced wing were studied, fabricated in test sections, and tested to failure in the lab. The most effective of these was a 3-tube design using a large tube for torsion, with small sparcap tubes top and bottom to take tension and compression loads from bending. The sparcaps prevented local buckling of the spar from ovalization under load of the large, thin torsion tube.

CS6.2.6.5 Weight Models

The structural weight was a critical design parameter as it strongly drove the power required. Estimates for the weight of the structure [6] used component weight data from two previous MIT human-powered aircraft, Chrysalis (1979) and Monarch (1983/4). The independence of spar weight and aspect ratio greatly simplified wing weight estimation. Spar weight was

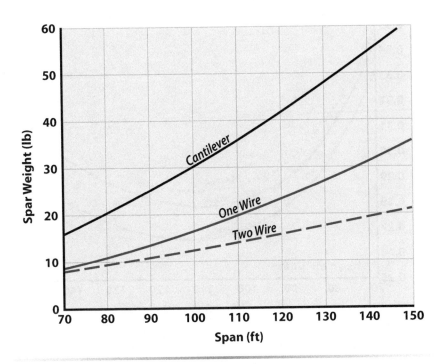

Figure CS6.9 Weight trends for wing spar vs span at 2 g load factor for 3 bracing schemes (from [3]).

a function of load factor, gross weight, wingspan and bracing scheme, and included the weight of the bracing wire. The component weights for the Mylar covering, foam LE (which covered a fixed fraction of the chord) and ribs (with fixed web thickness and spaced as a fraction of chord) were proportional to wing area. The trailing edge structure was modelled with constant weight per unit span. The fuselage was modelled with fixed weights for the cockpit, avionics, drive train, prop and other systems. The weight of the relatively lightly loaded tail surfaces were estimated with a fixed weight per area.

CS6.2.6.6 Sizing and Optimization

A computer program for HPA sizing was developed by Robert Parks using models for the pilot specific power (as a function of flight duration, see Fig. CS6.21), vehicle weight and aerodynamic drag as a function of wing area and aspect ratio. This was used to explore variations of the design parameters to select configurations with the greatest range [3].

A study of power required was made for wing areas between 100 and 1,000 ft² with pilot weight, wing strength and design lift coefficient held constant while wing area, aspect ratio and bracing configuration were varied. Figure CS6.10 shows the power required from the pilot (allowing

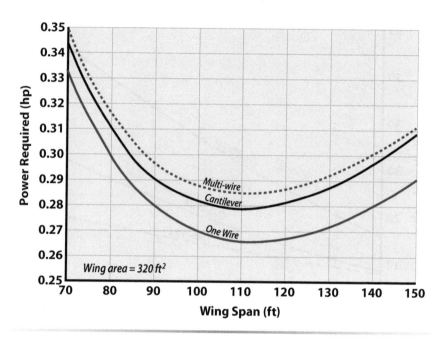

Figure CS6.10 Power required as a function of wing span for three bracing schemes for a wing area of 320 ft^2 (from [3]).

for estimated propeller and drivetrain efficiencies) as a function of wing-span for the three bracing configurations for a wing area of 320 ft^2 (roughly corresponding to an airspeed of 15 mph). Minimum power occurred for spans near 110 ft and aspect ratios of 38 for this wing area.

Optimizers and sizing programs, despite the best intentions of their creators, never include all the "real world" factors that will shape a design. While slavish attention to an optimizer is usually a prescription for folly, they are useful as navigational aids to the design space. While the optimum span for minimum power was found to be roughly 110 ft, this optimum was reasonably broad. This implies that small changes to reduce span to get a more compact aircraft will have modest performance penalties. The sizing program was then used to examine aircraft that were slightly smaller, trading off 2% in flight power for a more buildable and transportable design.

The sizing program was used to compare families of optimal aircraft for a range of wing areas with the three bracing schemes. Pilot weight and cruise lift coefficient were fixed and wing area was varied so that speed changed with area. For each wing area, the optimum span was found and reduced until a 2% power penalty was incurred. The power required, wing span and cruise speed were then compared. The sizing tradeoffs between the wing bracing schemes are shown in Fig. CS6.11, which plots typical results for a 140 lb pilot and cruise lift coefficient of 1.0. The multi-wire approach was

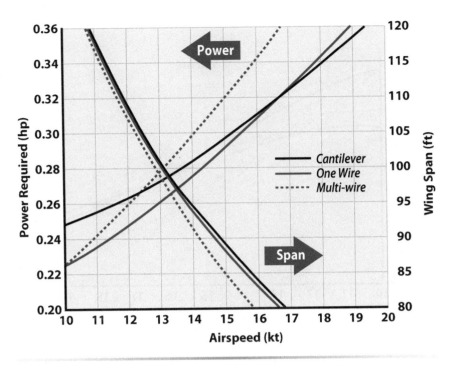

Figure CS6.11 Power and span for optimal aircraft (with 2% power penalty) for three bracing schemes (from [3]).

superior only at very low airspeed. The pure cantilever wing was best at speeds over 17 kt (20 mph), at power levels above 0.32 hp. The optimal aircraft power and span are both strongly dependent on cruise airspeed.

These results were combined with the pilot power model (power vs duration, see Fig. CS6.22) to estimate vehicle range as a function of cruise speed and wing bracing scheme. Typical output is shown in Fig. CS6.12 (for no wind) and in Fig. CS6.13 (for a 4 kt headwind). The results show that the one-wire concept is superior to the multi-wire for all speeds and is better than the cantilever wing at airspeeds less than 17 kt. While very large range can be attained by large, slow vehicles, they were very sensitive to wind and even a slight headwind reduced the range below the 60 mi requirement for all bracing concepts. This finding shows that, above a headwind speed of 2–3 kt, there are no optimal aircraft solutions that can meet the mission range goal, emphasizing the importance of selecting appropriate weather and winds for a flight.

Optimum speed was a function of the frequency distribution of acceptable weather conditions. In order to be able fly on 10% of the days in the weather window, a speed of 13 kt (15 mph) was required since the 10% probability level corresponded roughly to 4-hr long calm periods. The choice of 13 kt airspeed and the results of the range calculations led to the selection of the one-wire bracing configuration for the Daedalus aircraft.

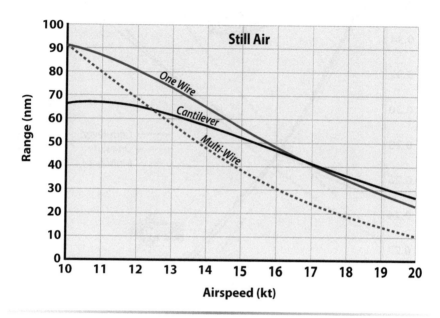

Figure CS6.12 Maximum still-air range vs airspeed for three bracing schemes (from [3]).

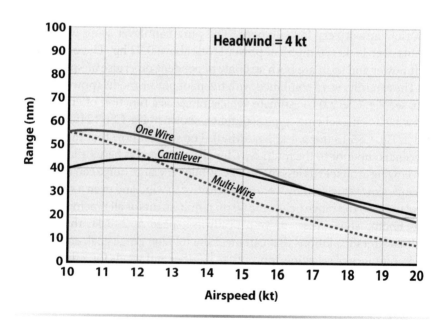

Figure CS6.13 Maximum range vs airspeed for three bracing schemes with a 4 kt headwind (from [3]).

CS6.2.7 Concept for the Daedalus Prototype

At the end of the feasibility study a nominal configuration was proposed for a prototype aircraft, as shown in Fig. CS6.2. This design was somewhat smaller, sized using a 3% penalty from the optimum power to limit the wingspan for a more compact, transportable vehicle. The design used a very high aspect ratio wing with a single-wire wing structure supporting a pilot in a semi-recumbent cycling position in a streamlined pod and boom arrangement with conventional aft tail surfaces. A pusher propeller was located behind the pod on the tailboom. This vehicle had a span of 30 m, a weight of 30 kg (68 lb) and an area of 30 m^2 and, if built as planned, would have the lowest power of any manned aircraft ever developed (3.0 W/kg specific power).

CS6.3 Phase II Design, Construction and Testing of the Prototype Aircraft

CS6.3.1 Phase II Overview

The second phase of the Daedalus project began in early 1986 and was focused on the detailed design and construction of the prototype aircraft proposed at the end of the feasibility study. The building and testing of this prototype was used as a risk reduction measure before committing to a record attempt in Greece.

The prototype aircraft was named (as appropriately as possible) the Michelob Light Eagle (or MLE, nicknamed Emily, or more formally the Light Eagle), after products sold by the primary Phase II sponsor Anheuser Busch. Detail design work was started in the spring of 1986 and construction of the Light Eagle began in June 1986. The aircraft was flown in flight tests at Hanscom Field, Massachusetts four months later and in winter flight testing at NASA Dryden. Phase II ended with a series of FAI records in early 1987, as shown in the project timeline in Fig. CS6.1.

The material that follows builds on the work in the Phase I study that largely defined the layout, aerodynamics, materials and structural concepts for the prototype. This section discusses the detailed design of the Light Eagle, including changes to sizing and structure, design of the wing, fuselage and tails as well as the drivetrain and controls. The remainder of the Phase II coverage reviews the flight testing and record flights of the Light Eagle prototype.

CS6.3.2 Design Evolution

The starting point for the prototype design was the configuration recommended in the Phase I feasibility study, shown in Fig. CS6.2. This was similar to the final Daedalus aircraft but somewhat smaller than the optimal span owing to a decision by the design team that a slightly lower wingspan aircraft would have practical advantages (e.g. transport, robustness and maneuver capability). The smaller aircraft, designated initially as the 30/30/30 configuration (for 30 m span, 30 m^2 wing area and 30 kg empty weight), gave up about 3% in power. This initial design was targeted for the 3.0 W/kg pilot power level that had been validated in tests with project pilot Lois McCallin.

The engineering process is rarely simple, and is often a bit messy involving unplanned changes and digressions that change the product. As the 30/30/30 preliminary design matured through detail design the span and weight of this configuration trended upwards as a result of myriad design trades and compromises, coming to rest at 31 m (102 ft) span and 31 kg (68 lb) with 30 m^2 (323 ft^2) wing area. It is interesting to note that the trend over the remainder of the program, from design and testing of the prototype to the final Daedalus configuration, was to drive the span up (initially to 114 ft on the long-tip version of the Light Eagle, finally to 112 ft on Daedalus), closer to the optimal span from the sizing program. At the same time the wing aspect ratio grew from 32.5, briefly to 38.4 and settled finally at 37.5 for the final aircraft. The weight target of 31 kg was exceeded substantially on the prototype (42 kg) but the final Daedalus aircraft ended up remarkably close to this target (just over 31 kg).

While minor changes were made to the aircraft design over the course of its evolution into Daedalus, the sizing work done in Phase I was checked and referenced for sensitivity effects but never wholly revised. The Phase I assumptions for weights, aerodynamics and pilot power proved to be fairly accurate, despite the guesswork involved. Regardless, by the end of the program these factors were certainly far better understood.

CS6.3.3 Detailed Design

The basic conceptual layout and sizing of the Light Eagle was taken from the Phase I feasibility study. The detailed design process, which started in April 1986 at MIT, refined the initial configuration and made minor changes to span and weight, to 102 ft and 31 kg. During construction the weight of the aircraft grew to 40 kg and, based on limited initial testing in the fall of 1986, appeared to have higher flight power than anticipated, due to the weight overrun. A longer set of wingtip panels was then constructed, to stretch the span to 114 ft to reduce induced drag and lower flight power.

The 3-view of the final version of the Light Eagle (with long tips) is shown in Fig. CS6.14. The basic dimensions, design parameters and performance of the long tip version of Light Eagle are compared with the Gossamer Albatross and the Daedalus 88 aircraft in Table CS6.3.

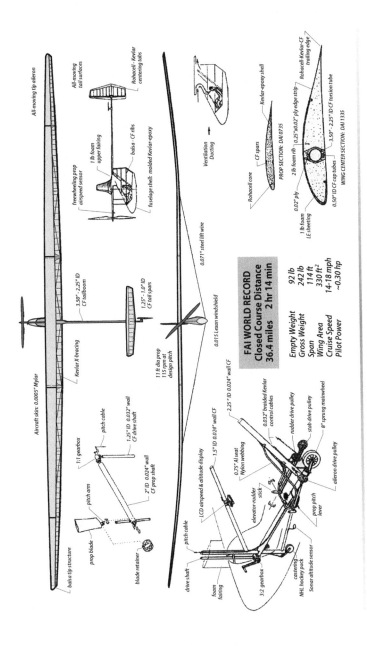

All-moving tip aileron

All-moving tail surfaces

Rohacell-Kevlar centering tabs

freewheeling prop airspeed sensor

1 lb foam upper fairing

balsa-CF ribs

fuselage shell: molded Kevlar-epoxy

Ventilation Ducting

Aircraft skin: 0.0005" Mylar

Kevlar X-bracing

3.50"-2.25" ID CF tailboom

1.25"-1.0" ID CF tail spars

11 ft dia prop 115 rpm at design pitch

0.071" steel lift wire

0.015 Lexan windshield

1:1 gearbox

pitch cable

1.25" ID 0.032" wall CF drive shaft

2" ID 0.024" wall CF prop shaft

LCD airspeed & altitude display

1.5" ID 0.024" wall CF

2.25" ID 0.024" wall CF

0.032" braided Kevlar control cables

rudder drive pulley

stab drive pulley

8" sprung mainwheel

pitch arm

0.75" Al seat Nylon webbing

prop blade

elevator-rudder stick

blade retainer

balsa tip structure

pitch cable

drive shaft

foam fairing

3:2 gearbox

castering NHL hockey puck

Sonar altitude sensor

prop pitch lever

aileron drive pulley

Rohacell core

CF spars

PROP SECTION: DAI 0735

Rohacell-Kevlar-CF trailing edge

Kevlar-epoxy shell

2 lb foam rib

0.25"x0.02" ply edge strip

0.02" ply

1 lb foam LE sheeting

3.50"-2.25" ID CF torsion tube

0.50" ID CF cap tubes

WING CENTER SECTION: DAI 1335

FAI WORLD RECORD
Closed Course Distance
36.4 miles   2 hr 14 min

Empty Weight	*92 lb*
Gross Weight	*242 lb*
Span	*114 ft*
Wing Area	*330 ft²*
Cruise Speed	*14-18 mph*
Pilot Power	*~0.30 hp*

Figure CS6.14 Layout and 3-view of Michelob Light Eagle prototype.

Table CS6.3 Comparison Between Distance-Record Human-Powered Aircraft (from [17])

	Gossamer albatross	Light eagle	Daedalus 88
Span ft (m)	95.1 (29.0)	113.85 (34.7)	111.88 (34.1)
Area ft^2 (m^2)	536 (49.8)	333.7 (31.0)	333.7 (31.0)
AR	19.0	38.4	37.5
Weights lb (kg)			
Empty	72.8 (33.0)	92.6 (42.0)	68.6 (31.1)
Pilot	135.6 (61.5)	151.1 (68.5)	159.9 (72.5)
Misc	4.4 (2.0)	8.8 (4.0)	13.2 (6.0)
Gross	212.8 (96.5)	252.5 (114.5)	241.7 (109.6)
Cruise conditions			
Specific power hp/lb (W/kg)	0.00219 (3.60)	0.00201 (3.30)	0.00177 (2.90)
Abs power hp (W)	0.297 (221)	0.303 (226)	0.282 (210)
V kt (m/s)	10.6 (5.4)	14.3 (7.3)	13.4 (6.9)
q lb/ft^2 (N/m^2)	0.378 (18.1)	0.687 (32.9)	0.603 (28.9)
Aero coefficients			
C_L	1.05	1.10	1.20
C_{D_0}	0.014	0.010	0.011
$C_{DS(parasite)}$	0.31	0.19	0.18
$L/D_{(typical\ wing\ 2-D)}$	75	112	118
$L/D_{(overall\ 3-D)}$	25.8	40.4	40.4
Drag lb (N)			
Induced	4.16 (18.5)	2.53 (11.2)	2.73 (12.2)
Profile	2.84 (12.6)	2.29 (10.2)	2.11 (9.4)
Parasite	1.26 (5.6)	1.42 (6.3)	1.14 (5.1)
Total	8.25 (36.7)	6.24 (27.8)	5.98 (26.6)
Distance nm (km)	19.3 (35.8)	31.7 (58.7)	63.0 (116.6)

The following sections discuss the design of the wing, fuselage, tails and drive system (pedals, gearboxes and propeller) as well as the controls and autopilot system.

CS6.3.4 Wing Structure

Although design load factors of 2 g were used in wing structural optimization in the feasibility study, the project mission profile had provision for a cliff launch for a 60 mi powered glide to the mainland so the Light Eagle was designed with a higher 3 g load factor for safety. The structural concept was based on a single bracing wire, from the fuselage at the landing gear to the half-span of the wing to reduce the root bending moment.

CS6.3.4.1 Wing Loads

The design conditions for the structure of an HPA differ from those used for conventional aircraft [15]. Traditional V-n diagrams were not directly applicable to the Daedalus aircraft, and limit and ultimate load scaling with a constant factor of safety was inefficient. The approach that was used considered the limited flight envelope of HPAs and also accounted for flexibility and dynamic effects on loading.

Several important flight parameters affected the loading cases: stall speed, maneuver speed, maximum dive speed and sideslip angle. The maneuver speed, Vm, is defined as the highest airspeed at or below which full control deflections can be made without risk of structural failure. This was assumed to be the highest speed achievable in level flight under human power. The aerodynamic design load factor was defined by the ratio between the selected V_m and the stall speed V_s

$$N_{des} = (V_m/V_s)^2$$

This criterion was used to size the Daedalus wing structure (at 1.75 g) but for Light Eagle, with its high altitude flight option, a higher load factor of 3 g was assumed for safety.

The dive speed, V_d, is defined as the maximum operating airspeed, roughly equivalent to V_{ne} on conventional aircraft. The dive speed V_d should ideally be only slightly higher than the maneuver speed for a low-altitude, energy limited aircraft. For the Light Eagle and Daedalus aircraft, which flew at altitudes in excess of 30 m, a higher dive speed of 11.3 m/s was used. Flutter was not a factor within this speed range. The torsional stiffness of the wing was designed so that the divergence speed for the Light Eagle was well above V_d. Torsional deformation of the wing at higher speeds unloaded the wingtips so that critical loadings for wing bending occurred closer to the maneuver speed, not the dive speed.

During the Light Eagle design the load cases considered moderate levels of sideslip (initially thought to be limited to 7.5 deg). Higher sideslip levels were found later in Phase II and Phase III flight testing, up to 30 deg, which were used in the final Daedalus aircraft design.

Load analysis was done with an early version of Drela's ASWING aero-elastic analysis code [16]. This considered nonlinear, deformed geometry including the effects of the compression imposed by the lift wire, and also included aerodynamic damping and inertial relief due to real and apparent mass. These flight-dynamic effects reduced peak loads far below those for a static clamped wing, especially for sideslip cases.

CS6.3.4.2 Wing Structural Design

The wing spar was a triple-tube design by structural designer Juan Cruz, as shown in Fig. CS6.7. A large central tube was constructed of ±40 deg. carbon fiber plies, and was optimized for shear and torsion loads. Two smaller tubes, made from 0 deg, and ±10 deg. plies, were used (top and bottom) for axial loads from wing bending This design was 4.4 lb lighter than an equivalent single tube design and was resistant to section ovalization under load without interior reinforcement.

The wing was constructed in five panels for ease of construction and transport (Fig. CS6.15). The three inboard sections were 28 ft in length while the tips were originally designed with 9 ft span, for a total span of 102 ft. The span was later increased to 114 ft with longer tip panels, although this reduced the structural margin somewhat, below the 3 g design condition. Although the wing was built to withstand a download equal to its own weight plus a slight gust load, it was supported by an additional download wire from a removable mast while on the ground and during short flights. The download wire was removed for only for very long flights or performance testing, at which times extra care was taken to support the wing for launch and recovery.

The wing structure used a single main spar with a LE shell of hot-wire cut beadboard foam, 2 lb/ft^3 density foam ribs with 0.02 in plywood capstrips and a Kevlar, carbon fiber and Rohacell trailing edge. To reduce compression loads on the trailing edge (which ended up plaguing the Light Eagle periodically with local compression failures from covering and flight loads) a forward in-plane bracing wire and struts were added in front of the spar to tension the trailing edge. This structural arrangement for the wing was stiff for normal bending loads but somewhat compliant for in-plane bending. Although adequate for flight loads this structure had a low in-plane bending frequency that showed up as fore-aft wing movement during pedalling.

Turning flight and roll control for slow-flying, long-span aircraft like Daedalus present special challenges. The very long wings are prone to such severe adverse yaw effects from ailerons that the rudder becomes the primary control for the pilot in lateral maneuvers. At these slow speeds the rudder can generate strong rolling moments due to yaw rate (from differential velocity across the span) which is more effective than ailerons for rapid changes to bank angle. Ailerons are used primarily in opposition to the turn to prevent overbanking (e.g. left aileron held in a steady right turn). Despite their drawbacks, the high-altitude flight requirement for Light Eagle suggested that ailerons were needed for the prototype.

Previous HPAs utilized either conventional ailerons or wing warping for roll control. Due to the single wire design of the Light Eagle, wing warping was not practical. Conventional hinged ailerons would have imposed a significant weight penalty due to the increase in torsional stiffness required to prevent aileron reversal as well as the additional structure to support the aileron hinge. The solution was to rotate the entire tip panel

Figure CS6.15 Center wing structure during assembly, with wing mount, spar, ribs, TE and compression struts showing use of Kevlar lashing to join components (from Daedalus 87). (Photo by Peggy Scott, 1987.)

of the wing for aileron control, imposing essentially no weight penalty on the structure itself. Additional weight consisted of 2 sets of bearings on the tip spars and the components of the aileron control linkage. The total weight of the system was approximately 2 lb.

The wing was initially designed using a high modulus prepreg—Thornel P75. As the project moved further into detailed design the P75 was not available. Without a suitable high modulus alternative, a standard lower modulus fiber, Thornel T300 was substituted, and new spars were designed around the existing tooling. Since much of the wing was designed to meet a stiffness constraint, the wing weight increased in nearly exact proportion to the modulus decrease, from close to 20 lbs to slightly more than 30 lbs. The spar weight increase alone (10 lb) represented a 17% growth in empty weight [6]. Table CS6.4 shows a comparison of the target weights and actual weights for the Light Eagle components. Although the weight overrun was principally in the wing spar many of the other components also experienced significant weight growth. The total 22 lb weight overrun had serious implications for the Phase II pilots, representing a 13% increase in the pilot's specific power (from 3.0 W/kg to 3.4 W/kg).

Experiencing this weight growth was a valuable lesson for the MIT students as it frequently happens in aircraft development programs. Draconian design measures and an experienced, talented building crew later in Phase III kept the weight for the Daedalus aircraft on target (so miracles do happen).

Table CS6.4 Light Eagle Target Versus Actual Weights (data from [3])

	Predicted (lb)	Actual (lb)
Fuselage	32.2	31.5
Wing spar	23.9	32.0
LE sheet	2.6	6.0
Wing ribs	2.2	4.6
Wing covering	1.6	3.0
Wire bracing	0.4	1.0
Tails	1.1	3.8
Misc/reinforcement	0.0	3.7
Total	66.6	88.1

CS6.3.5 UltraLight Construction

One of the challenges for a vehicle like Light Eagle or Daedalus concerns bonding or joining components such as struts to a wing spar or tubes in the fuselage structure. The key is to use minimal bonding agent to control weight growth. This requires discipline as the weight growth is subtle and insidious.

Tube joints were made by using lightweight lashings of Kevlar or carbon fiber tow and epoxy. This technique was developed for thin-walled aluminum tubes during earlier Chrysalis and Monarch HPA projects at MIT, and was used for all the thin-wall carbon fiber tubes in the Light Eagle and Daedalus. The butting tube end was closed with a shallow plug of Rohacell-31 foam and mitered to match the other tube. To form the joint, the tubes were first tack-glued to fix the joint geometry and tow lashings were wound under tension around the joint, first along the directions of principal stresses (across the joint), and then circumferentially to prevent peeling of the tows. Epoxy was applied during this wrapping process to fully saturate the tows, and the excess removed to give a high fiber/resin ratio. Failure of a joint was never observed in any of the HPA projects at MIT, instead buckling would occur in the tubes. Fig. CS6.16 shows a cutaway of the construction for a wing panel and assembly joint (for the Daedalus wing).

CS6.3.6 Fuselage

One of the most challenging parts of a successful HPA design is the fuselage and landing gear. The structure must be very light yet it has to handle flight and landing loads from a pilot that weighs more than two times the weight of the aircraft! Although the temptation for a long distance design would be to skimp on this structure, the reality is that these vehicles need to survive a long development and training flight program involving hundreds of major and minor takeoffs and landings.

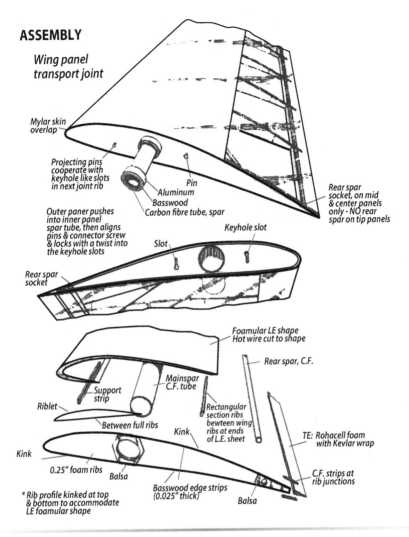

Figure CS6.16 Construction details for wing panel, showing assembly joint and wing components (from Daedalus wing) (from [4]).

The configuration recommended in the feasibility study used a "pusher" propeller located on the tail boom behind the wing. For the Light Eagle this was changed to a tractor configuration on the basis of weight, reliability and simplicity. The height of the fuselage was optimized, considering fuselage structural weight, wire weight, overall drag, as well as propeller efficiency (set by propeller diameter).

The fuselage consisted of an airfoil cross section pod to enclose the semi-recumbent pilot, highly swept in lateral planform to reduce wetted area. The swept planform minimized interference drag with the wing and was more resistant to flow separation when flown at a moderate sideslip angle, a very common occurrence for an HPA. The pod consisted of a thin molded Kevlar lower shell, with balsa/carbon fiber ribs with a 0.015 in. Lexan windshield. Polystyrene foam sheeting formed the upper fairing and cooling ducts. The aft portion of the pod was covered with aluminized Mylar to reduce solar heating.

Using results of the pilot cooling experiments (see Sec. CS6.2.4.1) a diffusion controlled inlet and ducting system was built into the upper fuselage pod to blow cooling air over the pilot's head and shoulders. Drainage holes in the fuselage below the pilot eliminated pilot sweat accumulation (added after a large pool accumulated during the long Light Eagle record flight).

The main forward fuselage structural elements consisted of seven carbon fiber tubes defining a minimal structure to support the pilot, landing gear, wing wire loads and propeller. The fuselage tubes were joined by lashing with Kevlar tow, as described earlier. The fuselage primary structure assembly weighed only 9.5 lb and the complete fuselage, tailboom and systems weighed 31.5 lb [6]. The main landing gear used an 8" diameter plastic wheel mounted in an aluminum fork that compressing a rubber block acting as spring and shock absorber. The seat's main mounting struts attached directly to the wheel-fork axle, so that much of the pilot's weight (twice the airframe weight!) bypassed the fuselage structure during landing impact. The nose gear was also sprung and was castered to allow ground steering via the rudder. This compact integration of the pilot support structure to the landing gear is essential for an HPA to absorb landing loads without high loads elsewhere in the structure.

CS6.3.7 Tail Surfaces

The horizontal and vertical tail surfaces used tubular carbon fiber spars with a CNC hot-wire foam LE shell, foam ribs and a Kevlar and Rohacell trailing edge. The tails were fully pivoting surfaces with 10% chord anti-servo tabs at the trailing edge. Anti-servo tabs deflect in the same direction as the surface, creating a restoring moment and were used for control force "feel" (from the rudder, particularly). The tails were hinged on "V" mounts bonded to the fuselage structure, giving a completely clean leading edge for laminar flow to reduce drag. The tail surfaces were optionally controlled by the autopilot via small servos that actuated the tabs such that the pilot could override the flight control system at all times.

While the tab provided stick feel to the pilot it also increased the lift slope and maximum lift of the surface. Overall, the use of the tab allowed for the use of slightly smaller rudder, saving weight. The disadvantage of the anti-servo tab on the rudder was increased stretch in the control lines, a factor discovered late in Phase III. Figure CS6.17 shows the business end of the Daedalus tailboom with the vertical tail installed in its V-mount.

Figure CS6.17 A view of the vertical tail and mounting structure at the end of the tailboom. This shows flight test hardware, installed for the Dryden testing, to measure tail deflection. (Photo by Peggy Scott, 1988.)

CS6.3.8 Drive Train

Many previous successful human-powered aircraft have used some version of a chain and sprocket drive system to connect the pedals to the prop shaft. These systems have the advantage of being light, efficient, and relatively easy to design and construct. However, the chain must twist to accommodate the 90 deg angle between the pedal and propeller axes, which makes it prone to jumping sprockets. For reliability the Light Eagle used a bevel gear and shaft system developed by Robert Parks [6]. A step-up gearbox with ratio 1:1.5 was used at the pedals to a high-speed vertical shaft and a 1:1 ratio gearbox to the propeller was used to match the required prop speed.

The gearboxes were fabricated out of aluminium with hardened steel gears. Standard lightweight bicycle crank arms and pedals attached to the lower gearbox. The lower gearbox weighed 2 lb, the upper unit 1.5 lb, while the cranks and standard pedals weighed 1.25 lb. The carbon fiber vertical driveshaft weighed about 8 oz and used a tailored layup to provide high torsional stiffness while retaining flexibility in bending to handle shaft mis-alignments. The gearboxes required care to set the gear mesh (using thin shims in the casing) but proved to be very reliable.

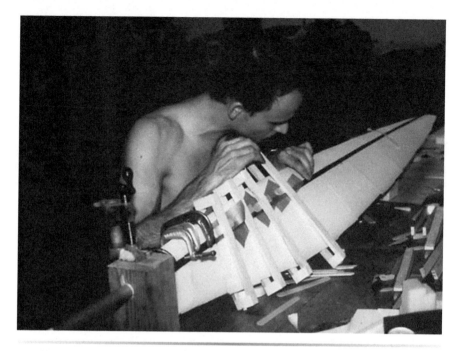

Figure CS6.18 Project aerodynamicist Mark Drela, intent on construction of the Light Eagle propeller out of hot-wire cut acrylic foam blocks and graphite spar. (Photo by Peggy Scott, 1986.)

CS6.3.8.1 Propeller

The propeller was designed using a minimum induced loss algorithm described by Larrabee [11], implemented in the MIT XROTOR code. The airfoil profile drag was also modeled to give a minimum overall loss design. The DAI 1238 airfoil used on the wing tips was used for the propeller blade sections. The 11 ft diameter propeller was designed to provide 7.5 lb of thrust at a speed of 15 mph and 115 rpm with an efficiency of 89% [6].

The propeller structure consisted of Rohacell acrylic foam core, hot wire cut in radial segments that were joined with unidirectional carbon fiber spar caps and a carbon fiber tube hub, as shown in Fig. CS6.18. The blades were covered with Kevlar and weighed 13 ounces apiece. A pilot-controllable variable pitch hub was used to match desired pilot pedalling rate. A very low pitch could also be selected during the initial takeoff roll to prevent prop stall at low advance ratios, significantly reducing the extra power needed for takeoff acceleration.

CS6.3.9 Controls

The Light Eagle was designed with three-axis control using rudder, elevator and ailerons. Although most flights used only rudder and elevator, the ailerons were used in steady state turns to prevent overbanking. Since normal rudder pedals were not practical in a pedal-powered airplane the Light Eagle used two wrist-actuated control sticks mounted on each side of and below the seat [6]. The right hand stick controlled the rudder and elevator, while the left stick controlled the ailerons. The ailerons were implemented as all-moving wing tip panels and required higher control forces. The propeller pitch control lever was similar to an old-style bicycle shifter, mounted directly in front of the pilot. Instruments included a pilot display of airspeed (using a small free-wheeling propeller sensor) and altitude above ground (using a sonar range sensor). A yaw string on the windshield was used as a sideslip indicator for turns.

CS6.3.9.1 Autopilot/Flight Control

The initial meteorological studies found that durations of more than 4 hr could be required for the Daedalus flight and that much of the calm wind conditions occurred during the night. A flight control/autopilot system to allow Daedalus to fly at night or in limited visibility was developed by Steve Finberg and Bryan Sullivan, engineers at the C.S. Draper Lab [9] and [10]. The autopilot provided wing levelling, airspeed hold and heading hold. A flight computer processed data from sensors in the wing and fuselage and displayed flight attitude to the pilot via an artificial horizon display or functioned as an autopilot to control the aircraft. Servos, connected to the cockpit computer via fiber-optic links drove small tabs on the rudder and elevator to actuate the tails. The altitude sensors were a Polaroid sonar range finder (accurate to 50 ft) and a barometric altimeter for high altitude flight. The entire autopilot system weighed approximately 2 lb [6].

This system was installed in the Light Eagle and tested during January 1987. The fly-by-light system worked well, and several flights were made with the aircraft controlled by a joystick in the cockpit. Neither the artificial horizon display nor the sensors displayed the reliability necessary for autopilot operation. The initial roll sensors (electrostatic sensors on the wings and fuselage) were plagued by noise problems. An alternative system using small gyroscopes was unable to achieve sufficient accuracy for use in the full autopilot system.

CS6.3.10 Flight Testing the Light Eagle Prototype

The Light Eagle was first flown on October 4, 1986 at Hanscom Air Force Base near Boston. Thirty-eight flights were conducted by several pilots including the primary project pilot at the time, Lois McCallin. The Hanscom flying was done with the original (31.1 m/102 ft span) configuration and were short, limited to the 1 mi length of the runway. In January 1987, flight operations were transferred to the NASA Dryden Flight

Research Facility at Edwards Air Force Base in California. A total of 48 flights, including long duration tests, were performed over the dry lakebed during this period. The Light Eagle, with extended wingtips (34.7 m/114 ft span), also set several new distance and duration records including a closed course distance record of 58.7 km (37.2 mi) on January 22, 1987 by Glenn Tremml and three feminine distance and duration records by Lois McCallin.

In addition to record flights that were necessary for pursuing sponsorship for Phase III, the primary goals of the flight test program were

1. Determine the physiological and manual control workloads imposed on the pilot.
2. Measure the aircraft performance, particularly the amount of mechanical power required to maintain level flight.
3. Evaluate the performance of the flight control system.

CS6.3.10.1   Flight Test Configurations

Flight testing was done in the early morning, normally from pre-dawn until breakup of the ground inversion layer with lakebed heating from the sun. In typical flight sessions, as the aircraft was prepared outside the hangar, frost would form on the flight surfaces (which would spoil laminar flow and drastically raise flight power). The frost was removed using a host of hot air guns as part of the morning pre-flight ritual and generally did not reform as the morning testing continued. Problems were encountered in the early flights such as failure of the landing gear and jamming of the gearboxes. The project team worked through these and other minor problems and flight testing continued in California. The Light Eagle proved to be a very capable, reliable workhorse and was used for a wide variety of experiments, measurements, training, and flight testing throughout the remainder of the program.

Very early in the flight test program with the original 102 ft span Light Eagle configuration it was realized that the power required to fly the airplane was too high (roughly 13% too high), principally due to the weight increase from 66 lb to 88 lb during construction. In an attempt to reduce flight power by lowering induced drag the wingspan was increased with longer wingtip panels (from 9 ft to 15 ft). This change increased the wingspan of the Light Eagle to 114 ft and the weight to almost 92 lb. The longer span version of the Light Eagle was used for most of the testing and for the record flights at Dryden.

CS6.3.10.2   Flying Qualities

The Light Eagle was first flown on October 3, 1986 by two engineering test pilots (including the author). The original shorter span (102 ft/31.1 m) version of the aircraft was short-coupled in pitch, initially with very neutral control feel and a tendency for pilot-induced pitch oscillations, probably exacerbated by the flexibility of the structure. Adding springs to the control stick and removing elevator mass balance alleviated this problem. This version of the Light Eagle proved to have reasonably good handling qualities (for a human-powered aircraft) and felt in its control response somewhat like a large, high-performance sailplane but with much slower roll response. Both athlete-pilots received flight instruction in high performance two-place sailplanes and this training transferred quite well to the Light Eagle. During longer flights over Rogers Dry Lake, a slight pitch phugoid with a period of 3–4 sec was observed but this was easily eliminated with attention to pitch attitude. The longer wing version of the Light Eagle also exhibited noticeable fore-aft in-plane motion of the wing tips at roughly 1.5 hz in response to pedalling loads. Its effect on the flight power is unknown.

Turning flight posed a greater challenge to the pilots due to the aircraft's initial lack of good lateral handling qualities. The high stiffness of the carbon fiber-epoxy wing spar in the original shorter-span configuration resulted in very little load-induced dihedral (roughly 0.5 m). The introduction of a slight bank to initiate a turn would lead to the development of large sideslip angles and loss of altitude. Strict attention to the yaw string and bank angle were required for smooth turns. When the wingspan was extended to 114 ft (34.7 m) to lower the flight power, the higher load-induced dihedral (to roughly 1.0 m) significantly increased roll-yaw coupling. The pilots reported a subjective reduction in the effort required to maintain level flight—an effect that was interpreted initially as the lower airframe power predicted by the design team. However, subsequent in-flight physiological measurements and tow testing were unable to verify a measurable power reduction. Transition problems on the outboard wing sections (described in the following section) may have produced additional drag that negated most of the benefits of increased span. The reduction in effort reported by the pilots was likely due to an improvement in handling qualities. With the enhanced lateral stability with the longer tips, turning flight posed much less of a problem and turns of 200 m radius were achieved during the two closed-course world-record flights.

CS6.3.10.3 Physiological Measurements

The original design power for the Light Eagle was 3.0 W/kg of pilot weight (0.3 hp for a 150 lb pilot). Validation of flight power relied on measurements of the pilot's heart rate so lightweight heart rate recorders were worn by the athlete pilots during all of the long duration testing in California. Figure CS6.19 shows the heart rate for the Light Eagle record flight by Glenn Tremml, the second Phase II "engine." Glenn's average cruise heart rate of 170 beats per minute, correlated with ergometer calibrations (see

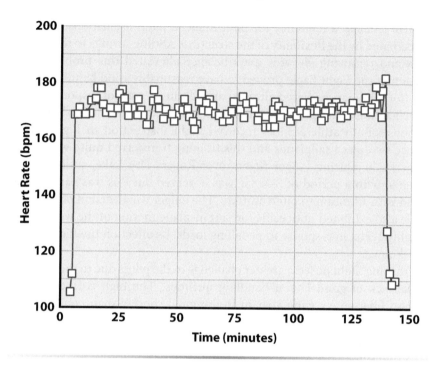

Figure CS6.19 Heart rate for Glenn Tremml during the Light Eagle record flight (from [8]).

Fig. CS6.6), gives a flight power of 3.3 to 3.4 W/kg. The peaks in heart rate correspond to pilot workload increase in turns at the ends of the triangular course. The official pilot power for the Light Eagle, obtained with heart rate and unpowered glide sink rate measurements, was estimated at 228 watts (at 108.8 kg gross weight), or 3.3 W/kg for the 68.5 kg pilot.

CS6.3.10.4 Tow and Glide Testing

In addition to the physiological methods described previously, direct measurements of aircraft power were attempted with tow testing and glide testing in still, early morning conditions [8]. Tow testing of airframe drag on the initial, shorter span Light Eagle proved to be unworkable due to tension variations from pilot inputs. Measuring sink rate in glide tests after towing to altitude proved similarly noisy. Power derived from gliding measurements of energy state, using height and velocity gave somewhat more consistent results but was sensitive to vertical atmospheric gradients where an air movement of 2 in/s can result in 30% error in *L/D*! The best results showed *L/D*s in the range 34 to 38 when corrected to out of ground effect performance. Although statistically inconclusive, these average figures agree roughly with the physiological power data. There was also a pronounced increase in power when the wing was contaminated with frost or dew, though this was never quantified.

CS6.3.10.5 Laminar Flow Testing

In-flight flow visualization of the flow on the Light Eagle wing, done at the end of the flight operations at Dryden, proved to be a pivotal, decisive test that helped shape the final Daedalus aircraft. Flow visualization was used to measure the extent of laminar flow on the wing in flight in order to understand the "missing" power reduction from the longer tips. It was also done to diagnose possible problems with Drela's untested wing airfoils. Visualization was done during gliding test flights by applying a mixture of kerosene and a fine black powder colorant to the upper and lower surfaces of the wing at several stations. The aircraft (without propeller) was quickly towed to altitude and descended at a constant speed. The kerosene mixture flowed along the wing chord in response to varying skin friction forces on the wing surface, making the transition from laminar to turbulent flow clearly visible after the flight. The transition locations compared favorably to design predictions except for the tips, which were almost fully turbulent. This at least partially explains why the longer tips for the Light Eagle did not reduce flight power. The polystyrene beadboard foam leading edge sheeting was very thin on the tips, on the order of the thickness of the expanded polystyrene "beads." Visual inspection of this area showed that the edges of the foam "beads" were imprinting slightly through the Mylar covering, effectively roughening the surface. The laminar separation bubbles on the wing were generally very pronounced (relatively well defined with long chordwise extent) indicating that there may be some additional drag associated with transition. Figure CS6.20 shows an area of the mid wing with laminar flow (smooth dark layer on forward portion of LE) and laminar bubble (area with mixing pattern) and downstream region of turbulent flow.

CS6.3.10.6 Record Flights

As part of the flight testing at Rogers Dry Lake several FAI record flights were made. On January 22, 1987 the Light Eagle set a new closed-course distance record of 58.7 km (37.2 mi) by Glenn Tremml. During that month, Lois McCallin set 3 additional World Records for human-powered flight: Distance around a closed course (feminine pilot, 15.4 km); Duration (feminine pilot, 37.5 min); and Straight Line Distance (feminine pilot, 6.8 km). Unfortunately the power required for the Light Eagle was about 10% too high (above her AT, anaerobic threshold) for her to maintain flight power aerobically. Lois remained with the project but retired from pilot status, a victim of the unforgiving physics of flight. The irony of this is that the final Daedalus aircraft, which had flight power levels of 2.9–3.0 W/kg, would have put Lois under her AT and permitted flights of several hours.

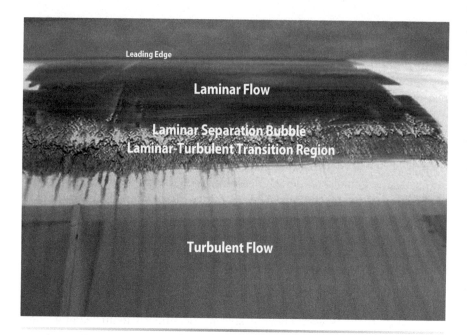

Figure CS6.20 Flow visualization of laminar flow and transition with separation bubble on Light Eagle mid-wing upper surface. (Photo by Peggy Scott, 1987.)

CS6.3.10.7 Overwater Testing

One additional test flight of the prototype was conducted after the Dryden testing, off the shore of Ninigret, Rhode Island on August 26, 1987. This flight lasted approximately 21 min and allowed the pilot to familiarize himself with flying over water. It also allowed the ground crew to practice assembling the airplane in outdoor, nightime conditions and the support boat crews to practice navigating alongside the airplane. This was a very committing test and emphasized the seriousness of overwater flights and the preparation required to execute them safely.

CS6.3.11 The Crucial Question

Was the prototype capable of completing the flight from Crete? With a specific power of roughly 3.3 W/kg (1.5 watt/lb) the early physiological testing predicted a duration limit for Olympic trained cyclists of around 3.25 hr or a range of just over 50 mi. On this basis it was decided that the prototype was unlikely to successfully complete the flight. Accordingly, in February 1987 the decision was made to proceed with design and construction of the Daedalus aircraft, to be optimized for lower flight power and to recruit a new set of higher power pilots for the record flight.

CS6.4 Phase III—Design, Construction and Testing of Daedalus

CS6.4.1 Phase III Overview

The final phase of the Daedalus project began in the spring 1987, following the Light Eagle record flights and flight testing at Dryden. Once the decision was made to proceed with Phase III the project was on a fast timeline (see Fig. CS6.1) to put two tested Daedalus vehicles in Crete, ready to fly, by late February 1988 to meet the spring weather window. This timeline also included recruitment, testing and training of a new pilot crew as well as planning and preparation of logistics that would support flight operations in Crete. The program also had another very important priority—the search for new sponsors for Phase III development, testing and the record attempts.

Preliminary design work on the new Daedalus aircraft started in the Spring of 1987 as Phase II flight testing wound down. By then the best estimates of flight power requirements for the Light Eagle prototype were 3.3–3.4 W/kg, a level deemed too high for a successful flight of over 4 hr in Crete. The design goal for Daedalus was clear—reduce the flight power to 3.0 W/kg or less. Although the external form of Daedalus was almost identical to the prototype, changes were made to virtually every part of the vehicle. The benefits of having built the Light Eagle as a prototype showed up in the extraordinary refinement of the Daedalus aircraft structure and systems.

The continuing meteorological studies had prompted a change to a low-altitude flight profile that allowed significantly lower structural margins. This, together with the extensive redesign, contributed to the reduction of empty weight from 92 lb for the Light Eagle to 68 lb for Daedalus. Also, based on the Phase II flow-visualization tests at Dryden the wing surface quality was improved with new wing LE materials and the wing airfoil family was redesigned for lower drag.

Significant upgrades were also made to the "engine" side of the flight power equation with a new crop of four athletes recruited from the ranks of Olympic-level amateur cyclists. These new athletes did not have flying experience but were trained over the fall and winter of 1987 to fly in sailplanes, in the Light Eagle and later in the Daedalus airframes. The new pilots were joined by Phase II veteran, Glenn Tremml, in a five-pilot rotation training hard for the flight from Crete.

Construction of two aircraft for the flight record attempts (Daedalus A and B, later rechristened Daedalus 87 and 88) began at MIT in June of 1987 at the Draper Lab Flight Facility in Bedford, MA. The two airframes were identical, to provide vehicle or component redundancy in case of damage. In a 72-hr final assembly marathon, the Daedalus 87 airframe was rolled out in October 1987 with the announcement of the financial sponsorship of the United Technologies Corporation for Phase III.

Daedalus 87 was delivered to NASA-Dryden for flight testing in November and the first flight was made on December 2, 1987. Initial flight testing went smoothly and the flight power appeared to be very low, in line with the 3.0 W/kg specific power goal at 13 kt airspeed. Unfortunately, due to a long string of poor weather, progress on flight testing and pilot familiarization flights moved slowly. By February 1988 the aircraft had accumulated only about an hour of flight testing. At this point the project was getting close to its scheduled departure for Greece to meet the 1988 spring weather window. This schedule went out the window when, on February 7, Daedalus 87 crashed on Rogers Dry Lake due to loss of control in thermal-induced gusts. Faced with possibly disastrous impact on the project, program manager John Langford wisely opted for delaying the departure for Greece by a month. The second airframe, fortunately very near completion, was brought up to flying status and rushed to California in late February while the damaged aircraft was repaired in Massachusetts. In a fast-paced flight testing program the Daedalus 88 airframe was modified to fix the control problem and completed qualification testing. The Daedalus airframes were transported to Crete in a Hellenic Air Force C-130 on March 26, 1988 and operations in Heraklion commenced so that the team would be ready for a record flight when the weather arrived.

CS6.4.2 Pilot Selection and Training

Meteorology studies and possible changes to the route of flight had led to concerns that the record flight could last more than the 4 hr originally envisioned. The physiological research had indicated that, for durations of 4 hr or more, countermeasures such as "carbohydrate loading" would be required to ensure that the pilot could sustain these long flights. In this approach the pilot, who has to maintain a high level of training, tapers off his workouts and "carbo loads" before a maximum effort to preload his muscles and liver with glycogen. Unfortunately the benefits of this glycogen loading only last for a few days, after which normal training must continue to sustain fitness. This type of loading strategy is used by competitive amateur cyclists who structure their training schedules around weekend races. This approach implied that at least four pilots in a rotation schedule would be required for flight readiness on any given day.

With additional sponsorship from the Shacklee Corporation a new crop of four athletes were recruited amongst the ranks of Olympic-level cyclists. These included Americans; Frank Scioscia, Eric Schmidt and Greg Zack and the Greek National cycling champion, Kanellos Kanellopoulos. These new "engines" were fantastic athletes, cycling up to 400 mi per week and significantly raising the bar in terms of specific power levels—but they had no flying experience. They were trained to fly over the course of the winter

Figure CS6.21 The five pilot-athletes in the Daedalus record flight rotation, under the wing of Daedalus 87. L-R, Greg Zack, Frank Scioscia, Glenn Tremml, Eric Schmidt, and Kanellos Kanellopoulos. (Photo by Peggy Scott, 1987.)

of 1987 in sailplanes and each made many flights in the Light Eagle in preparation for flying Daedalus. The new pilots joined Glenn Tremml, the experienced project pilot from Phase II who was able to train to approach their fitness level, in a pilot roster of five on continuous rotation. This photogenic group of athletes is shown in Fig. CS6.21.

The pilots were qualified initially with 4-hr ergometer tests at power levels of 3.1 W/kg, slightly above the nominal flight power (closer to the flight power when carrying 6 kg of drinking water). With concerns about longer flights due to headwinds, pilot qualification also included a gruelling 6-hr ergometer test on the Daedalus flight simulator as well as a 2-hr flight in the Light Eagle on the lakebed at NASA-Dryden.

CS6.4.3 Meteorology and Flight Route

The initial weather study undertaken by Jonathan Wyss [3] predicted the highest probability of acceptable conditions would occur during April of any given year. This analysis was reinforced throughout the project. This analysis suggested that, although during this period some 30–40% of the days might have flyable conditions, most of these windows would occur at night. The incidence of daylight flight windows was much lower, perhaps only a few days per year.

The historical analyses were reinforced by data obtained by the project's automatic weather stations that were deployed along the proposed western flight route in March 1986. Analysis of data from the stations and continued surveys of takeoff and landing sites led to increasing concern about micro-meteorology involved in operating (a) near the cliffs at Akra Spatha, (b) near the island of Kithira, and (c) in the channel between Kithira and the mainland, which was windy and heavily trafficked by large ships.

In December 1987 a special advisory panel of weather experts was convened. This panel reviewed the accumulated data and recommended that while the project's synoptic analysis was sound (i.e., selection of April as the prime flight window), the flight route should be restructured to avoid all possible land/sea interfaces, since these were the primary sources of temperature discontinuities that lead to local winds. The flight route was switched from one that began in western Crete near Maleme, paralleled the Akra Spatha Peninsula, and passed Kithera on the way to Neapolis, to one that began near Heraklion, passed the small island of Dia, and proceeded to Perissa Beach on the island of Santorini (see Fig. CS6.5).

Throughout the actual flight window, the project benefited from the combined support of analysts in both the U.S. and Greece. Hellenic Air Force meteorologists at Heraklion cooperated extensively in tracking weather conditions, launching radiosondes, and applying their local experience. The team deployed local observers on Santorini and in boats along the route. Steven Bussolari, the project's Director of Flight Operations, integrated all this data into a final flight decision.

During the spring 1988 weather window the project observed four days suitable for flight operations: two prior to our establishment of flight readiness, the record day on April 23, and one subsequent day in early May. This confirmed almost precisely the predictions of the feasibility study.

CS6.4.4 Engine Physiology

The pilot power prediction model used to size the Daedalus aircraft was based on the initial physiological testing done in the Phase I feasibility study. This model, which gives the maximum available specific power as a function of duration, is shown in Fig. CS6.22 along with a revised power model based on the additional physiological testing data obtained during the Daedalus pilot selection in 4-hr and 6-hr pilot qualification tests. Also plotted on this chart are the data from the Daedalus record flights for pilots Lois McCallin, Glenn Tremml (in the Light Eagle) and Kanellos Kanellopoulos in Daedalus 88, as well as estimated data for Bryan Allen in the Gossamer Albatross. Allen's flight in the Gossamer Albatross was at the

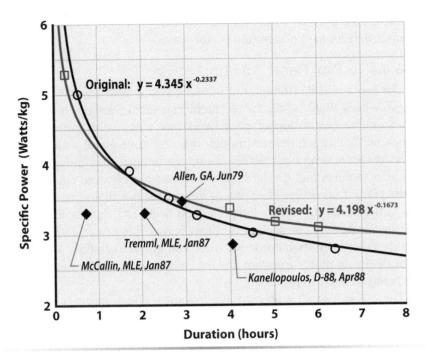

Figure CS6.22 Data for specific power vs duration for Daedalus, Light Eagle, and Gossamer Albatross with original and revised specific power models (from [17]).

nominal physiological power limit, while the Daedalus flights all occurred at somewhat lower power levels. The revised model is useful for preliminary design of human-powered vehicles but should be interpreted with caution as it is based on a small dataset for elite athletes trained specifically for a long duration task and will not, in general, apply to any given athlete.

As introduced in Sec. CS6.2.4, the key duration performance predictor for an individual athlete is how close they are operating to their AT, as measured by their oxygen uptake rate (VO2) as a percentage of their maximum rate (VO2max). From the data gathered in Phase I, it was concluded that for power levels requiring more than about 80% of VO2max, only a short duration (measured in minutes) can be expected. Below about 70% of VO2max, durations of 4 hr or more are possible, with duration limited primarily by a person's training and the availability of fuels and coolants. The Daedalus Project demonstrated durations of 6 hr at 60 to 65% of VO2max. The conclusions from these tests were (1) that specific power is the appropriate measure for HPA pilot duration performance; (2) VO2 versus specific power defines efficiency; and (3) the percentage of VO2max required for an individual determines the bounds of that individual's duration performance.

The VO2max percentages given earlier are useful as guidelines for selection of athletes for long-duration power. Testing at the required power levels (on an ergometer), with attention to hydration and cooling, is still the ultimate determinant of duration performance.

CS6.4.4.1 Pilot Heart Rate and Power

The most reliable (though not necessarily the most accurate) method for estimating flight power for the Light Eagle and Daedalus aircraft was based on the correlation of pilot heart rate and power from previous ergometer calibration runs. Although this was done on all long-duration flights there are several problems with this technique. First, the takeoff and climbout requires high power levels that must be produced anaerobically by the pilot. It takes time (five min or longer) for the pilot to recover from this initial transient so only long duration heart rate data is an accurate measure of power. The second problem is that this power correlation is subject to wide variation due to state of training and hydration. Thus, to be accurate, calibration runs were needed prior to and after a flight, which was rarely possible in practice.

There was also an unknown psychological effect on heart rate data. Initially it was thought that the physiological response (i.e., heart rate due to exertion) would override any psychological response (i.e., elevated heart rate due to excitement or fear). This appears to be true as the pilot approaches his anaerobic threshold but below that threshold psychological response becomes more significant. Pilot heart rates were typically much higher than expected during early Daedalus flights, even though the pilots reported subjectively lower power than they were accustomed to on the Light Eagle. Pilot heart rates also increased during turning flight due to an increase in the required flight power and also due to increased piloting workload. The variability of pilot power output as a function of heart rate due to training variation and psychological response makes it difficult to put absolute numbers on the power required for these vehicles.

CS6.4.4.2 Heart Rate During Record Flights

Heart rate histories for pilots Glenn Tremml (in the Light Eagle, in Phase II) and Kanellos Kanellopoulos (in Daedalus) during their record flights are shown in Fig. CS6.23. Also plotted are heart rates recorded for their 6-hr ergometer tests at 3.1 W/kg power. Note that Tremml's baseline heart rate was significantly higher than Kanellopoulos and his test was terminated at 5 hr. The heart rate for Tremml's record flight in the Light Eagle shows a heart rate of 170–180 bpm (corresponding to the higher power required for the Light Eagle at roughly 3.3–3.4 W/kg). The multiple peaks in heart rate during this flight coincide with turns around the triangular course, supporting the observation that piloting workload stresses affect heart rate.

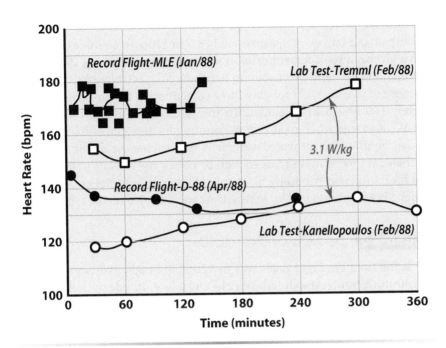

Figure CS6.23 Heart rate data for Daedalus record flights. Heart rates for Tremml's 1987 record flight in the Light Eagle and Kanellopoulos' 1988 record flight to Santorini in Daedalus 88 are compared with long-duration ergometer test data (from [17]).

The heart rate data for Kanellopoulos on the Daedalus record flight started off very high (for him) at 146 bpm after takeoff at Heraklion. The high heart rate was a mild concern to the crew monitoring from the boats but his heart rate soon fell to under 140 bpm and continued to fall for several hours, rising slightly towards the end of the flight in the approach to Santorini. It is interesting to note that throughout his flight Kanellopoulos flew very high, between 10–30 m altitude, well out of low ground effect. Based on this heart rate data it seems credible that, had all other factors remained constant, Kanellopoulos could have continued flying for at least two more hours.

CS6.4.4.3 HPA Fuel—Drink Development

Early in the project it was assumed that sufficient water would be carried during the flight for the pilot to offset fluid loss (due to sweating, about 2 lb/hr) and avoid dehydration. It also appeared that the pilots would carry sufficient glycogen stores in their muscle tissue and liver, especially if a "carbohydrate loading" strategy was adopted prior to the flight, so that fuel would not be a problem for the record flight. Subsequent long-duration tests indicated that depletion did occur (though it varied widely between individuals), which led to marked decline of blood glucose levels and accompanying fatigue at approximately 3.5 hr [13]. To counter this depletion a glycogen replacement drink was developed, containing 10% glucose polymer and 18 mg/liter of sodium (to maintain electrolyte balance in the presence of sweating). At roughly 1 l/hr consumption this provided approximately 100 grams/hr of glucose to the pilot, enough to produce about 900 W of energy when oxidized. Accounting for the 25% efficiency of the human engine, this provided sufficient fuel to sustain about 225 W of mechanical power, more than enough to power the aircraft in level flight. The fuel drink was promptly nicknamed "Ethanol" after its creator, Ethan Nadel. Six kilograms (13.2 lb) of this drink were carried on the record flight.

CS6.4.5 Design of Daedalus

At the time the Light Eagle was designed the flight profiles included a high altitude option (mountain launch) and a low altitude option (sea level launch). The high altitude option mandated a higher load factor for safety and initiated the development of an autopilot system. As the meteorology for the record flight was better understood the high altitude option was deemed too risky and the longer sea level flight profile from the airport at Maleme (later Heraklion) was adopted. The aircraft and structure design for Daedalus were accordingly optimized for low-level flight and the ultimate load factor was reduced from 3.0 g to 1.75 g on Daedalus.

While there were numerous changes to improve the aircraft the most significant single change was the reduction of the empty weight from 92 lb to less than 69 lb. This was accomplished by (1) the reduction of load factor, (2) use of higher modulus carbon fiber, (3) elimination of non-essential systems and components (automatic flight control system and ailerons), and (4) a comprehensive weight reduction program of all aircraft components— gram by gram.

CS6.4.5.1 Overall Design

Although in overall form the Daedalus aircraft closely resembled the Light Eagle prototype, significant changes were made to the materials, structure and aerodynamics to reduce weight and drag. A drawing with the three-view layout of the Daedalus aircraft is shown in Fig. CS6.24. Visually, the differences between the Daedalus and the Light Eagle (shown in a similar drawing in Fig. CS6.14) are subtle. Apart from the higher, curved dihedral of the wing on Daedalus, the changes appear largely in the details. The overall characteristics of the Daedalus 88 aircraft are compared to other record-holding HPAs, the Gossamer Albatross and Light Eagle, in Table CS6.3.

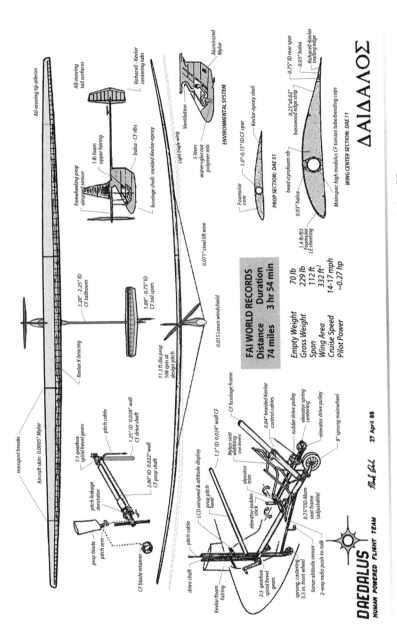

Figure CS6.24 Layout and 3-view of the Daedalus 88.

Since the project was going to build two Daedalus airframes more effort was spent early in Phase III on tooling, molds, and assembly jigs. This proved to be a wise investment in effort as it reduced construction time and ensured accurate assembly of the wing and flying surfaces.

CS6.4.5.2 Aerodynamics

The overall aerodynamic configuration for Daedalus was nearly identical to the Light Eagle in terms of layout and sizing of the wing, tails, and fuselage. Despite the similarity, changes were made to the design of almost every component to further improve performance. The most significant changes were made to the wing to eliminate the ailerons and to the wing airfoils to enhance laminar flow and reduce drag. Modest changes were also made to the tails to reduce weight and improve performance. A new propeller was optimized for the lower thrust required and used improved airfoils.

The wing area for Daedalus was the same as the Light Eagle (334 ft²) with slightly larger tip chords while the span was reduced slightly, to 112 ft. The aspect ratio for this wing was 37.5, slightly lower than the Light Eagle but still one of the highest ever built and flown. The vertical tail and horizontal tail were the same size as those on the Light Eagle but were built lighter, with leading edges of the smoother extruded foam used on the new wing to promote laminar flow. The anti-servo tab on the rudder was retained to enhance vertical tail effectiveness and to provide stick feel for the pilot but the servo tab was removed from the horizontal tail to save weight, and replaced with a pilot-adjustable spring-centering mechanism at the control stick.

It was decided, based on flights in California with the Light Eagle, that ailerons were not needed for broad turns or wing levelling on long, straight flights. Instead, Daedalus would use the strong yaw-roll coupling effects on slow-flying HPAs to turn using the rudder. The wing dihedral was correspondingly increased to 5.5 ft (from the 3.3 ft used on the Light Eagle) to increase yaw-roll coupling. This change saved approximately three lb and greatly simplified the mechanical complexity of the control system. The aileron system had been the only part of the Light Eagle control system to fail in flight, and it had failed twice. The lack of ailerons may have been a contributing factor to the Daedalus 87 flight test accident in February 1988 but this is unclear as the aircraft was also flown with inadequate dihedral (roughly ½ of the design dihedral) due to a shortened bracing wire.

The wing sections and wing LE structure were redesigned for Daedalus based on results of the Light Eagle flow visualization transition experiments and performance measurements. The wingtips for the Light Eagle had been found to be largely turbulent at the cruise C_Ls, which was traced to roughness in the LE foam shell. A new, more homogeneous, extruded polystyrene foam was selected to replace the rougher white beadboard foam for the leading edge shells, producing a much smoother wing surface. The use of extruded foam involved a minor weight penalty but it helped ensure that the wing would operate with laminar flow, as designed.

CS6.4.5.2.1 Wing Airfoil Design

A new family of low Reynolds number airfoils was developed by Mark Drela for the Daedalus wing in Phase III. The design methodology and process was similar to that used in the feasibility study for the Light Eagle wing airfoils to minimize drag using extensive laminar flow and to reduce losses in transitional separation bubbles.

The updated Daedalus airfoils, the DAE11, DAE21 and DAE31, were thinner than those on the Light Eagle and were designed with steeper, longer transition "ramps" (by shaping the pressure distributions) to reduce separation bubble losses to a greater degree than the Light Eagle airfoils. This was a bold and somewhat risky design move for an otherwise untested series of airfoil sections but further increased the performance of the Daedalus wing (roughly by 5%) so long as the surfaces could be made sufficiently smooth to get laminar flow. For example, the maximum L/D for the middle wing (DAI1336 vs DAE21) increased from 112 on the Light Eagle wing to 118 for the Daedalus wing.

CS6.4.5.2.2 Propeller Design

The Daedalus propeller was an evolution of the Light Eagle design, optimized for a lower cruise thrust. It was designed using the same methodology used for the Light Eagle propeller but used a new DAE51 low Reynolds number airfoil designed by Drela. The propeller was optimized for an advance ratio of 0.34, and featured a design point efficiency of nearly 90% with 6 lb (26.6 N) of thrust at 19.7 f/s (6.7 m/s). The 1.7 m blades were fabricated with an extruded foam (1.4 lb/ft^3) core over a tubular carbon fiber spar, with an outer shell of Kevlar. A variable pitch hub mechanism similar to that on the Light Eagle was used to allow the pilot to select a comfortable pedalling cadence or reduce pitch for takeoff.

CS6.4.5.3 Materials

There were two major changes in the materials used in the Daedalus. The first was the carbon fiber. Whereas the Light Eagle used standard carbon fiber with a prepreg modulus of 20 msi (Amoco T-300, Hercules AS-4), the Daedalus aircraft were built out of high-stiffness T50 fibers with a prepreg modulus of 35 msi provided by Amoco Performance Products. The materials used in the construction of Daedalus were detailed in an OSTIV paper by Cruz [14].

The second major materials change was in the CNC-cut polystyrene foam used for the leading edge of the wing and tail sections. The Light

Eagle used white beadboard foam with density slightly below 1 lb/ft^3. On Daedalus this was replaced with an extruded foam product, Foamular-150, with a density of 1.4 lb/ft^3. Thinner shells, made possible by its greater modulus and strength, were used to reduce the weight penalty to 1.5 lb overall. Foamular's advantage was its much smoother surface, which promoted laminar flow. Its trademark pink color gave the Daedalus wing and tails their distinctive appearance.

CS6.4.5.4 Structure

Although in external form the Daedalus was a near twin of the Light Eagle, the internal details differed significantly. In fact all the structural components were redesigned in some minor or major way to reduce weight. The primary wing spar ultimate load was reduced from 3 g to 1.75 g, and redesigned with higher-modulus material. This resulted in a simplification of the spar geometry, from a three-tube configuration (large center tube, small upper/lower cap tubes) to a single-tube geometry with integral top and bottom caps. The wing spar was reinforced internally with light Rohacell foam/balsa disks every 18 in. to prevent ovalization under load. The structure of the wing was also improved with addition of a smaller diameter aft tube spar (See Fig. CS6.8) and in-plane Kevlar cross-bracing. This main spar/aft spar configuration substantially decreased construction time and greatly increased in-plane stiffness of the wing [8], avoiding resonance with the pedalling frequency and virtually eliminating compression failures of the Kevlar trailing edge.

While the Light Eagle ribs were made from 2 lb/ft^3 extruded foam, the Daedalus wing ribs were made from 1 lb/ft^3 beadboard foam with bassword cap strips and balsa reinforcements at the spars. This change was made to reduce weight and increase the peel resistance of the capstrips using a lighter water-based adhesive that penetrated into the bead foam. The ribs were sliced from CNC hot-wired airfoil-shaped foam blocks laminated with basswood sheets, speeding up the mass production of hundreds of wing ribs for the two airframes.

The fuselage structure was also simplified, using a molded Kevlar upper fairing and air intake to replace the Light Eagle's shaped foam inlet system. All mechanical systems were redesigned, including the gearboxes, with a resultant weight savings of 2 lb. Further details of construction may be found in an article by McIntyre in AeroModeller [4].

Cruz and Drela have discussed the structural sizing and design conditions for the Daedalus wing in [15]. The Daedalus wing was sized using the same methodology as the Light Eagle wing but, due to the adoption of a lower altitude flight profile, used a lower design load factor of 1.75 g. The critical speeds for Daedalus were 11.3 m/s for dive speed, 7.8 m/s for maneuvering speed (the maximum level flight velocity under human power) and the stall speed of 5.8 m/s. The maneuver and stall speeds defined the design load factor of N_{des} = 1.75 g (see Sec. CS6.3.4.1). The

Figure CS6.25 Proof test for Daedalus wing structure in the hangar at Hanscom Field, MA. (Photo by Peggy Scott, 1987.)

critical design case for the Daedalus wing spar assumed, somewhat arbitrarily, a 1.25 g pull-up combined with a 30 deg sideslip angle. The same aeroelastic analysis approach developed for the Light Eagle was used for the loads cases to size the carbon fiber spars, simulating the wing deformations and inertial load relief so that strong non-linear effects on the structure were included. Figure CS6.25 shows the static proof test of the Daedalus wing structure at 1.5 g (without LE or covering).

Table CS6.5   Summary of Component Weights for Daedalus Aircraft, from Cruz [14]

	Kilograms	Pounds
Wing primary structure	11.382	25.09
Wing secondary structure	7.218	15.91
Wing miscellaneous	0.385	0.85
Wing total	1.895	41.85
Elevator total	0.538	1.19
Rudder total	0.623	1.37
Fuselage primary structure	3.307	7.29
Power system	2.798	6.17
Landing gear	0.728	1.60
Cockpit & pylon fairings	2.067	4.56
Control system	0.236	0.52
Seat & attachments	0.565	1.25
Water system (6 liter capacity)	0.212	0.47
Emergency tow system	0.029	0.06
Instruments	0.120	0.26
Radio	0.300	0.66
Miscellaneous	0.565	1.24
Fuselage total	10.927	24.09
Aircraft empty weight	31.073	68.5
Pilot weight (Kanellopoulos)	72.560	160
Consumables	6.000	13.2
Gross takeoff weight	109.633	241.7

The reduction in load factor, the higher modulus of the carbon fiber, a systematic redesign for reduced weight, and careful attention to detail reduced weight from 92 lb on the Light Eagle to 68.5 lb on the Daedalus. Table CS6.5 summarizes the Daedalus 88 airframe weights that are detailed, along with parametric weight models [14].

CS6.4.6 Phase III—Flight Testing

Flight testing of the Daedalus airframes took place at the NASA Dryden Flight Test Research Facility (Edwards AFB) in California. Daedalus 87 was shipped to Dryden in November 1987 and began flight tests early in December. Limited initial flight testing indicated that the power level was much lower than the Light Eagle. Based on pilot subjective feedback and heart rate data the power was roughly 3.0 W/kg, on track with the design

Figure CS6.26 Daedalus 87 before the February 7 crash on Rogers Dry Lake. Rigging with a shortened lift wire resulted in a very flat wing with only 3 ft of dihedral. (Photo by Peggy Scott, 1987.)

goals. Due to a long run of bad weather this airframe had accumulated only an hour of flight time before it was damaged in a crash on February 7, 1988.

CS6.4.6.1 The Daedalus 87 Crash

Due to a mistake in setting the length of the wing bracing wire the dihedral of the Daedalus 87, as it was rigged for the Dryden testing, was too low (only about 3 ft, much lower than the nominal design dihedral of 5.5 ft). Figure CS6.26 shows the Daedalus 87 in its flight test configuration with low dihedral. While this aircraft flew well in calm conditions, lateral control in turbulence greatly increased the piloting difficulty, to the point where pilot heart rates went up substantially due to workload and psychological stress. For example, pilot Eric Schmidt's heart rate before the Daedalus 87 crash was 170–180 bpm (his normal flight heart rates were in the 130 s)! Although the dihedral problem was identified before the accident, no action was taken before the crash. Given the rather dramatic improvement in the lateral characteristics of the Light Eagle when the dihedral was increased with the long tips and the reports of poor lateral handling qualities for the low dihedral Daedalus 87 in turbulent test sessions, it is surprising that the dihedral was not corrected earlier so that the crash would be avoided. Unfortunately, at the time, the project was in a headlong rush to wrap up the flight testing, head to Crete and be ready for the weather window. The

Figure CS6.27 Daedalus 87 crash on Rogers Dry Lake due to loss of control in thermal gust. The pilot, Eric Schmidt, was unhurt. (Photo by Peggy Scott, 1987.)

Dryden flight team was under pressure to complete testing and was reluctant to make changes to the aircraft, while the Massachusetts building team was busy rushing to finish Daedalus 88 and pack up to leave for Crete.

The Daedalus 87 crash occurred during a flying session that extended later in the morning than typical. Thermal convection was starting on the lakebed, and the aircraft encountered a thermal updraft under one wing that produced an uncommanded bank angle. Despite full corrective action, pilot Eric Schmidt, was unable to stop the aircraft from overbanking, and, in an ever-increasing slip, the wing contacted the ground and the aircraft ground looped, see Fig. CS6.27.

Investigation of the accident showed scratches in the wingtip that contacted the ground at a sideslip angle over 40 deg, a value initially thought to be too high to be representative of flight conditions. The immediate reaction after the accident was to question the decision to remove the ailerons from the Daedalus aircraft. The crashed airframe was shipped back to MIT for rebuilding while the second airframe, Daedalus 88, was rushed to completion.

In the interim, the Light Eagle, which had been setup for pilot training and flight control development, was modified for higher dihedral (6 ft) and was tested extensively for lateral control authority and turns. These tests showed that with higher dihedral, the handling characteristics of the

Figure CS6.28 Daedalus 88 with lengthened lift wire and 6.5 ft dihedral in flight tests on Rogers Dry Lake in March 1988. The aircraft was equipped with a data computer and sensors (see sideslip probe on mast) to validate control and handling. (Photo 1987 by Steve Finberg.)

Light Eagle improved markedly and rudder-only control was quite satisfactory for extended flights with turns.

The Daedalus 88 was shipped to Dryden and rigged with 6.5 ft of dihedral. It was equipped with sensors and a flight recorder for comprehensive flight testing to determine causes for the accident and qualify the vehicle (once the problems were fixed) for the record flight. Figure CS6.28 shows the Daedalus 88 in flight at Dryden in March, 1988 with increased dihedral, flight data computer and sensors.

Daedalus 88, to the relief of the flight test team, proved to be "in the groove" from its first flight. Rigged with higher dihedral (slightly higher than the design dihedral of 5.5 ft) the flight characteristics were vastly better than the Daedalus 87 and lateral control proved to be no problem for the pilots who reported much lower workload for maneuvering and turns. The pilots each made long familiarization flights in Daedalus 88 in training for the upcoming record flight. These flights were also used to gather data on handling in gusts and in turns. In the flight test program that lasted for several weeks two significant discoveries were made; first, that transient sideslip angles when maneuvering in gusts, even with the higher dihedral, were much greater than previously suspected, sometimes exceeding 30 deg. This showed that the 40 deg slip angle found on the crashed wingtip likely corresponded to the flight condition during the accident, and second, that stretch was found in the rudder cables that reduced the maximum in-flight rudder deflection to 7.5 deg, less than half the static deflection. The extra stretch in the rudder cables was due largely to the anti-servo tab, which produced high pitching moments opposing the rudder deflection. The cable stretch problem was corrected by re-tensioning the control stick and the Kevlar cabling. If uncorrected this problem could have shown up on the record flight and, combined with the low dihedral, might have led to serious loss of control. The crash, while hardly a lucky event, provided the impetus and time to find and fix these problems.

CS6.4.6.2 Lateral Control for HPAs

The problem of roll control for long-span, slow-flying aircraft like Daedalus is unique. The very long wings are prone to severe adverse yaw effects from ailerons. These adverse yaw effects are exacerbated by the low Reynolds numbers and the operation of the wing at high C_L, relatively near stall. While ailerons can be used to generate the rolling moments required for steady turns, it is not clear that they are the best means to do this under these flight conditions.

On the Light Eagle, for example, a turn could not be easily initiated with the ailerons—full application of right aileron (roughly −7.5 deg tip incidence change) caused the aircraft to yaw strongly to the left, with almost no immediate effect in roll. Instead, turns were initiated with the rudder using the strong coupling between rolling moment and sideslip (through dihedral) and rolling moment and yaw rate (which produced strong differential velocities across the span for these long wings and slow flight speeds). The combination of the initial yaw into the turn and yaw rate produced strong rolling moments that rolled the aircraft into a shallow bank. Once turning, ailerons were applied largely in the direction opposite the turn to control overbanking and rudder was used to control sideslip.

This turning strategy is reminiscent of the techniques used by 1970s soaring pilots to get the first generation of 22 m span super-sailplanes (such as the Nimbus I) to turn with full rudder and opposite application of ailerons (to use adverse yaw to get the wing to yaw and roll into the turn, assisting the rudder) then reversing the aileron into the turn to control banking.

The Daedalus aircraft were designed to turn without ailerons by increasing the coupling between sideslip and roll with added dihedral. In this approach the turning process was much simpler. The rudder is used to initiate the turn, rolling the aircraft with both yaw and yaw rate into the turn. As the aircraft rolls into the turn, sideslip (yaw out of the turn) is developed naturally by the rolling wing (yaw due to roll rate). If the dihedral is sufficiently high the steady state sideslip will be moderate, generating enough rolling moment to control overbanking, requiring the pilot to hold only a small amount of rudder into the turn to maintain yaw rate. Problems with this approach occur when the dihedral is too low so that very large sideslips are developed. Measurements of turns with the Light Eagle [10] with 6 ft dihedral showed that a relatively quick 3 deg/s semi-coordinated turn was established with application of full rudder for 4 s. Sideslip angle ramped up to 12 deg at 16 s into the maneuver at which point a steady state turn was established with only 3 deg of rudder. No aileron was required. This steady turn, at a bank angle of 2 deg, had a radius of 420 ft with the inside tip operating at 60% of the dynamic pressure of the outside tip. Recovery to straight flight was similarly easy using only rudder.

Such use of large dihedral for roll control has the side effect of giving an aircraft positive spiral stability, so that it may be flown "hands-off." This, combined with removal of ailerons and reliance on the rudder with yaw-roll coupling, significantly simplified the piloting task, reducing the number of controls that must be manipulated simultaneously by a pilot who must also keep pedalling to supply flight power.

CS6.4.6.3 Flight Testing and Training

During this period of intense flight testing the Light Eagle became the workhorse for the project. By the end of the Phase III flying the Light Eagle had accumulated more than 20 hr of flight time—probably an unofficial record for an HPA. Each pilot had to fly a 2-hr flight in the Light Eagle to qualify for the Daedalus record flight. The Light Eagle, equipped with a full set of sensors and a flight recorder, also flew with increased dihedral in instrumented flight tests of rudder-only turns and control. During this period comprehensive flight dynamics testing was done with the Light Eagle to characterize the flexible dynamics and control of the vehicle (work done by Siegfried Zerweckh at MIT and Jim Murray at NASA-Dryden [10]).

Finally, special hardware was installed in the Light Eagle (and later Daedalus 88) to directly measure flight power using driveshaft torque and rpm. A series of flights were made to measure flight power at a range of speeds and altitudes to validate the physiological power measurements. Unfortunately calibration errors in the sensors (not discovered until well after the tests) resulted in low flight power levels (157 W, or 0.21 hp), which do not line up with either theory or other measurements at 230 W. Nevertheless the driveshaft data forms a self-consistent set of power measurements that show the effects of altitude on flight power, in effect measuring the "actual" ground effect on the airframe. The results, shown in Fig. CS6.29, are not definitive but show that the benefits of flight in low ground effect are less pronounced than classical theory would predict. This data matches the subjective observations by the Daedalus pilots who typically flew at

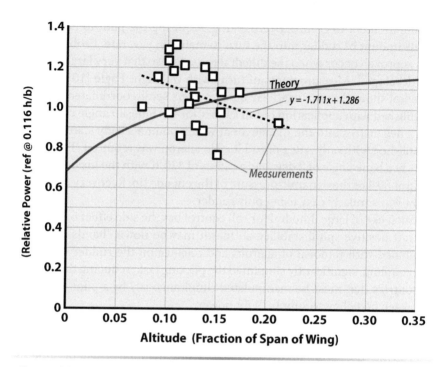

Figure CS6.29   Relative power in ground effect, recorded for Light Eagle at various speeds and altitudes, compared to classical ground effects. The data does not support the power reduction predicted for low altitudes (data from [10]).

higher, not lower, altitudes. It also matches the experience of Gossamer Albatross pilot, Bryan Allen, who, flying low and out of power late in his flight, signalled for a tow by his chase boat. Climbing several meters to allow hook-up, Allen found lower pedalling effort at higher altitude, enough for him to continue the flight and land in France. This effect is due to an additional power penalty near the ground or sea surface, it is not a fault with classical ground effects theory. Sullivan and Zerweckh [10] conjecture that higher turbulence levels in the lower Earth boundary layer could be a possible mechanism for increased wing parasite drag near the ground, raising flight power.

While Daedalus 88 was flown for many hours in the March flight test program, its gearboxes had been misbehaving throughout, making noise and wearing or breaking gear teeth, requiring constant re-adjustment to achieve quiet, low-wear operation. On the final test flights of Daedalus 88 at Dryden, just before the aircraft was shipped East for transport to Crete, the gears finally meshed into perfect, quiet, alignment. This was interpreted by the flight crew as a sign that Daedalus was ready at last for its long flight.

CS6.5 The Flight From Crete

Shortly after arrival in Crete, before the project was ready for a record attempt, Daedalus 88 was assembled and flown up and down the runway at Heraklion in a public relations and press event that was also a shakedown exercise for the airframe. On this sunny morning the wind was dead calm, and, gazing out to the north over the Aegean, the record flight seemed eminently possible. The day remained calm and the sea quiet all day. That was the last time for weeks that the team saw these conditions, instead the winds were often so strong they threatened to shred the portable hangar used to house the airplane and the sea was so rough the chase boats could rarely venture out. The improbability of reaching Santorini, over more than 70 mi of open sea in a gossamer airplane with a human engine, became very real.

While the project waited for weather the crew prepared, adding an emergency towline to the aircraft in case of pilot exhaustion, practicing rescue and lifesaving, getting familiar with the chase boats and forecasting weather. The aircraft was assembled and ready for launch at dawn on two occasions (with Frank Scioscia as pilot and again with Greg Zack as pilot) when the weather looked favorable. Each time the winds at Heraklion and along the flight route (as reported by team observers) became too strong near dawn and the flight was scrubbed.

On Saturday April 23, the weather, which had been watched closely for several days, looked very promising along the whole route of flight. By luck Kanellos Kanellopoulos, the Greek pilot, was up in the rotation. Kanello-poulos's takeoff just after dawn, involved a previously unpracticed, very committing maneuver flying over a 40–50 m cliff at the end of the runway. The takeoff, went smoothly and Kanellos, clearing the runway lip, headed out to sea, aiming to fly to the west of Dia, an island 7 mi off the coast of Crete, see Fig. CS6.30.

Pushed by a mild tailwind, averaging 3 mph, Kanellos flew very high, 10 m to 30 m above the sea, where he felt power was lowest. At one point, after several hours, near the middle of the flight, his ultrasonic altimeter failed. He kept to a strict schedule for his fluid "fuel" consumption (he had taken off with 6 liters, enough for a 6-hr flight before risking dehydration). Pursued by a small flotilla of boats Kanellos pedalled on, his heart rate dropping throughout the flight until he approached Santorini after almost 4 hr of flying, see Fig. CS6.31.

The landing site was Perissa Beach on the south coast of Santorini, a black volcanic sand beach, which by nearly 11am, was getting very warm. The local winds at the beach were high, near the flight velocity, running along the beach. Kanellos lined up for a landing approach into the wind, parallel to the shore, intending to make a shallow right turn and land in the

Figure CS6.30 Daedalus leaving Crete, after takeoff from runway at Heraklion as chase boats move into position. (Photo by Peggy Scott, 1988.)

Figure CS6.31 Daedalus approaching the south coast of Santorini at Perissa after a 4-hr flight. (Photo by Steve Finberg, 1988.)

Figure CS6.32 Daedalus trying to land into the wind at Perissa Beach, rebuffed by shoreline turbulence. (Photo by Peggy Scott, 1988.)

sand. As the right wingtip got close to the beach it was lifted by a gust and the aircraft was pushed offshore, see Fig. CS6.32.

Kanellos tried several times to cross the shoreline and on the third try, with a more forceful piloting effort, the tailboom snapped just ahead of the vertical stabilizer attachment point causing the control cables to pull the horizontal stabilizer to a full-up position. The aircraft pitched up and the main wing spar failed just inboard of the panel transport junction. At 10:58 am Daedalus 88 hit the water just 7 yards from shore after flying flawlessly for 4 hr and 72 mi. Kanellos dived out of the fuselage and swam to shore where he was hailed by a large crowd of crew members and onlookers. The Daedalus project was officially over and the pilot and the team were the holders of two FAI World records for human-powered aircraft—the absolute world distance record at 71.54 mi and the duration record at 3 hr 55 min.

The flight up to the approach for landing was amazingly turbulence-free over the entire route of flight above the sea, even with the tailwind. The high wind and turbulence at the shoreline was localized, generated by the "land/sea interface" between the heated sand and the cool sea. In retrospect a landing approach that crossed the interface rather than paralleled it might have allowed a more graceful ending, though it would have involved a crosswind landing. In any case the turbulence at the beach was worse than any encountered in flight testing on the dry lake in California and certainly put extreme stresses on the aircraft. Juan Cruz (project structural designer) later wryly observed that the structure must have been well designed because everything broke at once.

CS6.6 Epilogue

With the arrival of Daedalus 88 at Santorini, the project came to a palpable end—the end of three years of dreams, planning, training, and preparation. After a party to celebrate the flight the Daedalus crew began to drift off, back to MIT, off to tour Europe, back to their normal lives. For some the project was a launching point for new ventures—Langford, with Zerweckh, Tom Clancy and a few other Daedalus alumni, founded Aurora Flight Sciences, a company that focused initially on the design of unmanned aircraft for high altitude atmospheric research. Most of the project participants remained close and still get together once a year (near the end of April) to celebrate the record flight, an event always involving Greek food and wine.

Although Daedalus never flew again, the project workhorse, the Light Eagle, did. In 2009, Aurora Flight Sciences refurbished the Light Eagle and flew it to test solar powered airplane concepts, renamed as the Sun Light Eagle (SLE). A purely electric propulsion system was installed. A motor about 2 in. in diameter and 2 in. long, weighing 0.4 kg, with a specific power of over 3.0 kW/kg was geared to drive the original propeller, running off lithium batteries. Initial electric test flights were piloted by Lois McCallin. Then the plane was converted for unmanned operation by a remote pilot, and light weight solar cells were added to the wings. The plane made a few flights in New Mexico in support of a DARPA program. The SLE is currently in storage at Aurora Flight Sciences.

The pieces of the broken Daedalus 88 airframe were gathered up and shipped back to MIT, but the aircraft was never repaired and reassembled (unlike the Gossamer Albatross, which broke just after landing in France and was rebuilt to hang in the Smithsonian). The Daedalus 88 remains were transferred to the Smithsonian, where they still reside. Daedalus 87, the spare but identical airframe, hung for 20 years in the Boston Museum of Science, part of a long-term exhibit, gracing the entrance hall with its long, delicate wing. More than once the Daedalus alumni have looked up at it and wished they could take it down and fly it again. In December 2009 the Daedalus 87 airframe was moved for display at Dulles airport, Terminal B where it hangs today.

As to the ultimate limits of human-powered flight—Daedalus could have flown farther and longer, all the pieces were in place except an alternate, more distant destination. In the end, endeavours like this depend almost as much on logistics as they do on hardware and sweat. The Daedalus records will eventually be broken, and it will likely involve a very similar journey over the sea, hopefully avoiding those land/sea interfaces.

Acknowledgements

A team of over fifty people contributed to the ultimate success of Daedalus, and all deserve recognition. First and foremost is the project manager, John Langford and the senior engineering team: Juan Cruz, Mark Drela, Steve Finberg, Bob Parks, Bryan Sullivan, and Harold Youngren. Steve Bussolari and Ethan Nadel assembled and trained a team of six pilots: Kanellos Kanellopoulos, Lois McCallin, Frank Scioscia, Erik Schmidt, Glenn Tremml and Greg Zack. A team of dedicated undergraduates led by Dari Shalon, Tom Clancy and Tom Schmitter helped build and support the project's three aircraft. This passionate, inspired crew included: Jim Alman, Kevin Brown, Mary Chioschios, Jean Cote, Steve Darr, Jim Hilbing, Joice Himawan, T.C. Lau, Peter Neirinckx, Kelvin Phoon, Claudia Ranniger, Marc Schafer, Grant Schaffner, Louis Toth, Matt Thompson, Tim Townsend, Craig Wanke, Jim Wilkerson and Siegfried Zerweckh. Additional engineering talent, support for facilities, logistics, organization and everything else came from Ed Becotte, Jamie Dakoyannis, Debbie Douglas, Eunice Knight, Theo Korakianitis, Sarah Morris, Jim Murray, Bob Nesson, Mike O'Brian, Konstantin Pavlou, Joanna Prentiss, Christine Scott, Al Shaw, Mike Smith, John Tylco, Dave Watson and Jonathan Wyss. Finallly, Jack Kerrebrock (MIT Dean of Engineering), Brian Duff, and Peggie Scott helped garner the resources the project needed and give it the protection from outside interference that a small Skunk-Works style operation needs to be successful.

References

[1] Dorsey, G., *The Fullness of Wings: The Making of a New Daedalus*, Viking Press, 1990.

[2] *The Light Stuff*, NOVA Documentary, Mark Davis, WGBH, 1988.

[3] Langford, et al., *The Feasibility of A Human-powered Flight Between Crete and the Mainland of Greece*, MIT Department of Aeronautics, National Air & Space Museum, April 1986.

[4] McIntyre, J., "Man's Greatest Flight," *Aeromodeler*, August, 1988.

[5] Drela, M., "Low Reynolds Number Airfoil Design for the MIT Daedalus Prototype: A Case Study," *AIAA Journal of Aircraft*, Vol. 25, No. 8, Aug. 1988.

[6] Shalon, T., Langford, J.S., and Parks, R.W., "Design and Sizing of the MIT Daedalus Prototype," AIAA 87-2907, Aug. 1987.

[7] Burke, "The Gossamer Condor and Albatross: A Case Study in Aircraft Design," Pasadena, CA, Aerovironment #AV-R-80/540, 1980.

[8] Bussolari, S.R., Langford, J.S., and Youngren, H.H., "Flight Research with the MIT Daedalus Prototype," SAE Paper #87-1350, June 1987.

[9] Finberg, S.R. and Sullivan, R.B., "The Implementation of a Flight Control System in the Daedalus Human-powered Aircraft," *Royal Aeronautical Society Symposium on Human-powered Flight*, December 1986.

[10] Sullivan, R.B. and Zerweckh, S.H., "Flight Test Results for the Daedalus and Light Eagle Human-powered Aircraft," NASA Grant Report, October 1988.

[11] Larrabee, E.E., *The Daedalus Propeller: A Modern Application of Classical Propeller Theory*, Fall 1988.

[12] Bussolari, S.R. and Nadel, E.R., "The Physiological Limits of Long Duration Human Power Production," *Human Power*, in press, 1989.

[13] Nadel, E.R. and Bussolari, S.R., "The Daedalus Project: Physiological Problems and Solutions," *American Scientist*, July–August 1988.

[14] Cruz, J.R., "Weight Analysis of the Daedalus Human-powered Aircraft," *Proceedings of 21st OSTIV Conference*, Wiener Neustadt, Austria, Spring, 1989.

[15] Cruz, J.R. and Drela, M., "Structural Design Considerations and Load Determination for Human-powered Aircraft," *Proceedings of 21st OSTIV Conference*, Wiener Neustadt, Austria, Spring, 1989.

[16] Drela, M., "Method for Simultaneous Wing Aerodynamic and Structural Load Prediction," *Journal of Aircraft*, Vol. 27, No. 8, Aug. 1990.

[17] Langford J., "The Daedalus Project: A Summary of Lessons Learned," AIAA 83-2048, 1989.

Skyhawk: A Cessna Legend

Conrad F. Newberry

- Overview of the Skyhawk
- Biography of Clyde Cessna and Emergence of the Cessna Aircraft Company
- Skyhawk Predecessor Aircraft
- Skyhawk Need, Performance, Design Attributes, and Evolution
- Skyhawk Production, Cost, and Flight Test

Old and new Skyhawk fly in formation over the southeast Kansas countryside. The dark aircraft is the 1955 N5000A (the first production C-172) and the light aircraft is a 2006 C-172S model, N1665H. (Courtesy of Michael Fizer/AOPA)

The Cessna 172 makes flying like driving. You drive it into the air–and back down onto the runway.
Aviation writers (Thompson)

T he Cessna C-172 "Skyhawk" aircraft was designed in 1955 for a single mission—basic cross-country flight, a continuation of the C-170 "Family Car of the Air" concept. The design was revisionist in character. Cessna wanted an inexpensive (first cost and operational), safe, reliable, and easily maintained aircraft to compete with similar products of other general aviation manufacturers. The Skyhawk was the result of excellent engineering, production, and marketing collaboration. Some features, such as the wing and fuselage design, were products of Cessna's long-time corporate experience. Other features, such as the tricycle landing gear, were the result of product improvement and competitive product pressure. Cessna understood customer demands and desires, and was determined to meet those needs. The result was an inexpensive, robust, and dependable (as demonstrated by the C-172 endurance records of Heth/Burkhart set in September 1958 and the current record set by Timm/Cook in February 1959) aircraft. Proof of concept is that some 43,000 "Skyhawks" (more than any other aircraft model ever produced by any company) have been produced over the past sixty plus years, and the aircraft is still in production. Aircraft companies make their money by building airplanes, not by conducting research. *Cessna obviously got something very, very right!*

CS7.1 Introduction

The Cessna 172 Skyhawk is arguably the most popular aircraft ever built. Some 43,000 C-172 examples have been produced—more than any other aircraft model—for commercial, military, or general aviation. People have voted with their pocketbook, and hyperbole aside, have, by the sheer numbers purchased, elected the Skyhawk as their favorite aircraft [1].

The longevity of the 172 is remarkable. The basic design was first lofted in 1955; the first prototype (N41768, c/n 612) flight was on June 12, 1955 with Cessna Engineering Test Pilot, E. B. "Fritz" Feutz, at the controls. The first production aircraft (N5000A, c/n 28000) rolled off the Wichita, Kansas assembly line in 1955 and first flew on October 6, 1955 with Morton Brown, Cessna Chief Production Test Pilot, at the controls. Except for the ten-year interval between 1986 and 1996, when Cessna chose not to build single-engine aircraft due to public liability issues and regulations that were unfriendly to the general aviation community, the Skyhawk has been in continuous production for nearly half a century. This long-lived aircraft and its achievements were featured in the April 2006 issue of AOPA [Aircraft Owners and Pilots Association]. The C-172R and the C-172S are the current models of the Skyhawk being produced in Independence, Kansas; the end of production is nowhere in sight—the model may even be going green [1, 2, 3].

During the past century a number of great aircraft designs have emerged, some have become legends. Some of these legends include the DC-3 (C-47, R4D) "Dakota" or "Gooney Bird" (18,155 produced), "Spitfire" (20,351), P-51 "Mustang" (16,766), Bf 109 (33,984), B-17 "Flying Fortress" (12,731), B-24 "Liberator" (18,188), and the F-86 "SabreJet" (9860). With some 43,000 aircraft produced over a span of fifty years, the C-172 joins the ranks of these legends. The reader has probably flown as a pilot or passenger in one or more models of the Skyhawk or has an acquaintance that has. With its production numbers, longevity, and general aviation usage, the C-172, Skyhawk, is an icon in its own right. You can probably find a Skyhawk at any municipal airport. That said, just what is the C-172 [4, 5, 6, 7, 8, 9, 10, 11]?

The Cessna Skyhawk is (typically) a four-place general aviation aircraft manufactured in accordance with Federal Aviation Regulation (FAR) 23 specifications. It is slow, some say ungainly, reliable, dependable, operationally economic, relatively inexpensive to purchase, a single-wing-strut braced monoplane, pilot forgiving, and easy to operate. It evolved, in part, from the earlier C-140 and C-170 models. The C-172 helped solidify the international reputation of Cessna as a premier airplane manufacturer. Although the C-172 is primarily a business or family aircraft, variants have been used as military trainers and utility aircraft.

The standard C-172 model was introduced in 1955 as a 1956 model. A deluxe version of the standard model was unveiled, as the Skyhawk, in 1961. Skyhawk remained the deluxe version of the C-172 model until 1976, when all models of the C-172 became known as *Skyhawks*; subsequently, including this case study, all C-172 aircraft typically receive the "Skyhawk" appellation. As a corporation, Cessna has built nearly 200,000 aircraft of all types. The 50,000th (1963) and the 100,000th (May 27, 1975) aircraft to come off the production line were Skyhawks. It is noted that the 75,000th Cessna aircraft came off the line in 1967 [12, 13].

CS7.1.1 Clyde V. Cessna

The Cessna Aircraft Company (CAC) was founded by Clyde Vernon Cessna. Of French heritage (DeCessna), Clyde was born in Hawthorne, Iowa on December 5, 1879 to James William and Mary Vandora Skates Cessna. Clyde had an older brother, Roy Clarence Cessna. At some point prior to early 1881, Clyde's father filed "claim" to a quarter section (160 acres) of land in the middle, southern part of Kansas. The 1862 Homestead Act required individuals to live on a claim (quarter section) for a period of five years, improving the claim, at which time the individual could pay a filing fee of ten dollars to acquire title to the land. Most of the state of Kansas was settled as a consequence of this act [14, 15, 16].

Clyde and his family moved via train to their claim near (what would become) Rago, Kansas (Canton Township, Kingman County) when Clyde was a year old. The family's first Kansas residence was a sod house. James and Mary Cessna had five additional children (for a total of seven) after their move to Kansas—all were needed to turn the prairie farm into a profitable enterprise. Typical for that time and place, Clyde received only a fifth grade education in a one-room schoolhouse in nearby Raymond, Kansas. However, he was mechanically gifted and, as a teenager, became a much sought after mechanic for repairing farm machinery—and eventually automobiles and aircraft [14, 16].

On June 6, 1905, Clyde married Europa Elizabeth Dotzour, and they had two children: Eldon Wayne (May 5, 1907) and Wanda Delores (1909). In 1907, Clyde took a job as a mechanic with the Overland Farm dealership in Harper, Kansas (some 12 mi south of Rago). Clyde Cessna discovered that he was a born salesman. Soon, Clyde was selling automobiles as well as repairing both cars and farm implements. Clyde had ventured into the world of automotive dealerships, sales, and service [14].

Clyde developed a partnership with J. Watson in the Overland dealership in Enid, Oklahoma. Clyde's sales abilities were such that, in 1909, the dealership became the Cessna Automobile Company. On July 25, 1909, Louis Blériot flew across the English Channel from Calais to Dover and collected the 1000-lb prize offered by the *London Daily Mail* for the achievement. Upon reading of Blériot's epic flight, Clyde Cessna had an epiphany—he would build an airplane of his own [14].

1911 was a watershed year for Clyde Cessna. Featuring examples of the Blériot monoplane, the (John B.) Moisant International Aviators sponsored a widely advertised air circus in Oklahoma City on January 14, 1911. Clyde attended, and at age 31 saw his first powered flight, discovered that the French aviators earned $10,000 for their aerial demonstrations, and determined that he was in the wrong business. He gathered his life savings and went to New York City, where he spent several weeks working in the Queen Aeroplane Company assembly line making replicas of the Blériot monoplane. There he learned all he could about aircraft manufacturing, and bought an example of the Blériot monoplane (sans engine) and shipped it to Enid. On his way home, he bought an 80-hp V-8 engine in St. Louis. Clyde named his aircraft, "Silver Wings," and, due to difficulties with his V-8 engine, bought an Elbridge Aero Special engine (also 80-hp) to power the Blériot replica. Clyde still had to master the art of flying. On May 12, 1911, he moved his family to Jet, Oklahoma (some 40 mi north and a bit west of Enid) and the nearby salt plains to pursue his aviation activities [14, 15].

Soon after (self taught) Clyde felt he could fly without damaging the aircraft, he and brother Roy started to schedule exhibition flights, which would provide them with a chance to recoup financial losses. Clyde rebuilt "Silver Wings" during the winter of 1911, incorporating features he felt would improve handling and performance of the aircraft. Thus, he began what turned into an almost annual practice of using the winter months to rebuild and improve the aircraft to increase its effectiveness for the next year's exhibition circuit. In the years between 1911 and 1925, Clyde built or rebuilt 15 aircraft, mostly for exhibition flight—the area of aviation he felt offered the best chance for financial profit. In 1916, Clyde informally created the Cessna Aircraft Company [14, 15].

In 1925, Walter Beech and Lloyd Stearman, former Swallow Airplane Manufacturing Company executives, joined with Clyde Cessna and Walter Innes, Jr. in forming the Travel Air Manufacturing Company. Clyde contributed $25,000 (some say $5000), Beech invested $5000, Stearman contributed $700 and the plans for a new biplane design. Innes became president and treasurer, Cessna was vice-president, Beech became secretary, and Stearman was chief engineer of the new company. The biplane featured a tubular steel framework for the fuselage and tail assembly, a wood framed wing, and a Curtiss OX-5 engine. The aircraft was fabric (linen) covered and was designated the Model A. It sold for some $3500 (a bit more than a Ford Model A: $385–$1400) and was immediately successful in general aviation and in racing circles [14, 15].

Whereas Beech and Stearman both liked biplanes (the conventional design type for that time or era), Cessna considered the monoplane to be a superior design configuration. Due to differences in design philosophy, Cessna sold his interest in Travel Air to Walter Beech in 1927 (for $16,110) and left to start his own company [13, 14, 15].

CS7.1.2 Cessna Aircraft Company

The modern Cessna Aircraft Company began unofficially on April 19, 1927, in Wichita, to pursue the development of full cantilever winged monoplane designs—the Cessna (3-place) All Purpose and the Cessna (5-place) Common. The concept of a fully cantilevered wing was so radical for the time that Clyde paid Joseph S. Newell, S.B. (a Professor of Aeronautical Structural Engineering at the Massachusetts Institute of Technology) the sum of $800 to validate the structural integrity of his designs [14, 15].

The All Purpose first flew on August 13, 1927. After its first flight, the All Purpose was renamed the Phantom. The Phantom received some minor modifications and became the Cessna Series A (first production series airplane). The Series A configuration was a snug (small cabin), four-place, high-wing (fully cantilevered) aircraft, with a gross takeoff weight of some 2260 lb. A tubular steel framework was used for the fuselage and the empennage, while the wing featured a framework of wood. There were five models in the series, each had a two-letter designation; the first letter was the series designation, the second letter denoted the power plant. Thus, the model AA was powered by a 120-hp Anzani engine, the AW by the 125-lhp Warner engine (the only American powerplant in the group), AS by the 125-hp Siemans Halske, AC by the 130-hp Comet, and AF by the 150-hp Floco [31, 32].

Nineteen hundred twenty eight was a good year for Clyde Cessna. The Cessna Aircraft Company produced 46 aircraft, had a backlog of 96 more, and a production rate of 2.5 aircraft per week. Times were good. During the year Clyde developed the 2435 lb, 150 mph Model BW [B (second) Model Series, a super "A"] powered by a Wright J-5 220-hp engine. The three-door, six-place CW-6 [C (third) Model Series] powered by a 225-hp Wright J-5 also saw the light of day. The year 1929 looked even better: the Curtiss Flying Service signed a contract with Cessna on February 19, 1929 by which it was to purchase the entire production of Cessna aircraft [14, 15].

In October 1929, the Great Depression reared its ugly head: the Curtiss-Wright Flying Service went into bankruptcy and Cessna struggled for survival when the general aviation market became a shadow of its former self. Interestingly, there seemed to remain a small rising market for training gliders [14, 15].

In 1929, Cessna produced the single-place CG-1 (Cessna Glider, Model 1)—one prototype was built—designed by Eldon Cessna; Clyde's and Europa's son was now working for Clyde. This aircraft was immediately followed by the single-place CG-2. Some 300 examples of the CG-2 were built but only some 54 were sold. The glider had a sales price of $398. Overall, sales were not enough for Cessna to meet the payroll and service the corporate debt acquired largely by recent manufacturing capability expansion [14, 15].

The impact of the Depression on aviation in general and the light plane market in particular was such that the Cessna Aircraft Company produced no aircraft during the 1931–1933 period. In March 1931, the CAC Board of Directors voted to close the factory. The C. V. Cessna Aircraft Company (comprised of Clyde, his son Eldon, and engineer Garland Peed) was formed in late 1931 or early 1932 to build specialty, essentially hand-crafted, small production aircraft, such as the successful CR (Cessna Racer) racing series of aircraft (CR-1, -2, -3) during the early 1930s [14, 15].

Clyde's youngest sister, Grace Opal, married Dr. Eugene Wallace. The union produced three sons: Dwight, Dwane, and Deane. In May 1933, Dwane graduated from the Wichita University (WU) aeronautical engineering program (WU's fourth aeronautical engineering graduating class). One of his goals was to build airplanes with his uncle Clyde. Largely due to the efforts of the Wallace brothers, Dwight and Dwane, a revitalized Cessna Aircraft Company emerged from inactivity on January 17, 1934 at a stockholders meeting. The Cessna factory was reopened and soon started to produce the four-place C-34 (Cessna 1934 Model)—the "Worlds Most Efficient Airplane." This appellation was bestowed upon the aircraft because, at least in part, Cessna pilots won the Detroit News Trophy three consecutive races in years in which the races were held [1931 (Eldon Cessna flying an AW), 1935 (George Harte—Cessna Chief Test Pilot—flying a C-34), and 1936 (Dwane Wallace flying a C-34)]. The Detroit News Trophy race was a four-event contest in overall aircraft efficiency: economy, landing and takeoff capabilities, speed, and passenger safety and comfort [14, 15, 18]. The C-34 had a maximum speed of 162 mph, could cruise at 143 mph, average 16.9 mpg on long trips, and had a 1935 sales price (per aircraft) of $4995. With the C-34 configuration, Cessna, for the first time, incorporated a flap on a cantilevered wing [14, 15].

Selling his interest in the company (in 1935) to his two nephews, Dwane L. and Dwight S. Wallace, Clyde V. Cessna retired on October 28, 1936. He returned to Rago, Kansas, where he purchased a whole section (640 acres) west of his original 40-acre farm and resumed his earlier agrarian lifestyle. Dwane L. Wallace, at age 25, became President of the Cessna Aircraft Company [14, 15, 19].

CS7.1.3 The T-50

During the 1935–1938 timeframe, Dwane Wallace became convinced that a limited commercial market existed for a five-place, twin-engined, low-cost, aircraft that could support short haul/feeder airline operations. To fulfill this epiphany, he nurtured a low-winged, light twin-engined, five-place, fabric-covered, monoplane configuration constructed primarily of wood. The aircraft was given the T-50 designation by Cessna, and, later, Approved Type Certificate (ATC) 722 by the Civil Aeronautics Authority (CAA). Without realizing the full import of this epiphany, Wallace was positioning Cessna for its role in World War II [14, 15, 20].

The T-50 fuselage was framed with 4130X (chromium-molybdenum) tubular steel, Warren truss braced, and fabric-covered. The 41XX steels saw aviation applicability because they were relatively inexpensive, ductile, weldable, and enjoyed good deep-hardening attributes. Twin-vertical-tails were initially considered, but, due to flutter considerations, were rejected early in favor of a more conventional single wood-framed vertical fin [14, 15, 20, 21].

In the production aircraft, the 225-hp (takeoff and rated) R-755-7 engines in the prototype were replaced by 245-hp (takeoff, rated at 225-hp) L-4MB R-755-9 Jacobs (aka Shaky Jake) engines, and the fixed-pitch Curtiss-Reed propellers were replaced with constant speed metal Hamilton Standard propellers. The aircraft had a maximum speed of 175 mph at 7500 ft, a payload of 1400 lb, and a sales price of $29,675. Forty-three of the aircraft were built before production was interrupted by World War II [14, 15, 20, 22].

The tail-dragging T-50 had a wingspan of 41.92 ft, a height of 9.92 ft, and a length of 32.75 ft. The handcrafted wooden wing featured an NACA 23014 airfoil section at the root, tapering to an NACA 23012 section at rib number 22 near the wing tip; the aspect ratio was nearly six (5.94) [14, 15, 20].

The T-50 was christened the "Bobcat" by Cessna employees, but was sometimes, indelicately, called the "Bamboo Bomber" because of its particular configured use of steel, fabric, and the abundance of wood in its construction. Other terms of endearment for the T-50 were the "Rhapsody in Glue," "Useless-78," "Wichita Wobbler," and "San Joaquin Beaufighter." On May 16, 1940 President Roosevelt ask for the production of 50,000 airplanes annually (later increased to 60,000)—production numbers attained in November 1942. The United States Army Air Corps (USAAC) placed an order with Cessna in July 1940, for 33 examples of a military variant of the Bobcat to be used for transition and multi-engine pilot training. This military variant was given the USAAC AT-8 designation and featured 290-hp Lycoming R-680-9 engines instead of the initial Jacobs engines selected by Cessna for the aircraft. In September 1940, Canada ordered 180 T-50 aircraft and gave them the Crane Mk I designation. These Canadian T-50s were powered by the 245-hp Jacobs engines and wooden Hartzell propellers. A recent computerized cutaway model of the Bobcat is shown in Fig. CS7.1 [13, 14, 15, 23, 24].

Figure CS7.1   Cessna T-50 computer generated cut-away. (Courtesy of mikejamesmedia.com.)

Additional USAAF orders for the T-50 utilized the 245-hp Jacobs R-755-9 engines and were given an AT-17 designation. During the period from 1942 through 1944, some 3356 T-50s were ordered by the USAAF for use as light cargo and/or transport aircraft and were given the designation UC-78. A total of 5399 Bobcats were built by Cessna during World War II. Cessna management correctly concluded that after the war, these military aircraft would become surplus on the civilian market and that there would be no demand for new civilian T-50s. Cessna management correctly believed that after the war there would be many military pilots returning to civilian life and many more individuals who would want to become pilots and that both groups would provide a market for new general aviation aircraft. Perhaps inspired by the euphoria of victory, a Department of Commerce official told Congress that general aviation aircraft production could reach a postwar rate of 200,000 annually. Postwar surplus Bobcats sold for $1500 to $3000, depending upon their overall condition [14, 15].

CS7.1.4 Cessna Gliders

Cessna's contribution to the war effort was not limited to military variants of the T-50. The company's experience with gliders was exploited also. The USAAF had a requirement for several thousand assault troop gliders. They determined that the Waco Aircraft Company (Troy, Ohio) CG-4A (Hadrian) design met the requirements for the necessary 15-place assault, utility vehicle (Jeeps, 75-mm howitzers, field hospital, gasoline, and other cargo). The demand exceeded the production capacity of the Waco Company. A total of sixteen American companies (including some normally non-aviation corporations) served as prime contractors for World War II glider production (some 13,909 Hadrians were produced for World War II). Cessna was selected as the prime contractor for the construction of 1500 CG-4As; other Wichita firms, including Beech and Boeing, subcontracted glider work from Cessna. Cessna produced the outer wing panels and coordinated the production and assembly of the rest of the components. Beech produced the inner wing panels, the empennage surfaces, and all of the glider castings and forgings. Boeing (Wichita) provided final assembly [14, 15, 20].

CS7.2 The Skyhawk Lineage

The frequency of new models and model changes would suggest that within the conscious or unconscious Cessna corporate culture there is, or was, a penchant for frequent, if not annual, model changes (nurtured by marketing) that dates to the Clyde Cessna annual new aircraft models of the 1911–1925 era. By 2011, the company had produced a wide-ranging variety of some 200,000 aircraft (more aircraft than any other company in the world had produced), including some 43,000 Skyhawks. Although the company has produced a number of innovative "clean sheet of paper" designs, it also employs extensive revisionist design practices.

CS7.2.1 Cessna C-120/140

During World War II, Cessna initiated a highly publicized "Family Car of the Air" project for an assumed postwar demand created by veterans interested in a "family" (i.e., four-place) airplane. Aviation enthusiasts assumed a postwar reality that would include an airplane in every family garage. However, at war's end, with the G.I. Bill providing unexpected flight training funds for returning veterans, the market demand was for inexpensive trainers. Thus, the "Family Car of the Air" project was put on hold and plans were developed for a two-place aircraft with a training mission [2, 14, 15].

The result was the Cessna C-120/140 aircraft [Approved Type Certificate (ATC) 768], with a tail-dragging landing gear configuration. The C-120 was a less expensive C-140 with a bit less flight capability. For example, the C-140 had flaps and a rear window, whereas the C-120 had no flaps and no rear window. Both aircraft were created for the postwar training market, i.e., short range, local flying. Both models were successful in the short term [2, 15, 25].

First flight of the C-140 was on June 28, 1945, while the initial flight of the C-120 occurred on December 31, 1945. Eventually, 2164 examples of the C-120 were sold, compared to 4881 (plus 525 C-140As) examples of the C-140. Both models had a top speed of 120 mph, a gross weight of 1450 lb, a sea level rate of climb of 680 f/m, and a service ceiling of 15,500 ft, using an 85-hp Continental C-85-12/12F engine. The first C-140s were priced at $2995 per copy, while the C-120 had a cost of $2495 per copy. The C-120/140 project went from almost a "clean sheet of paper" (January 1945) to a production rate of 30 aircraft per day (September 1946) in a little over a year. A picture of the C-140 is shown in Fig. CS7.2 [2, 14, 15].

As shown in Table CS7.1, the C-120/140 aircraft had a height of 75 in., length of 21 ft, 6 in., and a span of 32 ft, 10 in. The wing featured a constant chord from the root to WS (wing station) 161, an NACA 2412 airfoil section throughout the constant chord portion of the wing, and a rib spacing that ranged from 8 to 15 in.; from WS 161 to WS 197 (tip), the airfoil section was symmetric with 12% thickness [2, 15, 26].

In May 1949, the fabric-covered rectangular wing of the C-140 was replaced by an all-metal wing with a 14-ft long (spanwise) rectangular center panel and straight-tapered outboard panels to create the C-140A configuration. The wing featured a 58.69-in. mac, an increase in aileron

Figure CS7.2 Cessna C-140. (Courtesy of CAC.)

area, and better flap efficiency. The new metal C-140A wing improved aircraft performance, stability, control, and appearance. Many of the geometric and performance characteristics of the C-140 can be seen in Tables CS7.1 through CS7.7. The new tapered wing provided a link to the future C-170 [2, 15, 27].

CS7.2.2 Cessna C-170

The 1945 "Family Car of the Air" concept was envisioned as a four-place aircraft whose sales price would be in the $1000–1500 range. This concept was designated project P-370 (Cessna used the letter P to identify experimental projects); the price was intended to be competitive with postwar automobile sedans. Postwar, this price range could not be attained, and the initial general aviation market was for a trainer type aircraft, not for a "Family Car of the Air." However, after the success of the C-120/140 trainer models, the "Family Car of the Air" concept resurfaced like the Phoenix in the form of the C-170. Some have referred to the C-170 as a stretched C-140 [2, 14, 15, 19].

The object of the C-170 design effort was to provide the customer with a low-cost, operationally affordable, comfortable (adequate legroom for all passengers), four-place aircraft. Essentially this meant creating a small,

Table CS7.1 Cessna Skyhawk Model Performance Data Evolution

Model attribute	Model									
	C-140	C-170	C-170B	C-172	C-172A	C-172A seaplane	C-172N	C-172R	C-172S	T-41A (C-1 72F/H/K)
Name	C-140 (std)	C-170 (std)	Business-liner	C-172 (std)	C-172 (std)	C-172 (std)	Skyhawk	Skyhawk	Skyhawk	Mes-calero
Date of 1st prototype flight	6/28/1945	11/5/1947	1952	6/12/1955	1960	1960	1977	4/19/1995	1998	1965
Crew	1	1	1	1	1	1	1	1	1	1
Capacity (crew + passengers)	2	4	4	4	4	3	4	4	4	2
Height (ft, in.)	6,3	6,5	6,7	8,6	8,11	9,11	8,9.5	8,11	8,11	8,9.5
Length (ft, in.)	21,6	25,0	24,11.5	25,0	26,4	27,0	26,11	27,2	26,11	26,11
Span (ft, in.)	32,10	36,0	36,0	36,0	36,0	35,10	35,10	36,1	36,1	36,0
Wing Area (ft^2)	159.3	175	174	175	174	174	174	174	175.5	174
Span Loading (psf)	44.16	61.11	61.11	61.11	61.11	61.95	64.19	67.90	69.3	63.59
Aspect Ratio	6.77	7.45	7.45	7.41	7.45	7.38	7.38	7.48	7.48	7.52
Main Gear Tread (ft, in.)	6,5	7,2	7,2	7,2	7,2	7,1.25	8,4.5	7,2	7,2	7,2
Empty Weight (lbf)	890	1260	1205	1260	1252	1415	1379	1620	1644	1260
Useful Load (lbf)	560	980	995	940	948	805	921	830	914	1040
Maximum Gross Weight (lbf)	1450	2200	2200	2200	2200	2106	2300	2450	2550	2000

Fuel Capacity (gal)	25	37.5	42	42	42	42	43	56	56	42
Wing Loading (psf)	9.1	12.6	12.64	12.6	12.6	12.7	13.2	14.1	14.4	13.2
Takeoff Distance (ft)*	1850	1820	1820	1525	1370	2390	1440	1685	1630	1525
Landing Distance (ft)*	1530	1145	1145	1250	1115	1345	1250	1295	1335	1250
Never Exceed Spd (mph/kt)	140/122	160/140	160/140	176/153	160/140	160/140	158/137	187/163	163/142	174/151
Maximum Speed (mph/kt)	120/104	140/122	140/122	135/117	140/122	108/94	125/109	141/122	145/126	139/121
@ Altitude (feet)	Sea Level	Sea Level	Sea Level	Sea Level	Sea Level	Sea Level	Sea Level	Sea Level	Sea Level	Sea Level
Cruising Speed (mph/kt)	105/91	120/104	121/105	124/108	131/114	106/92	122/106	140/122	143/124	130/113
@ Altitude (feet)	Sea Level	Sea Level		6000	8000	6500	8000	8000	8000	5000
Rate of Climb (fpm)	680	690	690	660	730	580	770	720	730	840
@ Altitude (feet)	Sea Level	Sea Level	Sea Level	Sea Level	Sea Level	Sea Level	Sea Level	Sea Level	Sea Level	Sea Level
Stall Speed flaps up (mph/kts)	49/43	58/50	58/50	58/50	58/50	59/51	50/43	51/44	53/46	53/46
flaps down	45/39	53/46	52/45	52/45	52/45	52/45	44/38	47/41	48/42	46/40
Range (miles /nm)	475/413	592/514	590/513	620/539	790/686	500/434	485/421	790/686	702/610	595/517
Service Ceiling (feet)	15,500	15,500	15,500	13,100	15,100	12,000	14,200	13,500	14,000	13,100
Engine	C-85-12/12F	C-145-2	C-145-2	C-O-300-A	C-O-300-C	C-O-300-C	L-O-320-H2AD	L-IO-360-L2A	L-IO-360-L2A	C-O-300-D
Maximum Power (bhp)	85	145	145	145	145	145	160	160	180	145
Power Loading (lbf/hp)	17.1	15.2	15.2	15.2	15.2	14.8	14.4	15.3	14.2	15.9

* Takeoff and landing distances are for 50 ft obstacles.

(continued)

Table CS7.1   Cessna Skyhawk Model Performance Data Evolution (continued)

Model attribute					Model					
	C-140	C-170	C-170B	C-172	C-172A	C-172A seaplane	C-172N	C-172R	C-172S	T-41A (C-1 72F/H/K)
Name	C-140 (std)	C-170 (std)	Business-liner	C-172 (std)	C-172 (std)	C-172 (std)	Skyhawk	Skyhawk	Skyhawk	Mes-calero
Propeller Diameter (ft, in.)	6,7	6,4	6,4	6,4	6,4	6,8	6,3	6,3	6,3	6,4
Propeller Manufacturer	Sen/McC fp	McCauley fp	McCauley fp	Sen/McC fp	Sen/McC fp	McC fp	McC fp	McC fp	McC fp	McC fp
Propeller Model (McCauley)	CM7148	MDM7655	MDM7655	1A170	1C172/ EM	1A175/SFC 8040	1C175/ MTM	1C235/LFA 7570	1A170/ JHA 7660	1C172/ EM
Base Price (US$)	2995	5475	7245	8750	9450	9450	22,300	124,500	190,600	negoti-ated
As Delivered Price (US$)	5003	7075	11,020	11,751	12,206		30,050	135,700	241,000	negoti-ated
Examples Produced	4881	730	2907	3757	994		6421			242
References	2, 3, 26, 32	2, 3, 26, 58	2, 3, 26, 58	2, 3, 26, 58	2, 3, 26, 58	2, 3, 26, 29, 58, 59	2, 3, 26, 58, 59	2, 29, 58, 59, 60	2, 29, 58, 59, 60	2, 3, 61, 62, 63

Notes: Since these nominal data are from several sources, the data given for a given model are not necessarily internally consistent. Fixed pitch is denoted by fp; Sen is Sensenich. McC denotes McCauley. US dollars are then year dollars.

Model attribute	C-140	C-170	C-170B	C-172	C-172A	C-172A (float-plane)	C-172N	C-172R	C-172S	T-41A (C-172F/H/K)
Name	C-140 (std)	C-170 (std)	Business-liner	C-172 (std)	C-172 (std)	C-172 (std)	Skyhawk	Skyhawk	Skyhawk	Mes-calero
Year of First Flight	6/28/1945	11/5/1947	/1952	6/12/1955	/1960	/1960	/1977	4/19/1995	/1998	/1965
Length (ft, in.)	21,6	25,0	24,11.5	25,0	26,4	27,0	26,11	27,2	26,11	26,6
Height (ft, in.)	6,3	6,5	6,7	8,6	8,11	9,11	8,9.5	8,11	8,11	8,11
Span (ft, in.)	32,10	36,0	36,0	36,0	36,0	35,10	35,10	36,1	36,1	36,2
Wing Area (ft^2)	159.3	175	174	175	174	174	174	174	175.5	174
Length root chord (ft, in.)	5,0.5									
Length tip chord (ft, in.)	0									
Maximum fuse depth (ft, in.)	4, 4.26*				5, 1.42*	5, 1.42*				
Maximum fuse width (ft, in.)	3, 4.99*				3, 5.21*	3, 5.21*				
Load factor (ultimate)	+4.57/-2.26	+3.8/-1.52	+3.8/-1.52	+3.8/-1.52	+3.8/-1.52		+3.8/-1.52	+3.8/-1.52	+3.8/-1.52	+3.8/-1.52
Normal gross weight	1450	2200	2200	2200	2200	2220	2300	2450	2550	2000
References	2,12, 15, 25	2,12, 24, 28	2,12, 24, 29	2,12, 14, 26	3,12, 19, 26	12,26, 103	3, 12, 26, 60	26, 40, 45	26,38,45	2, 26, 59, 99

Notes: Load factors are for flaps-up, at gross weight, normal category;
* denotes value based on scaled data from drawing.

Table CS7.3   Wing Data

Model attribute	C-140	C-170	C-170B	C-172	C-172A	C-172A (float-plane)	C-172N	C-172R	C-172S	T-41A (C-172F/H/K)
Name	C-140 (std)	C-170 (std)	Business-liner	C-172 (std)	C-172 (std)	C-172 (std)	Skyhawk	Skyhawk	Skyhawk	Mes-calero
Area [incl flaps, aileron] (ft²)	159.3	175	174	174	174	174	174	174	175.5	174
Span (ft, in.)	32,10	36,0	36,0	36,0	36,0	35,10	35,10	36,1	36,1	36,2
NACA Airfoil Section (root)	2412	2412	2412	2412	2412	2412	2412	2412	2412	2412
NACA Airfoil Section (tip)	0012	0012	0012	0012	0012	0012	0012	0012	0012	0012
Aspect Ratio	6.77	7.45	7.45	7.41	7.45	7.38	7.38	7.48	7.48	7.52
Chord, root (ft, in.)	5,0.5						5,4	5,4	5,4	5,4
Chord, tip (ft, in.)	0						3,8.5	3,8.5	3,8.5	3,8.5
mac length (ft, in.)	4,11.02				4,10.4	4,10.4	4,10.8	4,10.8	4,10.8	
Incidence, root (deg, min)	+1°0'		−1°30'	+1°30'	+1°30'	+1°30'	+1°30'		+1°30'	+1°30'
Incidence, tip (deg, min)	+1°0'		−1°30'	−1°30'	−1°30'	−1°30'	−1°30'		−1°30'	−1°30'
Wing Twist (degrees)			0	3.0° washout	3.0° washout	3.0° washout	3.0° washout		3.0° washout	3.0° washout
Max rib spacing—center section (in.)	15									
Max rib spacing—outbd panel (in.)	n.a.									
Dihedral LE ctr sect (deg, min)	+1°0'		+2°8'	+2°8'	+1°44'	+1°44'	+1°44'	+1°44'	+1°44'	+1°44'
Lift Strut Location (WS, in.)	105.78									
Total Aileron Area (ft²)	12.98		18.30	18.30	18.30	18.30	18.30		18.30	18.30
Total Flap Area (ft²)	8.74		21.23	21.23	21.23	21.23	21.20	21.20	21.20	21.23
References	2,3,15,25	2,3,12,28	2,3,24,29	2,3,24,36	2,3,24,39	2,3,24,103	3,12,26,60	26,40,45	26,38,45	2,59,98,99

Notes: n. a. denotes "not applicable".

Table CS7.4 Useful Load Data

Model attribute										
	C-140	C-170	C-170B	C-172	C-172A	C-172A (float-plane)	C-172N	C-172R	C-172S	T-41A (C-172F/H/K)
Name	C-140 (std)	C-170 (std)	Business-liner	C-172 (std)	C-172 (std)	C-172 (std)	Sky-hawk	Sky-hawk	Sky-hawk	Mes-calero
Crew & Passengers (lbf)	2	4	4	4	4	3	4	4	4	2
Oil (quarts)	4.5	8	8	8	8	8	6	8	8	8
Oil (weight at 7.5 lbf/gal)	8.43	15	15	15	15	15	11.25	15	15	15
Fuel (gal)	25	37.5	42	42	42	42	43	56	56	39
Fuel (weight at 6.0 lbf/gal)	150	225	252	252	252	252	258	336	336	234
Fuel (wt trapped in system, lbf)			30	30	30	30	18	18	18	18
Baggage (lbf)	80	120	115	120	120	120	120	120	120	120
Max Gross Weight (lbf)	1450	2200	2200	2200	2200	2106	2300	2450	2550	2000
Empty Weight (lbf)	890	1260	1205	1260	1252	1415	1379	1620	1644	1260
Useful Load (lbf)	590	980	980	910	910	805	897	857	900	1040
References	3, 24, 25, 26	2, 3, 26, 28	3, 24, 26, 58	2, 3, 26, 36	2, 3, 24, 39	3, 12, 103	3, 12, 26, 60	26, 40, 45	26, 38, 89	26, 59, 98, 99

Table CS7.5 Empennage Data

Model attribute	Model									
	C-140	C-170	C-170B	C-172	C-172A	C-172A (floatplane)	C-172N	C-172R	C-172S	T-41A (C-172F/H/K)
Name	C-140 (std)	C-170 (std)	Business-liner	C-172 (std)	C-172 (std)	C-172 (std)	Skyhawk	Skyhawk	Skyhawk	Mes-calero
Wing area (ft^2)	159.3	175.0	174.0	174.0	174.0	174.0	174.0	174.0	175.5	174.0
Wing mac (ft, in.)	4, 11.04	4, 11			4, 10.4	4, 10.4	4, 10.8	4, 10.8	4, 10.8	
Wing span (ft, in.)	32, 10	36, 0	36, 0	36, 0	36, 0	35, 10	35, 10	36, 1	36, 1	36, 0
Horizontal tail										
Airfoil section @ root (NACA)	0006		0009	0009	0009	0009				0009
Airfoil section @ tip (NACA)	0006		0006	0006	0006	0006				0006
Total Horiz Tail Area (ft^2)	23.3		35.22	35.22	34.67	34.67	36.09	36.09	36.09	36.09
Aspect Ratio	0.84		0.748	0.64	0.64		0.934			
Stabilizer Area (ft^2)	13.52		19.8	19.8	19.8	19.8	21.56	21.56	21.56	21.56
Elevator area (ft^2)	9.78		15.42	15.42	14.87	14.87	14.53	14.53	14.53	14.53
mac (ft, in.)				2, 11.3	2, 11.3		3, 6.5			

Incidence angle (deg,min)	-2°	-4.0°	-3°,30'							
l_H (ft, in.)	12,6	14,7.2	10,6.5	13,5.3*	13,5.3*	13,5.3*	13,5.3*	13,4.7*	13,4.7*	13,4.7*
Span (ft, in.)	8,10	10,6.5	10,8	10,8	10,8	10,8	10,8	11,4	11,4	11,4
Horizontal Tail Volume Coeff	0.374	0.580		0.546*	0.546*	0.546*	0.546*	0.544*	0.544*	0.544*
Vertical tail										
Airfoil section @ root (NACA)	0006		0009	0009	0009	0009	0009	0009	0009	0009
Airfoil section @ tip (NACA)	0006		0006	0006	0006	0006	0006	0006	0006	0006
Total Vertical Tail Area (ft^2)	11.5	18.42	18.42	18.04	18.04	18.67	18.67	18.67	18.67	18.67
Stabilizer Area (ft^2)	5.75	9.0	9.0	9.0	10.76	10.76	11.24	11.24	11.24	11.24
Rudder Area, (ft^2)	5.75	9.42	9.42	9.42	7.28	7.28	7.43	7.43	7.43	7.43
mac (ft, in.)				3,3.6*			3,10.1*			
Aspect Ratio	1.37*		1.43*	1.17*			0.934*			
Sweep of 25% chord				35°	35°	35°	35°	35°	35°	35°
l_V (ft, in.)	12,9.6	15,4.8		15,1.6*	15,1.6*	15,1.6*	4,8.24			5,0.04
Height (ft, in.)	4,0.34		4,0.876	5,0	5,0	4,8.24				
Vertical Tail Volume Coeff	0.028	0.045		0.0435*	0.0435*					
References	2,3,25,104	2,3,28,105	2,3,29,106	2,3,36,44	2,3,39,42	2,3,26,103	3,26,60,86	26,40,45	26,38,89	26,42,59,99

Notes: * Denotes value based on scaled data from drawing.

Table CS7.6 Propeller Characteristics

Model attribute	C-140	C-170	C-170B	C-172	C-172A	C-172A (floatplane)	C-172N	C-172R	C-172S	T-41 C-172F/H/K
Name	C-140 (std)	C-170 (std)	Business-liner	C-172 (std)	C-172 (std)	C-172 (std)	Skyhawk	Skyhawk	Skyhawk	Mescalero
Engine	C-85-12/12F	C-145-2	C-145-2	C-O-300-A	C-O-300-C	C-O-300-C	L-O-320-H2AD	L-IO-360-L2A	L-IO-360-L2A	C-O-300-D
Maximum Power (bhp)	85	145	145	145	145	145	160	160	180	145
Propeller Diameter (ft, in.)	6,7	6,4	6,4	6,4	6,4	6,8	6,3	6,3	6,3	6,4
Type	fixed pitch	fixed pitch	fixed pitch	fixed pitch	fixed pitch	fixed pitch	fixed pitch	fixed pitch	fixed pitch	fixed pitch
Number of Blades	2	2	2	2	2	2	2	2	2	2
Tip Speed (fps)	888	895	895	895	895	943	885	785	895	895
Critical Speed (mph/kt)							138/120			
Activity Factor							170	183.8	169.8	
t/c @ 0.75R							0.085	0.082	0.084	
Airfoil Section	RAF-6	RAF-6	RAF-6	RAF-6	RAF-6	RAF-6	RAF-6	RAF-6	RAF-6	RAF-6
Tip Sweep (deg)							0	0	0	
Rpm	2575	2700	2700	2700	2700	2700	2700	2400	2700	2700
Material	wood/Al	aluminum	aluminum	aluminum	aluminum	aluminum	aluminum	aluminum	aluminum	aluminum
Weight (lb)							30.1	35.0	35.0	
Manufacturer	Sen/McCaul	McCauley	McCauley	McCauley	McCauley	McCauley	McCauley	McCauley	McCauley	McCauley
Model Number (McCauley)	CM7148	MDM 7655	MDM 7655	1A170/ 7651	1C172/ EM76	1A175/ SFC8040	1C175/ MTM	1C235/ LFA7570	1A170/ JHA7660	1C172/ EM7653
References	15, 25	24, 28, 105	3, 29, 106	3, 26, 36, 44	3, 26, 39, 42	3, 26, 42, 103	26, 60, 86, 107	26, 40, 45, 89	26, 38, 45, 89	2, 59, 98, 99

Table CS7.7   Reciprocating Engine Characteristics

Engine attribute engine	SFC[1] (lb/hp-hr)	Horse-power/ rpm	Torque (in.-lb)	TBO (hr)	No. of cylin-ders	Displace (in.³)	Bore (in.)	Stroke (in.)	Compr Ratio	Fuel (Octane)	Oil press (psi)	Dry Wt (lb)	Cost ($)
Jacobs L-4 (R-755-7) c	0.53	225/2000			7 (R)	757.0	5.25	5.00	5.4:1	65	70-90	505	
Jacobs L-4 (R-755-9) c	0.53	245/2200			7 (R)	757.0	5.25	5.00	5.4:1	73	70-90	510	
C-85-12/12F (carb)	0.51	85/2575		1800	4 (HO)	188.0	4.0625	3.625	6.3:1	73	30-60	184	
C-145-2 (carb)	0.50	145/2700	330	1800	6 (HO)	301.37	4.0625	3.875	7.0:1	80/87	30-45	268	
C-O-300-A (carb)	0.45	145/2700	330	1800	6 (HO)	301.37	4.0625	3.875	7.0:1	80/87	30-60	268	
C-O-300-C (carb)	0.45	145/2700	330	1800	6 (HO)	301.37	4.0625	3.875	7.0:1	80/87	30-60	268	
C-O-300-D (carb)	0.45	145/2700	330	1800	6 (HO)	301.37	4.0625	3.875	7.0:1	80/87	30-60	268	
L-O-320-E2D (carb)	0.53	150/2700		2000	4 (HO)	319.75	5.125	3.875	7.0:1	80/87	60-90	269	42,000
L-O-320-H2AD (carb)	0.53	160/2700	390	2000	4 (HO)	319.75	5.125	3.875	9.0:1	100LL	60-90	269	45,000
L-O-320-D2J (carb)	0.50	160/2700	390	2000	4 (HO)	319.75	5.125	3.875	8.5:1	91/96	60-90	275	42,000
L-O-360-A4N (carb)	0.50	180/2700		2000	4 (HO)	360.0	5.125	4.375	8.5:1	100/100LL	50-90	296	44,000
L-IO-360-L2A (f inj)	0.51	160/2400		2000	4 (HO)	360.0	5.125	4.375	8.5:1	100/100LL	50-90	278	50,000
L-IO-360-L2A (f inj)	0.51	180/2700		2000	4 (HO)	360.0	5.125	4.375	8.5:1	100/100LL	50-90	278	50,000
References	22,63, 108	12, 22, 65	12, 26	26, 63	26, 65	12, 26, 63,108	12, 22, 63	12, 22, 65	12, 22, 63	12, 22, 108	12, 22, 65	12, 22	67

Notes: (1) at cruise conditions is radial; HO is horizontal opposed; LL is low lead; c/carb is carbureted ; f inj is fuel injected.

large cabined aircraft resulting in a c.g. travel much larger than usual for a four-place aircraft with control surfaces sized to provide adequate stability and control [2, 3, 14].

To lower the initial first cost, the aircraft was given a modified C-140 wing, maintaining the C-140 maximum wing chord (roughly) and expanding the center of the wing to increase the span sufficiently to achieve the required wing area. The result was that the C-170 employed ailerons sized for the smaller C-140, producing somewhat sluggish aileron control at slow speeds, such as those required during landing. Aileron control at flight speeds was adequate. However, elevator control forces for the C-170 were noticeably greater than those experienced with the C-140. The first prototype of the C-170 had its initial flight on November 5, 1947 [2, 14].

CS7.2.3 The C-170 Configuration

The comfortable, 4-place, C-170 (Type Certificate 799) was a conventional tail dragger with a high wing supported by two steel dual or V-struts with jury braces, similar to that found on the C-140. A picture of the C-170 is shown in Fig. CS7.3 [2, 15].

Figure CS7.3 Cessna C-170. (Courtesy of CAC.)

Table CS7.1 provides geometric, engine, and performance data on the C-170. It had a height of 6 ft, 5 in., a length of 25 ft, and a wingspan of 36 ft. The wing planform had an area of 175 ft^2 and an aspect ratio of 7.45. Its maximum speed was 140 mph (at sea level), its initial rate of climb was 690 f/m, and the service ceiling was 15,500 ft [24, 26, 28].

The main landing gear was manufactured from chromium-vanadium steel that had been heat treated and shot peened to improve its resistance to fatigue. Each strut was a single tapered spring leaf, eliminating the several components of more conventional shock strut gears [19, 28].

The C-170 featured a metal fuselage and empennage and a metal-framed, fabric covered wing—the fabric covered wing weighing less than an all-metal wing. The aircraft featured the extensive use of Alclad 24ST aluminum. The wing was covered with Grade A predoped fabric covered with a cellulose nitrate dope for weather protection. The floor mounted flap handle was located between the two front seats. The two front seats were rail-mounted and were capable of fore and aft adjustment. The rear seat was a bench seat with a back that was hinged at the bottom allowing some fore and aft adjustment at the top. The one-piece windscreen was made of Plexiglas. The C-170 had a steerable tail wheel with a solid rubber tire. A cigar/cigarette lighter was one of the cabin amenities [3, 26, 28].

CS7.2.4 L-19—The Bird Dog

In the late 1940s, the U.S. Army wanted to retire its veteran L-4s and L-5s. This meant that they needed a replacement aircraft. The new aircraft would be used for artillery spotting and utility work. The Army also wanted an all-metal aircraft because earlier fabric covered spotting and liaison aircraft had such a short service life. In response to the Army's request-for-proposal, Cessna designed the C-305, a military variant of the C-170 [14].

Army specifications required that the two-place, tandem seating, C-305, aka the L-19 [later it carried the O-1 (O for observation) designation] or "Bird Dog" have STOL capabilities. Thus, it could land over a 50-ft obstacle in less than 600 ft, using a unique Cessna flap design featuring 60-deg deflected slotted flaps. The L-19 could takeoff over a 50-ft obstacle in 560 ft; the aircraft could climb to 22,900 ft at a rate of 1290 f/m [14].

CS7.2.5 Cessna C-170B "Businessliner"

Cessna called the C-170B (Type Certificate 799) model the "Businessliner" in an attempt to interest businessmen in its potential as an executive type aircraft. There was a two-pronged approach to the Cessna sales strategy. First, convince businessmen that aircraft usage could enhance the productivity of their particular business. Second, convince those same businessmen that Cessna aircraft could provide this increased productivity better than other aircraft. A picture of the C-170B is shown in Fig. CS7.4 [3, 14, 29].

Early models of the C-170 featured plain flaps. The C-170B was fitted with high-lift single-slotted flaps, based upon the "Bird Dog" flaps, to lower landing speeds and shorten the ground run during takeoff. These flaps had Fowler action in that they traveled rearward and down when deflected. The unique flap track design provided an obstruction free lower wing surface so that passengers and crew entering the aircraft would not bump their heads on such obstructions when passing under the wing. These Para-Lift flaps had a maximum deflection of 40 deg [2, 3, 14].

The C-170B aileron (wing) profile transitioned from its NACA 2412 inboard shape to a symmetric section at the tip to accommodate inter-changeable wing tips [2].

A single, questionably streamlined, oval-shaped extruded hollow lift strut replaced the V-strut of earlier models. These struts had internal flat bearing plates on each end to effect attachment to the wing and fuselage. The empty weight of 1205 lb represented an increase of 20 lb over the previous model empty weight [2, 3].

Many of the physical characteristics of the C-170B are tabulated in Tables CS7.1 through CS7.7. The aircraft had a useful load of 980 lb. In 1955, the Continental C-145-2H powerplant was replaced with the Continental O-300-A. A total of 2907 C-170B Businessliners were produced from 1952 through January 1957 [2, 3].

Figure CS7.4 Cessna C-170B "Businessliner." (Courtesy of CAC.)

CS7.3 Cessna C-172 Skyhawk

In 1955, the markedly different (from earlier models) C-170C model became the C-172. In 1961, a deluxe version of the C-172B received the designation "Skyhawk." The Skyhawk name was initially applied only to deluxe versions of the C-172 model. However, after 1976, all C-172 models were called "Skyhawks." Herein, "Skyhawk" is the designation applied to all C-172 aircraft [2].

The original C-172 engineering, production, and flight test was performed in Cessna facilities in Wichita. General aviation liability issues caused Cessna to curtail production of a number of models, including the C-172, on May 28, 1986. The General Aviation Revitalization Act (GARA) of 1994 represented a resolution of these issues and enabled Cessna to restart their general aviation single engine production line in 1996 [14, 30, 31, 32].

With their facilities in Wichita being utilized for the production of other aircraft models, Cessna management looked elsewhere for their single piston-engine production venue. Independence, Kansas was selected as the restart site. This new Cessna facility shown in Fig. CS7.5 was opened

Figure CS7.5 Cessna facility in Independence, Kansas. (Courtesy of Michael Fizer/AOPA Pilot.)

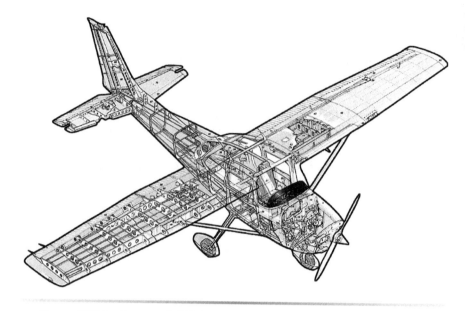

Figure CS7.6 Cessna C-172 Skyhawk. (Courtesy of Flight International.)

on July 3, 1995. The selection of Independence for the restart facility was a boon for the local economy. Many now consider Independence to be the "light aircraft capital of the world." Production at the new facility created some 1000 jobs and a (1996) payroll of some $20 million annually [14, 31, 32].

Several Cessna models are currently produced in Independence, including two versions of the C-172—the C-172R and the C-172S. Figure CS7.5 shows an aerial view of the Independence, Kansas Cessna plant [30, 31].

The Skyhawk is a very user-friendly aircraft. Simplicity is one of its virtues.

Production of the Skyhawk involves some 200 vendors or subcontractors. The lead-time for the engine and other parts is about six months.

A cutaway of the C-172 Skyhawk is shown in Fig. CS7.6. Geometric, physical and performance characteristics of some predecessor models (C-140, C-170, and C-170B) and selected models of the C-172, including the first and last two models, are shown in Table CS7.1.

CS7.3.1 The Need for the Skyhawk

In the second quarter of the 1950s, Cessna management was mulling over at least three factors related to their single-engine product lines. First, the C-180 Skywagon was Cessna's high performance (225-hp; 161 mph cruising speed, 1150 f/m rate of climb, takeoff in 995 ft, 20,000 ft service ceiling) four-place model (at $12,950). Cessna was not unaware that its customers appreciated high performance in their products [2, 26, 33].

Second, competitive pressures from other light plane manufacturers were forcing Cessna to consider the possible introduction of a fixed tricycle landing gear on some of their single-engine aircraft. Compared to the conventional taildragger configuration, the tricycle landing gear provides much improved straight ahead visibility over the nose during landing and takeoff. Further, a tricycle landing gear greatly improves takeoff, landing, and ground-handling characteristics compared to those same characteristics in a conventional taildragger (e.g., safety). The tricycle landing gear was not new to the aviation community in the early 1950s. Glenn Curtiss had used the concept as early as 1910. The concept had been used sporadically between 1910 and the mid-1930s. Bell Aircraft attracted considerable notice when they used a tricycle gear on the P-39 in 1937—the concept was not conventional in the mid-20th Century, particularly with general aviation aircraft. Most fighter aircraft circa 1930–1940 were tailgdraggers. However, the tricycle gear concept was not new to Cessna. They had used a tricycle gear on the C-310 (1953, a fast 5-place twin) and in the T-37 (Tweety Bird—1954, USAF trainer for transitioning from reciprocating engine powered to jet powered aircraft) [2, 34, 35].

The disadvantages associated with a fixed tricycle gear included the extra drag of the exposed gear and the related degradation in cruising speed. Additionally, Cessna management was concerned with the operational capability of the tricycle gear on soft fields and handling in high winds [2].

Third, the four-place C-170B (at a price of $8450—some $4500 less than the C-180) was in need of more elevator power to improve its 3-point landing performance. From a pilot perspective, more elevator power meant a reduction in stick force during full-flaps, 3-point landings with the c.g. at its forward position [2, 3].

Thus, in the quest for improved performance for the C-170B, the short version of the story is that the C-170B was to receive a new empennage to improve stability and control; a more powerful new engine, the C-O-300-A (an upgraded version of the C-145-2), and a fixed tricycle landing gear. The result was not a new C-170C variant but a new airplane— the C-172 [2].

CS7.3.2 Skyhawk Design Philosophy

For a number of very good reasons, most aircraft design textbooks imply (or state) that aircraft design starts with a set of requirements and a "clean sheet of paper." The clean sheet of paper approach enables the textbook writer to easily incorporate all of the salient features of the overall design process into a systematic paradigm. However, the veteran designer, when feasible and viable, will often incorporate prior subsystem or system design concepts into a new design in order to reduce cost and overall system risk. The use of such prior components or design concepts can result in a reduction in the time required to introduce the new aircraft into the operational inventory or market, increased reliability, and reduced maintenance-manhours-per-flight-hour (MMH/FH). Excessive use of new technologies can result in schedule delays, increased system cost, a reduction in examples produced, and, in worst cases, program cancellation.

CS7.3.2.1 General

Niles and Newell indicate that a new aircraft design is generally either a revision of an old design or a completely new design developed from a "clean sheet of paper." It is noted that a completely new design requires considerable data that must be either computed and/or estimated. In contrast, in a revisionist design, generally the only data that need to be computed and/or estimated are associated with the changes in the revision [17].

Revisionist designs typically result from small changes in technology, whereas completely new designs often incorporate significant new technologies such as monoplanes vs biplanes, jet propulsion vs reciprocating engine power, or all-metal vs wood manufacturing technologies. Revisionist designs typically are associated with incremental changes.

CS7.3.2.2 Singleness of Purpose

Niles and Newell note that size, requirements [performance (e.g., speeds, rate of climb), useful load, airworthiness requirements (e.g., load factors), specific requirements (e.g., limiting dimensions), and miscellaneous requirements (e.g., cost)], singleness of function, and power plant selection are issues that must be addressed by the designer. In particular, singleness of function should be of paramount importance.

> Every airplane is designed for some particular purpose, and the more limited this purpose is made, the greater is the likelihood of its successful attainment. This should be borne in mind by the designer, and everything should be subordinated to the attainment of the purpose desired.
>
> —*Niles and Newell*

Furthermore, the temptation to create a design that satisfies more than one function should be stringently resisted since the results are likely to be a design that does not really satisfy any function well. The aerospace landscape of the 20th century is littered with multi-function designs that, for a variety of reasons, never fully met expectations (i.e., requirements), e.g., the McDonnell Douglas/General Dynamics A-12. Too frequently, both customers and designers are guilty of not heeding the prophetic advice of Niles and Newell, with the result that limited resources are wasted [17].

The Skyhawk was designed for the quadriplace family/business cross country mission. The success of the design is due, in large part, to the fact that the single purpose of this design allows it to fulfill successfully the closely related family/business functions of pilot training, utility, and transport.

The design philosophy for the Cessna C-172 Skyhawk is basic revisionist. Each year usually finds some change in the model design (much like the automotive industry—Cessna the automobile salesman). Overall, one might sum up the Cessna philosophy as "make something the customer wants, keep it simple to reduce costs, make it reliable to enhance safety, make it easy to maintain, and make it cost-effective." The Skyhawk was designed to "Drive 'er down the runway and into the air!" [24].

CS7.3.2.3 Airworthiness Standards

The original C-172 (ATC 3A12) was designed to meet the airworthiness requirements of the Civil Aeronautics Administration (CAA), i.e., Civil Air Regulations (CAR) 3—Airplane Airworthiness; Normal, Utility, And Acrobatic Categories. Current Skyhawk models are designed to meet the requirements of the Federal Aviation Administration (FAA), i.e., Federal Aviation Regulations (FAR) Part 23—Airworthiness Standards: Normal, Utility, and Acrobatic Category Airplanes. However, the C-172 is not a purely aerobatic aircraft and is therefore designed to meet only the normal and utility portions of the Part 23 regulations [36, 37, 38].

CS7.3.3 Skyhawk Configuration

Figure CS7.6 shows a cutaway of the C-172 (courtesy of Flight International). In particular, one can see the full span front spar and the partial span rear spar. One can see the elegantly simple Land-O-Matic tricycle landing gear. The ribs are clearly seen as are the rectangular fuselage formers and bulkheads. It is noted that both the vertical and horizontal stabilizers have two spars. Multiple stringers are noted.

The 3-views shown in Figs. CS7.7 (1956 C-172), CS7.8 (1963 C-172D), CS7.9 (1973 C-172M) and CS7.10 (2006 C-172R) respectively illustrate

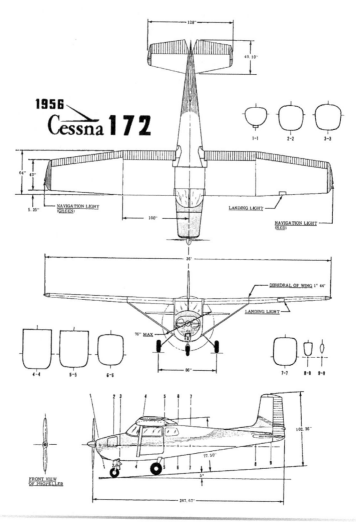

Figure CS7.7   Cessna C-172, 1956. (Courtesy of CAC.)

the evolution of the Skyhawk. The 1956 C-172 (Fig. CS7.7) shows the straight tapered outer wing panels, the "square" vertical tail feathers with little dorsal fin, the small (128-in. span) horizontal stabilizer, and the ubiquitous single lift strut. Figure CS7.8 shows the 1963 C-172D with the large (136-in. span) horizontal stabilizer, and the "Flight Sweep" vertical tail. It should be noted that the "Flight Sweep" tail was first used with the 1960 C-172A. The extended dorsal fin is shown in Fig. CS7.9 on the

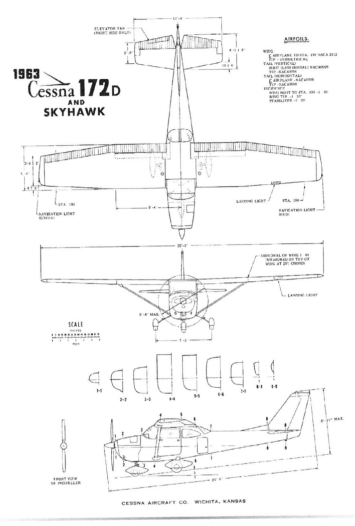

Figure CS7.8 Cessna C-172D, 1963. (Courtesy of CAC.)

1973 C-172M. The extended dorsal fin was introduced on the C-172L model. Figure CS7.10 shows the configuration of the C-172R, one of two current models being produced at the Independence Division. Table CS7.8 identifies a number of significant changes in the Skyhawk configuration over time. The basic dimensions did not change much with time as shown in these Figures and in Tables CS7.1, CS7.2, CS7.3, and CS7.5. Some of the overall dimension changes occurred as a result of

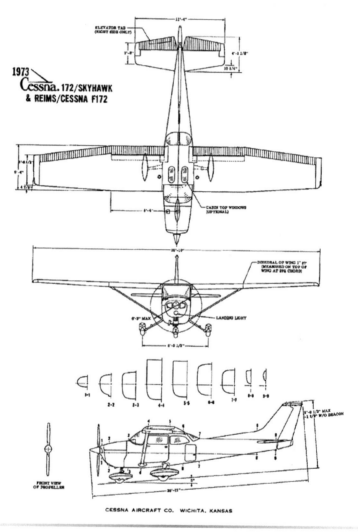

Figure CS7.9   Cessna C-172M, 1973. (Courtesy of CAC.)

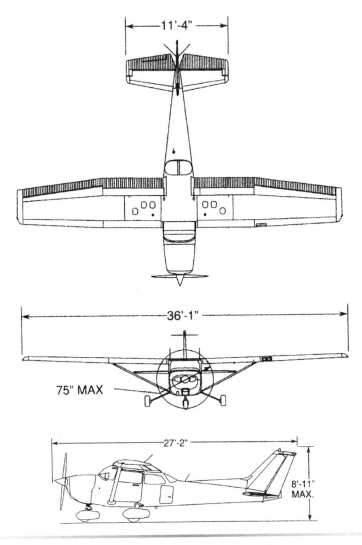

Figure CS7.10 Cessna C-172R, 2006. (Courtesy of CAC.)

different wing tip usage or the deletion or addition of navigational lights [26, 36, 39, 40].

Four of the many prominent features of current Skyhawks consist of triple corrosion protection for the aircraft, improved maintainability (compared to earlier models), the crashworthiness (safety) of the aircraft (FAR 23), and the G1000 avionics package. However, the G1000 package is optional (but selected by most owners). Further, the customer has the

Table CS7.8   C-172 Skyhawk Model Changes/Improvements

Model	Intro year	Model selected changes/improvements	References
C-172	1956	C-170B with Land-o-Matic tricycle landing gear; Clements' "square" vertical tail; six-cylinder C-O-300-A 145 hp engine introduced; retention of the C-170B Para-Lift flaps; steerable nose wheel +/– 8 deg from neutral.	1, 2, 3, 26, 36
C-172A	1960	Flight Sweep vertical tail introduced; the first C-172 model to be certified for floats; six-cylinder C-O-300-C engine replaces C-O-300-A engine; lighter, more efficient 1C172/EM propeller.	2, 3, 26, 39, 103
C-172B	1961	Main landing gear height shortened by 3 in. for better cabin ingress/egress; engine mounts were lengthened three inches to maintain acceptable propeller clearance; first use of Skyhawk designation; pointed spinner for drag reduction; new control wheel introduced.	3, 26, 28, 29, 58
C-172C	1962	Fiberglas wingtips for weight reduction; gross weight creeped up to 2250 lb.	2, 3, 26
C-172D	1963	Omni-Vision wraparound rear window made standard to provide passengers with a roomy cabin feeling; larger aft fuselage side windows; tailplane span increased by 8 in.; gross weight creepage to 2300 lb.	2, 3, 26
C-172E	1964	Center-mounted avionics in the instrument panel; electrical system fuses were replaced by circuit breakers.	2, 3, 12, 26
C-172F	1965	Better instrument panel lighting; manually operated flaps were replaced by electrically operated flaps; the elevator and elevator trim tab rigging was changed; the military version (back seat was taken out, six-cylinder C-O-300-D engine installed) was given the T-41A designation.	2, 3, 26, 59
C-172G	1966	Rotary door latches were introduced; the propeller spinner length was extended; the military version was given the T-41A designation.	2, 3, 12, 26
C-172H	1967	Nose gear oleo stroke reduced by 3 in.; the generator was replaced by a 60 amp alternator; an 11-point shock mount cowling was introduced; a split buss electrical system was introduced; the electrically powered stall warning horn was replaced by a pneumatically powered horn; the Rams Horn control wheel was introduced; basic "T" configured instrument panel; the military version was given the T-41A designation.	2, 3, 12, 26
C-172I	1968	The L-O-320-E2D "Blue Streak" 150-hp engine replaced the C-O-300-C engine; the fuel capacity was increased by three gallons to a total of 42 gallons (38 gallons usable).	2, 3, 12, 26

Model	Year	Description	Notes
C172J		No Skyhawk models with this designation.	
C-172K	1969	Control system lock nuts were replaced by Castellated nuts and cotter keys; aft side window area was increased by 50 %; a rudder tab was made ground adjustable; use of conical cambered wingtips decreased the wing span to 35 ft, 10 in.; a dorsal fin was added to the vertical tail; the military version was given the T-41A designation.	2, 3, 12, 26
C-172L	1971	Tubular steel main gear replaces steel spring leaf main gear to improve taxiing over soft ground; gray instrument panel; bonded cabin doors helped reduce drag; the wing flap drive was redesigned; the vertical stabilizer dorsal fin was extended forward to the rear window of the cabin, eliminating a side-slip pitch motion with large flap settings that occurred under some conditions; the propeller diameter was reduced by 1 in. to 75 in.; the main gear tread was increased by 13.5 in.	2, 3, 12, 26
C-172M	1973	Increased wing (cuffed, drooped) leading edge camber (increased leading edge radius) due to costumer demand, plastic fuel caps were utilized; rocker switches were incorporated into the electrical system; improved cabin sound proofing; improved door seals; increased avionics capacity; inertial reel shoulder harnesses introduced.	2, 3, 12, 26
C-172N	1977	The enigmatic L-O-320-H2AD 160-hp engine was installed in 1977; a 28-volt electrical system replaced the prior 14-volt system; a "notched" or pre-select flap control was introduced; urethane paint became the standard coating; front seats moved aft 4 in.	2, 3, 12, 26, 60
C-172O		No Skyhawk models with this designation.	
C-172P	1981	Rear seat shoulder harness became standard; the thickness of the windscreen and door windows was doubled; the gross weight grew to 2400 lb; the L-O-320-D2J became the standard engine in 1981.	2, 3, 12, 26, 43
C-172Q	1983	The L-O-360-A4N 180-hp engine installation was introduced; shoulder harnesses became standard for all seats; featured the 300 Series IFR avionics; air conditioning was introduced; the weight grew to 2550 lb.	2, 3, 12, 26
C-172R	1996	The L-IO-360-L2A 160 hp engine installation was introduced; the McCauley 1C235/LFA7570 propeller was introduced; advanced avionics; integral fuel tanks; gross weight dropped to 2450 lb.	26, 30, 40, 45, 89
C-172S	1998	The L-IO 360-L2A 180 hp engine installation was introduced; the McCauley 1A170E/JHA7660 propeller was introduced; gross weight grew to 2550 lb; Garmin G1000 glass cockpit (optional).	26, 30, 38, 45, 89

Notes: Variants of the basic models have not been considered in this table.

choice of the C-172 R (160-hp) or the C-172S (180-hp), if more performance is desired. The C-172S is sometimes referred to as the C-172SP (special performance) model [60].

CS7.3.4 Aerodynamics

Wing: The Skyhawk wing has its origins in the wings of the C-140 and C-170. The original C-172 wing was a C-170B wing. The original C-172 wing featured an NACA 2412 airfoil section. Section data for the NACA 2412 are presented in Table CS7.9. This section profile was maintained from the wing root to wing station (WS) 190. From WS 190 to WS 216 (wing tip) the wing section transitioned from the NACA 2412 profile to a 12% symmetric profile. As shown in Table CS7.3 the C-172 wing area is typically 174 ft^2 with a mean aerodynamic chord of 58.8 in. [3, 41].

Customer pressure caused Cessna to modify the leading edge of the wing with the C-172M model in 1973. More camber was added to the wing leading edge giving the leading edge a drooped look. The wing leading edge radius was also increased. The intent of the change was to provide better low speed aerodynamic effects i.e., near stall conditions. This resulted in the so-called "cuffed" or "cambered" leading edge. The original C-172 stalled in the neighborhood of 15 to 18 deg angle of attack; with the cuffed leading edge, the stall angle of attack may be a bit higher [2].

The unswept inner panel of the original C-172 wing was of constant 64-in. chord from the wing root to WS 100, with a straight taper from WS 100 to roughly WS 206, where the chord was 43 in.; the outer panel taper ratio was $\lambda = 0.6719$. The leading edge of the outer panel of the wing was swept (aft) some 2.75 deg. The actual wing mean aerodynamic chord varies between 58.4 and 58.8 in., depending upon the type of wing tip. As shown in Table CS7.3, these values have remained essentially constant over the lifespan of the Skyhawk [3].

Table CS7.9 Sectional Aerodynamics Attributes of Selected Airfoil Sections

Airfoil section	$c_{\ell max}$	c_{do}	c_{mac}	$\alpha_{\ell o}$ (deg)	$c_{\ell\alpha}/deg$	$\alpha_{c\ell max\ (deg)}$	a.c. (%chord)
RAF-6 (mod)	1.215			-3.3	0.0965	9.6	
NACA 0006	0.92	0.0052	0	0	0.098	9.0	0.243
NACA 0009	1.32	0.0056	0	0	0.098	13.4	0.240
NACA 2412	1.68	0.0060	-0.047	-2.0	0.098	16.8	0.245
References	41, 47	41, 47	41, 47	41, 47	41, 47	41, 47	41, 47

Note: RAF-6 data are for a maximum thickness of 8%.

The Skyhawk wing uses a Fowler type flap whereby the flaps extend rearward before they deflect. Prior to the C-172P, flaps could be deflected 10°, 20°, 30°, or 40°. Beginning with the C-172P model, the flaps cannot be deflected beyond 30°. The flap span extends from approximately WS 21.78, the wing-fuselage junction, to WS 100. The flap span is approximately 6.52 ft with a flap chord of 1.63 ft; the undeflected (exposed) flap chord is approximately 1.08 ft. The undeflected flap area is thus 14.12 ft^2; with flaps deflected, the total flap area is 21.23 ft^2. Deflected flaps add about 7.11 ft^2 to the wing area [3, 26, 41, 42].

The typical C-172 wing has an aspect ratio of roughly $AR = 7.45$. The inboard wing root airfoil section (WS 0) has an incidence of +1°30' and this incidence is constant to WS 100. The wing tip airfoil section has an incidence of −1°30'. Thus, over the outer wing panel, the incidence varies from +1°30' at WS 100 to −1°30' near the tip (say WS 206) with the incidence varying linearly over the outer wing panel. Thus, the wing outer panel has a total washout of 3°. This washout is provided, to some degree, in order to prevent the wing tips from stalling before the inboard section of the wing. The wing has utilized several wingtips throughout its life span. The wing has, however, pretty much retained a planform area of 174 ft^2 over time. The Skyhawk wing characteristics are shown in more detail in Table CS7.3 [2, 3, 41, 42].

The aileron span runs from WS 100 to roughly WS 206, approximately 106 in. The top planform aileron chord is approximately 9.35 in. in chordwise width, yielding a total aileron area of some 13.76 ft^2. Jane's gives the aileron area as 18.3 ft^2. It is known that aileron design often results in the bottom aileron chord length being greater than the top aileron chord length. Thus, a couple of additional inches in the C-172 bottom aileron chord length and/or a few inches in aileron span length could account for the difference between the scaled aileron area and the aileron area given by Jane's. The trailing edge portions of the wing along the aileron span were made symmetric in order to make the left and right ailerons interchangeable; this reduced production costs and enhanced aircraft maintainability. Beginning with the (1960) C-172A model, modified Frise ailerons were utilized by the Skyhawk [2, 3, 42].

Vertical: The original C-172 vertical tail (fin and rudder) utilized a symmetric NACA 0009 section at the root, excepting the dorsal fin (as applicable), tapering to a symmetric NACA 0006 section at the tip. Section lift, drag, and pitching moment characteristics for the 0006 and 0009 airfoil sections are presented in Table CS7.9. The rudder is horn-balanced to reduce pilot control effort. Symmetric sections are typically used for aircraft empennage control surfaces because both up and down vertical loads and left and right side loads need to be employed from time to time for vehicle control. The original vertical fin area was 9 ft^2 with a rudder area of 9.42 ft^2 [3, 41, 44].

The Flight-Sweep vertical tail of the 1960 C-172A had a vertical fin area of 10.76 ft^2, a rudder area of 7.28 ft^2, and a 25% fin chord line sweep of 35°. A dorsal fin was added in 1969 to the C-172K model. The dorsal fin was extended forward to the Omni-Vision rear window on the 1972 C-172L model. The large Fowler-type action flaps of the C-172 created a nose down pitching moment during sideslips when the flaps were deflected at large angles (e.g., during landing); the extended dorsal fin on the C-172L helped eliminate this problem. At the same time, a ground adjustable rudder tab was incorporated into the rudder. The C-172R and C-172S models have a vertical fin area of 11.24 ft^2 and a rudder area of 7.43 ft^2. Additional vertical tail data is presented in Table CS7.5 [2, 42, 44, 45].

Horizontal: The Skyhawk horn-balanced horizontal tail (fixed tail-plane plus elevator) utilized, like the vertical tail, a symmetric NACA 0009 section at the root tapering to a NACA 0006 section at the tip. The horn-balance tends to reduce pilot control effort. Cessna wanted its Skyhawk customers to feel that it took no more effort to fly the aircraft than it did to drive an automobile. Cessna's mantra was "drive it up, drive it down, you can turn and park it easier than an automobile." The original Skyhawk horizontal tail had a span of 128 in. The horizontal span was increased by 8 in. to 136 in. in the 1963 C-172D model. The original fixed tailplane area was 19.8 ft^2 and the total elevator area was 15.42 ft^2. The starboard (only) elevator has a trim tab of some 1.26 (scaled) ft^2. It should be noted that the horizontal tail had an incidence of −3.5° [2, 3, 44, 46].

With the increase in the span of the horizontal tail on the 1963 C-172D, there was an increase in stabilizer (tailplane) area to 20.16 ft^2. The corresponding total elevator area was 16.15 ft^2 (an increase of about a 5%). The C-172R and C-172S models have a stabilizer area of 21.56 ft^2 and a total elevator area of 14.53 ft^2—a very small loss in total horizontal tail area of less than 1%. Additional horizontal and vertical empennage data are presented in Table CS7.5 [45, 46].

Table CS7.9 presents several significant sectional aerodynamic characteristics (c_{l_α}, $c_{l_{max}}$, c_{mac}, etc.) of the four primary airfoils employed in the design of the C-172. Although the RAF (Royal Aircraft Factory)—6 dates to 1912 and is the oldest section of the four, its modified sectional characteristics are less well known than the other three (NACA 2412, 0006, 0009) because, today, it is largely limited to use in propeller design [43, 47].

Drag Polar: Loftin and Roskam both report that the C-172 has a zero-lift drag coefficient of $C_D = 0.0319$, an aspect ratio of $AR = 7.32$ (a bit low compared to the aspect ratios presented in Table CS7.1), and a wing span efficiency of $e = 0.75$. A simplified drag polar equation can be written as

$$C_D = C_{D_0} + (\pi\, eAR)^{-1} C_L^2.$$

With the values of C_{D_0}, e, and AR noted above, this simplified drag polar for the C-172 can be written as

$$C_D = 0.0319 + 0.05798\, C_L^2.$$

This drag polar is shown on Fig. CS7.11. The maximum L/D of 11.59 from Fig. CS7.11 agrees well with the published value of 11.6. In this era of computer worship, it should be remembered that until well into the late 50s and early 60s of the last century, the slide rule and graphical methods provided the answers to most of technology's questions. Graphical methods still provide a quick sanity check for engineering estimates based on more accurate analytical methods [48, 49, 50].

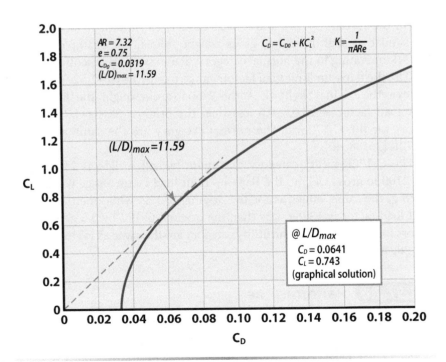

Figure CS7.11 C-172 Skyhawk drag polar.

CS7.3.5 Load Factors

Limit maneuvering load factors are the maximum load factors that one might expect to occur in a normal flight. These limit maneuvering load factors can be expected to be different depending upon aircraft usage in the normal, utility, or aerobatic category. The Skyhawk is designed only for normal and utility operation. In the normal category, the original C-172 was limited to a gross takeoff weight of 2200 lb and flaps-up limit load factors ranging from +3.8 to −1.52 (or +3.8 g to −1.52 g). At the 2200-lb gross weight, the flaps-down limit load factor was +3.5. Aerobatic maneuvers are not approved for Skyhawk normal category operation [36].

Operating in the utility category, the maximum C-172 gross takeoff limit was 1950 lb for the original C-172. The corresponding flaps-up limit maneuvering load factor range was +4.4 to −1.76. At this 1950 lb gross weight, the flaps-down limit load factor was +3.5. Although the Skyhawk is not designed for aerobatic flight, some maneuvers are permitted (i.e., chandelles, lazy eights, steep turns, and slow deceleration spins and stalls—no whip stalls) when the aircraft is operated in the utility category. It should be noted that when used in the utility category only the pilot and co-pilot seats should be occupied (maximum)—no back seat passengers and no baggage [36].

For the current C-172R and C-172S models, the limit load factor values are similar at slightly higher gross weight values. For example, for the C-172R operating in the normal category, at a maximum takeoff weight of 2450 lb, the flaps-up limit load factors range from +3.8 to −1.52—the same values as the original Skyhawk. At this 2450 lb gross weight, the flaps-down limit load factor is +3.0, −0.5 less than the original Skyhawk. The load factors for the C-172S, in the normal category are the same as for the C-172R [36, 38, 40].

The C-172R utility category maximum takeoff weight is 2100 lb. At this 2100 lb gross weight, the flaps-up limit load factor range is +4.4 g to −1.76 g, the same values as for the original C-172, while the flaps-down limit load factor is +3.0 g, less than the original C-172 by 0.5 g. The load factors for the C-172S, in the utility category are the same as for the C-172R [36, 38, 40].

CS7.3.6 Performance

Performance refers to such attributes as maximum speed attainable, cruise speeds, takeoff distance, landing distance, climb rates, lift-to-drag ratios, stall speeds, range, service ceiling, and endurance. Many of these attributes are tabulated in Table CS7.1 for selected models of the C-172 and some of their antecedents. Unless otherwise noted, the data in Table CS7.1

are for landplane versions of the selected aircraft. It should be noted that over time, the brake hp for the Skyhawk has increased from 145 to 180. The reference area for the C-172 is typically 174 ft^2. Wing loading (W/S) and power loading (W/bhp) have a strong influence on aircraft performance; S is the reference (wing planform) area, W is the aircraft weight, and bhp is the brake hp of the piston engine. For selected models of the Skyhawk, Table CS7.1 shows that the wing loading (at maximum gross weight) varies from 12.6 to 14.4 psf. For any given flight, the wing loading decreases with fuel burn. The power loading for the selected models of the C-172 shown in Table CS7.1 varies from 14.2 to 15.9; the power loading will decrease with fuel burn. It should be noted that the floatplane data for the C-172A are based on the installation of Edo 89-2000 floats; other floats could produce slightly different results. The Edo 89-2000 floats had an overall length of 16 ft, total weight of 240 lb (including installation accessories), normal draft of 13.5 in., and a fresh water displacement of 2000 lb per float [2, 51].

It should be noted that the values of the various attributes reported in Table CS7.1 should be taken as nominal values. Further, since the data in Table CS7.1 are from various sources, the attribute values may not be internally consistent.

Takeoff: Takeoff distance over a barrier can be considered to consist of four phases: ground roll, rotation, transition, and climb out. Table CS7.1 indicates that the C-140 and C170 required over 1800 ft for takeoff over a 50-ft barrier. For selected C-172 models, the landplane takeoff distance ranged from 1370 (C-172A) to 1685 (C-172R) ft. It should be noted that for the C-172A, the landplane takeoff distance was a little more than half the takeoff distance for the floatplane (Edo 89-2000 floats) variant (2390 ft) [2, 52].

Landing: The landing distance over a 50-ft obstacle can be considered a three-phase event: air distance, free roll distance, and braking distance. Table CS7.1 indicates that whereas the landplane version of the C-140 required some 1500 ft for landing, the selected landplane models of the C-170 and C-172 required landing distances ranging from some 1100 to 1300 ft. It should be remembered that the C-172 has large Fowler-action type flaps. Further, the floatplane version of the C-172A requires some 230 additional feet for landing than does the landplane version [52].

Speeds: Velocities in excess of the never-exceed speed or redline speed may result in aerodynamic instabilities and/or aircraft structural failure. From Table CS7.1, it can be seen that the never-exceed speed for the C-172 is generally in the 158–187 mph range, depending upon the model of interest.

As indicated in Table CS7.1, the maximum velocity for the landplane version of the Skyhawk ranges from 125 mph (C-172N) to 145 mph (C-172S). Compared to the landplane version of the C-172A, it should be noted that floats on the C-172A cause an approximate 30 mph drop in the maximum velocity of the aircraft [53].

Typical cruising flight is performed at 50 to 70% of normal rated engine bhp. As indicated in Table CS7.1, Skyhawk cruising speeds at reduced power settings are in the 122–143 mph range. For the C-172A model, the additional drag associated with floats causes a reduction of some 25 mph in cruising speed. The best cruise altitude for the Skyhawk is typically in the 6000–8000 ft range in altitude [53].

Climb: The rate of climb depends upon the excess power available at any given speed and altitude (between V_{max} and V_{min}). For the Skyhawk models and its predecessors (C-140, C-170), the maximum rate of climb occurs at sea level. It should be remembered that the selected aircraft in Table CS7.1, utilized a number of different engines and a number of different propellers [54, 55].

As indicated in Table CS7.1, the maximum rate of climb for the C-140 was nominally 680 f/m; for the C-170 and C-170B, it was nominally 690 f/m. For the Skyhawk, the maximum rate of climb ranges from 580 to 840 f/m, depending upon the selected model of interest. For the C-172A, the effect of floats is to reduce the maximum rate of climb by some 150 f/m.

Service Ceiling: Service ceiling is defined as the altitude at which the aircraft rate of climb is equal to 100 f/m. The service ceiling represents a practical upper altitude limit to controllable unaccelerated flight (i.e., static performance). Table CS7.1 indicates that the service ceiling for the Skyhawk's predecessor models was 15,500 ft. For the C-172, the service ceiling varies between 13,100 and 15,100 ft, depending upon the Skyhawk model of interest [55, 56, 57].

Range: As indicated in Table CS7.1, the range of the Skyhawk predecessors ranged from 475 to 592 mi. The range for the selected C-172 models in Table CS7.1 varies from 485 (C-172N) to 790 (C -172R) mi, depending upon the model of interest.

The equilibrium glide angle (power off) can be written as

$$\tan\theta = D/L,$$

where θ is the equilibrium glide angle, L is the aircraft lift, and D is the aircraft drag. The smallest gliding angle corresponds to the largest gliding range. Since the C-172 has a maximum lift-to-drag ratio of about 11.6, the smallest glide angle is about 4.93° [50, 54, 55].

Unrefueled Endurance: The maximum unrefueled endurance of the basic Skyhawk models is approximately 4.5 hr. The early Skyhawk models had a maximum endurance of some 4.5 hr. However, for later models, the maximum endurance varies from roughly 4.1 to 6.7 hr, depending upon the takeoff gross weight, rpm, cruise altitude and the fuel capacity [15, 29, 36, 59, 60].

Stall: Table CS7.1 gives the power-off stall speeds for the Skyhawk and its predecessors (C-140 and C-170). With flaps up, the stall speeds can be seen to range from 49 to 59 mph. With flaps down, the stall speeds are seen to range from 44 to 53 mph. It should be noted that power-on stall velocities tend to be less than power-off stall velocities [55].

CS7.3.7 Propulsion System

The Skyhawk tractor propulsion system consists of the engine, the propeller, and related accessories. The C-172 is a tractor in the sense that the propeller pulls the aircraft through the atmosphere.

Engines: A schematic of a typical flat horizontally opposed engine is shown in Fig. CS7.12. Flat horizontally opposed Continental engines

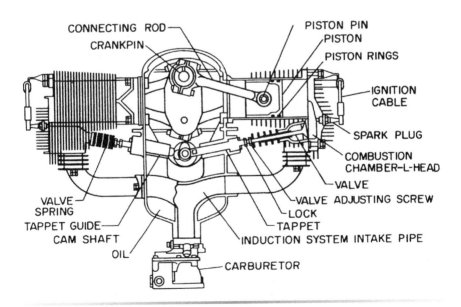

Figure CS7.12 Representative flat-horizontally opposed engine schematic. (From Borden & Cake.)

Figure CS7.13 Continental C-85-12 engine. (From Paul H. Wilkerson.)

were used in the early C-172 models and their antecedent designs. Model designations for these engines have been based both on horsepower and displacement. For example, the C-140 model utilized the four-cylinder C-85-12F/12 engine, which was rated at 85 hp at 2575 rpm and had a piston displacement of 188 in.^{3}. The C-85 engine is shown in Fig. CS7.13. Figure CS7.14 shows the engine performance curves for the Continental C-75, C-85, and C-90 engines [15, 62, 63, 64].

In Fig. CS7.14, the C-85 (ATC 233) full throttle brake hp curve indicates the full throttle hp produced by the engine at a given crankshaft rpm. The C-85 prop load curve indicates the power that can be absorbed by the propeller at a given rpm. The full throttle curve and the prop load curve are seen to intersect at 2575 rpm. This means that the C-85 engine is rated at 85 hp at 2575 rpm. Figure CS7.14 also shows the brake mean effective pressure (bmep) as a function of rpm for the C-85 (C-75 and C-90) engines. The bmep is the average pressure in the cylinder during the power stroke; it is a measure of the power produced by the engine. For the C-85 engine, the maximum bmep is approximately 140 psi (lb per square in.) at 2450 rpm. Additional C-85 engine characteristics are presented in Table CS7.7 [43, 62].

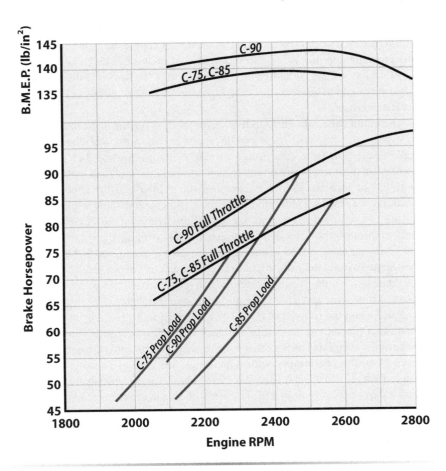

Figure CS7.14 Continental Motors C-75, C-85, and C-90 engine performance chart for sea level. (From Christy.)

SkyHawk Sets Single Engine Endurance Record

Refueled Endurance: The standard Skyhawk does not have the capability for aerial refueling. However on several occasions the aircraft has been fitted with a jury-rigged ground-to-air refueling subsystem. Perhaps the most noteworthy occasion occurred in 1958–59 at McCarran Field in Las Vegas, NV when Robert Elgin Timm and John Wayne Cook took off in a Cessna 172 (N9172B) on 4 December 1958 and landed on 7 February 1959 . . . a total time in the air of 64 days, 22 hr, 19 min, and 5 sec (almost 65 days). The Timm/Cook record setting flight began as a publicity stunt, gained momentum as a charity fundraiser, and stretch-finished as a still standing single engine

(continued)

endurance record. Before the flight the aircraft had approximately 1500 hr on the airframe and 450 hr on the standard continental C-145-2 engine. The C-145-2 has a TBO (time between overhaul) of 1800 hr (see Table CS7.7). Thus at the end of the record setting flight the engine had 2008 hr on it since its last overhaul.

The modifications to N9172B to prepare for the grueling flight took nearly two years to complete. These modifications included the addition of a 95 gallon Sorensen belly tank (total aircraft fuel capacity was 142 gallons), removal of the interior furnishings (except for the pilot's seat) and replacement with a four inch thick, four ft^2 square foam rubber pad for sleeping, the addition of a small stainless steel basin for washing and shaving, replacement of the co-pilot's side door with an accordion style door for easier access/ egress, through-the-firewall-plumbing to enable changing the engine oil and oil filters, an electric winch, and an electric pump to transfer fuel from the belly tank to the standard wing fuel tanks. The electric winch was installed to facilitate twice-a-day fueling. In addition to the standard instruments, a MkII Narco Omnigator and a Mitchell autopilot were installed.

Fueling and the transfer of other consumables was achieved by lowering a hook to a speeding ground support truck where it was attached to the fuel hose or device for delivering other consumables. The ground support truck (a Ford pickup) was equipped with a fuel tank, fuel hose and related support equipment. The fuel and other consumable supplies were typically transferred in the daytime on a remote straight stretch of road near Blythe, CA where the aircraft was flown over the truck at a speed matching the speed of the truck at an altitude of approximately 15 ft AGL. The winch hook was lowered and attached to the fuel hose or other delivery device and then winched into the aircraft where they were transferred. Needless to say considerable coordination between the aircraft and truck crews was essential. Near the end of the mission the fatigue of the air crew was making this maneuver very dangerous.

The flight was taxing on both the men and equipment. Carbonization of the engine combustion chambers and spark plugs resulted in the engine producing less than its maximum power, making it difficult to climb after resupply late in the mission. The flight did much to identify equipment weaknesses and maintenance issues. Overall, the flight demonstrated the safety, reliability and dependability of the Skyhawk design in particular and general aviation in general. It should be noted that the Timm/Cook flight broke the previous Heth/Burkhart flight time of 50 days, 19 min set 123 days earlier in another C-172 [61].

The six-cylinder, horizontally opposed, Continental C-145-2 powered the C-170 and C-170B Models. Figure CS7.15 shows the sea level engine performance curves for the C-145 (and the C-125) engine. These curves show the absolute dry manifold pressure, specific fuel consumption, and

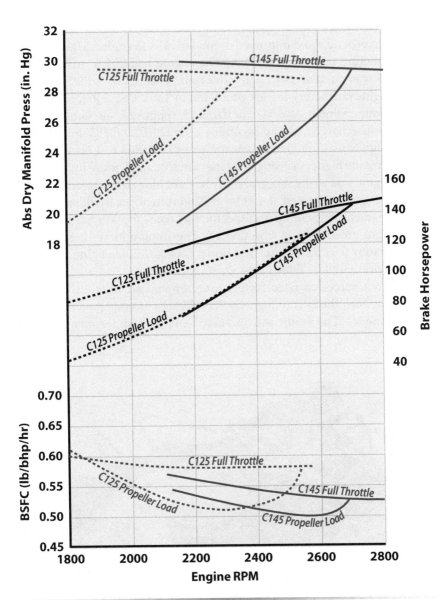

Figure CS7.15 Continental Motors C-125 and C-145 engine performance chart for sea level. (From Christy.)

brake hp as a function of engine rpm. The C-145 full throttle load and C-145 propeller load curves indicate that this engine was rated at 145 bhp at 2700 rpm. Further, this engine had a displacement of 301.37 in.[3] [62].

The original C-172 aircraft was powered by the six-cylinder, 4-cycle, air-cooled, direct drive, naturally aspirated (unsupercharged), horizontally opposed Continental C-O-300-A engine (TC 253). The C-O-300-A could be considered a new and improved or modern C-145 engine. Figure CS7.16 shows a picture of the C-O-300-A. The engine had 301.37 in.^{3} displacement and was rated at 145 hp at 2700 rpm at sea level. Additional C-O-300-A engine characteristics are presented in Table CS7.7 [62, 65].

The four cylinder, 4-cycle, horizontally opposed, air cooled, direct drive, unsupercharged, 150 hp Lycoming L-O-320-E2D (319.7 in.^{3} displacement) "Blue Streak" engine was installed in the 1968 C-172I aircraft, in part, to provide higher cruising velocities (and/or add 5 more hp). The Blue Streak engine was easily recognized by its light blue paint décor. The L-O-320-E2D engine was rated at 150 hp at 2700 rpm at sea level; the sea level engine characteristics of hp and specific fuel consumption (sfc) as a function of rpm are shown in Fig. CS7.17. The minimum sfc is seen to be approximately 0.53 lb/bhp-hr. Additional L-O-320-E2D engine characteristics are presented in Table CS7.7 [2, 26, 62, 65, 66].

Due to a perceived need for ten more hp and an engine that burned 100-octane fuel, the Lycoming L-O- 320-H2AD was employed in the 1977 C-172N model. The result was increased engine maintenance costs and

Figure CS7.16   Continental Motors C-O-300-A engine.
(From Paul H. Wilkerson.)

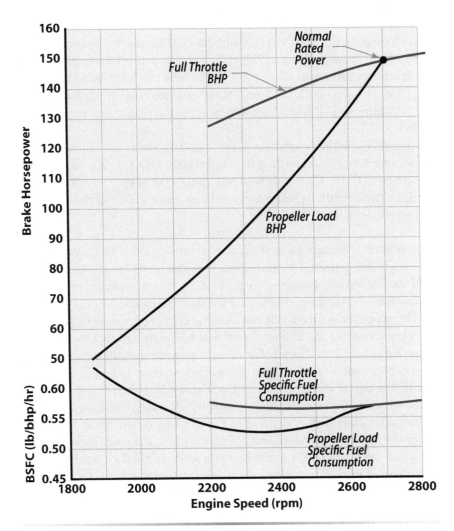

Figure CS7.17 Lycoming Motors L-O-320 engine performance chart for sea level. (From Christy.)

declining Skyhawk sales due, at least in part, to an insufficient lubrication problem associated with the engine tappets and camshaft lobes. This condition was corrected by switching to the Lycoming L-O-320-D2J engine in the 1981 C-172P model [2, 26].

The C-172R model was launched in 1996 when the product liability problems in general aviation were resolved. With the restart, Cessna moved production from Wichita, Kansas to Independence, Kansas, employed a fuel injection engine, improved the avionics, and added a second Skyhawk model, the C-172S/SP. It should be noted that all standard Skyhawk engines were carbureted until fuel-injected engines were introduced with the R and S models. Both the C-172R and the C-172S use the Lycoming L-IO-360-L2A for power. The C-172R uses a derated L-IO-360-L2A producing 160 hp at 2400 rpm while the C-172S uses the L-IO-360-L2A engine rated at 180 hp at 2700 rpm. Both engines use 100/100LL-octane fuel. Additional L-IO-360-L2A engine characteristics are presented in Table CS7.7 [26].

Table CS7.7 indicates that, in 2011 US dollars, the L-IO-360-L2A engine costs some $50,000. According to data in Table CS7.1, this is roughly 20% of the cost of the C-172S and about 36% of the cost of the C-172R. Depending upon the aircraft, a powerplant can be as much as 50% of the cost of the entire vehicle [67].

Propellers: Although most Skyhawks are powered by McCauley propellers, many are powered by propellers manufactured by other companies such as Sensenich. However, regardless of the manufacturer, all basic Skyhawk models utilize propellers of the fixed pitch ilk.

The propeller of an aircraft converts the energy produced by the engine into thrust to drive the aircraft forward. It has been said that the optimum propeller should have blades that are not so wide that they decrease efficiency, but wide enough to be structurally sound, and long enough to absorb the power of the engine without exceeding the critical tip speed. Thus, within limits, the larger the propeller diameter, the more efficient it is. Propeller efficiency is defined as the ratio of thrust hp to torque hp; under the best of conditions, this efficiency is less than 92% (few commercial fixed-pitch propellers exceed an efficiency in the low eighties). For a discussion of propeller theory and design, see Chapter 5 [68, 69, 70].

Propeller pitch is often defined as the distance the airplane travels in one propeller revolution. For a fixed-pitch propeller, the geometric pitch (the distance a blade element would advance in one revolution if it were moving along a helix/spiral having an angle equal to its blade angle) is constant across the blade length [43, 70, 71].

Throughout its production and operational life, the basic Skyhawk models have been powered by a variety of two-bladed, fixed-pitch propellers, notably manufactured by McCauley or Sensenich. These propellers have been typically manufactured in one piece, generally of forged metal or laminated wood. The blade angles are often chosen to maximize a single performance feature: takeoff, climb, cruise, or high-speed flight. A fixed-pitch propeller can also be designed as a compromise to satisfy two or more performance characteristics such as climb and cruise. Most propellers are designed as a compromise to provide the highest cruising speed in the neighborhood of 7500 ft altitude MSL. For example, the C-140 land plane was equipped with a McCauley CM7148 (48-in. pitch) propeller as standard, but could also be equipped with a CM7150 (50-in. pitch) to optimize cruise or the CM7146 (46-in. pitch) to enhance climb performance—all with a 71-in. diameter [26, 68, 70, 72].

Propellers are not inexpensive. For example, the Sensenich 76EM series propeller, compatible with 180 hp engines, retails for some $4000 per propeller. The Sensenich 74DM series propeller, matched with 150/160 hp engines, retails for approximately $3800 per propeller.

CS7.3.8 Structure

Except for furnishings, the Skyhawk is essentially an all-metal aircraft of semimonocoque construction. This means that the external skin or shell of the aircraft is supported by longitudinal stiffening members (e.g., stringers) and transverse supporting members [e.g., bulkheads (fuselage frames or wing ribs)]. The C-172 structure is illustrated in Fig. CS7.6.

As the Skyhawk has evolved, so have structural materials, manufacturing processes, and standardization. The current standard aluminum alloy designation system was established in 1954 to enhance production and utilization. As a result, 24S aluminum became 2024 aluminum. Early C-172 models used 24ST aluminum extensively, current models make extensive use of the high strength and good fatigue resistant 2024 aluminum alloy (the same alloy). The 2024 alloy features copper (3.8–4.9%), magnesium (1.2–1.8%), and manganese (0.3–0.9%) as the principal alloying materials. The alloy also has trace amounts of iron, chromium, zinc, titanium, and other elements. The ultimate tensile strength of the 2024 alloy is approximately 70,000 psi; the shear strength is approximately 41,000 psi; with a yield strength of some 50,000 psi. The material can be heat treated and, since the copper alloying element can somewhat degrade the corrosion resistance of the alloy, many 2024 products are often cladded to improve the corrosion resistance of the material [74, 75].

Compared to the weight of a cantilever wing, lift struts or wing braces can be used to reduce the weight of the wing. The struts of the C-120/140 were made of steel; those of the C-170 were made of steel, while those of the C-170B were made of aluminum. All Skyhawk struts were fabricated of aluminum. The C-172 lift struts (wing braces) are hollow aluminum tubes and have a tapered, questionably streamlined, oval cross-section (the radius of the leading edge is larger than the radius of the trailing edge). The upper ends of the Skyhawk struts are attached to the front wing spar. The lower ends of the struts are attached to the bottom of a forward fuselage bulkhead; the bulkhead that forms the A-pillar between the windscreen and the cabin door—one bulkhead aft of the firewall [17, 76, 77, 78].

Control surfaces are made of sheet metal and are corrugated (inverted V-shaped) for extra strength. Such corrugated control surfaces have been used on several Cessna models including the C-120/140, C-170, and the C-172, and may be considered a standardized design feature. These corrugations are spaced some three in. apart; the height of each corrugation is perhaps 1/8 to 3/16-in. If corrugations were not used, additional material, with corresponding extra weight, would have to be used to provide adequate control surface strength [76].

Weight creep is the curse of most aircraft model evolution. Whatever is done to improve each model of a given aircraft, the result is usually a weight increase due to added equipment or added structure—or both. During the model evolution of the C-172, Cessna was very successful in controlling its gross takeoff weight. During the first six years (1956–1962) the gross weight increased by only 50 lb. By 1963, the gross weight had escalated from 2200 to 2300 lb. It took another eighteen years (1981) for the gross weight to reach 2400 lb. The C-172S has a gross weight of 2550 lb. By any measure, this weight growth has been slow [2, 26].

The rivet is the typical Skyhawk sheet metal fastener of choice. Flush rivets are used extensively in the cowl (engine covering) area. Round headed rivets are used throughout the rest of the aircraft. The flush rivets are used over perhaps a quarter of the aircraft surface; round headed rivets are used over the remaining surface. Flush rivets are typically used to reduce surface protuberances and the resulting drag. However, the C-172 drag increment due to the round headed rivets should not be a big aerodynamic problem at the speeds flown by the Skyhawk [76, 79].

A typical Skyhawk weight statement is given in Table CS7.10. These data are for the 1961 C-172B model, whose gross weight was 2200 lb. Since weight creep is small with the Skyhawk, these weights may be representative for several C-172 models. The weight groupings are fairly standard. The useful load, including the fuel, is roughly 41.7% of the gross takeoff weight. The largest structural weight group is the wing group, which comprises some 21.5% of the total, closely followed by the propulsion group, which accounts for some 17% of the gross weight. The fuselage structure represents some 11.5% of the gross takeoff weight. The other weight groups represent small percentages of the gross takeoff weight. The C-172B paint scheme was somewhat limited to trim, which had a weight of about 3 lb, however, the current models have a complete topcoat of paint. The topcoat weighs approximately 20 lb [37].

Many of the Skyhawk parts are produced by outsource partners. However, the non-structural components of the seats (e.g., upholstery), and some of the special small parts are sub-assembly produced at the Independence plant.

Table CS7.10 Cessna Model C-172B Summary Weight Statement [37]

Weight item	Component weight (lb)	Group weight (lb)	Group weight as a % of gross weight	Weight item	Component weight (lb)	Group weight (lb)	Group weight as a % of gross weight
Wing group		**472.8**	**21.49**	**Fuselage group**		**252.9**	**11.49**
Basic structure	165.2			Basic structure	188.7		
Wing Tips (2)	4.0			Floor boards	14.7		
Access covers	1.2			Instrument panels	2.8		
Fairings & Fillets	3.0			Door – baggage compartment	3.0		
Struts (2)	20.8			Door – cabin, left	17.4		
Ailerons (2) [incl. bal. wt.]	21.6			Door – cabin, right	13.3		
Flaps (2)	16.4			Side windows 7 frames	2.8		
Misc. – nuts, bolts, etc.	3.8			Windscreen	10.2		
Empennage group		**60.7**	**2.76**	**Furnishings & equipment group**		**99.0**	**4.50**
Horizontal stabilizer	21.2			Front seats	24.8		
Elevator [incl. bal. wt.]	16.4			Rear seats	21.7		
Vertical fin & dorsal fin	11.4			Misc. accommodations	4.8		
Rudder	10.9			Furnishings	47.7		
Misc. – nuts, bolts, etc.	0.8			**Alighting gear group**		**121.9**	**5.54**
Propulsion group		**373.4**	**16.97**	Main Gear springs	51.6		
Basic engine	275.1			Main wheels & tires	28.6		
Air induction subsystem	2.8			Brake assembly & discs	6.4		
Exhaust subsystem	14.4			Nose gear assembly	19.1		
Cooling subsystem	6.2						

(continued)

Table CS7.10   Cessna Model C-172B Summary Weight Statement [37] (continued)

Weight item	Component weight (lb)	Group weight (lb)	Group weight as a % of gross weight
Fuel system	24.5		
Engine controls	1.8		
Starting system	16.0		
Propeller installation	32.6		
Nacelle group		31.2	1.42
Cowling assembly	18.9		
Engine mount assembly	12.3		
Air conditioning & anti-icing group		4.5	0.2
Cabin heating & defrosting subsystem	3.3		
Fresh air vents – front	1.2		
Surface controls group		31.2	1.42
Cockpit controls	9.9		
System controls	21.3		
Miscellaneous group		21.0	0.95
Standard paint scheme	3.0		
Unusable fuel	18.0		
Total weight – standard airplane (empty, dry, and unpainted)	1262.0	1262.0	57.36
Gross Weight of Standard C-172B	2200.0 lb		
Nose wheel & tire	8.5		
Steering	2.0		
Misc. – nuts, bolts, etc.	5.7		
Instruments & navigational equip. group		6.8	0.31
Navigational equipment	0.7		
Instruments	6.1		
Electrical group		41.4	1.88
Power supply equipment	34.3		
Power distribution & control	6.5		
Lights	0.6		
Hydraulic & pneumatic group		2.6	0.12
Brake system	2.6		
Licensed empty weight – standard airplane	1283.0	1283.0	58.3
Useful Load	917.0	917.0	41.7

The C-172 is a typical three-control aircraft i.e., the ailerons, elevators, and rudder are activated separately. (A two-control aircraft will have, for example, two controls such as the ailerons and rudder interconnected to a control wheel) [80, 81].

Longitudinal Stability: Elevator power is the ability of the elevator to produce a pitching moment. It can also be described by the derivative, $\partial C_m / \partial \delta_e$, where C_m is the pitching moment coefficient and δ_e is the elevator deflection angle. Elevator power is directly proportional to the horizontal tail volume coefficient, V_H. Horizontal tail volume ratios ($V_H = S_H \ell_H / S_{ref} \bar{c}$) are given for selected Skyhawk models in Table CS7.5. These tail volume ratio values are based on values of ℓ_t scaled from published three-view drawings, or taken from published data as noted in Table CS7.5. For example, the C-172A has a horizontal tail volume ratio of $V_H = 0.546$ (compared with $V_H = 0.490$ for the C-140A and $V_H = 0.760$ for the C-170A) [26, 82].

From a pilot's (and passenger's) perspective, longitudinal stability is probably the most important of the three aspects of static stability. Longitudinal stability refers to stability about the lateral axis, usually taken positive out the right wing (in the pilot's seat, looking forward) and perpendicular to the aircraft plane of symmetry. Longitudinal stability about the lateral axis is known as pitch stability. Pitch motion can be considered as uncoupled or independent of the aircraft's motion about the roll (longitudinal) axis or yaw (vertical) axis. For the elevator angle required to trim an aircraft, static stability in pitch requires that C_{m_0} be positive, and that $C_{m_\alpha} < 0$, i.e., that the derivative of the pitching moment coefficient with respect to angle of attack (or lift coefficient) be negative; the greater the negativity, the greater the stability [81, 83, 84].

One important design problem is the determination of the range of c.g. travel that provides satisfactory aircraft flying qualities. The axial coordinate of the center of gravity is typically located with respect to some longitudinal reference axis or fuselage reference line (FRL). The positions along this axis are typically called fuselage stations (FS, with numerical values in inches). Further, the c.g. travel range is often referred to its location on the wing mac. In the case of the C-172N, R, and S, the longitudinal datum origin (FS 0) is coincident with the location of the front face of the lower portion of the engine firewall—aft of this datum, fuselage station distance is positive and forward of the datum the fuselage station distance is negative [38, 40, 60].

The C-172R and C-172S models have a wing mac of 58.80 in., the leading edge of which is 25.90 in. aft of the datum. Thus, the mac of these two models extends from FS 25.90 to FS 84.70. The aft allowable center of gravity location is FS 47.3 while the forward center of gravity location (utility category) is at FS 35, a permissible travel of 12.3 in. or 20.9% of the wing mac. The forward c.g. limit is at 15.5% of the wing mac, while the aft limit is at 36.4% of the wing mac. It should be noted that the C-172N model has the most forward c.g. location at FS 35 and the most aft c.g. location at FS 47.3, values that are the same as for the C-172R and C-172S models [38, 40, 60].

The aft c.g. limit is typically set by the location of the neutral point of the airplane (i.e., the aerodynamic center of the airplane). If the aircraft is to have positive static stability, the aft c.g. limit should never be aft of the stick-fixed neutral point of the aircraft. Typically, the aft c.g. limit is just ahead of the stick-fixed neutral point. As the c.g. moves forward (e.g., due to fuel burn), the stability increases. The forward limit to the c.g. travel is, in part, determined by the ability of the elevator (up-elevator pitching moment) to balance the nose down pitching moment at $C_{L_{max}}$. Thus, the forward limit is based on controllability and is primarily determined by the elevator power [57, 84].

It should be noted that prior to the C-172D model, the span of the horizontal stabilizer was 10 ft 8 in. Beginning with the 1963 C-172D model, the horizontal span was increased to 11 ft 4 in., an increase of 8 in. This 8-in. increase in the span increased the horizontal tail area from 34.57 ft^2 to 36.31 ft^2 (reduced to 36.09 ft^2 for the C-172N, P, R, and S models). The horizontal stabilizer span for the C-172P model was 11 ft 3 in. The 11 ft 4 in. span returned with the C-172R and S models. Table CS7.5 shows many of the horizontal tail geometry characteristics for the selected models [26, 45, 46, 85, 86].

Directional Stability: Directional stability refers to stability about the vertical axis. Unlike longitudinal stability, which is uncoupled from the other two coordinates, directional stability and lateral stability are coupled by motions due to yaw (sideslip). Directional stability is also known as weathercock stability and is characterized by the fact that $C_{n_\beta} > 0$, i.e., static directional stability requires that C_{n_β} be positive. This means that when an airplane is at an angle of sideslip the yawing moment produced by such action is sufficient to return the aircraft to zero sideslip. The biggest contributors to directional stability are the vertical tail and the projected side area of the fuselage behind the center of gravity [49, 84].

Rudder power can be defined as the ability of the rudder to produce a yawing moment. Mathematically, rudder power can be described as $C_{n_{\delta_r}} = \partial Cn/\partial \delta_r$, which is directly proportional to the vertical tail volume coefficient, V_V. Vertical tail volume ratios ($V_V = S_V \ell_V / S_{ref} b$) are given for selected Skyhawk models in Table CS7.5. These tail volume ratio values are based on values of ℓ_V scaled from published three-view drawings, or values taken from published data as noted in Table CS7.5. For example, the C-172A has a vertical tail volume ratio of $V_V = 0.0435$ (compared with $V_V = 0.037$ for the C-140A and $V_V = 0.055$ for the C-170A; Pazmany gives a value of $V_V = 0.028$ for the C-140). For this value of the C-172A vertical tail volume ratio, the scaled parameter ℓ_V (moment arm) was taken to be the distance from an assumed c.g. location (quarter chord of the mac) to the aerodynamic center of the vertical tail [82, 87].

In 1960, Cessna made a decision to provide the Skyhawk with a rakish vertical tail. The quarter chord of the C-172A vertical tail was swept aft 35°. Customers approved! Since there is no free lunch, the price of the new vertical tail was slightly reduced rudder power, a slight decrease in C_{n_β}, and a slight increase in the weight of the vertical tail [2, 26].

Large single-slotted flaps with Fowler-type action were featured on the C-170 and C-172 aircraft. When these flaps were deployed at flap settings greater than 30° during sideslips, a nose down pitching moment was experienced. This nose down pitching moment phenomena was not completely eliminated until a larger dorsal fin was added to the C-172L in 1972. Further, 30° is the largest flap setting available on current models of the Skyhawk [2, 38].

A larger vertical tail, rather than the addition of a dorsal fin, would have been one option for Cessna. It should be noted that, for the same vertical fin area, a lower vertical fin with a dorsal appendage will typically create less parasite drag than a higher vertical fin with no dorsal appendage, and provide greater directional stability [54].

Lateral Stability: Aileron power is the ability of the ailerons to produce a rolling moment. This aileron power can be expressed by the derivative, $\partial C_\ell / \partial \delta_a = C_{\ell_{\delta_a}}$, where C_ℓ is the rolling moment coefficient and δ_a is the aileron deflection angle. With differential ailerons, the aileron angle is the average of the displacement angles of the two ailerons. Typically, $C_{l_{\delta_a}} < 0$, consistent with right aileron down producing greater lift on the starboard wing (conversely, left aileron up diminishing the lift on the port wing) creating a roll to the left. Strictly speaking, to date, no one has found a way to provide a purely aerodynamic means for providing an airplane with static stability, about the roll axis. That said, the dihedral effect contributes to a negative C_{ℓ_β} ($C_{\ell_\beta} = \partial C_{\ell_\beta} / \partial \beta < 0$) which will aid lateral stability. The rolling moment due to sideslip is sometimes referred to as the dihedral effect, but is not a static stability property in the strict sense of the word [57, 84].

The dihedral angle is the angle between the chordal plane of the wing and the horizontal plane. If the tip chord of the wing lies above the root chord of the wing, the aircraft is said to have a positive dihedral angle [55, 83, 84].

In general, one desires some dihedral effect but not too much. Some dihedral effect provides some stable C_{ℓ_β}, and minimizes spiral divergence (the tendency to depart from wings level equilibrium and roll off into a spiral). However, too much dihedral can make roll control difficult in gusty air. High wing aircraft typically require less dihedral that low wing aircraft. It should be noted that all early Skyhawk models have a dihedral angle range of $1° \leq \gamma \leq 2° \, 8'$, with most of the later Skyhawk models having a dihedral angle of $\gamma = 1° \, 44'$, see Table CS7.3 [54, 57, 81, 83, 84].

Active Controls: Cables, pulleys, and cranks (i.e., mechanical control systems) comprised the control surface systems of early aircraft and much of current general aviation aircraft, including the several models of the Skyhawk. Although active controls (fly-y-wire with stability augmentation) typically require less pilot effort than those with mechanical rigging, they do not necessarily represent the best design approach for all aircraft. It seems likely that active controls have not been employed in the Skyhawk design due to a number of factors. Cable, pulley, and crank control system rigging typically requires less weight, is less complicated, is as (or more) reliable, requires less maintenance, and is less costly than corresponding active control systems.

It should be noted that mechanical control subsystems generally become lighter, less expensive, and less complex than active control subsystems as one moves from high performance military and/or large transport aircraft to the smaller general aviation aircraft. As aircraft size and/or performance are reduced, the optimization of aircraft control subsystems increasingly tends to favor mechanical rather than active control subsystem utilization. Cost effectiveness is, to some degree in the eyes of the beholder. However, if the requirement is to design an aircraft that is easy to fly, robust in performance, easily maintained, practical, inexpensive to operate, and with a low first cost, chances are that it will not feature active controls. Just because one can do something (e.g., employ active controls) does not mean that it should be done. Any design feature should be efficacious to the aircraft mission. The designer constantly needs to be reminded of these axioms. Designers who do not pay attention to these axioms tend to promote the fulfillment of Augustine's Ninth Law and thereby price the product out of the market [88].

CS7.4 Skyhawk Components and Subsystems

Many of the Skyhawk components and subsystems were borrowed from predecessor aircraft: the C-120, C-140, and C-170. It was this revisionist approach that helped keep the cost and risk low for the C-172 Skyhawk.

CS7.4.1 Cabin

One can define the Skyhawk cabin as the volume of the aircraft fuselage between the firewall and the aft end of the baggage compartment. This volume has changed some over time. It is instructive to remember that the cabin of the first or original C-172 was identical to the cabin of the C-170B [2, 3, 36, 58].

The maximum useable cabin volume (excepting the volume for the pilot's seat) for the C-172B/C-172 was obtained with the baggage compartment shelf, the rear seat, and the co-pilot's seat removed. The removal of these items reduced the empty weight by some 40 lb. The maximum cabin volume was 55 ft^3. The C-170B/C-172 cabin extended from the engine firewall aft to the rear of the baggage compartment, from fuselage station (FS) 0 to roughly FS 91 [36, 58].

In the C-172R model, the cabin extends from FS 0 to roughly FS 142, the rear of the second baggage area. The end of the first baggage area is roughly FS 108. The aft second baggage area is some 34 in. in length with a vertical height of only 10.25 in. [40].

The Skyhawk cabin doors are true quadrilaterals in shape, no two sides parallel. There is one door on each side of the cabin. The front height of each door is 40.5 in. and the maximum width is 37 in. These doors are large enough to provide easy access into and out of the aircraft. These dimensions have changed only fractionally over time [36, 40].

The C-170B owner's manual says, "The windshield is a single piece, full floating, 'free blown' unit of 'Longlife' plastic." However, there was a center bar (WS 0) down the middle of the windshield, so the windshield was actually in two pieces. The C-172 models through C-172C retained this two-piece windscreen. A true one-piece windscreen was introduced in 1963 with the C-172D model [36, 39, 58].

C-170B crew/passenger visibility was provided by the front windscreen; a window in each door; and two windows aft of the doors, one on either side of the cabin. Excepting the left door window, all C-170B cabin windows were fixed (unopenable). The left door window was hinged at the top, which allowed it to be opened outward and upward. This window configuration was carried over to the initial C-172 model design and continued through the C-172C model [36, 58].

In 1963, a rear window (Omni Vision, with a center bar) was added to the C-172D cabin. At the same time, the aft side windows were increased in size. In 1969, the aft side windows were enlarged by 50%. Windscreen and window thicknesses were doubled for safety considerations in the 1983 C-172P model [40].

The front seats of the C-170B were individually mounted on tracks on the cabin floor. This allowed the individual front seats to be moved fore or aft as required. The single bench-type back seat was built to accommodate two people. This back seat was hinged at the bottom and was adjustable. The hinge allowed the back seat to be folded forward to provide access to the baggage area. This cabin seat configuration was retained by the initial C-172 model and continued essentially through current models. In the

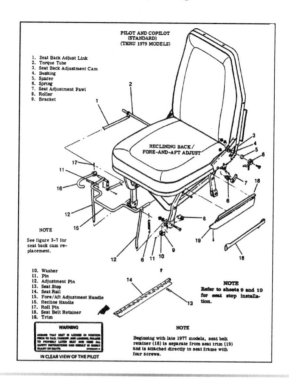

Figure CS7.18 C-172P pilot and co-pilot adjustable front seat assembly. (Courtesy of CAC.)

C-172R/C-172S models, the front seats are adjustable up and down as well as fore and aft; the seat back angle for each seat is also adjustable. Front seat shoulder harnesses were made standard in the 1971 C-172L model. Inertial reel shoulder harnesses were introduced (as an option) in the 1975 C-172M. Shoulder harnesses for the back seat were made standard in the 1984 C-172P model. The current models have inertia reel harnesses for all crew members and passengers [40, 58, 89].

Although the seat frames are currently an outsourced item, the seats themselves are made in the Independence Cessna plant. The contoured, energy absorbing front and rear seats of the current C-172R and C-172S models are dynamically tested to demonstrate the sustainability of a 26 g force. Figure CS7.18, courtesy of the Cessna Aircraft Company, is an assembly sketch of the standard C-172P Cessna single seat construction [89, 90].

A glove compartment was located on the right hand side of the instrument panel. The ubiquitous cigar or cigarette lighter was also located in the instrument panel. Four ash receivers (trays) were installed: two in front, one on either side of the cabin near the windscreen, and two located just aft of the rear door post bulkhead on either side of the cabin for use by passengers occupying the back seat. Increased cabin soundproofing was introduced in the 1976 C-172M model [36, 58].

Center-mounted avionics were added to the C-172E instrument panel in 1964. The basic "T" style instrument panel was added in 1967 to the C-172H model. The Garmin G1000 "glass cockpit" was introduced (as an option) in the C-172S cabin in 1998 [30].

CS7.4.2 Instruments

The instrument panel for the C-172 is shown in Fig. CS7.19 (part a) while the instrument panel for the C-172S is shown in Fig. CS7.19 (part b). As can be seen, with time, more instruments have been added as standard equipment [36, 38].

In the 1956 C-172, with the exception of the outside air temperature (OAT) all instruments were located on the instrument panel. The airspeed indicator (left side) and altimeter (left center) were both part of the pitot-static system. The pitot tube was located (then and now) near the leading edge of the port wing close to the lift-strut-wing junction (the stall warning transmitter is just a few inches away on the leading edge). The static port was located on the left forward side of the fuselage. The OAT gage is located on the right-hand side of the cabin in the upper corner of the windscreen in the ventilator. It should be noted that on the original C-172, the outside air temperature gage and the turn and bank indicator were considered optional equipment [36].

The basic "T" instrument panel was introduced with the 1967 C-172H to accommodate the growing number of instruments. The "T" refers to the configuration of the basic flight control instruments on the instrument panel. The console or pedestal below the center of the panel (between the two front seats) also contains a number of instruments, including the elevator trim control, the fuel shutoff valve, and the fuel selector valve handle in the C-172S model [38].

(a) C-172   (b) C-172S

Figure CS7.19   C-172/C-172S Instrument panel comparison.
(Courtesy of CAC.)

The instrument panel for the C-17S is comprised of subpanels of metal construction to allow related groups of instruments to be removed for servicing, repair, or other reasons without disturbing the entire panel. Flight instruments are located on one panel segment in front of the pilot. To the right of the flight instruments subpanel is a subpanel containing the engine tachometer and assorted navigational instruments. A center panel contains avionics equipment. The right hand side of the instrument panel contains avionics equipment and room for future instrument expansion. The center pedestal contains a hand-held microphone (and bracket) and other controls [38].

CS7.4.3 Landing Gear

Pazmany has stated the obvious, that takeoff is optional but landing is mandatory. The Cessna patented Safety Landing Gear used on the C-140 and the C-170 was coupled with a steerable nose wheel at the terminus of an air-oil shock strut (incorporating a shimmy dampener) to create the Land-O-Matic tricycle gear developed for the first C-172s. The Cessna patented Safety Landing Gear consisted of a single tapered sprung chrome vanadium steel leaf for each main gear i.e., a two-piece gear. The steel leaf was heat treated and shot peened to enhance its fatigue resistance [36, 87].

Fixed Gear: The steel Land-O-Matic main (leaf) gear requires very low maintenance; it may need to be painted occasionally to prevent rust. To achieve the maximum cushioning effect of the landing gear, correct tire pressure needs to be maintained. In the initial Skyhawks, the correct main gear tire pressure was 24 psig, while the correct nose wheel tire pressure was 20 psig. The correct main gear tire pressure for the C-172S is 38 psig, for the nose gear tire, a pressure of 45 psig should be maintained. To demonstrate the reliability of the main Land-O-Matic gear, a device was designed by Cessna to simulate constant rough ground landings. Initially displayed to the public at a conference in Cleveland, Ohio in November 1946, and later at the St. Louis Air Fair, 4,525,165 landings had been simulated by May 1947 with no damage to the gear or its attachment points [15, 28, 36, 38].

The Sylvester J. "Steve" Wittman designed Cessna steel spring Safety Landing Gear was advertised by Cessna (e.g., for the C-140) to "... effectively combat ground looping tendencies ... glides the plane smoothly and easily over rough fields. It's high on endurance, low on drag." The gear was also advertised to have built-in cushion action. Cessna paid Wittman royalties or annual fees for 17 years before buying the patent from him. Figure CS7.20 shows an assembly sketch of the sprung-steel, low-maintenance, and single-leaf main Wittman gear used on the early Skyhawks. In 1971, the C-172L introduced a tubular steel landing gear with a somewhat streamlined cross section. Figure CS7.21 illustrates the tubular

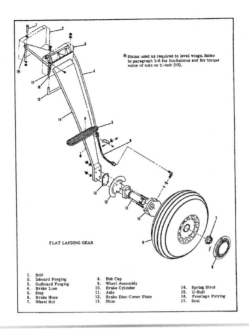

Figure CS7.20   C-172 single leaf main landing gear assembly.
(Courtesy of CAC.)

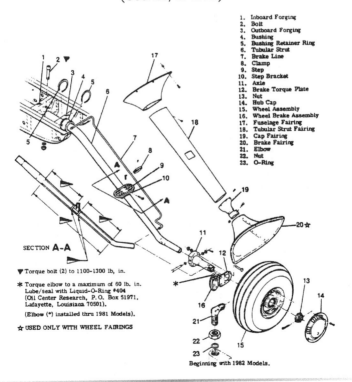

Figure CS7.21   C-172P tubular strut main landing gear assembly.
(Courtesy of CAC.)

steel landing gear assembly for the C-172P model. Figure CS7.22 illustrates the C-172P model nose gear assembly. Figure CS7.23 is an assembly illustration of the C-172P model nose gear strut. In 1958, the C-172 main gear was swept aft by 3 in. [2, 15, 90].

It should be noted that there are kits to convert the C-172 landing gear from a tricycle configuration to a conventional taildragger configuration. There are some aviation aficionados who believe that any real airplane has a taildragger landing gear configuration [12].

Retractable Gear: The Cutlass RG, a Cessna C-172 variant with a retractable landing gear (RG), was introduced in 1980. In one sense, the Cutlass filled a light aircraft retractable gear market gap left by the production discontinuation of the C-177 Cardinal RG in 1978. The 1980 Cutlass also featured a 76.5 in, variable pitch, constant speed McCauley propeller, and a 180 hp L-O-360-F1A6 (360 in.^{3}) engine. The 172 RG had a top speed of 145 kt and a gross takeoff weight of 2658 lb. The Cutlass

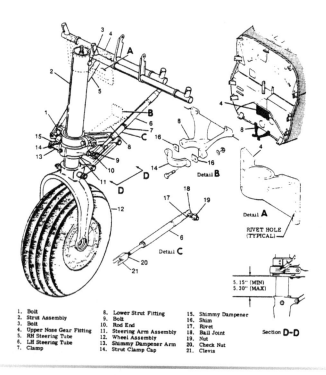

1. Bolt	8. Lower Strut Fitting	15. Shimmy Dampener	
2. Strut Assembly	9. Bolt	16. Shim	
3. Bolt	10. Rod End	17. Rivet	
4. Upper Nose Gear Fitting	11. Steering Arm Assembly	18. Ball Joint	
5. RH Steering Tube	12. Wheel Assembly	19. Nut	
6. LH Steering Tube	13. Shimmy Dampener Arm	20. Check Nut	
7. Clamp	14. Strut Clamp Cap	21. Clevis	

Figure CS7.22   C-172P nose wheel landing gear assembly.
(Courtesy of CAC.)

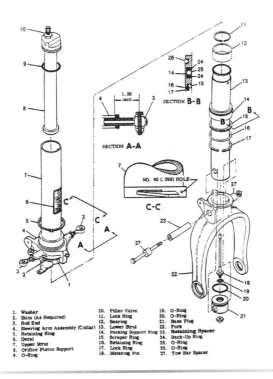

Figure CS7.23 C-172P nose wheel landing gear strut assembly. (Courtesy of CAC.)

weighed 351 lb more, was 20 kt faster (maximum), and cost $18,302 (average) more than the standard 1980 Skyhawk, the C-172N, with the 160 hp L-O-320-H2AD engine. However, only 1159 Cutlass RG aircraft (with annually declining sales) would be sold through 1985—an average of only some 193 per year; only 14 were built in 1985. The retractable gear, more powerful engine, greater weight, and 20 kt (18 kt in cruise) greater speed were not worth the incremental cost of $18,302—the market wasn't interested [26].

However, as sometimes happens with the Law of Unintended Consequences, the Cutlass RG is a useful, if not ideal, aircraft for "complex training" (advanced training) required of flight instructors and commercial pilots. "Complex training" requires "complex aircraft." According to § 61.31(e) of the Code of Federal Regulations, Title 14, Aeronautics and Space, "complex aircraft" are required to have a controllable pitch propeller, flaps, and retractable landing gear [91].

CS7.4.4 Electrical Subsystem

The original C-172 utilized a 12-volt, direct-current energy subsystem. The subsystem was powered by an engine-driven generator. Stand-by electrical power was provided by a 12-volt battery when the generator was not operating. Most electrical circuits were protected by fuses; however, an automatic resetting circuit breaker protected the stall warning, and turn and bank indicators. A schematic of the electrical subsystem for the C-172A is shown in Fig. CS7.24 [36, 39].

The 1965 C-172F model featured an electrical subsystem powered by a 14-volt, direct-current engine-driven generator. The clock was protected by a fuse; the rest of the electrical circuits were protected by circuit breakers [59].

The 14-volt system in earlier models was replaced by a 28-volt subsystem in the 1978 C-172N model. The C-172R and C-172S models continue to be equipped with a direct-current 28-volt electrical subsystem. The electrical subsystem for current Skyhawk models is powered by a belt-driven, 60-amp alternator, with a 24-volt battery as a stand-by energy source. A split primary bus bar is used to distribute electrical energy for most circuits [38, 40, 60].

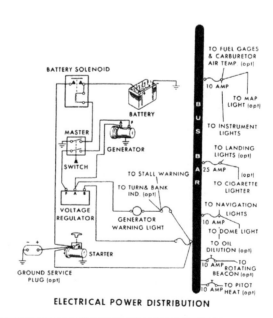

Figure CS7.24   C-172A electrical power distribution schematic. (Courtesy of CAC.)

CS7.4.5 Fuel Subsystem

The original C-172 had two 21-gallon aluminum fuel tanks. One tank each installed in the inboard, forward internal portion of each wing. This was a safety design decision. If the wings were to become detached during a crash landing, the fuel tanks would tend to stay with the wings, minimizing cabin exposure to fuel hazards. Only 18.5 gallons in each tank was usable fuel, the remaining 2.5 gallons of fuel were absorbed by the fuel plumbing subsystem (i.e. trapped fuel, ullage). Fuel was gravity fed through a fuel selector valve and fuel strainer to the engine carburetor. Highly leaded fuels were not recommended. A schematic of the fuel subsystem for the C-172A is shown in Fig. CS7.25 [36, 39, 92].

Each fuel tank has a sump drain plug on the underside of the wing on a line perpendicular to the rear edge of the cabin door and several inches from the side of the fuselage. These plugs enable one to drain any sediment or water that may have accumulated in the tank. The owner's manuals for the C-172 and the 172A state that the wing fuel tank sumps should be drained during 100-hr inspection. However, draining the sumps should be part of every pre-flight inspection [36, 39, 40].

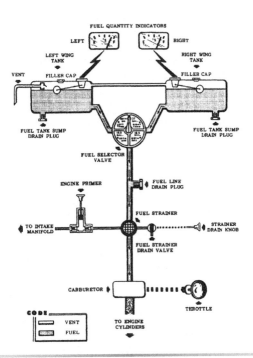

Figure CS7.25   C-172A fuel subsystem schematic. (Courtesy of CAC.)

The C-172R and C-172S Skyhawk models utilize two (one in each wing) vented fuel tanks constructed in an integral or "wet" wing configuration. This type of installation eliminates the weight of a separate fuel tank and/or bladder. A "wet" wing fuel cell is a fuel cell whose sides are part of the actual wing structure, i.e., no fuel bladders. Each tank holds 28 gallons when full; 1.5 gallons of which is unusable. Skyhawk fuel cell integrity is demonstrated during manufacture by cell pressurization in a water tank—if there are no bubbles, there is no leak. To accelerate the curing process, fuel tank cell seals are oven cured in two cycles, one cycle of six hr and one cycle of eight hr. For each cycle, the temperature is maintained at 125° F and the humidity is held within the range of 65–70% [38, 40, 76].

The fuel-injected engines used in the C-172R and C-172S use 100/100LL-octane gasoline. The other Skyhawk models have carbureted engines, most of which use 80/87 octane gasoline. Table CS7.7 indicates the recommended fuel grade for the several engines used for the Skyhawk and it close predecessors. It is recommended that the fuel tanks be refilled immediately after landing in order to minimize moisture condensation in the tanks [36, 38, 40].

CS7.4.6 Hydraulic Subsystem

The Skyhawk brake subsystem consists of a single-disc, hydraulically actuated brake for each of the main gear wheels. A brake master cylinder is mechanically attached to each pilot's rudder pedals—toe pressure to the top of either set of rudder pedals activates the brakes. A hydraulic line connects the master cylinder to the disc brake. The hydraulic line is shown in main gear assembly drawing, Fig. CS7.20 and CS7.21 [36, 40].

CS7.4.7 Flight Control Subsystems

The flight control subsystem for the four-place C-172 includes flaps, ailerons, rudder, elevators, and assorted tabs, control stops, chains, cables, bellcranks, and turnbuckles. The Skyhawk and its antecedents have utilized a rather simple control system for the past 60 years—one that has not seen appreciable change [36, 40].

CS7.4.7.1 Flap Subsystem

The Skyhawk flaps are hinged near the trailing edge of the inboard section of the wing. The single-slotted flap subsystem with Fowler-action (rearward movement, downward deflection) was activated by a flap handle (lever) mounted between the two front seats on early Skyhawks. Over time, this flap handle was replaced by a switch lever located near the junction of the instrument panel and the center pedestal on the C-172R/S models. There are mechanical stops at 0°, 10°, 20°, 30° and 40° on the C-172 and C-172A, but stops only at 0°, 10°, 20°, and 30° on the C-172R/S. Initially, the flap travel was 0–38°, with +2°, –1° tolerance. The 40° flap deflection capability was removed with the 1981 C-172P model [36, 40].

Flaps provide added lift for short-field takeoff, but perhaps more importantly, create significant drag. This added drag during landing can provide for an increase in the glide angle, enabling the pilot to bring the aircraft in over a barrier and land in a shorter distance than would otherwise be possible. The 10° flap setting is recommended for unusually short-field takeoff. The use of flaps is not recommended for cross wind takeoffs. The flaps may be activated (lowered or raised) at flight speeds less than 100 mph [36, 40].

The Skyhawk flaps are inboard of the ailerons and extend on either side of the fuselage from roughly WS 21.78 to WS 100. The flap chord is approximately 12.96 in. undeflected (approximately 19.56 in. deflected), with a span of roughly 78.22 in. The total flap area for the Skyhawk is some 21.23 ft^2 as indicated in Table CS7.3. The flap control subsystem is shown schematically in Fig. CS7.26 for the C-172 model [26, 36, 43, 93].

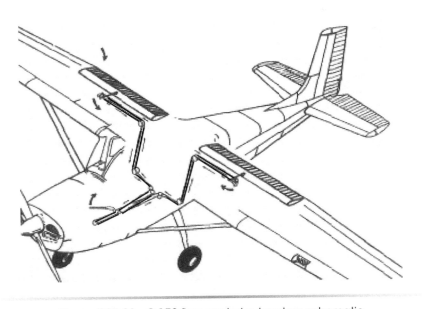

Figure CS7.26 C-172 flap control subsystem schematic.
(Courtesy of CAC.)

CS7.4.7.2   Aileron Subsystem

As the ailerons are deflected the aircraft rolls toward the up aileron. The down aileron creates more drag than the up aileron causing the aircraft to turn (or yaw) opposite to the direction the pilot wants to go. This yawing tendency is known as adverse yaw and can be corrected or controlled by some combination of aileron design and rudder usage. The use of rudder with aileron to produce a coordinated turn is a basic piloting technique.

One type of aileron design that can be used to mitigate adverse yaw is the Frise aileron. The Frise aileron typically employs aerodynamic balance in its design. Beginning with the C-172A through the C-172S models, the Skyhawk has successfully used modified Frise ailerons [36, 40, 45, 57, 93].

Aileron travel on the original C-172 was 20° up and 14° down, with a tolerance of ±1°. Total travel of the ailerons is limited to 34° by mechanical stops in the aileron bellcrank. The aileron bellcrank is located at WS 151.38 on the starboard wing and at WS 151.44 on the port wing—roughly at the spanwise center of the aileron—and transfers cable commands to the aileron push/pull tubes [36, 76].

The undeflected aileron chord (top planform) is roughly 9.35 in. with an aileron span of some 106 in. Frise aileron design often has a lower surface chord width greater than the upper surface chord width. The total aileron area is some 18.30 ft^2 as indicated in Table CS7.3. The aileron control subsystem is shown schematically in Fig. CS7.27 for the C-172N model [36, 40].

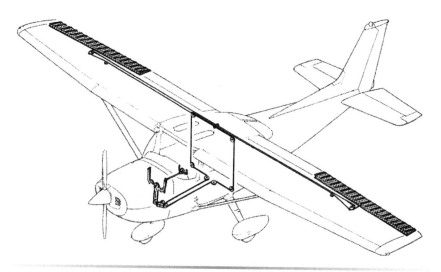

Figure CS7.27   C-172N aileron control subsystem schematic. (Courtesy of CAC.)

CS7.4.7.3 Horizontal Tail—Elevator Subsystem

The horizontal tail is part of the empennage and typically consists of three (one or two in an all-moving horizontal tail) parts: a fixed horizontal stabilizer, a moveable elevator, and a moveable elevator trim tab. The task of the fixed horizontal stabilizer is to keep the nose of the aircraft from pitching by providing basic pitch stability. The elevator is hinged at the trailing edge of the horizontal stabilizer and controls the wing angle-of-attack (pitch). The elevator tab provides control wheel pressure relief (reduction in pilot control forces) during flight [83, 94].

The C-172S horizontal stabilizer is constructed of fore and aft spars, ribs, stiffeners, and wrap around skin panels. The elevator is of the shielded horned (or paddle) balance design and is constructed of a forward spar, a rear channel, ribs, bellcrank, corrugated surface skins, formed leading edge skins, and a torque tube. On the original Skyhawk, elevator travel is 28° up and 26° down, with +1°, −0° tolerances. Turnbuckles are used to maintain a cable tension of some 30 lb. The elevator trim tab actuator (screwjack, in the C-172) is located internally in the starboard portion of the horizontal stabilizer [36, 40].

A small elevator trim tab, an auxiliary movable control surface, is located on the trailing edge of the starboard elevator. The tab is used to relieve control wheel pressure, i.e., pilot force, during flight and is controlled by a relatively small, vertically mounted tab control wheel near the flap lever. Aft movement of the tab control wheel trims nose up and forward movement trims nose down. The elevator trim tab is rectangular in shape and is hinged just forward of the center of the starboard elevator training edge. The C-172F trim tab has an area of some 1.46 ft^2 (scaled). The elevator trim tab in the original C-172 has an area of some 1.21 ft^2. The elevator control subsystem for the C-172N is shown in Fig. CS7.28. Over time, the horizontal stabilizer and elevator areas have varied a bit. For example, as indicated in Table CS7.5, the horizontal stabilizer area has varied from 19.8 ft^2 for the C-172, to 21.56 ft^2 for the C-172S. The elevator area has varied from 15.42 ft^2 for the C-172 to 16.15 ft^2 for the T-41A (C-172F). Other empennage data are provided in Table CS7.5 [36, 40, 59].

CS7.4.7.4 Vertical Tail—Rudder Subsystem

The vertical tail provides weathercock stability and directional control (around the yaw axis) for an airplane in a number of flight conditions, including spin recovery, adverse yaw, slipstream rotation, cross wind landings and takeoffs, and, for multi-engine aircraft, asymmetric thrust. The vertical stabilizer is the forward, fixed portion of the vertical tail and helps control the direction of flight. The moveable rudder is hinged to the trailing edge of the vertical stabilizer. The rudder provides control (together with

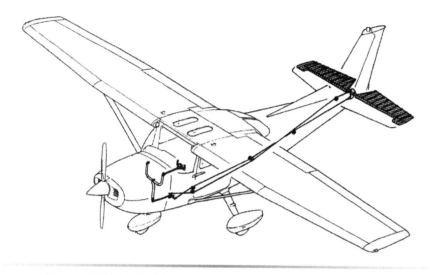

Figure CS7.28 C-172N elevator control subsystem schematic. (Courtesy of CAC.)

the ailerons) for turning flight. The rudder must be large enough to maintain zero degrees sideslip under the most adverse case turning flight and/or asymmetric thrust (multi-engine operation) condition—this requirement typically sizes the rudder [54, 83, 84].

The vertical tail of the C-172S consists of a fixed fin constructed of a spar, sheet metal ribs, wraparound skin panel, formed leading edge skin, and a dorsal. The rudder consists of formed leading edge skin, a center spar, ribs, spar with hinge brackets, wraparound skin, and a trailing edge ground adjustable (thin pie shaped) trim tab located at the rudder base. The tab is attached to the rudder along one of the long sides of its configuration; the tab is approximately 19.2 in.^{2} in size. Portions of this structure can be seen in the C-172 cutaway in Fig. CS7.7. The height of the vertical tail may change a bit from model to model depending upon the particular navigation light or beacon affixed to the top of the tail [38].

The rudder control subsystem for the C-172 is shown in Fig. CS7.29. Table CS7.5 indicates that the C-172A vertical tail area is 18.04 ft^2, thus, the C-172A vertical tail aspect ratio is approximately, $AR_V = 1.18$. The rudder area for the C-172A is 7.28 ft^2 and the fin area is 10.76 ft^2. For the C-172S, the rudder area is 7.43 ft^2 and the fin area is 11.24 ft^2; thus, the C-172S total vertical tail area is 18.67 ft^2, a bit larger than the corresponding area of the C-172A [26, 38, 45, 93].

The original dorsal fin added 0.872 ft^2 to the vertical tail area. The later, larger dorsal fin added approximately 3.125 ft^2 to the vertical tail area, an increase of 2.25 ft^2 in area over the original dorsal fin.

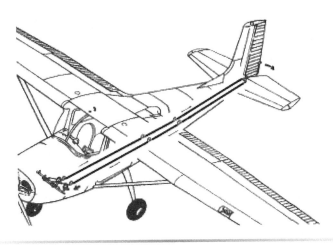

Figure CS7.29 C-172 rudder control subsystem schematic. (Courtesy of CAC.)

For the Skyhawk, pedals located just aft of the firewall and directly in front of the pilot's seat are used to operate the C-172 rudder. Rudder travel is 16°, with ±1° of tolerance, to either the left or right. On the original Skyhawk, rudder travel was limited by stops (bolts) located on the furthermost aft bulkhead [26, 36, 40].

CS7.5 Skyhawk Operation and Flight Test

CS7.5.1 Cost

In the mid-1940s, it was widely believed that the "Family Car of the Air" could be built for $1000–$1500. More realistic estimates turned out to be in the $2000–$2500 price range and above. Note that the basic cost of the C-140 (see Table CS7.1) was $2995 but the as delivered cost was $5003 [15, 81].

The unit cost of a Skyhawk depends upon a number of factors. There is a basic cost. The components of the basic cost include the cost of the engine, instruments, and the airframe. The cost increases with the number of options a customer chooses to purchase e.g., additional instruments, air conditioning, wheel pants, and/or larger gas tanks. Table CS7.1 shows a basic cost and an as delivered cost. For example, the 1956 C-172 had a basic cost of $8750 (see Table CS7.1) and an as delivered cost of $11,751. The difference in these numbers is, in large part, due to the options desired by the customer [26].

Over time, the Skyhawk engine hp has increased from 145 hp to 180 hp; as engine hp increases so does engine cost. As instruments evolved from "round" analog gauges to the glass cockpit, the number and cost of instruments has increased.

Cessna provides each new (and returning) customer with alternative choices. The customer can stay at a base price or select from a number of available options (e.g., floats, instruments, wheel pants) to obtain a model configuration that meets his/her flight needs and affordable desires. Purchasing options obviously increases the aircraft price. For example, Peter Bedell reports that the C-172SP (special performance S-model, type certificate for the C-172S model) he flew had a base price of $172,500 but the as-flown version cost $241,000. Thus, for any given model, the price is flexible, depending upon the needs and desires of the customer [26, 30].

From 1956 through 1974, the average price of the C-172 stayed under $20,000. From 1974 through 2006, the mean cost of the Skyhawk escalated some $4600 per year. Part of this cost escalation is due to inflation. However, with some options, a current Skyhawk can cost in excess of $300,000. Figure CS7.30 shows the escalation in the average price of the Skyhawk over time. For the first twenty-five years, the price was below $40,000. However, for the past thirty years there has been a steady upward climb in the cost of the Skyhawk [3, 26, 30].

In 2011, the typical price of the C-172R is reported to be in the neighborhood of $269,500, while that of the C-172SP is around $301,000. The operating costs for these aircraft are in the neighborhood of $1.00 (C-172R) to $1.02 (C-172SP) per mile or roughly $105 (C-17R) to $111 (C-172SP) per hour [30].

It isn't just that there is a demand for new models of the C-172 Skyhawk, there is great market demand for older versions of the Skyhawk. For example, the original C-172 model (circa 1956) sold for an average price of $11,751; in 2005, the average market price for this model was $24,500. Similarly, the 1977 C-172N sold for $30,050 new, but commanded a price of $52,000 in 2005 [26].

CS7.5.2 Production

The major Skyhawk subassemblies consist of the aft fuselage, cabin, power plant cowling, both wings, horizontal stabilizer, vertical stabilizer, and flight controls. The fabrication of a new Skyhawk begins with the wing rib drillouts—a wing rib is held in a fabrication tool while the prospective rivet holes are marked and drilled. As the fabrication progresses, the jigs or tools tend to get larger and heavier. Where appropriate, the part being fabricated can be rotated to enable the worker to more conveniently perform the particular fabrication task. Assembly tools are positioned

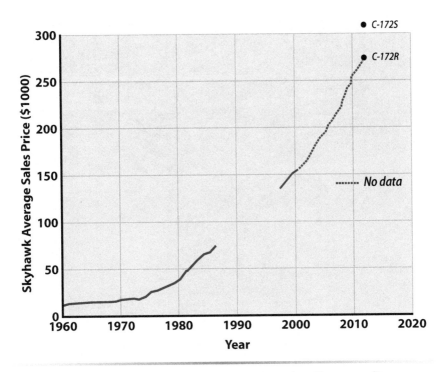

Figure CS7.30 Average Skyhawk sales price (then year $).

and hand-tools are arranged at each station to maximize worker comfort, i.e., the tools are ergonomically constructed. The tools or jigs tend to be heavy metallic structures—the bigger, the heavier—whose major control points, using a laser system, can be accurately held to tolerances on the order of 0.0001 in.

In 1996, at the restart of C-172 production, the production rate was 10–12 aircraft per week. In mid-2010, the production rate was four aircraft per week. The Skyhawk production line in Independence, Kansas is shown in Fig. CS7.31.

Some of the Skyhawk jigs currently in use have been used (occasionally with some modification) since they were first installed on the original production assembly line in 1955. These jigs or tools are re-mastered (checked for accuracy) at least on an annual basis, more often if the assemblers think there may be a problem. Additionally, wear checks are made periodically. Wear portions of the jigs are repaired or replaced as needed.

During high volume Skyhawk production, the assembly line may consist of as many as 24 different assembly stations. Each worker on such a line may perform only one task or a group of related tasks. As the production

Figure CS7.31 Skyhawk production line at Independence, Kansas. (Courtesy of CAC.)

rate decreases, the number of assembly line stations can decrease to as few as sixteen, with each worker performing as many as four separate tasks or groups of related tasks.

Most assemblies are moved along the assembly line by workers pushing roller-mounted assemblies to the next station. In some isolated instances, an overhead crane may be used to move a very large assembly. When production rates are large, only one aircraft type may be produced by a single assembly line. However, when production rates are lower, a given assembly line may be mixed in the sense that different (maybe two or three) aircraft models may be moving down the line in some staggered fashion. In such cases, the workers have to be very flexible and capable in completing their assembly tasks at a given station.

Assembly parts are concentrated in a "parts supermarket." Material handlers assemble a "parts kit" consisting of all the parts required to perform a particular assembly station operation and transfer the kit to the appropriate assembly line station. In an effort to improve assembly line efficiency, "toolboxes" are currently (2010) being assembled so that all the tools necessary to complete the fabrication process at any given assembly line station will be readily available at that particular station. Thus, the "parts kit" and the assembly line "toolbox" provide the worker or workers at each station along the line with the parts and tools required to complete the necessary assembly work. In the absence of a station "toolbox," tools may be moved from one station to another as the need arises. Rapid prototyping is used to develop new parts and new subassemblies.

Corrosion protection is a major consideration in the operating life of any Skyhawk model. Weather and neglect nibble at the structural integrity of any structure exposed to the atmosphere—including aircraft. Air and moisture have a debilitating structural affect on metallic surfaces. Cessna builds corrosion protection into each Skyhawk model via a triple corrosion protection scheme. This triple corrosion protection scheme includes (1) a corrosion resistant chromate primer applied to all metallic parts prior to assembly, (2) application of an alodine conversion coating to exposed surfaces after assembly, and (3) application of a full coat of corrosion primer prior to the final topcoat of paint.

Environmental issues are given serious consideration in the manufacturing processes. Lead in the fuel is addressed above. Paint primers typically contain both chromium and lead. Many of the fasteners contain cadmium (via electroplating). Lead, chromium, and cadmium are hazardous materials and their use is monitored accordingly. Floor sweepings, which may contain some of these materials, are handled in an appropriate fashion.

CS7.5.3 Flight Test

Every production aircraft is tested on one or two (more, if necessary for some reason) flights to ascertain that the aircraft operates as intended. These tests are performed to demonstrate that the aircraft functions properly. Every Skyhawk is flight tested to ensure that the aircraft is rigged to fly straight and level in coordinated flight without aileron or rudder control force required. Rigging adjusts the aileron, flaps, elevator, and rudder control surface angles and alignment.

Initial wing rigging is achieved by setting two eccentric bushings, one on each rear spar attachment, at the neutral position. These two bushings should always be rotated together whenever a setting change is required. The rotation of these two bushings changes the wing angle of attack (incidence) [2, 36].

The fuselage is flexed in a "wing set" maneuver. This maneuver essentially requires turning the aircraft through two 360-degree turns (once to the left and once to the right) at a constant altitude and a bank angle of 70 deg. This maneuver induces up to a 3 g load on the aircraft and tends to tighten up the structure in at least a semi-permanent "set" [76].

Every tenth aircraft is flight tested for acceptable spin recovery characteristics. A two-turn spin, in each direction, is conducted to demonstrate that recovery can be affected within a fraction of a turn with normal control application [76].

Every C-172 is flight tested to check stall warning function and the adjustment of the fixed rudder tab. The Stall Horn is checked to demonstrate that its activation is within acceptable limits (typically within a few mph of stall) depending upon the aircraft configuration, e.g., flaps up, flaps down, power-on, and/or power-off. The electrically powered harsh sounding horn and stall warning vane used on the early Skyhawks was replaced by a unique pneumatic stall warning device on the C-172H (1967) model (the device is also used on other Cessna aircraft). This device consists of an opening in the port wing leading edge near the lift-strut-wing junction wing station. This opening is located beneath the local airflow stagnation point. In the normal operating speed range above stall, this opening is located in a region of positive pressure, which forces an inboard flow of air through a small reed horn producing no noise. However, as speed is reduced to within some five to ten mph of stall, the inlet area of the device moves into a strong negative (suction) pressure region created by the local airflow. This causes an outward flow of air through the reed horn producing what is sometimes a two-tone warning: a soft initial sound followed by a louder, harsher, and higher-pitched sound as the stall velocity is approached [2].

To insure that the fuel selector switch is working properly, a fuel selector switch functional check is made by turning the fuel selector to the shut-off position in flight; this should cause the engine to stop. The fuel selector switch is then turned back to the "on" position. Since the propeller is still windmilling, if everything is working properly, the engine re-starts [76].

Full break stalls are performed on each aircraft to ensure that normal aircraft control exists up to the point of stall, with stall recovery affected with the normal use of controls. Power-off stalls are performed with the engine at idle power, not with the engine shut-off [76].

CS7.5.4 Maintenance Considerations

Periodic maintenance should be considered in the design of any aircraft. The construction materials, manufacturing processes, subsystem selection, and on-board equipment packaging all contribute to maintenance expense. For example, the instrument panel of the C-172R/S models is segmented so that, if a particular instrument requires attention, only a small segment of the overall panel needs to be removed to affect a repair or replacement. Typically, customers do not want to purchase a product that requires high levels of maintenance. Not only does the customer have to pay for the maintenance in such cases, the customer must also be without the use of the product while the maintenance procedure is being performed. Keep it simple is an excellent design mantra, but do so in the context of keeping the customer happy.

Certain inspections and maintenance procedures should be followed if a given aircraft is to retain the performance, dependability, and reliability it possessed when new. The 1956 Skyhawk Owner's Manual states that "Airplanes are built to be used and regular use tends to keep them in good condition." A systematic schedule of lubrication and routine maintenance should be followed [36].

Annual aircraft inspections may be required by law. More frequent inspections and service have been demonstrated to be cost effective. Cessna has suggested that some maintenance items be checked on a daily (per flight), 25-flight hr, 100-hr, and 500-hr interval, depending upon the maintenance item. For example, the C-172 engine oil level should be checked before every flight; the nose gear torque links should be lubricated every twenty-five hr; according to the C-172A owner's manual, the gyro instrument air filters should be replaced every 100 hr; and the wheel bearings should be repacked every 500 hr. Some maintenance items can be performed on as needed basis [39].

Depending upon the maintenance and/or repair procedure to be performed, it may be necessary to lift the aircraft off the ground. When such elevation is required, slings about the C-172 engine mount fuselage attachment and the aft section of the fuselage may be used, or the aircraft may be elevated by jacking (see Fig. CS7.32). Jacking tends to be the lifting method of choice. The appropriate service manuals provide specific details of the jacking or lift-sling procedure to be followed. Wheel repairs and brake relining are examples of instances when such lifting or elevation may be required [36, 60].

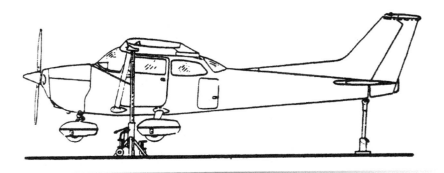

Figure CS7.32 C-172P proper jacking for maintenance and repair. (Courtesy of CAC.)

CS7.5.5 Skyhawk Safety

In 1949, as a result of crash worthiness or survivability research by Crash Injury Research (CIR) and other agencies during the preceding six years, more people were walking away from lightplane crashes than was previously the case. This beneficial development occurred because most manufacturers of the leading lightplane models had begun to incorporate crashworthiness or survivability features into their aircraft design [95].

By 1949, Crash Injury Research, then administered by the Cornell University Medical College, had studied some 600 lightplane accidents. The results of these studies indicated that maximum aircraft occupant protection could be achieved by designing aircraft with survivability features including high energy absorbing structures, progressively collapsing structures, more rugged cockpits, cockpit tubing which buckled outward, stronger safety belts, stronger control wheels, moving instrument panels forward or crew and passengers aft, more rugged landing gear, and impact circuit breakers to reduce fire hazards. Shoulder harness and seat belts were shown to be effective in providing occupant protection from head injuries. Since this early work, research has continued to be done in order to constantly improve occupant survivability from all aircraft accidents and mishaps [95].

Increasing engine reliability promotes aircraft safety. A study of 1947 data indicated that 94% of all engine failures (for engines with less than 200 hp) resulted in forced landings. Further, one accident in twelve resulted in either serious or fatal injury to one or more of the occupants. This 1947 study indicated that powerplant failures could be grouped into the failure categories of engine structure, fuel subsystem, ignition subsystem, propeller assembly, lubrication subsystem, engine accessories, control subsystem, and miscellaneous. Almost 70% of all powerplant failures were due to some fuel subsystem failure. Next in line was engine structure failure at roughly 4%. Suggested solutions included subsystem duplication, change in basic methods (e.g., switching from carburetion to fuel injection), better fuel subsystem warning methods, and improved subsystem design [96].

Table CS7.11 shows (circa 1986) NTSB Cessna 172 safety data compared with similar data for other general aviation aircraft. The accident rates shown in Table CS7.11 are for 100,000 flying hr. The Skyhawk was the safest lightplane in the categories of fatal accident rate and engine failure. Of the comparable aircraft, only the C-150 had fewer in-flight airframe accidents than the C-172. The C-172 accident rates are among the lowest in the remaining categories in this study [12].

The AOPA Air Safety Foundation looked at all C-172 accidents occurring between 1982 and 1988—some 1600 accidents. The C-172

Table CS7.11 Cessna 172 Safety Record (Accidents per 100, 000 Flying Hours)—NTSB Data

Accident type	Frequency comparison range	Cessna 172 frequency
Mean Fatal Accident Rate	4.84–1.65	1.65*
Engine Failure	12.36–1.41	1.41
In-Flight Airframe Failure	1.49–0.02	0.03
Stall	22.47–0.36	0.77
Hard Landing	3.5–0.19	0.71
Ground Loop	22.6–0.17	1.00
Undershoot	2.41–0.10	0.26
Overshoot	2.35–0.23	1.00
Reference	12	

* All single-engine Cessna aircraft.

safety record was compared with that of several other four-place general aviation aircraft comprising the bulk of the training and entry-level transportation fleet. In terms of accidents per 100,000 flying hr, the Skyhawk had fewer accidents per 100,000 flying hr than other comparable aircraft. The C-172 had fewer serious accidents as well. Of these serious accidents, maneuvering was the leading phase of flight for the accident occurrence. In summary, the Skyhawk is one of the safest aircraft one can fly [97].

CS7.6 T-41 Mescalero

The first C-172 (N5000A) came off the production line in 1955. However, prior to 1964 only the Austrian Luftstreitkräfte seemed (for certain) to recognize the military potential of the Skyhawk. Even then, Austria possessed only one example of the C-172. In 1964, the USAF conducted an industry wide competition for an off-the-shelf, fixed-pitch, propeller-driven primary trainer. Specifically, the competition was conducted to select an aircraft that would be used for the initial phase of primary flight training, screening pilot candidates for further training. The competition was a two-step process: step one was product qualification, step two was cost. Collectively, Beech (submitted two proposals), Piper (submitted three), and Cessna (submitted three) submitted a total of eight proposals. All three Cessna proposals met the qualification requirements. However, Cessna bid only the lower cost C-172 variant design [2, 30, 98].

CS7.6.1 T-41A

The USAF selected the proposal that featured the variant C-172, Skyhawk, design and assigned the T-41A designation to the aircraft. This variant design seated two (instead of the usual four—the back seat was removed), utilized essentially civilian avionics, incorporated the naturally aspirated Continental O-300-D engine, and otherwise was a standard C-172F. Cessna and the USAF signed a contract on July 31, 1964 for 170 aircraft for delivery between September 1964 and July 1965; first delivery was within 60 days. Eventually, the USAF bought 216 T-41As from the production runs of the Cessna C-172F, C-172G, C-172H, and C-172K models—reconfigured to USAF requirements. An additional 26 T-41As (C-172Gs) were purchased for delivery to the Peruvian Air Force under the US Military Assistance Program (MAP). Eight T-41As were purchased by the Ecuadorian Air Force. The aircraft has also been used by the Greek, Chilean, and El Salvadorian Air Forces [2, 3, 30, 98, 99].

The T-41A was in production from 1964 to 1970, in total, approximately 237 (or 242) were built: C-172F (170), C-172G (26), C-172H (34), and C-172K (7). Thus, most T-41A aircraft were basic, off-the-shelf, reconfigured C-172F models. Although the price of the T-41A was a negotiated one, the cost per aircraft (for the initial 170 aircraft) is said to have been less than $7000. Geometric, performance, and related characteristics of the T-41A are given in Tables CS7.1 thru CS7.7. It should be noted that at one time USAF student pilots had some 30 hr in the T-41A before transitioning to the T-37 "Tweety Bird" ("Screaming Mimi," or "Tweet") primary jet trainer also produced by Cessna [2, 3, 30, 98, 99].

CS7.6.2 T-41B

In 1966, the U. S. Army Training Command determined that it needed a primary/advanced pilot training aircraft that could operate in high ambient temperatures from small, unimproved airfields, e.g., Fort Rucker in Alabama and Fort Stewart near Savannah, Georgia, and was looking essentially for an off-the-shelf solution. The Army was aware of the success of the T-41A as a trainer for the Air Force but determined that it needed a more powerful aircraft. AVCOM held a flight demonstration competition, with priority given to takeoff and climb capability, at Parks Bi-State airport in St. Louis. This competition was won by Cessna with a modified R172E entry. Accordingly, the U. S. Army contracted with Cessna for 255, basically off-the-shelf but modified, R172E aircraft for delivery in 1967. These aircraft were given the T-41B designation, named the Mescalero, and differed from the T-41A in several respects. Whereas the T-41A was powered by a carbureted Continental O-300-D 145-hp engine with a fixed-pitch propeller, the T-41B was powered by the fuel-injected Continental IO-360-D 210-hp engine matched with a McCauley two-bladed D2A34C67/76S constant-speed 76-in. propeller. Further, the T-41A was (basically) a reconfigured C-172F, G, H, or K model, whereas the T-41B was essentially a reconfigured R172E, a variant of the C-172 [2, 99, 100].

CS7.6.3 T-41C, D

In 1967, USAF studies supported the need for a two-seat pilot training aircraft that could operate from the USAF Academy (USAFA) airfield in Colorado Springs (6572 ft), Colorado in summertime temperatures of some 100°F. This aircraft would be scheduled for primary pilot training and the preliminary selection of USAFA cadets for later assignment to a more advanced pilot training program. One requirement for the selected aircraft was that it have a fixed-pitch propeller. Cessna's solution was a modified R172E model with a 14-volt electrical subsystem and a fixed-pitch propeller (an example of give the customers what they want). This aircraft was given the T-41C designation. Fifty-two T-41C aircraft were eventually purchased by the USAF: 45 in 1968 and 7 in 1969 [2, 3, 99, 101].

The last Mescalero, the T-41D, was purchased by the USAF primarily for distribution through the MAP during the years 1968–1972. It was the most widely produced variant of the T-41 model series. In total, 843 Mescaleros (all variants) were produced [2, 3, 99, 101].

CS7.7 The Skyhawk Future

If you want to market a product that has broad appeal in a particular segment of the market (e.g., general aviation market), keep it simple. Arguably, the low-speed, high-wing, strut-mounted, four-place, cost-effective C-172 Skyhawk with its fixed tricycle landing gear and the fixed pitch propeller is about as simple as a general aviation aircraft can get. The Skyhawk is recognizable, is technically uncomplicated, has a reasonably affordable first cost, is relatively inexpensive to operate, has excellent handling qualities, and is easy to fly. You can find at least one at almost every municipal airfield. Almost every aviation enthusiast has had at least one ride in a Skyhawk, and, with the sometimes aggressive help of the marketing department, some 43,000 individuals (pairs and groups) have purchased one. It has an international reputation as a safe, efficient, and reliable aircraft.

The C-172, in one version or another, has successfully performed a wide variety of civilian and military functions. In civilian guise, it has performed as a trainer, businessliner, family van, pleasure craft, executive transport, and utility aircraft. Militarily, it has served as a pilot trainer, utility transport, artillery spotter, and in some cases has carried weapons pylons.

As new technologies have become available (e.g., the G1000 glass cockpit), Cessna has incorporated them into a standard model or offered them as options to a standard model to interested customers. Improvements in performance (e.g., more powerful engine) and handling qualities (e.g., addition of the dorsal fin) have been effected as flight experience has suggested or customer response has dictated. After some 60 years, the final epithet regarding the ubiquitous Skyhawk remains to be written.

CS7.7.1 The Greening of the Skyhawk

The high operating costs (e.g., five dollar a gallon avgas) of general aviation aircraft have inspired some to look for ways to reduce these costs. One in particular, George Bye, of Bye Aerospace, Inc. and its subsidiary, Bye Energy, Inc., is working to show that an electric-motor lithium-ion-battery propulsion system can produce, for the same weight, the same performance, including range, as a reciprocating engine powered aircraft. A modified 1978 C-172N with an off-the-shelf electric motor installed and a six-bladed composite McCauley propeller has been selected as the prototype vehicle [102].

While the ultimate goal of this Bye Green Flight Project is to cut operating costs by some 50%, eliminate pollutant emissions, and lower engine maintenance costs, current technologies will only provide an electric powered Skyhawk with a two-hr endurance. This two-hr flight time is more than adequate for pilot training functions. Greater endurance will require, among other things, further advancement in battery technologies. According to Bye, the rate of electric power technology advancement is such that these ultimate goals of this project should be attainable in the near future [102].

The COO of Bye, Charles B. Johnson (former President and COO of Cessna, circa 2002–2004), has estimated that an FAA-approved supplemental type certificate for the Green Skyhawk is attainable in roughly two years from the initiation of development. The target cost of converting a reciprocating engine propulsion subsystem to an electric propulsion subsystem is that of a complete reciprocating engine overhaul [14, 102].

The future might see Cessna offering the electric powered propulsion subsystem as an option with Bye Energy offering reciprocating engine power conversion to electric power for existing Skyhawks. Proof-of-concept could lead to cost and environmental issue changes in related fields of the aerospace industry [102].

What Dwane Wallace once said of the general aviation industry, could easily be said of Skyhawk's future: "It's early in the morning and the sun is shining" [14].

Acknowledgements

The author appreciates the time, effort, and energy that the following individuals (named and unnamed) have spent in helping me with this project. It has improved the quality and accuracy of my work. However, at this point all the errors are mine. At Cessna, I thank Andrew Woodward, Doug Oliver, Terry Clark, Larry Taylor, Brian Richardet, Charles Pate, Bob Wethington, Joe Latas, and particularly David Levy for taking time to help me and offering me some of their expertise. Eleanor S. Uhlinger and her staff at the Dudley Knox Library at the Naval Postgraduate School, particularly Erma Fink, have found numerous books and other documents I thought necessary for the completion of this project—thank you. I thank Steve Boser at Sensenich for his telephonic and electronic comments and help. Mark Maughmer, Wally Fowler, Bill Mason, and Bob Ball took time to review the manuscript and offer constructive suggestions—thank you. A vote of appreciation to Miriam Stoner at AOPA Pilot and Michael Fizer for allowing me to use some of the pictures from their April 2006 C-172 50th anniversary issue. I thank Pat Pierce at Teledyne Continental for his patience and care in answering my questions regarding Continental engines. A big thank you to Debbie Roberts and Dawn Hartwell at Flight International (FlightGlobal) for allowing me to use their cutaway of the Skyhawk. Many thanks to Lee Nicolai and Grant Carichner for their wise counsel and guidance. A number of other people have helped in a variety of other ways (especially those who have given me permission to use specific photographs or figures) to make this work possible—I thank them for their time and effort.

About the Author

Conrad F. Newberry received his AA degree from Independence (Kansas) Junior College, BEME (Aeronautical Sequence) degree from the University of Southern California, MSME and MAEd degrees from California State University, Los Angeles, and his Doctor of Environmental Science and Engineering (D.Env.) degree from UCLA. He is or has been a registered engineer in California, Kansas, Texas, and North Carolina; and is an AAEE certified Air Pollution Control Engineer. Conrad has held aerospace industry engineering or engineering related positions supporting L-5, B-47, F-86, X-15, F-5, F-18, Athena (missile), and Space Shuttle design, development, and/or manufacture. He is Professor Emeritus at California State Polytechnic University, Pomona and at the Naval Postgraduate School, and is the recipient of the ASEE/AIAA John Leland Atwood Outstanding Educator Award, the ASEE Fred Merryfield Design Award [creative excellence in teaching engineering design], and a U.S. Navy Meritorious Civilian Service Award. He is a Fellow of both AIAA and ASEE.

References

[1] Robb, David W., "A 172 Reunion," *AOPA Pilot*, Vol. 49, No. 4, April 2006, pp. 80–84.

[2] Thompson, William D., *Cessna-Wings for the World: The Single-Engine Development Story*, Maverick Publications, Inc., Bend, Oregon, 1991, pp. 1–4, 21–44, 52–53, 70–76.

[3] Phillips, Edward H., *Wings of Cessna, Model 120 to the Citation III*, Flying Books, 1986, pp. 14–16, 18, 22, 36.

[4] Morgan, Len, *Famous Aircraft: The Douglas DC-3, Famous Aircraft Series*, Aero Publishers, Inc., Fallbrook, California, 1980.

[5] Darling, Kev, *Merlin Powered Spitfires*, WarbirdTech Series, Vol. 35, Specialty Press Publishers and Wholesalers, North Branch, Minnesota, 2002.

[6] Johnsen, Frederick A., *North American P-51 Mustang*, Vol. 5, WarbirdTech Series, Specialty Press Publishers and Wholesalers, North Branch, Minnesota, 1996.

[7] *F-51D Mustang Handbook*, Aviation Books, Flying Enterprise Publications, Dallas, Texas.

[8] Shacklady, Edward, *Messerschmitt Bf109*, Tempus Publishing, Inc., Charleston, South Carolina, 2000.

[9] Drendel, Lou, *B-17 Flying Fortress*, Walk Around Series, Walk Around No. 16, Squadron/Signal Publications, Inc., Carrollton, Texas, 1998.

[10] Bowman, Martin W., *The B-24 Liberator: 1939–1945*, Patrick Stephens, Limited, Northamptonshire, United Kingdom, 1979, 1989, p. 11.

[11] Hughes, Kris and Walter Dranem, *North American F-86 SabreJet Day Fighters*, WarbirdTech Series, Vol. 3, Specialty Press Publishers and Wholesalers, North Branch, Minnesota, 1996.

[12] Clarke, Bill, *The Cessna 172*, 2nd ed., TAB Books (division of McGraw-Hill, Inc.), Blue Summit, Pennsylvania, 1993, pp. 14–20, 38, 267–274.

[13] Hays, John D., Chad Kannady, Ben Matthaei, Sandra Reddish, Benjamin Hruska, Charles J. Lawrence, Jay M. Price, and Theresa St. Romaine, *Images of America: Wichita's Legacy of Flight*, Arcadia Publishing, Charleston, South Carolina, 2003 (The AIAA—Wichita Section with Jay M. Price), pp. 50, 61, 111.

[14] Rodengen, Jeffrey L., *The Legend of CESSNA*, Write Stuff Enterprises, Inc., Fort Lauderdale, Florida, 2007, pp. 22–40, 46–80, 85–102, 106–116, 123, 135, 232–235.

[15] Abel, Alan, Drina Welch Abel, and Paul Matt, *Cessna's Golden Age*, Wind Canyon Books, Inc., Niceville, Florida, 2001, pp. 1–17, 28–43, 69, 89, 52–55.

[16] Isely, Bliss and W. M. Richards, *Four Centuries in Kansas*, The State of Kansas, Topeka, Kansas, 1941, pp. 215, 227–228.

[17] Niles, Alfred S. and Joseph S. Newell, *Airplane Structures*, 3rd ed., Vol. 1, John Wiley & Sons, Inc., 1943, pp. 11–15.

[18] Snyder, Melvin, "History of Department of Aerospace Engineering at Wichita State University," *Aerospace Engineering Education During the First Century of Flight*, edited by Barnes McCormick, Conrad Newberry, and Eric Jumper, AIAA, Reston, Virginia, 2004, pp. 112–126.

[19] Phillips, Edward H., *Cessna: A Master's Expression*, Flying Books, Eagan, Minnesota, 1985, pp. 115.

[20] Mayborn, Mitch and Bob Pickett, *Cessna Guidebook*, Vol. 1, Flying Enterprise Publications, Dallas, Texas, 1973, pp. 3, 30, 36, 76–80, 100–101.

[21] Clark, Donald S. and Wilbur R. Varney, Physical Metallurgy for Engineers, D. Van Nostrand Company, Inc., New York, 1952, pp. 176, 388–393.

[22] Wilkinson, Paul H., *Aircraft Engines of the World 1944*, Paul H. Wilkinson, New York, 1944, pp. 70–71.

[23] http://www.mikejamesmedia.com/3d_catalog_info_t50.html, March 8, 2011.

[24] Christy, Joe, *The Complete Guide to the Single-Engine Cessnas*, 3rd ed., TAB Books, 1979, pp. 10, 12–17, 28, 42.

[25] *Operation Manual for 120 Cessna 140*, Cessna Aircraft Company, pp. 123–13, Wichita, Kansas, 1946, pp. 7–26.

[26] Cavanaugh, Jim, *Standard Catalogue of Cessna Single Engine Aircraft*, 3rd ed., (revised by Randall A. Augustinian), Jones Publishing, Inc., Iola, Wisconsin, 2006, pp. 164–237, 248–251, 312–313.

[27] "Cessna Trades Fabric for Metal," *Aviation Week*, Vol. 50, No. 21, May 23, 1949, p. 39.

[28] Bridgman, Leonard (compiler and editor), *Jane's All the World Aircraft: 1950–51*, 38th ed., McGraw-Hill Book Company, Inc., New York, 1950, pp. 211c–212c.

[29] *Cessna 170 '52, '53, '54, and '55 [C-170B] Models Owner's Manual*, Cessna Aircraft Company, pp. 213–13, Wichita, Kansas, 1988, pp. 31, 40.

[30] Bedell, Peter A., "The Skyhawk Turns 50," AOPA Pilot, Vol. 49, No. 4, April 2006, pp. 70–79.

[31] Boatman, Julie K., "A Decade of Independence," AOPA Pilot, Vol. 49, No. 4, April 2006, pp. 85–87.

[32] Smith, Allen, "Cessna Came to Town 15 Years Ago," *Independence Daily Reporter*, Vol. 130, No. 210, pp. 1 (lower half, columns 1–5), 5 (upper half, columns 1–3), dated July 3, 2011.

[33] Cook, LeRoy, "Cessna Development Through the Golden Years," *Standard Catalogue of Cessna Single Engine Aircraft*, 3rd ed., (revised by Randall A. Augustinak), Jim Cavanagh (Iola, Wisconsin, Jones Publishing, Inc.), 2006, pp. 18–23.

[34] Bell, Lawrence D., "Fighter Philosophy," *Aviation*, Vol. 40, No. 7, July 1941, pp. 74–75, 90, 98.

[35] Matthews, Birch, "COBRA!," Bell Aircraft Corporation 1943–1946, Schiffer Publishing, Ltd., Atglen, Pennsylvania, 1996, p. 76.

[36] *1956 Cessna 172 Owner's Manual*, Cessna Aircraft Company, P130A-13, Wichita, Kansas, 1956, pp. iv, 10–14, 20–21, 35–36, 46–47, 49, 51–55, 66.

[37] Wood, K. D., *Aerospace Vehicle Design, Volume I: Aircraft Design*, Johnson Publishing Company, Boulder, Colorado, pp. A11, A13, A24–A25, A141-A150, A177–A178.

[38] *172S Skyhawk Information Manual*, Cessna Aircraft Company, 172SIM, Wichita, Kansas, 2004, pp. 6–4, 6–16, 7–9 to 7–12.

[39] *Your 1960 Cessna 172 (C-172A)*, Cessna Aircraft Company, P188–13, Wichita, Kansas, 1960, pp. 1–21.

[40] *172R Skyhawk Information Manual*, Cessna Aircraft Company, 172RIM, Wichita, Kansas, 1996, pp. 1–2, 6–4, 6–11, 6–16, 7–6 to 7–8, 7–14 to 7–15, 7–29.

[41] Abbott, Ira H. and Albert E. von Doenhoff, *Theory of Wing Sections; Including a Summary of Airfoil Data*, Dover Publications, Inc., New York, 1959, pp. 113, 410, 478, 479.

[42] Taylor, John W. R. (Editor and Complier), *Jane's All The World's Aircraft: 1961–62*, 49th ed., Sampson Low, Marston & Company, Ltd., London, United Kingdom, 1961, pp. 227–231.

[43] Lowry, John T., *Performance of Light Aircraft*, AIAA, Reston, Virginia, pp. 145–183.

[44] Bridgman, Leonard (Editor and Complier) and John W. R. Taylor (Assistant Compiler), *Jane's All The World's Aircraft: 1957–58*, 45th ed., Sampson Low, Marston & Company, Ltd., London, United Kingdom, 1957, pp. 263–266.

[45] Jackson, Paul (Editor-in-Chief), Kenneth Munson (Deputy Editor), Lindsay Peacock (Assistant Editor), and John W. R. Taylor, (Editor Emeritus), *Jane's All The World's Aircraft: 1999–2000*, 88th ed., Jane's Information Group, Inc., Alexandria, Virginia, 1999, pp. 624–627.

[46] Taylor, John W. R. (Editor and Complier), *Jane's All The World's Aircraft: 1963–64*, 50th ed., Sampson Low, Marston & Company, Ltd., London, United Kingdom, 1963, pp. 189–192.

[47] Weick, Fred E., *Aircraft Propeller Design*, McGraw-Hill Book Company, Inc., New York, 1930, pp. 29–35, 114–118, 252–256.

[48] Loftin, Laurence K., *Quest for Performance: The Evolution of Modern Aircraft*, *NASA SP-468*, National Aeronautics and Space Administration, Washington, D.C., 1985, p. 145.

[49] Anderson, John D. Jr., *Introduction to Flight*, 3rd ed., McGraw-Hill, Inc., New York, 1978, pp. 306–308.

[50] Roskam, Jan, *Airplane Design: Part I., Preliminary Sizing of Airplanes*, Roskam Aviation and Engineering Corporation, Ottawa, Kansas, 1985, p. 164.

[51] "EDO Floats," EDO-AIRE Seaplane Division, Melville, New York, pp. 8–9.

[52] Nicolai, Leland M. and Grant E. Carichner, *Fundamentals of Aircraft and Airship Design, Volume I*, AIAA, Reston, Virginia, 2010, pp. 260–267.

[53] Dwinnell, James H., *Principles of Aerodynamics*, McGraw-Hill Book Company, Inc., New York, 1949, pp. 299–300.

[54] Dole, Charles E., *Flight Theory and Aerodynamics: A Practical Guide for Operational Safety*, Wiley-Interscience Publication, John Wiley & Sons, New York, 1981, 169–171, 267–268, 270–271, 274–277.

[55] Dommasch, Daniel O., Sydney S. Sherby, and Thomas F. Connolly, *Airplane Aerodynamics*, 4th ed., Pitman Publishing Corporation, New York, 1667, pp. 211–243, 294–297, 349, 371–376, 499–501.

[56] Diehl, Walter Stuart, *Engineering Aerodynamics*, rev. ed., The Ronald Press Company, Inc., New York, 1936, p. 375.

[57] Perkins, Courtland D and Robert E. Hage, *Airplane Performance Stability and Control*, John Wiley & Sons, Inc., New York, 1949, pp. 161, 246–255, 343–359, 367–368.

[58] *1956 Cessna 170 (C-170B) Owner's Manual*, Cessna Aircraft Company, P130-13, Wichita, Kansas, 1956, pp. 14, 16–21, 42, 49, 56.

[59] *1965 Cessna Model 172 (C-172F) and Skyhawk Owner's Manual*, Cessna Aircraft Company, D615-13, Wichita, Kansas, 1984, pp. 2–3, 4–5.

[60] *Pilot's Operating Handbook, Cessna Skyhawk, 1977 Model 172N*, Cessna Aircraft Company, D1082-13, Wichita, Kansas, 1977, pp. 6–12.

[61] Ells, Steven W., "Endurance Test, Circa 1958—150,000 Miles Without a Landing in a Cessna 172," *AOPA Pilot*, Vol. 51, No. 3, March, 2008, pp. 141–147.

[62] Christy, Joe, *Engines for Homebuilt Aircraft & Ultralights*, TAB Books, Inc., Blue Ridge Summit, Pennsylvania, 1983, pp. 60–63, 74–79.

[63] Wilkinson, Paul H., *Aircraft Engines of the World 1956*, Paul H. Wilkinson, New York, 1956, pp. 208–211.

[64] Borden, Norman E., Jr. and Walter J. Cake, *Fundamentals of Aircraft Piston Engines*, Hayden Book Company, Inc., New York, 1971, pp. 90, 95.

[65] Wilkinson, Paul H., *Aircraft Engines of the World 1964/65*, Paul H. Wilkinson, Washington, D. C., 1965, pp. 206, 208, 225.

[66] Taylor, John W. R. (Editor and Complier), *Jane's All The World's Aircraft: 1975–76*, 62nd ed., Franklin Watts, Inc., New York, 1976, pp. 736, 754, 769, 771.

[67] *Aftermarket Engine Price List*, Lycoming Engines, Williamsport, Pennsylvania, July 5, 2011, pp. 10–11, 14, 16.

[68] Sunderland, Luther D., "Application of Wooden and Metal Propellers," *Sport Aviation*, Experimental Aircraft Association, Vol. 22, No. 11 Oshkosh, Wisconsin, pp. 15–23.

[69] Manly, G. B., *Aircraft Powerplant Manual*, Frederick J. Drake & Company, Chicago, Illinois, 1942, pp. 513–553.

[70] Chapel, Charles Edward, Ralph D. Bent, and James L. McKinley, *Aircraft Power Plants*, rev. ed., McGraw-Hill Book Company, Inc., New York, 1955, pp. 283–314.

[71] Houghton, E. L., and A. E. Brock, *Aerodynamics for Engineering Students*, St. Martins Press, New York, 1970, pp. 105–123.

[72] *McCauley Propeller System Technology Guide*, McCauley Propeller Systems, Wichita, Kansas, p. 1.

[73] Boser, Steven (Sensenich Vice-President), personal communication (e-mail), August 24, 2011, pp. 1–4.

[74] "Chemical Composition Limits for Wrought Aluminum Alloys," *Steel*, Vol. 135, No. 6, August 9, 1954, pp. 14.

[75] "Alloy 2024 Sheet and Plate," SPD-10-036, Alcoa Mill Products, Inc., Bettendorf, Iowa, March 26, 2011.

[76] Levy, David W., personal communication (e-mails) RE: C-172 Information, February 14; RE: More C-172 questions, March 9; FW: Skyhawk, March 22; RE: Skyhawk info, May 3; RE: C-172 info, May 18; RE: C-172 info, May 20; RE: a few more questions, July 18, 2011.

[77] Bhatia, Manav and William H. Mason, "Influence of Strut on Wing Structural Behavior of a Strut-Braced Wing," unpublished manuscript, June 21, 2011, pp. 1–21.

[78] Bhatia, Manav and William H. Mason, "Why a Strut-Braced Wing Weighs Less than a Cantilever Wing," unpublished manuscript, July 5, 2011, pp. 1–4.

[79] Bartholomew, Niles Clark, *Aircraft Inspection Methods*, Pitman Publishing Corporation, New York, 1940, pp. 20–25.

[80] "Title 14—Aeronautics and Space, Parts 1 to 59," *Code of Federal Regulations*, Office of the Federal Register, National Archives and Records Administration, Revised as of January 1, 1986; Part 23, pp. 100–231.

[81] Thurston, David B., *Design for Flying*, McGraw-Hill Book Company, New York, 1978, pp. 12, 60–63, 93.

[82] Corning, Gerald, *Supersonic and Subsonic, CTOL and VTOL, Airplane Design*, Gerald Corning, College Park, Maryland, 1976, p. 9:42.

[83] Van Deventer, C. N., *An Introduction to General Aeronautics*, American Technical Society, 1965, pp. 85, 89–100, 144–145.

[84] Etkin, Bernard, *Dynamics of Flight—Stability and Control*, 2nd ed., John Wiley & Sons, New York, 1982, pp. 3–9, 70–71, 75–81.

[85] Taylor, John W. R. (Editor and Complier), *Jane's All The World's Aircraft: 1962–63*, 49th ed., Sampson Low, Marston & Company, Ltd., London, United Kingdom, 1962, pp. 193–195.

[86] Taylor, John W.R. (Editor and Complier) and Kenneth Munson (Assistant Editor), *Jane's All The World's Aircraft: 1977–78*, 64th ed., Macdonald and Jane's Publishers, Ltd., London, England, United Kingdom, 1964, pp. 264–267.

[87] Pazmany, Ladislao, *Landing Gear Design for Light Aircraft*, Vol. 1, Pazmany Aircraft Corporation, San Diego, California, pp. iv.

[88] Augustine, Norman R., *Augustine's Laws*, AIAA, New York, 1983, p. 55.

[89] Jackson, Paul (Editor-in-Chief), Kenneth Munson (Deputy Editor), Lindsay Peacock (Assistant Editor), and John W. R. Taylor, (Editor Emeritus), *Jane's All The World's Aircraft: 1997–98*, 88th ed., Jane's Information Group, Inc., Alexandria, Virginia, 1997, pp. 609–612.

[90] *Model 172 1977–1986 Service Manual*, Cessna Aircraft Company, Wichita, Kansas, 1986, pp. 2–2, 3–14, 5–1, 5–22, 5–30.

[91] "Title 14—Aeronautics and Space, Parts 1 to 59," *Code of Federal Regulations*, Office of the Federal Register, National Archives and Records Administration, Revised as of January 1, 2012; Part 23, §61.1(b)(3).

[92] Thurston, David B., *Design for Safety*, Macmillan Publishing Company, New York, 1987, pp. 70–75, 89, 106, 110–111.

[93] Taylor, John W. R. (Editor and Complier), *Jane's All The World's Aircraft: 1960–61*, 48th ed., Sampson Low, Marston & Company, Ltd., London, United Kingdom, 1960, pp. 277–280.

[94] Klemin, Alexander, "Aerodynamics of the Airfoil," *Elements of Technical Aerodynamics*, National Aeronautics Council, New York, 1942, pp. 19–35.

[95] "Design Crash Protection Into Lightplanes," *Aviation Week*, Vol. 50, No. 18, May 2, 1949, pp. 19–22.

[96] Weick, Fred E., "Powerplant Failures—Causes and Cures," *Aviation Week*, Vol. 50, No. 10, March 7, 1949, pp. 21–25.

[97] Landsberg, Bruce, "Cessna 172 Safety Review," *AOPA Pilot*, Vol. 38, No. 12, December, 1995, pp. 117–118.

[98] Taylor, John W. R. (Editor and Complier), *Jane's All The World's Aircraft: 1965–66*, 52nd ed., Sampson Low, Marston & Company, Ltd., London, United Kingdom, 1965, pp. 204–207.

[99] Shiel, Walt, Jan Forsgren, and Mike Little, T-41 Mescalero: *The Military Cessna 172*, Slipdown Mountain Publications LLC, Lake Linden, Michigan, 2006, pp. 135.

[100] Taylor, John W. R. (Editor and Complier), *Jane's All The World's Aircraft: 1967–68*, 54th ed., Sampson Low, Marston & Company, Ltd., London, United Kingdom, 1967, pp. 232–235.

[101] Taylor, John W. R. (Editor and Complier), *Jane's All The World's Aircraft: 1969–70*, 50th ed., McGraw-Hill Book Company, New York, 1969, pp. 293–298.

[102] Garvey, William, "An Electrifying Cessna," *Aviation Week & Space Technology*, Vol. 172, No. 44, December 6, 2010, p. 47.

[103] *1960 Cessna Model 172 (C-172A) Seaplane Supplement to the Owner's Manual*, Cessna Aircraft Company, P188A-13, Wichita, Kansas, 1960, pp. 1–5.

[104] Bridgman, Leonard (Editor and Compiler), *Jane's All The World's Aircraft: 1948*, 36th ed., The Macmillan Company, New York, 1948, pp. 232c–233c.

[105] Bridgman, Leonard (Editor and Compiler), *Jane's All The World's Aircraft: 1949–50*, 37th ed., The McGraw-Hill Book Company, Inc., New York, 1949, pp. 203c–204c.

[106] Bridgman, Leonard (Editor and Compiler), *Jane's All The World's Aircraft: 1952–53*, 40th ed., Samson Low, Marston & Company, Ltd., London, United Kingdom, 1952, pp. 186–187.

[107] Roskam, Jan, "Airplane Design: Part III, Layout Design of Cockpit, Fuselage, Wing, and Empennage: Cutaways and Inboard Profiles," Roskam Aviation and Engineering Corporation, Ottawa, Kansas, 1986, p. 296.

[108] "U. S. Reciprocating Engines," *Aviation Week*, Vol. 50, No. 9, February 28, 1949, p. 30.

T-46A and Fairchild Republic Company

Leland Nicolai

- The NGT Proposal
- Performance Comparisons Between the T-46A and T-37B
- The Republic Aviation Company
- The Roll Out
- The End of Thunder

The two T-46A flight test aircraft are shown flying formation during flight test in 1986. The flight test went well and impressed those familiar with the T-37 with its performance and maintainability features. But it was clear to everyone that the T-46A was not going anywhere.

If you want to work for me, don't surprise me.
And when you tell me . . . tell me everything.
General Colin Powell

CS8.1 Introduction

I n 1981 the Air Force issued a request for proposal (RFP) for the Next Generation Trainer (NGT). The NGT was to replace the Cessna T-37B used for undergraduate pilot training in the Air Force. The Fairchild Republic Company on Long Island, New York won the competition for the Development, Test, and Evaluation (DT&E) of two flight test aircraft and one static test article. Their bid was $104M. Garrett Turbine Engines in Arizona was given a $115M contract to develop a high bypass turbofan engine (F 109) for the trainer.

Fairchild Republic had done their homework and turned in an outstanding proposal. The company had designed and built a two-thirds scale flying NGT "concept demonstrator" and performed flight testing on its own prior to submitting the proposal, allowing key aerodynamic and handling quality data to be collected to substantiate detailed numerical analysis and simulation.

The T-46A (shown in Figs. CS8.1 and CS8.2) was a very fine design sporting the trademark Republic twin vertical tail like the A-10. The twin tail empennage promised rock solid spin recovery. The cockpit featured side-by-side seating and standard round analog gauges with the control layout designed for optimum ease of operation for the student in terms of

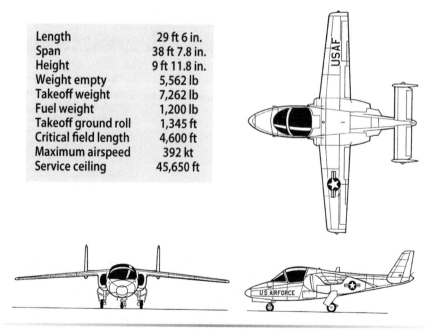

Length	29 ft 6 in.
Span	38 ft 7.8 in.
Height	9 ft 11.8 in.
Weight empty	5,562 lb
Takeoff weight	7,262 lb
Fuel weight	1,200 lb
Takeoff ground roll	1,345 ft
Critical field length	4,600 ft
Maximum airspeed	392 kt
Service ceiling	45,650 ft

Figure CS8.1 T-46A performance summary.

Figure CS8.2 The Fairchild Republic/USAF T-46A.

visibility and reach. The aircraft also had many built-in maintainability features that enhanced ground personnel access to different areas of the aircraft for servicing. The front part of the nose could be unlatched and hinged forward, allowing access to avionics equipment located inside. The engine access doors were at shoulder level allowing a change of the F 109-GA-100 turbofan engine in only thirty min. The T-46A with the fuel efficient F 109 engine was projected to use only 1/3 of the fuel of the aircraft it was replacing. The F 109 engine had an SL TSFC of 0.41 that compared to the J 69-T-25 engine in the T-37B, which had a SL TSFC of 1.14. It was projected that the T-46A would pay for its development in 10 years with the fuel it would save.

Another major improvement of the Republic trainer was that the aircraft would be pressurized for operation at 42,500 ft permitting the aircraft to fly above most of the bad weather. This had been a significant shortcoming of the unpressurized T-37B which was limited to flight altitudes below 25,000 ft. The T-46A also had the zero-zero capability ACES II ejection seats.

A performance comparison of the T-46A and the T-37B is shown on Table CS8.1.

The T-46A in flight test made the cover of the 25 November 1985 Aviation Week. The future looked bright for the T-46A.

Table CS8.1 Performance Comparison of the T-46A and T-37B

Item	RFP	T-46A	T-37B
Critical field length (ft)	5000	4600	7500
Engine-out take-off climb gradient (%)	3.5	3.62	Negative
Normal landing distance (ft)	5000	5110	7500
Landing approach speed (KIAS)	90–100	100	100
Engine-out go-around climb gradient (%)	2.0	2.71	Negative
Cruise speed at 25,000 ft (KTAS)	300	370	326
Rate of climb at 25,000 ft (FPM)	2000	2263	900
Sustained load factor at 25,000 ft (g)	2.5	2.63	1.7
Sustained load factor at 15,000 ft (g)	–	3.5	2.5
Cruise altitude (ft)	35,000	42,000	25,000
Ferry range (nm)	–	1160	576
Service ceiling (ft)	–	45,650	25,000
Maximum speed at 35,000 ft (KTAS)	–	392	NA
Takeoff gross weight (lb)	–	7262	6625
Empty weight (lb)	–	5562	4056
Fuel (lb)	–	1197	2010

CS8.2 The Fairchild Republic Company

The Republic Aviation Company was born in the late 1930s as Seversky Aircraft located in Farmingdale, NY on Long Island. It became Republic Aviation shortly before WWII and produced more than 26,000 military aircraft that fought in all major wars of the 20th century, such as the P-47 Thunderbolt (WWII), F-84 Thunderjet (Korean war), F-105 Thunderchief (Vietnam war), and the A-10 Thunderbolt II. Republic was always a "one trick pony" operation having only one major program at a time. This made for great peaks and valleys in employment. In 1964 Republic delivered the last of 833 F-105s and the future looked bleak for the company. Fairchild Industries took over Republic in 1965 in order for the company to survive.

In 1970 the Air force issued an RFP for a low cost (less than $3M unit cost), easy to maintain, close-air-support aircraft. The new aircraft was to be designed around a 30-mm GAU-8 gatling gun. The GAU-8 was as large as a Volkswagen and was perfect for close-air-support. Northrop and Fairchild Republic won contracts to build two prototypes each of the YA-9 and YA-10 respectively. The Air Force selected Fairchild Republic to build 715 A-10As in 1973 and the future looked good for the company. In 1981 the production of the A-10 ended and the NGT program was considered a "must win" for the company.

As mentioned earlier Fairchild Republic did their homework and prepared well for the NGT competition. They rolled up their cost to build two flight test aircraft and a static test article and the cost was $132M. Competitive intelligence revealed that North American Rockwell was going to bid $120M. So Fairchild Republic dropped their price to $104M and won the fixed price DT&E contract in July 1982. The T-46A team was in trouble right from the start as they tried to do $132M worth of work for $104M, and because it was a fixed price contract any cost over $104M was borne by the company.

In order to save money they got their suppliers to "bet on the come of a projected 650 unit production run" and invest their own money during DT&E. As a result suppliers had knowingly signed on to the program aware of the fact that they would be providing free parts for the first two aircraft along with associated testing and documentation that was also performed at their cost. This approach was flawed because without financial incentives, many vendors gave the parts low priority resulting in parts that were late and of low quality [1].

The T-46A team tried to cut corners to save money and ran into trouble resulting in rework and more time and money. The late parts and rework caused the schedule to start slipping. When the Air Force and Fairchild Industries management asked how things were going the program team would withhold information, provide half truths and try to conceal the problems.

One of the milestones in the contract was a roll-out event. At roll-out the first aircraft was to be ready for ground tests and system checkouts with all the equipment items installed. The roll-out was scheduled for February 1985 but at that date the aircraft needed another eight months to be ready. The T-46A team chose not to inform the Air Force of the situation and proceeded with the roll-out.

The roll-out was a splendid affair with smoke, dancing lights, live music and an invitee list that included all the bigwigs from Fairchild and the Air Force, along with members of the New York State congressional delegation. The center of attention was test article #1 and from fifty ft the aircraft looked great as shown on Fig. CS8.3, but as you got up close it was apparent that all the equipment items were not installed and parts of the airplane were made of plastic, cardboard, and wood.

The roll-out embarrassed Fairchild Industries and the Air Force. Slowly the full disclosure of the behind schedule and over budget status became known, which infuriated the Air Force.

Retribution came quickly. The Air Force advised the Secretary of Defense about the sorry state of the T-46A program and the deliberate withholding of important information by the program personnel. The Secretary of Defense ordered a contractor operations review (COR) for June 1985. The Air Force COR team was at Farmingdale from June 4 through

Figure CS8.3   T-46A test article #1 at the Feb. 1985 roll-out.

June 13 and wrote up everything they saw wrong from the level of company management down to safety violations in the cafeteria. Every single item that could be written up was, in apparent retaliation for the botched roll-out. The result was the Air Force reducing the monthly progress payments from $8M to $4M and deferring the approval of the first production lot of T-46A aircraft. The reduction of the progress payments further aggravated the financial situation of the company. In addition the company, already strapped for cash, had to fix the problems identified by the COR.

The T-46A team continued to prepare the aircraft for first flight but by this time they had blown past the $104M contract award and were spending their own money.

In September 1985 things went from bad to worse when the Secretary of Defense recommended cancelling the T-46A program as part of a budget trimming exercise [2]. Shortly thereafter the management of Fairchild Industries announced that it was no longer interested in building airplanes and was looking for a buyer of the Farmingdale plant.

The first DT&E aircraft flew on 15 October 1985 (see Fig. CS8.4) and the second about nine months later. Both aircraft conducted a normal flight test program at Edwards AFB, Ca (500 hr) and impressed those familiar with the T-37B with its performance and maintainability features [3, 4]. But it was clear to everyone that the T-46A was not going anywhere. The

Figure CS8.4 First flight of the T-46A on October 15, 1985.

real winner of the flight test was the Garrett F 109-GA-100 turbofan engine which performed great.

The end finally came on Friday the 13th in March 1987 when the Air Force formally terminated the T-46A program [5]. The two DT&E aircraft are still at Edwards on display.

The sustaining engineering for the A-10A was transferred to Grumman in the fall of 1987. Fairchild Industries auctioned off the plant equipment and old tooling (most of it being sold for scrap) and closed the plant. Today the plant area is a movie complex and shopping center with very little indication that a major piece of U.S. military aviation history ever existed there.

CS8.3 The Message Takeaway

The decision by the T-46A program team to "low ball" the proposal price and to enter into a supplier "buy-in" subcontract could be viewed as poor program judgement and reconciled by the Air Force and Fairchild Industries management. But the deliberate with holding of information that led to the embarrassing roll-out was inexcusable and a violation of ethical behavior and mutual trust [6].

Sometime in the early 1950s Clarence "Kelly" Johnson wrote down 14 Operating Rules for the Lockheed Skunk Works. These 14 Rules were written around the Skunk Works motto of "Be Quick, Be Quiet, and Be On Time." Sixty years later these rules are still relevant. These 14 Rules are shown on color plate 1. A careful review of Kelly's 14 Rules for good project management reveals that the T-46A program team violated the following two rules:

6. There must be a monthly cost review covering not only what has been spent but also projected costs to the conclusion of the program. Don't have the books 90 days late and don't surprise the customer with sudden overruns.

12. There must be mutual trust between the military project organization and the contractor with very close cooperation and liason on a day-to-day basis. This cuts down misunderstanding and correspondence to an absolute minimum.

References

[1] Neubeck, Ken, "End of the Thunder," *Wings and Air Power*, Vol. 14, pp. 14–29.
[2] Keller, Bill, "Pentagon Backing Proposal To Drop A Fairchild Plane," *New York Times*, p. 1, Sept. 1985.
[3] "USAF/Fairchild T-46A Begins 22-Month Flight Test Program," *Aviation Week and Space Technology*, p. 2, Oct. 1985.
[4] "T-46A Trainer Testing Accelerates Despite USAF Bid To Cancel Program," *Aviation Week and Space Technology*, pp. 66–69, Mar. 1986.
[5] Bernstein, James, "AF Confirms Fairchild's Fears," *Newsday*, Mar. 1986.
[6] Fink, Donald E., "T-46A—Dead or Alive" (Editorial), *Aviation Week and Space Technology*, p. 13, Aug. 1986.

Case Study 9 — Foot-Launched Glider Design and Performance

Paul Dees

Ryan Vo

- History of Foot-Launched Gliding
- Technical Summary of the Key Design Aspects of the Flex-Wing
- Analyzing Performance of Several Wings
- Review of Rigid Wing Hang Glider Design
- Overview of Paraglider Design and Performance

Hang gliding is an aerial sport that comes the closest to humankind mimicking the free flight of birds. This sport experienced a rebirth in popularity during the 1970s with the advent of new, simple wings developed by pioneers like John Dickensen based on a wing concept developed from NASA engineer Francis Rogallo.

The sensation of flying through the air is so delightful that the operators immediately desire to make another glide.
Octave Chanute, 1897

Nomenclature

α	= angle of attack, degrees
C_D	= drag coefficient
C_{D_i}	= inviscid induced drag coefficient due to wing planform and twist
$C_{D_{misc}}$	= parasitic drag coefficient due to miscellaneous items
$C_{D_{profile}}$	= profile drag coefficient
C_L	= lift coefficient
$C_{L_{max}}$	= maximum lift coefficient
D	= drag, pounds
γ	= flight path angle, degrees
L	= lift, pounds
L/D	= lift/drag
ρ	= air density, slugs/ft^3
q	= dynamic pressure, pounds/ft^2
S_{wing}	= wing reference area, ft^2
θ	= pitch angle, degrees
V	= velocity, miles per hour
V_g	= velocity over the ground, miles per hour
VG	= variable geometry
V_s	= sink rate velocity, miles per hour or ft/min
W	= weight, pounds

CS9.1 Introduction

P ilots who fly hang gliders and paragliders believe these aircraft come as close as is humanly possible to mimicking the flight of birds. What follows is a brief history of foot-launched gliding, followed by a technical summary of the key design aspects of the most popular style of hang glider, the "flex-wing," which evolved from the original work of Francis Rogallo and John Dickensen, during the 1950s and 1960s. Next, the performance of several wings will be analyzed with some comparisons to data provided by the manufacturers of these gliders. Rigid wing hang glider and paraglider design and performance will also be discussed.

CS9.2 The Rebirth of Hang Gliding

The first heavier-than-air gliding flight likely took place in 1849 when Sir George Cayley launched his coachman across a valley in England. During the late 1800s, the first hang gliders were built and flown by famous pioneers such as the German engineer Otto Lilienthal, Percy Pilcher of Scotland, and Octave Chanute and John Montgomery of the United States. Three of these four died piloting their experimental craft, evidence of the dangers faced by the pioneers of flight. Octave Chanute's biplane glider was the first successful human-carrying glider that set the pattern for biplane aircraft structure for decades to follow. Shortly thereafter, Orville and Wilbur Wright experimented with gliders that improved upon Chanute's design in 1900–1902. The Wright brothers perfected their control system and basic turns with their 1902 glider before applying power to invent the airplane in 1903. The Wrights also experienced the first well-documented, true soaring flight in October 1911 at Kitty Hawk, with a flight that lasted 9 min, 45 sec. References [1–10] document many of these pioneering efforts.

After powered flight became a reality, gliding activity became more of interest for recreation. After World War I, Germany was prohibited by the armistice from flying powered aircraft, and gliding thrived as a sport. The first competition was held in Rhön in 1920 and pioneer Willy Pelzner won in a biplane hang glider. The following years saw the advent of gliders that eventually evolved into the efficient sailplanes of today. Figure CS9.1 illustrates the chronology of these early gliding aircraft.

Hang gliding experienced a rebirth in the 1960s and 1970s when John Dickensen of Australia developed a practical pilot support and weight shift control method coupled with the wing design invented by Francis and Gertrude Rogallo of the United States. Barry Palmer flew a foot-launched Rogallo wing prior to John Dickensen, but it was not practical enough to be widely imitated. John Dickensen's wings were used by Australians Bill Moyes and Bill Bennett to promote hang gliding across the world, and it really caught on in the early 1970s. By 1974 there were about forty manufacturers of "Standard Rogallo" hang gliders in the United States. One could buy a ready-to-fly hang glider for about $400, but its design deficiencies combined with with little or no instruction on proper flying technique led to many fatalities. Eventually the design technology matured and safety improved greatly with the advent of helmet usage, backup emergency parachutes, and industry standards developed for standardized full-scale strength and pitch stability testing, standardized instruction, and pilot ratings. The Hang Glider Manufacturers Association (HGMA) oversees the full-scale testing of hang gliders in the United States and the Deutscher Hangegleiter Verband, which is the German Hang Gliding and Paragliding Organization (DHV), does this in Europe [11, 12]. The sport has greatly improved safety its record and been successfully self-regulated in the United States for nearly three decades by the United States Hang Gliding and Paragliding Association [13].

The gliders evolved into two basic types of aircraft: The more sophisticated, advanced Rogallo/Dickensen wings that are now called flex-wings; and gliders with rigid structure more like private aircraft that are known as rigid wings. In the 1980s and 1990s paragliders greatly increased in popularity and their design technology continues to evolve rapidly. Figure CS9.2

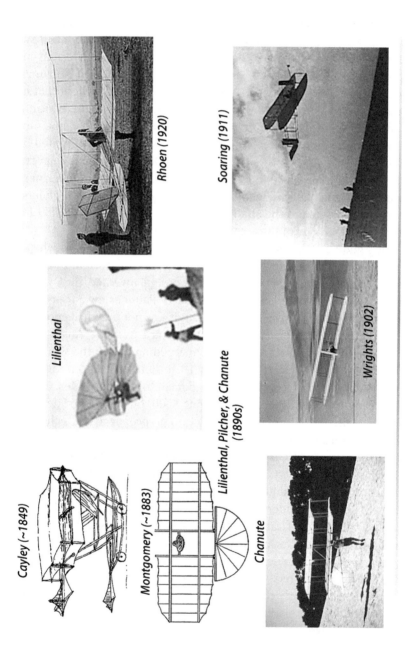

Figure CS9.1   Early gliding aircraft through the 1920s.

Rigid Wings

Paragliders

'Standard' Rogallo

Flex Wing

Francis Rogallo

Barry Palmer

John Dickensen

Figure CS9.2 Hang gliders and paragliders since their rebirth in popularity.

illustrates the chronology of foot-launched gliders since their rebirth of popularity.

CS9.3 Flex-Wing Hang Glider Design Features and Geometry

The typical flex-wing hang glider has structure and aerodynamic design features that differ significantly from typical lightweight private or ultra-light aircraft. Figure CS9.3 shows some of the design nomenclature.

Key design features include:

- A tailless design that has adequate pitch stability, enabled by a combination of the low pilot position, wing sweep, and washout twist.
- Pitch control is by pilot weight shift fore and aft.
- Roll/yaw control is by pilot weight shift side to side, causing passive differential sail twist.
- Lift loads are carried by the sail leading edge (LE), side wires, keel, and crossbar.
- Battens slide in pockets in the sail, held in tension at the trailing edge using special hardware or ties.
- A tight sail held in place by LE tube bending stiffness is used to control spanwise wing twist.
- Passive load relief occurs at high load factors by sail twisting and the outboard LE bending aft. There are complex aerodynamic and structural interactions that are known by active pilots and the few glider designers that work for the glider manufacturers but it would be difficult to get quantifiable measurements of them.

The pilot takes off by donning the helmet and harness, clipping it into the glider and double checking they are safely hooked in, then walking up

Figure CS9.3   Flex-wing hang glider parts. [Courtesy of Kay Dees]

to the launch area. After confirming conditions are safe, the pilot yells, "Clear!" to notify all in the area of the launch. The pilot lifts the glider by the down tubes and begins an aggressive run down hill into the wind, keeping the wings level, maintaining an optimum angle of attack, and transitioning into safe flight. Once away from the hill, the pilot retracts their legs into the prone harness and zips it closed. Typical harnesses are very comfortable and enclose the pilot's body, except for the head and arms, and are warm at higher altitudes, accommodating multi-hour journeys. Flights often involve circling in thermal lift, followed by straight glides.

Towing aloft using a variety of techniques is commonplace in flat terrain lacking hill or mountain launches. The current world distance records originated from tow flights.

An emergency hand-deployed parachute is carried in the harness and if used, the pilot stays attached to the glider. Safe deployments have occurred very close to the ground with insufficient time to bail out, so the parachute enables the possibility of a life-saving descent of both the pilot and glider wreckage as a unit.

Landings approaches are typically accomplished using either an aircraft or figure eight approach to a designated landing field. The pilot unzips the harness and extends their legs to maintain safe airspeed and retain control with their hands still on the base tube. They transition their hands to the down tubes and descend to a few feet above the ground, gradually increasing angle of attack as airspeed bleeds off. The pilot then deliberately stalls the glider through an aggressive flare maneuver by abruptly pushing out and pitching up the nose of the glider to arrest forward motion. Most skilled pilots can do this and land on their feet within a few feet of a designated target. This landing flare creates additional lift due to the pitch rate and enables landings at nearly zero ground speed if done properly.

Flex-wing hang gliders make up the vast majority of gliders being flown today due to their lighter weight, lower cost, and more robust design than rigid wings. There are only a handful of hang glider manufacturers worldwide and they include Wills Wing, Moyes, Icaro, Aeros, Airborne, and Northwing [14–19]. Wills Wing gliders will be discussed and their product line is typical of the gliders available from the other manufacturers as well.

Flex-wings vary from easy to fly and inexpensive trainers (Fig. CS9.4), to high performance wings capable of very good glide capability and cross country flights of hundreds of miles (Fig. CS9.5). Flex-wings are built in a variety of wing area sizes to accommodate pilots of various body weights. The Wills Wing Falcon 3 trainer of Fig. CS9.4 is a "single-surfaced" wing. The Falcon 3 has a convenient setup that includes an exposed crossbar that creates more drag, but also enables the glider to be very light weight and have superb handling qualities. Many advanced pilots enjoy Falcons due to these qualities despite the glider being designed originally as a trainer.

Figure CS9.4 Falcon 3 training glider. [Darren Darsey]

Figure CS9.5 Talon 2 advanced flex-wing. [Oz Report]

The Talon 2 (also called T2) of Fig. CS9.5 features a fully enclosed crossbar and has a tighter wing of smaller area, requiring an advanced pilot rating to fly it and fully enjoy its capabilities. The Talon 2 eliminates the upper rigging present on the other gliders to reduce drag. This results in a weight penalty in order for the structure to still meet strength requirements at negative load factors. The Talon weighs and costs more than the Falcon. The advanced gliders have a variable geometry (VG) feature where, with the pull of a line, the pilot can tighten crossbar tension, reducing sail twist, and effectively trimming at a higher airspeed with lower drag.

Figure CS9.6 shows wing planforms and basic data for the entire spectrum of gliders available from Wills Wing as of 2011 and also includes intermediate gliders such as the Sport 2 and U2. These two wings offer the superb handling like the Falcon and almost reach the high speed glide performance of the Talon 2, but with less glider weight. The U2 has a higher wing loading, higher aspect ratio, and higher performance than the Sport 2. Both have the variable geometry feature of the Talon 2 for improved *L/D* and reduced trim forces at high speed. In Fig. CS9.6 a single wing size of glider of each model is shown, typically the closest size appropriate for a pilot weighing 160 to 170 lb.

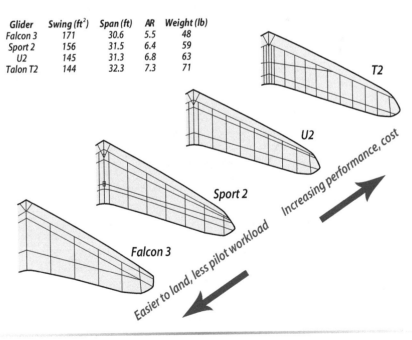

Glider	Swing (ft^2)	Span (ft)	AR	Weight (lb)
Falcon 3	171	30.6	5.5	48
Sport 2	156	31.5	6.4	59
U2	145	31.3	6.8	63
Talon T2	144	32.3	7.3	71

Figure CS9.6 Wills Wing hang glider product line.

CS9.4 Equations of Straight Gliding Flight

The equations for gliding flight are as follows with the geometry shown in Fig. CS9.7.

α = Angle of attack, (+) as shown

θ = Pitch angle, (−) as shown, $\theta = \alpha + \gamma$

γ = Flight path angle, (−) as shown

$$L - W \cos\gamma = 0$$

$$D - W \sin\gamma = 0$$

$$\tan\gamma = D/L = V_S/V_G = C_D/C_L$$

$$C_L = L/q\,S_{ref} \approx W/q\,S_{ref} \text{ (for small } \gamma)$$

$$C_D = D/q\,S_{ref}$$

$$q = \tfrac{1}{2}\rho V^2$$

$$V = \sqrt{2L/(\rho S C_L)} = \sqrt{2W\cos\gamma/(\rho S C_L)} \approx \sqrt{2W/(\rho S C_L)}$$

$$V = \sqrt{V_S^2 + V_G^2}$$

$$V_{S_{ft/min}} = 88\,V_{MPH}\sin\gamma$$

$$L/D = \left(88/V_{S_{ft/min}}\right)\sqrt{V_{MPH}^2 - \left(V_{S_{ft/min}}/88\right)^2}$$

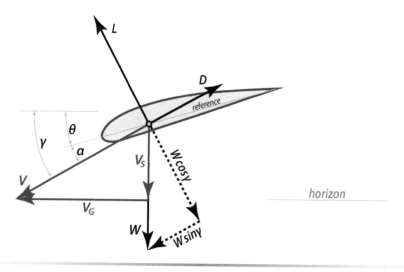

Figure CS9.7 The geometry of gliding flight.

CS9.5 **Drag and Performance of Four Flex-wing Hang Gliders**

Wills Wing publishes performance results for their wings in the public domain, providing an opportunity to reverse engineer the individual drag components and attempt to match their data with a family of total drag polars, each being relative to each other. The generic glide and sink rate polars [20] are shown in Fig. CS9.8 and are for pilot weights 130% greater than the minimum weight rating for each glider. They assume the pilot is flying with excellent form, with wings level using proper technique to control yaw excursions. These polars are provided in a generic sense so pilots can calculate their glider performance using the specific weights and sizes of a given model of glider. Data were read from the curves of Fig. CS9.8 and by assuming a given glider size and pilot weight, the values of C_L and C_D could be calculated to generate plots of L/D versus C_L (to be shown later in the study).

The drag polars of all four glider models were estimated using classical aircraft design drag methods where the drag was broken down into three fundamental parts: Parasitic drag ($C_{D_{misc}}$), induced drag due to lift

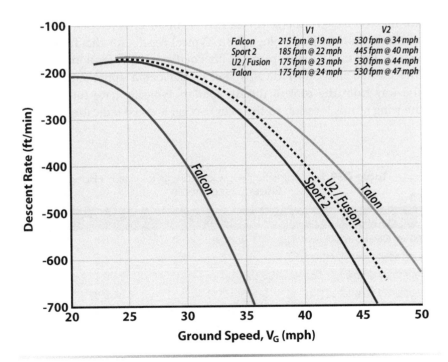

Figure CS9.8 Glide performance of Wills Wing gliders. [20]

(or $C_{D_{induced}}$), and profile drag ($C_{D_{profile}}$). These three parts were then added together in a build up of total drag. Lift curves were not required to estimate glide performance but $C_{L_{max}}$ levels were needed and were estimated assuming no dynamic $C_{L_{max}}$ benefit due to landing flare pitch rate. Actual landing speeds at flare would suggest an additional delta $C_{L_{max}}$ due to pitch rate on the order of 0.5 to 0.8.

$C_{D_{misc}}$ is the miscellaneous parasitic drag originating from items such as the exposed pilot, cables, control bar, wheels, etc. Estimates were done using sources such as Hoerner [21]. It is typically expressed as the product of $C_{D_{misc}}$ and swing which is equivalent to the product of the drag coefficient based on frontal area and the frontal area itself. The pilot plus harness drag was derived from Kilkenny [22] and adjusted using engineering judgment to more contemporary values. Tables CS9.1a and b show the assumed values used for this analysis.

C_{D_i} or $C_{D_{induced}}$ is the induced drag due to lift creation and driven by the wing planform and twist (washout). Hang gliders have large amounts of twist, with intermediate gliders having nearly 20 deg of twist change from the root to the tip airfoil sections. This is due to the structural arrangement. Flex-wings have no rear spar and the only way to minimize sail twist is through a stiff sail and LE, which adds weight and stiffens handling qualities. With VG activated the twist can be reduced to approximately half that amount on the Talon 2. C_{D_i} was estimated using a public domain vortex-lattice method called Athena Vortex Lattice, or AVL, developed at Massachusetts Institute of Technology [23]. Typical results for the Talon 2 are shown in Fig. CS9.9. The spanwise lift coefficient distribution is shown by the dashed line and indicates stall would begin approximately 25% of the way from the root to the tip. Gliders typically have tufts of yarn in the wing at this location so the pilot can get a visual indication in flight

Table CS9.1a Parasitic Drag Coefficients for Various Hang
Glider Components

C_D assumptions, based on frontal area	
Exposed round crossbar near wing lower surface	0.8
Round downtubes, basetube (F3)	1.0
Streamlined downtubes, basetube (S2, U2)	0.13
Streamlined downtubes (T2)	0.09
Wheels	1.0
Kingpost	0.3
Exposed cables	0.7–1.0

Table CS9.1b Miscellaneous Parasitic $C_D \times S_{frontal}$ for Various Hang Gliders

Bottoms up buildup of $C_D \times S_{frontal}$ of misc drag items, ft^2				
Components	F3	S2	U2	T2
Pilot and harness	1.70	1.70	1.50	1.50
Exposed crossbar	2.05	0	0	0
Downtubes	0.97	0.11	0.11	0.07
Basetube	0.42	0.04	0.04	0.04
Wheels	0.04	0.04	0.04	0
Corner fittings	0.06	0.02	0.02	0.02
Kingpost	0.09	0.09	0.09	0
Lower side wires	0.17	0.17	0.17	0.17
Fore/aft lower wires	0.18	0.18	0.18	0.18
Top landing wires	0.15	0.15	0.15	0
Top fore/aft wires	0.06	0.06	0.06	0
Reflex wires	0.14	0.07	0.07	0
Keel skin friction	0.01	0.01	0.01	0.01
Total	6.04	2.64	2.44	1.99
Round-off total	6.0	2.6	2.4	2.0

when flow separation first begins. The spanload distribution is shown in blue and is close to elliptic, showing the level of refinement achieved through decades of incremental development.

Estimates of twist were made based on analysis of photographs adjusted after consultation with Steve Pearson, the glider designer at Wills Wing. Measurement of in-flight glider geometry would make an interesting study as from personal observation, sail twist and camber changes with changes in pilot weight, load factor, and airspeed. By intent, flex-wing hang gliders have selective geometric flexibility of their structure. The subtle nuances of how gliders have stiffness to take flight loads and perform well at high speed, yet have flexibility for pleasing handling qualities, are only understood by a few successful hang glider designers. Rigorous engineering studies have not been conducted to document or further optimize these factors.

For this study, AVL was used to analyze the gliders with VG on and off to model two different twist distributions. A transition from VG off (e.g., sail with more twist) at low speed to VG full on (e.g., sail with less twist) was assumed based on the author's flight experience. In the case of the Sport 2, the AVL predicted C_{D_i} levels seemed excessively high and were adjusted

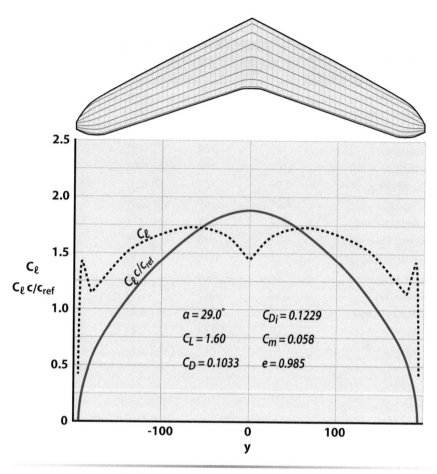

Figure CS9.9   T2 induced drag estimate using AVL.

downward to family with the other three glider models (Fig. CS9.10). Curves were fit through these data so values could be calculated at any C_L. Figure CS9.11 shows the difference between estimated C_{D_i} via AVL and the ideal induced drag.

$C_{D_{profile}}$ included the airfoil profile drag, including the effects of boundary layer growth and flow separation across various parts of the wing at either high or low C_Ls. It was estimated by first taking total drag polars derived from the Wills Wing Web site shown in Fig. CS9.8, and then subtracting the parasitic and induced drags from the total drag. This yielded a residual drag coefficient versus C_L that was assumed to include the profile drag. The raw data were in need of adjustment to assure that they formed a family of curves that represented a progression across the product line of Wills Wing gliders. Figure CS9.11 shows these data after adjustment to

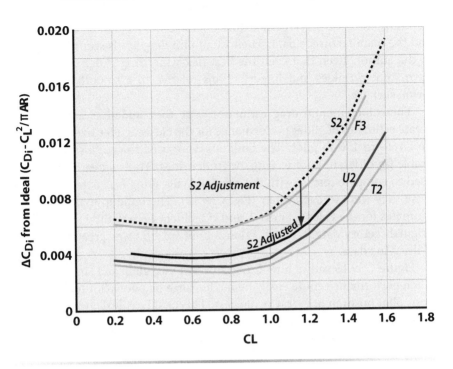

Figure CS9.10 Induced drag estimates.

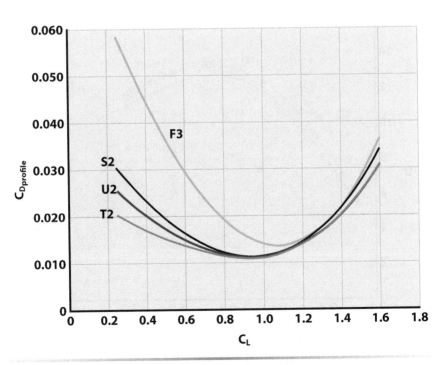

Figure CS9.11 Adjusted profile drag estimates.

minimize the glide performance differences between the built-up estimates and the manufacturer's data. Their minimum drag levels seemed very low to the author, given that even the best flex-wing hang glider sails tend to have some wrinkles and imperfections relative to a smooth theoretical airfoil surface.

These incremental drag estimates were then added as a buildup to create estimates of glider performance for the closest glider size that would fit a 160- to 170-lb pilot. The buildups were used to create spreadsheets usable for different pilot weights or air densities. These spreadsheets were used to compare results with the original Wills Wing *L/D* data to see how close they were to each other (Fig. CS9.12). The buildup maximum *L/D* estimates for the Falcon, Sport 2, and U2 gliders were slightly less than the original factory data, but the Talon 2 built-up *L/D* was greater than the Wills Wing data.

Figure CS9.13 shows how *L/D* translated into glide and sink rate data. Glide performance could easily be made worse by poor pilot technique or poor pilot position or a draggy harness. This could easily degrade *L/D* by a point or more. Typical flight data using a GPS and flown in calm air at constant airspeed in back-to-back flights can easily yield much larger uncertainty bands than the differences between the buildup data and the Wills

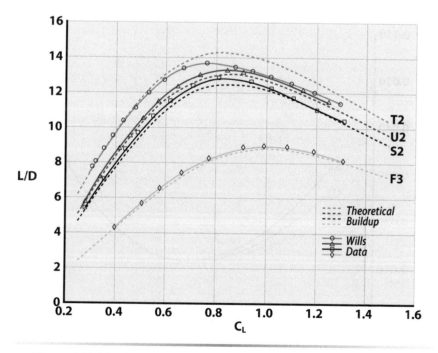

Figure CS9.12   Comparison of *L/D* estimates to manufacturer's data.

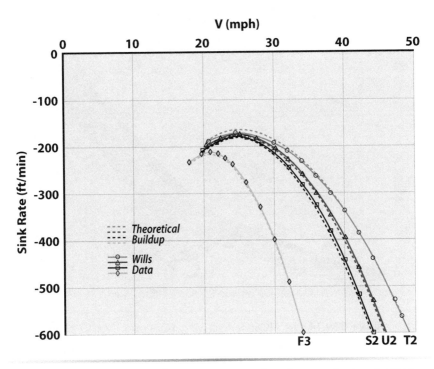

Figure CS9.13   Glide performance comparison to manufacturer's data.

Wing data. The buildup can be considered reliable regarding increments between the Wills Wing models. What is lacking is a rigorous and disciplined, independent flight test comparison amongst the various popular glider models. It should also be noted that typical hang glider pilots who analyze their in-flight GPS data do not seem to achieve these high of *L/D* and low of sink rates.

Once the drag estimates were completed for the rigid wing gliders and for the paraglider, a new family of profile drag curves were created and the glide performance was reworked and the glide performance will again be compared with the manufacturer's data. These new estimates put sink rate performance closer to what typical pilots have measured in long, still air, straight line glides.

CS9.6   Rigid Wing Hang Glider Design Features and Performance

Rigid wings provide a further improvement in glide performance over topless flex-wings like the Talon, although there is some debate regarding how much or if their cost, weight, and greater fragility are worthwhile.

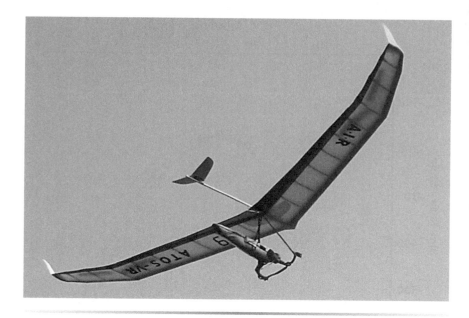

Figure CS9.14   Air ATOS VR rigid wing. [Oz Report, Vol 10, #39]

A typical rigid wing, the Air ATOS VR, is shown in Fig. CS9.14. Rigid wings typically have greater wing span, higher wing aspect ratio, lower sweep wing planforms, and some designs also include winglets for a further boost in effective span.

The ATOS VR design [24] represents state of the art rigid wing design that yields an *L/D* in the high teens, which is impressive considering the drag of an exposed pilot. Design features of the ATOS VR that differentiate it from the flex-wings include:

- Fully cantilevered wing structure with carbon fiber D-spar LE, carbon ribs attached to the LE and covered with sailcloth. It is fragile and must be carefully maintained.
- Less wing sweep so a small tail improves pitch damping and pitch stability.
- Roll/yaw control is by spoilers and is actuated by pilot weight shift side to side. Note that the weight shift by itself provides negligible roll effectiveness.
- Inboard simple trailing edge flaps of triangular planform.
- Ribs fold against D-spar when glider is disassembled. The airfoil is typically constrained to not have undercamber or chordwise curvature reversals.
- Small winglets.

Glide polar data of the ATOS VR were available at the manufacturer's Web site, and built-up estimates were performed in a similar manner to the flex-wing estimates. Additional data on profile drag were available in Ruhle [25], and were used to better estimate the drag impact of having a smooth airfoil over the D-spar region and fabric covering aft of that. This profile drag proved invaluable for creating a profile drag family of curves for all the gliders of this study. As mentioned, the flex-wing profile drag polars were then revised and glide performance updated.

Also, because the ATOS VR utilized flaps, an adjustment was made to induced drag due to a spanload that departed from the optimum. Figure CS9.15 shows a comparison between the built-up drag estimate and the manufacturer's data at an all-up weight of 140 kg.

The Brightstar SWIFT design, unlike the ATOS VR, integrates the pilot into the wing in a cage-like structure with the pilot in a supine rather than prone position (Fig. CS9.16). It has a higher L/D in the mid 20s if a hang-cage pilot fairing is used. Brightstar sold the manufacturing rights to Aeriane, which currently also produces a powered version [26]. It is heavier than the ATOS line of gliders but the glide performance is improved. Design features of the SWIFT that differentiate it from other designs include:

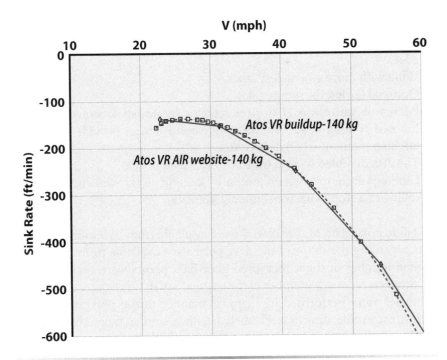

Figure CS9.15 Glide performance comparison to manufacturer's data.

Figure CS9.16 Brightstar SWIFT high performance glider.
[Courtesy Dr. Stephen Morris]

- Fully cantilevered composite wing structure from leading to trailing edge.
- Wing with more sweep, less taper.
- Optional fairing for supine pilot.
- Pitch, roll control via TE devices actuated by side stick control.
- Inboard trailing edge flaps that by careful design provide a nose-up pitching moment when deflected.
- The disassembled wing is transported in a large box.
- Landings accomplished on foot or by landing on the skid and aft wheel, rolling to a stop like a conventional sailplane.

Glide polar data of the SWIFT were available from two sources. Trend data regarding the shape of the drag polar were available from [27], and a limited number of flight-measured glide data points were available from Dr. Stephen Morris, a member of the original SWIFT design team. Built-up estimates were performed in a similar manner to the previously shown estimates. Profile drag was estimated with a similar polar shape as the ATOS VR, but with lower drag due to the smooth surface over the entire airfoil and the presence of some laminar flow as predicted by the SWIFT design team [28]. Figure CS9.17 compares the built-up drag estimate to the Morris-provided flight data at an all-up weight of 335 lbs.

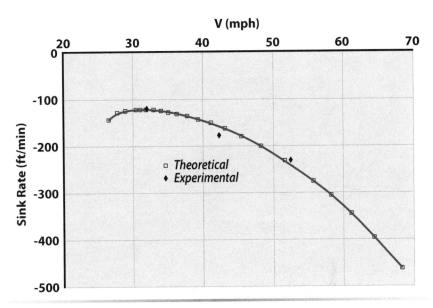

Figure CS9.17 SWIFT glide performance.

CS9.7 Paraglider Design Features and Performance

Paragliders have increased in popularity among free flight pilots due to their convenience features. Unlike hang gliders, paragliders can be easily transported on the pilot's back in a specialized back pack. This opens up a range of launch sites that are accessible by foot. Paragliders typically have a lower wing loading than hang gliders and fly much slower, enabling them to work smaller areas of lift and land in smaller landing fields that are not suitable for hang gliders. A typical paraglider is shown in Fig. CS9.18. Paragliders do not have hard structure and maintain their shape in flight by ram air forces created by air entering openings at the stagnation point on the LE. The pilot is suspended below the wing by a series of thin lines. Because the wing shape is held by lift forces, the wing has a distinctive arch shape when viewed from the front. This anhedral angle is destabilizing in roll, but the pilot is suspended so far below the wing that the pendulum stability overcomes the instability. It is possible for paragliders to collapse in extreme conditions, though most are very resistant to collapse and easily re-inflated if one does occur.

Typical design features include:

- Very compact package that fits into a special backpack.
- Wing shape held by aerodynamic forces in the wing cells rather than rigid structure.

Figure CS9.18   Paraglider pilot on final approach to land. [Courtesy Kim Fujishige Rosman]

- Pilot suspended far below the wing via multiple, thin lines.
- Pilot is in supine rather than prone position like on hang gliders.
- No tail surfaces.
- Control via pilot pulling on brake lines that attach to wing at the trailing edge. Pulling down on one of the lines pulls down the trailing edge of that wing, increasing aft camber.
- Pitch control via pulling both brake lines, roll control via pulling one or the other, combined with weight shift.
- Airspeed remains fairly constant with some variation available through a "speed bar" foot stirrup feature on some wings where the pilots weight can be shifted forward to increase airspeed through reducing the length of the front "A" lines.
- Trim airspeed is defined as when there is no brake pull force, and is above the speed for maximum L/D.
- Sink rate can be temporarily increased by deliberately inducing a symmetric, partial collapse of the wing tips, commonly referred to as "big ears."

Takeoffs are initiated by laying out the lines and the uniflated wing carefully on the ground, and then getting in the harness. When it is safe to launch the pilot pulls the wing upward so the cells can inflate in the wind, yells "Clear!" as previously discussed, and then runs into the wind and down the hill into the air. Some pilots prefer facing downhill and some prefer a reverse launch where they face the wing uphill and then carefully turn back downhill for the launch run once the wing is safely inflated and overhead. Launches can be challenging in windy conditions. Landings are accomplished similar to hang gliders, only the pilot pulls on the both brake lines simultaneously to pitch up the wing and flare before landing on their feet.

Performance was estimated for a typical intermediate paraglider, the UP Makalu 2 [29] (see Fig. CS9.19). Drag was estimated using a similar process as used for the hang gliders with drag of all the lines estimated through use of a line diagram. Profile drag was higher for the hang gliders because the airfoils typically have a base area at their trailing edge and there is pillowing between the cells of the wing that generates additional drag. It was also estimated to increase significantly at C_Ls above and below the C_L for

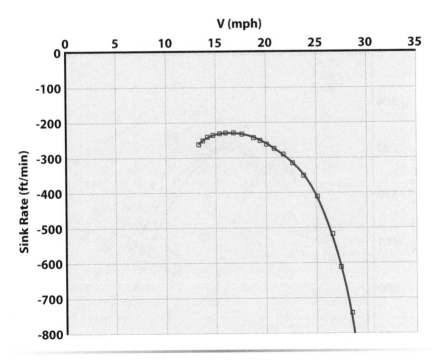

Figure CS9.19   Makalu 2 paraglider glider performance.

maximum L/D. This is because at high C_Ls the brakes are being used and drag increases and at low C_Ls the air inlets typically deform causing drag. Flight test data were not available but the manufacturer's Web site provided minimum, trim, and maximum speeds for a given pilot weight, providing clues. Also discussions with a Makalu 2 owner provided additional insights on performance. A popular book on paragliding [30] also provided some performance clues.

CS9.8 Overall Performance Summary

The broad spectrum of foot-launched glider performance can be seen by comparing all the wings together once the flex-wing profile drag polars were adjusted. Glide performance is shown in Fig. CS9.20. A summary of the pertinent geometric and performance characteristics for all the gliders considered herein is shown on Table CS9.2, all for a pilot weight of 170 lb at sea level on a standard day. The L/D is shown in Fig. CS9.21.

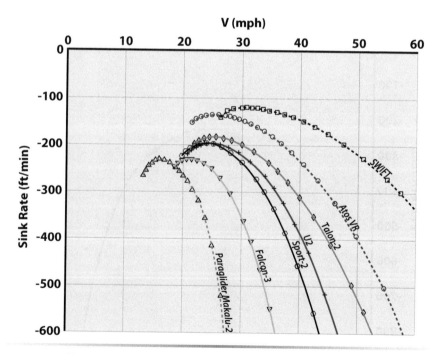

Figure CS9.20 Glide performance summary—all types.

Table CS9.2 Geometric and Performance Summary—All Types

	Span (ft)	S_{wing} (ft^2)	AR	Glider Wt (lb)	Total Wt (lb)	W/ S_{wing} (lb/ft^2)	V_{stall} (mph)	$V_{S_{min}}$ (ft/min)	V for $V_{S_{min}}$ (mph)	L/D	V for L/D (mph)
gid Wings											
WIFT	39.0	140	10.9	155	325	2.32	26	121	31	24.3	36
os VR	45.1	153	13.3	97	292	1.91	22	137	26	17.8	31
ex-Wings											
lon 2	32.3	144	7.2	71	266	1.85	21	185	26	13.1	30
2	31.3	145	6.8	63	258	1.78	21	197	25	12.0	30
oort 2	31.5	156	6.4	59	254	1.63	20	198	24	11.4	28
alcon 3	30.6	171	5.5	48	243	1.42	19	231	22	8.7	25
araglider											
akalu 2	31.5	269	3.7	14	195	0.72	13	230	16	6.6	19

CS9.9 Design Considerations and Requirements

Foot-launched gliders have unique design considerations that must be taken into account to properly develop a new wing.

These include the following factors:

- $C_{L_{max}}$ and wing area must be sufficient to enable launching and landing on foot. A stall speed not accounting for dynamic $C_{L_{max}}$ of no greater than 20 mph is desirable for safe, easy landings.
- Wing loadings higher than about 1.9 lb per square foot will drive the need for high lift devices.
- Weight must be kept very low, limiting wing span and to some degree, wing surface quality. An empty glider weight of no more than 75 lb is desirable to enable safe launches in no-wind conditions.
- Pilots outweigh their gliders so vertical location of center of gravity is very important, with a pilot location below the wing offering pendulum stability advantages.
- Positive static pitch stability is a must. With the pilot located below the wing, the glider will be the least stable right at stall. A static margin of +10% of the mean aerodynamic chord is good.
- Flexible airframes cannot be developed using only conventional aircraft design analysis techniques. Most have evolved through trail and error validated by testing. Full scale truck-based testing has successfully been used to assure pitch stability.

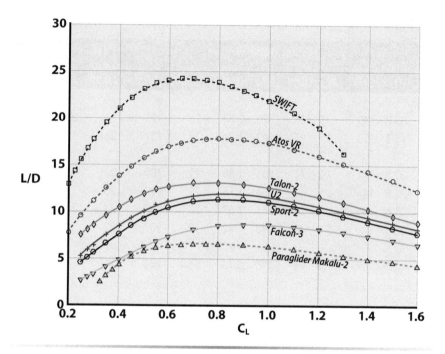

Figure CS9.21   *L/D* summary—all types.

- Glider performance must be optimized and balanced between climb performance circling in thermal lift and glide performance between thermals, sometimes into the wind.
- Sufficient roll rate to quickly capture a bank angle in thermal lift is most desirable.
- The ability to penetrate into a headwind is very helpful to make a designated landing field.
- Ground handling is an important factor. Wings swept aft are safer for advanced cliff launches and wings above the pilot are superior when launching near vegetation.
- Gliders are often assembled in windy, undeveloped areas and glider structure must be robust to handle non-flight loads.
- Static balance of the glider is important, so an empty glider center of gravity near the pilot's shoulders is desirable. This consideration tends to favor tailless configurations, or those with a small tail.
- Portability is an important consideration. If the wing requires a trailer one may as well fly a sailplane and get the superior glide performance. At a minimum, the wing should fold into a long, thin package that can be carried on a truck rack with several other wings.
- Glider cost must be kept low and the business case for wings made of laid-up composite materials is challenging at best.

CS9.10 Conclusions

- Foot-launched gliders have evolved into safe, efficient, inexpensive, and extremely fun sport aircraft.
- Flex-wing glider performance varies in L/D from 9 to 13 depending on complexity and degree of pilot skill required.
- Drag components can be estimated based on factory-provided glide data, then adjusted to family better with each other.
- There is still a need for rigorous flight testing to better quantify in flight hang glider geometry and glide performance.
- Rigid wing gliders have achieved L/Ds in the 20s though at a weight penalty.
- Paragliders have greatly increased in popularity and fly at lower airspeeds and L/Ds than hang gliders, but are often able to work smaller areas of lift.

Acknowledgments

The author is grateful to Steve Pearson at Wills Wing for providing background information on their gliders and their data; to Kay Dees, Darren Darsey, Ryan Voight, Davis Straub, and the Oz Report for the use of photos; and to Ilan Kroo of Stanford University and to Stephen Morris for SWIFT data and images. Thanks to Len Baron for an excellent peer review and to Rick Rosman for providing paraglider information. Thanks also to the United States Hang Gliding and Paragliding Association.

About the Author

Paul Dees has been flying hang gliders since the early 1970s and is an advanced-rated pilot by the United States Hang Gliding and Paragliding Association. He is also a Boeing Associate Technical Fellow in product development and has been a part of the configuration design, test, or aerodynamics analysis teams on aircraft programs including the Northrop F-20 Tigershark, the McDonnell Douglas AV-8B Harrier and T-45 Goshawk, and the Boeing 737NG, 747-8, 757, 767-400, 777 Product Development, and 787 commercial airplane programs. He also worked on the High Speed Civil Transport and Sonic Cruiser development programs.

References

[1] Spearman, A., John, *Joseph Montgomery, Father of Basic Flying*, University of Santa Clara, CA, 1967.
[2] Nitsch, S., *Vom Sprung zum Flug*, Brandenburgisches Verlagshaus, 1991.
[3] Means, J., *The Aeronautical Annual*, 1897, W. B. Clarke and Co., 1897.
[4] Dees, P. W., "The 100-Year Chanute Glider Replica, An Adventure in Education," presented at the 1997 World Aviation Congress, Anaheim, CA, October, 1997, AIAA Paper No. 1997-5573.

[5] Jarrett, P., *Another Icarus: Percy Pilcher and the Quest for Flight*, Smithsonian Institution Press, 1987.

[6] McFarland, M.W., *The Papers of Wilbur and Orville Wright*, Vol. I, Ayer Company Reprint, Salem, NH, 1990.

[7] Crouch, T., *The Bishop's Boys*, W.W. Norton & Co., New York, 1989.

[8] Kochersberger, K., Ash, A., Sandusky, R., Hyde, K., "An Evaluation of the Wright 1901 Glider Using Full Scale Wind Tunnel Data," presented at the 40th AIAA Aerospace Sciences Meeting, Reno, NV, January, 2002, AIAA Paper No. 2002-1134.

[9] Dees, P., "How Gliders Helped the Wright Brothers Invent the Airplane," presented at the 41st AIAA Aerospace Sciences Meeting, Reno, NV, January, 2003, AIAA Paper No. 2003-0095.

[10] Kochersberger, K., Player, J., Ash, A., "Performance Analysis of the Wright 1902 Glider Using Full Scale Wind Tunnel Data," presented at the 41st AIAA Aerospace Sciences Meeting, Reno, NV, January, 2003, AIAA Paper No. 2003-0096.

[11] HGMA Web site: http://www.hgma.net

[12] DHV Web site: http://www.dhv.de/typo/Home-English.3.0.html

[13] USHPA Web site: http://www.ushpa.aero

[14] Wills Wing Web site: http://www.willswing.com

[15] Moyes Web site: http://www.moyes.com.au or http://www.moyesusa.com

[16] Icaro Web site: http://www.icaro2000.com

[17] Aeros Web site: http://www.aeros.com.us

[18] Airborne Australia Web site: http://www.airborne.com.au

[19] Northwing Web site: http://www.northwing.com

[20] Wills Wing polar article: http://www.willswing.com/Articles/Article.asp?reqArticleName=PolarData

[21] Hoerner, Sighard, *Fluid Dynamic Drag*, self published, 1965.

[22] Kilkenny, E.A., "Full Scale Wind Tunnel Tests on Hang Glider Pilots," Cranfield College of Aeronautics Report 8416, April, 1984.

[23] Athena Vortex Lattice Web site: http://web.mit.edu/drela/Public/web/avl/

[24] ATOS Web site: http://www.a-i-r.de

[25] Ruhle, Felix, "Untersuchung der Aerdynamischen Eigenshaften eines fusstartfahigen Nurfluglers," Thesis for Institut Fur Aerodynamik und Gasdynamik, University of Stuttgart, July, 1994.

[26] SWIFT Web site: http://www.aeriane.com/

[27] Kroo, I, "Design and Development of the SWIFT: A Foot-Launched Sailplane," presented at the 18th AIAA Applied Aerodynamics Conference, Denver, CO, August, 2000, AIAA Paper No. 2000-4336.

[28] Conversations and email with Dr. Ilan Kroo and Dr. Stephen Morris, February, 2011.

[29] UP Paraglider web site: http://www.up-europe.com/index.php?id=133.

[30] Pagen, Dennis, *The Art of Paragliding*, Sport Aviation Publications, 2004, p. 201.

Appendix A Conversions

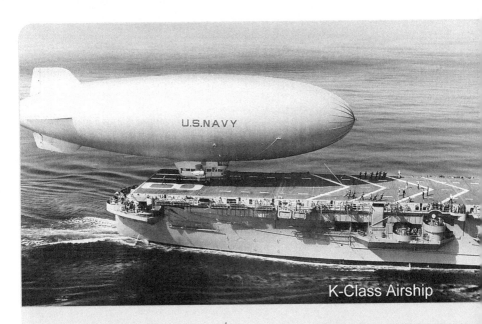

K-Class Airship

- Unit Conversions
- Temperature Conversions
- Gases and Liquids

In 1998, NASA sent a probe to Mars. The probe was part of the Stardust program and was to land on the Martian surface, collect some dirt samples, and return to Earth. The probe crashed into the Martian surface because the contractor that built the probe mixed up English and metric units on the drawings.

The solution to all the world's ills is to pick one set of units and force everybody to use them.

Leland Nicolai

A.1 Unit Conversions

A.1.1 Length

Multiply	By	To obtain
Centimeter (cm)	3.281×10^{-2}	Feet
	3.938×10^{-1}	Inches
	1.000×10^{-5}	Kilometers
	1.000×10^{-2}	Meters
	1.094×10^{-2}	Yards
Foot (ft)	30.48	Centimeters
	12.00	Inches
	3.048×10^{-4}	Kilometers
	3.048×10^{-1}	Meters
	1.894×10^{-4}	Miles
	3.333×10^{-1}	Yards
Inch (in.)	2.540	Centimeters
	8.333×10^{-2}	Feet
	2.540×10^{-2}	Meters
	2.778×10^{-2}	Yards
	1.000×10^{-3}	Miles
Meter (m)	1.000×10^{2}	Centimeters
	3.281	Feet
	39.37	Inches
	1.000×10^{-3}	Kilometers
	6.214×10^{-4}	Miles
	1.094	Yards
Statute mile (mile or mi)	5.280×10^{3}	Feet
	1.609	Kilometers
	1.760×10^{3}	Yards
	0.868976	Nautical miles
Nautical mile (nmile)	6.076×10^{3}	Feet
	1.852×10^{3}	Meters
	1.15078	Miles
Yard (yd)	91.44	Centimeters
	3.000	Feet
	36.00	Inches
	9.144×10^{-1}	Meters
	5.682×10^{-4}	Miles

A.1.2 Area

Multiply	By	To obtain
Acre	4.356×10^4	Square feet
	4.047×10^3	Square meters
	1.562×10^{-3}	Square miles
Square centimeter (cm^2)	1.076×10^{-3}	Square feet
	1.550×10^{-1}	Square inches
	1.000×10^{-4}	Square meters
	1.000×10^2	Square millimeters
Square foot (ft^2)	2.296×10^{-5}	Acres
	1.440×10^2	Square inches
	9.290×10^{-2}	Square meters
	3.587×10^{-8}	Square miles
	1.111×10^{-1}	Square yards
Square inch (in.^{2})	6.4516	Square Centimeters
	6.944×10^{-3}	Square feet
	6.452×10^{-4}	Square meters
Square kilometer (km)	2.471×10^2	Acres
	1.076×10^7	Square feet
	3.861×10^{-1}	Square miles
Square meter (m^2)	2.471×10^{-4}	Acres
	1.000×10^4	Square centimeters
	10.76	Square feet
	1.550×10^3	Square inches
	3.861×10^{-7}	Square miles
Square mile	6.40×10^2	Acres
	2.778×10^7	Square feet
	2.590	Square kilometers
	2.590×10^6	Square meters
	3.0976×10^6	Square yards

A.1.3 Volume

Multiply	By	To obtain
Cubic centimeter (cm³)	3.531×10^{-5}	Cubic feet
	6.1024×10^{-2}	Cubic inches
	1.000×10^{-6}	Cubic meters
	1.308×10^{-6}	Cubic yards
	3.381×10^{-2}	Fluid ounce
Cubic foot (ft³)	2.832×10^{4}	Cubic centimeters
	1.728×10^{3}	Cubic inches
	2.832×10^{-2}	Cubic meters
	28.317	Liters
	7.481	Gallons
Cubic inch (in.³)	16.39	Cubic centimeters
	5.787×10^{-4}	Cubic feet
	1.639×10^{-5}	Cubic meters
Cubic meter (m³)	1.000×10^{6}	Cubic centimeters
	35.31	Cubic feet
	6.102×10^{4}	Cubic inches
	1.308	Cubic yards
Gallon (U.S.) (gal)	1.3368×10^{-1}	Cubic feet
	3.78542	Liters
	3.785×10^{-3}	Cubic meters
	231	Cubic inches
	128	Fluid ounces
	8.000	Pints
	4.000	Quarts
Imperial gallon	2.774×10^{2}	Cubic inches
	1.201	Gallons (U.S.)
	4.546	Liters
Liter	3.532×10^{-2}	Cubic feet
	0.2642	Gallons
	1.000×10^{-3}	Cubic meters
	2.113	Pints
	1.05669	Quarts
	33.8142	Fluid ounces
Pint (U.S.) (pt)	1.671×10^{-2}	Cubic feet
	1.250×10^{-1}	Gallons
	4.732×10^{-1}	Liters
	0.5	Quarts
	28.875	Cubic inches
	16	Fluid ounces
Quart (U.S.) (qt)	3.342×10^{-2}	Cubic feet
	2.500×10^{-1}	Gallons
	9.463×10^{-1}	Liters
	2	Pints

A.1.4 Velocity

Multiply	By	To obtain
Centimeter per second (cm/s)	3.281×10^{-2}	Feet per second
	3.937×10^{-1}	Inches per second
	1.000×10^{-2}	Meters per second
Foot per second (fps or ft/s)	30.48	Centimeters per second
	1.097	Kilometers per hour
	5.921×10^{-1}	Knots
	3.048×10^{-1}	Meters per second
	6.818×10^{-1}	Miles per hour
Inch per second (ips)	8.333×10^{-2}	Feet per second
	2.540	Centimeters per second
Kilometer per hour (km/h)	9.113×10^{-1}	Feet per second
	5.396×10^{-1}	Knots
	6.214×10^{-1}	Miles per hour
Knot (kt)	1.689	Feet per second
	1.151	Miles per hour
	1.000	Nautical miles per hour
	1.852	Kilometers per hour
Meter per second (m/s)	3.281	Feet per second
	3.600	Kilometers per hour
	1.943	Knots
	2.237	Miles per hour
Mile per hour (mph)	1.467	Feet per second
	1.609	Kilometers per hour
	0.8684	Knots
	0.4470	Meters per second

A.1.5 Acceleration

Feet per second^2 (ft/s^2)	30.48	Centimeters per second^2
	0.6818	Miles per hour-second

A.1.6 Angular Rate and Frequency

Multiply	By	To obtain
Radians per second (rad/s)	0.1592	Revolutions per second
	9.549	Revolutions per minute
	57.296	Degrees per second
Revolutions per minute (rpm)	0.01667	Revolutions per second
	0.10472	Radians per second
	6	Degrees per second
Cycle per second (cps)	1.000	Hertz
	2π	Radians per second

A.1.7 Mass

Multiply	By	To obtain
Kilogram (kg)	1.000×10^3	Grams
	6.854×10^{-2}	Slugs
Slug	1.459×10^4	Grams
	14.59	Kilograms

A.1.8 Weight

Multiply	By	To obtain
Gram (g)	3.528×10^{-2}	Ounces
	2.205×10^{-3}	Pounds
Pound (lb)	4.536×10^2	Grams
	16	Ounces
Short ton	2000	Pounds
	907.185	Kilograms
Metric tonne	2205	Pounds
	1000	Kilograms

A.1.9 Force

Multiply	By	To obtain
Dyne	1.020×10^{-3}	Grams
	1.000×10^{-5}	Newtons
	2.248×10^{-6}	Pounds
Gram (g)	3.528×10^{-2}	Ounces
	2.205×10^{-3}	Pounds
	9.807×10^2	Dynes
	9.807×10^{-3}	Newtons
Kilogram (kg)	2.205	Pounds
	9.807	Newtons
	70.93	Poundals
Pound (lb)	4.536×10^{-1}	Kilograms
	4.448	Newtons
	32.17	Poundals
Poundal	1.410×10^{-2}	Kilograms
	1.383×10^{-1}	Newtons
	3.108×10^{-2}	Pounds

A.1.10 Pressure

Multiply	By	To obtain
Atmosphere (atm)	29.92	Inches of mercury (0°C)
	760	Millimeters of mercury (0°C)
	1.0133	Bars
	14.70	Pounds per square inch
	1.01325×10^6	Dynes per square centimeter
	1.01325×10^5	Newtons per square meter
Bar	9.870×10^{-7}	Atmospheres
	1.000	Dyne per square centimeter
	1.0×10^5	Newtons per square meter
	7.501×10^2	Millimeters of mercury (0°C)
	1.451×10^{-5}	Pounds per square inch
Dyne per square centimeter (dyne/cm²)	2.952×10^{-5}	Inches of mercury (0°C)
	1.020×10^{-2}	Kilograms per square meter
	7.501×10^{-4}	Millimeters of mercury (0°C)
	1.450×10^{-5}	Pounds per square inch
Inch of mercury (in. Hg)	3.342×10^{-2}	Atmospheres (0°C)
	3.388×10^{-2}	Bars
	3.388×10^3	Dynes per square centimeter
	13.60	Inches of water
	25.40	Millimeters of mercury
	3.388×10^3	Newtons per square meter
	70.73	Pounds per square foot
	4.912×10^{-1}	Pounds per square inch
Inch of water (in. H₂O) (4°C)	2.458×10^{-3}	Atmospheres
	7.355×10^{-2}	Inches of mercury
	1.868	Millimeters of mercury
	2.491×10^2	Newtons per square meter
	3.613×10^{-2}	Pounds per square inch
	5.203	Pounds per square foot
Kilogram per square meter (kg/m²)	9.678×10^{-5}	Atmospheres
	98.07	Bars
	2.896×10^{-3}	Inches of mercury
	9.807	Newtons per square meter
	6.588	Poundals per square foot
	2.048×10^{-1}	Pounds per square foot
	1.422×10^{-3}	Pounds per square inch

(continued)

Multiply	By	To obtain
Millimeter of mercury (0°C) (torr or mm Hg)	1.333×10^3	Dynes per square centimeter
	3.937×10^{-2}	Inches of mercury
	5.354×10^{-1}	Inches of water
	1.333×10^2	Newtons per square meter
	1.934×10^{-2}	Pounds per square inch
Newton per square meter [pascal (Pa)] (N/m²)	9.869×10^{-6}	Atmospheres
	10	Dynes per square centimeter
	2.953×10^{-4}	Inches of mercury
	1.020×10^{-1}	Kilograms per square meter
	2.089×10^{-2}	Pounds per square foot
	1.450×10^{-4}	Pounds per square inch
Pound per square foot (psf)	4.725×10^{-4}	Atmospheres
	4.788×10^{-4}	Bars
	4.788×10^2	Dynes per square centimeter
	1.414×10^{-2}	Inches of mercury
	4.882	Kilograms per square meter
	47.88	Newtons per square meter
	6.944×10^{-3}	Pounds per square inch
Pound per square inch (psi)	6.804×10^{-2}	Atmospheres
	6.895×10^4	Dynes per square centimeter
	2.036	Inches of mercury
	7.031×10^{-2}	Kilograms per square meter
	6.895×10^3	Newtons per square meter
	1.44×10^2	Pounds per square foot

A.1.11 Density

Multiply	By	To obtain
Pound per cubic foot (lb/ft³)	5.787×10^{-4}	Pounds per cubic inch
	16.018	Kilograms per cubic meter
	1.6018×10^{-2}	Grams per cubic centimeter

A.1.12 Work and Energy

Multiply	By	To obtain
British thermal unit (Btu)	2.530×10^2	Calories
	7.783×10^2	Foot pounds
	3.927×10^{-4}	Horsepower hours
	1.055×10^3	Joules
	1.055×10^3	Newton meters
	2.930×10^{-4}	Kilowatt hours
	1.055×10^3	Watt seconds
Foot pound (ft·lb)	1.285×10^{-3}	British thermal units
	5.050×10^{-7}	Horsepower hours
	1.356	Joules
	3.766×10^{-7}	Kilowatt hours
	1.356	Newton meters
Horsepower hour (hp·h)	2.545×10^3	British thermal units
	1.980×10^6	Foot pounds
	2.684×10^6	Joules
	7.457×10^{-1}	Kilowatt hours
Joule	9.486×10^{-4}	British thermal units
	2.389×10^{-1}	Calories
	1.000×10^7	Dyne centimeters (ergs)
	7.376×10^{-1}	Foot pounds
	1.000	Newton meter
	1.000	Watt second
Kilowatt hour (kWh)	3.415×10^3	British thermal units
	2.655×10^6	Foot pounds
	1.341	Horsepower hours
	3.600×10^6	Joules
	3.670×10^5	Kilogram meters
	3.600×10^6	Watt seconds
Dyne centimeter	7.3756×10^{-8}	Foot pounds
	1.000×10^{-7}	Newton meters

A.1.13 Power

Multiply	By	To obtain
British thermal unit per minute (BTU/min)	3.969×10^6	Calories per second
	12.97	Foot-pounds per second
	2.357×10^{-2}	Horsepower
	17.58	Joules per second
	2.987×10^{-2}	Kilogram meters per second
	17.58	Watts
Foot-pound per second (ft·lb/s)	7.713×10^{-2}	British thermal units per minute
	3.239×10^{-1}	Calories per second
	1.818×10^{-3}	Horsepower
	1.356	Joules per second
	1.383×10^{-1}	Kilogram meters per second
	1.356	Watts
Horsepower (hp)	42.42	British thermal units per minute
	550	Foot-pounds per second
	33,000	Foot-pounds per minute
	7.457×10^2	Joules per second
	76.04	Kilogram-meters per second
	7.457×10^2	Watts
Kilogram-meter per second	33.47	British Thermal Units per minute
	7.233	Foot-pounds per second
Watt (joule per second) (W)	5.689×10^{-2}	British thermal units per minute
	2.388×10^{-1}	Calories per second
	7.376×10^{-1}	Foot-pounds per second
	1.341×10^{-3}	Horsepower
	1.020×10^{-1}	Kilogram-meters per second

A.2 Temperature Conversions

- $T(^{\circ}C) = (5/9)\,[T(^{\circ}F) - 32]$
- $T(^{\circ}C) = (5/9)\,[T(^{\circ}R) - 491.67]$
- $T(^{\circ}C) = T(^{\circ}K) - 273.15$
- $T(^{\circ}F) = (9/5)\,T(^{\circ}C) + 32$
- $T(^{\circ}F) = (9/5)\,[T(^{\circ}K) - 273.15] + 32$
- $T(^{\circ}F) = T(^{\circ}R) - 459.67$

A.3 Gases and Liquids

A.3.1 Standard Values for Air at Sea Level

- $p_0 = 2116.22$ psi $= 1.01325 \times 10^5$ N/m^2 = 29.92 in. Hg = 760 mm Hg
- $T_0 = 518.67°$R = 59.0°F = 288.15°K = 15.0°C
- $g_0 = 32.174$ ft/s^2 = 9.80665 m/s^2
- $\rho_0 = 0.002377$ slug/ft^3 = 0.12492 kg·s^2/m^4
- $\nu_0 = 1.5723 \times 10^{-4}$ ft^2/s $= 1.4607 \times 10^{-5}$ m^2/s
- $\mu_0 = 1.2024 \times 10^{-5}$ lb/ft·s $= 1.7894 \times 10^{-5}$ kg/m·s
- $\mu_0 = 3.737 \times 10^{-7}$ slug/(ft·s)

A.3.2 Specific Weights of Other Gases at One Atmosphere and 0°C

- Carbon dioxide = 0.12341 lb/ft^3
- Helium = 0.01114 lb/ft^3
- Hydrogen = 0.005611 lb/ft^3
- Nitrogen = 0.07807 lb/ft^3
- Oxygen = 0.089212 lb/ft^3

A.3.3 Specific Weights (Specific Gravity) of Some Liquids at 0°C

- Alcohol (methyl) = 50.5 lb/ft^3 (0.810)
- Gasoline = 44.9 lb/ft^3 (0.72)
- JP1 = 49.7 lb/ft^3 (0.80)
- JP3 = 48.2 lb/ft^3 (0.775)
- JP4 = 49.0 lb/ft^3 (0.785)
- JP5 = 51.1 lb/ft^3 (0.817)
- JP7 = 48.6–50.3 lb/ft^3 (0.779–0.806)
- JP8 = 55.81 lb/ft^3 (0.894)
- JP10 = 58.62 lb/ft^3 (0.939)
- Kerosene = 51.2 lb/ft^3 (0.82)
- Sea water = 63.99 lb/ft^3 (1.025)
- Water = 62.43 lb/ft^3 (1.000)

- 1976 U.S. Standard Atmosphere
- MIL-210A Atmospheric Data
- MIL-210A 10% & 1% Probability Atmospheric Data
- MIL-210A Navy Non-Standard Atmospheric Data

This serene setting is a reminder that the fundamental shape of a hot air balloon is the same as it has always been. Better and more vibrant materials are available and, when combined with light-weight burners, the performance of these recreational balloons increases.

Science can amuse and fascinate us all, but it is engineering that changes the world.

Isaac Asimov

The U.S. military has definitions for atmospheres that are different from a standard day, whether "standard day" is defined by MIL-210A or the 1976 U.S. Standard Atmosphere (Table B.1). Tables B.2–B.4 summarize the temperature profiles of those nonstandard atmospheres as defined by the U.S. Air Force and the U.S. Navy. The standard day temperature profile is shown on each plot (Figs. B.1 and B.2) for reference. The 1% and 10% hot and cold days represent temperature profiles that would occur only 1% or 10% of the time, respectively. The unique temperature profiles defined by the U.S. Navy are also included for completeness. All temperature profiles are plotted, and the plotted data are included in the tables.

Table B.1   1976 U.S. Standard Atmosphere

Alt.	Press.	P/P_0	Density	ρ/ρ_0	Temp.	Temp.	T/T_0	Viscosity	V_{sonic}
(ft)	(lb/ft^2)	(δ)	(slug/ft^3)	(σ)	°R	°F	(θ)	(slug/ft-s)	(ft/s)
0	2116.2	1.0000	0.002377	1.0000	518.7	59.00	1.0000	3.737E-07	1116.4
1000	2040.9	0.9644	0.002308	0.9711	515.1	55.43	0.9931	3.717E-07	1112.6
2000	1967.7	0.9298	0.002241	0.9428	511.5	51.87	0.9862	3.697E-07	1108.7
3000	1896.6	0.8962	0.002175	0.9151	508.0	48.30	0.9794	3.677E-07	1104.9
4000	1827.7	0.8637	0.002111	0.8881	504.4	44.74	0.9725	3.657E-07	1101.0
5000	1760.8	0.8320	0.002048	0.8617	500.8	41.17	0.9656	3.637E-07	1097.1
6000	1695.9	0.8014	0.001987	0.8359	497.3	37.60	0.9587	3.616E-07	1093.2
7000	1632.9	0.7716	0.001927	0.8106	493.7	34.04	0.9519	3.596E-07	1089.3
8000	1571.9	0.7428	0.001868	0.7860	490.1	30.47	0.9450	3.575E-07	1085.3
9000	1512.7	0.7148	0.001811	0.7620	486.6	26.90	0.9381	3.555E-07	1081.4
10,000	1455.3	0.6877	0.001755	0.7385	483.0	23.34	0.9312	3.534E-07	1077.4
11,000	1399.7	0.6614	0.001701	0.7156	479.4	19.77	0.9244	3.513E-07	1073.4
12,000	1345.9	0.6360	0.001648	0.6932	475.9	16.21	0.9175	3.493E-07	1069.4
13,000	1293.7	0.6113	0.001596	0.6713	472.3	12.64	0.9106	3.472E-07	1065.4
14,000	1243.2	0.5875	0.001545	0.6500	468.7	9.07	0.9037	3.451E-07	1061.4
15,000	1194.3	0.5643	0.001496	0.6292	465.2	5.51	0.8969	3.430E-07	1057.3
16,000	1146.9	0.5420	0.001447	0.6090	461.6	1.94	0.8900	3.409E-07	1053.3
17,000	1101.1	0.5203	0.001400	0.5892	458.0	−1.62	0.8831	3.388E-07	1049.2
18,000	1056.8	0.4994	0.001355	0.5699	454.5	−5.19	0.8762	3.367E-07	1045.1
19,000	1013.9	0.4791	0.001310	0.5511	450.9	−8.76	0.8694	3.345E-07	1041.0
20,000	972.5	0.4595	0.001266	0.5328	447.3	−12.32	0.8625	3.324E-07	1036.8
21,000	932.4	0.4406	0.001224	0.5150	443.8	−15.89	0.8556	3.303E-07	1032.7
22,000	893.7	0.4223	0.001183	0.4976	440.2	−19.46	0.8487	3.281E-07	1028.6
23,000	856.3	0.4046	0.001142	0.4807	436.6	−23.02	0.8419	3.259E-07	1024.4
24,000	820.2	0.3876	0.001103	0.4642	433.1	−26.59	0.8350	3.238E-07	1020.2

It.	Press.	P/P_0	Density	ρ/ρ_0	Temp.	Temp.	T/T_0	Viscosity	V_{sonic}
't)	(lb/ft²)	(δ)	(slug/ft³)	(σ)	°R	°F	(θ)	(slug/ft·s)	(ft/s)
25,000	785.3	0.3711	0.001065	0.4481	429.5	−30.15	0.8281	3.216E−07	1016.0
26,000	751.64	0.3552	0.001028	0.4325	425.95	−33.72	0.8212	3.194E−07	1011.7
27,000	719.15	0.3398	0.000992	0.4173	422.38	−37.29	0.8144	3.172E−07	1007.5
28,000	687.81	0.3250	0.000957	0.4025	418.82	−40.85	0.8075	3.150E−07	1003.2
29,000	657.58	0.3107	0.000923	0.3881	415.25	−44.42	0.8006	3.128E−07	999.0
30,000	628.43	0.2970	0.000889	0.3741	411.69	−47.98	0.7937	3.106E−07	994.7
31,000	600.35	0.2837	0.000857	0.3605	408.12	−51.55	0.7869	3.084E−07	990.3
32,000	573.28	0.2709	0.000826	0.3473	404.55	−55.12	0.7800	3.061E−07	986.0
33,000	547.21	0.2586	0.000795	0.3345	400.99	−58.68	0.7731	3.039E−07	981.7
34,000	522.12	0.2467	0.000765	0.3220	397.42	−62.25	0.7662	3.016E−07	977.3
35,000	497.96	0.2353	0.000737	0.3099	393.85	−65.82	0.7594	2.994E−07	972.9
36,000	474.71	0.2243	0.000709	0.2981	390.29	−69.38	0.7525	2.971E−07	968.5
37,000	452.43	0.2138	0.000676	0.2843	389.97	−69.70	0.7519	2.969E−07	968.1
38,000	431.20	0.2038	0.000644	0.2710	389.97	−69.70	0.7519	2.969E−07	968.1
39,000	410.97	0.1942	0.000614	0.2583	389.97	−69.70	0.7519	2.969E−07	968.1
40,000	391.68	0.1851	0.000585	0.2462	389.97	−69.70	0.7519	2.969E−07	968.1
41,000	373.30	0.1764	0.000558	0.2346	389.97	−69.70	0.7519	2.969E−07	968.1
42,000	355.78	0.1681	0.000531	0.2236	389.97	−69.70	0.7519	2.969E−07	968.1
43,000	339.09	0.1602	0.000507	0.2131	389.97	−69.70	0.7519	2.969E−07	968.1
44,000	323.17	0.1527	0.000483	0.2031	389.97	−69.70	0.7519	2.969E−07	968.1
45,000	308.01	0.1455	0.000460	0.1936	389.97	−69.70	0.7519	2.969E−07	968.1
46,000	293.56	0.1387	0.000439	0.1845	389.97	−69.70	0.7519	2.969E−07	968.1
47,000	279.78	0.1322	0.000418	0.1758	389.97	−69.70	0.7519	2.969E−07	968.1
48,000	266.65	0.1260	0.000398	0.1676	389.97	−69.70	0.7519	2.969E−07	968.1
49,000	254.14	0.1201	0.000380	0.1597	389.97	−69.70	0.7519	2.969E−07	968.1
50,000	242.21	0.1145	0.000362	0.1522	389.97	−69.70	0.7519	2.969E−07	968.1
51,000	230.85	0.1091	0.000345	0.1451	389.97	−69.70	0.7519	2.969E−07	968.1
52,000	220.01	0.1040	0.000329	0.1383	389.97	−69.70	0.7519	2.969E−07	968.1
53,000	209.69	0.0991	0.000313	0.1318	389.97	−69.70	0.7519	2.969E−07	968.1
54,000	199.85	0.0944	0.000299	0.1256	389.97	−69.70	0.7519	2.969E−07	968.1
55,000	190.47	0.0900	0.000285	0.1197	389.97	−69.70	0.7519	2.969E−07	968.1
56,000	181.53	0.0858	0.000271	0.1141	389.97	−69.70	0.7519	2.969E−07	968.1
57,000	173.01	0.0818	0.000258	0.1087	389.97	−69.70	0.7519	2.969E−07	968.1
58,000	164.90	0.0779	0.000246	0.1036	389.97	−69.70	0.7519	2.969E−07	968.1
59,000	157.16	0.0743	0.000235	0.0988	389.97	−69.70	0.7519	2.969E−07	968.1
60,000	149.78	0.0708	0.000224	0.0941	389.97	−69.70	0.7519	2.969E−07	968.1
61,000	142.75	0.0675	0.000213	0.0897	389.97	−69.70	0.7519	2.969E−07	968.1
62,000	136.05	0.0643	0.000203	0.0855	389.97	−69.70	0.7519	2.969E−07	968.1
63,000	129.67	0.0613	0.000194	0.0815	389.97	−69.70	0.7519	2.969E−07	968.1

(continued)

Alt.	Press.	P/P_0	Density	ρ/ρ_0	Temp.	Temp.	T/T_0	Viscosity	V_{sonic}
(ft)	(lb/ft²)	(δ)	(slug/ft³)	(σ)	°R	°F	(θ)	(slug/ft·s)	(ft/s)
64,000	123.59	0.0584	0.000185	0.0777	389.97	−69.70	0.7519	2.969E−07	968.1
65,000	117.79	0.0557	0.000176	0.0740	389.97	−69.70	0.7519	2.969E−07	968.1
66,000	112.26	0.0530	0.000168	0.0705	390.18	−69.49	0.7523	2.970E−07	968.3
67,000	107.00	0.0506	0.000160	0.0671	390.73	−68.94	0.7533	2.974E−07	969.0
68,000	101.99	0.0482	0.000152	0.0639	391.28	−68.39	0.7544	2.977E−07	969.7
69,000	97.22	0.0459	0.000145	0.0608	391.83	−67.84	0.7554	2.981E−07	970.4
70,000	92.68	0.0438	0.000138	0.0579	392.37	−67.30	0.7565	2.984E−07	971.1
71,000	88.36	0.0418	0.000131	0.0551	392.92	−66.75	0.7576	2.988E−07	971.7
72,000	84.25	0.0398	0.000125	0.0525	393.47	−66.20	0.7586	2.991E−07	972.4
73,000	80.33	0.0380	0.000119	0.0500	394.02	−65.65	0.7597	2.995E−07	973.1
74,000	76.60	0.0362	0.000113	0.0476	394.57	−65.10	0.7607	2.998E−07	973.8
75,000	73.05	0.0345	0.000108	0.0453	395.12	−64.55	0.7618	3.002E−07	974.4
76,000	69.67	0.0329	0.000103	0.0432	395.67	−64.00	0.7628	3.005E−07	975.1
77,000	66.45	0.0314	0.000098	0.0411	396.22	−63.45	0.7639	3.009E−07	975.8
78,000	63.38	0.0300	0.000093	0.0392	396.76	−62.91	0.7650	3.012E−07	976.5
79,000	60.46	0.0286	0.000089	0.0373	397.31	−62.36	0.7660	3.016E−07	977.1
80,000	57.67	0.0273	0.000084	0.0355	397.86	−61.81	0.7671	3.019E−07	977.8
81,000	55.02	0.0260	0.000080	0.0338	398.41	−61.26	0.7681	3.023E−07	978.5
82,000	52.50	0.0248	0.000077	0.0322	398.96	−60.71	0.7692	3.026E−07	979.2
83,000	50.09	0.0237	0.000073	0.0307	399.51	−60.16	0.7703	3.030E−07	979.8
84,000	47.79	0.0226	0.000070	0.0293	400.06	−59.61	0.7713	3.033E−07	980.5
85,000	45.61	0.0216	0.000066	0.0279	400.60	−59.07	0.7724	3.037E−07	981.2
86,000	43.52	0.0206	0.000063	0.0266	401.15	−58.52	0.7734	3.040E−07	981.9
87,000	41.54	0.0196	0.000060	0.0253	401.70	−57.97	0.7745	3.043E−07	982.5
88,000	39.65	0.0187	0.000057	0.0242	402.25	−57.42	0.7755	3.047E−07	983.2
89,000	37.84	0.0179	0.000055	0.0230	402.80	−56.87	0.7766	3.050E−07	983.9
90,000	36.12	0.0171	0.000052	0.0219	403.35	−56.32	0.7777	3.054E−07	984.5
91,000	34.48	0.0163	0.000050	0.0209	403.90	−55.77	0.7787	3.057E−07	985.2
92,000	32.92	0.0156	0.000047	0.0200	404.44	−55.23	0.7798	3.061E−07	985.9
93,000	31.43	0.0149	0.000045	0.0190	404.99	−54.68	0.7808	3.064E−07	986.5
94,000	30.01	0.0142	0.000043	0.0181	405.54	−54.13	0.7819	3.068E−07	987.2
95,000	28.66	0.0135	0.000041	0.0173	406.09	−53.58	0.7829	3.071E−07	987.9
96,000	27.36	0.0129	0.000039	0.0165	406.64	−53.03	0.7840	3.074E−07	988.5
97,000	26.13	0.0123	0.000037	0.0157	407.19	−52.48	0.7851	3.078E−07	989.2
98,000	24.96	0.0118	0.000036	0.0150	407.74	−51.93	0.7861	3.081E−07	989.9
99,000	23.84	0.0113	0.000034	0.0143	408.29	−51.38	0.7872	3.085E−07	990.5
100,000	22.77	0.0108	0.000032	0.0136	408.83	−50.84	0.7882	3.088E−07	991.2

Unit conversions: m = 0.3048(ft); N/m² = 47.88(lb/ft²); kg·s²/m⁴ = 0.01903(slug/ft³); Temp°K = 5/9(Temp°R); q/M² = 0.7P; Temp°C = 5/9(Temp°R−491.67); kg/m·s = 0.02088(slug/ft·s); m/s = 0.3048(ft/s); knots = 0.5921(ft/s)

Table B.2 MIL-210A Nonstandard Atmospheres

Cold day		Polar day		STANDARD DAY		Tropical day		Hot day	
Alt. (ft)	T (°R)	Alt. (ft)	T (°R)	Alt. (ft)	T (°R)	Alt. (ft)	T (°R)	Alt. (ft)	T (°R)
0	399.7	0	444.0	0	518.7	0	549.5	0	562.7
3311	444.7	3243	453.9	5000	500.8	5000	530.1	5000	543.4
5000	444.7	5000	453.0	10,000	483.0	10,000	510.7	10,000	523.6
10,000	444.7	9882	450.3	15,000	465.2	15,000	491.3	15,000	504.6
10,744	444.7	10,000	450.0	20,000	447.3	20,000	472.0	20,000	485.2
15,000	430.6	15,000	435.9	25,000	429.5	25,000	452.7	25,000	466.4
20,000	413.6	20,000	421.7	30,000	411.7	30,000	433.4	30,000	447.4
25,000	395.8	25,000	407.4	35,000	393.9	35,000	414.1	35,000	429.6
30,000	376.9	30,000	393.0	36,089	390.0	40,000	395.1	39,400	414.7
30,715	374.7	30,065	392.7	40,000	390.0	45,000	376.9	39,500	414.7
35,000	374.7	35,000	391.4	45,000	390.0	50,000	359.6	40,000	414.9
40,000	374.7	40,000	390.1	50,000	390.0	53,595	347.7	45,000	417.1
42,377	374.7	45,000	388.8	55,000	390.0	55,000	350.7	50,000	419.5
45,000	361.1	50,000	387.5	60,000	390.0	60,000	361.7	50,400	419.5
50,000	336.8	55,000	386.2	65,000	390.0	65,000	373.0	50,500	419.5
50,583	334.7	60,000	385.0	65,617	390.0	69,620	383.7	55,000	420.6
55,000	334.7	65,000	383.7	70,000	392.4	70,000	384.2	60,000	421.5
60,000	334.7	70,000	382.4	75,000	395.1	75,000	390.7	65,000	422.5
61,087	334.7	75,000	381.1	80,000	397.9	80,000	397.4	70,000	425.0
65,000	346.5	80,000	379.8	85,000	400.6	85,000	404.2	75,000	428.5
70,000	359.2	85,000	378.7	90,000	403.3	90,000	410.9	80,000	432.0
73,055	365.7	86,092	378.3	95,000	406.1	95,000	417.6	85,000	435.9
75,000	365.2	90,000	378.3	100,000	408.8	100,000	424.3	90,000	439.9
80,000	363.8	95,000	378.3					95,000	443.8
85,000	361.9	100,000	378.3					100,000	448.1
90,000	360.0								
95,000	358.0								
100,000	355.8								

Table B.3   MIL-210B Nonstandard Days

Alt. (ft)	COLDB (1%) T (°R)	COLDB (10%) T (°R)	HOTB (10%) T (°R)	HOTB (1%) T (°R)
0	381.7	394.7	572.7	579.7
5000	403.7	419.7	550.7	553.7
10,000	409.7	422.7	527.7	533.7
15,000	401.7	411.7	509.7	517.7
20,000	389.7	395.7	495.7	503.7
25,000	375.7	381.7	479.7	487.7
30,000	363.7	373.7	465.7	474.7
35,000	359.7	369.7	450.7	463.7
40,000	359.7	369.7	436.7	449.7
45,000	357.7	363.7	427.7	437.7
50,000	347.7	355.7	421.7	427.7
60,000	334.7	343.7	421.7	427.7
70,000	339.7	343.7	421.7	427.7

Table B.4 MIL-210B Navy Nonstandard Days

Coastal		COLDC		HOTC	
Alt. (ft)	T (°R)	Alt. (ft)	T (°R)	Alt. (ft)	T (°R)
0	572.7	0	445.7	0	545.7
1000	563.7	1000	447.7	1000	549.7
2000	555.7	2000	447.7	2000	549.7
2500	551.7	2500	447.7	2500	549.7
5000	538.7	5000	442.7	5000	538.7
7500	527.7	7500	437.7	7500	527.7
10,000	518.7	10,000	431.7	10,000	518.7
12,500	509.7	12,500	426.7	12,500	509.7
15,000	500.7	15,000	421.7	15,000	500.7
17,500	491.7	17,500	415.7	17,500	491.7
20,000	483.7	20,000	409.7	20,000	483.7
22,500	474.7	22,500	403.7	22,500	474.7
25,000	465.7	25,000	397.7	25,000	465.7
27,500	456.7	27,500	391.7	27,500	456.7
30,000	447.7	30,000	384.7	30,000	447.7
32,500	439.7	32,500	377.7	32,500	439.7
35,000	429.7	35,000	374.7	35,000	429.7
37,500	421.7	37,500	372.7	37,500	421.7
39,500	415.7	40,000	368.7	39,500	415.7
40,000	416.7	42,500	365.7	40,000	416.7
42,500	419.7	45,000	360.7	42,500	419.7
45,000	421.7	47,500	355.7	45,000	421.7
47,500	421.7	50,000	349.7	47,500	421.7
50,000	421.7			50,000	421.7
52,500	421.7			52,500	421.7

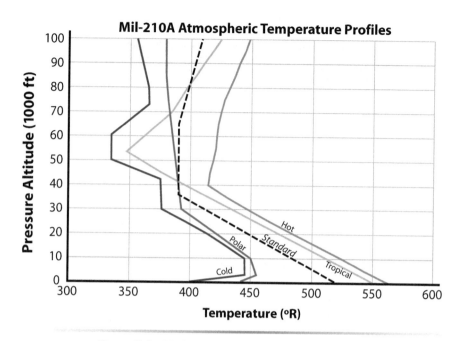

Figure B.1 Mil-210A atmospheric temperature profiles.

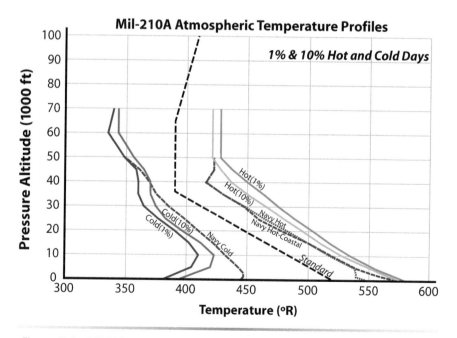

Figure B.2 Mil-210A atmospheric temperature profiles 1% and 10% hot and cold days.

USS Macon near Moffett Field Hangar One

- Defining and Understanding "Added Mass"
- Overview of Flight Conditions that Experience Added Mass
- Displaced Mass Ratio
- Review of Prolate Spheroid and its Added Mass and Moment of Inertia

An overview of the Moffett Field Facility, which was a world-class airship hangar during the 1930s. Built for the USS Macon in 1933, Hangar One was used only for large airships until 1935 when the Macon crashed. Since then the facility has had multiple uses it but its final utility is uncertain in 2013.

If you can't do it with brainpower, you can't do it with manpower working overtime.

Clarence "Kelly" Johnson

C.1 Added, Apparent, or Virtual Mass

One of the more misunderstood effects in all of aerodynamics is *added mass*. This term is sometimes referred to as apparent mass or virtual mass. These terms do not accurately convey its true character since they imply that there is something mystical about them. In this book the term *added mass* will be used. Another misconception is that added mass is only present for buoyant airships. Added mass affects the motion of every object that is accelerating or decelerating in a surrounding fluid, whether or not there is any appreciable buoyancy. However, we will see that the effects of added mass are only significant when the mass of the object is similar to that of the displaced surrounding fluid.

Although the Imlay Paper [2] gives the easiest conceptualization of added mass it is not perfect. This write-up will use the nomenclature from [2] as well as the time derivative of energy approach, which gives the clearest visualization of the source of added mass. The resulting equations represent a complete set of equations of motion including potential flow aerodynamics. However, some terms in these equations are not strictly due to added mass. This can present problems when these equations of motion are combined with typical aerodynamic static and dynamic coefficients derived from wind tunnel testing, water tunnel testing, or CFD. Care must be taken to ensure that aerodynamic forces and moments that are captured in the aerodynamic model are removed from the potential flow equations of motion without affecting the added mass terms.

When a body accelerates, decelerates, or changes direction while moving in a fluid, it behaves as though it has more mass than it actually does. The apparent increase in mass and distribution of this added mass varies with the nature of the motion. Generally, the added mass phenomenon is poorly understood because it isn't significant enough to impact airplane performance. Therefore, added mass is not included in aircraft design, not because it does not exist, but because it is extremely small compared to other forces and moments. However, added mass is always present for any body accelerating/decelerating through a fluid irrespective of its buoyancy. Only vehicles whose displaced external fluid mass is similar to the vehicle's mass may experience a significant effect from these external fluid forces. As will be seen later, when the equations of motion are developed, these forces are the result of the time rate of change of kinetic energy being added to the flow field. Vehicle accelerations/decelerations add work to the surrounding fluid, and it is this effect that will be captured in these equations.

> What the added mass phenomenon is, and what terms should be used to represent it in equations of motion, do not appear to be understood clearly by many persons.
>
> —*Frederick Imlay*

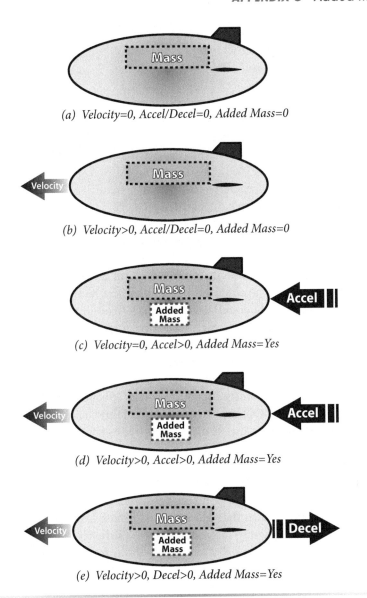

(a) Velocity=0, Accel/Decel=0, Added Mass=0

(b) Velocity>0, Accel/Decel=0, Added Mass=0

(c) Velocity=0, Accel>0, Added Mass=Yes

(d) Velocity>0, Accel>0, Added Mass=Yes

(e) Velocity>0, Decel>0, Added Mass=Yes

Figure C.1   Added mass for various flight conditions.

Figure C.1 summarizes at a top level those flight conditions that experience an added mass. Notice that the velocity of the vehicle is not relevant to the existence of added mass. It is purely an acceleration/deceleration effect. To get some perspective on the magnitude of added mass, Table C.1 compares this effect for various well known objects. Table C.1 introduces the term *displaced mass ratio* (*DMR*) that determines the magnitude of the *added mass* term. *DMR* is defined as the ratio of the mass of the displaced fluid volume (usually air for an airship or water for a submarine) to the actual mass of the vehicle including its internal gases. The low *DMR* for the Boeing 747 means that there is no measurable *added mass* because the mass of air displaced by the volume of the B-747 is trivial compared to its actual mass. However, objects such as an airship or air bubble have significant *DMR* values that create modest to large added mass terms. Based on *BR* note that the beach ball in air is less buoyant than an aluminum ball bearing in water!

> The air bubble is an interesting example in that its large *BR* would suggest that there is a force equivalent to 770 g's that would make the bubble accelerate rapidly to impossible speeds. In fact, its very large added mass term slows the bubble to its nominal speeds and reduces the force to about 2 g's.

As a first step in developing the equations for added mass, the force F_1 is defined as the *force that the fluid exerts on the body*. Then, F_1 is broken into its cartesian components X_1, Y_1, Z_1, in Eq. (C.1) and the fluid moment, FM_1, is made up of components K_1, M_1, N_1, and is expressed in a likewise fashion as Eq. (C.2). The assumptions are that the body is rigid and refers to a set of moving orthogonal Cartesian axes fixed in the body. The total force F_1 is the vector sum of each force component along the *x*, *y*, and *z* axes. Likewise, the total fluid moment FM_1 is the vector sum of each moment component along the *x*, *y*, and *z* axes.

There are a number of papers that discuss added mass but the most noteworthy ones are by Imlay [2] and Munk [3]. Shapes without symmetry about any of the 3 axes that are traveling in a real fluid have 3 force equations and 3 moment equations that contain 36 terms, as shown in Eqs. (C.3) and (C.4). If the fluid is ideal (irrotational and frictionless) then the 36 terms reduce to 21. Bodies with complex shapes and/or little symmetry must have their added mass estimated by CFD codes capable of modeling complex geometries or their added mass can be measured in a water tunnel test facility. Experience has shown that full potential flow (ideal fluid) solvers do a reasonable job of predicting the added mass for simple body shapes.

Using the unit vectors *i, j, k* allows the fluid force on the body to be written as

Table C.1 Buoyancy and Displaced Mass Ratios of Various Objects [1]

Object	DMR	BR
Boeing 747-400	0.01	−0.01
Aluminum Ball Bearing (in water)	0.36	−0.64
20 in. Beach Ball	0.13	−0.87
Goodyear Airship (Spirit of America)	1.25	0.98
Air Bubble (in water)	~770	~770
DMR = Displaced Mass Ratio BR = Buoyancy Ratio		

$$F_1 = X_1\boldsymbol{i} + Y_1\boldsymbol{j} + Z_1\boldsymbol{k} \qquad (\textit{fluid force on body}) \qquad (C.1)$$

Using the unit vectors $\boldsymbol{i}, \boldsymbol{j}, \boldsymbol{k}$ allows the fluid moment on the body to be written as

$$FM_1 = K_1\boldsymbol{i} + M_1\boldsymbol{j} + N_1\boldsymbol{k} \qquad (\textit{fluid moment on body}) \qquad (C.2)$$

These equations are based on the fact that the added mass force, F_1, is the mathematical result where X_1 is the force component due to a change in the fluid kinetic energy in the x-direction and likewise for Y_1 and Z_1.

FORCES $\hfill$ (C.3)

$$
\begin{aligned}
X_1 = \; & X_{\dot{u}}\dot{u} + X_{\dot{w}}(\dot{w} + uq) + X_{\dot{q}}\dot{q} + Z_{\dot{w}}wq + Z_{\dot{q}}q^2 && \text{(longitudinal)} \\
& + X_{\dot{v}}\dot{v} + X_{\dot{p}}\dot{p} + X_{\dot{r}}\dot{r} - Y_{\dot{v}}vr - Y_{\dot{p}}rp - Y_{\dot{r}}r^2 && \text{(lateral)} \\
& - X_{\dot{v}}ur - Y_{\dot{w}}wr && \text{(mixed terms)} \\
& + Y_{\dot{w}}vq + Z_{\dot{p}}pq - (Y_{\dot{q}} - Z_r)qr && \text{(mixed terms - usually small)} \\
Y_1 = \; & X_{\dot{v}}\dot{u} + Y_{\dot{w}}\dot{w} + Y_{\dot{q}}\dot{q} && \text{(longitudinal)} \\
& + Y_{\dot{v}}\dot{v} + Y_{\dot{p}}\dot{p} + Y_{\dot{r}}\dot{r} + X_{\dot{v}}vr - Y_{\dot{w}}vp + X_{\dot{r}}r^2 + (X_{\dot{p}} - Z_{\dot{r}})rp - Z_{\dot{p}}p^2 && \text{(lateral)} \\
& - X_{\dot{w}}(up - wr) + X_{\dot{u}}ur - Z_{\dot{w}}wp && \text{(mixed terms)} \\
& - Z_{\dot{q}}pq + X_{\dot{q}}qr && \text{(mixed terms - usually small)} \\
Z_1 = \; & X_{\dot{w}}(\dot{u} + wq) + Z_{\dot{w}}\dot{w} + Z_{\dot{q}}\dot{q} - X_{\dot{u}}uq - X_{\dot{q}}q^2 && \text{(longitudinal)} \\
& + Y_{\dot{w}}\dot{v} + Z_{\dot{p}}\dot{p} + Z_{\dot{r}}\dot{r} + Y_{\dot{v}}vp + Y_{\dot{r}}rp + Y_{\dot{p}}p^2 && \text{(lateral)} \\
& + X_{\dot{v}}up + Y_{\dot{w}}wp && \text{(mixed terms)} \\
& - X_{\dot{v}}vq - (X_{\dot{p}} - Y_{\dot{q}})(pq - X_{\dot{r}})qr && \text{(mixed terms - usually small)}
\end{aligned}
$$

MOMENTS $\hfill$ (C.4)

$$
\begin{aligned}
K_1 = \; & X_{\dot{p}}\dot{u} + Z_{\dot{p}}\dot{w} + K_{\dot{q}}\dot{q} - X_{\dot{v}}wu + X_{\dot{r}}uq - Y_{\dot{w}}w^2 - (Y_{\dot{q}} - Z_{\dot{r}})wq + M_{\dot{r}}q^2 && \text{(longitudinal)} \\
& + Y_{\dot{p}}\dot{v} + K_{\dot{p}}\dot{p} + K_{\dot{r}}\dot{r} + Y_{\dot{w}}v^2 - (Y_{\dot{q}} + Z_{\dot{r}})vr + Z_{\dot{p}}vp - M_{\dot{r}}r^2 - K_{\dot{q}}rp && \text{(lateral)} \\
& + X_{\dot{w}}uv - (Y_{\dot{v}} + Z_{\dot{w}})vw - (Y_{\dot{r}} + Z_{\dot{q}})wr - Y_{\dot{p}}wp - X_{\dot{q}}ur && \text{(mixed terms)} \\
& + (Y_{\dot{r}} + Z_{\dot{q}})vq + K_{\dot{r}}pq - (M_{\dot{q}} - N_{\dot{r}})qr && \text{(mixed terms - usually small)} \\
M_1 = \; & X_{\dot{q}}(\dot{u} + wq) + Z_{\dot{q}}(\dot{w} - uq) + M_{\dot{q}}\dot{q} - X_{\dot{w}}(u^2 - w^2) - (Z_{\dot{w}} - X_{\dot{u}})wu && \text{(longitudinal)} \\
& + Y_{\dot{q}}\dot{v} + K_{\dot{q}}\dot{p} + M_{\dot{r}}\dot{r} + Y_{\dot{p}}vr - Y_{\dot{r}}vp + K_{\dot{r}}(p^2 - r^2) + (K_{\dot{p}} - N_{\dot{r}})rp && \text{(lateral)} \\
& - Y_{\dot{w}}uv + X_{\dot{v}}vw - (X_{\dot{r}} + Z_{\dot{p}})(up - wr) + (X_{\dot{p}} - Z_{\dot{r}})(wp + ur) && \text{(mixed terms)} \\
& - M_{\dot{r}}pq + K_{\dot{q}}qr && \text{(mixed terms - usually small)} \\
N_1 = \; & X_{\dot{r}}\dot{u} + Z_{\dot{r}}\dot{w} + M_{\dot{r}}\dot{q} + X_{\dot{v}}u^2 + Y_{\dot{w}}wu - (X_{\dot{p}} - Y_{\dot{q}})uq - Z_{\dot{p}}w_q - K_{\dot{q}}q^2 && \text{(longitudinal)} \\
& + Y_{\dot{r}}\dot{v} + K_{\dot{r}}\dot{p} + N_{\dot{r}}\dot{r} - X_{\dot{v}}v^2 - X_{\dot{r}}vr - (X_{\dot{p}} - Y_{\dot{q}})vp + M_{\dot{r}}rp + K_{\dot{q}}p^2 && \text{(lateral)} \\
& - (X_{\dot{u}} - Y_{\dot{v}})uv - X_{\dot{w}}vw + (X_{\dot{q}} + Y_{\dot{p}})up + Y_{\dot{r}}ur + Z_{\dot{q}}wp && \text{(mixed terms)} \\
& - (X_{\dot{q}} + Y_{\dot{p}})vq - (K_{\dot{p}} - M_{\dot{q}})pq - K_{\dot{r}}qr && \text{(mixed terms - usually small)}
\end{aligned}
$$

Typically, airship bodies and submarines have a vertical plane of symmetry that is closely approximated by prolate spheroids with aft fins. These shapes and planes of symmetry apply to conventional airships as well, but less so for a non-axisymmetric shape that is typical of a hybrid airship. Applying the symmetry relationships in Eq. (C.5) to Eq. (C.3) and Eq. (C.4) [2] leads to the 6 equations (3 forces and 3 moments) in Eq. (C.6).

For the following equations $X_{\dot{v}} = \partial X/\partial \dot{v}$ $Y_{\dot{w}} = \partial Y/\partial \dot{w}$ etc.

$$X_{\dot{v}} = X_{\dot{w}} = X_{\dot{p}} = X_{\dot{q}} = X_{\dot{r}} = 0$$

$$Y_{\dot{w}} = Y_{\dot{q}} = 0$$

$$Z_{\dot{p}} = Z_{\dot{r}} = 0 \qquad \textbf{Body of revolution}$$

$$K_{\dot{q}} = 0 \qquad \textbf{with symmetric fins} \qquad \text{(C.5)}$$

$$M_{\dot{r}} = 0 \qquad \text{due to symmetry}$$

$$Y_{\dot{v}} = Z_{\dot{w}}, \ Y_{\dot{r}} = Z_{\dot{q}}, \ M_{\dot{q}} = N_{\dot{r}} \qquad \text{due to symmetry}$$

Symmetry Relationships

The problem becomes one of calculating each of the force and fluid moment components along each of the unit vector axes. When a body is accelerated/decelerated its work on the surrounding fluid produces a force of the fluid back on the body that, when divided by the body's acceleration, gives a value referred to here as *added mass*.

Reviewing the results in Eq. (C.6) some simplifications become clear. Most obvious is that for rotational motions, $K_{\dot{p}}$ would be negligible compared to $N_{\dot{r}}$ for M_1 and N_1. If there were no fins, $K_{\dot{p}}$ would be exactly zero but even with the usual size of tail fins it can be assumed zero as noted in Eq. (C.6).

$$X_1 = X_{\dot{u}}\dot{u} - Y_{\dot{v}}(rv - qw) - Y_{\dot{r}}(q^2 + r^2)$$
$$Y_1 = X_{\dot{u}}ru + Y_{\dot{v}}(\dot{v} - pw) + Y_{\dot{r}}(\dot{r} + pq)$$
$$Z_1 = -X_{\dot{u}}uq + Y_{\dot{v}}(\dot{w} + pv) - Y_{\dot{r}}(\dot{q} - pr)$$
$$K_1 = K_{\dot{p}}\dot{p} \qquad \text{\{zero for no fins, generally small\}}$$
$$M_1 = (X_{\dot{u}} - Y_{\dot{v}})uw - Y_{\dot{r}}(\dot{w} + pv - qu) + N_{\dot{r}}\dot{q} - (N_{\dot{r}} - K_{\dot{p}})pr$$
$$N_1 = -(X_{\dot{u}} - Y_{\dot{v}})vu + Y_{\dot{r}}(\dot{v} + ru - pw) + N_{\dot{r}}\dot{r} + (N_{\dot{r}} - K_{\dot{p}})pq$$

$$\left. \begin{array}{l} \textbf{Body of revolution} \\ \textbf{with symmetric fins} \end{array} \right\} \text{(C.6)}$$

where a sample definition of terms is shown below -

$X_{\dot{u}}$ is the force in the x-direction due to acceleration along the x-axis.

$Y_{\dot{v}}$ is the force in the y-direction due to acceleration along the y-axis.

$Y_{\dot{r}}$ is the force in the y-direction due to rotational acceleration about the z-axis.

$N_{\dot{r}}$ is the moment about the z-axis due to rotational acceleration about the z-axis, etc.

It is instructive to take another look at Fig. C.1. Here, there are five different motions that exhibit varying amounts of added mass. For the two cases where the body is at rest or has a constant velocity there is no additional external force F_1 beyond normal aerodynamic forces. In other words, F_1 does not include forces on the body due to velocity, such as lift and drag; rather it accounts for the added forces generated by the fluid on the body because the body was accelerated or decelerated. These added fluid forces reduce the body's acceleration, which makes the body behave as though it has more mass than it really does. Calling this effect *added mass* is merely an algebraic convenience since the body's mass is the same but its behavior is as though its mass was greater.

The remaining three motions in Fig. C.1 do experience added mass because they all involve accelerations or decelerations. Imlay's [2] approach describes new motions that are the result of body accelerations moving an otherwise stationary fluid because the fluid must move aside and then close behind as the body passes through it. This body motion imparts a kinetic energy to the fluid that it did not have in its original quiescent state—in other words, the body has done work on the fluid. Complete equations of motion for the body must include this kinetic energy given to the fluid by the body and immediately returned to the body as added mass and added moment of inertia. However, when body motion is steady (acceleration = 0) the fluid kinetic energy is constant for an ideal fluid and there is no net work done on the fluid so the *added mass* and *added moment of inertia* terms are zero. Figure C.1 should be studied carefully to understand all of the situations where *added mass* comes into play.

The next step in this discussion is the relationship between body geometry and the *added mass* derivatives, which are not functions of the body mass at all but are a result of the mass of displaced fluid of the body volume. We will now examine the theoretical values of these derivatives for the simplified geometry of prolate spheroids (no tails or fins).

Equation (C.7) shows the equation for an ellipsoid with elliptical cross sections in all three planes of symmetry that have the principal axes go through the origin at the center of the ellipsoid.

$$\frac{x^2}{a^2} + \frac{y^2}{b^2} + \frac{z^2}{c^2} = 1 \text{ where } a, b, \text{ and } c \text{ are the semiaxes.} \qquad \text{(C.7)}$$

The simplified geometry of an ellipsoid results in the following terms being zero -

$$X_{\dot{v}} = X_{\dot{w}} = X_{\dot{p}} = X_{\dot{q}} = X_{\dot{r}} = Y_{\dot{w}} = Y_{\dot{p}} = Y_{\dot{q}} = Y_{\dot{r}} = 0$$
$$Z_{\dot{p}} = Z_{\dot{q}} = Z_{\dot{r}} = K_{\dot{q}} = K_{\dot{r}} = M_{\dot{r}} = 0$$

where $X_{\dot{v}}$ is a force in the x-direction due to an acceleration in the y-direction and $Z_{\dot{q}}$ is the moment about the z-axis due to an angular acceleration about the y-axis.

Using nomenclature from [2], α_0 and β_0 are constants that describe the relative geometric proportions of an ellipsoid. From the added mass equations for an ellipsoid it is evident that the added mass derivatives are merely functions of body shape and size as well as the fluid density.

Prolate Spheroid

Looking at the special case of an ellipsoid called a prolate spheroid, where $b = c$, $a > b$, the ellipsoidal restrictions on Eqs. (C.3) and (C.4) simplify to Eq. (C.8) while recognizing that because of symmetry $Y_{\dot{v}} = Z_{\dot{w}}$, $M_{\dot{q}} = N_{\dot{r}}$, and $K_{\dot{p}} = 0$. Using these identities for a prolate spheroid, the ellipsoidal equations are reduced to the following-

$$X_{\dot{u}} = -\overbrace{\frac{\alpha_0}{2-\alpha_0}}^{k_1}\overbrace{\frac{4}{3}\pi\rho ab^2}^{mass}$$

Prolate Spheroids

$$Y_{\dot{v}} = Z_{\dot{w}} = -\overbrace{\frac{\beta_0}{2-\beta_0}}^{k_2}\overbrace{\frac{4}{3}\pi\rho ab^2}^{mass}$$

$$K_{\dot{p}} = 0$$

$$N_{\dot{r}} = M_{\dot{q}} = -\frac{1}{5}\overbrace{\frac{(b^2-a^2)^2\,(\alpha_0-\beta_0)}{2(b^2-a^2)+(b^2+a^2)\,(\beta_0-\alpha_0)}}^{k'}\overbrace{\frac{4}{3}\pi\rho ab^2}^{mass}$$

$$(C.8)$$

Combining Eqs. (C.9) – (C.11) for a prolate spheroid and using Lamb's [3] k factors, yields the final simplified set of equations shown as Eq. (C.12). Factors k_1 and $k_2 = k_3$ are plotted in Fig. C.2.

For a prolate spheroid where ε is the eccentricity or ellipticity of the meridian elliptical section and is defined as

$$\varepsilon^2 = 1 - (b/a)^2 \qquad (C.9)$$

and α_0 and β_0 terms are defined as -

$$\alpha_0 = \frac{2\left(1-\varepsilon^2\right)}{\varepsilon^3}\left(\frac{1}{2}\log\frac{1+\varepsilon}{1-\varepsilon}-\varepsilon\right) \qquad \beta_0 = \frac{1}{\varepsilon^2}-\frac{1-\varepsilon^2}{2\varepsilon^3}\log\frac{1+3}{1-\varepsilon} \qquad (C.10)$$

Lamb's k Factors -

$$\text{Let } k_1 = \frac{\alpha_0}{2-\alpha_0} \qquad (C.11a)$$

$$k_2 = \frac{\beta_0}{2 - \beta_0} \tag{C.11b}$$

$$k' = \frac{\varepsilon^4 (\beta_0 - \alpha_0)}{(2 - \varepsilon^2)\left[2\varepsilon^2 - (2 - \varepsilon^2)(\beta_0 - \alpha_0)\right]} \tag{C.11c}$$

$$X_{\dot{u}} = -k_1 (4/3) \pi \rho a b^2 \qquad \text{(force in x-direction)} \tag{C.12a}$$

$$Y_{\dot{v}} = Z_{\dot{w}} = -k_2 (4/3) \pi \rho a b^2 \qquad \text{(force in y or z-direction)} \tag{C.12b}$$

$$N_{\dot{r}} = M_{\dot{q}} = -k' (4/15) \pi \rho a b^2 (a^2 + b^2) \qquad \text{(moment around y or z axes)} \tag{C.12c}$$

where, the mass of the displaced volume of the surrounding fluid is

$$\text{mass of displaced fluid} = (4/3) \pi \rho a b^2$$

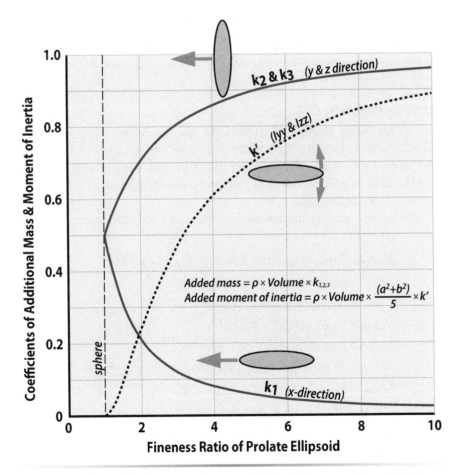

Figure C.2 Added mass and moment of inertia for a prolate ellipsoid.

and the moment of inertia of the displaced mass is

$$\text{moment of inertia of displaced mass} = (4/15)\pi\rho ab^2 (a^2 + b^2) \quad \text{(C.13)}$$

Equation (C.13) can be split into its two fundamental parts,

$$\text{mass of displaced volume for an ellipsoid} = (4/3)\pi\rho ab^2$$
$$\text{and the moment term for an ellipsoid} = (a^2 + b^2)/5$$

All of the remaining *added mass* derivatives are zero for the *prolate ellipsoid*. The factors k_1, k_2, and k' are determined solely by the fineness ratio (a/b) of the spheroid. These results are generally true for any body, therefore the *added mass* derivatives for any body can always be expressed as some proportionality factor (k_1, k_2, k') times either the mass of fluid displaced by the body or the moment of inertia of that displaced fluid mass. A common misconception is that the *added mass* proportionality factors are multipliers on the mass of the object. In fact the *added mass* factor applies to the displaced mass of the surrounding fluid (usually air for airships or water for submarines) and not the mass of the vehicle itself.

Sample Problem C.1

The following example illustrates the magnitude of *added mass* and *added moment of inertia* for a prolate ellipsoid whose envelope is considered to be a thin shell encapsulating the lifting gas. Equations (D.10) and (D.11) will be needed to calculate the total moment of inertia for the envelope and its contained gas.

The following are examples of added mass and added moment of inertia calculations:

Prolate ellipsoid ($a = 100$ ft, $b = 25$ ft, and $c = 25$ ft)

FR = length/diameter = 200/50 = 4.0

Mass of envelope = 2800 lb (given)

Volume = 261,800 ft^3 = $4/3\pi ab^2$

Mass of helium in envelope = 2765 lb

$\quad\quad\quad${261,800 × 0.01114 × (491.7/518.7)} @SL (see Table 2.3)

Mass of displaced air = 20,020 lb @SL

Ellipsoid total mass = 2800 lb + 2765 lb = 5565 lb

Envelope moment of inertia = 8,750,000 lb-ft^2 [Eq. (D.11)]

Helium moment of inertia = 2765 × ($100^2 + 25^2$)/5

$\quad\quad\quad\quad\quad\quad\quad$ = 5,875,625 lb-ft^2 [Eq. (D.10)]

Total moment of inertia (envelope + gas) = 14,625,625 lb-ft^2

Moment of inertia of displaced air $= 20{,}020 \times (100^2 + 25^2)/5$
$$= 42{,}542{,}500 \text{ lb-ft}^2$$

$k_1 = 0.082$, $k_2 = 0.860$, $k' = 0.61$ (see Fig. C.2)

Thus, adding all contributions to mass and moment of inertia results in -

Added mass in x-direction $= 0.082 \times 20{,}020 = 1642$ lb

Added mass in y or z-direction $= 0.86 \times 20{,}020 = 17{,}217$ lb

Total mass (x-direction with accel/decel) $= 5{,}565 + 1{,}642 = 7{,}207$ lb

Total mass (y or z-direction with accel/decel) $= 5{,}565 + 17{,}217$
$$= 22{,}782 \text{ lb}$$

Added moment of inertia (with accel/decel) $= 0.61 \times 42{,}542{,}500$
$$= 25{,}950{,}925 \text{ lb-ft}^2$$

Total moment of inertia (with accel/decel) $= 14{,}625{,}625 + 25{,}950{,}925$
$$= \mathbf{40{,}576{,}550 \text{ lb-ft}^2}$$

Notice that the added mass and moments of inertia can be substantial in the y- or z-directions but much less in the x-direction. These calculations show that masses and moments of inertia with added terms have no value by themselves. They become important only when total forces or moments are applied to an airship to assess the controllability and handling qualities of an airship in various flight conditions.

A non-intuitive aspect of added mass and moment is that the *factors* are not functions of the amount of acceleration. The factors are only influenced by body geometry and surrounding fluid density. However, the amount of the added mass *is* used to calculate the dimensional value of a body force for a given body mass and acceleration. In this case the greater the acceleration the greater the force.

Most treatments of *added mass* and/or *moment of inertia* assume that this effect happens instantaneously. This is not so. In fact, the effect of added mass may lag the acceleration/deceleration by a second or more. When flight control characteristics are used to develop flight control laws, this time delay should be included. However, the impact of this time lag on the handling qualities of an airship is outside the scope of this book.

References

[1] Atkinson, C.J. and Urso, R.G., "Modeling of Apparent Mass Effects for the Real-Time Simulation of a Hybrid Airship," AIAA 2006–6619, New York, 2006.

[2] Imlay, F.H., "The Complete Expressions for 'Added Mass' of a Rigid Body Moving in an Ideal Fluid," TMB Report 1528, July 1961.

[3] Lamb, H., *Hydrodynamics*, Sixth Edition, Dover Publications, pp. 160–169, 1932.

[4] Munk, M., "Some Tables of the Factor of Apparent Additional Mass," NACA TN-197, July 1924.

Aerostat 7

- Using Properties of Ellipsoids to Approximate Characteristics of a Hybrid Airship
- The Effect of Lobes on Surface Area and Perimeter; Calculating Surface Area of Lobed Shapes

The 71M is a modern aerostat. These aerostats provide cost-effective ISR information for areas where the airspace is controlled. They are perfect vehicles for monitoring border activity. They have been used in many parts of the world including Iraq and Afghanistan.

It is better to light one candle than curse the darkness.
Chinese Proverb

D.1 Useful Design Information

The geometric and physical properties of ellipsoids can be found in Eqs. (D.1) thru (D.11). From these relationships many important design parameters can be calculated.

Although these equations are for ellipsoids they are also used to help define characteristics of hybrid airships that are not bodies of revolution. Notice that the calculations for surface area and perimeter for ellipsoids are approximate as there are no exact equations. However, the equation for volume [Eqs. (D.1) and (D.2)] are exact. Notice also that there are a few exceptions where planform area is used as the reference value instead of $Vol^{2/3}$. This is only done when trying to merge airplane and airship data together. Otherwise, $Vol^{2/3}$ is always used as the reference quantity.

Figure D.1 shows how the concept of equivalent diameter is used to estimate some of the characteristics of a hybrid airship. The length of the ellipse is the length of the hybrid airship. The equivalent diameter, d_e, is computed as it usually is in aerodynamics where the frontal area of the body and the circle with diameter $= d_e$ have the same areas. This also introduces the concept of aspect ratio, AR. Aspect ratio is computed the same way for ellipsoids as for hybrid lobed configurations.

Properties of Ellipsoids (a, b, c, are semi-axis lengths of an ellipsoid)

$$\text{Volume} = \frac{4}{3}\pi abc = \pi \, \ell_B \, width \, height/6 \tag{D.1}$$

$$\text{Volume} = \frac{4}{3}\pi ab^2 = \pi \, \ell_B \, d_e^2/6 \qquad (body \ of \ revolution) \tag{D.2}$$

$$\text{Surface Area} \approx 4\pi \left[\frac{a^p b^p + a^p c^p + b^p c^p}{3}\right]^{\frac{1}{p}} \text{ where } p = 1.6075 \tag{D.3}$$

$$\text{Surface Area} \approx 3.88 \, Vol^{2/3} \, FR^{1/3} \qquad (bodies \ of \ revolution \ only) \tag{D.4}$$

$$\text{Ellipse Perimeter} \approx \pi \left[3(a+b) - \sqrt{10ab + 3(a^2 + b^2)}\right] \tag{D.5}$$

$$\text{Aspect Ratio, } AR = span^2/area = width^2/S_{plan} \tag{D.6}$$

$$\text{Planform Area, } S_{plan} = \pi \ell_B d_e/4 \tag{D.7}$$

$$\frac{x^2}{a^2} + \frac{y^2}{b^2} + \frac{z^2}{c^2} = 1 \qquad (scalene \ ellipsoid) \tag{D.8}$$

$$mass = m = \rho \, Vol \tag{D.9}$$

$$I_{xx} = m\frac{b^2 + c^2}{5} \quad I_{yy} = m\frac{a^2 + c^2}{5} \quad I_{zz} = m\frac{a^2 + b^2}{5} \quad (solid) \tag{D.10}$$

$$I_{yy} \text{ or } I_{zz} = m\left[a^2 + b^2 + \frac{2ab^3 + 2a^2b^2}{2ab + b^2}\right] \qquad (thin \ shell) \tag{D.11}$$

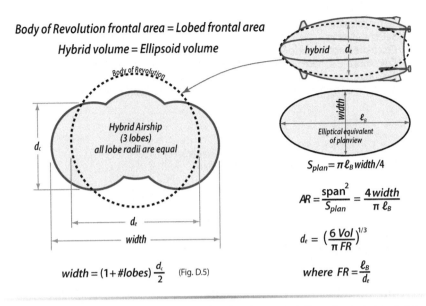

Figure D.1 Use of ellipsoid to approximate characteristics of a hybrid airship.

Figure D.2 summarizes how lobes change the surface area and perimeter of a hybrid airship relative to an ellipsoid. This detailed figure goes through all of the steps and derivations that contrast the geometries of lobed configurations to ellipsoids. Notice the condition for the hybrid geometry. Assume that lobes are placed one R from each other. This is a good assumption as real designs will turn out to be very close.

Figure D.3 gives the value of the ratio equivalent diameter/circular diameter of lobe, d_e/d_c. Given d_e for a configuration and the number of lobes yields the diameter of the lobes, d_c. For a given internal pressure this allows the load/inch to be calculated, which gives the envelope weight for a specific material. Based on this figure it should be clear that volume, FR, and #lobes fully describes the geometry of a hybrid airship body.

Figure D.4 is included as a quick check for calculating the surface area. The equation for calculating surface area of an ellipsoid is unwieldly and it is easy to make a mistake. Use this to check initial calculations.

Figure D.5 shows how the equivalent flat plate skin friction coefficient varies for given amounts of laminar flow. This figure is also good for checking initial calculations. The amount of skin friction reduction is quickly known. Although this figure is for 4000 ft these data should still be good for relative changes in C_f at other altitudes as well.

Process for calculating Surface Area of lobed shape:
1) Select # of lobes
2) Calculate length, width, & height of ellipsoid for the given volume & FR
3) Calculate the surface area using ellipse length, width, & height (Eq. D.3)
4) Calculate the lobed shape surface area by multiplying ellipse surface area by the ratio (Perimeter$_{lobes}$/Perimeter$_{ellipse}$) in the table below.

Assumes additional lobes are offset by R relative to the one next to it.

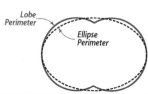

R = radius of lobe
All lobed geometries are 2R in height
Ellipses have the same height and width as lobed shapes

Front View	# Lobes	Lobes Perimeter	Lobes Width	Ellipse Area	Ellipse Perimeter	Perimeter$_{lobes}$ / Perimeter$_{ellipse}$
●	1	$6\pi R/3$	$2R$	$2\pi R^2/8$	$2\pi R$	1.0
●	2	$8\pi R/3$	$3R$	$3\pi R^2/8$	$2.525\pi R$	1.057
●	3	$10\pi R/3$	$4R$	$4\pi R^2/8$	$3.084\pi R$	1.081
●	4	$12\pi R/3$	$5R$	$5\pi R^2/8$	$3.663\pi R$	1.092
●	5	$14\pi R/3$	$6R$	$6\pi R^2/8$	$4.254\pi R$	1.098

$$\frac{Perimeter_{lobes}}{Perimeter_{ellipse}} = 1.122 - 0.1226 \left(\frac{1}{\# lobes}\right)$$

Figure D.2   Effect of lobes on surface area and perimeter.

Sample Problem D.1: Surface Area of Lobed Shape

For $Vol = 1{,}000{,}000$ ft^3, $FR = 3$, and #lobes = 3
For a body of revolution $d_e = (6\ Vol/\pi\ FR)^{1/3} = 86.0$ ft, $\ell_B = 258.1$ ft
Lobe diameter is calculated (Fig. D.3) as $d_c = 86.0/1.50 = 57.3$ ft
The lobe radius is $d_c/2 = 28.7$ ft
Width of the lobed configuration (Fig. D.2) $= (1 + 3)(28.7) = 114.6$ ft

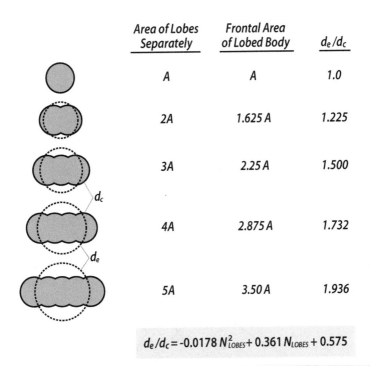

	Area of Lobes Separately	Frontal Area of Lobed Body	d_e/d_c
	A	A	1.0
	2A	1.625 A	1.225
	3A	2.25 A	1.500
	4A	2.875 A	1.732
	5A	3.50 A	1.936

$$d_e/d_c = -0.0178\, N_{LOBES}^2 + 0.361\, N_{LOBES} + 0.575$$

Figure D.3 Frontal areas and equivalent diameters of lobed bodies.

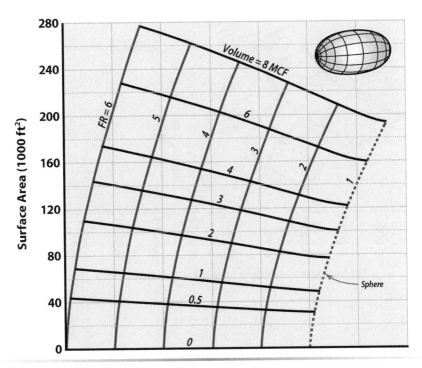

Figure D.4 Surface area of ellipsoidal bodies for varying *FR* and volume.

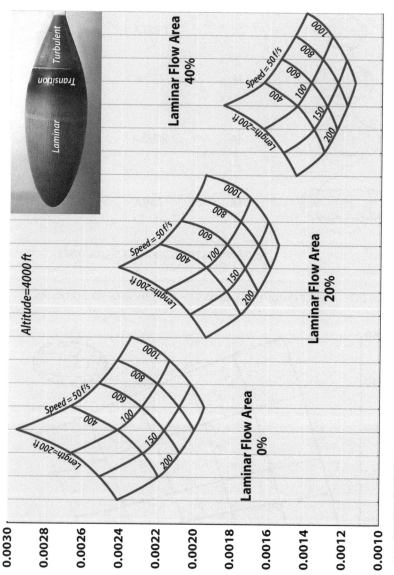

Figure D.5 Equivalent coefficient of friction for bodies of revoltution.

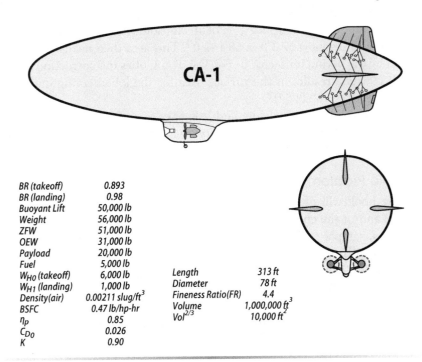

BR (takeoff)	0.893
BR (landing)	0.98
Buoyant Lift	50,000 lb
Weight	56,000 lb
ZFW	51,000 lb
OEW	31,000 lb
Payload	20,000 lb
Fuel	5,000 lb
W_{H0} (takeoff)	6,000 lb
W_{H1} (landing)	1,000 lb
Density(air)	0.00211 slug/ft^3
BSFC	0.47 lb/hp-hr
η_P	0.85
C_{D_0}	0.026
K	0.90

Length	313 ft
Diameter	78 ft
Fineness Ratio(FR)	4.4
Volume	1,000,000 ft^3
Vol$^{2/3}$	10,000 ft^2

Figure D.6 Typical layout of a conventional airship (CA-1).

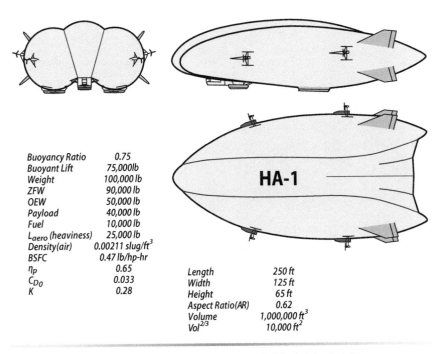

Buoyancy Ratio	0.75
Buoyant Lift	75,000 lb
Weight	100,000 lb
ZFW	90,000 lb
OEW	50,000 lb
Payload	40,000 lb
Fuel	10,000 lb
L_{aero} (heaviness)	25,000 lb
Density(air)	0.00211 slug/ft^3
BSFC	0.47 lb/hp-hr
η_P	0.65
C_{D_0}	0.033
K	0.28

Length	250 ft
Width	125 ft
Height	65 ft
Aspect Ratio(AR)	0.62
Volume	1,000,000 ft^3
Vol$^{2/3}$	10,000 ft^2

Figure D.7 Typical layout of a hybrid airship (HA-1).

Using Eq. (D.3) the surface area of an ellipsoid with $\ell_B = 258.1$, width = 114.6 ft, and height = 57.3 ft is 58,182 ft^2. This area then needs to be corrected by the perimeter factor in Fig. D.4. For 3 lobes the correction factor is 1.081, which results in the surface area for the lobed configuration of $(1.081)(58,182) = 62,895$ ft^2

Since Eq. (D.3) is unwieldly and it is easy to make a mistake using it Fig. D.4 is included as a quick sanity check for the surface area of an ellipsoid.

Sample Problem D.2: Vehicle Characteristics

The book will often use the characteristics of a conventional airship (CA-1) and/or the characteristics of a hybrid airship (HA-1) in sample calculations, sample problems, or general discussions. The following data is presented as being representative of each airship type but is not for any specific vehicle. Flight will always be assumed to be at 4000 ft on a standard day.

CA-1 is an ellipsoidal body of revolution with the characteristics shown in Fig. D.6, which shows a 3-D view of CA-1 and its assumed dimensions and geometry. It is powered by (2) reciprocating engines driving propellors.

HA-1 is a 3-lobe hybrid design that assumes it has an Air Cushion Landing System (ACLS). Figure D.7 shows the 3-view of HA-1. It is powered by (4) reciprocating engines driving propellors.

Historical Airship Database

- Overview and Review of Airship Characteristics and Performance

Built in 1918 for military purposes, the R-31 had a to speed of 70 mph, which wa faster than any other airshi then in service. She was powered by six 275 hp engines. As a fuel-saving design change one engine was removed, reducing the airship's speed to 65 mph. The rigid structure was ma of wood.

A scientist discovers that which exists . . .
an engineer creates that which never was.
Theodore Von Kármán

The following database of airship characteristics and performance is the aggregation of information from several sources [1, 2]. Sometimes differing data from more than one source was available and in those cases the authors chose the data that seemed the most credible.

This database has been expanded to include several parameters that have been calculated or measured from reliable three-view drawings. First, envelope volume has been added for rigid designs that quote gas cell volumes but rarely include the displacement volume of the external hull. Calculations were made using the maximum diameter and length along with actual nose, mid-section, and aft end shapes.

Next, horizontal and vertical tail size data is included for numerous configurations and is the source of the tail sizing plots in Chapter 7. Some of this data comes from old drawings and wind tunnel models while the rest of the data comes from tracing airship plan views and side views, and estimating the horizontal and vertical tail areas. The generalized parameters, horizontal and vertical *tail volume coefficients*, are included where tail sizing information was available. For those configurations that had known tail sizes but an unknown tail moment arm, the moment arm was assumed to be 38% of the body length. While C_{HT} is computed conventionally, C_{VT} is not. Normally, the vertical tail volume coefficient is referenced to span. For airships this is not an important parameter so body length has been substituted instead. This results in the following relationships.

$$C_{HT} = S_{HT}\, \ell_{HT}/\ell_B\, Vol^{2/3}$$
$$C_{VT} = S_{VT}\, \ell_{VT}/\ell_B\, Vol^{2/3}$$

Calculating the drag coefficient was attempted for those configurations where the maximum speed and horsepower were known. However, the results were unacceptable given the many assumptions that had to be made for this calculation.

Table E.1   Summary of Performance Characteristics for Historical Airships

rship	Vol (ft³)	Year	Type	Gross Wt (lb)	Empty Wt (lb)	Heavi-ness (lb)	Alt ceil (ft)	Max speed (kt)	# Eng	Eng (hp)
4	95,000	1919	NR	6424	4200	2224	8000	40.0	1	90
C A-60+	68,000	1988	NR	4270	3070	1200	7300	47.8	2	80
ROS 40A	60,035	1997	NR	1200	800	400	10,000	36.7	2	68
ROS 40B	88,570	2002	NR	5710	4220	1490	9840	44.3	2	125
Class	84,000	1917	NR	5680	3600	2080	8500	40.8	1	100
Class	181,000	1918	NR	12,239	7600	4639	8600	52.1	2	300
Class	190,000	1919	NR	12,239	7900	4339	8800	50.4	2	250
Class	95,000	1918	NR	6424	4541	1883	8000	48.7	1	150
-1	183,000	1929								
oodyear GZ-20	202,700	1969	NR	12,840	9840	3000		43.4	2	210
ifford Steam Airship	88,287	1852	NR						1	3
-1	174,800	1922								
-3-8	416,000	1941								
-x	456,000									
Class	123,000	1942								
ghtship A-150	150,000	2003	NR	9689	6305	3384	10,000	54.7	2	180
1-2	202,200	1929	NR							
A Military	180,000	1921	NR	12,172	7651	4521	10,500	56.5	2	260
1 "Norge"	670,980	1924	SR	53,500	24,200	29,300		60.8	3	250
avy A (DN)1	110,000	1916	NR	7438	5834	1604	6000	30.4	2	140
orth Sea Class	360,000	1918	NR	27,400	13,900	13,500		45.3	2	250
)-1	127,000	1919								
arseval PL 27	1,105,344	1918	NR							
2100	5,156,000	1929	R			236,365	350,610		6	700
2-23	1,040,000	1917	NR	79,500	39,900	39,600		45.0	4	250
2-31	1,610,000	1918	R	123,000	68,300	54,700		61.0	6	300
2-33	2,100,000	1919	R	160,000	81,700	78,300		52.7	5	250
2-9	930,000	1916	R	71,000	42,100	28,900		39.1		600

(continued)

Airship	Vol (ft³)	Year	Type	Gross Wt (lb)	Empty Wt (lb)	Heaviness (lb)	Alt ceil (ft)	Max speed (kt)	# Eng	Eng (hp)
Republique Class	130,000	1908	SR							
Sentinel 1000	353,100	1991	NR			5962	8000	60.0	2	300
Sentinel 8500	3,000,000	1995	NR					87.0	3	1800
Skyship 600	235,400	1984	NR	15,653	10,488	5165	7000	56.5	2	255
TC-14	357,000	1924	NR	11,584	7511	4073		50.4	2	380
TE-1	80,200	1926	NR	4630	3296	1334		39.1	2	80
TF-1	52,290	1926	NR	2870	1937	933		34.8	1	40
US-LTA 138-S	138,000	1987	NR	8900	5883	3017	9000	56.5	1	300
WDL 1B	254,275	1997	NR	16,243	11,243	5000	6561	58.0	2	210
Zep LZ1,3	442,398	1900	R	28,880	22,707	6173	2135	14.8	2	14
Zep LZ2	442,398	1906	R	26,565	20,392	6173	2790	21.4	2	80
Zep LZ4,5,6	584,810	1908	R	38,360	28,108	10,251		30.1	2	105
Zep LZ7,8	739,768	1910	R	49,383	34,392	15,873		32.5	3	120
Zep LZ9, 10,12	699,626	1911	R	45,525	29,982	15,542	8040	40.8	3	145
Zep LZ11,13	740,378	1912	R	47,840	33,400	14,440		40.8	3	145
Zep LZ14	892,278	1912	R	57,540	39,462	18,078				
Zep LZ15,16	801,954	1912	R	50,044				41.2	3	180
Zep LZ17, 19,20	790,621	1912	R		39,462	10,585		41.2	3	180
Zep LZ18	1,108,411	1913	R	69,114	44,643	24,471		40.8	4	180
Zep LZ21	835,811	1913	R	53,461	34,061	21,605		39.8	3	180
Zep LZ22,23	880,859	1913	R	56,658	37,147	19,511		38.9	3	180
Zep LZ24,25, 27-35,37	892,330	1914	R	57,540	37,257	20,282	9185	45.5	3	180
Zep LZ26	1,049,394	1914	R	63,933	37,037	26,896		43.7	3	180
Zep LZ36,39	1,050,717	1915	R	63,713	39,241	24,471		45.9	3	180
Zep LZ38, 40-58, 60,63	1,413,214	1915	R	81,570	45,855	35,714	9185	51.9	4	210
Zep LZ59, 61,64-69, 71,73, 77,81	1,437,632	1915	R	91,601	52,138	39,462		51.5	4	240

rship	Vol (ft³)	Year	Type	Gross Wt (lb)	Empty Wt (lb)	Heavi-ness (lb)	Alt ceil (ft)	Max speed (kt)	# Eng	Eng (hp)
ep LZ62, 72,74–76, 78-80, 82–90	2,593,357	1915	R	140,653	69,224	71,429	12,795	43.5	6	240
ep LZ91-99	2,574,191	1917	R	142,747	59,303	83,444	17,060	54.0	5	237
ep LZ-96 (~L-49)	2,538,744	1917	R	160,000	58,200	101,800		57.5	1	1200
ep LZ100, 101,103, 105–111	2,574,191	1917	R	143,298	55,115	88,183	21,325	58.1	5	240
ep LZ102, 104	2,997,663	1917	R	175,265	60,406	47,840	21,655	55.6	5	240
ep LZ112	2,793,181	1918	R	159,171	54,453	104,718	20,340	70.8	7	240
ep LZ113, 114	2,997,058	1918	R	175,176	54,453	120,723	6235	70.8	7	240
ep LZ120-Bodensee	1,053,363	1919	R	51,367	29,101	22,046		71.5	4	245
ep LZ126 (ZR-3, LA)	4,141,794	1924	R	179,233	93,034	86,199		63.6	5	400
ep LZ127-Graf Zep	4,469,439	1928	R	191,799	147,928	43,871		69.2	5	530
ep LZ129-Hindenburg	8,076,393	1936	R	511,464	286,596	143,298		72.9	4	1050
ep LZ130-Graf Zeppelin II	8,076,393	1936	R	511,464	286,596	143,298		72.9	4	1050
EPPELIN NT (N 07)	306,360	1997	SR	23,567			8530	67.0	3	200
odiac D6	324,944	1918	NR	25,000	11,900	13,100		43.0	2	256
PK2 (ZSG-2)	527,000	1951	NR							
P4K (ZSG-4)	527,000	1955	NR							
P5K (ZS2G-1)	650,000	1955	NR							
NPM-x	725,000		NR	45,900	30,700	15,200				
PG-2W (ZP2N-1)	975,000	1954	NR							
PG-3W (EZ-1C)	1,465,000	1958	NR							
PN-1 (ZPG-1)	875,000	1952	NR							

(continued)

Airship	Vol (ft³)	Year	Type	Gross Wt (lb)	Empty Wt (lb)	Heaviness (lb)	Alt ceil (ft)	Max speed (kt)	# Eng	Eng (hp)
ZR-1 Shenandoah (~L49)	2,289,861	1923	R	175,000	80,200	94,800		54.5	6	357
ZR-2 / R-38	2,960,000		R							
ZRS-4,5 (Akron, Macon)	7,400,000	1931	R	461,207	244,700	216,490		73.0	8	570

Source: [1], [2], and Wikipedia

Table E.2 Summary of Geometry Characteristics for Historical Airships

Airship	Type	Envelope vol (ft³)	Gas vol (ft³)	Ballonet vol (ft³)	Len (ft)	Diam (ft)	H,V arm (ft)	Horiz area (ft²)	Horiz vol coeff	Vert area (ft²)	Vert vol coeff
A-4	NR	95,000	95,000		162	33.5	62	354	0.0646	266	0.2348
ABC A-60+	NR	68,000	68,000	13,600	128	36					
AEROS 40A	NR	60,035	60,035	16,810	122	31					
AEROS 40B	NR	88,570	88,570	19,396	143	34.75					
B-Class	NR	84,000	84,000	19,250	163	31.5	62	375	0.0743	208	0.2132
C-Class	NR	181,000	181,000	55,250	196	42	74	530	0.0629	432	0.2394
D-Class	NR	190,000	190,000	55,250	198	42	75	495	0.0569	460	0.2493
E-Class	NR	95,000	95,000	16,500	162	33.5	62	458	0.0836	229	0.2021
G-1	NR	183,000	183,000	39,200	187	42.8					
Goodyear GZ-20	NR	202,700	202,700		192	50	73			280	0.1184
Gifford Steam Airship	NR	88,287	88,287		144						
J-1	NR	174,800	174,800	53,450	168	45	64	464	0.0564	404	0.1833
K-3-8	NR	416,000	416,000	114,000	249	57.9					
K-x	NR	456,000	456,000	104,400	250	60.0	95	992	0.0636	815	0.2178
L-Class	NR	123,000	123,000	29,400	148	39.5					
Lightship A-150	NR	150,000	150,000	38,558	165	46					
M-2	NR	648,000	648,000		302	69.5	115	1270	0.0644	1130	0.2492
MA Military	NR	180,000	180,000		169	48	64	546	0.0651	348	0.1460
N 1 "Norge"	SR	670,980	670,980		348	85.3	132			834	0.1686
Navy A (DN)1	NR	110,000	110,000		175	35					
North Sea Class	NR	360,000	360,000	128,000	262	57	100	1124	0.0844	323	0.1115
O-1	SR	127,000	127,000		178	35.4					
Parseval PL 27	NR	1,105,344	1,105,344		515	64					

rship	Type	Envelope vol (ft³)	Gas vol (ft³)	Ballonet vol (ft³)	Len (ft)	Diam (ft)	H,V arm (ft)	Horiz area (ft²)	Horiz vol coeff	Vert area (ft²)	Vert vol coeff
00	R	5,156,000	5,156,000		720	133.4					
23	NR	1,040,000	1,040,000		535	53	203	2280	0.0844	1880	0.7025
31	R	1,610,000	1,610,000		615	65.5	234	2191	0.0606	2060	0.5351
33	R	2,100,000	2,100,000		643	79					
9	R	930,000	930,000		526	53	200	2620	0.1045	1676	0.6634
publique Class	SR	130,000	130,000	25,650	200	35.8					
ntinel 1000	NR	353,100	353,100	91,806	222	54.7	84			732	0.2260
ntinel 8500	NR	3,000,000	3,000,000	1,260,000	452	111.5					
yship 600	NR	235,400	235,400	61,200	194	50	74		0.0000	437	0.1687
-14	NR	357,000	357,000		236	53.9					
-1	NR	80,200	80,200		136	34	52	332	0.0678	290	0.2370
-1	NR	52,290	52,290		106	30.9	40	328	0.0891	229	0.2135
S-LTA 138-S	NR	138,000	138,000	36,000	160	41.3					
DL 1B	NR	254,275	254,275	69,750	197	53.75					
ep LZ1,3	R	442,398	399,054	None	420	38.25					
ep LZ2	R	442,398	367,270	None	420	38.25					
ep LZ4,5,6	R	584,810	529,715	None	446	42.667					
ep LZ7,8	R	739,768	681,570	None	486	46					
ep LZ9,10,12	R	699,626	629,000	None	459	46					
ep LZ11,13	R	740,378	660,380	None	486	46					
ep LZ14	R	892,278	803,050	None	518	48.9					
ep LZ15,16	R	801,954	688,630	None	466	48.9					
ep LZ17, 19,20	R	790,621	688,630	None	459	48.9					
ep LZ18	R	1,108,411	953,490	None	518	54.5					
ep LZ21	R	835,811	737,015	None	486	48.9					
ep LZ22,23	R	880,859	781,860	None	512	48.9					
ep LZ24,25, 27–35,37	R	892,330	793,515	None	518	48.9					
ep LZ26	R	1,049,394	882,860	None	529	52.5					
ep LZ36,39	R	1,050,717	879,330	None	530	52.5					
ep LZ38, 40–58, 60,63	R	1,413,214	1,126,530	None	537	61.33					
ep LZ59,61, 64–69,71, 73,77,81	R	1,437,632	1,264,260	None	586	61.25					
ep LZ62,72, 74–76,78– 80,82–90	R	2,593,357	1,949,360	None	650	78.5					

(continued)

Airship	Type	Envelope vol (ft³)	Gas vol (ft³)	Ballonet vol (ft³)	Len (ft)	Diam (ft)	H,V arm (ft)	Horiz area (ft²)	Horiz vol coeff	Vert area (ft²)	Vert vol coeff
Zep LZ91–99	R	2,574,191	1,970,000	None	645	78.5					
Zep LZ-96 (~L-49)	R	2,538,744	1,975,500	None	644	78	245	2456	0.0501	1864	0.3142
Zep LZ100, 101,103, 105–111	R	2,574,191	1,977,610	None	645	78.5					
Zep LZ102,104	R	2,997,663	2,419,040	None	743	78.5					
Zep LZ112	R	2,793,181	2,196,560	None	693	78.5					
Zep LZ113,114	R	2,997,058	2,419,040	None	743	78.5					
Zep LZ120-Bodensee	R	1,053,363	796,345	None	429	61.25					
Zep LZ126 (ZR-3, LA)	R	4,141,794	2,472,000	None	656	104.7	249	2510	0.0370	2485	0.2294
Zep LZ127-Graf Zep	R	4,469,439	3,708,020	None	776	100					
Zep LZ129-Hindenburg	R	8,076,393	7,062,895	None	804	135.2					
Zep LZ130-Graf Zeppelin II	R	8,076,393	7,062,895	None	804	135.2					
ZEPPELIN NT (N 07)	SR	306,360	290,450	77,700	246	46.5	93		0.0000	317	0.1402
Zodiac D6	NR	324,944	324,944		263	59	100	838	0.0674	881	0.3151
ZPK2 (ZSG-2)	NR	425,000	425,000	121,800	249	57.8	95	992	0.0667	815	0.2360
ZP4K (ZSG-4)	NR	527,000	527,000	121,800	266.5	68.5	101	1108	0.0645	888	0.2012
ZP5K (ZS2G-1)	NR	650,000	650,000	169,500	282	67.6	107	1720	0.0871	860	0.1817
ZNPM-x	NR	725,000	725,000	203,000	311	73.3	118	1280	0.0603	1137	0.2270
ZPG-2W (ZP2N-1)	NR	975,000	975,000	247,300	342.7	75.4	130	1511	0.0584	1511	0.2654
ZPG-3W (EZ-1C)	NR	1,465,000	1,465,000	383,000	403	85.1					
ZPN-1 (ZPG-1)	NR	875,000	875,000	220,800	321	73.5	122	1518	0.0631	1518	0.2754
ZR-1 Shenandoah (~L49)	R	2,488,906	2,115,174	None	680.3	79.7	259	2870	0.0594	2335	0.4124
ZR-2 / R-38	R	3,125,745	2,724,000	None	695	85.5					
ZRS-4,5 (Akron, Macon)	R	8,542,981	6,850,000	None	785	133.0	298	6980	0.0635	7170	0.3848

Source: [1], [2], and Wikipedia

References

[1] Chant, C., *The Zeppelin*, Barnes & Noble Books, 2000.
[2] Shock, J.R., *U.S. Navy Airships 1915–1962*, Atlantis Productions, 1992.

Appendix F

Helium Availability— Now and in the Future

Brad Baughman

- The History of Helium, Including Its Origins, Current Supply, and Sales
- Review and Speculation on Helium's Supply and Pricing Impact on the Production Growth of New Airships
- Future of Helium

The Graf Zeppelin II was designed to use helium after the Hindenburg's accident. However, Germany was unable to get helium from the United States so LZ-130 flew 30 flights using hydrogen. In 1940 "Graf 2" was grounded and along with the partially built LZ-131 used for scrap to build airplanes for the Luftwaffe.

It is better to beg forgiveness ... than ask permission.
Skunk Works Motto

Helium is a natural resource, and like oil, coal, and natural gas is available as the laws of supply and demand will dictate. There is a common misconception that the world's supply of helium will be depleted in the near future. This is far from the truth. It is perhaps a perception created by the misunderstanding of a number of factors including the U.S. government selling off its helium reserves by 2015, unexpected changes in consumption over the past decade, and temporary shortages created by factory repairs. These factors, combined with sensationalistic journalism, have created a confusing picture of the availability of helium both now and in the future.

F.1   Where Helium Comes From

Helium is the second most abundant element in the universe. It is second on the chemical periodic table of elements and is the smallest and lightest of the noble or inert gases. Terrestrial helium is trapped beneath the Earth's surface along with the abundant supply of natural gas we use to cook and heat our homes. Helium is extracted by fractional distillation from natural gas that contains up to 7% helium. Since helium has a lower boiling point than any other element, low temperature and high pressure are used to liquefy nearly all the other gases (mostly nitrogen and methane). The resulting crude helium gas is purified by successive exposures to lowering temperatures, in which almost all of the remaining nitrogen and other gases are precipitated out of the gaseous mixture. Activated charcoal is used as a final purification step, usually resulting in 99.995% pure Grade-A helium, with neon being the remaining principal impurity. In a final production step, most of the helium that is produced is liquefied via a cryogenic process. This is necessary for applications requiring liquid helium and also allows helium suppliers to reduce the cost of long distance transportation, as the largest liquid helium containers have more than five times the capacity of the largest gaseous helium tube trailers. Diffusion of crude natural gas through special semi-permeable membranes and other barriers is another method used to recover and purify helium. Helium is commercially available commodity in either liquid or gaseous form and can be purchased just about anywhere in the world. As a liquid, it can be supplied in small containers called Dewar's that hold up to 35.3 ft^3 (1000 liters) or in large ISO containers that have nominal capacities as large as 1483 ft^3 (around 11,000 U.S. gallons). In gaseous form, small quantities of helium are supplied in high-pressure cylinders holding approximately 282 ft^3, while large quantities of high pressure gas are supplied in tube trailers that have capacities of up to approximately 172,000 ft^3.

Today, the world's helium resources are estimated at 1.87 trillion ft^3 (1870 billion ft^3) with 43% of those resources inside the United States [1]. The breakdown of helium resources worldwide is shown in Fig. F.1. The

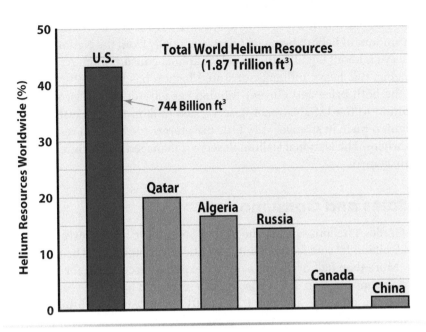

Figure F.1   Known world helium resources.

greatest U.S. reserves of helium are in Hugoton and the nearby gas fields of southwest Kansas and the panhandles of Texas and Oklahoma. It is estimated that the resource base for yet-unproven helium in natural gas in the U.S. is 1000 to 2000 trillion ft^3 (1–2,000,000 billion ft^3), about 500–1000 times the proven reserves [2]. Around the world these numbers are likely greater since less exploration has been done. While most of the helium trapped within Earth has been there for a long time, it is interesting to note that Earth is actually creating helium on an annual basis. Terrestrial helium is a byproduct of the natural radioactive decay of heavy radioactive elements (thorium and uranium), as the alpha particles emitted by such decays consist of helium-4 nuclei. This radiogenic helium is trapped with natural gas in concentrations up to 7% by volume generating an estimated 3000 metric tons (594 million ft^3) of helium per year. That's about 25% of the U.S. annual helium consumption.

In 1925, the U.S. government established the National Helium Reserve at Amarillo, Texas, with the goal of supplying military airships in time of war and commercial airships in peacetime. By 1995, 35.3 billion ft^3 of the gas had been collected at the National Helium Reserve, and the reserve was $1.4 billion in debt, prompting Congress to phase it out. The resulting "Helium Privatization Act of 1996" (Public Law 104–273) directed the United States Department of the Interior to start emptying the reserve in

2005 and bleed it down to 600 million ft³ by 2015. Owing to this directive, the amount of helium being withdrawn annually from the National Helium Reserve is about equal to the total U.S. annual consumption. As one might imagine, this heavy influence of stored helium has had significant effects on the both price and current refining capacity of helium producers. In addition to the U.S. National Reserve, 466 million ft³ of privately owned helium remain in storage. Very little privately owned helium reserves exist elsewhere. The National Helium Reserve is managed by the Bureau of Land Management.

F.2 Sales and Consumption

Grade-A helium sales by the United States have been slowly increasing to 4.9 billion ft³ sold last year but domestic consumption has been decreasing. More than half of today's sales are exported accounting for 73% of the world's consumption. U.S. consumption increased up to the year 2000 and has decreased since (see Fig. F.2). In 2011, approximately 6 billion ft³ of helium was extracted from natural gas or withdrawn from helium reserves with approximately 76% from the United States, 11% from Algeria, and most of the remainder from Russia, Poland and Qatar, which opened a new helium plant and now produces 8% of the world's helium needs. In 2010, 45% of the world's needs were met by withdrawing from the U.S. National Reserve (see Fig. F.3).

In 2011, 28.5 trillion ft³ of natural gas was produced in the United States [3]. Conservatively estimating the helium percentage to be 1% (can be as high as 7%), there could have been 285 billion ft³ or more of helium produced yet only 2.9 billion ft³ were produced meaning only about 1% of

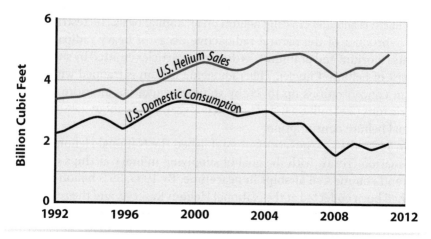

Figure F.2   U.S. helium sales and consumption.

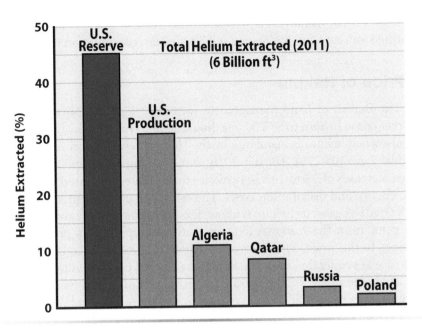

Figure F.3 2011 World helium extracted.

the helium streaming out of the ground in natural gas facilities across the United States was captured and depicted in Fig. F.4. The percentage captured worldwide is even less.

While helium's use in airships remains limited, its use in cryogenics, pressurizing, purging, welding, leak detection, and the ever popular "party balloon" applications account for most of today's helium consumption. Helium consumption grew as the fiber optic industry boomed in the 1990s being the second largest consumable in cable fabrication. As competition grew, manufacturers looked for ways of reducing production costs. Helium recovery systems were designed to collect the helium used in the production.

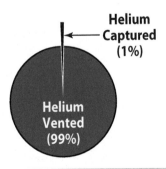

Figure F.4 U.S. natural gas production.

Up to 80% of the helium used to transfer heat from the fiber to chiller is now captured for reuse and has helped stabilize helium consumption rates.

F.3 Price of Helium

The Bureau of Land Management raised the FY 2011 price of open-market crude helium to $75.00 per thousand ft^3. The price for the government-owned helium is mandated by the Helium Privatization Act of 1996 (Public Law 104-273). During 2010, some helium suppliers announced price increases of 5% to 10% in response to continued increased raw material, energy, and distribution costs. The estimated price for private industry's Grade-A gaseous helium is about $160 per thousand ft^3, about double the price from the National Reserve, with some producers posting surcharges to this price. The price of pure helium is expected to continue to increase as production costs, including the price of crude helium, increase. As illustrated in Fig. F.5 the private industry price of helium has increased about 14% per year since 2001, yet the price from the National Reserve has only grown 4% per year over the same time period.

Adding to the variability in the marketplace was the recent maintenance shutdown of a major helium production facility. The Shute Creek Gas Plant in western Wyoming is the world's largest helium facility, producing more than 20% of the U.S. helium supply. Recent maintenance at the Exxon Mobil Corp. plant took longer than expected, about six weeks instead of four, creating a temporary shortage of helium and problems for users who were not prepared. Similar maintenance actions are common, such as repairs of a pipeline in Texas, a major plant in Wyoming running below capacity, and

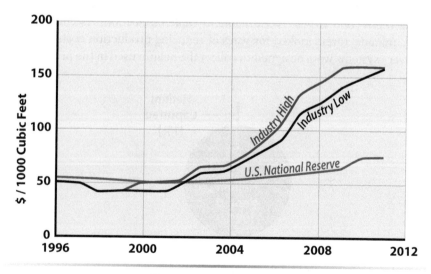

Figure F.5 Helium price.

facility repairs in Algeria, Poland, and Australia. This temporary shortage was not enough to spur new helium facility construction, but when the production demand and price of helium rises after 2015, new plants will be built to increase production capacity. Worldwide, eight new helium plant projects are scheduled for start up between 2011 and 2017, including a new helium plant that started up in Australia in March 2011. Two projects are scheduled for start up in the United States during 2011–14 near Riley Ridge, WY, and St. Johns, AZ. Other plants are planned for Algeria, China, India, Indonesia, Qatar, and Russia.

While the United States is reducing its National Reserve at prices far below commercial market rates, the price of helium is artificially suppressed, reducing economic incentives to expand production and distribution capabilities in the United States and around the world. When the National Reserve stops reducing its stockpile of helium in 2015, the world price will rise accordingly and producers around the world will find a reason to expand their capacity.

F.4 Impact of Hybrid Airships

Of critical importance to the budding airship market is how the supply and pricing will affect the production growth of new airship hulls. Will this significantly affect the operating or production costs? As an example, let's use a notional 3-million-cubic-foot cargo airship. At the current private industry price of $160 per thousand ft^3, it will cost $480,000 to fill this cargo airship. At an estimated produced value of $60M per vehicle, the helium cost accounts for just 0.8% of the vehicle cost. If the cost of helium doubles,

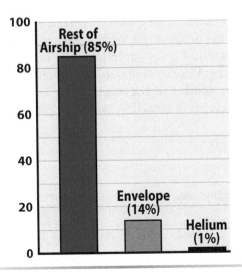

Figure F.6 Hybrid airship cost breakdown.

it would increase the vehicle production cost by less than 1%. With the envelope costing about $8M and the rest of the vehicle costing another $50M, helium is not the driving production cost as shown in Fig. F.6. Compared to most aircraft materials, helium remains a very inexpensive structural material. To further put this into perspective, the amount of helium needed for one 3-million-cubic-foot cargo airship is 0.07% of the 4.4 billion ft^3 sold annually by the United States. It would take a production rate of about 15 cargo airships a year to increase the annual U.S. helium sales by 1%. Even with hundreds of cargo airships in service, the replenishment rates of the entire fleet would struggle to approach even the 1% of current annual production figures. Airships have the further benefit of not requiring the higher grade helium needed by medical equipment and scientific applications (99.9999%), so the more commonly available industrial grade (98%+) helium is suitable for airships allowing us to shop for best price.

F.5 Conclusion

While Grade-A helium can be scarce at times, as it was in late summer 2011 when the plant in Shute Creek was shut down for repairs, there are ample supplies of unrefined industrial grade helium in the world to meet current needs for hundreds of years to come. The addition of cargo airship production has at most a few percentage point impact on helium production, which at this point is only capturing 1% of the helium coming out of the ground in natural gas. Selling off large quantities of helium from the National Reserve at cut rate prices has temporarily taken away the incentive for private industry to maintain or upgrade their production capability. As that effect fades in 2015, the laws of supply and demand will again self regulate production rates and costs, and new plants, already planned, will come on line and support demand. Despite what you may have read, the world's supplies of helium are not disappearing for hundreds of years, and the advent of new cargo airships is not going to dramatically impact the availability of helium. At the current U.S. consumption, there is enough helium in the U.S. reserves and U.S. resources to last more than 150 years; more if we stop exporting and thousands of years if we count the yet-unproven reserves.

References

[1] Mineral Commodity Summaries, U.S. Geological Survey, Jan. 1996–2012.
[2] "The Impact of Selling the National Helium Reserve," Committee on the Impact of Selling the Federal Helium Reserve, Commission on Physical Sciences, Mathematics and Applications, Commission on Engineering and Technical Systems, National Research Council, 2000.
[3] Annual Energy Review, U.S. Energy Information Administration, Sept. 2012.

INDEX

SUPPORTING MATERIALS

A complete listing of titles in the AIAA Education Series is available from ARC. Visit ARC frequently to stay abreast of product changes, corrections, special offers, and new publications.

AIAA is committed to devoting resources to the education of both practicing and future aerospace professionals. In 1996, the AIAA Foundation was founded. Its programs enhance scientific literacy and advance the arts and sciences of aerospace. For more information, please visit www.aiaafoundation.org.

1. The Skunk Works manager must be delegated practically complete control of his program in all aspects. He should report to a division president or higher.
2. Strong but small project offices must be provided both by the military and industry.
3. The number of people having any connection with the project must be restricted in an almost vicious manner. Use a small number of good people (10% to 25% compared to the so-called normal systems).
4. A very simple drawing and drawing release system with great flexibility for making changes must be provided.
5. There must be a minimum number of reports required, but important work must be recorded thoroughly.
6. There must be a monthly cost review covering not only what has been spent and committed but also projected costs to the conclusion of the program. Don't have the books 90 days late, and don't surprise the customer with sudden overruns.
7. The contractor must be delegated and must assume more than normal responsibility to get good vendor bids for subcontract on the project. Commercial bid procedures are very often better than military ones.
8. The inspection system as currently used by the Skunk Works, which has been approved by both the Air Force and Navy, meets the intent of existing military requirements and should be used on new projects. Push more basic inspection responsibility back to subcontractors and vendors. Don't duplicate so much inspection.
9. The contractor must be delegated the authority to test his final product in flight. He can and must test it in the initial stages. If he doesn't, he rapidly loses his competency to design other vehicles.
10. The specifications applying to the hardware must be agreed to well in advance of contracting. The Skunk Works practice of having a specification section stating clearly which important military specification items will not knowingly be complied with and reasons therefore is highly recommended.
11. Funding a program must be timely so that the contractor doesn't have to keep running to the bank to support government projects.
12. There must be mutual trust between the military project organization and the contractor with very close cooperation and liaison on a day-to-day basis. This cuts down misunderstanding and correspondence to an absolute minimum.
13. Access by outsiders to the project and its personnel must be strictly controlled by appropriate security measures.
14. Because only a few people will be used in engineering and most other areas, ways must be provided to reward good performance by pay not based on the number of personnel supervised.

Kelley Johnson's 14 Rules of Management (Skunk Works)

P-791 in final assembly

P-791 starts rollout prior to first flight

P-791 first flight at Palmdale, California (Jan 31, 2006)

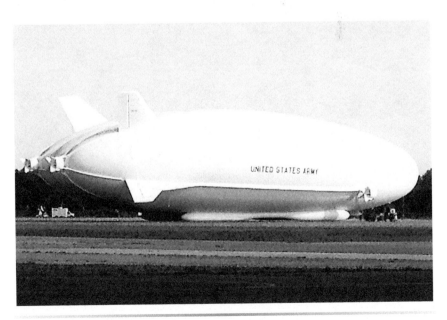

NGC/Army LEMV hybrid airship readies for first flight (August 10, 2012)

Boeing X-32 and Lockheed Martin X-35 aircraft (JSF competition)

Lockheed Martin X-35 performing a vertical landing

SR-71 showing Mach diamonds from a high power setting

Lockheed Martin X-35 performing maneuvers during flight test phase

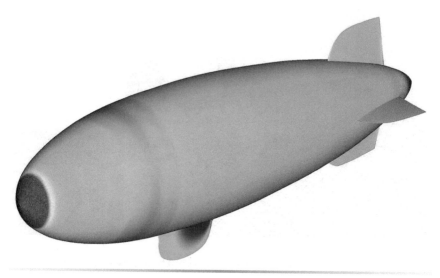

ZP4K pressure distribution

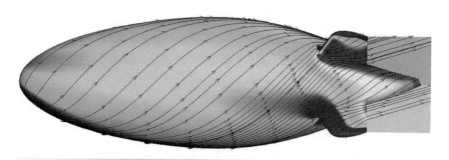

ISIS pressure distribution

HondaJet

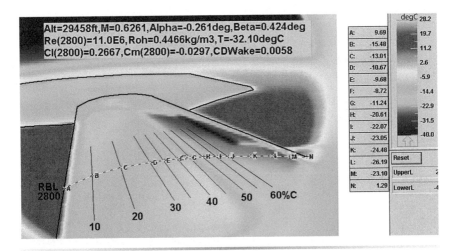

Flight test data of boundary layer transition
using wing surface temperatures on the HondaJet

Ryan Voight enjoys the miracle of flight over Utah in a
Wills Wing T2C hang glider

The Swift high performance hang glider (shown without pilot fairing)

Cessna 172S (forground) and Cessna 172 (background)

Cessna 172 Skyhawk—classic design

ZLT Airship in Botswana

A-170 Airship moored

Hindenburg exploding and burning at Lakehurst landing site (May 6, 1937)

Hindenburg flying over Manhattan

USS Shenandoah (ZR-1) internal structure for rigid shell

Daedalus enroute to Santorini with pilot Kanellos Kanelopoulos (April 23, 1988)

Daedalus enroute to Santorini near the end of the flight

Launch apparatus for a high altitude scientific
balloon with its payload module

The Ultra-Long Duration Balloon (ULDB) launched by NASA
(pumpkin balloon design)

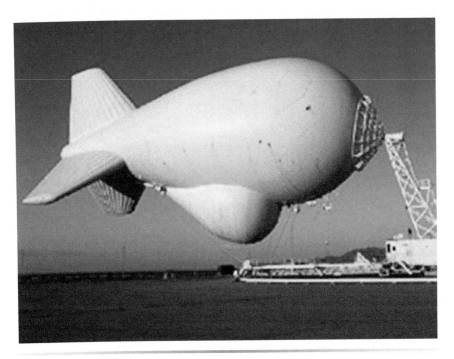

Tethered Aerostat Radar System (TARS)

Lockheed Martin ISIS high altitude airship—final design

T-46A trainer aircraft during flight test at Edwards AFB (1986)

SR-71C